Spectroscopy for Materials Characterization

Spectroscopy for Materials Characterization

Edited by

Simonpietro Agnello
Department of Physics and Chemistry – Emilio Segrè
University of Palermo
Palermo
Italy

WILEY

This edition first published 2021

Published by John Wiley & Sons, Inc., Hoboken, New Jersey
Published simultaneously in Canada.

Registered Office
John Wiley & Sons, Inc., 111 River Street, Hoboken, NJ 07030, USA

Editorial Office
111 River Street, Hoboken, NJ 07030, USA

For details of our global editorial offices, customer services, and more information about Wiley products visit us at www.wiley.com.

Wiley also publishes its books in a variety of electronic formats and by print-on-demand. Some content that appears in standard print versions of this book may not be available in other formats.

For general information on our other products and services or for technical support, please contact our Customer Care Department within the United States at (800) 762-2974, outside the United States at (317) 572-3993 or fax (317) 572-4002.

Wiley also publishes its books in a variety of electronic formats. Some content that appears in print may not be available in electronic formats. For more information about Wiley products, visit our web site at ww.wiley.com.

Library of Congress Cataloging-in-Publication Data Applied for

ISBN: 9781119697329

Cover design by Wiley
Cover image: © khalus/gettyimages

Set in 9.5/12.5pt STIXTwoText by Straive, Pondicherry, India

10 9 8 7 6 5 4 3 2 1

MIX
Paper from responsible sources
FSC® C013604
FSC www.fsc.org

Contents

Preface

Ten years of experience in a basic laboratory course of spectroscopy for master's students in physics as well as some courses for PhD students in physics and materials science together with more than 20 years of experimental work in physics and materials science convinced me of the opportunity to join the different basic aspects and technical approaches for the characterization of materials in a book. My contact with scientists in physics, chemistry, engineering, biology, and many other research fields emphasized the need to collect accurate and broad knowledge on spectroscopic techniques both from the fundamental point of view and from the technical perspective to impart the skills required to think about experiments, to interpret their results, and last but not least to advance theoretical knowledge. Motivated by these aspects, I started the project by inviting different experts in the various fields of spectroscopy, usually employed in the study of materials from the nanoscale up to the macroscopic scale and also in the field of astrophysics. In general, any atom, molecule, or condensed phase up to a solid needs to be characterized, from the species identification to structural definition and electron dynamics. Consequently, this practice-oriented handbook collects the presentation of different spectroscopic techniques nowadays used to carry out experiments and is thought for master's students, PhD students, and researchers who are not yet experts in the field of spectroscopy. Each chapter is a standalone contribution dedicated to a different technique and is composed of a basic theory introduction at a level adapted to students, plus more advanced information for experts. Then, details about the characteristic instrumental features and apparatus are considered, aiming to give hints for the appropriate arrangement of a typical experiment and, in some cases, for the understanding of the choice of an instrumental setup and the selection of technical features. Each chapter is completed by case studies that can be reproduced by employing commercially available or laboratory-prepared samples, or by examples of applications mainly from the above-cited fields of materials science, chemistry, and engineering. Furthermore, hints for the deepening one's understanding of the topic with more specialist books are given along the way.

The first chapter is dedicated to the basic aspects of radiation–matter interaction in the visible-ultraviolet range and the fundamentals of absorption and emission (fluorescence and phosphorescence) processes of atoms and molecules. This chapter is followed by two chapters on time-resolved spectroscopy at the nanosecond scale and the more recent frontier approach of ultrafast spectroscopy at the femtosecond scale. These chapters detail the aspects of the electron dynamics in the excited states after pulsed excitation through emission and absorption processes including the Raman effect. Then, a chapter dedicated to fluorescence microscopy imaging at the nanoscale domain is presented with confocal and two-photon techniques, where aspects of spatial resolution and dynamics are explored. The molecular vibrational spectroscopy in the infrared range is tackled in Chapters 5 and 6 by both the absorption and the Raman scattering effects. These chapters report on the fundamental aspects of radiation-molecular vibration interaction and consider the traditional experimental methods, detailing the most recent applications of microscopy and joined vibrational–atomic force microscopy. The following two chapters consider spectroscopy related to irradiation effects. The first is focused on the thermoluminescence phenomenon of light emission induced by heating a sample after it has been exposed to ionizing radiation environment. The spectroscopic aspects are introduced and the basics of the phenomenon together with the interpretation of trap states and release thermodynamics are deepened in view of the application in many emerging fields such as dosimetry. In a similar context, Chapter 8 reports on the phenomena of radiation-induced absorption, emission, and compaction with a focus on the novel online (in situ) measurements, particularly useful for the spectroscopic study of the materials' modifications during their exposure to ionizing radiation as well as their recovery processes. Chapters 9 and 10 treat the magnetic spectroscopy illustrating the electron and nuclear phenomena. In particular, electron paramagnetic resonance in the static and dynamic domains are dealt with in the former chapter, considering hyperfine interaction, transient dynamics, and the $S = 1$ state paramagnetism. In the latter chapter, the standard as well as the non-conventional nuclear magnetic resonance phenomena are introduced including the Fourier transform, the Magic Angle Spinning, and the time domain studies. The book then reports on high-energy spectroscopy with three chapters on X-rays and ultraviolet radiation. Chapter 11 reports on the absorption process of X-rays in near edge and pre-edge, also including fine structure determination and in situ spectroscopy. Chapter 12 introduces photoelectron emission illustrating the surface maps and the depth and angle resolved profiles. Successively, Chapter 13 covers vacuum ultraviolet excitations considering the most widely available sources. Finally, two structural spectroscopy chapters on transmission electron spectroscopy (TEM) and atomic force microscopy (AFM) are included. They report on the structural investigation at the atomic and nanometric scale including those spectroscopic aspects pertaining to high-energy excitation in the former technique and to the novel interpretation of spectroscopy in the latter.

The book, as stated above, is the result of the contribution of different scientists mainly working in the field of physics, chemistry, surface science, and nanoscience. Their different points of view assure a multidisciplinary approach to the techniques and, as a consequence, appeal to a broad readership of researchers and students dedicated to the studies on materials.

The Editor
Simonpietro Agnello

List of Contributors

Simonpietro Agnello
Department of Physics and
Chemistry – Emilio Segrè
University of Palermo
Palermo
Italy

Raffaele Giuseppe Agostino
Dipartimento di Fisica
Università della Calabria
Rende
CS
Italy

Antonino Alessi
Fraunhofer Institute for
Technological Trend Analysis
Appelsgarten 2
Euskirchen
Germany

Laboratoire des Solides Irradiés
CEA/DRF/IRAMIS,
CNRS, Ecole polytechnique
Institut Polytechnique de Paris
Palaiseau
France

Gianpiero Buscarino
Department of Physics and
Chemistry – Emilio Segrè
University of Palermo
Palermo
Italy

Marco Cannas
Department of Physics and
Chemistry – Emilio Segrè
University of Palermo
Palermo
Italy

Pellegrino Conte
Department of Agriculture,
Food and Forestry Sciences
University of Palermo
Palermo
Italy

Vincenzo De Michele
Laboratoire Hubert Curien
Université Jean Monnet
Saint-Etienne
France

Giuliana Faggio
Department of Information Engineering
Infrastructure and Sustainable Energy
University Mediterranea of Reggio Calabria
Reggio Calabria
Italy

Tiziana Fiore
Department of Physics and
Chemistry – Emilio Segrè
University of Palermo
Palermo
Italy

Vincenzo Formoso
Dipartimento di Fisica
Università della Calabria
Rende
CS
Italy

Franco Gelardi
Department of Physics and
Chemistry – Emilio Segrè
University of Palermo
Palermo
Italy

Francesco Giannici
Department of Physics and
Chemistry – Emilio Segrè
University of Palermo
Palermo
Italy

Sylvain Girard
Laboratoire Hubert Curien
Université Jean Monnet
Saint-Etienne
France

Rossella Grillo
Department of Information Engineering
Infrastructure and Sustainable Energy
University Mediterranea of Reggio Calabria
Reggio Calabria
Italy

Alessandro Longo
The European Synchrotron
Grenoble
France
ISMN-CNR
Palermo
Italy

Fabrizio Messina
Department of Physics and
Chemistry – Emilio Segrè
University of Palermo
Palermo
Italy

Giacomo Messina
Department of Information Engineering
Infrastructure and Sustainable Energy
University Mediterranea of Reggio Calabria
Reggio Calabria
Italy

Adriana Morana
Laboratoire Hubert Curien
Université Jean Monnet
Saint-Etienne
France

Federico Moretti
Lawrence Berkeley National Laboratory
Berkeley
CA
USA

Claudia Pellerito
Department of Physics and
Chemistry – Emilio Segrè
University of Palermo
Palermo
Italy

Christoph J. Sahle
The European Synchrotron
Grenoble
France

Giuseppe Sancataldo
Department of Physics and
Chemistry – Emilio Segrè
University of Palermo
Palermo
Italy

Alice Sciortino
Department of Physics and
Chemistry – Emilio Segrè
University of Palermo
Palermo
Italy

Michelangelo Scopelliti
Department of Physics and Chemistry – Emilio Segrè
University of Palermo
Palermo
Italy

Alberto Spinella
Advanced Technologies Network Center (ATeN Center)
University of Palermo
Palermo
Italy

Lavinia Vaccaro
Department of Physics and Chemistry – Emilio Segrè
University of Palermo
Palermo
Italy

Valeria Vetri
Department of Physics and Chemistry – Emilio Segrè
University of Palermo
Palermo
Italy

Anatoly F. Zatsepin
Russian Academy of Sciences
Ural Division
Yekaterinburg
Russia

Institute of Physics and Technology
Ural Federal University
Yekaterinburg
Russia

Dmitry A. Zatsepin
Russian Academy of Sciences
Ural Division
Yekaterinburg
Russia

Institute of Physics and Technology
Ural Federal University
Yekaterinburg
Russia

1

Radiation–Matter Interaction Principles

Optical Absorption and Emission in the Visible-Ultraviolet Region

Simonpietro Agnello

Department of Physics and Chemistry – Emilio Segrè, University of Palermo, Palermo, Italy

1.1 Empirical Aspects of Radiation–Matter Interaction

The basic everyday experience of the colors around us inspires the knowledge of the phenomenon of radiation–matter interaction. Sun, the source of daylight, emits a large quantity of electromagnetic field frequencies (ν), or wavelengths (λ). When these rays impinge on matter, they can be reflected (scattered) or absorbed by the constituent atoms and molecules. These phenomena give rise to the appearance of colored objects since the light arriving at our eyes (a light sensor) depends on the emitted radiation or the reflected one, having fewer frequencies than the original sunlight due to the absorption effect of matter around us. In this chapter, some empirical aspects of absorption and emission phenomena will be introduced.

1.1.1 Optical Absorption: The Lambert–Beer Law

An introductory experiment that highlights the effect of absorption of light is represented in Figure 1.1.

A parallel beam of lightwave with wavelength λ and intensity $I_0(\lambda)$ impinges perpendicularly on a face of a parallelepiped specimen of matter. It is useful to recall that λ and ν are connected by the speed of light c (2.9979×10^8 m·s^{-1} in vacuum): $\lambda = c/\nu$ [1]. Passing through the sample, the light intensity could be reduced, and at the exit, the amount I measured at λ could be diminished to the value $I_t(\lambda)$ [2–4]. The eventual intensity reduction inside the specimen increases continuously on increasing the size L. If a portion of thickness dx of the sample is considered, it is expected that passing through it, I decreases by a quantity dI (for simplicity, λ will be omitted henceforth). This effect depends on the presence of absorbing centers in the volume considered, on the one side, and on their physical properties, on the other. These features are taken into account by the concentration of centers, N, and by their cross section, σ. Larger is the concentration of centers, larger absorption will take place. On increasing the probability of radiation–matter interaction, σ increases too, as well as the absorption effect. If a homogeneous and isotropic distribution of absorbing centers inside the volume of the sample is considered, it can be assumed that [3]:

$$\mathrm{d}I = -N\sigma I\mathrm{d}x \tag{1.1}$$

Spectroscopy for Materials Characterization, First Edition. Edited by Simonpietro Agnello.

where the units are J $(\text{cm}^2\cdot\text{s})^{-1}$ for I, centers cm^{-3} for N, cm^2 for σ, and cm for dx. By considering the entire sample, Eq. (1.1) can be integrated to obtain:

$$\int_{I_0}^{I_t} \frac{dI}{I} = -\int_0^L \sigma N dx \quad (1.2)$$

and determine the solution

$$\ln \frac{I_0}{I_t} = \sigma NL. \quad (1.3)$$

In general, from solution (1.3), it is found that at a position x inside the sample

$$I(x) = I_0 e^{-\sigma Nx} \quad (1.4)$$

this is the *Lambert–Beer* law that expresses the attenuation of light intensity as a function of the thickness of the sample traversed [2, 3]. A typical expected profile of light intensity in traversing a sample is reported in Figure 1.1. It is useful to introduce some quantities commonly associated to the absorption effect. The empirical one is the *absorption coefficient* defined by the experimental macroscopic measurement of attenuation:

$$I(x) = I_0 e^{-\alpha x}. \quad (1.5)$$

It is easy to show that $\alpha = \sigma N$, connecting the macroscopic quantities α and N to the microscopic one σ (see Section 1.2 to find the relation to atomic and molecular properties). Then, we report the instrumental quantity, the *optical density* (OD), also called *absorbance* (A) [5, 6]:

$$\text{OD} = A = \log_{10} \frac{I_0}{I_t} = 0.4343 \ln \frac{I_0}{I_t} = 0.4343\alpha L = 0.4343\sigma NL \quad (1.6)$$

and the *transmittance*, T:

$$T = \frac{I_t}{I_0} = 10^{-\text{OD}}. \quad (1.7)$$

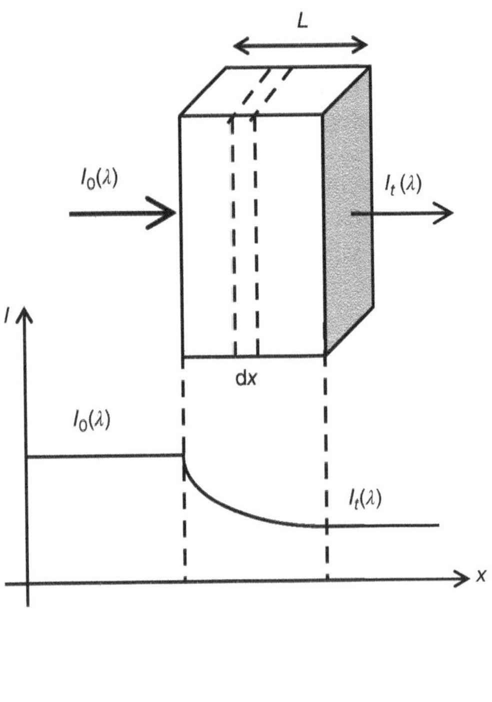

Figure 1.1 Schematic representation of a beam of light at wavelength λ passing through a parallelepiped of matter. In the bottom, the qualitative decrease of intensity is reported. I_0 is the impinging intensity and I_t the transmitted one.

It is worth noting that when OD ≪ 1, $1 - T = 1 - 10^{-\mathrm{OD}} \approx 1 - (1 - \mathrm{OD}) = \mathrm{OD} = A$, so the absorbance and the optical density can be derived directly from the transmittance [6]. Finally, it is also useful to introduce a quite diffuse alternative to Eq. (1.6):

$$\mathrm{OD} = A = \log_{10}\frac{I_0}{I_t} = \varepsilon CL \tag{1.8}$$

where ε is the *molar extinction coefficient*, or *molar absorption coefficient*, having units liter/(mole·cm) [$\mathrm{M^{-1}\cdot cm^{-1}}$], and C is the concentration of absorbing centers, in mole/liter [M]. By equating (1.6) and (1.8), it is shown that

$$0.4343\sigma N = \varepsilon C \tag{1.9}$$

having used centers cm^{-3} for N and cm^2 for σ. Then, considering the Avogadro's number, $N_\mathrm{A} = 6.022\ 10^{23}$ centers mole^{-1}, we obtain the conversion formula

$$\sigma(\mathrm{cm}^2) = \frac{\varepsilon(\mathrm{M^{-1}cm^{-1}})C(\mathrm{M})}{0.4343N(\mathrm{centers/cm^3})} = \frac{\varepsilon(\mathrm{M^{-1}cm^{-1}})C(\mathrm{M})}{0.4343\frac{N(\mathrm{centers\cdot L^{-1}})}{1000}} = \frac{\varepsilon(\mathrm{M^{-1}cm^{-1}})C(\mathrm{M})}{0.4343\frac{N_\mathrm{A}\left(\frac{\mathrm{centers}}{\mathrm{mole}}\right)C(\mathrm{M})}{1000}}$$
$$= \frac{1000\varepsilon(\mathrm{M^{-1}cm^{-1}})}{0.4343\cdot 6.022\cdot 10^{23}} = 3.82\cdot 10^{-21}\varepsilon(\mathrm{M^{-1}cm^{-1}}) \tag{1.10}$$

these quantities are related to the electronic states of absorbing centers, as will be shown later.

Concluding, the Lambert–Beer law states that the optical density is proportional to the concentration of absorbing centers and to their electronic properties. All of the above considerations can be extended to any λ and the study of absorption as a function of the wavelength impinging on the sample gives origin to the absorption spectrum.

It is worth noting that some physical phenomena can influence the experimental evaluation of the optical density. The light scattering (both elastic process, Rayleigh scattering, and anelastic process, Raman scattering [7, 8]) can deviate the beam and avoid its exit in the detection direction. This effect could give origin to an inexact estimate of OD and can be evidenced by a λ^{-4} background dependence of absorbance [1, 7]. In particular, it could be erroneously concluded that photons have been absorbed whereas only their path has been deviated by the matter without any energy transfer from the electromagnetic field to the atoms. A second physical effect is the emission of light from the sample caused by the return of the electron to its thermal equilibrium state after the absorption phenomenon, promoting it to an excited state (see further). The emission is usually at a wavelength different to the impinging one, but if the light exiting from the sample is not recorded identifying the λ, as usually done in a single monochromator spectrometer, a wrong estimate of the optical density can be done. The latter effect could be relevant if absorption is large and photons of impinging light are highly reduced in number through the sample and the exiting counted photons mainly coincide with those emitted. The latter effect can be instrumentally avoided by using a double monochromator setup. Neglecting instrumental effects like stray light, that is parasitic light arriving at the detector not passing through the sample, and signal-to-noise limits [2, 3], another physical effect to be taken into account is reflection [1, 4]. When the parallel beam reported in Figure 1.1 impinges perpendicularly on the sample surface, the mismatch of refractive index between the medium (n_1) and the sample (n_2) induces a transmitted and a reflected beam [1, 4]. Introducing the reflectivity r for normal incidence of light:

$$r = \left(\frac{n_1 - n_2}{n_1 + n_2}\right)^2 \tag{1.11}$$

the light entering the sample has the intensity

$$I_e = I_0(1-r). \tag{1.12}$$

This light is attenuated, according to the Lambert–Beer law (1.4). Furthermore, before exiting the sample, the light is reflected again on the exit surface. It is found that the transmitted light for single reflection path is given by

$$I_t = [I_0(1-r)]e^{-\sigma NL} \cdot (1-r) = I_0(1-r)^2 e^{-\sigma NL} \tag{1.13}$$

and the absorbance estimation is

$$\mathrm{OD} = A = \log_{10}\frac{I_0}{I_t} = -2\log_{10}(1-r) + 0.4343\sigma NL. \tag{1.14}$$

This result shows that, due to the reflection effect, an absorption different from zero is experimentally observed even in the absence of absorbing centers, that is when $N = 0$. Taking the more accurate multiple reflections effect between the two surfaces with refractive index mismatch between the sample and the medium, it is found that [4]:

$$I_t = pI_0 e^{-\sigma NL} \approx \frac{(1-r)^2}{(1-r^2)} I_0 e^{-\sigma NL} = \frac{2n_1 n_2}{n_1^2 + n_2^2} I_0 e^{-\sigma NL} \tag{1.15}$$

where p is the reflection factor. When this factor, or the refractive index dependence on the wavelength, is not known, the "parasitic" effect of reflection cannot be estimated. A technical solution is to take the measurement of the same material using two different thicknesses, if possible. In fact, considering two samples of thickness L_1 and L_2, respectively, we obtain:

$$T_1 = \frac{I_{t_1}}{I_0} = pe^{-\sigma NL_1} \qquad T_2 = \frac{I_{t_2}}{I_0} = pe^{-\sigma NL_2} \tag{1.16}$$

$$\frac{T_1}{T_2} = e^{-\sigma N(L_1 - L_2)} \qquad \sigma N = \alpha = -\frac{1}{L_1 - L_2}\ln\frac{T_1}{T_2}. \tag{1.17}$$

This way, it is shown that the two measurements enable to find the experimentally relevant features related to the absorbing centers: the cross section and the concentration, or the absorption coefficient.

To conclude this paragraph, in Figure 1.2, a typical absorption spectrum is reported with the absorbance in the vertical axis and the wavelength (in nanometer) in the horizontal axis.

It can be observed that the amount of absorbance (or optical density) changes by changing the wavelength, with a profile depending on the specific features of the investigated material. To carry out a meaningful interpretation of the spectrum, taking in due account the spectral profile and the electronic state distribution, the wavelength axis has to be changed into an axis of energy, E (usually reported in electronvolt, eV; $1\,\mathrm{eV} = 1.602 \cdot 10^{-19}$ J). To achieve this aim, it is useful to refer to the Planck–Einstein relation [9]: $E = h\nu$, where h is the Planck's constant ($6.626 \cdot 10^{-34}$ J·s $= 4.136 \cdot 10^{-15}$ eV·s). To convert the axis from wavelength to energy, one can use the formula:

$$E = h\nu = h\frac{c}{\lambda} \tag{1.18}$$

and the conversion equation

$$E(\mathrm{eV}) \cdot \lambda(\mathrm{nm}) = hc = 1240\,\mathrm{eV} \cdot \mathrm{nm} \tag{1.19}$$

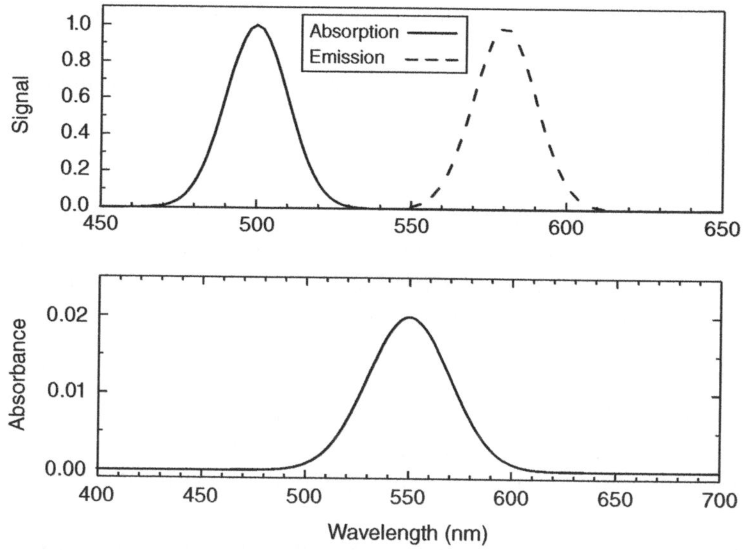

Figure 1.2 Bottom: Typical absorption spectrum registered as a function of wavelength. Top: Representative experimental absorption (continuous line) and emission (dashed line) spectra registered as a function of wavelength.

that enables to correlate a value of energy with the wavelength and vice versa. Another useful quantity in spectroscopy is the wavenumber, $\bar{\nu}$. This is defined by (wavelength)$^{-1}$, $1/\lambda$, and, using Eq. (1.18), it is shown that

$$\bar{\nu} = \frac{1}{\lambda} = \frac{E}{hc}. \tag{1.20}$$

The wavenumber is usually reported in units of cm^{-1}. Combining Eqs. (1.19) and (1.20), it is found that

$$\bar{\nu}(\text{cm}^{-1}) = \frac{E(\text{eV})}{h(\text{eV} \cdot \text{s})c\left(\frac{\text{cm}}{\text{s}}\right)} = \frac{E(\text{eV})}{1240 \cdot 10^{-7}(\text{eV} \cdot \text{cm})} = \frac{8066\, E(\text{eV})}{1\,\text{eV} \cdot \text{cm}}. \tag{1.21}$$

Concluding this paragraph, it is worth mentioning that the absorption phenomenon is one of the basic processes of the radiation–matter interaction and it is extended in a wide range of energy of the electromagnetic spectrum. The underling physical process is related to the specific atomic or molecular species absorbing the energy from the electromagnetic wave [8, 9]. The frequency range of interest for this chapter includes the visible (Vis) radiation and goes from the near infrared (NIR) to the ultraviolet (UV). In particular, the visible range in vacuum extends in frequency from about $3.8 \cdot 10^{14}$ to $7.5 \cdot 10^{14}$ Hz, in wavelength from 800 to 400 nm, and in energy from 1.6 to 3.1 eV [1].

1.1.2 Emission: Fluorescence and Phosphorescence

The absorption of light at a given wavelength λ by a sample is physically associated to an energy transfer from the radiation electromagnetic field to the electrons of the matter constituting the sample. In this phenomenon, the electrons are typically promoted from one energy level (ground state) to another level of higher energy (excited state). In the process, the electron system is put out of thermal equilibrium; so, each electron spontaneously tends to return to its initial energy level, releasing the acquired energy. In an ideal experiment, a stationary state can be observed in which by continuously illuminating the sample with a radiation at λ, in a given spatial direction, another radiation is emitted by the sample isotropically in space with wavelength $\lambda' > \lambda$. This phenomenon is known as photoluminescence and can sometimes be observed also by naked eye, illuminating a sample with high-energy photons, typically in the UV part of the spectrum, and revealing an emission in the visible range (an example of this effect is observed in the luminescence emergency panels, emitting light when they are in the dark after being illuminated by electric lamps or sunlight). It is worth underlining that typically, as shown in Figure 1.2, the absorbed radiation wavelength (energy) and the emitted wavelength (energy) are related by

$$\lambda_{abs} < \lambda_{em} \qquad (E_{abs} > E_{em}) \tag{1.22}$$

due to some internal processes subtracting energy to the electrons when they return to their ground state. The difference between the photon energy at which the absorption maximum amplitude occurs and the energy where the emission maximum amplitude occurs is called Stokes shift [2, 5].

In considering the phenomenology of the emission process, we can distinguish two kinds of phenomena. Both of them are related to the emission of light, but their time dependence is very different. In particular, one emission process lasts for a long time (more than μs) after the removal of exciting light (like in the luminescence emergency panels) and is called ***phosphorescence***. The other rapidly (less than 0.1 μs) decreases in amplitude as the exciting light is turned off and is known as ***fluorescence***. An experimental procedure to distinguish these processes consists in recording the emission amplitude at a given wavelength as a function of time after the excitation light is turned off. In Figure 1.3 is reported the result of a typical time-resolved photoluminescence experiment. The excitation light impinges on the sample up to the time 1 ms, to let the system reach a stationary state, and the emission is recorded at a selected wavelength λ_{em}. The emission intensity amplitude recorded is constant, in accordance to the stationary state. At the time $t_0 = 1$ ms, the excitation is rapidly removed by turning it off or by an opportune shutting system. The signal is

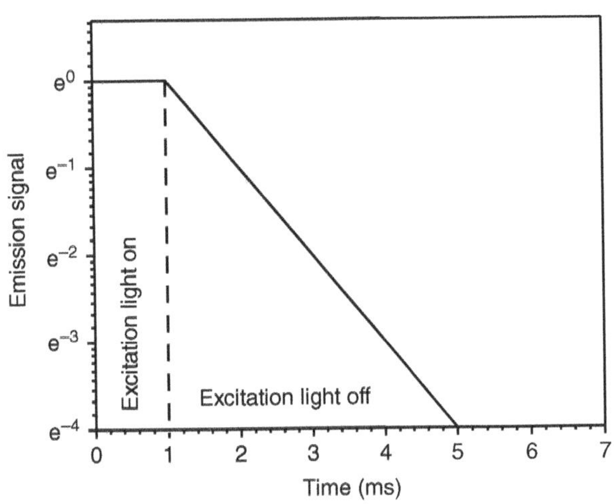

Figure 1.3 Typical decay curve as a function of time of the emission. For $t < 1$ ms, the exciting light is illuminating the sample, for $t \geq 1$ ms it is turned off.

then continuously recorded at $t > 1$ ms at λ_{em} and it is found that its amplitude decreases. In the figure is reported a typical decay with single exponential law:

$$I_{\lambda_{em}}(t) = I_{\lambda_{em}}(t_0)e^{-\frac{t-t_0}{\tau}}. \tag{1.23}$$

By this procedure, it is possible to record the characteristic time τ that defines the lifetime, or the decay time, of the photoluminescence, and obtain the time needed to reach a value of the emission amplitude 1/e of its stationary value in a given experiment [2, 10]. Values of τ up to ~10 ns are typical of fluorescence phenomena and larger lifetimes, up to 10^3 s, are characteristic of phosphorescence, enabling empirically to distinguish them [2, 5, 10]. In the following, a microscopic interpretation of the reported phenomena is given. It is worth mentioning that specific instrumentations are needed to carry out time-resolved photoluminescence [2, 5].

1.2 Microscopic Point of View

The empirical observations of absorption and emission phenomena contain very important information on the electronic and molecular properties of matter. In this view, it is fundamental to understand what kind of knowledge can be obtained from such experiments. In this paragraph, the theoretical bases that enable to determine microscopic features about the electronic states from the macroscopic measurements will be deepened.

1.2.1 Einstein Coefficients

A simplified model of the atom constituted by two nondegenerate energy levels is assumed to evaluate the interaction with radiation. As reported in Figure 1.4, the lower energy level is E_1 and the higher energy level is E_2. This system interacts with the thermal equilibrium radiation field at temperature T.

According to Planck's theory, the energy distribution of the radiation is given by [8, 11, 12]

$$\varrho(\nu) = \frac{8\pi h\nu^3}{c^3}\frac{1}{e^{\frac{h\nu}{kT}} - 1} \tag{1.24}$$

where $\rho(\nu)$ is the density of energy for unit volume and unit frequency interval, h the Planck's constant, ν the radiation frequency, c the speed of light, k the Boltzmann's constant ($1.38 \cdot 10^{-23}$ J K^{-1}), and T the absolute temperature.

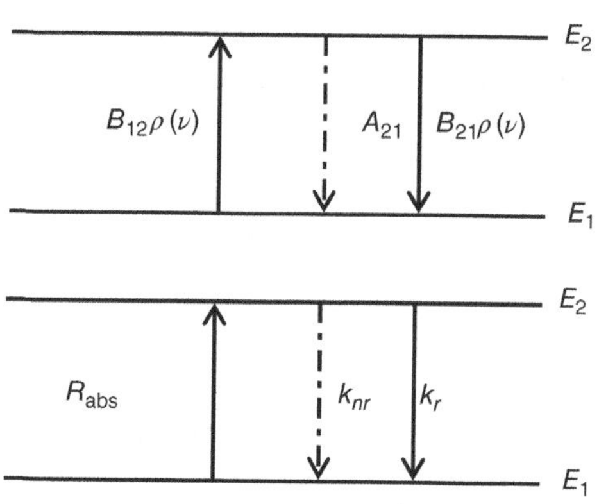

Figure 1.4 Top: Schematic representation of two energy-level atom. The arrows represent the absorption and emission transition processes whose probability is given by Einstein's coefficients A_{21}, B_{12}, B_{21} and the density of radiation $\rho(\nu)$. Bottom: Schematic representation of two energy levels of an atom. The arrows represent the excitation and relaxation transitions. The absorption rate is R_{abs}, the radiative emission rate is k_r, and the non-radiative transition rate is k_{nr}.

The interaction between a density N of atoms for unit of volume and the radiation field causes an exchange of energy with those electromagnetic waves' modes having frequency related to the atom's energy levels' separation by the equation $E_2 - E_1 = h\nu$. As a consequence, based on Einstein's treatment, the following processes can occur [8, 13]:

- Transition from the state E_1 to the state E_2, stimulated by the absorption of a photon; based on Einstein's theory, the rate of this process is given by

$$N_1 B_{12} \varrho(\nu) \tag{1.25}$$

where N_1 is the density of atoms (population) in the lower energy state.
- Transition from the state E_2 to the state E_1, stimulated by the emission of a photon; the rate is given by

$$N_2 B_{21} \varrho(\nu) \tag{1.26}$$

where N_2 is the density of atoms in the upper energy state.
- Spontaneous transition from the state E_2 to the state E_1 with a rate

$$N_2 A_{21}. \tag{1.27}$$

The Einstein's coefficients A_{21}, B_{12}, and B_{21} have been used. In particular, it is worth observing that B_{12} and B_{21} are related to the presence of the field (stimulated processes of absorption and emission, respectively), whereas A_{21} is present also without electromagnetic field and is related to spontaneous emission. This term is related to the radiative emission lifetime introduced in the previous paragraph and, in detail, it is the reciprocal of the lifetime at low temperature, $A_{21} = 1/\tau$ [13]. At thermal equilibrium, the population of atomic states should reach a stationary condition and it is expected that

$$\frac{\mathrm{d}N_1}{\mathrm{d}t} = -\frac{\mathrm{d}N_2}{\mathrm{d}t} = 0 \tag{1.28}$$

and, based on the above reported processes, one obtains

$$\frac{\mathrm{d}N_1}{\mathrm{d}t} = -N_1 B_{12} \varrho(\nu) + N_2 [B_{21} \varrho(\nu) + A_{21}] = -\frac{\mathrm{d}N_2}{\mathrm{d}t} = 0 \tag{1.29}$$

and the relation

$$\frac{N_2}{N_1} = \frac{B_{12} \varrho(\nu)}{[B_{21} \varrho(\nu) + A_{21}]}. \tag{1.30}$$

The Boltzmann distribution at thermal equilibrium in a two-level system without degeneracy predicts that [14]

$$\frac{N_2}{N_1} = \mathrm{e}^{-\frac{h\nu}{kT}}. \tag{1.31}$$

Equating (1.30) and (1.31) and solving with respect to $\rho(\nu)$, it is found that

$$\varrho(\nu) = \frac{A_{21} \mathrm{e}^{-\frac{h\nu}{kT}}}{\left[B_{12} - B_{21} \mathrm{e}^{-\frac{h\nu}{kT}}\right]}. \tag{1.32}$$

This distribution of energy density in the electromagnetic field should coincide with the Planck's law at thermal equilibrium. As a consequence, by equating (1.32) and (1.24), it is found that

$$\frac{B_{12}}{A_{21}} = \frac{B_{21}}{A_{21}} = \frac{c^3}{8\pi h\nu^3} \tag{1.33}$$

and, finally

$$A_{21} = \frac{8\pi h\nu^3}{c^3} B_{21} \tag{1.34}$$

$$B_{12} = B_{21}. \tag{1.35}$$

In the case of degenerate energy levels with degeneracy g_1 and g_2, it is shown that (1.35) transforms into

$$g_1 B_{12} = g_2 B_{21} \tag{1.36}$$

whereas (1.34) remains unchanged [8, 13].

Using the quantum mechanical treatment of the interaction between radiation and matter and, in particular, neglecting any magnetic contribution and considering the electric dipole approximation, the atom can be described by a dipole moment

$$\boldsymbol{\mu} = e\boldsymbol{r} \tag{1.37}$$

where e is the electron charge ($1.602 \cdot 10^{-19}$ C) and $\boldsymbol{r}$ is its position vector with respect to the atomic nucleus. The time-dependent perturbation theory enables to show that the probability to populate the higher energy level of the atom E_2 (multiplied by unit frequency interval), starting from the level with energy E_1, is given by [8, 9, 13]:

$$P_2(t) = \frac{2\pi}{\hbar^2} |\langle E_1 \mid V \mid E_2 \rangle|^2 t \tag{1.38}$$

where V is the interaction energy between the electric field and the electric dipole moment:

$$V = -\boldsymbol{\mu} \cdot \boldsymbol{\mathcal{E}} \tag{1.39}$$

and $\hbar = h/2\pi$. Considering a linearly polarized lightwave with electric field of amplitude $\mathcal{E}_0$, wavevector $\boldsymbol{k}$, and angular frequency $\omega = 2\pi\nu$

$$\boldsymbol{\mathcal{E}} = \mathcal{E}_0 \cos(\omega t - \boldsymbol{k} \cdot \boldsymbol{r}) \tag{1.40}$$

the probability of population of the excited state per unit of time, coinciding with the transition rate, is then given by

$$R_{12} = \frac{2\pi}{\hbar^2} |\langle E_1 \mid V \mid E_2 \rangle|^2 = \frac{2\pi}{\hbar^2} |\langle E_1 \mid \boldsymbol{\mu} \cdot \boldsymbol{\mathcal{E}} \mid E_2 \rangle|^2 = \frac{2\pi}{\hbar^2} \mathcal{E}_0^2 |\langle E_1 \mid \mu \mid E_2 \rangle|^2 = \frac{2\pi}{\hbar^2} \mathcal{E}_0^2 \mu_{12}^2 \tag{1.41}$$

in which the electric dipole matrix element μ_{12} relative to the considered atomic states in the direction of the external field has been introduced. Using (1.29), it is possible to find that

$$B_{12}\varrho(\nu) = R_{12} = \frac{2\pi}{\hbar^2} \mathcal{E}_0^2 \mu_{12}^2. \tag{1.42}$$

This result shows a connection between the macroscopic empiric quantities and the microscopic ones related to the quantum mechanical states of the electron in the atom. In particular, it is shown that the transition probability is related to the electric dipole matrix element μ_{12}.

Considering that in vacuum [8, 13]

$$\varrho(\nu) = \varepsilon_0\, \mathcal{E}_0^2 \tag{1.43}$$

ε_0 being the permittivity of free space (8.854×10^{-12} kg^{-1} m^{-3} s^4 A^2), it is possible to find that

$$B_{12} = \frac{2\pi}{\hbar^2 \varepsilon_0}\, \mu_{12}^2. \tag{1.44}$$

Furthermore, since the intensity of radiation and the energy density are related by [1, 8, 13]

$$I(\nu) = \varrho(\nu)c \tag{1.45}$$

using (1.42), it is found that the transition rate between the atom's energy states is given by

$$R_{12} = B_{12}\varrho(\nu) = B_{12}\frac{I(\nu)}{c} \tag{1.46}$$

a connection with the intensity of radiation is made explicit now. The rate of energy absorbed per unit of volume by the atom from the electromagnetic field can then be written as

$$N_1 R_{12} h\nu = N_1 B_{12}\frac{I(\nu)}{c} h\nu. \tag{1.47}$$

By assuming that all the atoms reside in the N_1 state, this is the energy lost by the radiation field. If a sample of thickness dx is considered, the energy lost for unit area by the electromagnetic wave is then

$$-N_1 B_{12}\frac{I(\nu)}{c} h\nu \mathrm{d}x. \tag{1.48}$$

By recalling the Lambert–Beer law in differential form from (1.1), it is shown that

$$\mathrm{d}I(\nu) = -N_1\sigma(\nu)I(\nu)\mathrm{d}x = -\alpha(\nu)I(\nu)\mathrm{d}x = -N_1 B_{12}\frac{I(\nu)}{c} h\nu \mathrm{d}x \tag{1.49}$$

where the frequency dependence has been inserted, and finally one obtains

$$\alpha(\nu) = N_1 B_{12}\frac{h\nu}{c}. \tag{1.50}$$

This is a more direct connection between experimental parameters and the microscopic ones. In fact, by using (1.44), it is found that

$$\alpha(\nu) = N_1 \frac{2\pi}{\hbar^2 \varepsilon_0}\mu_{12}^2 \frac{h\nu}{c} = N_1 \frac{8\pi^3 \nu}{hc\varepsilon_0}\mu_{12}^2. \tag{1.51}$$

Integrating (1.51) over the entire frequency range pertaining to the given atomic (or molecular) species gives

$$\int \alpha(\nu)\mathrm{d}\nu = N_1 \frac{8\pi^3 \hat{\nu}}{hc\varepsilon_0}\mu_{12}^2 \tag{1.52}$$

where the central absorption frequency $\hat{\nu}$ has been introduced that is usually related to the maximum of the absorption band. This expression shows that absorption measurements give information on the electric dipole matrix element μ_{12} once the concentration of absorbing centers N_1 is known. The connection reported in (1.41) with the electronic states' wave functions enables to obtain information about them and vice versa, i.e. once the dipole matrix element is known, from the integral of the absorption band, the concentration of absorbing centers can be found.

1.2.2 Oscillator Strength, Lifetime, Quantum Yield

In the previous paragraph, we have determined theoretical quantities relating the atomic wavefunctions of the energy levels to spectroscopic observables. In the simplified model of the atom with a single electron, it is considered that this latter can oscillate in a harmonic potential well. The atomic system is then a charged harmonic oscillator. The wavefunctions of this system enable to evaluate the electric dipole matrix element μ_{12} and to determine the theoretical integrated absorption reported by (1.52). It can be shown that the expected value of integrated absorption is [5, 11]:

$$\int \alpha(\nu)\mathrm{d}\nu = \frac{N_{\mathrm{A}}e^2}{1000\mathrm{c.m}} \ \mathrm{sec}^{-1}\mathrm{cm}^{-1}\mathrm{mol}^{-1}\mathrm{L} \tag{1.53}$$

where N_{A} is Avogadro's number, c is the speed of light, and m is the electron mass ($9.109 \cdot 10^{-31}$ kg). Equation (1.53) gives a numerical value that can be compared with experimental results. This comparison gives origin to the quantity called *oscillator strength* and usually given by f:

$$f = \frac{\left[\int \alpha(\nu)\mathrm{d}\nu\right]_{\mathrm{experimental}}}{\frac{N_{\mathrm{A}}e^2}{1000\mathrm{c.m}} \ \mathrm{sec}^{-1}\mathrm{cm}^{-1}\mathrm{mol}^{-1}\mathrm{L}}. \tag{1.54}$$

The oscillator strength is a dimensionless quantity characterizing the transition between the two considered energy levels E_1 and E_2. By introducing [5, 11]

$$\frac{\int \alpha(\nu)\mathrm{d}\nu}{c} = \int \alpha(\overline{\nu})\mathrm{d}\overline{\nu} \tag{1.55}$$

it is shown that

$$f = 4.33 \cdot 10^{-9} \left[\int \alpha(\overline{\nu})\mathrm{d}\overline{\nu}\right]_{\mathrm{experimental}}. \tag{1.56}$$

Expected values of f are lower equal than unity and on decreasing of the probability of the absorption process decreases too. This feature is linked to the selection rules that highlight those quantum transitions between energy levels giving an electric dipole matrix element different from zero [5].

Another form of the oscillator strength is [8, 15]

$$f = \frac{8\pi^2 m\nu}{3he^2}\mu_{12}^2 \tag{1.57}$$

which, on the basis of (1.34), (1.35) and (1.44), can be written

$$f = \frac{m\varepsilon_0 c^3}{24\pi^2 e^2 \nu^2} A_{21}. \tag{1.58}$$

Considering that $A_{21} = 1/\tau$ [13], as stated above, it is shown that

$$f = \frac{m\varepsilon_0 c^3}{24\pi^2 e^2 \nu^2}\frac{1}{\tau} = \frac{m\varepsilon_0 c\,\lambda^2}{24\pi^2 e^2}\frac{1}{\tau} \tag{1.59}$$

connecting the oscillator strength to the radiative emission lifetime at low temperature. Furthermore, using (1.52) to determine μ_{12}, Eq. (1.57) becomes

$$f = \frac{8\pi^2 m\nu}{3he^2}\left[N_1\frac{8\pi^3\nu}{hc\varepsilon_0}\right]^{-1}\int\alpha(\nu)\mathrm{d}\nu \tag{1.60}$$

giving

$$\int\alpha(\nu)\mathrm{d}\nu = \frac{3\pi e^2}{mc\varepsilon_0}N_1 f. \tag{1.61}$$

This formula relates the integrated absorption coefficient to the concentration of absorbing centers through the oscillator strength. In particular, given a concentration N_1, the area of the absorption curve is higher at larger oscillator strength. Furthermore, both the oscillator strength and the concentration of absorbing centers can be experimentally determined from this formula once one of the two parameters is known. This result shows the relevance of the absorption measurements and once more the exploitability of this experimental technique to determine microscopic information about the matter.

In general, the relation between absorption and emission processes can be described by

$$I_{\mathrm{em}}(\nu) = \eta I_0(\nu)\left(1-10^{-\varepsilon CL}\right) = \eta I_0(\nu)\left(1-\mathrm{e}^{-\alpha L}\right) \tag{1.62}$$

where the absorption effect is reported using (1.5) and (1.8) and differences in absorption and emission frequency ν have been neglected, assuming that one is the inverse process of the other. The above formula shows that the number of emitted photons by a sample of thickness L depends on the number of photons impinging on the sample, I_0, and the number of them giving absorption effect, as evidenced by the terms in parenthesis. The efficiency of emission is determined by η, which is called (external) radiative quantum efficiency or quantum yield [2, 10, 16]. It is worth noting that in the case of low absorption effect, $\varepsilon CL \ll 1$, $\alpha L \ll 1$, the emission is proportional to the concentration of absorbing species. The quantum efficiency of photoluminescence is defined as the ratio of the number of emitted photons to the number of absorbed photons

$$\eta = \frac{\text{number of emitted photons}}{\text{number of absorbed photons}}. \tag{1.63}$$

In the interaction process between radiation and a two-level atom, in which the absorption drives the electron from the starting state E_1 to the final state E_2 at higher energy, as reported in Figure 1.4, not all of the absorbed photons give rise to an emitted photon since other processes could drive the excited electron back to the E_1 state. Based on the above considerations, it is clear that $\eta \leq 1$. To go deeper into the connection between η and other physical parameters, it is useful to consider the simplified energy level scheme of the two-level atom reported in Figure 1.4. The absorption process, in which the electron is promoted to the E_2 state, is represented by the vertical arrow connecting E_1 and E_2 states. This process has a rate, probability per unit of time, R_{abs}. The system in the excited state is out of thermal equilibrium because typically $E_2 - E_1 > kT$. As a consequence, the electron returns to the lower energy level. This relaxation process could occur by the emission of a photon with frequency $\nu = (E_2 - E_1)/h$, denoted as radiative process with a rate k_{r}. Furthermore, the electron could relax without emission of photons, not radiatively, by exchanging energy with its environment with a rate k_{nr}. The rate equation for the variation of the population N_1 of the state with energy E_1 can be written as

$$\frac{\mathrm{d}N_1}{\mathrm{d}t} = -R_{\mathrm{abs}}N_1 + (k_{\mathrm{r}} + k_{\mathrm{nr}})N_2 \tag{1.64}$$

where the population of the excited state N_2 has been introduced. This equation is a balance between absorbed photons and relaxation from the excited state through radiative (emitted photons) and non-radiative processes. Under stationary equilibrium conditions, there will be no change in the populations with time and it is found that

$$\frac{dN_1}{dt} = 0 = -R_{abs}N_1 + (k_r + k_{nr})N_2; \qquad R_{abs}N_1 = (k_r + k_{nr})N_2. \tag{1.65}$$

It is reasonable to assume that the number of absorbed photons is proportional to the number of atoms being excited by the radiation, so it can be written

$$I_{abs} \propto R_{abs}N_1 = (k_r + k_{nr})N_2. \tag{1.66}$$

On the other hand, the number of emitted photons is proportional to the number of atoms relaxing radiatively from the E_2 state, and it is possible to assume

$$I_{em} \propto k_r N_2. \tag{1.67}$$

From the above considerations, it is found that the quantum efficiency is given by

$$\eta = \frac{I_{em}}{I_{abs}} = \frac{k_r}{k_r + k_{nr}}. \tag{1.68}$$

Introducing now the rate equation for the excited state

$$\frac{dN_2}{dt} = +R_{abs}N_1 - (k_r + k_{nr})N_2 \tag{1.69}$$

it is found that after the stationary regime is attained, when the population N_2 has stabilized, if the radiation field is suddenly removed at the time t_0 and, as a consequence, R_{abs} is suddenly put to zero, the rate equation of the excited state becomes

$$\frac{dN_2}{dt} = -(k_r + k_{nr})N_2 \tag{1.70}$$

whose solution for $t \geq t_0$ is

$$N_2(t) = N_2(t_0)e^{-(k_r + k_{nr})(t - t_0)} \tag{1.71}$$

$N_2(t_0)$ being the stationary population of the excited state. Comparing Eq. (1.71) with (1.23), it is found that

$$\frac{1}{\tau} = (k_r + k_{nr}) \tag{1.72}$$

and finally, it is found that

$$\eta = \frac{k_r}{k_r + k_{nr}} = k_r \cdot \tau. \tag{1.73}$$

It is worth considering that, typically, the non-radiative processes are active at high temperature when the two-level system is coupled to other degrees of freedom of the atomic system. At variance, at very low temperature, $k_{nr} = 0$. On this basis, it is found that at $T = 0\,\text{K}$ (or at low enough temperature)

$$\frac{1}{\tau_0} = k_r \tag{1.74}$$

$$\eta = 1 \tag{1.75}$$

and a measure of the emission lifetime at low temperature, τ_0, is a direct estimate of k_r. It is expected that the quantum efficiency decreases on increasing the temperature due to the activation of the non-radiative relaxation processes [2, 5, 10]. In addition, the intrinsic or natural lifetime, $\tau_0 = 1/k_r$, can in general be determined at any temperature by the measure of the quantum efficiency and of the emission lifetime using (1.73)

$$\tau_0 = \frac{1}{k_r} = \frac{\tau}{\eta} \tag{1.76}$$

this shows that typically the lifetime decreases on increasing the temperature. In general, the lifetime concept and the quantum efficiency are relevant aspects in the dynamic study of emitting centers and in the characterization of optoelectronic devices. They give useful information on the coupling of the system to its neighborhood.

1.2.3 Vibronic States: Homogeneous and Inhomogeneous Lineshape

The absorption and emission processes have usually a dependence on the frequency ν of the exciting electromagnetic field. In particular, the amplitude of the absorption coefficient or the intensity of the emitted radiation changes by changing ν. This effect is related to the specific features of the centers interacting with the electromagnetic field through the distribution of the density of their electronic states as a function of ν or of the electron's energy [8, 9]. In a given experiment, where the intensity of transmitted or emitted light is measured as a function of the frequency, a profile of I (ν) is recorded. This profile is usually called lineshape of the spectrum. In this context, it is possible to distinguish two contributions to the lineshape: homogeneous and inhomogeneous. This classification is related to the origin of the physical process contributing to the spectral width of the lineshape [2, 6, 11, 13]. It is now useful to consider a wider class of systems interacting with the radiation: the molecules. In this case, in addition to the degrees of freedom of the electrons, the degrees of freedom of the nuclei are present. In particular, not only the electrons can move and change their energy, but also the nuclei can move relative to each other, adding a contribution to the total energy of the system. The features affecting the lineshape can now be listed, collecting them in the above-cited two classes [8, 13].

Homogeneous contributions

1) Once in the excited state, the electron should return to the starting state in a time that corresponds to the emission lifetime. This time, based on the Heisenberg uncertainty principle, is linked to an indetermination in the energy value of the excited state. Such effect gives rise to a spread in the energy of the transition and, as a consequence, a width of the lineshape. In addition, in the case of a gas of atoms or molecules, the collisions between the particles give rise to changes in the permanence time in the excited state and, therefore, impose changes to the lifetime and width of the lineshape.
2) When the system is a molecule, the vibrational degrees of freedom affect the distribution of energy levels and give rise to specific lineshapes (see Eq. 1.92).

Inhomogeneous contributions

1) When the atoms or molecules of a given species are embedded in an environment, the different spatial distributions of the other atoms and molecules originate local electric fields that cause a different energy levels' separation and distribution among the centers under investigation.

2) When atoms or molecules move in space, their thermal velocity with respect to an observer fixed in space is randomly spread according to the Maxwell–Boltzmann distribution, giving rise to frequency variation of the absorbed or emitted light due to the Doppler effect.

Typically, the homogeneous effect is "intrinsic" to a given lineshape and the inhomogeneous effect induces a replica of the intrinsic lineshape, overall giving a broader lineshape that is the envelope of the many replicas. Neglecting the vibrations, the more common homogeneous lineshape is the Lorentzian, whereas the inhomogeneous lineshape is the Gaussian. The intermediate lineshape that is a Gaussian convolution of Lorentzian lineshapes is known as Voigt lineshape. The analytic forms of these functions are:

$$L(\nu) = \frac{1}{\pi}\frac{A}{(\nu - \nu_0)^2 + A^2} \qquad \text{Lorentzian} \tag{1.77}$$

this lineshape is centered at ν_0 and is characterized by a full width at half maximum (FWHM) of the amplitude equal to $2A$;

$$G(\nu) = \frac{1}{\sqrt{2\pi}\sigma} e^{-\frac{(\nu-\nu_0)^2}{2\sigma^2}} \qquad \text{Gaussian} \tag{1.78}$$

this is centered at ν_0 and has FWHM $= 2\sqrt{2\ln(2)}\sigma$; finally, the Voigt function can be defined as

$$V(\nu) = \int_{-\infty}^{\infty} G(\nu')L(\nu - \nu')\mathrm{d}\nu' \qquad \text{Voigt} \tag{1.79}$$

where the Gaussian and Lorentzian contributions have been introduced. The final shape of the Voigt function depends on the balance between the FWHM of the two composing contributions.

To go deeper into the homogeneous lineshape features, the case of a molecular system has to be considered. It is then necessary to determine its electronic states in detail. First of all, the molecule is a many-body system constituted by the electrons, the nuclei (it is useful to consider this unit for spectroscopic aims, associating the constituent parts: protons and neutrons, motion) and their motion and interaction. The more general Hamiltonian for this system is

$$H = K_\mathrm{e} + K_\mathrm{n} + U_\mathrm{ee} + U_\mathrm{en} + U_\mathrm{nn} \tag{1.80}$$

where the subscripts label electrons' (e) and nuclei's (n) kinetic and potential energies, in order: the electrons kinetic energy; the nuclei kinetic energy; the electron–electron interaction potential energy; the electron–nuclei interaction potential energy; the nuclei–nuclei interaction potential energy. This Hamiltonian is really complex and it can be simplified considering that the proton mass is $>10^3$ times larger than the electron mass. As a consequence, it can be assumed that the electrons' motion is much faster than the nuclei's motion and that each electron explores a slowly varying configuration of the nuclei. Such "adiabatic" approximation is usually known as Born–Oppenheimer approximation [5, 8, 15, 17]. It assumes that the nuclei are practically fixed in space during the electrons' motion, influencing the latter's energy statically. On the other side, the nuclei experience the electrons' motion as an average value since the latter instantaneously adapt themselves to the given nuclear configuration. Such rapid rearrangement imposes that the nuclear energy is essentially related to the instantaneous nuclei positions. This enables to define a nuclear configuration and associate to it the overall electrons–nuclei system energy. Under this approximation, a configurational energy can be introduced

$$U_\mathrm{conf} = \overline{K_\mathrm{e} + U_\mathrm{ee} + U_\mathrm{en}} + U_\mathrm{nn} \tag{1.81}$$

where the first three terms, depending on electrons' motion, are an averaged (over the electrons' coordinates and motion) contribution for a given spatial configuration of nuclei. The modified Hamiltonian of the overall system is then

$$H = K_{\mathrm{n}} + U_{\mathrm{conf}} \tag{1.82}$$

where the configurational potential energy and the kinetic energy of the nuclei are explicitly reported. In the harmonic approximation, U_{conf} is simplified as a quadratic function of the generalized coordinate Q [15, 18]. This is the internuclear distance in the most simple diatomic molecule or a normal coordinate in the case of a polyatomic molecule [5, 7]. It can be written

$$U_{\mathrm{conf}} = \frac{1}{2}m\omega^2Q^2 + U_{\mathrm{g}} \tag{1.83}$$

where the equilibrium position (minimum of the potential energy) $Q = 0$ has been chosen for simplicity and m and ω are the mass and frequency of the oscillation mode whereas U_{g} is a constant representing the minimum energy of the given electronic configuration (related to spin and orbital motion of electrons). The complete Hamiltonian then becomes

$$H = K_{\mathrm{n}} + \frac{1}{2}m\omega^2Q^2 + U_{\mathrm{g}} \tag{1.84}$$

that has the harmonic oscillator form [9]. The solution of the Schrodinger equation using this Hamiltonian gives eigenvalues and eigenfunctions of the nuclear coordinate [9]. The total energy of the system based on (1.81) will include the electrons' energy. A representation of this energy is shown in Figure 1.5.

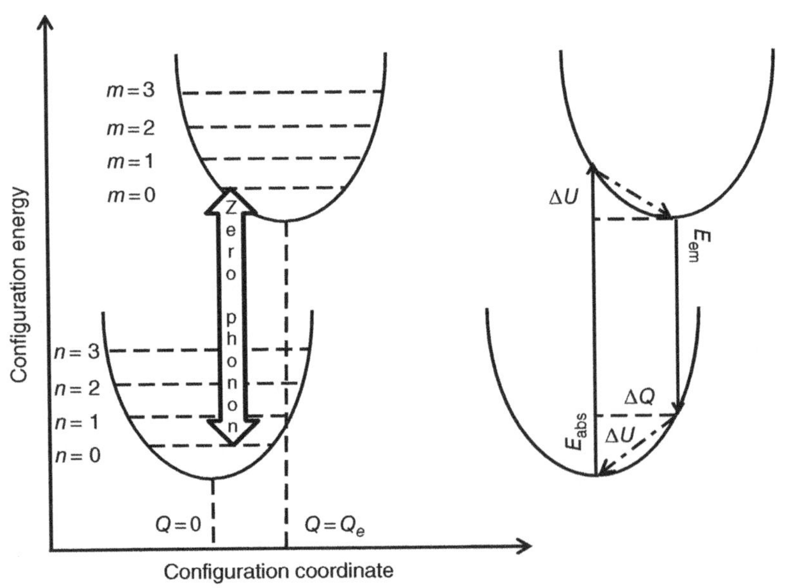

Figure 1.5 On the left: Schematic representation of the configuration total energy for the ground (lower) and the excited (upper) electronic states of a molecule. The continuous lines refer to the total energy, the horizontal dashed lines refer to the nuclei vibrational energy levels, marked by their quantum numbers. The equilibrium configuration coordinates are $Q = 0$ and $Q = Q_{\mathrm{e}}$ for the ground and excited states, respectively. The zero-phonon line is highlighted by the double arrow. On the right: The excitation (E_{abs}) and emission (E_{em}) pathways are reported by continuous arrows; ΔQ and ΔU highlight the change in configuration coordinate and potential energy, respectively; the dash-dotted arrows mark the nuclear relaxation processes.

Each parabola schematizes the energy of the electronic configuration. In the lower parabola, representing the ground state of the system, the electrons have given spin and orbital angular momenta. In the upper parabola, schematizing the first excited state, the spin and orbital angular momenta have in general changed. The parabolic approximation gives the shape of the total energy, and the solutions of the harmonic oscillator define the possible quantized energy for the nuclear motion (represented by dashed lines in the figure). This means that not all the energy values are possible but just those marked by the dashed lines. Anyway, the energy of vibration is much lower than the energy of electrons' interaction, and almost a continuous change can be considered within the given parabola, schematizing the electron configuration.

Considering the excited energy level, it is characterized, in general, by a configuration coordinate of equilibrium $Q = Q_e$ different with respect to the electronic ground state. Assuming again a harmonic approximation, the total energy of the excited state can be written as

$$U'_{conf} = \frac{1}{2}m\omega^2 Q^2 + U_g + E - FQ \tag{1.85}$$

where E is the absorption energy and the linear electron–phonon coupling approximation has been applied, represented by the term with F [15, 18]. The parameter F shows that the configuration coordinate Q of the minimum potential energy in the ground state and that in the excited state are different due to the connection between this energy and the electronic configuration (analytically, (1.85) is the formula of a parabola whose vertex position in the energy-Q space is different from that of the parabola in (1.83) but they have the same concavity). By changing the electronic configuration, the potential energy changes and also the nuclear configuration adapts itself to a novel equilibrium position. By finding the minima of (1.83) and (1.85), it is demonstrated that the difference in the abscissa of minima is

$$\Delta Q = Q_e = \frac{F}{m\omega^2}. \tag{1.86}$$

In the linear electron–phonon approximation, the oscillation frequency ω of nuclei is the same in the ground and in the excited state. This is described by parabolic potentials with the same concavity in both states. As reported in Figure 1.5, this also implies that moving by ΔQ far from the minima, the same energy change ΔU is found in both states. By the scheme reported in the above figure, it is found that

$$\Delta U = \frac{E_{abs} - E_{em}}{2} \tag{1.87}$$

where E_{abs} and E_{em} are the absorption and emission energies and the energy difference ($E_{abs} - E_{em}$) between the maximum of the absorption profile and that of the emission is called *Stokes shift*. Usually, it is considered that the electronic transition occurs in a much shorter time than the nuclear motion, so the nuclear configuration coordinate is unchanged during both the absorption and emission processes [5]. This statement is known as *Franck–Condon principle* [5]. From (1.83), it is found that for $Q = Q_e$ the configuration energy change of the ground state is

$$\Delta U = \frac{1}{2}m\omega^2 {Q_e}^2 = \frac{1}{2}\frac{F^2}{m\omega^2}. \tag{1.88}$$

The same absolute value is found for the excited state. This value corresponds to the relaxation energy after the absorption or the emission processes. In particular, based on the Franck–Condon principle, in any transition, the nuclei's coordinate and momentum are unchanged and a sudden electronic configuration change occurs [5]. Once the electronic state has changed, the nuclei relax toward the minimum energy of vibration compatible with the system's temperature. At 0 K, this transition is toward the minimum vibrational energy, that is the minimum of the parabolas describing the energy of nuclei (dash-dotted arrows in Figure 1.5). The *Stokes shift* marks the presence of a non-null electron–phonon coupling. If the associated energy difference is zero, the two electronic states have the minima of the potential curves for the same value of Q, the two parabolas are vertically aligned, and transition between the same vibrational levels occurs without energy difference between absorption and emission. To go deeper into this aspect, the Born–Oppenheimer approximation is considered again. The wavefunction ψ that solves the Schrodinger equation can be factorized, separating the nuclei's and the electrons' contributions

$$\psi(\mathbf{r}, \mathbf{Q}) = \varphi(\mathbf{r}, \mathbf{Q})\vartheta(\mathbf{Q}) \tag{1.89}$$

where φ, ϑ refer to the electronic and nuclear wavefunctions, respectively, the first being parametrically dependent on $\mathbf{Q}$, the nuclear coordinate, and the latter being independent on the electronic coordinate $\mathbf{r}$. Furthermore, the electrons' wavefunction, on the basis of the Condon approximation, depends on the average value of the nuclear coordinate [5, 15]

$$\varphi(\mathbf{r}, \mathbf{Q}) \approx \varphi(\mathbf{r}, \mathbf{Q}^{\text{avg}}). \tag{1.90}$$

To evaluate the probability of transition between the two states reported in Figure 1.5, the dipole matrix element introduced in (1.41) should be considered. In particular, the value

$$\mu_{12}^2 = |\langle E_1 \mid \mu \mid E_2\rangle|^2 = |\langle \psi_2(\mathbf{r}, \mathbf{Q}) \mid er \mid \psi_1(\mathbf{r}, \mathbf{Q})\rangle|^2 \tag{1.91}$$

has to be determined between the ground state designated by the energy E_1 and the excited state E_2. By considering (1.90) and the separation of nuclear and electronic coordinates, it is found that [5, 15]:

$$\mu_{12}^2 = \left|\left\langle \varphi_2\left(\mathbf{r}, \mathbf{Q}_2^{\text{avg}}\right) \mid er \mid \varphi_1\left(\mathbf{r}, \mathbf{Q}_1^{\text{avg}}\right)\right\rangle\right|^2 \cdot \left|\left\langle \vartheta_2^m(\mathbf{Q}) \mid \vartheta_1^n(\mathbf{Q})\right\rangle\right|^2 = |D_{12}|^2 \cdot |M_{nm}|^2 \tag{1.92}$$

where the indices 1 and 2 refer to the lower and higher electronic energy levels and the indices n and m refer to the vibrational levels of the nuclei. It is worth underlining that the overall energy of the considered molecular system is the combination of the electrons' and nuclei's interactions. The latter is determined by the vibrational state marked by the quantum numbers reported in Figure 1.5. Overall, the transition involves electronic states and nuclear vibrational quantum states; so, the transition is called vibronic transition [5]. Equation (1.92) shows that the first factor gives the amplitude of the probability, being linked to the oscillator strength, and the second factor, $|M_{nm}|^2$, is responsible for the shape of the band for the given transition of absorption or emission. This is known as the *Franck–Condon* factor [5, 15]. In particular, since it is related to the harmonic oscillator solutions of the Schrodinger equation, this factor is null if $n \neq m$ for a given oscillator [9]. But because the solutions considered pertain to different equilibrium configuration of oscillators, with the same frequency, the orthonormal wavefunctions of the harmonic oscillators $\vartheta_1^n(\mathbf{Q}), \vartheta_2^m(\mathbf{Q})$ are eigenstates of the "same" oscillator but are referred to the different equilibrium (central) positions $Q = 0$, $Q = Q_e$, respectively; so their integrals M_{nm} could in general differ from zero. Furthermore, once a wavefunction $\vartheta_1^n(\mathbf{Q})$ is chosen, with fixed n, it can be decomposed by the full set of $\vartheta_2^m(\mathbf{Q})$ considering all possible values of m, because the latter set is a basis for the space of wavefunctions [5]. It is then found that the overall transition probability from a starting state $\varphi_1\left(\mathbf{r}, \mathbf{Q}_1^{\text{avg}}\right) \vartheta_1^n(\mathbf{Q})$ to any of the excited vibronic states $\varphi_2\left(\mathbf{r}, \mathbf{Q}_2^{\text{avg}}\right) \vartheta_2^m(\mathbf{Q})$ is given by

$$\left(\mu_{12}^2\right)_n = \sum_{m=0}^{\infty} |D_{12}|^2 \cdot |M_{nm}|^2 = |D_{12}|^2 \sum_{m=0}^{\infty} |M_{nm}|^2 = |D_{12}|^2 \tag{1.93}$$

the summation being equal to 1 due to the orthonormality condition of used wavefunctions [5, 9, 15]. This finding explains that the transition probability from a given vibrational state is dependent on the electronic part of (1.92) whereas the nuclear part is responsible for the shape. This result gives origin to the lineshape of the absorption or emission band, since the same considerations can be done inverting the initial state and because $|M_{nm}|^2 = |M_{mn}|^2$. In particular, the homogeneous lineshape for a molecular species is dependent on the electron–phonon coupling and on its strength through $|M_{nm}|^2$. It is also found that absorption and emission lineshapes are symmetric with respect to the transition energy individuated by the M_{00} element, known as the *zero-phonon line* (ZPL, reported symbolically in Figure 1.5) [5, 15]. This transition is from the electronic ground state without vibration excitation to the excited electronic state without vibration excitation and vice versa, and it coincides for absorption and emission. In general, the vibration quantum $\hbar\omega$ is much larger than the thermal energy kT at room temperature; so, only the ground vibrational level, $n = 0$, is populated in the ground electronic state. In this case, the most relevant terms for the evaluation of the transition are $|M_{0m}|^2$ and for the absorption process at energy $E = E_{00} + m\,\hbar\omega$, one can write

$$A(E) = A(E_{00} + m\,\hbar\omega) \propto |M_{0m}|^2 \tag{1.94}$$

giving the amplitude of absorption for the transition from the ground vibrational level of the electronic ground state to the mth vibrational level of the excited electronic state, where E_{00} is the zero-phonon line energy. Analogously, for the emission process, one can write

$$L(E) = A(E_{00} - n\,\hbar\omega) \propto |M_{n0}|^2 \tag{1.95}$$

considering that at room temperature, after the absorption process, the system relaxes to the lowest vibrational level in the excited state, as shown by the dash-dotted arrow in Figure 1.5, and then it goes back to the electronic ground state, occupying one of the many vibrational levels depending on the $|M_{n0}|^2$ factor. On the basis of the wavefunctions of the harmonic oscillator, it can be demonstrated that [5, 15, 18, 19]

$$|M_{0m}|^2 = \frac{S^m}{m!}\ \mathrm{e}^{-S} \tag{1.96}$$

where the Huang–Rhys factor S has been inserted [19]

$$S = \frac{\Delta U}{\hbar\omega} \tag{1.97}$$

connecting the Stokes shift of the absorption and emission transitions to the vibrational energy of the oscillator considered. Equation (1.96) is a Poisson distribution with variance S, standard deviation $\sqrt{S}$, and average value S [20]. The Huang–Rhys factor measures the nuclear relaxation energy ΔU in units of the vibrational quantum of energy $\hbar\omega$. In particular, the larger relaxation occurs in the presence of a strong electron–phonon coupling, implying that when the electron is excited, a large modification of the equilibrium position of nuclei occurs, with the ensuing relative shifts of the parabola describing the energy of the ground and of the excited state, as schematized in Figure 1.5 [15, 18]. It is interesting to observe that on increasing the Huang–Rhys factor, the lineshape of the considered transition, given by the distribution of $|M_{0m}|^2$, changes from strongly asymmetric ($0 < S < 1$) with predominance of the zero-phonon line, to gently asymmetric ($1 < S < 6$) with residual of the ZPL, to symmetric and almost Gaussian ($S > 10$) with small contribution of the ZPL [15]. The Huang–Rhys factor can be determined from the Stokes shift once the vibration frequency is known using (1.87) and (1.97). Furthermore, S can be determined from the area of the ZPL and the total area of the absorption (or emission) bands. In fact, starting from (1.96), it can be shown that

$$\frac{\text{Area ZPL}}{\text{Total area}} = \frac{|M_{00}|^2}{\sum_{m=0}^{\infty} |M_{0m}|^2} = \mathrm{e}^{-S}. \tag{1.98}$$

Finally, for large values of S, the variance of the Poisson distribution is equal to S and the spectral variance in terms of $\hbar\omega$ will be $S(\hbar\omega)^2$. From the Gaussian profile reported by (1.78), it is then determined that

$$\text{FWHM} = 2\sqrt{2\ln(2)}\sigma = 2\sqrt{2\ln(2)}\hbar\omega\sqrt{S} = 2.35\hbar\omega\sqrt{S} \tag{1.99}$$

and the Huang–Rhys factor could be experimentally estimated. To conclude these considerations on the homogeneous lineshape, a more detailed treatment should include the many possible vibration degrees of freedom of a polyatomic molecule and replicas of the considered features have to be inserted with different Huang–Rhys factors for each mode [8, 18, 21].

1.2.4 Jablonski Energy Level Diagram: Permitted and Forbidden Transitions

In the previous paragraph, the simplest molecular model using the Born–Oppenheimer approximation enabled to determine the dipole moment matrix element (1.92). The first factor is related to the electronic wavefunction and the second factor is due to the nuclear wavefunction. It is usual to consider two contributions in the electronic wavefunction, the first due to the orbital motion and the second due to the spin degrees of freedom. The dipole moment matrix element can then be written

$$\mu_{12}^2 = |\langle \varphi_2{}^l \varphi_2{}^s \mid er \mid \varphi_1{}^l \varphi_1{}^s \rangle|^2 \cdot |\langle \vartheta_2^m \mid \vartheta_1^n \rangle|^2 = |\langle \varphi_2{}^l \mid er \mid \varphi_1{}^l \rangle|^2 \cdot |\langle \varphi_2{}^s \mid \varphi_1{}^s \rangle|^2 \cdot |\langle \vartheta_2^m \mid \vartheta_1^n \rangle|^2 \tag{1.100}$$

where the first factor accounts for the spatial dependence of the electron motion (orbital contribution), the second factor for the spin contribution, and the third factor for the nuclear vibration (Franck–Condon factor). Each of these factors contributes to the evaluation of the dipole moment matrix element, and they give rise to the selection rules for the vibronic transition [5, 8]. A given transition is usually called spin forbidden if

$$|\langle \varphi_2{}^s \mid \varphi_1{}^s \rangle|^2 = 0 \tag{1.101}$$

this is typically the most limiting rule. The dipole moment matrix element is related to the oscillator strength by Eq. (1.57) and to the experimental absorption through (1.56) and the molar extinction coefficient through (1.8); so, it is observed that the latter parameter is in the range $(10^{-5} < \varepsilon < 10^{0})\ \mathrm{M^{-1}\ cm^{-1}}$ for spin forbidden transitions. When the orbital factor is null

$$|\langle \varphi_2{}^l \mid er \mid \varphi_1{}^l \rangle|^2 = 0 \tag{1.102}$$

the transition is called orbitally forbidden and the range $(10^{0} < \varepsilon < 10^{3})\ \mathrm{M^{-1}\ cm^{-1}}$ is found for the molar extinction coefficient. Finally, values $(10^{3} < \varepsilon < 10^{5})\ \mathrm{M^{-1}\ cm^{-1}}$ typically pertain to allowed transitions. It is worth observing that these rules could not be strictly respected since in some cases the vibronic states could have a not pure spin or orbital angular momentum contribution, being instead a mixture of states [5, 8, 9].

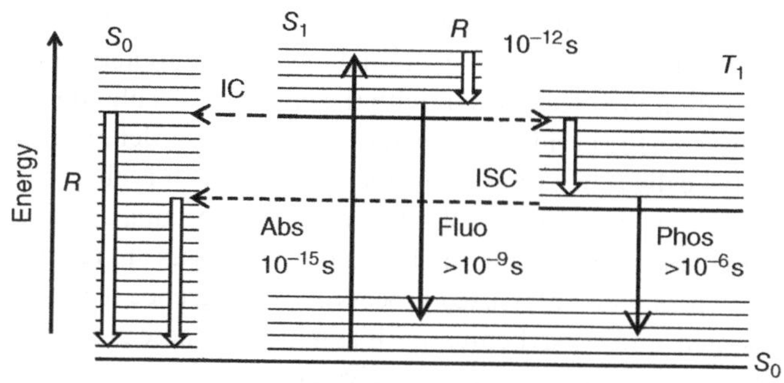

Figure 1.6 Jablonski diagram for the transition processes of electrons among vibronic states. The electronic levels are labeled by S_0, S_1, T_1 and thick horizontal lines, thinner horizontal lines mark the vibrational levels. Continuous arrows represent photon-related (radiative) transitions; white arrows mark relaxation transitions among vibrational levels; short-dashed arrows represent the intersystem crossing process (ISC); dashed line the internal conversion process. Typical times of the processes of absorption (Abs) fluorescence (Fluo), phosphorescence (Phos), and vibrational relaxation (*R*) are inserted.

The overall sequence of transitions occurring among the energy states of a molecule can now be described in more detail. Figure 1.6 shows a schematic representation of the processes connecting the ground state S_0 and the excited state S_1, assumed to be spin singlet states, and the excited spin triplet state T_1.

This scheme is known as the Jablonski diagram [2, 5]. The system is assumed to be at a temperature T characterized by a thermal energy much lower than the vibrational energy of nuclei: $kT \ll \hbar\omega$, so only the lowest vibrational levels could be populated in thermal equilibrium; as an extreme case, it could be assumed that $T = 0$ K. The interaction with an electromagnetic field of opportune frequency gives rise to the absorption process in which a photon is lost by the field and the electron is promoted from the S_0 to the S_1 state. This effect is in a typical timescale much faster than any nuclear motion and can be assumed to occur in 10^{-15} s [5]. As stated above, the nuclei are in a fixed position and with given momentum during this process; as a consequence, if the starting vibrational state at low T is the ground vibrational state, the arrival vibrational state, in the electron excited state S_1, is not necessarily the lowest vibrational energy level (this depends on the Franck–Condon factor). This is sketched by the tip of the absorption arrow pointing to a high vibrational level in S_1. Successively, the nuclei relax and release their vibrational energy, reaching the lowest vibrational level (marked by the white arrow *R*), in a time typical of nuclear vibration: 10^{-12} s [7, 14]. On reaching the lowest vibrational level in S_1, the system could relax to the S_0 state through coupling to its highly excited vibrational levels and the internal conversion (IC, dashed arrow) process and without energy release to the electromagnetic field (non-radiative relaxation, heating of the molecule). It is also possible that the system relaxes to the S_0 state by emitting a photon. This process is called fluorescence and the overall permanence in the S_1 state of an ensemble of similar molecules gives the lifetime of this emission process. A typical timescale starting from 10^{-9} s characterizes this process. This transition is spin allowed and the short lifetime is a characteristic feature. It is worth noting that also in this case the arrival vibrational state in S_0 is determined by selection rules and could not be the ground vibrational level. Another pathway from the S_1 state involves the molecular vibrations and the spin–orbit coupling [2, 5]. This process enables to change the spin state of the electrons, giving a change from singlet (S_1) to triplet (T_1) state. This interaction process is known as intersystem crossing (ISC, short-dashed arrow) and gives rise to a non-radiative relaxation between excited states. Once in T_1, where the vibrational state could be different from the ground state, a vibrational relaxation occurs to the ground vibrational level (white arrow). Successively, the system could reach the S_0 state through ISC relaxation to its high-energy vibrational states (short-dashed arrow). Furthermore, a transition to S_0 could occur with the emission of a photon by the phosphorescence phenomenon. In this case, the permanence in the excited state T_1 determines the lifetime of this emission process that typically is in a timescale larger than 10^{-4} s. It is observed that this transition is spin forbidden, since a passage from triplet to singlet state occurs, and the related lifetime is much longer than for fluorescence. Overall, the ISC and IC processes are related to vibrations of the molecule and are known as phonon-assisted processes. The presence of vibrations is related to the temperature of the system and an Arrhenius law is assumed for these processes [2, 5]. In detail, the rate of the intersystem crossing process, K_{ISC}, is given by

$$K_{\mathrm{ISC}} = K_0\ \mathrm{e}^{-\frac{\Delta E}{kT}} \tag{1.103}$$

where K_0 is a pre-exponential factor taking into account entropic-statistical factors, ΔE is the activation energy of the process, and k is the Boltzmann constant.

The Jablonski diagram is useful to describe the overall emission features of a system and to schematize the energy levels distribution and their dynamics aspects.

1.2.5 Excited States Rate Equations

To go deeper in the emission features, a simplified Jablonski diagram for the transition processes of electrons among vibronic states can be considered, joining the representations reported in Figure 1.4 (bottom) and Figure 1.6. The singlet energy levels S_0 and S_1 are connected by the absorption process and by the radiative fluorescence emission from S_1 with rate K_r^F, and the non-radiative decay with rate K_{nr}^F. The T_1 state is populated by the intersystem crossing with rate K_{ISC}, and it is connected to the S_0 state by the phosphorescence emission, with rate K_r^P, and by the non-radiative process with rate K_{nr}^P. Under light excitation, it is possible to describe the time-dependent population of the excited singlet, N^{S_1}, and triplet, N^{T_1}, states by the rate equations [15, 18]:

$$\frac{\mathrm{d}N^{S_1}}{\mathrm{d}t} = I_0\left[1-\mathrm{e}^{-\alpha L}\right] - \left[K_r^{\mathrm{F}} + K_{\mathrm{nr}}^{\mathrm{F}} + K_{\mathrm{ISC}}\right] N^{S_1} \tag{1.104}$$

$$\frac{\mathrm{d}N^{T_1}}{\mathrm{d}t} = K_{\mathrm{ISC}}N^{S_1} - \left[K_{\mathrm{r}}^{\mathrm{P}} + K_{\mathrm{nr}}^{\mathrm{P}}\right] N^{T_1} \tag{1.105}$$

where the absorbed light through a sample of thickness L, giving transitions from S_0 to S_1, has been introduced based on Eq. (1.5). It is observed that the emission from the excited states depends on their population; so, it can be stated that for the fluorescence the emitted light is given by

$$I^{\mathrm{F}}(E_{\mathrm{exc}}, E_{\mathrm{em}}, T) = K_{\mathrm{r}}^{\mathrm{F}}N^{S_1}(E_{\mathrm{exc}}, T)f(E_{\mathrm{em}}) \tag{1.106}$$

where the excitation, E_{exc}, and the emission, E_{em}, energies have been introduced together with the temperature dependence and a lineshape $f(E_{\mathrm{em}})$ including the homogeneous and inhomogeneous distributions of levels [18]. Analogously, for phosphorescence, it is found that

$$I^{\mathrm{P}}(E_{\mathrm{exc}}, E_{\mathrm{em}}, T) = K_{\mathrm{r}}^{\mathrm{P}}N^{T_1}(E_{\mathrm{exc}}, T)g(E_{\mathrm{em}}) \tag{1.107}$$

where the lineshape for the triplet to singlet emission of phosphorescence, $g(E_{\mathrm{em}})$, has been inserted. In the stationary state (ss), it is found that

$$\frac{\mathrm{d}N^{S_1}}{\mathrm{d}t} = 0; \qquad N_{\mathrm{ss}}^{S_1} = \frac{I_0[1-\mathrm{e}^{-\alpha L}]}{\left[K_{\mathrm{r}}^{\mathrm{F}} + K_{\mathrm{nr}}^{\mathrm{F}} + K_{\mathrm{ISC}}\right]} \tag{1.108}$$

$$\frac{dN^{T_1}}{dt} = 0; \qquad N_{ss}^{T_1} = \frac{K_{ISC} N_{ss}^{S_1}}{\left[K_r^P + K_{nr}^P\right]} \tag{1.109}$$

which can be substituted into (1.106) and (1.107) to obtain

$$I_{ss}^{F}(E_{exc}, E_{em}, T) = K_r^F \frac{I_0[1 - e^{-\alpha L}]}{\left[K_r^F + K_{nr}^F + K_{ISC}\right]} f(E_{em}) \tag{1.110}$$

$$I_{ss}^{P}(E_{exc}, E_{em}, T) = K_r^P \frac{K_{ISC}}{\left[K_r^P + K_{nr}^P\right]} \frac{I_0[1 - e^{-\alpha L}]}{\left[K_r^F + K_{nr}^F + K_{ISC}\right]} g(E_{em}). \tag{1.111}$$

Apart from the dependence on the rate of transitions, these equations show that the emission intensities of both fluorescence and phosphorescence depend on the absorption process. In this context, in the case of low absorption, it is found that $(1 - e^{-\alpha L}) \sim \alpha L$. Since the absorption coefficient is a function of the excitation energy, both Eqs. (1.110) and (1.111) enable to determine $\alpha(E_{exc})$ when the E_{em} is fixed and to reconstruct the absorption profile. This kind of measurement is known as excitation spectrum [2, 18] and enables to determine the connection between the spectra of phosphorescence and fluorescence, relating them to the same absorption pathway, and to reconstruct the Jablonski diagram.

Finally, in analogy to Eq. (1.70), when the excitation light is suddenly removed, Eqs. (1.104) and (1.105) enable to determine the lifetime of fluorescence and of phosphorescence and to demonstrate that they are, respectively [18]:

$$\tau_F = \left[K_r^F + K_{nr}^F + K_{ISC}\right]^{-1} \tag{1.112}$$

$$\tau_P = \left[K_r^P + K_{nr}^P\right]^{-1}. \tag{1.113}$$

Considering the temperature dependence of the non-radiative processes and their rate, these equations show that a lifetime dependence on temperature is present and that at low T the non-radiative rates are canceled, enabling to evaluate the true radiative rate [18].

1.3 Instrumental Setups

The instrumentation used and the opportune choice of instrumental features strongly determine the results of a spectroscopy measurement. This paragraph is devoted to summarize the main components of the instrumentation and those technical features that distinguish them aiming to give the basis for an aware choice of parameters during an experiment or the opportunity to choose a good instrumental configuration for carrying out an appropriate experiment.

1.3.1 Typical Block Diagram of Spectrometers

In spectroscopy experiments, radiation–matter interaction is the fundamental process that should be investigated by the instrument. In every instrument, a source of radiation is needed and then, to analyze the effect, the radiation should be selected in terms of energy, or wavelength, and the influence on the sample to be investigated should be detected. Typical block diagrams for simple absorption and emission instruments are reported in Figure 1.7.

In a typical absorption experiment, a light source generates the electromagnetic wave. This is directed to a monochromator that is able to select a given wavelength (energy) of the lightwave within an opportune interval (bandwidth), depending on the construction parameters. The light

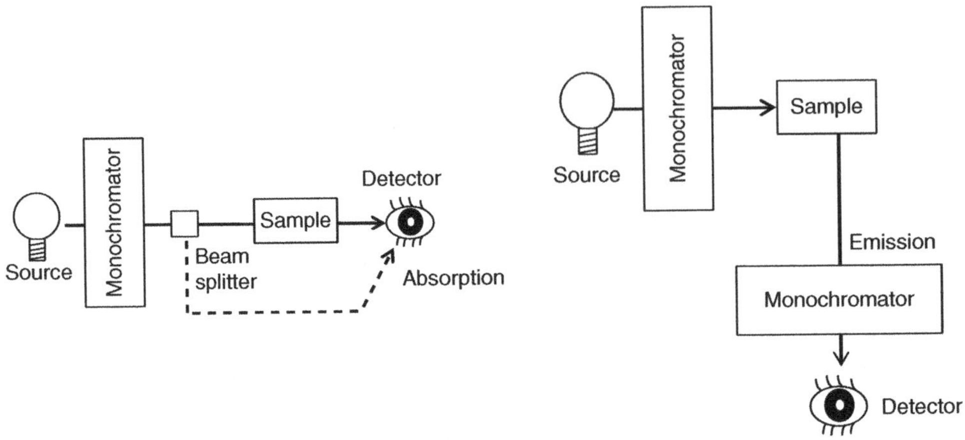

Figure 1.7 Pictorial representation of the instrumentations to carry out spectroscopy experiments. On the left, the components of an absorption spectrophotometer consisting of a light source, an entrance monochromator, the sample, and a detector are drawn. The double beam configuration is sketched by the beam splitter and the dashed arrow representing the reference beam. On the right, the scheme for an emission spectrofluorometer shows the additional exit monochromator and the side detector to record the emitted light.

impinges then on the sample that is opportunely inserted in its path. The light emerging from the sample is detected by a detector and its intensity is registered by a computer (not reported). As reported in Section 1.1.1, comparing the source intensity without the sample ($I_0(\lambda)$) and that after the passage through the sample ($I_t(\lambda)$), an absorption measurement can be carried out. This kind of setup is known as single-beam spectrophotometer and it needs a couple of measurements to record $I_0(\lambda)$ and $I_t(\lambda)$. In some systems, a beam splitter is inserted before the sample to split the lightwave in two identical rays (see dashed path in Figure 1.7). This system is known as double-beam spectrometer and uses the unperturbed ray as a reference to record $I_0(\lambda)$ and the ray passing through the sample to "contemporarily" detect $I_t(\lambda)$. In this case, a great advantage is usually obtained by using a single detector to avoid mismatches of the spectral response of the detecting system [2, 10].

In the photoluminescence experiment, a spectrofluorometer is used where the source of radiation is placed before one monochromator (entrance) to select the wavelength (energy) to excite the sample. When an emission process is active, this is typically isotropic in space and the emitted light can be recorded everywhere. Usually, to avoid any interference with the transmitted light, the emission is recorded in the 90° geometry shown in Figure 1.7 or, seldom, in backscattering geometry (the emitted light is recorded in the same direction as the exciting light, the impinging and emerging rays form 180° angle, not reported). To determine the characteristics of the emitted light, a second monochromator (exit) is inserted after the sample to select the wavelength (energy) of the light. Finally, a detector is inserted that is connected to a computer (not reported).

1.3.2 Light Sources

The first element of a spectrometer, as shown in Figure 1.7, is the light source. Different types of sources are usually employed in ultraviolet-visible-infrared (UV-Vis-IR) spectroscopy [2, 10, 22].

- Incandescent lamp: it consists in a metallic filament (typically tungsten) inside a transparent glass bulb filled by halogen gas to maintain the filament. The filament is traversed by an electric current heating it by Joule effect and increasing its temperature [10]. A black body-like emission occurs, giving a continuous spectrum comprising typically the Vis-IR range.

- Discharge arc lamp: a couple of electrodes (anode and cathode) are inserted in a transparent glass bulb filled by a low-pressure gas of a given element (e.g. H, Hg, Na, Xe). By applying a high voltage (~1000 V) to the electrodes, a discharge occurs through the gas ionizing it. The ionized gas emits characteristic lines by the transition of the electrons through the excited states of the element. The ensuing spectrum is a line spectrum with a continuous background related to the temperature and the pressure of the gas. It is usual, for spectroscopy, to use high-pressure discharge lamps to enhance the continuous background and employ them in a large spectral range. In particular, lamps with Xe are used for UV-Vis emission spectroscopy and with Deuterium for UV and up to vacuum-UV absorption and emission.
- Laser [10]: various types of laser light are employed for spectroscopy due to their precise wavelength, spatial collimation, high intensity and time resolution in the pulsed regime [23]. These systems are realized with opportune cavities with reflecting walls to obtain the lasing effect.

Gas lasers (e.g. He–Ne: 633 nm; Ar: 514.5 nm; Kr: 647 nm) employ an electric discharge to ionize a low-pressure gas and induce the population inversion. Typically, they are characterized by continuous wave emission.

Excimer lasers use mixture of gases [Xe, HCl, Ne (as buffer); Kr, F, He (as buffer)] at high pressure to obtain molecular complexes in the excited state (XeCl∗, KrF∗) through an electric discharge. During the excited state decay, a laser pulse is generated with duration in the window 5–20 ns. The population inversion is related to the disappearance of the ground state once the complex, which is unstable, dissociates.

Dye lasers use organic molecules in solution as laser active medium with emission in the range 300–1000 nm depending on the dye. The emission range is tunable up to 30 nm and can be pulsed by pumping it with an excimer laser.

Solid state lasers have been the first system used by Maiman in 1960 to prove coherent emission [24]. They include the first member of the family, the Ruby laser (Al_2O_3:Cr^{3+}, 694.3 nm), the Nd-YAG laser (1064 nm), and the Ti-sapphire laser (Al_2O_3:Ti^{3+}, ~800 nm, wide band). In these systems, a pumping lamp populates an excited state starting from which a long lifetime excited state is successively populated, inducing a population inversion and the following stimulated emission. The emission can be pulsed and a timescale of nanoseconds or down to femtoseconds can be reached. Furthermore, due to the high intensity of some of these systems, nonlinear effects can be obtained through nonlinear optics to generate harmonics or continuous very fast pulsed emission.

Diode lasers are light emitting systems using semiconductors [23, 25]. They are solid structures of layered semiconductors with opportune bandgap tuning. This way, it is possible to obtain recombination of electrons and holes in a direct bandgap region. The charge injection by an external electric potential enables the emission of photons in this active region. By opportune combination of the constituting elements of the semiconductors (for example changing the x balance in $Ga_{1-x}Al_xAs$), the emission can be engineered. The typical commercial diode lasers emit in the UV-Vis-IR range.

1.3.3 Dispersion Elements: Gratings and Resolution Power

In typical spectroscopy experiments, it is necessary to select the energy, or wavelength, of the radiation that impinges on the sample under study, to determine its effect, or to analyze the radiation emitted by the sample. For example, in the case of a continuous light source, it is mandatory to select those wavelengths that could induce an absorption effect; otherwise, it would not be possible to understand the microscopic features giving the process and to reconstruct the fundamental information of the energy levels' distribution. The typical systems used to disperse and select the wavelength of a source comprise a prism, a grating, or an interferometer [2, 7, 22]. Each of these systems enables to select a wavelength and an interval around it, called bandwidth. The choice of the specific dispersing tool depends on the kind of experiment and the spectral resolution needed, that means the bandwidth of the selected radiation, or the light throughput. For the UV-Vis spectroscopy, the most used system is the grating with the monochromator [2, 10]. The latter is an optical system with a couple of slits to let the light enter and exit it, some mirrors to drive the light inside it, and one or more gratings to spatially separate the wavelengths and select them. The core part of this system is the grating that is constituted by a series of N slits of equal width, parallel to each other and with distance d [1]. It is shown that for a parallel light beam impinging perpendicularly to a transmission planar grating, there is a maximum of intensity of transmitted light for those rays of wavelength λ deviated with respect to the normal to the grating plane by the angle θ

$$\mathrm{d}\sin(\theta) = \pm m\lambda \tag{1.114}$$

where m is an integer, giving the order of interference [1]. The presence of more slits implies that between two adjacent maxima, given by different values of m, there are minima of intensity for

$$N\mathrm{d}\sin(\theta') = \pm m\lambda \qquad \text{with} \quad m = 1, 2, \ldots, N-1. \tag{1.115}$$

These equations can be used to show that the angular distance between the maxima of interference of two wavelengths at distance $\Delta\lambda$ is given by

$$\frac{\Delta\theta}{\Delta\lambda} = \frac{m}{\mathrm{d}\cos(\theta)} \qquad \text{angular dispersion} \tag{1.116}$$

giving the angular dispersion of a grating. This depends on the distance between the slits, usually reported as (number of lines)/mm, and it increases on increasing the lines/mm constituting the grating (lower value of d) and on increasing the order m of interference. If a monochromator of focal length f is considered [2, 22], the linear dispersion is obtained

$$\frac{\Delta x}{\Delta\lambda} = \frac{m}{\mathrm{d}\cos(\theta)}f \qquad \text{linear dispersion} \tag{1.117}$$

yielding the spatial distance Δx, known as linear distance, on a planar screen of the interference maxima of two wavelengths separated by $\Delta\lambda$, and showing that this distance increases on increasing the focal length of the monochromator. Finally, in order to be able to separate the interference maxima of two adjacent wavelengths, the condition usually applied is that the minimum intensity of one wavelength coincides with the maximum intensity of the other. This condition gives the resolving power of a grating:

$$\frac{\lambda}{\Delta\lambda} = Nm \qquad \text{resolving power} \tag{1.118}$$

this is the minimum difference of wavelengths that a grating with N lines can resolve at the interference order m. This feature depends on the number N of slits illuminated and does not depend on their relative distance d. All of the reported features are fundamental for the good choice of a grating to be used in a monochromator.

Finally, in a plane reflection grating, it is usual to give a specific shape profile to the reflecting lines [22]. In this case, it is shown that a plane wave, impinging with the angle I with respect to the normal to the grating plane, is reflected at the angle D with respect to the same normal direction according to the equation [22]

$$d[\sin(I) + \sin(D)] = m\lambda \qquad \text{basic equation of grating} \tag{1.119}$$

where I and D have the same sign when referred to the same side with respect to the normal to the grating plane. It is found that for m = 0, for any λ, Eq. (1.119) is verified for $I = -D$, that is total reflection condition, the impinging and the reflected beam are symmetric with respect to the normal to the grating plane. This is known as zero order of the gratings and enables to determine a reference position inside the monochromator, the one that gives the absolute maximum of reflected light because all the wavelengths superimpose. Moving the grating with respect to this position, the angles can be measured and the reflected wavelength satisfying Eq. (1.119) can be determined. This procedure is usually applied in the spectroscopy instrumentations to measure the output wavelength from the monochromator. Finally, this kind of reflecting grating is characterized by a wavelength, or equivalently an angle, of maximum efficiency of reflection with dispersion (not the zero order), called blaze wavelength, and the efficiency of dispersion is near the intensity maximum for typically 100 nm around this wavelength [22].

1.3.4 Detectors: Photodiode, Photomultiplier, Charge Coupled Device

The fundamental part of a spectroscopy experiment is the detector. It should be sensitive in order to detect the photons also at very low number for unit of time, and avoid having noise signal due to electronics. The most used detectors in UV-Vis-IR spectroscopy are the photodiode, the photomultiplier (PMT), and the charge coupled device (CCD). The most sensitive among them is the PMT, which is also characterized by a fast enough time of detection, response time. The photodiode and the PMT are typically used in scanning spectroscopy system, where they are coupled to monochromators that select the wavelength to be revealed. The CCD is typically coupled to a grating and is able to contemporarily detect many wavelengths, thus enhancing the speed of recording of a spectrum. The working principle of these detectors is now briefly summarized [2, 10, 22, 25].

- The photodiode is a solid state detector based on the junction of two extrinsic semiconductors with n- and p-doping [14, 25]. In the junction, the charge depletion region is formed that is characterized by the presence of an intrinsic electric field [25]. When photons of energy larger than the bandgap of the semiconductors impinge in this region, an electron–hole couple is generated and is separated, ejecting these charges toward the neutral regions of the two semiconductors. This effect induces a current, if the photodiode is short-circuited or, equivalently, a voltage if the photodiode is open circuit. Both these signals are proportional to the number of impinging photons per unit of time, or equivalently to the intensity of radiation. A saturation effect for high radiation intensity could occur due to the recombination of electron–holes simultaneously generated in the depletion region, not giving a current signal. Typical spectral sensitivity of the photodiode depends on the semiconductor used and its bandgap. For Si, the spectral range is 200–1100 nm, for Ge 400–1800 nm, and for PbSnTe 2000–18 000 nm [10, 25]. Many different combinations exist nowadays, showing the extension of spectral response from far-IR to UV [25]. One limit of a photodiode is related to the generation of thermal couples of electron–holes in the depletion region giving rise to a noise signal. To contain this effect, low-temperature detectors are employed, cooling them with liquid nitrogen (77 K) [22].

- The photomultiplier is formed by a photocathode that is an emitting layer that when illuminated with photons ejects electrons by photoelectric effect [2, 10, 12]. A second electrode acts as an anode to accelerate the photoelectrons. Beneath this element is present an array of secondary emitters of electrons, each called dynode. In particular, each electron impinging with opportune energy on a dynode is able to release other electrons. Opportunely accelerating the photoelectron emitted by the photocathode toward the anode, the electron has kinetic energy large enough to release the secondary electrons from the first dynode. By applying the opportune voltage, the secondary electrons are accelerated toward the next dynode and they release other electrons. An avalanche process can be generated this way. In particular, the impinging photon effect is amplified by a factor δ^n by an array of n dynodes, each emitting δ electrons characterizing the photomultiplier by a high gain. Materials used for the photocathode are metallic leagues: $CsNa_2$, KSb, Cs_3Sb, KCsSb, and also semiconductors: GaAs, InGaAs, that are characterized by a low work function [10, 14]. For the dynode, other leagues are used: Be-Cu, GaP. A general property of the PMT, apart from its large gain, is that the spectral response depends on the employed material. It has the drawback of the dark current due to spontaneously emitted electrons by the dynodes that are accelerated by the high voltages usually employed to reduce the response time of the PMT. Indeed, this voltage is used to reduce the travel time of the electrons in the array of dynodes and the PMTs are very fast detectors often employed in time-resolved spectroscopy. These detectors are also used in the photon counting mode that enables to detect very low-intensity radiation (few events in time). In this procedure, the photoemitted electrons originate a high-intensity signal in short time, because they are correlated to the arrival event of a photon and they traverse the entire dynodes array, which can be distinguished from the low-intensity signal continuous in time due to thermal electrons generated with larger probability by few dynodes [10]. Typically, PMT cannot work in the low energy range of photo-detection since the work function of metals is not low enough.
- The CCD detector is constituted by an array of metal-oxide-semiconductor (MOS) capacitors [25]. These are typically p-type doped Si covered by SiO_2 insulator and with top metallic gate. Electric contacts on top metal and bottom Si enable the switching of each MOS between the accumulation and depletion or null state [10, 25]. By applying a positive voltage to the gate, the depletion regime is activated in the Si near to the Si–SiO_2 interface, implying the accumulation of negative charge. If photons impinge the depleted region in this regime, they increase the number of negative charge trapped. This number is proportional to the intensity of light. In a CCD, the array of MOS is opportunely wired and subjected to a sequence of voltages to let three nearby MOSs constitute a pixel where the central MOS is active and the two edges inactive when the detector is exposed to light. As a consequence, during light exposure, the charge is accumulated only in the central MOS of a pixel. Successively, an applied voltage time sequence opportunely transfers the charge from the central MOS to one of the nearby in a sequence to let the accumulated charge be read by an opportune system. In this way, the array of pixels "traps" the information from the impinging radiation synchronously. Typically, CCDs are formed by arrays of multiples of 256 pixels in a square or rectangular matrix with rows and columns of given spatial extension depending on the construction of MOS. The CCD is usually put after a grating so each pixel is hit by a different wavelength due to the spatial dispersion imposed to the light. The columns of pixels select the wavelength whereas the rows enhance the signal reading for each wavelength. The advantage of a CCD is the recording of all the wavelengths in a spectral range synchronously without the need to change the position of the grating.

1.4 Case Studies

Examples of absorption and photoluminescence measurements will be reported in this paragraph in order to give some hints on the settings to carry out a satisfying experiment and with the aim to let correlate information to reconstruct the electronic energy levels of a given chromophore.

1.4.1 Optical Absorption in Visible-Ultraviolet Range

A typical experiment of absorption consists in the investigation of the light transmitted by a given sample. The block scheme of the instrumentation has been presented in Section 1.3.1 and it includes a light source and a system to select the wavelength interacting with the sample. Two possible configurations are typically used. In the first, the wavelength of the source is selected by a monochromator and it is sent to the sample and the transmitted light is detected. In the second, all the radiation of the source is sent to the sample and the analysis of the wavelength is done after the sample by a grating and a CCD. Obviously, in each case, a blank or a reference spectrum of the source should be done without the sample to determine the intensity of light arriving at the sample before to traverse it and give rise to the absorption process. In Sections 1.4.1.1, 1.4.1.2 examples of the practice to apply to obtain good spectra are reported.

1.4.1.1 Scanning Device (Bandwidth and Scanning Speed Effects)

In a scanning spectrophotometer, the wavelength is selected before to send the light to the sample. As a consequence, it is necessary to take into consideration the time to select the wavelength and correctly connect it to the time necessary to record the signal passing through the sample by the detector. The latter is typically determined by the detector response time (RT). The time to select the wavelength is the scanning speed (SS) of the spectrophotometer. Both parameters can be adapted by the user, in particular larger response time typically warrants a high signal-to-noise ratio and a good analysis of the results. Considering that a monochromator selects a wavelength with a range around it, called bandwidth (BW), and that the BW determines the spectral resolution, it is a good practice to use a BW lower than the spectral width of the band (or bands) that is recorded. This spectral width can be experimentally measured by the full width at half maximum (FWHM), that is the spectral distance of the points at half amplitude of the band's signal maximum. It is opportune to set at least BW < 0.5 FWHM to avoid distortions. After this choice has been done, it is opportune to set RT < BW/SS to avoid other distortions. Indeed, this choice ensures that the time needed for the detector to record a signal is lower than the time necessary for the instrument to change the wavelength in an interval equal to its spectral resolution.

To show the effects that could be produced by a wrong choice of the instrumental parameters, a holmium glass filter has been employed [26, 27]. This sample is characterized by a series of narrow, well-defined peaks in the spectral range from the visible to the UV; is commercially available; and can be used for teaching-calibrating purposes. The spectrophotometer used enables to change the BW in the range 0.1–5 nm. Furthermore, it has SS in the range 10–1000 nm min^{-1} and RT in the interval 0.03–240 s. After an explorative spectrum, the band of the holmium filter centered at about 360 nm has been selected. The SS has been fixed to 20 nm min^{-1} and the RT to 0.25 s. The first experiment consists in changing the BW.

As reported in Figure 1.8, the different BWs give rise to spectral distortion evidencing that an inaccurate choice compromises the interpretation of the experimental result. In particular, larger BW widens the spectral shape. As reported in the inset of the figure, the effect of distortion can be quantitatively determined by measuring the FWHM. A linear dependence of the FWHM is found

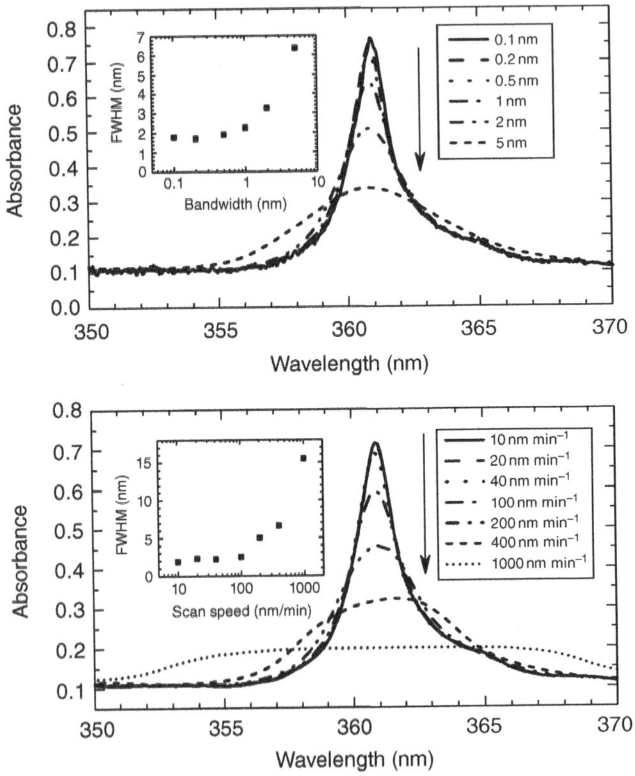

Figure 1.8 Top: Absorption spectra of holmium glass filter at various bandwidths of the spectrophotometer from 0.1 to 5 nm. The inset reports the dependence of the full width at half maximum on the various employed bandwidths. Bottom: Absorption spectra of holmium glass filter at various scan speeds of the spectrophotometer from 10 to 1000 nm min^{-1}. The inset reports the dependence of the full width at half maximum on the various employed scan speeds.

on the BW for large values of the latter, whereas for small BW the true width of the band is measured. In particular, the condition BW = 0.4 FWHM somehow guarantees a good estimation of the true width of the band.

In the second experiment, the SS has been changed, fixing the BW = 0.5 nm and the RT = 1 s. As reported in Figure 1.8, the fast scanning of the spectrum induces a strong distortion because of the delay in the response of the detector with respect to the change in the wavelength. These examples show that the instrumental parameters should be opportunely set in order to avoid distortion and at the same time guarantee a good signal-to-noise ratio to enable a good analysis of the results.

As a further case, the absorption spectrum of a high-purity fused quartz glass of commercial origin (Infrasil 301, by Hereaus [28]) is shown in Figure 1.9 with the optimized instrumental parameters BW = 2 nm, RT = 4 s, and SS = 10 nm min^{-1}. The spectrum is reported as a function of energy to clarify those physical features not correctly described by the wavelength, like the spectral shape. Furthermore, the absorption coefficient is reported taking into account the thickness of the experimental sample. It is possible to observe that the used glass features an absorption band peaked at about 5.14 eV superimposed to a larger absorption at higher energy. The lower energy peak is associated to an electronic transition and it is of interest for the determination of the radiative relaxation linked to the photoluminescence presented in Section 1.4.2.1.

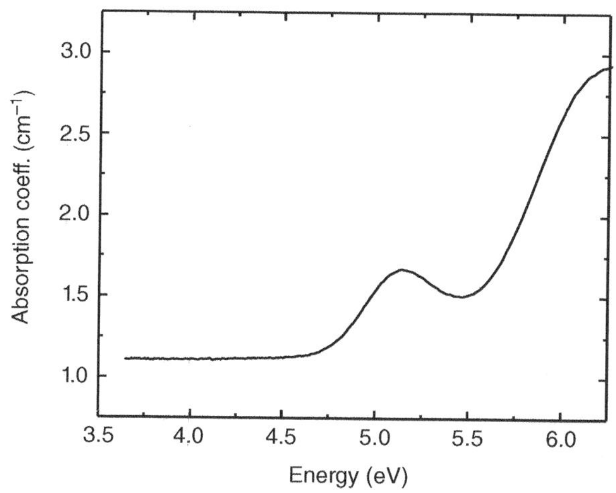

Figure 1.9 Absorption spectrum of a commercial fused quartz glass featuring an absorption band in the UV range.

1.4.1.2 CCD Fiber Optic Device

Many setups for absorption measurements are nowadays using fiber optics technology. These systems use compact light sources and CCD detectors coupled to gratings to obtain fiber optics spectrometers. The light from the source is driven by an optical fiber to the sample holder directing the light collimated by a lens perpendicular to the sample surface. The light exiting from the sample is collected by a second lens and is directed to another optical fiber that drives the light to the grating that disperses it and then is detected by a CCD. These systems are typically of small dimension and are portable, with many advantages for coupling them in various kinds of experiments like in situ measurements. The opportune choice of source, fibers, and detector enables to use these systems for UV-Vis-IR spectroscopy.

A fiber optic instrument equipped with two light sources: a deuterium lamp and a tungsten lamp, and a spectral resolution of 1.7 nm has been used to detect the absorption of quantum dots of CdSe/ZnS. In particular, commercial core–shell nanoparticles with nominal 4 nm core of CdSe and ZnS shell have been dispersed in solution of toluene to detect the absorption as a function of nanoparticles' concentration [29, 30]. A quartz cuvette of 1 cm has been used for the measurements. Due to quantum confinement effect, the nanoparticles feature an absorption band related to the size of the semiconducting core [29]. As reported in Figure 1.10, a prominent band at 585 nm can be observed for all the concentrations used ranging from 10^{-3} up to 10^{-1} mg ml^{-1}. This spectral position is compatible with literature data for this nanoparticles size. The inset of the figure reports the amplitude of absorbance for the maximum of the band as a function of the concentration. A linear trend is found in accordance with the Lambert–Beer law Eq. (1.6).

1.4.2 Photoluminescence

In a photoluminescence experiment, a sample is illuminated by a source to induce the absorption process and the excitation of the electron from the ground state to an excited state. Returning back to the ground state, the electron could emit light. This emission is spectroscopically studied. In a typical photoluminescence setup, as reported in the block scheme of Figure 1.7, two

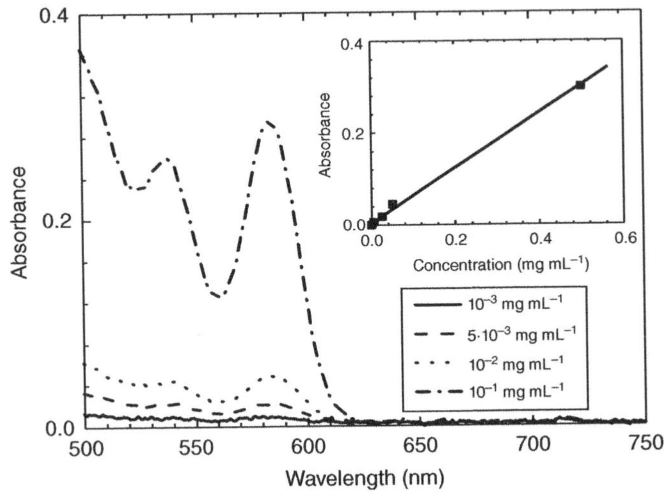

Figure 1.10 Absorption spectra of CdSe/ZnS core–shell nanoparticles at different concentrations in toluene solution. The inset shows the peak absorbance amplitude at 585 nm as a function of concentration; the straight line is a guide for the eye.

monochromators are used to select the excitation wavelength and the emission wavelength. In the following, the two kinds of measurements that can be done are illustrated.

1.4.2.1 Emission and Excitation Spectra: Energy Levels Reconstruction

As discussed above, the photoluminescence experiment consists in illuminating the sample with an opportune wavelength selected by the excitation monochromator. This wavelength is fixed inside the range of an absorption band recorded in a preliminary absorption experiment, as those illustrated in the previous paragraphs. The light emitted by the sample is recorded by a detector after passing through a second monochromator. The scan of the latter gives rise to the spectrum, usually called *emission spectrum*. As an example, the emission spectrum of the sample of high-purity fused quartz glass of commercial origin (Infrasil 301, by Hereaus [28]), reported in Section 1.4.1.1, has been recorded by a scanning spectrofluorometer using as light source a 150 W xenon arc lamp. To record this spectrum, the excitation has been fixed at 5.0 eV where the sample features an absorption band, as reported in Figure 1.9. The instrumental parameters have been fixed in order to not distort the spectra. Furthermore, the spectral response of the detector and exit monochromator has been considered to correct the spectra [2]. As reported in Figure 1.11, two emission bands can be resolved with maximum amplitude at 3.17 and 4.26 eV, respectively. These bands are marked, according to literature [18, 31, 32], β and α_E bands, respectively.

To complete the investigation of the stationary emission process, the excitation spectrum can be recorded. It consists in registering the amplitude of the emission at a given wavelength by scanning the excitation wavelength. Instrumentally, this means that the exit monochromator reported in Figure 1.7 is at fixed wavelength, whereas the entrance monochromator changes the wavelength of the source radiation that impinges on the sample.

As reported in Figure 1.11, the excitation spectrum in the case of the commercial fused quartz glass sample can be recorded in correspondence of the two peaks of emission marked β and α_E. The inset of Figure 1.11 reports the two obtained excitation spectra, fixing the emission energy at 3.17 and 4.26 eV. These spectra have very similar lineshapes and are centered at about the same

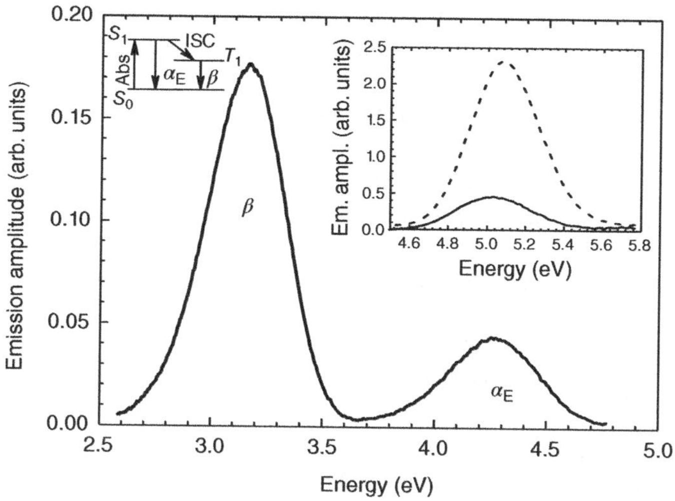

Figure 1.11 Emission spectrum of the commercial fused quartz glass with the absorption spectrum reported in Figure 1.9. The excitation energy has been fixed at 5.0 eV. The emission bands are marked by β, α_E according to literature [18, 31, 32] and a schematic Jablonski diagram is reported in the top left side of the panel. In the inset, the excitation spectra are reported for the emission peaked at 3.17 eV (dashed line) and 4.26 eV (continuous line).

energy 5.0 eV. This position strongly resembles the peak of the absorption band reported in Figure 1.9, proving that the two emission peaks can be related to electronic states populated through the same absorption transition. Based on these considerations, it is possible to draw the Jablonski diagram reported in Figure 1.11 for the chromophore present in the commercial fused quartz glass and responsible for the absorption band at about 5.14 eV and for the two emissions at 3.17 and 4.26 eV. In particular, the absorption process is due to the transition between the S_0 and S_1 state. The emission associated to the α_E band is the reverse of the absorption, with a Stokes shift due to the vibronic nature of the system consisting of a chromophore in a solid matrix with molecular vibrations. Finally, the emission associated to the β band is related to the transfer from the excited S_1 state to the T_1 state assisted by the intersystem crossing process (ISC) and the subsequent radiative transition to the S_0 state [15, 18, 31–33]. Even if the demonstration of this pathway is beyond the scope of this chapter, the commercial availability of the sample used for both the absorption and the emission experiments is a useful hint for training and comprehension of the physics of absorption and emission spectroscopy.

References

1 Rossi, B. (1957). *Optics*. Addison-Wesley Publ. Co.
2 Lakowicz, J.R. (2006). *Principles of Fluorescence Spectroscopy*. Springer.
3 Svanberg, S. (2004). *Atomic and Molecular Spectroscopy*. Springer.
4 Bach, H. and Neuroth, N. (1998). *The Properties of Optical Glass*. Springer Verlag.
5 Harris, D.C. and Bertolucci, M.D. (1978). *Symmetry and Spectroscopy: an Introduction to Vibrational and Electronic Spectroscopy*. Oxford University Press.
6 Sole', J.G., Bausa', L.E., and Jaque, D. (2005). *An Introduction to the Optical Spectroscopy of Inorganic Solids*. Wiley.

7 Ferraro, J.R., Nakamoto, K., and Brown, C.W. (2003). *Introductory Raman Spectroscopy*. Elsevier.

8 Bransden, B.H. and Joachain, C.J. (1983). *Physics of Atoms and Molecules*. Longman Scientific & Technical.

9 Cohen-Tannoudji, C., Diu, B., and Laloë, F. (1977). *Quantum Mechanics*. Wiley.

10 Vij, D.R. (ed.) (1998). *Luminescence of Solids*. Plenum.

11 Barrow, G.M. (1962). *Introduction to Molecular Spectroscopy*. McGraw Hill Book Company.

12 Eisberg, R. and Resnick, R. (1985). *Quantum Physics of Atoms, Molecules, Solids, Nuclei and Particles*. Wiley.

13 Steinfeld, J.I. (1985). *Molecules and Radiation: An Introduction to Modern Molecular Spectroscopy*, 2e. Dover Publications.

14 Kittel, C. and Kroemer, E. (1980). *Thermal Physics*, 2e. W.H. Freeman and Company.

15 Pacchioni, G., Griscom, D.L., and Skuja, L. (eds.) (2000). *Defects in SiO_2 and Related Dielectrics: Science and Technology. Springer.*

16 De Mello, J.C., Wittmann, H.F., and Friend, R.H. (1997). An improved experimental determination of external photoluminescence quantum efficiency. *Adv. Mater.* 9: 230–233.

17 Born, M. and Oppenheimer, R. (1927). Zur Quantentheorie der Molekeln. *Ann. Phys.* 389: 457–484.

18 Nalwa, H.S. (ed.) (2001). *Silicon-Based Materials and Devices*. Academic Press.

19 Huang, K. and Rhys, A. (1950). Theory of light absorption and non-radiative transitions in F-centres. *Proc. R. Soc. A* 204: 406–423.

20 Taylor, J.R. (1982). *An Introduction to Error Analysis: The Study of Uncertainties in Physical Measurements*. University Science Books.

21 Herzberg, G. (1966). *Molecular Spectra and Molecular Structure: III Electronic Spectra and Electronic Structure of Polyatomic Molecules*. Van Nostrand Reinhold Company.

22 Bass, M., Van Stryland, E.W., Williams, D.R., and Wolfe, W.L. (1995). *Handbook of Optics*, 2e. McGraw-Hill, Inc.

23 Milonni, P.W. and Eberly, J.H. (1988). *Lasers*. Wiley.

24 Maiman, T.H. (1960). Stimulated optical radiation in Ruby. *Nature* 187: 493–494.

25 Sze, S.M. and Lee, M.K. (2012). *Semiconductor Devices – Physics and Technology*, 3e. Wiley.

26 Hellma GmbH & Co (2020). KG Klosterrunsstraße 5 Müllheim. www.hellma.com.

27 Allen, D.W. (2007). Holmium oxide glass wavelength standards. *J. Res. Natl. Instr. Stand. Technol.* 112: 303–306.

28 Heraeus Quarzglas Bitterfeld GmbH & Co (2020). KG, Heraeus Comvance Heraeusstr. 06803 Bitterfeld-Wolfen. www.heraeus.com.

29 Jasieniak, J., Smith, L., van Embden, J. et al. (2009). Re-examination of the size-dependent absorption properties of CdSe quantum dots. *J. Phys. Chem. C* 113: 19468–19474.

30 Sigma-Aldrich, Inc (2020). www.sigmaaldrich.com.

31 Agnello, S., Boscaino, R., Cannas, M. et al. (2003). Temperature and excitation energy dependence of decay processes of luminescence in Ge-doped silica. *Phy. Rev. B* 68: 165201. (1–5).

32 Cannizzo, A., Agnello, S., Boscaino, R. et al. (2003). Role of vitreous matrix on the optical activity of Ge-doped silica. *J. Phys. Chem. Sol.* 64: 2437–2443.

33 Skuja, L. (1998). Optically active oxygen-deficiency-related centers in amorphous silicon dioxide. *J. Non-Cryst. Sol.* 239: 16–48.

2

Time-Resolved Photoluminescence

Marco Cannas and Lavinia Vaccaro

Department of Physics and Chemistry – Emilio Segrè, University of Palermo, Palermo, Italy

2.1 Introduction to Photoluminescence Spectroscopy

Luminescence is a paramount property for several optical applications including lighting, laser sources, detectors, and, recently, modern nanotechnologies (bioimaging, optoelectronics). To this aim, the research is currently active toward the development of production methods successful to finely control the physical and chemical characteristics of materials, thus tailoring their emission. This challenge strongly stimulates the luminescence spectroscopy to be more and more performing by new solutions that improve the efficiency of excitation sources and detectors used in the experimental setup. A crucial part for the study of luminescence properties is covered by the time-resolved technique that allows to characterize in detail the properties of excited state from which the photon emission originates. Time-resolved luminescence provides indeed the measure of spectroscopic parameters (lifetime, quantum yield, oscillator strength, and electron–phonon coupling) useful to fully describe the optical cycle excitation/emission.

The purpose of this chapter is to provide a theoretical background of the luminescence properties related to color centers in wide band-gap insulators. This is a very important issue that has been dealt with in several textbooks (see, for example, Refs. [1–7]). In particular, we will focus on specific features arising from the electron–phonon coupling (Zero-Phonon Line and vibrational sidebands), which are fundamental for the interpretation of luminescence experiments in solids.

2.1.1 Photoluminescence Properties Related to Points Defects: Electron–Phonon Coupling

Point defects are usually defined in the context of a crystalline network: if the regular array of atoms is interrupted, the lattice site is occupied differently than in the ideal crystal and it is called *point defect*. Defects include unoccupied sites (vacancies); occupied sites that in the perfect crystal are unoccupied (interstitial); impurities at sites that in the crystal lattice either are occupied by atoms of the pure material (substitutional impurities) or are unoccupied (interstitial impurities). The concept of defect may be also extended to amorphous materials even if the lack of regularity (long-range order) introduces differences respect to the crystal, where a defect has fixed orientation and symmetry. The presence of defects in a crystalline or amorphous matrix may drastically modify

Spectroscopy for Materials Characterization, First Edition. Edited by Simonpietro Agnello.

the optical properties of the host material. In fact, they exist in different localized electronic states that cause optical transitions as absorption and luminescence with lower energies than the fundamental absorption edge of the material, from valence to conduction band. For these reasons, point defects are also called color-centers or chromophores. Even if these transitions are localized at the defect site, the optical spectra will be influenced, to a greater or a lesser extent, by the fact that the color-center is embedded in a solid matrix, either crystalline or amorphous. Indeed, it is closely surrounded by neighboring atoms with which it interacts; then, the description of its optical properties requires the defect-matrix complex to be considered.

To compute the electronic states involved in the optical transitions, such a complex is treated as a system of n electrons (mass $m \cong 9.109 \times 10^{-31}$ kg, coordinate r) and N nuclei (mass M_α, coordinate R) which interact by Coulomb forces. In some ways, the defect-matrix complex could be considered as an oversimplification of a molecule, whose energy levels and optical transitions are treated in the previous chapter. For the case considered here, the Hamiltonian is given by:

$$H = \frac{1}{2m}\sum_{i=1}^{n} p_i^2 + \frac{1}{2}\sum_{\alpha=1}^{N} \frac{P_\alpha^2}{M_\alpha} + V(r,R), \tag{2.1}$$

where the first and the second terms are the kinetic energy of electrons and nuclei, respectively, and $V(r, R)$ is the interaction potential energy given by:

$$V(r,R) = \frac{1}{2}\sum_i\sum_j \frac{e^2}{|r_i - r_j|} + \frac{1}{2}\sum_\alpha\sum_\beta \frac{e^2 Z_\alpha Z_\beta}{|R_\alpha - R_\beta|} - \sum_\alpha\sum_i \frac{e^2 Z_\alpha}{|R_\alpha - r_i|}, \tag{2.2}$$

where e is the electron charge ($e \cong 1.602 \times 10^{-19}$ C) and Z is the atomic number.

The Schrödinger's equation $H\psi(r, R) = E\psi(r, R)$, where $\psi(r, R)$ and E are the wave function and the energy eigenvalue of the defect-matrix complex, is a many-body problem that cannot be exactly solved. To this purpose, it is usual to adopt the *Adiabatic Approximation* based on the substantial difference between the electron mass m and the nuclear mass M_α ($M_\alpha/m \geq 10^3$) so that electrons move much faster than nuclei, namely, the nuclei are almost fixed. According to this approximation, the wave function $\psi_{l,n}(r, R)$ which describes the stationary state of the system is given by the product:

$$\psi_{l,n}(r,R) = \varphi_l(r,R)\varphi_{l,n}(R), \tag{2.3}$$

where $\varphi_l(r, R)$ and $\varphi_{l,n}(R)$ are the electronic and nuclear wave functions and are solutions of the two equations:

$$-\frac{\hbar^2}{2m}\sum_j \nabla_j^2 \varphi_l(r,R) + V(r,R)\varphi_l(r,R) = W_l(R)\varphi_l(r,R) \tag{2.4}$$

$$-\frac{\hbar^2}{2}\sum_\alpha \frac{\nabla_\alpha^2}{M_\alpha}\varphi_{l,n}(R) + W_l(R)\varphi_{l,n}(R) = E_{l,n}\varphi_{l,n}(R), \tag{2.5}$$

where $\hbar = h/2\pi$ is the reduced Planck's constant ($h \cong 6.626 \times 10^{-34}$ J·s), and the indices l and n represent the electronic and nuclear states, respectively. Equation (2.4) describes the stationary states of the electrons moving in the field of fixed nuclei and experiencing a potential energy $V(r, R)$. For different nuclear positions, $V(r, R)$ changes and both $\varphi_l(r, R)$ and $W_l(R)$ depend parametrically on R. The motion of the nuclei is governed by the second equation (Eq. 2.5), where $W_l(R)$ plays the role of the potential energy and $E_{l,n}$ represents the eigenvalue of the total energy of the defect-matrix complex.

If the system is in a stable state, the nuclear motions reduce to small vibrations about the equilibrium positions R_{l0}. In the simplest case, when R is the distance between two nuclei, $W_l(R)$ can be expanded in a Taylor series up to quadratic terms:

$$W_l(R) = W_l(R_{l0}) + \frac{1}{2}\left(\frac{d^2W_l}{dR^2}\right)_{R=R_{l0}}(R-R_{l0})^2 + \cdots \equiv W_l(R_{l0}) + \frac{1}{2}a_l(R-R_{l0})^2. \tag{2.6}$$

The vibration frequency in the lth electronic state is related to the coefficient of quadratic terms in expansion (2.6) by $\omega_l = (a_l/\mu)^{1/2}$, where μ is the reduced mass of the system. The substitution of Eq. (2.6) into (2.5) transforms it into the Schrödinger equation of a harmonic oscillator. Its solution is well known: the energy levels are $E_{l,\,n} = W_l(R_{l0}) + \hbar\omega_l(n + 1/2)$, where n is the vibrational quantum number and $\varphi_{l,\,n}(R)$ are Hermite polynomials multiplied by Gaussian functions [3, 4].

Most generally, the defect-matrix complex consists of N nuclei with $f = 3N - 6$ degrees of freedom and the function $W_l(R)$ can be expanded in a series analogous to Eq. (2.6). In this case, it is appropriate to introduce new variables (*normal coordinates*), q_s ($s = 1,2,\ldots,f$), so that the problem reduces to find the energy levels and wave functions for the stationary states of a set of f independent harmonic oscillators (*normal modes*). For each normal coordinate, the nuclear potential curve takes the form of a parabola centered at $q_{l,\,s0}$:

$$W_{l,s}(q_s) = W_{l,s}(q_{l,s0}) + \frac{1}{2}\left(\frac{d^2W_{l,s}}{dq_s^2}\right)_{q_s=q_{l,s0}}(q_s-q_{l,s0})^2 + \cdots \equiv W_{l,s}(q_{l,s0}) + \frac{1}{2}a_{l,s}(q_s-q_{l,s0})^2. \tag{2.7}$$

The total energy is:

$$E_{l,n} = \sum_{s=1}^{f}\hbar\omega_{l,s}\left(n_s+\frac{1}{2}\right) = \hbar\omega_{l,1}\left(n_1+\frac{1}{2}\right) + \hbar\omega_{l,2}\left(n_2+\frac{1}{2}\right) + \cdots + \hbar\omega_{l,f}\left(n_f+\frac{1}{2}\right), \tag{2.8}$$

where the index s indicates the configurational coordinate with vibrational frequency $\omega_{l,\,s}$, and n denotes the set of vibrational quantum numbers n_s. The total wave function is therefore the product of the individual normal oscillators:

$$\varphi_{l,n}(q) = \prod_{s=1}^{f}\varphi_{l,n_s}(q_s) = \varphi_{l,n_1}(q_1)\varphi_{l,n_2}(q_2)\cdots\varphi_{l,n_f}\left(q_f\right). \tag{2.9}$$

The interplay between the normal modes of the defect and those of the whole solid is crucial to determine the optical lineshape. Indeed, it is known that the eigenfunction spectrum for the normal modes of a solid consists of alternating allowed and forbidden bands. When a defect, composed of a finite number of atoms undergoing vibrations, is introduced into the solid, we can distinguish two different cases [2].

Band vibrations: The vibrational frequency of the atoms included in the defect lies in one of the allowed bands of the solid matrix. The defect is in resonance with the eigenfrequencies of the host network and radiates elastic waves, thus losing energy. In this case, all N atoms, both belonging to the defect and to the host matrix, participate to the motion and share the finite energy of the normal mode. Then, each atom has energy depending on N^{-1} and its displacement is related to $N^{-1/2}$. Vibrations of this kind are called *band vibrations*.

Localized vibrations: The vibrational frequency of the atoms included in the defect lies in one of the forbidden bands of the solid. In this case, since the defect is not in resonance with any eigenfrequency of the unperturbed host matrix, it does not radiate elastic waves. The vibration amplitudes of the atoms in the solid drop off rapidly with increasing the distance from the defect; only the atoms of the defect environment participate in the vibration and their displacements are independent of N. Such vibrations are called *localized vibrations*.

2.1.2 Optical Transitions: The Franck–Condon Principle

To describe the optical transition between two electronic states, Figure 2.1 shows a configuration coordinate (q_s) diagram where the potential energy curves of the ground, $W_{I,\,s}(q_s)$), and excited, $W_{II,s}(q_s)$, states are represented together with the vibrational levels, and ε_0 is the energy difference between them. Since the electronic state changes in a time ($\sim 10^{-15}$ s) much shorter than the nuclear vibration ($\sim 10^{-12}$ s), it can be assumed that the nuclei do not move nor change their momenta during the electronic transition (*Franck–Condon Principle*) [8, 9]. Therefore, both the absorption (from the ground to the excited state) and the luminescence (from the excited to the ground state) are represented by vertical arrows.

According to quantum mechanics, optical absorption and luminescence processes are quite well described by the first-order time-dependent perturbation theory. Let us consider the electronic state I, $\psi_{I,\,n}(r,q) = \varphi_I(r,q)\varphi_{I,\,n}(q)$, with the set of vibrational levels $n \equiv \{n_1, n_2, \ldots, n_f\}$ thermally populated in accordance with the Boltzmann distribution; the absorption transition probability $W(\mathrm{I}, n \rightarrow \mathrm{II}, m)$ to the set of vibrational levels $m \equiv \{m_1, m_2, \ldots, m_f\}$ of the electronic state II, $\psi_{\mathrm{II},\,m}(r,q) = \varphi_{\mathrm{II}}(r,q)\varphi_{\mathrm{II},\,m}(q)$, is proportional to the absolute square of the matrix element of the perturbation operator. Since the light wavelength is much greater than the size of the defect (*electric dipole approximation*), the perturbation operator will be the dipole moment due to the electronic and nuclear charges:

$$P = D_{\mathrm{el}} + D_{\mathrm{nucl}}. \tag{2.10}$$

It is worth noting that in the description of electronic transitions, the contribution D_{nucl}, involving purely vibrational transitions within a single electronic state, can be neglected. The matrix element of P is therefore:

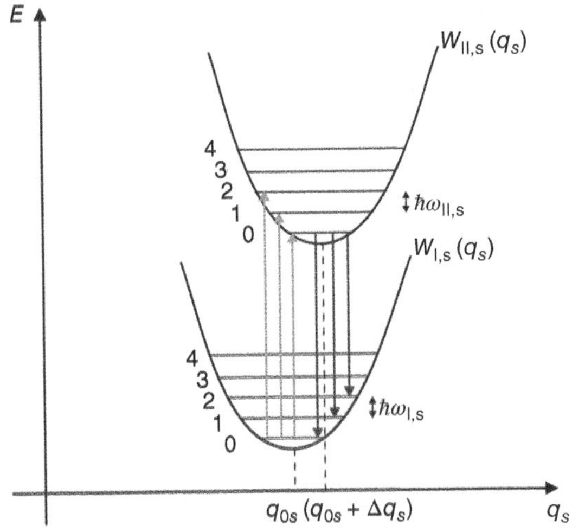

Figure 2.1 Configuration coordinate diagram. The potential energy of the ground $W_{I,\,s}$ and the excited $W_{II,\,s}$ electronic states are depicted together with the associated vibrational levels. For simplicity sake, the electronic transitions (vertical arrows) are supposed to take place from the lower vibrational level.

$$P_{\mathrm{I},n\to\mathrm{II},m} = \iint \mathrm{d}r\ \ \mathrm{d}q\,\varphi_{\mathrm{II}}^{*}(r,q)\,\varphi_{\mathrm{II},m}^{*}(q)\,D_{\mathrm{el}}\,\varphi_{I}(r,q)\,\varphi_{\mathrm{I},n}(q)$$
$$= \int \mathrm{d}q\left[\int \mathrm{d}r\,\varphi_{\mathrm{II}}^{*}(r,q)\,D_{\mathrm{el}}(r)\,\varphi_{I}(r,q)\right]\varphi_{\mathrm{II},m}^{*}(q)\,\varphi_{\mathrm{I},n}(q). \tag{2.11}$$

Hereafter, we will use the notation:

$$D_{\mathrm{I}\to\mathrm{II}}(q) \equiv \int \mathrm{d}r\,\varphi_{\mathrm{II}}^{*}(r,q)\,D_{\mathrm{el}}(r)\,\varphi_{I}(r,q) \tag{2.12}$$

to indicate the electronic matrix element. $D_{\mathrm{I}\to\mathrm{II}}(q)$ can be argued to be only weakly dependent on the nuclear coordinates, and in agreement with the *Condon Approximation*, it can be replaced by its value at the nuclear equilibrium position, $D^{0}_{\mathrm{I}\to\mathrm{II}}$.

Thus, the absorption transition probability is given by:

$$W(\mathrm{I},n\ \rightarrow\ \mathrm{II},m) \propto \left|D^{0}_{\mathrm{I}\to\mathrm{II}}\right|^{2}\left|\int \varphi_{\mathrm{II},m}^{*}(q)\,\varphi_{I,n}(q)\ \ \mathrm{d}q\right|^{2}. \tag{2.13}$$

The first factor is the electronic part and is proportional to the overall intensity of the absorption band. The second factor, called *Franck–Condon integral*, is the nuclear part; it measures the overlap between the vibrational functions of the ground and excited state and determines the band shape [10, 11].

After absorption, the nuclei relax toward the minimum energy configuration in a much shorter time (10^{-12}–10^{-11} s) than the luminescence lifetime ($\geq 10^{-9}$ s). This implies that at the time when light emission occurs, the vibrational levels of the electronic excited state are populated in accordance with a thermal distribution. The rate of light emission is given by the relationship between the *Einstein coefficients* of stimulated absorption (b) and spontaneous emission (a):

$$a = \frac{\hbar\omega^{3}n_{\mathrm{ref}}^{3}b}{\pi c^{3}}, \tag{2.14}$$

where n_{ref} is the refraction index of the medium. The emission transition probability is obtained by combining Eqs. (2.13) and (2.14). The indexes of the nuclear functions are different from those of Eq. (2.13) as a consequence of the displacement of the equilibrium position:

$$W(\mathrm{II},m'\ \rightarrow\ \mathrm{I},n') \propto \frac{\hbar\omega^{3}n^{3}}{\pi c^{3}}\left|D^{0}_{\mathrm{I}\to\mathrm{II}}\right|^{2}\left|\int \varphi_{\mathrm{I},n'}^{*}(q)\,\varphi_{\mathrm{II},m'}(q)\ \ \mathrm{d}q\right|^{2}. \tag{2.15}$$

A case of particular interest is that in which the vibration frequency does not change during the electronic transition ($\hbar\omega_{\mathrm{I},\,s} = \hbar\omega_{\mathrm{II},\,s}$), and, consequently, $W_{\mathrm{I},\,s}(q_s)$ and $W_{\mathrm{II},\,s}(q_s)$ have the same curvature; this is referred to as *linear electron–phonon coupling*. Consequently, the vibrational wave functions in the ground and excited electronic states are equal in pairs, thus leading to a symmetrical relationship:

$$\left|\int \varphi_{\mathrm{II},m}^{*}(q)\,\varphi_{\mathrm{I},n}(q)\ \ \mathrm{d}q\right|^{2} = \left|\int \varphi_{\mathrm{I},m}^{*}(q)\,\varphi_{\mathrm{II},n}(q)\ \ \mathrm{d}q\right|^{2}. \tag{2.16}$$

Since the absorption and luminescence transitions occur under temperature equilibrium, the vibrational nth level, both in the ground and in the excited states, are populated following the same distribution. Therefore, the symmetry property of the *Franck–Condon integral* (Eq. 2.16) applies to the absorption and luminescence lineshapes; as a consequence they are characterized by a *mirror symmetry* with symmetry plane at the energy of the transition between the lowest vibrational levels.

2.1.3 Zero-Phonon Line

Optical transitions in which the vibrational state of the system does not change are called *purely electronic transitions* or *zero-phonon transitions*, since they occur without creation or annihilation of phonons; in this case, absorption and luminescence are resonant [6]. An example of purely electronic absorption is shown in the configuration coordinate (q_s) diagram of Figure 2.2: E_{00}, E_{11}, and E_{22} represent the energies of the $0 \rightarrow 0$, $1 \rightarrow 1$, and $2 \rightarrow 2$ transitions. In general, $W_{I,\,s}(q_s)$ and $W_{II,\,s}(q_s)$ have different curvatures ($\hbar\omega_{I,\,s} \neq \hbar\omega_{II,\,s}$), then the purely electronic transitions do not occur at the same energy and are different from ε_0: $E_{00} \neq E_{11} \neq E_{22} \neq \varepsilon_0$. In the *linear electron–phonon coupling* approximation ($\hbar\omega_{I,\,s} = \hbar\omega_{II,\,s}$), all the zero-phonon transitions have the same energy ($E_{00} = E_{11} = E_{nn}$) coinciding with ε_0. In this case, the spectrum evidences a single line called *Zero-Phonon Line* (ZPL) of energy $E_{00} = \hbar\omega_0$, which is due to the superposition of a set of lines arising from transitions between different pairs of vibrational levels. Therefore, ZPL lies exactly at the energy of the purely electronic transition and is independent of temperature.

Ideally, ZPL has zero width and its shape is accounted for by the δ-function:

$$I_{ZPL}(E) \propto \delta(E - E_{00}). \tag{2.17}$$

In reality, ZPL has a finite width due to the lifetime τ of the excited electronic state; δ-function in Eq. (2.17) is, therefore, replaced by a Lorentzian curve with full width at half maximum (FWHM) of $(\hbar/\tau)$:

$$I_{ZPL}(E) \propto \frac{(\hbar/2\tau)^2}{(E - E_{00})^2 + (\hbar/2\tau)^2}. \tag{2.18}$$

The *zero-phonon transition* probability for each configuration coordinate q_s is:

$$W(\mathrm{I}, n_s \rightarrow \mathrm{II}, n_s) \propto \left| \int \varphi^*_{\mathrm{II},n_s}(q_s)\varphi_{\mathrm{I},n_s}(q_s)\ \mathrm{d}q_s \right|^2, \tag{2.19}$$

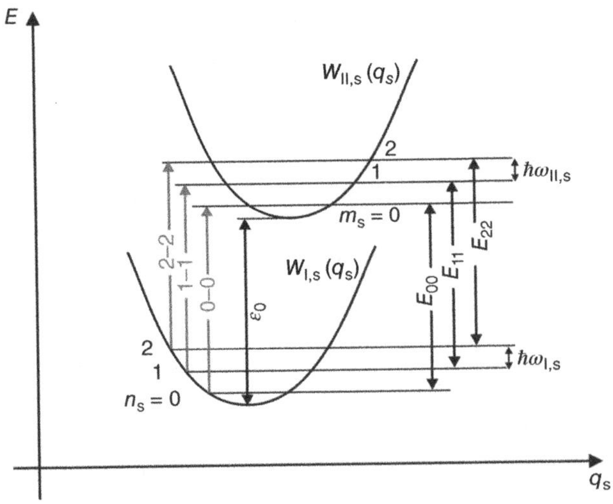

Figure 2.2 Configuration coordinate diagram where are depicted the *zero-phonon transitions* $0 \rightarrow 0$, $1 \rightarrow 1$, and $2 \rightarrow 2$.

where the vibrational levels of the electronic states I and II coincide ($n_s = m_s$). According to Eq. (2.9), the total probability is given by:

$$W(\mathrm{I}, n \to \mathrm{II}, n) \propto \prod_{s=1}^{f} \left| \int \varphi^*_{\mathrm{II},n_s}(q_s)\varphi_{\mathrm{I},n_s}(q_s)\ \mathrm{d}q_s \right|^2. \tag{2.20}$$

To calculate the temperature dependence of the ZPL intensity, $I_{\mathrm{ZPL}}(T)$, we have to average over the Boltzmann distribution to account for the thermal population of the vibrational levels n_s:

$$\begin{aligned} I_{\mathrm{ZPL}}(T) &= A\langle W(\mathrm{I}, n \to \mathrm{II}, n)\rangle_T \\ &= A|D|^2F(T)\sum_{[n]} \exp\left[-\frac{\hbar}{kT}\sum_{s=1}^{f}\left(n_s + \frac{1}{2}\right)\omega_{\mathrm{I},s}\right] \times \prod_{s=1}^{f}\left|\int \varphi^*_{\mathrm{II},n_s}(q_s)\varphi_{\mathrm{I},n_s}(q_s)\ \mathrm{d}q_s\right|^2. \end{aligned} \tag{2.21}$$

A is a constant independent of the frequency, $D \equiv D^0_{\mathrm{II}\to\mathrm{I}}$ is the electron matrix element in the Condon approximation, $F(T)$ is the normalizing factor of the Boltzmann's distribution function, k is the Boltzmann's constant ($k \cong 1.381 \times 10^{-23}$ J·K^{-1}), $\omega_{\mathrm{I},\,s}$ is the frequency of the sth normal mode in the electronic state I, and $[n]$ represents all the possible sets n_s.

First, only the contribution of the band vibrations is considered. As discussed above, the equilibrium positions of the electronic states I and II in the configuration coordinate q_s differ for an infinitesimal quantity Δq_s, whereas the oscillator frequencies do not change, ($\omega_{\mathrm{I},\,s} = \omega_{\mathrm{II},\,s} = \omega_s$). The relationship between the vibrational wave functions is therefore:

$$\varphi_{\mathrm{II},n_s}(q_s) = \varphi_{I,n_s}(q_s - \Delta q_s) = \varphi_{\mathrm{I},n_s}(q_s) - \frac{\mathrm{d}\varphi_{\mathrm{I},n_s}}{\mathrm{d}q_s}\Delta q_s + \frac{1}{2}\frac{\mathrm{d}^2\varphi_{I,n_s}}{\mathrm{d}q_s^2}\Delta q_s{}^2 + \cdots. \tag{2.22}$$

The overlap integral in Eq. (2.21) becomes:

$$\begin{aligned} J_{n_s} &= \int_{-\infty}^{+\infty} \varphi^*_{\mathrm{II},n_s}\varphi_{\mathrm{I},n_s}\mathrm{d}q_s \\ &= \int_{-\infty}^{+\infty} \varphi^*_{\mathrm{I},n_s}\varphi_{\mathrm{I},n_s}\mathrm{d}q_s - \Delta q_s\int_{-\infty}^{+\infty} \varphi^*_{\mathrm{I},n_s}\frac{\mathrm{d}}{\mathrm{d}q_s}\varphi_{\mathrm{I},n_s}\mathrm{d}q_s + \frac{1}{2}\Delta q_s{}^2\int_{-\infty}^{+\infty} \varphi^*_{\mathrm{I},n_s}\frac{\mathrm{d}^2}{\mathrm{d}q_s^2}\varphi_{\mathrm{I},n_s}\mathrm{d}q_s + \cdots \\ &\cong 1 + \frac{1}{2}\Delta q_s{}^2\int_{-\infty}^{+\infty} \varphi^*_{\mathrm{I},n_s}\frac{\mathrm{d}^2}{\mathrm{d}q_s^2}\varphi_{\mathrm{I},n_s}\mathrm{d}q_s, \end{aligned} \tag{2.23}$$

where we have account for the properties of orthonormality and symmetry of the vibrational functions. The remaining integral can be expressed in terms of the average kinetic energy:

$$\frac{1}{2}\int_{-\infty}^{+\infty} \varphi^*_{\mathrm{I},n_s}\frac{\mathrm{d}^2}{\mathrm{d}q_s^2}\varphi_{\mathrm{I},n_s}\mathrm{d}q_s = -\frac{M_s}{\hbar^2}\int_{-\infty}^{+\infty} \varphi^*_{\mathrm{I},n_s}\hat{T}\varphi_{\mathrm{I},n_s}\mathrm{d}q_s = -\frac{M_s}{\hbar^2}\langle\hat{T}\rangle_{\mathrm{I},n_s}, \tag{2.24}$$

where $\langle\hat{T}\rangle_{\mathrm{I},n_s}$ is the kinetic energy averaged over the nth state of the sth oscillator and M_s is the nuclear mass defined in the normal mode s. According to the virial theorem:

$$\langle\hat{T}\rangle_{I,n_s} = \frac{1}{2}\left(n_s + \frac{1}{2}\right)\hbar\omega_s, \tag{2.25}$$

the overlap integral is:

$$J_{n_s} = 1 - \frac{\Delta q_s{}^2 M_s}{2\hbar^2} \hbar\omega_s \left(n_s + \frac{1}{2}\right) \quad (2.26)$$

and its square is

$$J_{n_s}^2 \cong 1 - \frac{\Delta q_s{}^2 M_s}{\hbar^2} \hbar\omega_s \left(n_s + \frac{1}{2}\right), \quad (2.27)$$

where the terms of the order of Δq_s^4 are neglected.

To solve Eq. (2.21), because of the independence of normal modes, the states of the oscillator 1 are separated from the remaining set n and are averaged over them:

$$\begin{aligned} I_{\mathrm{ZPL}}(T) &= A|D|^2 F_1(T) \sum_{n_1=1}^{\infty} \exp\left[-\frac{\hbar\omega_s}{kT}\left(n_1 + \frac{1}{2}\right)\right] \times J_{n_1}^2 \\ &\cdot F^*(T) \sum_{[n^*]} \exp\left[-\frac{\hbar}{kT}\sum_{s=2}^{f}\left(n_s + \frac{1}{2}\right)\omega_s\right] \times \prod_{s=2}^{f} J_{n_s}^2. \end{aligned} \quad (2.28)$$

The asterisks on the indices n and on the normalizing factor $F(T)$ mean that the oscillator 1 is excluded from the set. The first factor in Eq. (2.28) becomes:

$$F_1(T) \sum_{n_1=1}^{\infty} \exp\left[-\frac{\hbar\omega_s}{kT}\left(n_1 + \frac{1}{2}\right)\right]\left[1 - \frac{\Delta q_s{}^2 M_s}{\hbar^2}\hbar\omega_s\left(n_s + \frac{1}{2}\right)\right] = 1 - \frac{\Delta q_s{}^2 M_s}{\hbar^2} E_1(T). \quad (2.29)$$

$E_1(T)$ is the average thermal energy of a linear harmonic oscillator:

$$E_1(T) = \hbar\omega_s\left(\overline{n}_1 + \frac{1}{2}\right) = \frac{1}{2}\hbar\omega_s \coth\frac{\hbar\omega_s}{2kT}, \quad (2.30)$$

where $\overline{n}_1$ is the average quantum number of the oscillator 1. The same algorithm is applied on the states of other oscillator numbers and after a bit of algebra, the integrated intensity of ZPL is found to be:

$$I_{\mathrm{ZPL}}(T) = A|D|^2 \exp\left[-\sum_{s=1}^{f} \frac{\Delta q_s{}^2 M_s}{\hbar^2} kT_s\right], \quad (2.31)$$

where T_s is the *effective temperature* of a harmonic oscillator:

$$kT_s \equiv E_s(T) = \hbar\omega_s\left(\overline{n}_s + \frac{1}{2}\right) \equiv \frac{1}{2}\hbar\omega_s \coth\frac{\hbar\omega_s}{2kT}. \quad (2.32)$$

It is useful to introduce the dimensionless parameter, called *Huang Rhys factor* [12], for the band vibrations:

$$S_s^B \equiv \left(\frac{1}{2} M_s \omega_s^2 \Delta q_s{}^2\right) / \hbar\omega_s. \quad (2.33)$$

The physical meaning of the Huang Rhys factor has been introduced in the previous chapter. According to Eq. (2.33), $2\hbar\omega_s S_s^B$ equals the difference between the absorption energy E_A (vertical distance from the minimum of $W_{I,s}(q_s)$ to $W_{II,s}(q_s)$) and the luminescence energy E_L (vertical distance from the minimum of $W_{II,s}(q_s)$ to $W_{I,s}(q_s)$). The difference between E_A and E_L is the so-called *Stokes shift*, S_s^B is therefore:

$$S_s^B = \frac{E_A - E_L}{2\hbar\omega_s} = \frac{\text{Stokes shift}}{2\hbar\omega_s}, \tag{2.34}$$

it gives the amount of the vibrational relaxation energy in units of the vibrational quantum.

After introducing S_s^B, the expression for $I_{ZPL}(T)$ can be rewritten as:

$$I_{ZPL}(T) = A|D|^2 \exp\left[-\sum_{s=1}^{f} S_s^B \coth\frac{\hbar\omega_s}{2kT}\right] \equiv A|D|^2 \exp\left[-\sum_{s=1}^{f} S_s^B \frac{2kT_s}{\hbar\omega_s}\right]. \tag{2.35}$$

At low temperature, namely $kT \ll \hbar\omega_s$, $\coth(\hbar\omega_s/2kT) \to 1$ or $T_s \to (\hbar\omega_s)/2k$ and $I_{ZPL}(T)$ becomes:

$$I_{ZPL}(T \to 0) = A|D|^2 \exp\left[-\sum_{s=1}^{f} S_s^B\right], \tag{2.36}$$

that increases with decreasing the *total Huang Rhys factor*, $\sum_{s=1}^{f} S_s^B$. At high temperature, namely $kT \gg \hbar\omega_s$, $\coth(\hbar\omega_s/2kT) \to 2kT/\hbar\omega_s$ or $kT_s \to kT$ and the expression for $I_{ZPL}(T)$ becomes:

$$I_{ZPL}(kT \gg \hbar\omega_s) = A|D|^2 \exp\left[-kT\sum_{s=1}^{f} \frac{2S_s^B}{\hbar\omega_s}\right]. \tag{2.37}$$

In this case, $I_{ZPL}(T)$ decreases exponentially with increasing temperature and drops faster the larger is the *Huang Rhys factor.*

2.1.4 Phonon Line Structure

In the following, we deal with the phonon coupled transitions to succeed in describing the shape of the whole band and its temperature dependence; the effects of band vibrations and localized vibrations will be separately discussed.

As introduced in Section 2.1.1, band vibrations correspond to the transitions in which phonons of the matrix are created or annihilated. They appear as broad continuous bands whose shape, $L_{vib}(\omega, T)$, depends on the spectral density of phonons and on the perturbation nearby the defect. As a consequence, $L_{vib}(\omega, T)$ cannot be derived exactly and we will limit ourselves to indicate it by its formal expression. In fact, the integrated intensity of the whole vibronic band is given by:

$$I_{vib}(T) = \sum_n v_n \sum_m W(\text{I}, n \to \text{II}, m). \tag{2.38}$$

The sum over n is an average over the vibrational levels in the ground electronic state, each level having a weight v_n. The sum over m corresponds to all the possible transitions from the ground to the excited state including the ZPL ($n = m$). Under the Condon approximation, the previous equation can be written as:

$$
\begin{aligned}
I_{\text{vib}}(T) &= \sum_n v_n \sum_m |D|^2 \left| \int \varphi^*_{\text{II},m}(q)\varphi_{\text{I},n}(q)\ \text{d}q \right|^2 \\
&= \sum_n v_n \sum_m |D|^2 \int \varphi^*_{\text{II},m}(q')\varphi_{\text{I},n}(q')\ \text{d}q' \int \varphi^*_{\text{I},n}(q)\,\varphi_{\text{II},m}(q)\ \text{d}q \\
&= \sum_n v_n |D|^2 \int\!\!\int \text{d}q\ \text{d}q' \varphi^*_{\text{I},n}(q)\,\varphi_{\text{I},n}(q') \sum_m \varphi^*_{\text{II},m}(q')\varphi_{\text{II},m}(q) \\
&= \sum_n v_n |D|^2 \int \text{d}q \varphi^*_{\text{I},n}(q)\,\varphi_{\text{I},n}(q) \\
&= \sum_n v_n |D|^2 \int \text{d}q |\varphi_{\text{I},n}(q)|^2,
\end{aligned}
\tag{2.39}
$$

where the properties of the complete set of functions $\varphi_{\text{II},\ m}(q)$ have been used:

$$
\sum_m \varphi^*_{\text{II},m}(q)\varphi_{\text{II},m}(q') = \delta(q-q'). \tag{2.40}
$$

The last term in Eq. (2.39) becomes:

$$
I_{\text{vib}}(T) = \sum_n v_n |D|^2 \int |\varphi_{I,n}(q)|^2\ \text{d}q = N|D|^2 = \text{const}, \tag{2.41}
$$

where N is the total number of defects in the solid matrix. Equation (2.41) indicates that I_{vib} does not depend on the distribution function v_n and, therefore, it does not change with temperature. This property leads to a constant value of the total area under the spectrum; the decrease of ZPL with increasing temperature is compensated by the increase of the other part of the spectrum (vibrational background).

Luminescence spectra have the same features provided that non-radiative transitions are absent so as to keep constant the total area. To single out the temperature dependence of the ZPL, it is useful to consider the ratio between its intensity and that of the whole band. This relative intensity is known as *Debye–Waller factor* and is given by:

$$
\alpha(T) = \frac{I_{\text{ZPL}}(T)}{I_{\text{vib}}(T)} = \exp\left[-\sum_{s=1}^{f} S_s^B \coth\frac{\hbar\omega_s}{2kT} \right]. \tag{2.42}
$$

The localized vibrations cause a change of the potential curve $W_{ls}(q_s)$ and of its minimum position during the electronic transition. In the simplest case of a single localized mode with frequency Ω and under the approximation $\hbar\Omega \gg kT$, the thermal excitation of the vibrational levels in the ground electronic state can be neglected. It is worth noting that, since the localized vibration frequencies are quite high ($\hbar\Omega \sim 10^{-1}$ eV), this condition is satisfied in a wide temperature range.

The configurational coordinate of the single localized mode is the nuclear distance R and the vibration levels are denoted as λ and λ' in the ground and excited electronic states. The overlap integral in Eq. (2.13) is therefore calculated between the wave function $\varphi_{\text{II},\lambda'}(R)$ in the excited state and $\varphi_{\text{I},0}(R)$ in the ground state where only the vibrational level λ=0 is populated:

$$
J_{0\lambda'} = \int_{-\infty}^{+\infty} \varphi^*_{\text{II},\lambda'}(R)\varphi_{I,0}(R)\ \text{d}R. \tag{2.43}
$$

For the nuclear harmonic oscillator with mass M, we can exploit the relationship $\hbar\Omega = M\Omega^2\overline{R}^2$, $\overline{R} \equiv (\hbar/m\Omega)^{1/2}$ being the nuclear average displacement from equilibrium position. Then, the wave functions $\varphi_{I,0}$ and $\varphi_{II,\lambda'}$ can be expressed as a function of the dimensionless parameter $\xi = R/\overline{R}$, and they assume the form of a Hermite polynomial multiplied by a Gaussian function:

$$\varphi_{I,0}(\xi) = \frac{1}{\sqrt{\pi}} e^{-\xi^2/2} \tag{2.44}$$

$$\varphi_{II,\lambda'}(\xi) = \frac{(-1)^{\lambda'}}{\sqrt{2^{\lambda'}\lambda'!\pi}} e^{\xi^2/2} \frac{d^{\lambda'}}{d\xi^{\lambda'}} e^{-\xi^2}. \tag{2.45}$$

Owing to the linear electron–phonon coupling approximation, the vibrational wave functions are in the following relation: $\varphi_{I,0}(\xi) = \varphi_{II,0}(\xi - \xi_0)$, where $\xi_0 = R_0/\overline{R}$ is the shift between the two potential curves, the equilibrium position of the oscillator in electronic state II being taken as the origin. Equation (2.43) becomes:

$$J_{0\lambda'} = C_{\lambda'} \int_{-\infty}^{+\infty} e^{-\xi\xi_0} \frac{d^{\lambda'}}{d\xi^{\lambda'}} e^{-\xi^2}\, d\xi, \tag{2.46}$$

where we have introduced the notation:

$$C_{\lambda'} \equiv \frac{(-1)^{\lambda'}}{\sqrt{2^{\lambda'}\lambda'!\pi}} e^{-\xi_0^2/2}. \tag{2.47}$$

After solving this integral by parts, we obtain the transition probability $W(I,0 \rightarrow II,\lambda')$:

$$W(I,0 \rightarrow II,\lambda') = J_{0\lambda'}^2 = e^{-\xi_0^2/2}\left(\frac{\xi_0^2}{2}\right)\frac{1}{\lambda'!} = e^{-S^L}\frac{(S^L)^{\lambda'}}{\lambda'!}, \tag{2.48}$$

that is given by the Poisson distribution, where $S^L \equiv \xi_0^2/2 \equiv M\Omega^2 R_0^2/2\hbar\Omega$ is the Huang–Rhys factor for the localized vibration.

Figure 2.3 displays representative patterns of Eq. (2.48) corresponding to [11]:

a) Weak electron–phonon coupling ($0 < S^L < 1$). In this case, the maximum transition probability is that connecting the lower vibrational levels in the ground and excited electronic states, which is the ZPL.
b) Medium electron–phonon coupling ($1 < S^L < 6$). In this case, the probability distribution is asymmetric and is peaked at vibrational levels with $\lambda' > 0$ and the ZPL transition is still probable.
c) Strong electron–phonon coupling ($S^L > 10$). In this case, the distribution is symmetric (Gaussian shape) and the ZPL transition probability vanishes.

2.1.5 Vibrational Structure

To derive the properties of the optical spectra, the overall effect of band and localized vibrations must be taken into account. Since the coupling to each mode is independent from all the others, the total transition probability can be factorized in the product of the transition probabilities for each mode. If we consider a single localized vibration of frequency Ω, and f band vibrations, the transition probability from (I,λ,n_s) to (II,λ',m_s) is:

$$W(I,\lambda,n_s \rightarrow II,\lambda',m_s) = W_{\lambda,\lambda'} \prod_{s=1}^{f} W_{n_s,m_s}. \tag{2.49}$$

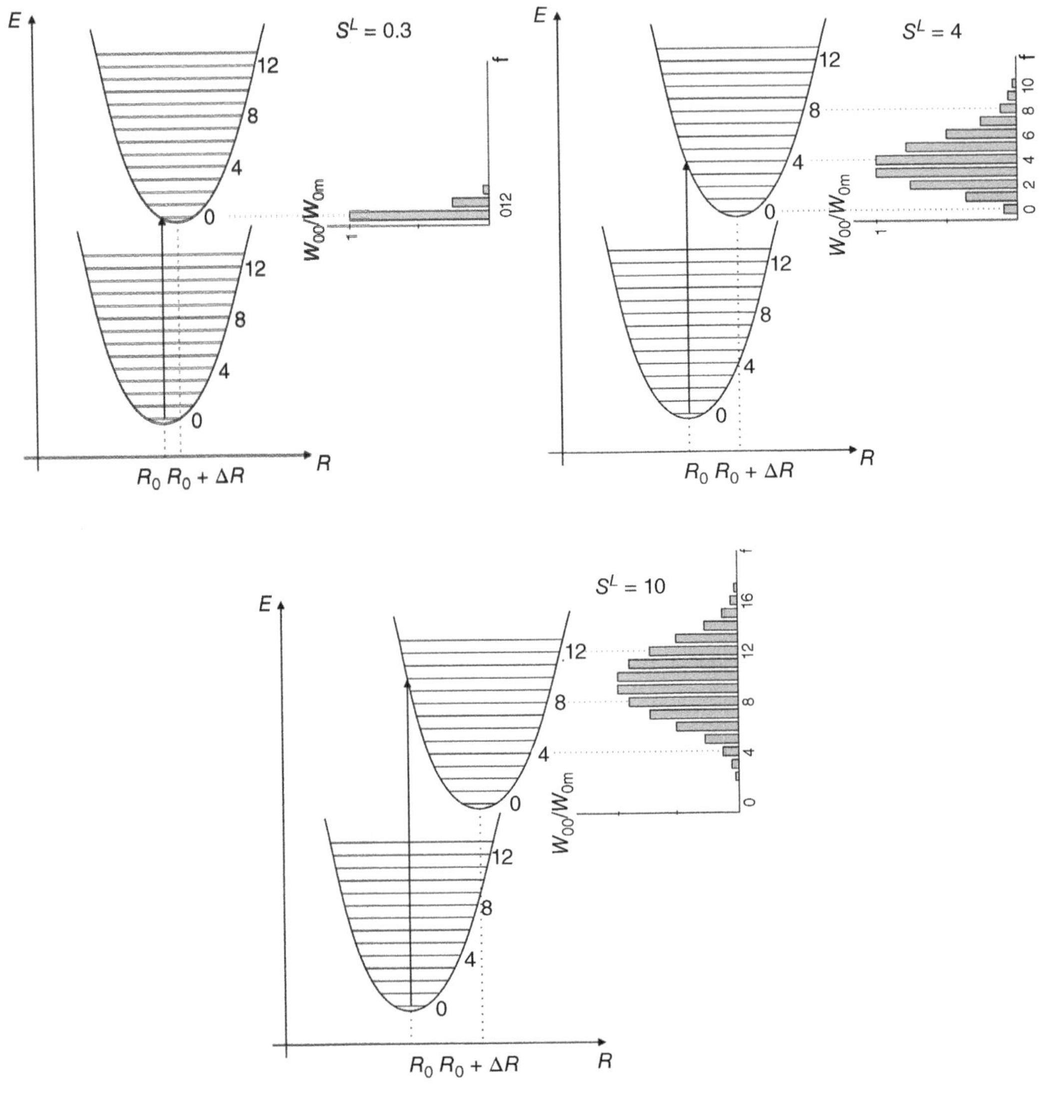

Figure 2.3 Configuration coordinate diagram showing the potential energy of the ground and excited electronic states for three values of the Huang–Rhys factor for the localized vibration, S^L. The distributions of the vibronic transition probability are also reported. Source: Modified from Skuja [11].

I and II are the electronic states, λ and λ' denote the localized vibration, n_s and m_s are the sets of the band vibrations, $W_{\lambda,\lambda'}$ is the transition probability between the localized vibration levels in the ground and excited electronic states, and W_{n_s,m_s} is the transition probability between the band vibration levels. In the approximation $kT \ll \hbar\Omega$, $\lambda = 0$ and the previous equation averaged over the band vibration states is:

$$\langle W(\mathrm{I}, 0, n_s \rightarrow \mathrm{II}, \lambda', m_s)\rangle_T = W_{0,\lambda'} \left\langle \prod_{s=1}^{f} W_{n_s,m_s} \right\rangle_T = W_{0,\lambda'} L_{\mathrm{vib}}(\omega, T), \tag{2.50}$$

where ω is the frequency of the transition $n_s \rightarrow m_s$.

Equations (2.49) and (2.50) determine the spectral features of the whole optical band. The intensity of transitions coupled to the localized vibration going from $\lambda = 0$ to λ' ($\lambda' = 0, 1, 2...$), $W_{0,\lambda'}$, is not influenced by the interaction with the band vibrations so that is distributed according to the Poisson distribution of Eq. (2.48). The band vibrations, in turn, give rise to an internal lineshape, $L_{\text{vib}}(\omega, T)$, featuring the properties of the vibronic band in the absence of a localized vibration: a ZPL and a vibrational background. Upon increasing temperature, the ZPL intensity decreases and the vibrational background increases so that the total intensity remains constant. $L_{\text{vib}}(\omega, T)$ follows a *similarity law*: the overall spectrum is a series of replicas of *vibronic lines* spaced apart by $\lambda'\Omega$ ($\lambda' = 0,1,2, ...$) from the electronic transition. The intensity of each replica is given by the factor $W_{0,\lambda'}$ governed by Eq. (2.48); therefore, the absorption transition has the expression:

$$\begin{aligned} I_{\text{hom}}^{\text{abs}}(\omega, T) &= W_{0,0}L_{\text{vib}}(\omega, T) + W_{0,1}L_{\text{vib}}(\omega - \Omega, T) + W_{0,2}L_{\text{vib}}(\omega - 2\Omega, T) \\ &\quad + \cdots + W_{0,\lambda'}L_{\text{vib}}(\omega - \lambda'\Omega, T) + \cdots = \sum_{\lambda'} W_{0,\lambda'}L_{\text{vib}}(\omega - \lambda'\Omega, T) \\ &= \sum_{\lambda'} \mathrm{e}^{-S^L} \frac{(S^L)^{\lambda'}}{\lambda'!} L_{\text{vib}}(\omega - \lambda'\Omega, T). \end{aligned} \tag{2.51}$$

For the luminescence transition, the *similarity law* also applies: the overall spectrum is a series of replicas of the vibronic bands $L_{\text{vib}}(\omega, T)$ spaced apart by $-\lambda\Omega$ ($\lambda = 0,1,2, ...$) from the electronic transition:

$$\begin{aligned} I_{\text{hom}}^{\text{lum}}(\omega, T) &= W_{0,0}L_{\text{vib}}(\omega, T) + W_{0,1}L_{\text{vib}}(\omega + \Omega, T) + W_{0,2}L_{\text{vib}}(\omega + 2\Omega, T) \\ &\quad + \cdots + W_{0,\lambda}L_{vib}(\omega + \lambda\Omega, T) + \cdots = \sum_{\lambda} W_{0,\lambda}L_{\text{vib}}(\omega + \lambda\Omega, T) \\ &= \sum_{\lambda} \mathrm{e}^{-S^L} \frac{(S^L)^{\lambda}}{\lambda!} L_{\text{vib}}(\omega + \lambda\Omega, T), \end{aligned} \tag{2.52}$$

$I_{\text{hom}}^{\text{abs}}(\omega, T)$ and $I_{\text{hom}}^{\text{lum}}(\omega, T)$ describe the spectral shape of a single defect and are therefore referred to as a homogeneous property.

We note that, when the thermal excitation of the localized vibrations is taken into account ($kT \sim \hbar\Omega, \lambda \geq 0$), two substantial differences emerge: the replicas $L_{\text{vib}}(\omega, T)$ appear in the anti-Stokes region of the spectrum and the factor $W_{0,\lambda'}$ is replaced by a thermally averaged one. In this case, since the localized vibrational states with $\lambda > 0$ contribute to the spectrum, both absorption and luminescence bands widen with increasing temperature. Given a single localized mode of frequency Ω linearly coupled with the electronic transition (Huang–Rhys factor S), the width of the optical band, measured as FWHM, is given by:

$$\text{FWHM}(T) = \sqrt{8(\ln 2)} \times \hbar\Omega[S \coth(\hbar\Omega/2kT)]^{1/2}. \tag{2.53}$$

2.1.6 Inhomogeneous Effects

Till now, we have dealt with the homogeneous spectral lineshape of a single defect that is expressed by Eqs. (2.51) and (2.52) related to absorption and luminescence, respectively, which apply to the coupling with a single localized mode. In fact, the homogeneous features can be measured in systems where all defects are absolutely identical and undergo precisely identical changes under the influence of the surrounding matrix. A class of solids that quite well reproduces this condition is that of the crystals containing defects embedded exactly in equivalent positions with respect to the matrix.

A completely different framework is represented by amorphous solids where, due to the disordered network, each defect is surrounded by a different local environment; this site-to-site nonequivalence results in inhomogeneous effects on the spectral features [10, 11]. In the simplest approximation, it is assumed that inhomogeneous fluctuations cause an *energy shift* of the homogeneous spectrum as a whole without any changes in its shape, whereas the other spectroscopic parameters (transition probability, phonon energy, ...) remain constant. In this case, it is convenient to introduce a one-dimensional inhomogeneous distribution function $w_{inh}(E_{00})$, so that $w_{inh}(E_{00})\Delta E$ represents the fraction of defects having their ZPL in the energy interval ΔE around E_{00}. The inhomogeneous distribution is usually described by a bell-shaped Gaussian function:

$$w_{inh}(E_{00}) = \frac{1}{\sqrt{2\pi}\sigma_{inh}} e^{-\frac{\left(E_{00}-\overline{E}_{00}\right)^2}{2\sigma_{inh}^2}} \tag{2.54}$$

peaked at the mean energy $\overline{E}_{00}$ whose width σ_{inh} does not depend on temperature. In the general case, the whole optical spectrum, both absorption and luminescence, is therefore given by the convolution of Eqs. (2.51, 2.52) and (2.54) in which the first is conveniently rewritten as a function of E_{00}:

$$I(E) = I_{hom}(E - E_{00}) \otimes w_{inh}(E_{00}). \tag{2.55}$$

Because of the inhomogeneity effects, the vibronic spectral features are smeared out; then, the whole optical band appears to be structureless and its total width is determined by the different weights of homogeneous and inhomogeneous broadening mechanisms.

2.2 Experimental Methods and Analysis

2.2.1 Time-Resolved Luminescence

To outline the time-resolved experiments, we consider the interaction between the exciting light, with photon energy E_{exc} and an ensemble of noninteracting defects contained in a solid sample with thickness d, as shown in Figure 2.4. Since we are dealing with electronic transitions, we assume that, regardless of temperature, all defects are in the ground state before interacting with the excitation light and indicate their concentration as N_0.

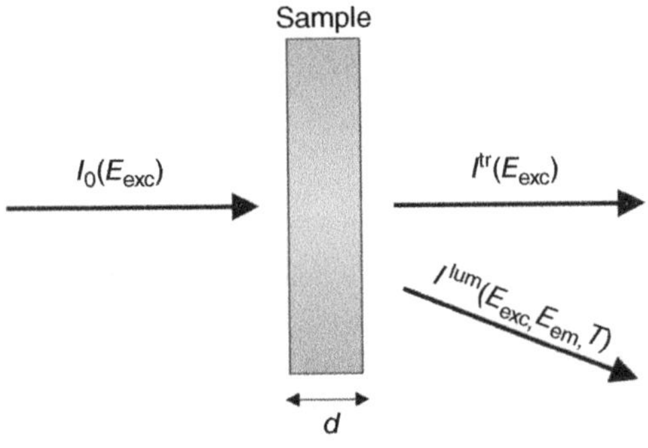

Figure 2.4 Scheme describing the absorption and luminescence of a sample of thickness d. I_0 is the incident intensity, $I^{tr}(E_{exc})$ is the transmitted intensity, and $I^{lum}(E_{exc}, E_{em}, T)$ is the emitted luminescence intensity.

If I_0 is the incident intensity, according to the *Lambert–Beer law* the transmitted intensity, I^{tr}, is given by:

$$I^{tr}(E_{exc}) = I_0(E_{exc})e^{-\alpha(E_{exc})d}, \tag{2.56}$$

where α is the absorption coefficient. This macroscopic parameter is proportional to the defect concentration, N_0, and is related to the transition probability between the ground and excited electronic states associated with the single defect, as described in the previous section. Its dependence on excitation energy, $\alpha(E_{exc})$, represents the absorption spectrum whose shape is due to homogeneous and inhomogeneous contributions, in agreement with Eq. (2.55).

The intensity absorbed by the sample is the difference between I_0 and I^{tr}:

$$I^{abs}(E_{exc}) = I_0(E_{exc})\left[1 - e^{-\alpha(E_{exc})d}\right]. \tag{2.57}$$

If the absorption coefficient and the concentration of the absorbing defects are known, the *absorption cross section* σE_{exc} can be obtained as:

$$\sigma(E_{exc}) = \frac{\alpha(E_{exc})}{N_0}. \tag{2.58}$$

Another spectroscopic parameter, frequently used to quantify the absorption probability, is the dimensionless *oscillator strength* f of an electric dipole transition of energy E between the initial, I, and final, II, electronic states:

$$f = \frac{2mE}{3\hbar^2 e^2}|D_{I \rightarrow II}|^2, \tag{2.59}$$

where $D_{I \rightarrow II}$ is the electric dipole matrix element defined in Eq. (2.12). The linear relation between the maximum value of $\alpha(E_{exc})$, α_{max}, and N_0 can be expressed by the *Smakula's equation* that, for a Gaussian bandshape, is:

$$N_0 f = 8.72 \times 10^{16} \frac{n}{(n^2+2)^2}\alpha_{max} \cdot (\text{FWHM}), \tag{2.60}$$

where n is the index of refraction of the medium.

Due to the absorption process, ($N_1 E_{exc}$) defects will be in the excited state during exposure of the sample to the excitation light, then a portion of them can decay radiatively, with a rate k_r, thus originating the photon emission (luminescence) with energy $E_{em} \leq E_{ex}$, while the remaining excited defects decay non-radiatively, with a rate k_{nr} that depends on temperature. The luminescence intensity is given by:

$$I^{lum}(E_{exc}, E_{em}) = k_r N_1(E_{exc}) I^{lum}(E_{em}), \tag{2.61}$$

where $I^{lum}(E_{em})$ is the emission lineshape determined by homogeneous and inhomogeneous contributions, in agreement with Eq. (2.55).

The variation rate of the excited state population, N_1, depends on the absorption and decay processes, both radiative and non-radiative, according to the following equation:

$$\frac{dN_1(E_{exc}, T)}{dt} = I_0(E_{exc})\left[1 - e^{-\alpha(E_{exc})d}\right] - (k_r + k_{nr}(T))N_1(E_{exc}, T). \tag{2.62}$$

As described in the previous chapter, steady-state luminescence experiments are performed when the system undergoes a continuous excitation; in this case $dN_1/dt = 0$ and combining Eqs. (2.62) and (2.61) we get:

$$I^{\text{lum}}(E_{\text{exc}}, E_{\text{em}}, T) = \frac{k_r}{k_r + k_{nr}} I_0(E_{\text{exc}}) \left[1 - e^{-\alpha(E_{\text{exc}})d}\right] I^{\text{lum}}(E_{\text{em}}), \tag{2.63}$$

where $k_r/(k_r + k_{nr}) = \eta$ is the *quantum yield* that is the ratio between the number of emitted and absorbed photons. From Eq. (2.63), we can define two types of luminescence spectra:

i) The *emission spectrum* (PL spectrum) in which the intensity is measured as a function of E_{em} for fixed E_{exc}. This type of spectrum measures the shape and intensity of the band emitted by the sample.
ii) The *excitation spectrum* (PLE spectrum) measures the luminescence intensity, monitored at a fixed E_{em}, as a function of E_{exc} and represents the excitation efficiency of the luminescence spectrum.

The time dependence of luminescence spectra is studied by using pulsed excitation. After the excitation of an ensemble of point defects with a light pulse, which produces a population of $N_1(0)$ of the excited state, the light source is switched off ($I_0 = 0$) and $N_1(t)$ decays according with:

$$N_1(t) = N(0)e^{-t/\tau}, \tag{2.64}$$

where $\tau = 1/(k_r + k_{nr})$. Combining with Eq. (2.61), we obtain the expression of the luminescence time decay:

$$I^{\text{lum}}(E_{\text{exc}}, E_{\text{em}}, T, t) = k_r I^{\text{lum}}(E_{\text{exc}}, E_{\text{em}}, T) N_1(0) e^{-t/\tau}. \tag{2.65}$$

Equation (2.65) defines the time-resolved luminescence, which can be experimentally tested by detecting PL spectra at fixed time delays from the excitation pulse.

The measure of the lifetime at low temperature, at which non-radiative rates can be neglected, is only dependent on the radiative rate ($\tau_r = 1/k_r$); it allows to determine the oscillator strength by using the following equation, derived from the relation between *Einstein's coefficient* for spontaneous and stimulated emission:

$$f = \frac{1}{E^2\tau_r}\frac{m\hbar^2c^3}{2e^2}. \tag{2.66}$$

2.2.2 Site-Selective Luminescence

The use of tunable lasers in the time-resolved luminescence setup is very advantageous to improve the sensitivity in the acquisition of emission spectra. On the other hand, it allows a selective excitation of defects, on which the site-selective luminescence is based. To provide a theoretical background on this technique, we consider the absorption and luminescence lineshapes discussed in Section 2.1.5 accounting for the homogeneous and inhomogeneous parts. In fact, each defect contributes to the absorption through a homogeneous lineshape governed by the coupling between the electronic transition and both the band vibration and the localized modes. Under the hypothesis of linear coupling with a single localized phonon $\hbar\Omega$, the homogeneous lineshape is expressed by:

$$I^{\text{abs}}_{\text{hom}}(E_{\text{exc}} - E_{00}) = \sum_{\lambda'} e^{-S^L} \frac{(S^L)^{\lambda'}}{\lambda'!} L_{\text{vib}}(E_{\text{exc}} - E_{00} - \lambda'\hbar\Omega), \tag{2.67}$$

where E_{00} is the energy of the ZPL and S^L is the Huang–Rhys factor associated with the localized mode. The inhomogeneity arises from the site-to-site nonequivalence of defects embedded in the amorphous network. It is well-founded to assume that only the ZPL features a statistical distribution $w_{inh}(E_{00})$; the total absorption lineshape, $I^{abs}(E_{exc})$, is therefore given by the convolution:

$$I^{abs}(E_{exc}) = \sum_{\lambda'} e^{-S^L} \frac{(S^L)^{\lambda'}}{\lambda'!} L_{vib}(E_{exc} - E_{00} - \lambda'\hbar\Omega), \otimes w_{inh}(E_{00}) \tag{2.68}$$

that makes explicit the dependence on the microscopic parameters (E_{00}, S^L, $\hbar\Omega$) in Eq. (2.57). Similarly, the luminescence lineshape is the convolution between a homogeneous function, which is mirror-symmetric of $I^{abs}_{hom}(E_{exc} - E_{00})$ against E_{00}, and the inhomogeneous distribution $w_{inh}(E_{00})$, while its amplitude is proportional to the absorbed intensity, scaled by the quantum yield factor $\eta \le 1$:

$$\begin{aligned} I^{lum}(E_{exc}, E_{em}) &= \eta I^{abs}_{hom}(E_{exc} - E_{00}) \otimes w_{inh}(E_{00}) \times I^{lum}_{hom}(E_{00} - E_{em}) \otimes w_{inh}(E_{00}) \\ &= \eta \sum_{\lambda'} e^{-S^L} \frac{(S^L)^{\lambda'}}{\lambda'!} L_{vib}(E_{exc} - E_{00} - \lambda'\hbar\Omega) \otimes w_{inh}(E_{00}) \\ &\times \sum_{\lambda} e^{-S^L} \frac{(S^L)^{\lambda}}{\lambda!} L_{vib}(E_{00} - E_{em} - \lambda\hbar\Omega) \otimes w_{inh}(E_{00}), \end{aligned} \tag{2.69}$$

that reproduces the lineshape function appearing in Eq. (2.63).

The attempt to single out homogeneous and inhomogeneous lineshapes by site-selective luminescence is successful only for a peculiar subclass of defects having a very low electron–phonon coupling ($S^L \approx 0$). In this case, Eq. (2.69) becomes:

$$I^{lum}(E_{exc}, E_{em}) = \eta L_{vib}(E_{exc} - E_{00}) \otimes w_{inh}(E_{00}) \times L_{vib}(E_{00} - E_{em}) \otimes w_{inh}(E_{00}). \tag{2.70}$$

Under this condition, absorption and luminescence spectra overlap and the ZPL lies in this region. It is, therefore, possible to site-selectively excite a specific defect subset within $w_{inh}(E_{00})$; the site-selective luminescence intensity is given by:

$$\begin{aligned} I^{S.S.}(E_{exc}, E_{em}) &= \eta L_{vib}(E_{exc} - E_{00}) \times L_{vib}(E_{00} - E_{em}) \otimes w_{inh}(E_{00}) \\ &= \eta \int L_{vib}(E_{exc} - E_{00}) L_{vib}(E_{00} - E_{em}) w_{inh}(E_{00})\ dE_{00}. \end{aligned} \tag{2.71}$$

At low temperature, $L_{vib}(E_{exc} - E_{00})$ as well as $L_{vib}(E_{00} - E_{em})$ vanish for negative arguments, so that the site-selective luminescence is detected with the emission resonant to excitation, $EE_{ex} = E_{em}$, namely the ZPL, and Eq. (2.71) reduces to:

$$I^{S.S.}(E_{exc} = E_{em}) = L^2_{vib}(0) w_{inh}(E_{00}) \tag{2.72}$$

thus allowing the measure of the inhomogeneous distribution $w_{inh}(E_{00})$.

2.2.3 Basic Design of Experimental Setup: Pulsed Laser Sources; Monochromators; Detectors

In this section, we describe a typical setup that can be conveniently used to perform time-resolved photoluminescence measurements. The main components of the experimental station are: an integrated *Tunable Laser System*, a *Spectrograph*, an *Intensified CCD Camera*.

2.2.3.1 Tunable Laser

The tunable laser is an integrated system that provides the excitation radiation with a wavelength that can be conveniently adjusted within a wide spectral range covering UV-Visible-IR. A very common example exploits the third harmonic of a Q-switched Nd:YAG laser (355 nm) to pump an optical parametric oscillator (OPO) that converts it into a tunable output.

The laser active medium is a crystal of yttrium-aluminum-garnet (YAG) doped with Nd^{3+} ions, which is pumped by a flashed Xenon lamp via the ~800 nm Nd^{3+} absorption transition. The Q-switching is triggered by an electro-optic crystal (Pockel cell, PC) placed within the laser cavity. When the PC is not polarized, the cavity Q-factor is low and the population inversion is high without any laser oscillations. When the PC is polarized, the Q-factor suddenly increases and the laser action starts with a strong initial inversion, thus resulting in the buildup of an intense pulse (~10^2 mJ, ~5 ns long).

The laser beam at $\lambda_0 = 1064$ nm (fundamental harmonic) produced by the Nd:YAG laser is directed on a second harmonic generator (SHG), where the wavelength is converted to $\lambda_0/2 = 532$ nm, and then on a third harmonic generator (THG), where it is converted to $\lambda_0/3 = 355$ nm. The SHG and THG are nonlinear KD*P (KH_2PO_4) birefringent crystals, cut at the proper angle for the required wavelength. The nonlinear conversion process critically depends on the relative orientation of the polarization axis of the incident beam and the axes of the nonlinear crystals. The maximum efficiency is obtained when the phase matching is verified; it requires that the phase velocities of the frequency-doubled and the incident waves are the same within the crystal.

Then, the third harmonic of frequency ω_p traverses the OPO, a birefringent nonlinear BBO (β-BaB_2O_4) crystal, and is converted into two beams, *signal* and *idler*, with frequencies ω_s and ω_i, respectively. Because of the energy conservation, the following condition applies:

$$\omega_p = \omega_s + \omega_i. \tag{2.73}$$

To achieve the phase matching condition within the OPO, the conservation of momentum has to be fulfilled as well:

$$\overleftarrow{k_p} = \overleftarrow{k_s} + \overleftarrow{k_i}. \tag{2.74}$$

The previous equation can be written in the form:

$$n_p\omega_p = n_s\omega_s + n_i\omega_i, \tag{2.75}$$

where n_p, n_s, and n_i are the refractive indices of the nonlinear crystal at the frequencies ω_p, ω_s, and ω_i, respectively. Since the refractive index depends on the polarization of the light and the angle of incidence with respect to the optical axis of the BBO crystal, the OPO output can be thus continuously tuned over a wide spectral range by varying the crystal orientation. The *idler* polarization is perpendicular to the optical bench while the *signal* polarization is parallel; then a polarizer is placed in front of the OPO to select one of them. Usually, the *signal* wavelength can be varied from 410 to 710 nm and the *idler* wavelength from 710 to 2400 nm. Moreover, the output wavelength range can be extended down to 210 nm by suitable UV modules, SHG nonlinear crystals, that halve the wavelength of the OPO beam. In such tunable laser systems, the beam intensity can reach tens of mJ pulse^{-1}, the linewidth is ~1 meV.

2.2.3.2 Time-Resolved Detection System: Spectrograph and Intensified CCD Camera

The time-resolved detection of luminescence spectra usually combines a *Spectrograph* and an *Intensified CCD Camera*. The light emitted by the sample enters a slit, whose width can be changed, then it is spectrally resolved by a spectrograph equipped with more gratings differing in the blaze wavelength, λ_{blaze}, and in the spectral resolution depending on number of grooves per mm. Then, the luminescence light hits a *photocathode*, placed at the entrance of the Intensified CCD Camera, that releases electrons that are accelerated into the *Micro-Channel Plate* (MCP). The MPC is an image intensifier consisting of a thin semiconductive glass plate, which is perforated by more than 10^6 small holes (channels) with a diameter in the range $10 \div 25$ μm. Since the inner surface has a high secondary emission coefficient, the electrons that hit the channel walls generate additional electrons with a gain depending on the voltage at the MCP output. The electrons are further accelerated by a high voltage ($5 \div 8$ kV) and strike the phosphor coating on the fluorescent screen causing it to release photons. These photons are transferred to the surface of CCD, by optical fibers and produce charges at the pixels they strike. These charges, which are proportional to the number of incident photons, are then converted to an analog voltage, that is input to a A/D converter where it is digitally encoded and transmitted to the interface of the computer. Therefore, due to the MCP gain, for each photon that strikes the photocathode surface, many photons ($>10^3$) are produced. Moreover, the possibility of varying the photocathode voltage allows to enable or to disable the CCD: in the Gate ON mode, the photocathode voltage is usually set at −200 V and the CCD sees the light; in the Gate OFF mode, the photocathode voltage is zero and the CCD does not see the light.

As shown in the diagram of Figure 2.5, the *Gate Width* Δt determines the amplitude of the time window during which the CCD is enabled to reveal the luminescence light (Gate ON mode); while the *Gate Delay* T_D regulates the temporal shift of the acquisition window with respect to the trigger signal (that is the arrival of the laser pulse). This setup allows the detection of time-resolved luminescence spectra synchronized with the laser excitation pulses. The measured PL intensity is the light emitted from the sample, given by Eq. (2.65), integrated from T_D to $T_D + \Delta t$, in agreement with:

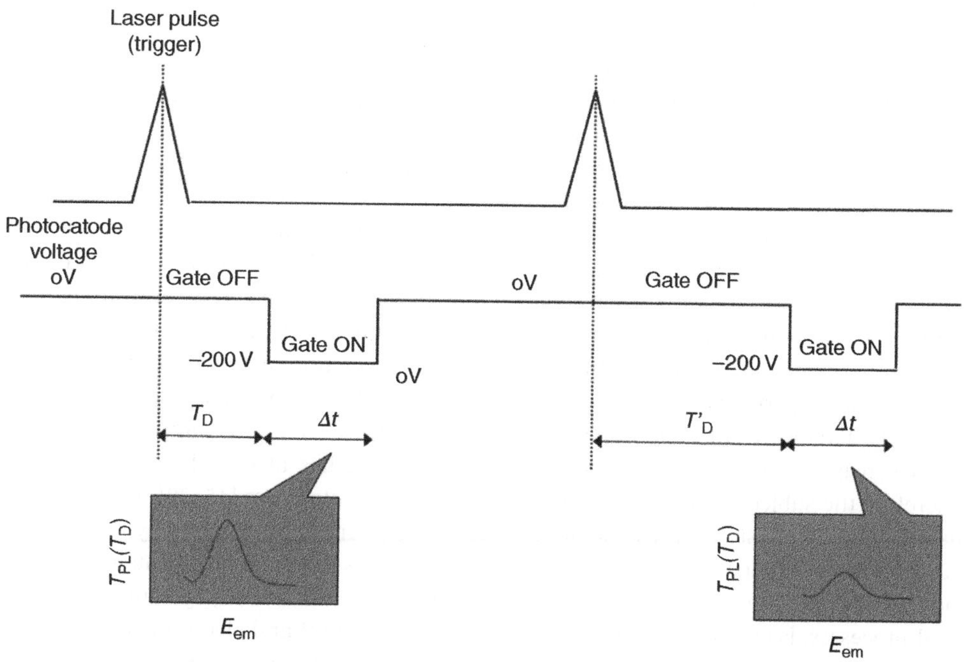

Figure 2.5 Diagram of the CCD timing: Gate ON and Gate OFF modes.

$$I_{\rm PL} = \int_{T_{\rm D}}^{T_{\rm D}+\Delta t} I^{\rm lum}(t)\ {\rm d}t. \tag{2.76}$$

To acquire the decay curve, from which the lifetime τ can be estimated, it is necessary to acquire a set of PL spectra with a fixed Δt and $T_{\rm D}$ ranging from zero to several τ, when the luminescence signal is extinguished. In fact, if ($\Delta t \ll \tau$), the integral in (2.76) reduces to:

$$\int_{T_{\rm D}}^{T_{\rm D}+\Delta t} I^{\rm lum}(t)\ {\rm d}t \simeq I_{\rm PL}(T_{\rm D}), \tag{2.77}$$

that well describes the luminescence decay on increasing $T_{\rm D}$.

2.3 Case Studies: Luminescent Point Defects in Amorphous SiO_2

Interest toward the optical properties of point defects in *amorphous* SiO_2 (silica) is a timely debated issue for its fundamental aspects in the science of amorphous solids and is constantly motivated by the key role of this material in high-tech devices, see, for instance, review papers [13, 14] and references therein. Silica is a model material both for its simple structure and for the possibility to compare its properties with those of its crystalline counterpart (α-quartz). Moreover, due to its excellent transparency, from the mid-IR to vacuum-UV, is indispensable for long-range low-loss optical communication fibers and is the best glassy material for high-power pulsed laser optics. Point defects are relevant because they determine a wide number of optical phenomena. They can be not only detrimental for the use of silica, as it is the case of the transmission losses, but they are also successfully exploited to build modern devices, such as fiber Bragg gratings based on the change of refractive index induced by radiation (photosensitivity). One of the most relevant optically active defects in the silica network is the oxygen dangling bond or *nonbridging oxygen hole center* (NBOHC), ($\equiv$Si—O—)$_3$Si—O$^{\bullet}$, which is characterized by absorption bands in the visible and UV spectral range and by a luminescence around 1.9–2.0 eV, the latter being considered its optical fingerprint. NBOHC has, indeed, exceptional characteristics. On the one hand, it is common to bulk and surface silica, thus influencing several applications (the transmission of optical fibers or the emission of silica nanoparticles characterized by a high specific surface). On the other hand, NBOHC is among the intrinsic defects in oxides with the smallest electron–phonon coupling; this feature allows the site-selective excitation/detection of the purely electronic transition or ZPL.

This section deals with the luminescence of the NBOHC at the silica surface: a model system to evidence the effectiveness of the time-resolved technique in the study of electronic transitions and their coupling with phonons. The reported results are set in an important and timely issue concerning a wide class of optical phenomena influenced by the size reduction of materials to nanoscale. Nanometer-sized silica particles (nanosilica) are, indeed, characterized by a large specific surface area that favors a large concentration of defects that have a crucial role in determining the high emissivity, which is surprising if compared to the optical properties of their bulk counterpart, and are therefore the subject of huge attention in the specialized literature [15–20]. It also worth noting that in these nanosystems, the generation of defects is strongly conditioned by the accessibility of surface sites for molecules of the environment [21]. This is particularly advantageous from a fundamental viewpoint because the possibility of stabilizing specific defects, by controlled thermochemical processes, is crucial to better understand their structural and electronic properties.

At silica surfaces, the structure of NBOHC is in fact determined by its generation occurring by the chemical reaction of an axially symmetric surface-E' center, $\equiv$Si$^{\bullet}$, with an oxygen. Then, its

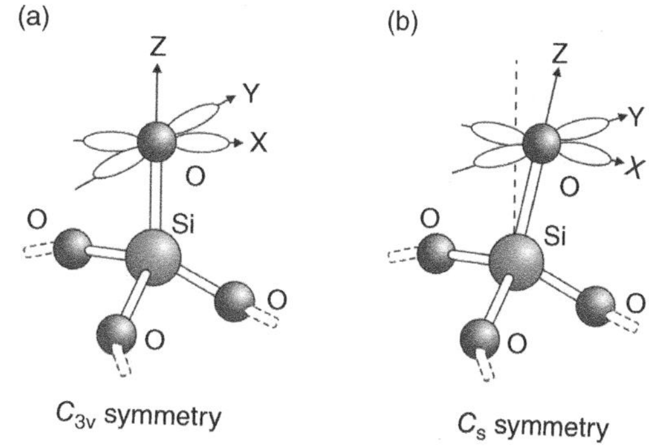

Figure 2.6 Structure of the NBOHC and p orbitals of the dangling oxygen in the (panel a) C_{3v} and (panel b) C_S symmetries.

symmetry is expected to be C_{3v}, in accordance with the structure of the surface-E' coordination sphere that is known by the electron spin resonance (ESR) properties: an axially symmetric g tensor and nearly coincident hyperfine constants due to the interaction with the ^{29}Si atoms bonded to the three basal O [22]. In Figure 2.6a, we report the structural model of the surface NBOHC with a C_{3v} symmetry, where the 2p orbitals of the dangling oxygen are also shown, $2p_x$ and $2p_y$. Quantum chemical calculations performed by Radzig [21] have shown that, due to the *Jahn–Teller effect*, the symmetry of the surface NBOHC deviates from C_{3v} and becomes C_S, as sketched in Figure 2.6b. This removes the degeneracy of the ground state between $2p_x$ and $2p_y$; the calculated energy difference between these states is ~0.1 eV.

2.3.1 Emission Spectra and Lifetime Measurements

Several studies have evidenced that NBOHC at the silica surface emits a luminescence band around 2.0 eV with a composite excitation profile consisting of a peak at 2.0 eV, nearly overlapping with the emission, and an UV broadband with peaks at 4.8 and 6.0 eV [23–25]. Time-resolved spectra have been performed in agreement with the experimental setup described in the previous section.

In Figure 2.7 are reported time-resolved PL spectra acquired under pulsed laser excitation at 2.07 eV (panel a) and at 4.77 eV (panel b) with $\Delta t = 4$ μs and T_D going from 1 to 248 μs, when the decay of the PL emission is almost completed.

To analyze the luminescence decay from the PL spectra reported in Figure 2.7, we have derived the decay kinetics of the two sub-bands ($E_{em} = 1.91$ eV and $E_{em} = 1.99$ eV) under visible ($E_{exc} = 2.07$ eV) and UV ($E_{exc} = 4.77$ eV) excitation (Figure 2.8a). As evident from the semilogarithmic scale, the PL decay curves deviate from a single exponential law. This behavior is common to color centers embedded in amorphous network and is consistent with a multiexponential curve with decay constants inhomogeneously distributed. The quantitative analysis of the PL time decay, *I(t)*, has been carried out by a *stretched exponential decay function*:

$$I(t) \propto \exp[-(t/\tau)^{\gamma}], \quad (2.78)$$

where τ is the lifetime and $\gamma \leq 1$ is a stretching parameter that measures the deviation from a single exponential decay. The results obtained from the fitting procedure are reported in Table 2.1.

To complete the lifetime study, in Figure 2.8b we show the temperature dependence of the decay kinetics of the PL band excited at $E_{exc} = 4.77$ eV. Since the two sub-bands have similar properties, we only display the curve related to one of them ($E_{em} = 1.99$ eV). The deviation from a single exponential law is maintained in the investigated temperature range and the lifetime slightly increases from $\tau = 41.2 \pm 0.5$ μs, at $T = 300$ K, to $\tau = 52.0 \pm 0.5$ μs, at $T = 10$ K. Combining the luminescence

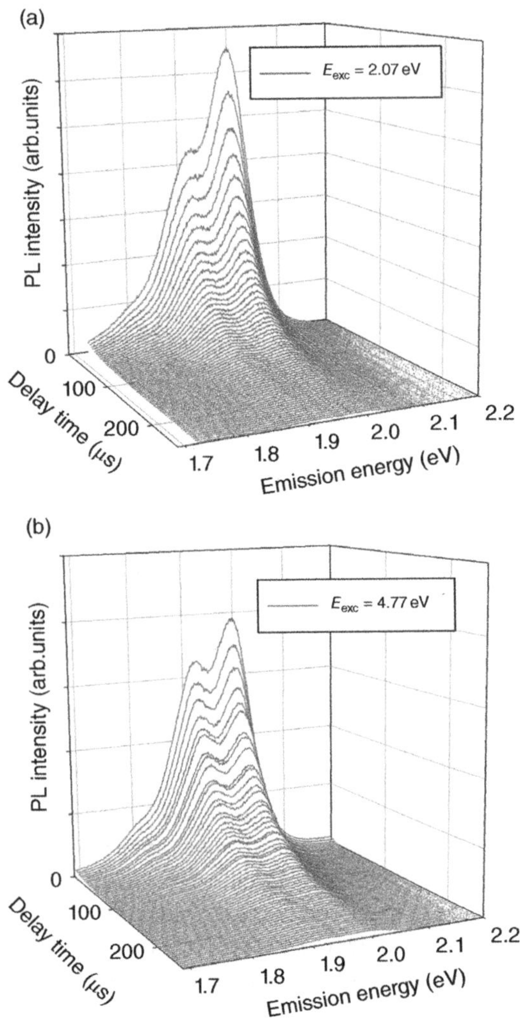

Figure 2.7 Time-resolved PL spectra acquired at different delays in the sample containing ($\equiv$Si—O—)$_3$Si—O$^\bullet$ surface-defects under laser excitation at 2.07 eV (panel a) and 4.77 eV (panel b).

lifetime at low temperature ($\tau \approx 50$ μs) and the Einstein coefficient relation (Eq. 2.66), it is possible to estimate the oscillator strength of the 2.0 eV absorption band for the surface-NBOHC: $f \approx 6 \times 10^{-5}$.

We observe that the spectral features of surface-NBOHC can be accounted for by an energy-level scheme, where two different pathways of excitation/emission can be distinguished. The first includes the cycle 2.0 eV (excitation)/1.9 eV (emission) occurring between two electronic states, whose small Stokes shift is consistent with a very weak electron–phonon coupling. The second cycle, UV and vacuum UV (excitation)/1.9 eV (emission), involves additional electronic states and the large Stokes shift is due to non-radiative electronic relaxations. The excited state from which the PL takes place is commonly associated with a lone pair in both nonbonding 2p orbitals of the dangling oxygen [26–28]. In contrast, conflicting models have been put forward to account for the states originating the excitation bands; one of the most accepted hypotheses that the 2.0 eV band is associated with the charge transfer from the Si—O$^\bullet$ bonding orbital to one of the nonbonding orbitals, while in the UV and vacuum-UV bands, the charge transfer originates from the nonbonding 2p orbitals of the basal oxygen [21, 26, 28]. In this framework, the time-resolved experiments

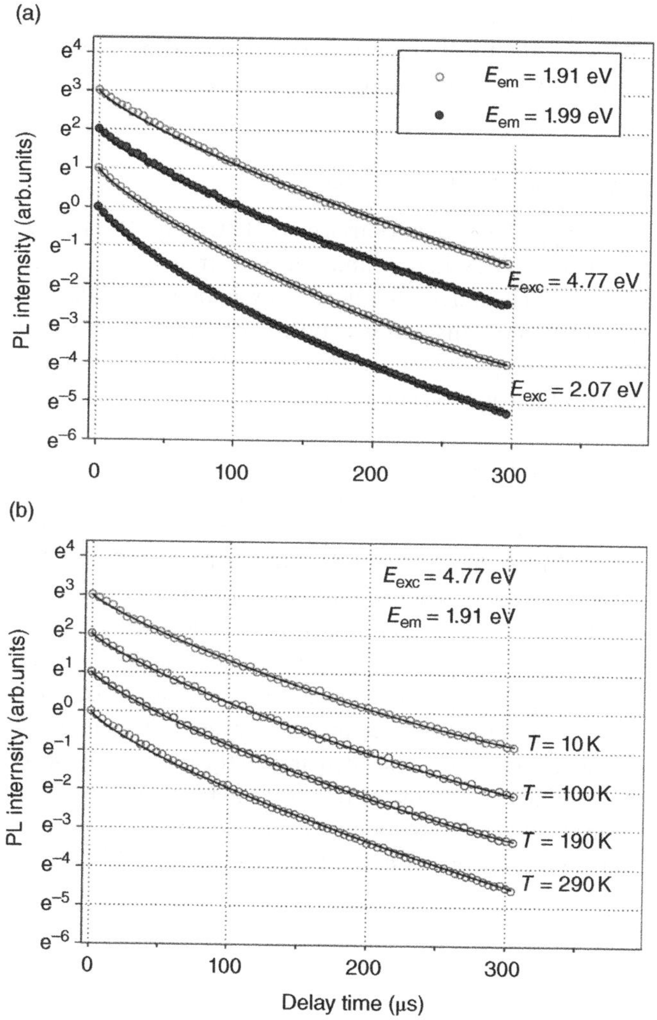

Figure 2.8 Panel (a): Semilog plots of the PL decay in surface-NBOHC (≡S—i—O—)$_3$Si—O$^\bullet$ detected at room temperature at E_{em} = 1.91 and 1.99 eV under laser excitation at E_{exc} = 4.77 and 2.07 eV. Panel (b): Semilog plots of the PL decay in surface-NBOHC (≡Si—O—)$_3$Si—O$^\bullet$ detected at different temperatures at E_{em} = 1.91 eV under E_{exc} = 4.77 eV. For viewing purposes, the initial values of the decay curves are arbitrarily scaled. Full lines plot the best fit curves of Eq. (2.78).

Table 2.1 Best-fitting parameters obtained by Eq. (2.78) for the decay kinetics of the two PL sub-bands measured in the sample containing the surface-NBOHC (≡Si—O—)$_3$Si—O$^\bullet$.

E_{exc} (eV)	E_{em} (eV)	τ (μs)	γ
2.07	1.92	35.2 ± 0.5	0.79 ± 0.02
	1.99	29.9 ± 0.5	0.75 ± 0.02
4.77	1.92	44.3 ± 0.5	0.78 ± 0.02
	1.99	41.2 ± 0.5	0.76 ± 0.02

The associated errors derive from the best-fitting procedures.

help in the interpretation since the long lifetime ($\tau \approx 50$ μs) points out the forbidden character of the PL transition due to the small overlap between the filled 2p orbitals of the nonbridging oxygen atom, which identify the excited state, and that where the charge transfer terminates.

2.3.2 Zero-Phonon Line Probed by Site-Selective Luminescence

The results reported in the previous section have evidenced that surface-NBOHC ($\equiv$Si – O–)$_3$Si – O$^\bullet$ is characterized by a small Stokes shift between its excitation and emission transitions peaked around 2 eV. This implies the possibility to detect, under site-selective excitation, the ZPL and the vibrational structures with which the electronic transition is coupled. The main purposes of this study are: (i) the measure of the stretching frequency of the Si—O$^\bullet$ bond in the ground and in the excited electronic state; (ii) the measure of the phonon coupling parameters; (iii) the measure of the inhomogeneous distribution of the ZPL.

Vibrational properties: Figure 2.9 shows the effects of temperature on time-resolved PL spectra measured with $E_{exc} = 1.997$ eV. At $T = 290$ K, the emission is characterized by two sub-bands peaked at 1.92 ± 0.01 and 1.99 ± 0.01 eV, and it extends over the anti-Stokes region. On lowering temperature, the PL amplitude increases, the anti-Stokes part vanishes and, below 150 K, the ZPL resonant with the excitation is increasingly evident together with a vibrational structure at 920 cm^{-1} apart from it. The origin of the 920 cm^{-1} line will be clarified in the following.

The temperature dependence of the ratio between the intensities of ZPL and the whole band, I_{ZPL}/I_{TOT}, namely the Debye–Waller factor $\alpha(T)$, is shown in Figure 2.10; it allows to quantify the thermal deactivation of the environment vibrational modes the PL transition is coupled to. Panel (a) illustrates the measure of I_{TOT} (shaded area) and I_{0L} (shaded area in the inset) in the spectrum detected at $T = 8$ K. Panel (b) evidences that $\alpha(T)$ decreases from 0.11 to 0.005 on increasing temperature from 8 to 137 K. As reported in the previous section, the expression of $\alpha(T)$ is derived under the straightforward approximation (homogeneous system of defects characterized by an

Figure 2.9 Time-resolved PL spectra of surface-NBOHC ($\equiv$Si—O—)^{3}Si—O$^\bullet$ under pulsed laser excitation at $E_{exc} = 1.997$ eV measured on decreasing temperature from 290K to 8 K. At lower temperature, the ZPL and the vibration 920 cm^{-1} apart from it are clearly visible.

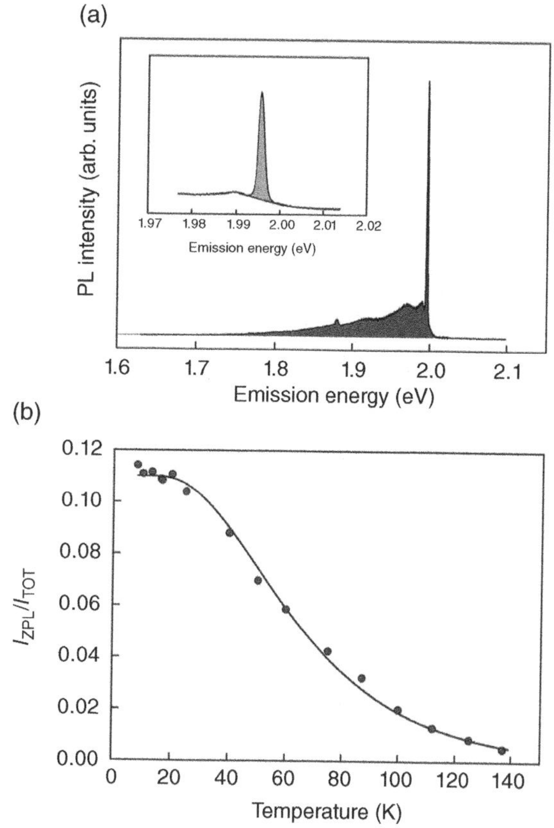

Figure 2.10 Panel (a): Time-resolved PL spectrum of surface-NBOHC ($(\equiv Si{-}O{-})_3Si{-}O^{\bullet}$) under pulsed laser excitation at E_{exc} = 1.997 eV measured at T = 8 K. The shaded area represents the total integrated intensity, I_{TOT}, the shaded area in the inset corresponds to the integrated intensity of ZPL, I_{ZPL}. Panel (b): Temperature dependence of the Debye–Waller factor; solid line is the best fit curve of Eq. (2.79).

electronic transition linearly coupled to a single mode of mean effective frequency ϖ), so that all defects are selectively excited, namely ZPL is in resonance with laser light:

$$\alpha(T) = \exp\left[-S_{\text{tot}}^{B} \times \coth\left(\frac{\hbar\varpi}{2k_{\text{B}}T}\right)\right], \tag{2.79}$$

where S_{tot}^{B}, the total Huang–Rhys factor, is the coupling strength averaged over the totality of phonons. Experimental results and the curve of Eq. (2.79) are in good agreement and the best-fit parameters turn out to be: $S_{\text{tot}}^{B} = 2.2 \pm 0.1$ and $\varpi = 89 \pm 7\ \text{cm}^{-1}$. However, as we will demonstrate in the following, the optical lineshape of surface-NBOHC has an inhomogeneous component. This implies that a fraction of NBOHC is not selectively excited (the ZPL is not in resonance with the laser light), so as to give a contribution to the spectrum independent of temperature; for this reason, $S_{\text{tot}}^{B} = 2.2$ represents an upper limit of the actual value. We also note that the calculated values lead to a Stokes shift smaller than $2S_{\text{tot}}^{B} \times \hbar\varpi \approx 0.05$ eV, in good agreement with the experimental results on the emission/excitation spectra of this surface-NBOHC variant.

The most significant features of the local vibrations, coupled to the electronic transition around 2 eV, are derived by the emission spectra reported in the upper and lower side of Figure 2.11. The emission excited at 1.997 eV shows the ZPL, whose FWHM is $\approx$1.4 meV (11 cm^{-1}), coincident with the laser line; that is, the ZPL originates from those centers located within the laser spectral linewidth in a much larger inhomogeneous distribution. At lower energies one observes two *phonon sidebands* centered at 923 ± 3 and $1840 \pm 10\ \text{cm}^{-1}$ apart from the ZPL.

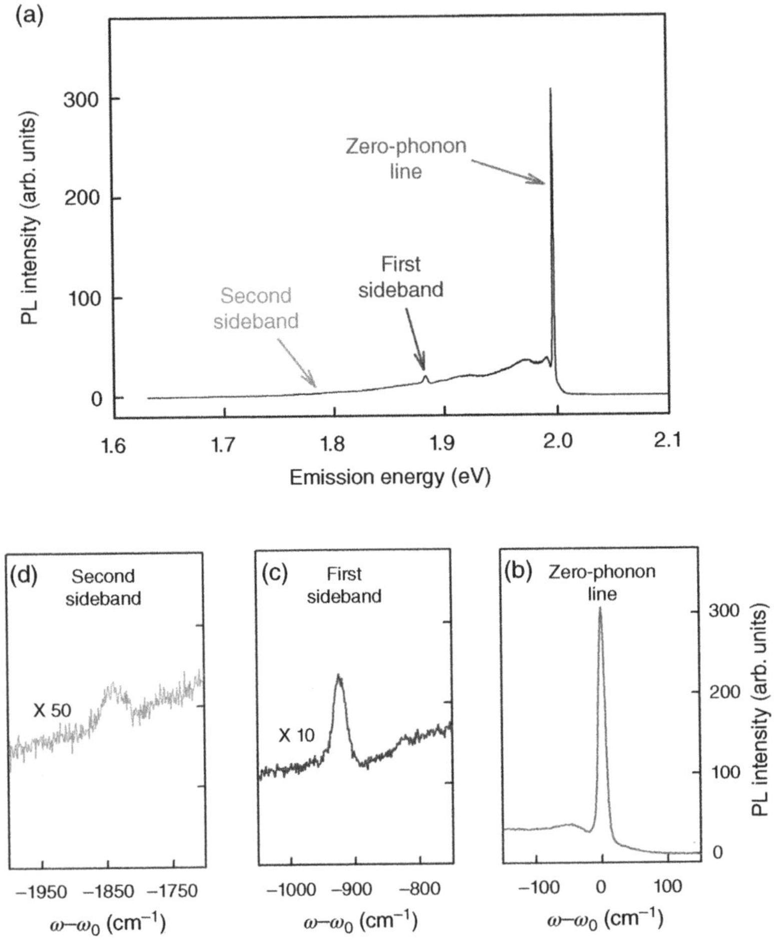

Figure 2.11 Panel (a): Time-resolved PL spectrum of surface-NBOHC ($\equiv$Si—O—$)_3$Si—O$^\bullet$ measured at $T = 8$ K under pulsed laser excitation at $E_{exc} = \hbar\omega_0 = 1.997$ eV. Panels (b–d): Zooms that show the ZPL profile (b), the first (c) and the second (d) sidebands plotted as a function of distance from the laser line ω_0.

In agreement with Figure 2.2, they represent the transitions from the lower vibrational level in the excited electronic state to the first and second vibrational levels in the ground electronic state, respectively. The measured values, therefore, identify the fundamental, ω_g, and the overtone, $2\omega_g$, frequencies of the nearly equally spaced vibrational levels of the surface-nonbridging oxygen in the electronic ground state. We note that these sidebands are more and more wider than the ZPL, their FWHM being ~20 and 40 cm^{-1} for the first and the second line, respectively; this spreading is due to an inhomogeneous distribution of the vibrational frequency of centers having the same ZPL. From these emission spectra, we also measure the ratio between the integrated intensities of ZPL and the first and second vibrational lines: $I_{0L}/I_{1L} = 13.5 \pm 0.5$, $I_{0L}/I_{2L} = 160 \pm 30$, the error being mainly due to the inaccuracy in the reference line subtraction to account for the overlapping with nonselectively excited luminescence. Based on the linear electron–phonon coupling, subsequently evidenced by the equal values of vibration frequency both in the ground and in the excited state, we compare these values with the Poisson's distribution, $I_{kL} = \exp(-S)^L \times (S^L)^k/k!$ and extract

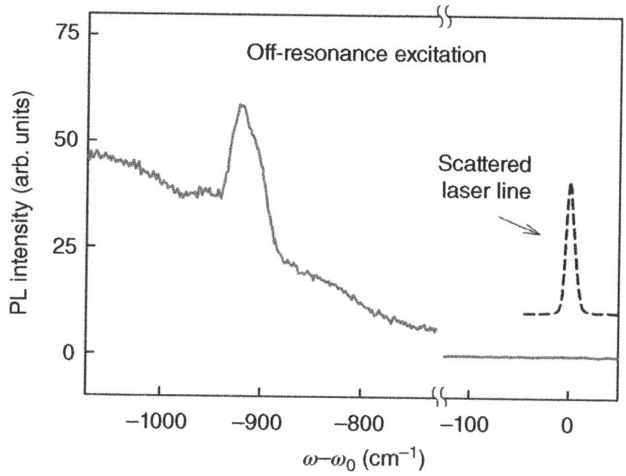

Figure 2.12 Time-resolved PL spectrum of the surface-NBOHC (≡Si—O—)$_3$Si—O$^•$ measured at T = 10 K under pulsed laser excitation at $E_{exc} = \hbar\omega_0 = 2.112$ eV and plotted as a function of the distance from the laser line ω_0. The laser lineshape, acquired by the scattered light, is also shown arbitrarily scaled respect to the emission spectrum.

the partial Huang–Rhys factor S^L: $I_{0L}/I_{1L} = 1/S^L$ yields $S^L = 0.074 \pm 0.003$ and $I_{0L}/I_{2L} = 2/(S^L)^2$ yields $S^L = 0.11 \pm 0.01$. We note that the difference between the values of S is larger than the experimental uncertainty; despite this incongruence, these results quantify the low coupling of the electronic transition with the local Si—O$^•$ stretching mode, namely, the nearly absent relaxation of the Si—O$^•$ bond after excitation.

We now return to the experimental results on the (≡Si—O—)$_3$Si—O$^•$ to complete the description with the vibrational properties of the excited electronic state. Figure 2.12 shows the spectrum obtained under excitation at $E_{exc} = 2.112$ eV, where no resonant luminescence appears. In contrast, a sharp line is detected at lower energies, shifted by $920 \pm 3\,cm^{-1}$ from the excitation; the accuracy being determined by the preliminary detection of the scattered laser line shown in the same figure. This sharp line is the off-resonance ZPL that, in agreement with Figure 2.2, is excited through the vibrational level 1 of the excited electronic state. Then, the value of $920 \pm 3\,cm^{-1}$ is the measure of the frequency ω_e of the surface Si—O$^•$ stretching in the excited electronic state that is almost coincident with that in the ground electronic state. This finding demonstrates the validity of the linear coupling between the optical transition around 2.0 eV and the localized mode. Moreover, on the basis of the reduced mass of the Si—O molecule ($m^* = 1.692 \times 10^{-26}$ kg), it is possible to calculate the force constant of Si—O$^•$ bond at the surface: $k = (\omega_{surf})^2 \cdot m^* \approx 508\,N\,m^{-1}$.

Inhomogeneous broadening measured by ZPL distribution: Finally, we report the study of the inhomogeneous properties of NBOHC at the surface of silica taking advantage of time-resolved experiments being able to detect ZPL under tunable laser excitation [29]. Figure 2.13a shows a series of time-resolved emission spectra measured at $T = 10$ K with the excitation energy stepwise incremented from 1.887 to 2.077 eV (minimum step 0.003 eV); each spectrum is displayed in the vicinity of the excitation energy thus evidencing the ZPL. From these spectra, we plot in Figure 2.13b the distribution of the ZPL intensity. The experimental data are best fitted by a Gaussian curve centered at 1.995 ± 0.003 eV with FWHM of 0.042 ± 0.005 eV ($340 \pm 40\,cm^{-1}$) that

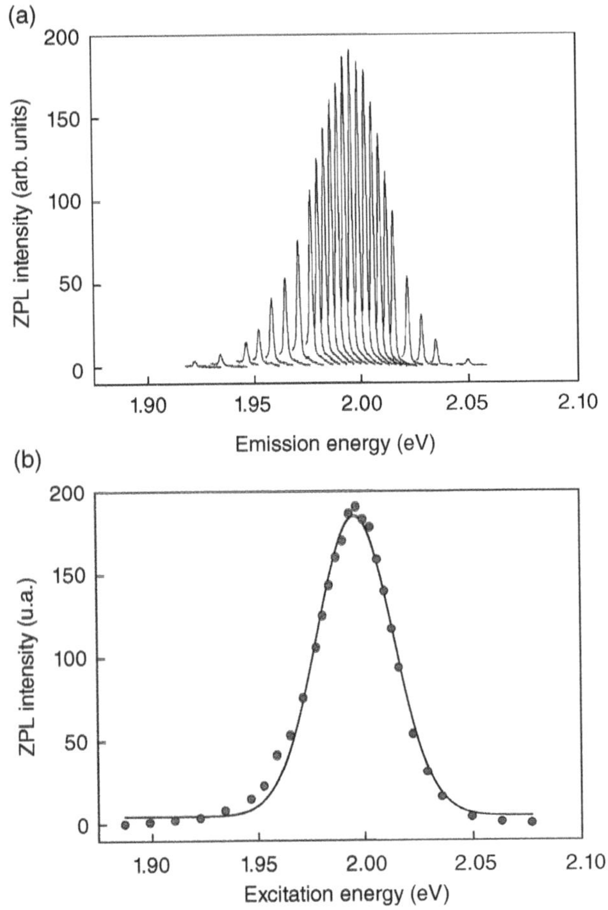

Figure 2.13 Panel (a): Time-resolved PL spectra of surface-NBOHC $(\equiv Si{-}O{-})_3Si{-}O^{\bullet}$, measured at $T = 10$ K, in the ZPL region at different excitation energies from 1.89 to 2.08 eV with step 0.01, 0.006 and 0.003 eV. Panel (b): Distribution of the ZPL intensity, obtained by the spectra reported in the panel (a) (symbols), superimposed to its Gaussian best fit curve.

represents the inhomogeneous distribution $w_{inh}(E_{00})$ of the electronic transitions, due to the different local environment surrounding the $(\equiv Si{-}O{-})_3Si{-}O^{\bullet}$ at the silica surface. Indeed, the inhomogeneity is related to the structural disorder of the silica network in the long- and local-range in comparison with the SiO_4 tetrahedron size. The first is intrinsic to the amorphous state, and is mainly accounted for by the statistical distribution of the Si—O—Si bond angle and the size of (Si—O) n ring structure [30, 31]. The second is due to the presence of point defects that introduce a local distortion into the surroundings, such distortion being dependent on the site [11].

We observe that the main experimental outcome is the detectability of the ZPL under site-selective excitation of inhomogeneously distributed centers, thus allowing the inhomogeneous curve to be drawn directly. The detection of the ZPL is therefore a probe of the silica structure near the NBOHC; this potential is precluded for other defects in silica, because of the stronger phonon coupling of their optical transitions. In those cases, the deconvolution between homogeneous and inhomogeneous broadening can be done only indirectly.

References

1 Herzberg, G. (1966). *Molecular Spectra and Molecular Structure*. New York: Van Nostrand Reinhold.
2 Rebane, K.K. (1970). *Impurity Spectra of Solids*. New York: Plenum Press.
3 Stoneham, A.M. (1975). *Theory of Defects in Solids*. Oxford: Clarendon Press.
4 Watts, R.K. (1977). *Point defects in Crystals*. New York: John Wiley & Sons.
5 Yen, W.M. and Selzer, P.M. (eds.) (1981). *Laser Spectroscopy of Solids*. New York: Springer-Verlag.
6 Sild, O. and Haller, K. (eds.) (1988). *Zero Phonon Lines and Spectral Hole Burning In Spectroscopy and Photochemistry*. New York: Springer-Verlag.
7 Vij, D.R. (ed.) (1998). *Luminescence of Solids*. New York: Plenum Press.
8 Franck, E.G. and Dymond, E.G. (1926). *Trans. Faraday Soc.* **21**: 536.
9 Condon, E.U. (1928). *Phys. Rev.* **32**: 858.
10 Cannas, M. (1998). Point defects in amorphous SiO_2: optical activity in the visible, UV and vacuum-UV spectral regions. PhD thesis, Dipartimento di Scienze Fisiche ed Astronomiche, Università di Palermo, Italy.
11 Skuja, L. (2000). *Defects in SiO_2 and Related Dielectrics: Science and Technology (eds. G. Pacchioni, L. Skuja and D.L. Griscom). USA: Kluwer Academic Publishers.*
12 Huang, K. and Rhys, A. (1950). *Proc. Roy. Soc.* **A204**: 406.
13 Skuja, L., Hirano, M., Hosono, H., and Kajihara, K. (2005). *Phys. Phys. Stat. Sol. (c)* **2**: 15.
14 Girard, S., Alessi, A., Richard, N. et al. (2019). *Rev. Phys.* **4**: 100032.
15 Vaccaro, L., Morana, A., Radzig, V., and Cannas, M. (2011). *J. Phys. Chem. C* **115**: 19476.
16 Bonacchi, S., Genovese, D., Juris, R. et al. (2011). *Angew. Chem., Int. Ed.* **50**: 4056.
17 Rimola, A., Costa, D., Sodupe, M. et al. (2013). *Chem. Rev.* **113**: 4216.
18 Carbonaro, C., Corpino, R., Ricci, P. et al. (2014). *J. Phys. Chem. C* **118**: 26219.
19 Tarpani, L., Ruhlandt, D., Latterini, L. et al. (2016). *Nano Lett.* **16**: 4312.
20 Spallino, L., Spera, M., Vaccaro, L. et al. (2014). *Phys. Chem. Chem. Phys.* **16**: 22028.
21 Radzig, V.A. (2000). Defects in SiO_2 and Related Dielectrics: Science and Technology. In: (eds. G. Pacchioni, L. Skuja and D.L. Griscom). USA: Kluwer Academic Publishers.
22 Radzig, V.A. (2001). *Chem. Phys. Rep.* **19**: 469.
23 Vaccaro, L., Cannas, M., Radzig, V., and Boscaino, R. (2008). *Phys. Rev. B* **78**: 075421.
24 Vaccaro, L., Cannas, M., and Radzig, V. (2008). *Phys. Rev. B* **78**: 233408.
25 Vaccaro, L., Cannas, M., and Radzig, V. (2009). *J. Non Cryst. Solids* **355**: 1020.
26 Suzuki, T., Skuja, L., Kajihara, K. et al. (2003). *Phys. Rev. Lett.* **90**: 186404.
27 Bakos, T., Rashkeev, S.N., and Pantelides, S.T. (2004). *Phys. Rev. B* **70**: 075203.
28 Giordano, L., Sushko, P.V., Pacchioni, G., and Shluger, A.L. (2007). *Phys. Rev. B* **75**: 024109.
29 Vaccaro, L. and Cannas, M. (2010). *J. Phys. Condens. Matter* **23**: 235801.
30 Hobbs, L.W. and Yuan, X. (2000). *Defects in SiO_2 and Related Dielectrics: Science and Technology (eds. G. Pacchioni, L. Skuja and D.L. Griscom). USA: Kluwer Academic Publishers.*
31 Mauri, F., Pasquarello, A., Pfrommer, B.G. et al. (2000). *Phys. Rev. B* **62**: R4786.

3

Ultrafast Optical Spectroscopies

Alice Sciortino and Fabrizio Messina

Department of Physics and Chemistry – Emilio Segrè, University of Palermo, Palermo, Italy

3.1 Femtosecond Spectroscopy: An Overview

Traditional steady state spectroscopies, addressed in Chapter 1 of this book, are essential tools to characterize a physicochemical system through the extensive mapping of its energy landscape and transitions. However, these classical methods cannot capture the dynamical aspects driving the response of the system after an initial perturbation. These aspects can only be addressed by time-resolved spectroscopic techniques, which are methods capable of following in time the evolution of a physicochemical system, after it has been initially brought out of equilibrium by a defined stimulus, such as photon absorption. Indeed, a thorough understanding of these *dynamics*, as they unravel in time, is often crucial to fully elucidate the behavior of a system of physical, chemical, or biological interest.

Photoexcitation of any physical system triggers a complex sequence of phenomena occurring over many temporal orders of magnitude after initial photon absorption, and responsible for the ultimate outcome of the photocycle. The primary and most fundamental dynamics, however, usually occur on picosecond (1 ps = 10^{-12} s) or femtosecond (1 fs = 10^{-15} s) timescales for many well-known processes of fundamental interest in photophysics and photochemistry [1]. In particular, it can be argued that femtoseconds are the "fundamental" timescale for any physicochemical process involving short-range atomic rearrangements, such as chemical reactions or molecular relaxations, because the vibrational period of nuclei always fall in the femtosecond time range ($\sim$10 fs for the OH vibration). Thereby, it is evident that very fast, or rather "ultrafast," spectroscopic techniques are compulsory to investigate these types of phenomena.

This chapter addresses a variety of experimental methods usually referred to as ultrafast or femtosecond spectroscopies. These techniques are capable of time resolutions reaching a few femtoseconds, which are essential to reconstruct in detail the course of events initiated by photoexcitation, achieving a comprehensive understanding of the photocycle of any physical system. Examples of typical phenomena which are addressed by femtosecond methods are molecular energy relaxations, such as internal conversion or intersystem crossing [2], solvation dynamics, electron and energy transfer events [3–5], fluorescence quenching [5], dynamics of charge carriers and excitons in semiconductors [6], and photochemical reactions [7]. For these reasons, femtosecond techniques are today well-established and considered a fundamental tool in spectroscopy.

Spectroscopy for Materials Characterization, First Edition. Edited by Simonpietro Agnello.

For example, whenever a molecule in solution phase is electronically excited by UV or VIS light, the surrounding solvent responds to the electronic redistribution within the molecule by a series of short- and long-range rearrangements, collectively named *solvation dynamics*. These dynamics typically occur on femtosecond and picosecond timescales [3, 4] and are a key component of the behavior of any system in solution phase. As another example, when charge carriers in a bulk or nano-sized semiconductor are promoted to a high energy state by photon absorption, their successive relaxation involves processes, such as electron–electron or electron–phonon scattering, or Auger recombination, which are of fundamental importance and equally take place on such, ultrafast, timescales [8, 9].

The general idea behind most femtosecond spectroscopies is to use at least two, or more, light pulses with very short time durations to follow in real time the undergoing dynamics. One of the pulses, for example, is resonant to an electronic transition of the investigated system, and its absorption by the system causes an injection of energy and a quasi-instantaneous redistribution of the electronic charge. Then, the successive dynamics are studied by taking spectroscopic snapshots of the excited systems at variable delays, by the use of a second light pulse. This can be done by using one of several possible spectroscopic observables, capable of retrieving different types of information, such as the UV/VIS absorption, reflectance or luminescence, or the infrared vibrational absorption. For example, transient absorption (TA) or pump/probe methods, discussed in Section 3.3, exploit a pump/probe approach, where the changes of visible or infrared absorption are detected and followed in time after excitation. Other methods rely on different observables, such as the spontaneous emission (ultrafast fluorescence, discussed in Section 3.4) or the Raman scattering (Section 3.5). From the technical point of view, femtosecond spectroscopies are characteristically nonlinear optical experiments, which make large use of a wide toolbox of laboratory techniques in nonlinear and laser optics, to manipulate, generate, detect, and control femtosecond-pulsed light beams in various spectral regions, as described in Section 3.2.

The development of the field of femtosecond spectroscopy has been characterized by a progressive and dramatic improvement in what can be practically achieved. With time, a femtosecond time-resolved version has been developed for almost any traditional spectroscopic technique, including, in recent times, photoemission spectroscopy [10], X-ray scattering [11], X-ray absorption [12], or optical microscopy methods [13]. The last frontier in the field is pushing even more the time resolution of these experiments, to the extent that the first attosecond ($1\text{ as} = 10^{-18}$ s) experiments have been emerging in the last years [10, 14].

When these methods are properly combined, it is often possible to literally track in real time the flow of charge and energy through time and space after photoexcitation. The potential of these methods is testified by their applications on a wide variety of different systems and nanosystems, such as semiconductor NPs [8], molecules and macromolecules in solution [15], carbon nanomaterials [16], and many others, always providing very useful insight on their photoinduced behaviors. Besides reconstructing the photocycle, a further capability of femtosecond experiments is disentangling homogeneous and inhomogeneous broadenings of the spectral lines, via methods like TA *hole burning* [17] or *four-wave mixing* [18], which do not have an equivalent in traditional spectroscopy.

The chapter is organized as follows. First, a general presentation of the characteristics of fs laser beams will be presented in Section 3.2, as needed to follow the rest of the chapter. Then, three well-established ultrafast spectroscopic methods will be described: transient absorption (Section 3.3), femtosecond-resolved fluorescence (Section 3.4), and femtosecond Raman (Section 3.5). In the final section (Section 3.6), four different case studies will be presented, to illustrate the utility of these techniques in selected real-case scenarios.

3.2 Ultrafast Optical Pulses

3.2.1 General Properties

An ultrafast optical pulse is an electromagnetic pulse characterized by a very short duration (from few femtoseconds to few hundreds of femtoseconds) and a broad spectral distribution (10–100 nm FWHM) in the near-infrared, visible, or UV spectral range. Generating ultrafast pulses relies on the use of a *mode-locked* laser, which exploits the amplification of a large number of laser modes oscillating in-phase within the laser cavity [19, 20]. The most widespread type of femtosecond mode-locked laser in modern spectroscopy is the Ti:sapphire laser. A typical Ti:sapphire oscillator emits laser pulses with a central wavelength tunable around 800 nm, typical duration of 10–100 fs, energy of 1–100 nJ pulse^{-1}, and repetition rate of ~80 MHz. By using an external amplifier, these pulses can be then amplified up to μJ or mJ per pulse with a proportional reduction of the repetition rate. Because of the very short duration, these numbers imply intensities as high as tens of GW per cm^2, which can easily be achieved even without focusing. These intense, amplified pulses are then available to feed a range of experiments in nonlinear optics and spectroscopy such as those described in this chapter. The details of mode-locking will not be further discussed here, and the rest of this section will be devoted to describing some general properties of propagating femtosecond light pulses.

The time dependence of the oscillating electric field in an amplified ultrashort pulse is described by $E(t) = \mathrm{Re}\left(E_0 \mathrm{e}^{-\gamma t^2 + i\omega_0 t}\right)$. The wave amplitude follows a Gaussian envelope, with the shape factor γ proportional to the squared inverse of the duration of the pulse: $\gamma \propto \Delta t^{-2}$. To obtain the spectral content of the pulse, one can calculate the Fourier transform of $E(t)$, which is shown to be again a Gaussian function, centered at the frequency ω_0 and with a frequency width $\Delta\omega$ proportional to $\gamma^{1/2}$ [19]. This entails a strict relation between the time duration of the pulse and its spectral width, according to Heisenberg's uncertainty principle $\Delta t \Delta\omega \geq \frac{1}{2}$. This leads to the fundamental consequence that ultrashort laser pulses are intrinsically non-monochromatic: in order to have pulse durations in the femtosecond range, it is compulsory to have a broad enough spectral distribution, typically in the tens of nanometers. When the equivalence of the uncertainty principle is verified, the pulse is named Fourier transform-limited. This condition can only be perfectly achieved by a Gaussian pulse, and guarantees the shortest possible pulse for a given spectral width. For example, a transform-limited Gaussian pulse with 10 fs duration, peaking at 800 nm, shows a spectral bandwidth of 94 nm. As discussed hereafter, one consequence of the broad bandwidth of femtosecond pulses is that their propagation is affected by strong dispersion effects [19, 20]. On the other hand, their extremely high peak intensities, due to the short duration, lead to intense nonlinear optical effects.

3.2.1.1 Dispersion Effect: Group Velocity Dispersion

When dealing with optical pulses with femtosecond pulse durations, it is important to consider the effects of group velocity dispersion (GVD). The latter affects the duration of a light pulse which traverses any media, because of the frequency dependence of the refractive index $n(\omega)$. GVD is defined as:

$$\mathrm{GVD} = \left(\frac{\mathrm{d}^2 k}{\mathrm{d}\omega^2}\right)_{\omega_0} = \frac{\mathrm{d}}{\mathrm{d}\omega}\left(\frac{1}{v_g(\omega)}\right)_{\omega_0} \tag{3.1}$$

where v_g is the group velocity. The latter can be written as:

$$v_g \approx \frac{c}{n(\omega)}\left(1 - \frac{\omega}{n(\omega)}\frac{\mathrm{d}n(\omega)}{\mathrm{d}(\omega)}\right). \tag{3.2}$$

GVD is measured in $\mathrm{fs}^2\,\mathrm{mm}^{-1}$ and, in a transparent region, is typically positive because of the characteristic dependence of n on frequency. During propagation, every spectral component of the pulse acquires a different delay, resulting in a temporal broadening of the pulse without any spectral changes [21]. To visualize the effect of GVD on a Gaussian pulse passing through a medium, a simulation is shown in Figure 3.1. From top to bottom, three pulses are shown, representing a transform-limited Gaussian with FWHM = 5 fs centered at 550 nm, and the same pulse after passing through 1 or 2 mm SiO_2, respectively. As evident from the figure, GVD substantially enlarges the pulse duration, increasing it to several hundreds of femtoseconds. The pulse duration Δt, that is the FWHM of the Gaussian intensity profile, broadens to Δt_b given by:

$$\Delta t_b = \Delta t\sqrt{1 + (4 \cdot \ln(2) \cdot \mathrm{GVD} \cdot L \cdot \Delta t^{-2})^2} \tag{3.3}$$

where L is the propagation length inside the material [19]. Besides broadening, GVD causes a frequency *chirp*, that is a time dependence of the instantaneous frequency of the pulse, given by $\omega(t) = \frac{\partial \phi}{\partial t}$. In Figure 3.1, the *chirp* is evident from the comparison of the frequencies of the two sides of the pulse, showing that the instantaneous frequency is redder in the front part of the pulse

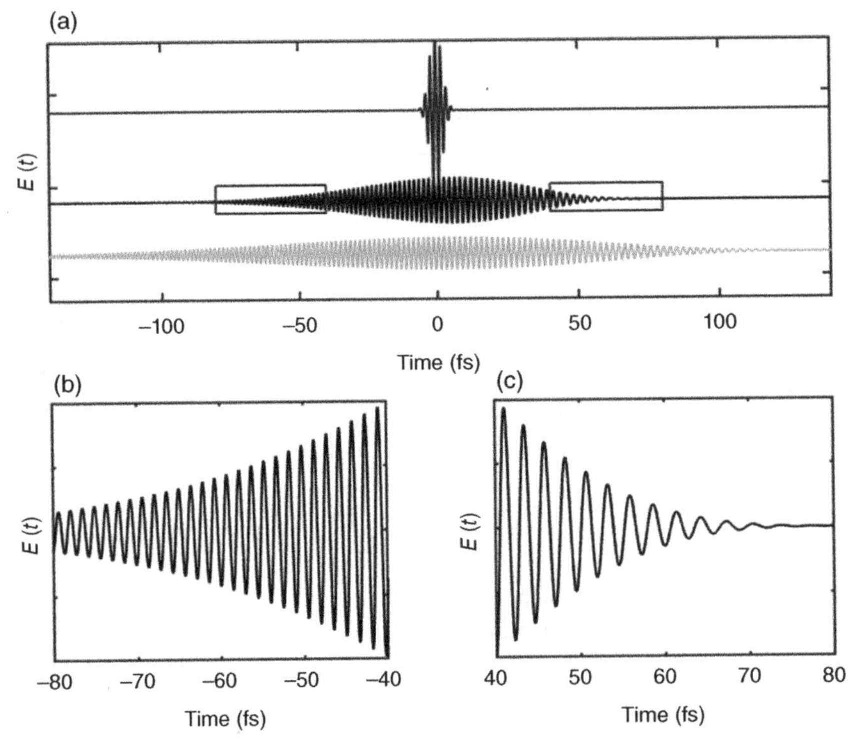

Figure 3.1 Panel (a): Simulation of a gaussian pulse centered at 550 nm with FHWM = 5 fs (first curve from the top) after propagation through a SiO_2 medium of 1 mm thickness (second curve) and 2 mm (third curve). Panel (b and c): zooms of the tails of the black pulse (squares). Each wavelength is delayed by a different phase, resulting in a longer pulse with a positive chirp (the redder frequencies are faster than the bluer).

and bluer in the back [19, 20], which is exactly the effect of a positive GVD. In particular, the instantaneous frequency acquires an approximately linear time dependence, $\omega(t) = \omega_0 + \alpha t$, because the phase of the wave acquires a quadratic time term.

Pulse broadening and chirp acquired by femtosecond pulses during their propagation in optical setups need to be put under control in order to preserve good time resolution. One way to do it is to limit the use of transparent optical components, preferring the use of reflective optics only. Some special methods exist to manipulate the chirp, such as what is called a pulse compressor, built by using a pair of prisms or gratings. In a pulse compressor, one can add negative GVD (redder part of a pulse propagates slower than the blue part) which compensates the effect of pulse broadening in a transparent media, recompressing the pulse [19, 20].

Last but not least, dispersion also affects the temporal overlap of two pulses centered at different wavelengths which pass through the same medium, because their group velocities are generally different: this effect is called *group velocity mismatch*, or GVM. Thereby, if the pulses are initially overlapping in time at a certain point in space, they overlap no more after some propagation distance within a dispersive medium. The GVM effect, for example, can be very important in the generation of pulses through nonlinear effects because it can limit the effective interaction length between the two pulses.

3.2.2 Nonlinear Optics: Basis and Applications

3.2.2.1 Second Harmonic Generation and Sum Frequency Generation

Second harmonic generation (SHG) is a nonlinear optical phenomenon in which two photons of the same frequency, interacting in a nonlinear material, are converted in a single photon with doubled frequency [22]. The polarization $\vec{P}$ of a medium excited by an electrical field $\vec{E}$ can be expressed as [19, 20]:

$$\frac{\vec{P}(\omega)}{\varepsilon_0} = \chi^{(1)}(\omega)\vec{E}(\omega) + \chi^{(2)}(\omega)\vec{E}(\omega)\vec{E}(\omega) + \chi^{(3)}(\omega)\vec{E}(\omega)\vec{E}(\omega)\vec{E}(\omega) + \cdots \tag{3.4}$$

where, in general, $\chi^{(n)}$ is a tensor. The first term of the equation describes the phenomena usually encountered in linear optics, while the other describes nonlinear effects at different orders in the electric field. Under certain conditions ($\chi^{(2)} \neq 0$, as generally occurs in a non-centrosymmetrical medium), two photons at the same frequency ω_1, passing through an appropriate medium, are combined to generate a new photon with a doubled frequency ($2\omega_1$). The process follows the laws of energy ($\omega_1 + \omega_1 = 2\omega_1$) and momentum conservation ($\vec{k}_1 + \vec{k}_2 = \vec{k}_3$) and in not-depleted pump condition (that is, negligible pump absorption) it is possible to describe the intensity of the new beam as [19, 20]:

$$I(2\omega_1) = \frac{2^7\pi^3\omega_1^2\chi_{\text{eff}}^2L^2}{n^3c^3}I^2(\omega_1)\left(\frac{\sin(\Delta kl/2)}{\Delta kl/2}\right)^2 \tag{3.5}$$

where n is the refractive index, L is the optical path within the nonlinear material, $\Delta k = k_2 - 2k_1$ is the so-called phase mismatch, and χ_{eff} is the effective susceptibility which is a certain combination of the components of the $\chi^{(2)}$ tensor, which depends on the material and on its orientation. The intensity of the new beam depends on the square of the incident beam intensity, on the length L (with a quadratic dependence if $\Delta k = 0$), and on the degree of phase mismatch. Fulfilling the condition $\Delta k = 0$, named *phase matching*, gives maximally efficient SHG, and corresponds to the conservation of momentum in the process. It can be seen as a situation in which first and second harmonic beams propagate in the medium with the same speed. In order to achieve this, the refractive index at ω and 2ω has to be the same. Although this is not generally possible in isotropic media, such a limitation can be overcome by using (uniaxial) birefringent media such as beta-barium borate (BBO). The latter display two different refractive indexes, ordinary (n_o), and extraordinary (n_e), for beams with two orthogonal polarizations, where the extraordinary index also depends on the angle θ between the $\vec{k}$ of the beam and the optical axis of the crystal. Thus, one can achieve phase matching, for example, if the beam at ω propagates as an ordinary beam, while the beam at 2ω propagates as an extraordinary beam. Then, changing the orientation of the crystal, it is possible to find an angle θ for which $n_e(2\omega, \theta) = n_o(\omega)$, fulfilling the phase matching condition.

Because femtosecond pulses are intrinsically broadband, another important parameter in SHG is the extent of spectral bandwidth which is effectively doubled (acceptance bandwidth), not necessarily coincident with the whole pulse bandwidth. In fact, the phase matching condition is exactly fulfilled only at a given wavelength, and therefore it cannot be exactly fulfilled across the entire pulse bandwidth. In practice, assuming that the first harmonic beam at λ_1 propagates as an ordinary beam, and a SHG beam is produced as an extraordinary beam at $\lambda_2 = \lambda_1/2$, the portion $d\lambda_1$ of the doubled pulse bandwidth is given by [23]:

$$d\lambda_1 = 0.88\frac{\lambda_1}{L}\left(\frac{\partial n_o}{\partial \lambda_1}\bigg|_{\omega} - 0.5\frac{\partial n_e}{\partial \lambda_2}\bigg|_{2\omega}\right)^{-1}. \quad (3.6)$$

Therefore, to increase the bandwidth of the doubled beam, it is necessary to decrease the crystal thickness L, at the cost of SHG efficiency (proportional to L^2). Therefore, according to the experimental requirements, one needs to find the right compromise between the two needs.

SHG is a specific nonlinear process which involves two photons with the same energy. Other second-order nonlinear processes are possible, such as sum and difference frequency generation (SFG and DFG), where two photons with different energies combine together into a third photon. As for SHG, these processes need to satisfy energy and momentum conservation, which for SFG are expressed by: $\omega_3 = \omega_1 + \omega_2$ and $\vec{k}_3 = \vec{k}_1 + \vec{k}_2$ (equations for DFG have a minus sign). Also here, phase matching can be achieved within a birefringent nonlinear crystal. As compared to the efficiency equation reported in Eq. (3.5), the intensity of the pulse obtained by SFG or DFG depends on the product of the two incoming intensities ($I(\omega_3) \propto I_{\omega_1} \cdot I_{\omega_2}$) [24].

3.2.2.2 Noncollinear Optical Parametric Amplifier

A noncollinear optical parametric amplifier (NOPA) is an optical device capable of producing tunable femtosecond radiation in the visible or near-infrared region. The output of a NOPA is obtained by the interaction of two beams in a nonlinear crystal: a strong pump (ω_p) and a weak and broadband *seed* ($\omega_s < \omega_p$). The pump is used to amplify the seed intensity, producing a strong output beam, named the *signal*, while creating another beam named *idler* at ω_i, where $\omega_p = \omega_s + \omega_i$ for energy conservation. In practice, the nonlinear process involved can be seen as difference frequency generation (DFG) between the pump and the seed. If the pump is the second harmonic at 400 nm of the Ti:sapphire beam, ω_i falls in the infrared region, and the output signal wavelength can be typically tuned from 490 to 760 nm. Considering that the seed is a broadband pulse, changing the orientation of the nonlinear crystal allows to amplify different wavelengths based on the particular phase matching condition fulfilled in a given orientation [25].

Optical parametric amplification can be obtained either in a collinear or noncollinear geometry. The NOPA configuration uses the latter, allowing to compensate for the group velocity mismatch (GVM) between the two pulses (Figure 3.2a), which limits the interaction length and, therefore, the amplification of the seed into the signal.

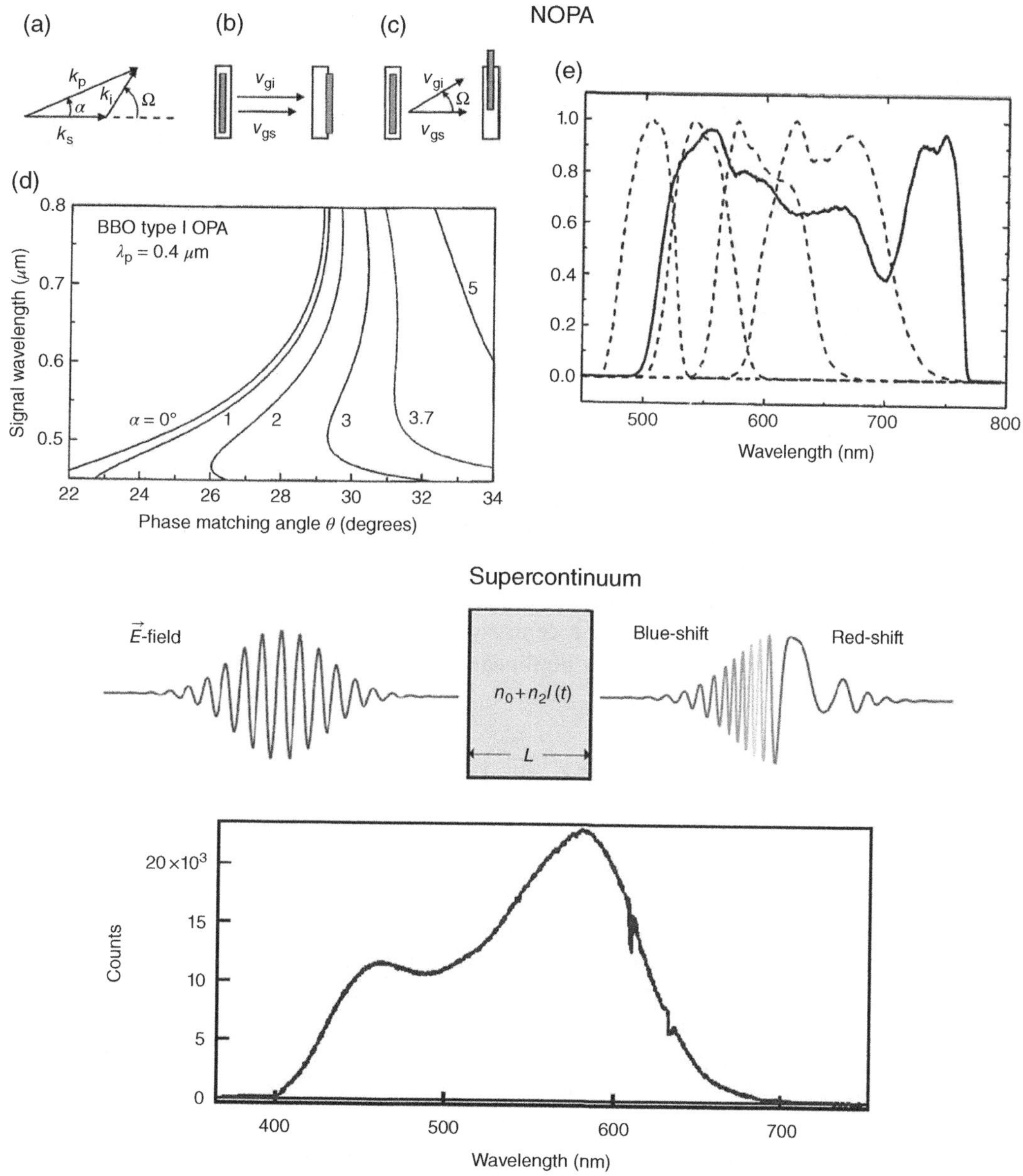

Figure 3.2 Top: (a) Wavevectors of pump ($\vec{k}_p$), signal ($\vec{k}_s$), and idler ($\vec{k}_i$) in the NOPA geometry; (b) group velocity mismatch of signal and idler pulses in collinear geometry and (c) in noncollinear geometry. The interaction length between the pulses is longer in (c) than in (b). (d) Phase matching curves for a NOPA pumped at 400 nm, as a function of the pump-signal angle. (e) Solid curve: NOPA spectrum under optimum alignment conditions and a compressed seed; dashed line: sequence of spectra obtained by changing the white light chirp. *Source:* Reprinted from [25], with the permission of AIP Publishing. Bottom: Generation of supercontinuum pulse (white light) from a red pulse propagating in a centrosymmetric medium and supercontinuum spectrum generated by 800 nm beam pulse focused in a 2 mm D_2O quartz cell filtered by a short-pass filter at 750 nm.

The advantages of a noncollinear configuration can be also explained in different terms. In a noncollinear configuration, the crystal orientation angle at which phase matching occurs for amplification of a given wavelength becomes dependent on the angle α between the pump and signal wavevectors propagating inside the crystal, as in Figure 3.2d. In particular, as demonstrated in Figure 3.2e, the phase matching condition is simultaneously achieved over a very broad wavelength range when $\alpha \sim 3.7°$. This allows to produce very broadband output pulses, which can be then compressed to very short temporal durations even below 10 fs. Usually, the output pulse is not broad as the near-vertical line in Figure 3.2e because the seed is temporally chirped and the interaction with the pump only involves a relatively narrow portion of the seed spectrum. Then, changing the alignment (spatially and temporally) between pump and signal allows to amplify different portions of the spectrum, as shown by the dashed lines in Figure 3.2e.

3.2.2.3 Supercontinuum Generation

The weak seed of a NOPA, usually, is a spectrally broad ultrashort light pulse. These pulses are called *white light* or *supercontinuum* pulses [26] and are generated by a third-order nonlinear phenomenon named *self-phase modulation* (SPM), combined to another phenomenon called *self-focusing* (SF). Besides the NOPA seed, these white light pulses are also used as probe pulses for transient absorption experiments discussed in the next section. To understand the generation of these supercontinuum pulses, consider a centrosymmetric material such as a liquid or glass ($\chi^{(2)} = 0$). Taking into account third-order nonlinear effects, the instantaneous polarization can be written as:

$$\frac{\vec{P}}{\varepsilon_0} = \chi^{(1)}\vec{E} + \chi^{(3)}\vec{E}\,\vec{E}\,\vec{E} = \vec{E}\left(\chi^{(1)} + \chi^{(3)}|E|^2\right) \tag{3.7}$$

which makes it possible to write the square of the refractive index n as:

$$n^2 = 1 + \chi^{(1)} + \chi^{(3)}|E|^2 = n_0^2 + \chi^{(3)}I. \tag{3.8}$$

Therefore, a propagating pulse will produce a change of the refractive index related to its instantaneous intensity. Considering the temporal and spatial distribution of the pulse intensity, and considering that $\chi^{(3)}$ is usually very small, the last expression is usually approximated in first order as $n \approx n_0 + n_2I(r, t)$, where the nonlinear refractive index n_2 is closely related to $\chi^{(3)}$. Considering a Gaussian beam, the intensity depends on position and time. Therefore, one may expect both spatial and temporal changes of the refractive index, which give rise, respectively, to SF and SPM [26].

The SPM is the effect related to the temporal dependence of beam intensity and, in particular, to the variations of the phase of the pulse. The applied field can be written as:

$$E(t,x) = E_0 e^{-i(\omega_0 t - kx)} \quad \text{where} \quad k = \frac{\omega_0}{c}n(t). \tag{3.9}$$

Therefore, the instantaneous phase of the pulse depends on time as:

$$\begin{aligned} \phi(t) &= \omega_0 t - \frac{\omega_0}{c}n(t)x \\ &= \omega_0 t - \frac{\omega_0 n_0}{c}x - \frac{\omega_0}{c}xn_2I(t). \end{aligned} \tag{3.10}$$

Considering the instantaneous frequency $\omega(t)$:

$$\omega(t) = \frac{\partial}{\partial t}\varphi(t) = \omega_0 - n_2\frac{\omega_0 x}{c}\frac{\partial I(t)}{\partial t} \tag{3.11}$$

the last term implies a change (self-modulation) of the frequency of the propagating light in a way controlled by the intensity envelope of the pulse itself. This effect causes the introduction of new frequencies in the spectrum of the pulse, symmetrically higher and lower than ω_0 according to the alternating signs of $\frac{\partial I(t)}{\partial t}$ in a Gaussian pulse (Figure 3.2).

A discussion of SPM, however, cannot be disentangled from the effects of SF. In fact, SF occurs in parallel to SPM because of the *spatial* dependence of the intensity in the pulse, which, for a Gaussian beam, is much stronger in the center of the beam than the sides. For this reason, the beam is capable of modifying the local refractive index, making it higher at the center of the beam if $n_2 > 0$. This causes a focusing of the beam along its path, because the region of modified refractive index behaves as a lens.

In laboratory practice, SF is easily observed by prefocusing by an ordinary lens an intense femtosecond laser beam (at least a few μJ per pulse) to a spot of a few tens of μm within a transparent medium with significant $\chi^{(3)}$, such as ordinary glass, a water cell, or crystals such as sapphire. Focusing the beam allows to reach a threshold intensity above which the onset of SF occurs. As a consequence, the beam spontaneously shrinks down to a *filament* with much smaller (a few μm) cross section, within a few millimeters of propagation length. In practice, SF stops when the diameter of the filament is so small (a few μm) that the diffraction is strong enough to balance the effect and prevent further self-focusing. Obviously, self-focusing causes a dramatic increase in the local intensity of the electric field. Thereby, SPM is strongly enhanced within the filament, strongly contributing to a dramatic broadening of the pulse spectrum and to the generation of an intense white light. Therefore, the formation of a stable filament is essential to have a stable and intense white light pulse. The final output of these processes is a spectral broad pulse as a consequence of combined SPM and SF, and it is also temporally broad and strongly chirped as a consequence of group velocity dispersion (GVD), as depicted in Figure 3.2. For instance, if the white light is generated from a 800 nm beam passing through a 2 mm cuvette of D_2O, the pulse covers a very broad range which is symmetric with respect to 800 nm, from which the visible part can be then selected by a filter (Figure 3.2).

3.3 Transient Absorption Spectroscopy

Ultrafast transient absorption (TA), or pump/probe, spectroscopy is a nonlinear spectroscopic method based on measuring the changes in the absorption spectrum of a system following an external excitation [8, 27–32]. In a TA experiment, the sample is photoexcited by a femtosecond pulse called *pump* and the variations of the absorption spectrum are measured by another, delayed, ultrafast pulse named *probe*. The probe is usually spectrally broad (400–700 nm) and this allows to record simultaneously the changes of the absorption spectrum in a wide spectral range. Moreover, the variations in the entire spectrum are recorded at different time delays between the two pulses, yielding kinetic traces of the time-dependent absorption coefficient at every wavelength.

In these experiments, the pump pulse is normally resonant to one of the electronic transitions of the sample, in order to bring it from the ground state to an upper energy state. Then, the instrument records the intensity of a probe light pulse which has traversed, at certain delay after the excitation, the excited spot on the sample, and compares this with the result of an identical measurement without the pump pulse. As explained hereafter, the TA signal is obtained from the ratio between the probe intensities recorded with and without excitation.

3.3.1 The Experimental Method

If we indicate with I_u and I_p the probe light intensities transmitted through the unexcited (u) and photoexcited (p) sample, and writing the number of absorbers in the system as $N_0 = N_g + N_e$, that is the sum of non-excited (N_g) and excited absorbers (N_e), the Beer–Lambert law can be used to express I_u and I_p in terms of the variation $\Delta\sigma$ of the attenuation cross section:

$$I_u(\omega) = I_0(\omega)e^{-\sigma_g(\omega)dN_0} \tag{3.12}$$

$$\begin{aligned} I_p(\omega) &= I_0(\omega)e^{-\sigma_g(\omega)d(N_0 - N_e) - \sigma_e(\omega)d(N_e)} \\ &= I_0(\omega)e^{-\Delta\sigma(\omega)dN_e - \sigma_g(\omega)dN_0} \\ &= I_u(\omega)e^{-\Delta\sigma(\omega)dN_e} \end{aligned} \tag{3.13}$$

$$\Delta\sigma(\omega) = -\frac{1}{dN_e}\ln\left(\frac{I_p(\omega)}{I_u(\omega)}\right) \tag{3.14}$$

where σ_g and σ_e are the attenuation cross sections in the ground and excited state, respectively, $\Delta\sigma = \sigma_e - \sigma_g$, and d is the sample thickness. In practice, the recorded TA signal is most commonly expressed as a differential optical density ΔOD, which is indeed proportional to the change in cross section: $\Delta OD = \Delta\sigma dN_e/2.303$. Thus, the TA signal is given by:

$$\Delta OD = OD_p - OD_u = -\frac{1}{2.303}\ln\frac{I_p}{I_u}. \tag{3.15}$$

This is the quantity that is experimentally obtained from I_u and I_p generally reported in a TA experiment. Considering that the variation of the absorption is normally very small, $I_p - I_u << I_u$, it is possible to linearize Eq. (3.15) and calculate the TA signal as:

$$\Delta OD = \frac{1}{2.303}\frac{I_u - I_p}{I_u}. \tag{3.16}$$

In general, ΔOD(ω) is the superposition of three types of contributions: ground state bleaching (GSB), stimulated emission (SE), and excited state absorption (ESA). Their nature is illustrated in Figure 3.3 using a molecular system with three electronic levels as an example. Similar considerations apply to molecules with more than three levels, or to other types of physical systems, such as bulk solids or nanoparticles.

GSB is the negative TA signal related to the depopulation of the ground state upon pumping. After pumping, photoexcited molecules are no more in the ground state and they cannot contribute anymore to the ground state absorption. Therefore, the probe is less absorbed in the spectral range of the 0 → 1 transition (see Figure 3.3) and this appears as a negative signal in the TA spectrum, resembling one or more of the steady state absorption bands.

Also, SE is a negative signal and it occurs because the probe pulse passing through the excited volume stimulates the emission of a photon from the excited state to the ground state. It produces a negative signal because the detector records an increase in intensity of the probe in the photoexcited system, in the spectral region of the 1 → 0 transition. The spectral shape of SE is very similar to the photoluminescence band except for its negative sign. However, it is worth noting that the SE cross section, and intensity, is proportional to the emission frequency, while the intensity of stationary fluorescence depends on the cube of emission frequency, as it is easy to see from Einstein's coefficients. In a perfect two-level system, GSB and SE would be perfectly degenerate and be indistinguishable, but in a typical molecular system, they are separated by a Stokes shift.

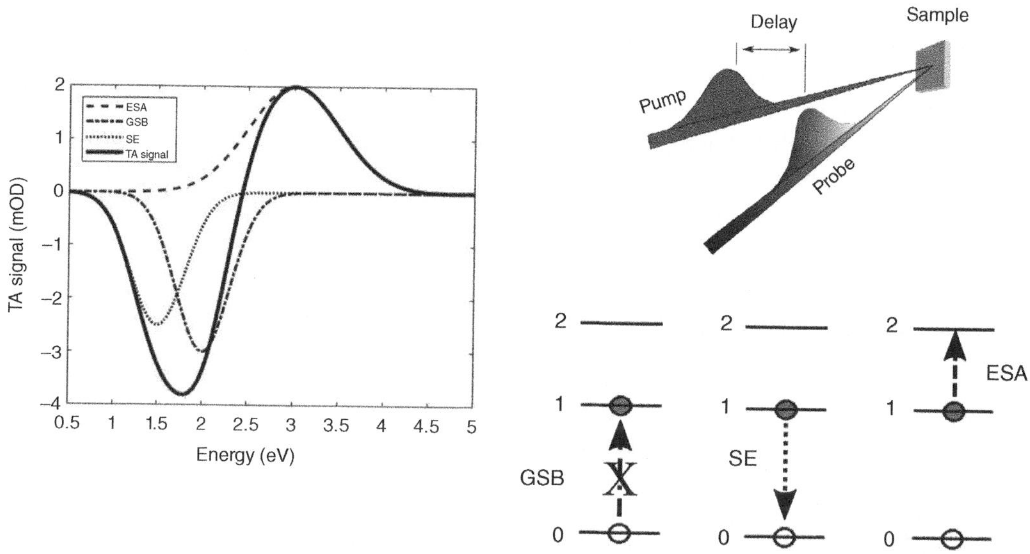

Figure 3.3 On the left panel, a hypothetical signal decomposed into three contributions. Diagram of the typical signals in a TA experiments: ground state bleaching (GSB), due to the decreasing absorption of the sample; stimulated emission (SE) due to the interaction with the probe which stimulates photon emission and the consequent decay of the excited system back to the ground state; and excited state absorption (ESA) related to the possibility of new transitions toward higher energy states. Upper right: simple scheme of a TA measurement: pump and the probe pulses overlap on the sample, and changing the delay between them allows to measure the changes in the absorption spectrum induced by photoexcitation and subsequent dynamics.

Last but not least, ESA is the only positive signal that is recorded in a TA experiment. It is due to the new electronic transitions, such as $1 \rightarrow 2$, that become possible from the excited state, because probe absorption can bring the system to other higher energy states [33].

Overall, a TA experiment is simultaneously sensitive to several electronic transitions accessible from the ground and excited state(s) of the system. Virtually any photoexcited system is expected to provide several observables that can be used to follow the dynamics, making the technique very versatile. For instance, suppose that, for a given system, the GSB is difficult to observe because it falls out of the probing window. Then, one can still resort to SE or ESA as observables to monitor the evolving system. While a singlet excited state will typically give rise to strong SE, a triplet state is not expected to, due to the very low radiative rate toward the ground state. However, it should still be detectable through an ESA transition toward some higher triplet state, which should remain strongly allowed. Overall, following the spectral position and the intensity of one or more of these signals allows to reconstruct in detail the undergoing relaxations of the system, provided that the time resolution is good enough to resolve all the dynamics of interest (usually <100 fs). The analysis of TA data provides a wealth of information on the relaxation dynamics of the photoexcited system, such as excited state population decay, internal conversion or intersystem crossing, charge and energy transfer, energy relaxation, solvation of the excited state, and more [3, 8, 29, 34].

To achieve femtosecond time resolution, it is necessary to have short pump and probe pulses (50–80 fs) which keep their duration along the setup (for example they should not traverse too much glass) and, moreover, it is necessary that the two pulses are synchronized down to the scale of a few femtoseconds. Synchronization on these timescales cannot be achieved by the use of electronics. The only way to achieve such a precision is to use all-optical strategies. In fact, a transient absorption setup typically exploits a *delay stage* in the optical path of one of the beams, which allows to change the relative delay between two pulses increasing or decreasing the path of one of them. It is easy to calculate that an increase in the path of 3 mm corresponds to a delay of 10 ps. In a standard setup, both beams derive from a single laser, the original beam being split in two portions which are used to generate separately the pump and the probe. In general, both pump and probe derive from the same laser source because it would be harder to synchronize with femtosecond accuracy two beams originating from two different lasers. The next section describes in more detail the main technical characteristics of a typical TA setup.

3.3.2 Typical Experimental Setups

A typical TA experimental setup involves the use of a laser that generates femtosecond pulses. Although the duration and the spectral characteristics of the pulses depend on the specific type of laser medium, many modern TA setups are based on amplified Ti:sapphire lasers, which typically produce pulses peaked at 800 nm with durations in the range 30–200 fs.

In the simplest configuration, the output of the laser is immediately split into two parts, which are used to generate the pump and the probe, respectively. The scheme of a typical setup is reported in Figure 3.4.

On the pump arm, the 800 nm beam passes through a series of optical elements used to control the characteristics of the beam, such as manipulating the spatial mode or the polarization, or altering the wavelength. For example, in Figure 3.4 the 800 nm beam is frequency-doubled (type I phase matching) by a β-BBO crystal in order to create a 400 nm beam (10–20% efficiency is easily achieved) meant to excite the sample. Alternatively, the fundamental Ti:sapphire beam could be sent to a NOPA, in order to fully tune the pump wavelength. Then, the beam is chopped, variably delayed, and focused on the sample. Pump spot on the sample should have a Gaussian spatial profile, with typical spot sizes of 50–200 μm. Importantly, the TA signal should be linear with the

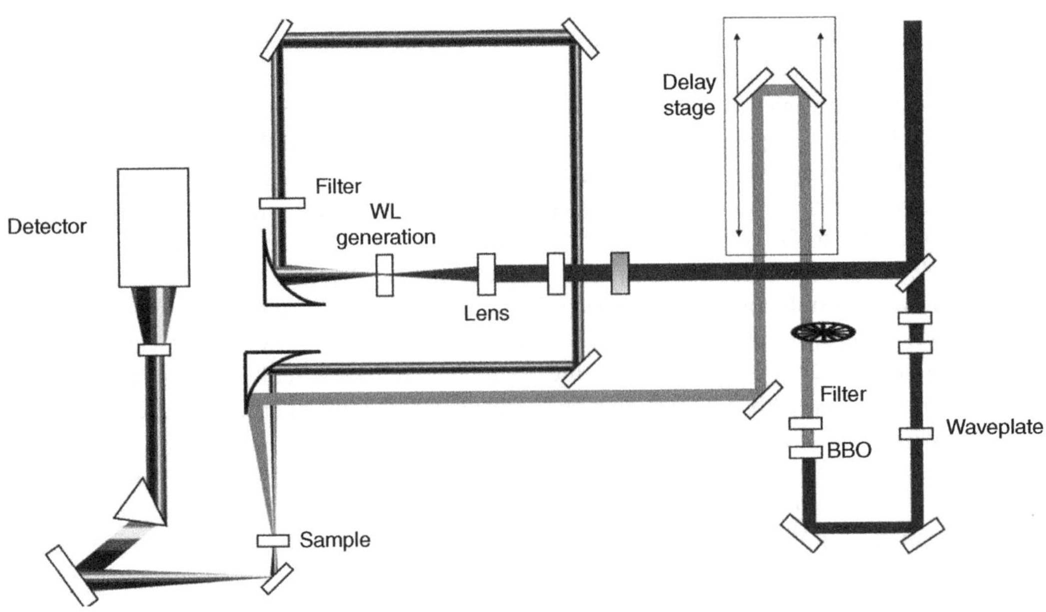

Figure 3.4 Scheme of a typical pump–probe setup.

intensity of the pump, which means that the photoexcited transition is not saturated and that only a small percentage (typically 10% or less) of the system is excited. Besides, using excessive pump power may give rise to other problems, such as multi-photon excitation contributions, white light generation in the sample, and so on. For all these reasons, typical pump energies per pulse in TA are usually limited to ten to hundred nJ pulse^{-1}.

On the second arm in Figure 3.4, supercontinuum generation is used to generate the probe pulse, a very common configuration in TA. The white light can be easily generated focusing a small portion of the 800 nm beam in a suitable material (sapphire crystal, water cuvette, etc.), generating a broadband pulse extending from 400 to 700 nm. The intensity of the 800 nm beam can be controlled and it is usually fixed at the value ~1 μJ, which creates a single stable filament in the media. The spectral profile of the white light can be controlled by changing the media and by various filters, which are also used to get rid of the intense residual light at the fundamental wavelength. After generation and filtering of the white light probe, the latter can be split into two (not shown in Figure 3.4) by a beam splitter, creating a second, reference probe beam, which does not pass through the excited part of the sample and can be used to correct any artifacts coming from probe intensity and spectral fluctuations. Because of the highly nonlinear nature of white light generation, which is intrinsically very noisy, its fluctuations can be very strong, making the probe reference beam very useful for noise reduction.

After generation, the white light beam needs to be collimated and then focused on the sample. Probe spot on the sample should be Gaussian, of identical or slightly smaller size than the pump spot. In the latter case, the system is less sensitive to possible misalignments in the pump/probe overlap. Several variants of this approach are possible. For example, one can double the 800 nm before generating the white light. The supercontinuum generated from 400 nm then extends deeper to the UV, allowing to probe in the ~300–600 nm spectral region. Another variant involves the use of two NOPAs, one used to produce the tunable pump pulse, and the other to produce a relatively broadband, and tunable probe [28].

As anticipated, the pump path is controlled by a motorized delay stage allowing a precise electronical control of pump–probe delay with precision of few femtoseconds. The probe and the pump are directed in such a way to overlap within the sample. Therefore, the absorption spectrum measured by the probe is collected from the region which has been previously photoexcited by the pump. If the sample is a liquid, it is generally made to continuously flow in a thin flow cell, or in a liquid jet, which strongly reduces photodamage. The flow speed should be regulated in order that every pump pulse hits a fresh portion of the sample. If the sample is a solid, it is usually moved by a motor stage in order to limit the excitation damage. The thickness of the sample is kept as low as possible (ideally, a few hundred μm) to reduce GVD and GVM effects which would tend to degrade time resolution.

After the sample, the probe beam is finally dispersed through a monochromator and sent to the detector, which measures its spectrum. Typically, the pump pulse is chopped, so that the detector alternates measures of I_p and I_u, allowing for a direct estimation of the TA signal according to Eq. (3.15). In case a reference probe beam is added, two independent detectors are used, in order to use the reference beam to correct the TA signal for probe fluctuations.

A full TA experiment consists in a temporal scan in which the delay between the two pulses is changed. The pump and probe pulses spatially overlap within the sample volume for the entire duration of the scan; in the time domain, the "zero time" of the experiment is identified when the pulses also overlap in time. Typically, time zero is found from the observation of the, so-called, cross-phase modulation (CPM) [35]. This is a signal related to the interaction of the probe with the variation of the refractive index induced by the pump inside the medium. It is responsible for a strong distortion of the data when the two pulses temporally overlap in the sample, so that it can be used to easily locate time zero. Once time zero is found, the TA scan is designed to cover from $t < 0$ to maximum delays of several hundreds of picoseconds, or even a few nanoseconds, in order to fully reconstruct the kinetics initiated by photoexcitation.

The ultimate time resolution of a TA experiment is controlled by the time duration of the two pulses. It is easily 100 fs or less, and can be as short as <10 fs in extreme cases [36, 37]. If the pulses are transform-limited ($\Delta\omega\Delta t = 0.5$), the time resolution is simply given by the cross-correlation between the two pulse intensity profiles. Therefore, the duration of the pump pulse should be made as short as possible to optimize time resolution, minimizing GVD effects. The requirements, however, are less severe for the probe pulse. In fact, the time resolution does not change even if the probe pulse is chirped by GVD. Although GVD temporally broadens the probe pulse, this only implies a different temporal overlap between the pump and every wavelength of the probe. This effect needs to be compensated during data analysis (see next section), but does not degrade time resolution, because the TA measurement is carried out separately on each spectral component of the probe, dispersed on a multichannel detector after interacting with the sample. Thus, even if the probe is chirped, the time resolution is identical to that which would be obtained by transform-limited pulses of the same total spectral width of the probe pulse [38]. In practice, probe pulses are often obtained by supercontinuum generation, yielding a very large bandwidth, hence a very short transform-limited duration. In this situation, the factor ultimately controlling the temporal resolution is the duration of the pump pulse only, which should be kept as short as possible.

3.3.3 Data Analysis and Interpretation

Before data analysis, TA data need to be corrected for the effects of GVD and CPM. Because of GVD-induced chirp, the probe pulse can be broadened to durations as long as ≈1 ps, and the time $t_0(\lambda)$ of the interaction with the pump depends on wavelength. This causes an uncertainty on the definition of *zero time*, which is a crucial point in time-resolved measurements. One way to measure $t_0(\lambda)$ is conducting a TA experiment in a reference sample (e.g. the pure solvent) where only the CPM signal is observed. Its narrow temporal width can be then used to estimate the time resolution of the setup, and the time at which CPM is observed for any λ yields the GVD correction curve $t_0(\lambda)$. Once this is known, kinetic traces from the sample data set are temporally shifted [33], so as to eliminate the wavelength dependence of the zero time, eliminating GVD effects. As for CPM, it is usually very difficult to compensate for its effects on the data. Therefore, after GVD corrections, the spectra collected in the temporal window around time zero where CPM is observed are typically removed from the data. As a consequence, the first useful spectrum is collected after a certain minimum delay from time zero, of the order of the time resolution of the experiment.

After GVD and CPM corrections, the following step is the extraction of the dynamics and the definition of the characteristic timescales of the sample dynamics. Whichever is the chosen data analysis method, its aim is to disentangle the various temporal dynamics contained within the signal and associate them with well-defined spectral features. There are several approaches to do this. At the qualitative level, directly inspecting the TA spectra at various time delays and comparing the kinetic traces at different wavelengths is often the first step to have an idea on what processes are observed, and their approximate timescales. In this respect, one should keep in mind that TA spectra can be generally read using the same rational approach that is used to read traditional steady state optical data, where intensity variations are usually due to depopulation, and spectral shifts are due to energy relaxations.

After a qualitative analysis of TA data, more sophisticated approaches are usually applied. One of them is singular value decomposition (SVD) followed by a global analysis (GA) [39]. SVD is a mathematical method by which data are decomposed to a minimum number of relevant kinetics and spectra, and at the same time, white noise is removed from the data. GA is combined with SVD in order to find time constants which can describe the entire data, providing a global model simultaneously fitting the kinetics at all wavelengths. From the SVD and the global analysis, decay associated spectra (DAS) are extracted, which describe relevant changes of TA signal for each time constant found during GA [40]. Assuming a physical model that includes only incoherent relaxations (not oscillations or phase variations), the GA typically leads to describe the signal as follows:

$$\mathrm{TA}(\lambda, t) = \sum_i \mathrm{DAS}_i(\lambda) \cdot \mathrm{e}_{t>0}^{-\frac{t}{\tau_i}} \otimes \mathrm{IRF}(K_\mathrm{b}, t_0 = 0, t) \tag{3.17}$$

where τ_i and DAS_i are the characteristic times and the relative amplitudes, and IRF is the instrumental response function that is supposed to be a Gaussian function with a width (≈FWHM/2.35) K_b and peaking $t_0 = 0$, and $\otimes$ is the convolution operator. Once DAS_i and related timescales τ_i are identified, specific knowledge of the system needs to be used to attribute to each of these processes a definite meaning, finally leading to an exhaustive model of the sequence of relaxation events initiated by photoexcitation.

3.4 Ultrafast Fluorescence Spectroscopies

Another group of powerful ultrafast spectroscopic techniques are the fluorescence-based techniques, such as fluorescence upconversion (FLUC) and Kerr gating-based time-resolved fluorescence. Both of them are time-resolved electronic spectroscopic techniques capable of selectively detecting the (spontaneous) emission, typically with 50–200 fs time resolution, without any GSB or ESA contaminations. Because these methods are only sensitive to excited state dynamics, the signals they produce are much easier to interpret than TA; at the same time, some information remains hidden, such as ground state dynamics, or the population of non-emissive states. For example, only fluorescent (singlet) excited states are usually measurable with these methods. In contrast, most (but not all) triplet excited states remain "dark" because their phosphorescence has too low a radiative rate to provide a detectable signal.

Both FLUC and Kerr gating experiments require the use of two pulses (called *excitation* and *gate*) and the possibility of changing the time delay between them. The excitation pulse directly interacts with the sample, initiating the fluorescence emission. Then, the gate pulse controls the temporal window in which the emitted fluorescence is sampled with femtosecond time resolution. In both cases, varying the delay between excitation and gate allows to follow the dynamics of the fluorescence. However, the two techniques achieve time resolution through the use of two different nonlinear optical effects. A FLUC experiment exploits SFG between the gate beam and the fluorescence, in order to sample the fluorescence signal within a time window defined by the cross-correlation between the excitation and the gate. The Kerr-gated ultrafast fluorescence is founded instead on the optical Kerr effect, i.e. on the modifications of refractive index of a medium induced by the exposure to a femtosecond pulse. A Kerr medium is used as an optical shutter activated by the gate beam, to select a temporal "slice" of the fluorescence signal to be measured. Details on the technical aspects of the two spectroscopies are given in the next subsections.

3.4.1 FLUC: The Experimental Method

A typical FLUC experiment involves the excitation of a sample by the pump beam and the collection of the largest possible fraction of the isotropically emitted fluorescence [2–4, 41–44]. This is achieved by large solid-angle collection optics, such as a parabolic mirror with short focal length and large diameter. The fluorescence is then collimated and refocused on a nonlinear crystal such as a BBO or KDP, by a pair of parabolic mirrors arranged in a telescope. In the BBO, the fluorescence spatially overlaps with the second beam, the gate, so that sum frequency generation occurs and a frequency upconverted pulse is generated and sent to detection. Because SFG occurs only within the temporal duration of the gate, SFG involves only a defined femtosecond temporal "slice" of the emission, even if the latter is nanosecond-lived. By measuring the intensity of the upconverted pulse as a function of excitation–gate delay, the entire emission kinetics can then be reconstructed.

3.4.2 FLUC: Typical Experimental Setups

A possible implementation of these concepts is reported in the bottom of Figure 3.5. Here, the gate wavelength is set at 800 nm, and the fluorescence is in the visible range, yielding an upconverted pulse in the near-UV, as can be obtained from the SFG equation: $\omega_{\mathrm{PL}} + \omega_{\mathrm{gate}} = \omega_{\mathrm{up}}$. Some technical upgrades allow to detect a fluorescence in the UV range (≈280 nm) [2, 42], which corresponds to generating and detecting the upconverted pulse in the deep-UV (≈210 nm). While the excitation

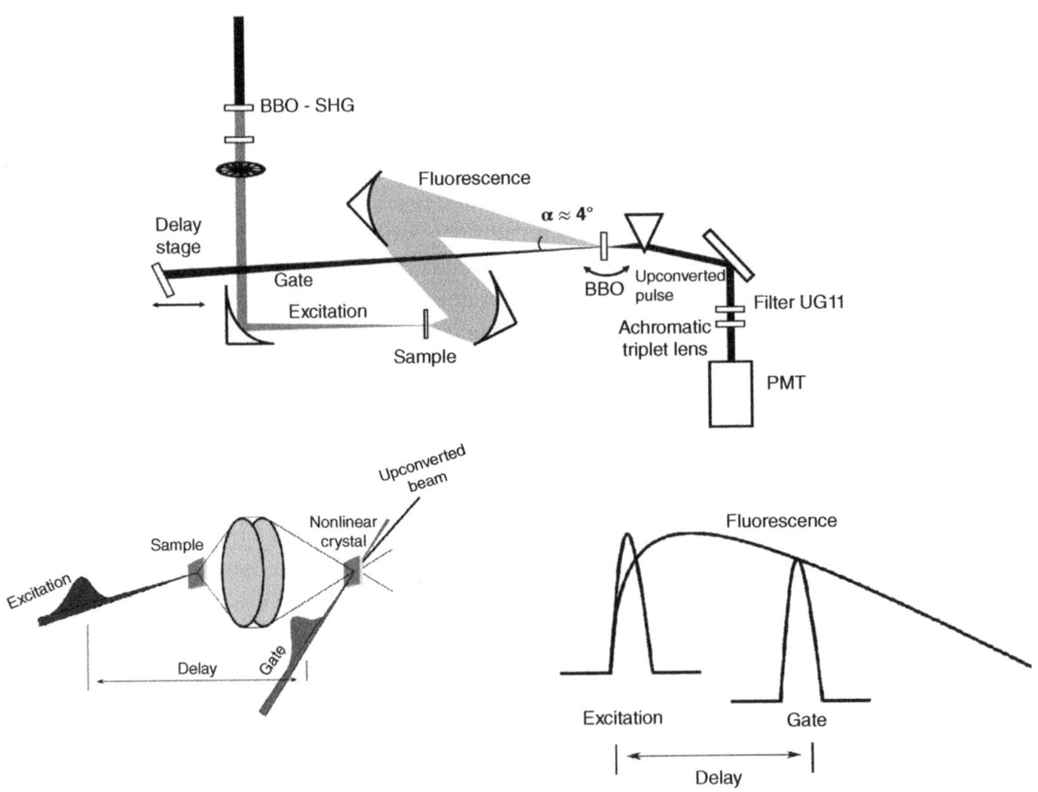

Figure 3.5 Scheme of a typical fluorescence upconversion setup.

pulse intensity is limited by linearity requirements, the intensity of the gate beam cannot usually overcome 1 μJ per pulse because of practical reasons (e.g. BBO damage). Moreover, the thickness of the BBO is limited by time resolution requirements (e.g. limiting GVD in the SFG process), and the number of fluorescence photons participating in SFG process at any given delay is only a tiny portion of the total emission, because of the very short duration of the gate pulse (tens of femtoseconds vs. nanosecond lifetime). Thereby, the upconverted signal is in practice very weak, even as low as <1 upconverted photon per excitation pulse. FLUC thus requires great care in isolating this signal, which is achieved by a combination of spatial and spectral filtering methods, and finally directing to a highly sensitive detection system, such as a single photon counter or a highly sensitive CCD [2].

Recording the upconverted pulse as a function of the gate delay allows to reconstruct the kinetics of the emission at a given wavelength, at variable delays from photoexcitation, with femtosecond time resolution. In particular, the temporal resolution of these experiments is ultimately determined, once detrimental effects are eliminated (e.g. GVM in the BBO), by the cross-correlation between the pump and the gate, which can be as low as a few tens of femtoseconds. Notably, efficient SFG can only occur if the *phase matching condition* ($\Delta k = 0$) is fulfilled, which only occurs at a precise orientation of the nonlinear crystal. Thus, the simplest approach to FLUC is acquiring single-wavelength fluorescence kinetics, as a function of excitation–gate delay, for a given crystal orientation. Rotating the nonlinear crystal changes the phase matching condition and allows to upconvert different spectral portions of the emission band. Then, repeating the measurement with different crystal positions allows to reconstruct the dynamics of the entire fluorescence band as a function of wavelength and time [45].

In Figure 3.5, the excitation beam is obtained by SHG of the fundamental Ti:sapphire beam, chopped, and finally focused on the sample by a parabolic mirror. After excitation of the sample, the spontaneous emission is collected and collimated by second parabolic mirror and, then, focused by a third parabolic mirror with the same diameter into the SFG crystal, where the delayed gate at 800 nm finally overlaps to the fluorescence spot. The setup uses a slightly noncollinear geometry to facilitate separation of the three beams after the nonlinear crystal, although the noncollinearity angle is kept very low in order to reduce GVM effects within the crystal, which would otherwise degrade time resolution. After the nonlinear crystal, the upconverted emission passes through a monochromator, and UV pass-band filters, and is finally sent to a photomultiplier (PMT), the output of which is read by a lock-in amplifier. In this configuration, the data are collected as single-wavelength traces. The detected wavelength is determined by rotating the nonlinear crystal, thus, determining a certain phase matching angle. More sophisticated setups use broad wavelength detection, which is obtained by substituting the PMT with a CCD camera, and by performing a computer-controlled rotation of the nonlinear crystal during data acquisition. By this method, the whole emitted spectrum can be acquired at any given delay [2, 42].

Interestingly, because of the phase matching condition, the SFG process is automatically polarization sensitive, meaning that only the fluorescence component that has the same polarization of the gate (type I phase matching) is upconverted in the BBO and detected by the setup. Therefore, by adding a waveplate in the pump arm, as in a TA experiment, the polarization of the latter can be controlled, and polarization-sensitive measurements can be easily achieved by a setup such as in Figure 3.5. To achieve this aim, one repeats each measurement in the two pump polarizations, parallel (*para*) or perpendicular (*ortho*) to the gate, respectively, and calculates the fluorescence anisotropy $r(t)$:

$$r(t) = \frac{I_{\text{para}}(t) - I_{\text{ortho}}(t)}{I_{\text{para}}(t) + 2 \cdot I_{\text{ortho}}(t)}. \tag{3.18}$$

Femtosecond fluorescence anisotropy measurements often provide important information regarding the occurrence of random rotational diffusion within the system, or any other phenomena (e.g. energy transfer and site-to-site excitation migration) capable of randomizing the orientation of the electric dipoles [46].

The actual design of a FLUC setup involves a relatively complex optimization problem involving multiple geometrical parameters to find the best compromise between signal intensity, noise rejection, and time resolution. While this topic cannot be fully covered here, a significant example is given by the geometry of the telescope directing the fluorescence into the nonlinear crystal. In fact, the focal lengths of the two parabolic mirrors used in Figure 3.5 must be chosen in such a way that the fluorescence enters the BBO within the acceptance angle for the SFG process. If n_o is the ordinary refractive index, and θ_m is the phase matching angle, the SFG acceptance angle is given by [47]:

$$\Delta\theta = \frac{2.78 n_{o,PL} \lambda_{PL}}{L\left[1 - \left(n_{o,PL}\lambda_{up}\right) / \left(n_{up}(\theta_m)\lambda_{PL}\right)\right]}. \tag{3.19}$$

The need to fulfill this condition imposes relatively tight limitations on the possible geometrical parameters of a setup such as in Figure 3.5. Similar considerations can be done for many other relevant geometrical parameters, the optimization of which constitutes an important aspect in the design of any FLUC setup [47].

3.4.3 FLUC: Data Analysis and Interpretation

Data collected in a FLUC experiments are usually single-wavelength traces. The treatment and correction of the data involve, as for the TA, the definition of a common zero time for all the kinetics, the GVD correction, and an assessment of the temporal resolution. A useful technique to measure the temporal resolution is the measurement of the temporal evolution of a Raman signal of a pure solvent. Because Raman scattering is essentially instantaneous, the temporal width of the Raman signal corresponds to the IRF of the experiment and it can be used to calibrate the experiment. After these preliminary procedures, the results can be treated by GA methods similar to what discussed above for TA, although FLUC data are generally much easier to interpret, inasmuch as they are much less affected by spectral congestion issues, and only contain one type of signal, that is spontaneous emission. In fact, many FLUC signals essentially consist in a single time-dependent fluorescence band, or a few with no mutual overlap. In this case, FLUC data are often simpler to analyze by direct band integration methods (e.g. moment analysis), leading to very straightforward interpretations. For example, one popular application of FLUC is probing solvent relaxations around a chromophore through the observation of picosecond and femtosecond dynamical Stokes shift induced on its fluorescence band. This approach has proven very valuable to analyze solvation dynamics in water [3], and hydration dynamics around proteins, a problem of great biological importance [4, 44].

3.4.4 Kerr-Based Femtosecond Fluorescence Spectroscopy

Kerr-based femtosecond fluorescence spectroscopy exploits the optical Kerr effect, through which a material becomes birefringent under the action of the intense electric field provided by a light pulse [48, 49]. As for upconversion method, the sample is pumped by a laser beam and the consequent fluorescence is collected. Unlike FLUC, the emission is focused into an isotropic Kerr medium (glass plate or solvent) arranged between two crossed polarizers intercepting the fluorescence path. In standard conditions, the output after the second polarizer is zero (closed shutter condition).

When the gate pulse passes through the Kerr media, it *opens the shutter*, by inducing a transient change in the refractive indexes of the medium that makes it birefringent. Thereby, the emission traversing the Kerr cell becomes elliptically polarized, a portion of it passes through the second polarizer, and reaches the detector. Changing the delay between the pump and the gate then allows to record fluorescence kinetics. In can be shown that the gate pulse should be polarized at 45° with respect to the excitation in order to obtain maximum output [48]. Considering that a phase matching condition is not required in this scheme, a broadband detection is easy and limited only by the optical absorption of the used components and of the detection device.

The temporal resolution is determined by whatever is shorter between the cross-correlation between excitation and gate and the cross-correlation between the excitation and the intrinsic response function of the Kerr medium. In fact, if the Kerr medium reacts slowly to the gate modifications, then the temporal resolution is dominated by the latter. For example, with an excitation of ≈50 fs at 475 nm and a gate of 40 fs, the time resolution is ≈120 fs using a 1 mm-thick fused silica plate as a Kerr medium [49].

3.5 Femtosecond Stimulated Raman Spectroscopy

Femtosecond time-resolved methods are not limited to electronic spectroscopies, but can also be extended to meet the advantages of vibrationally sensitive techniques, such as infrared absorption and Raman. In this regard, this section addresses ultrafast stimulated Raman spectroscopy, which is one of the most powerful methods to combine the structural sensitivity and chemical selectivity of vibrational methods with the power of femtosecond time resolution [50–55].

3.5.1 The Experimental Method

Classic Raman spectroscopy exploits the Raman effect, discussed in detail in Chapter 5 of this book, that is the inelastic scattering of light associated to the generation or annihilation of vibrational quanta of energy ω_V during the interaction of the light beam with the system under study. A typical steady state Raman experiment interrogates the sample by a laser beam at ω, and analyzes the spectral distribution of photons scattered in all directions. In the most common case (Stokes Raman scattering), part of the scattered radiation is found to be red-shifted to $\omega' = \omega - \omega_V$, because an energy ω_V has been spent to excite phonons of this frequency within the sample. Thus, spectral analysis of the scattered radiation reveals a series of peaks at various ω_V reporting on the vibrational mode structure of the system. Vibrational frequencies being much smaller than visible ones, the exciting beam must be narrow enough in frequency to avoid spectral overlap between ω and ω'. Therefore, the vibrational spectral resolution is of the order of the FWHM of the Raman pump beam. Raman spectroscopy is a powerful probe of the vibrational mode structure of any physical system and displays a sensitive dependence on both electronic and molecular structure. Thus, it can be used for accurate chemical recognition and to reveal, with high specificity, any structural changes of the system at study.

On these grounds, femtosecond time-resolved Raman (TRR) experiments can be conceived as a powerful probe of excited state dynamics of any photoexcited physical system, overcoming the lack of structural sensitivity typical of absorption and fluorescence spectroscopies. In particular, Raman spectra being characteristic of any given state of the system, one should expect strong changes of the vibrational mode pattern whenever the system transitions to an excited state, and subsequently evolves. Following these changes by TRR can be used to extract detailed information on any structural changes or photochemical processes initiated by photoexcitation.

Broadly speaking, TRR spectroscopy is conceptually similar to TA, except, in this case, the observable becomes the change in the Raman spectrum of the sample between a pumped and unpumped situation [52]. TRR is not the only example of femtosecond-resolved vibrational spectroscopy. An alternative is provided by mid-infrared TA experiments, which are the time-resolved counterpart of infrared absorption experiments [56, 57]. Similar to TRR, infrared TA can also probe the vibrational mode structure, but one advantage of TRR spectroscopy is the exclusive use of visible- or near-infrared-light pulses, which are easier to generate and detect, especially if broadband pulses are needed. Furthermore, due to the higher energy, visible-light pulses are inherently shorter in time, ultimately providing better time resolution [51]. Another alternative is given by time-domain techniques founded on TA, such as impulsive stimulated Raman scattering (ISRC) [58], which works best with low-frequency modes and generally displays lower signal-to-noise ratios than state-of-the-art TRR. For these reasons, TRR is generally recognized as the most powerful approach to achieve vibrational spectroscopy with femtosecond resolution, and will be discussed in detail in the rest of this section.

3.5.2 Typical Experimental Setups

The simplest and most straightforward path to implement a TRR experiment is to use a pump–probe approach very similar to TA experiment, except for the fact that the probe beam is used as a Raman pump. Thus, instead of measuring the change in probe transmittance, which is the basis of TA, one can achieve TRR by collecting the isotropic Raman signal produced by scattering of the delayed probe pulse, and analyzing it in a spectrograph as a function of pump–probe delay [54]. However, this method, named probe-induced Raman scattering, is made very difficult by the very small cross section of Raman scattering, and also brings inherent limitations in time resolution. In fact, considering the time–energy uncertainty relation, the probe beam used as a Raman pump beam must be relatively broad in time to be narrow enough in the spectral domain. For example, a pulse centered at 550 nm, with a FWHM of 1 nm will guarantee a ω_V spectral resolution of $\sim$30 cm^{-1}, which may be adequate to probe a Raman spectrum, but with a transform-limited pulse duration as long as 450 fs. This problem is fundamentally unsurmountable and limits the time resolution to the picosecond range.

In fact, one of the most effective approaches to carry out a TRR experiment is use a more complex scheme, initially proposed at the end of the 1990s [55], based on what is called the *stimulated* Raman process. The relation between spontaneous and stimulated Raman process is similar to that between spontaneous and stimulated emission: shortly, the Raman scattering cross section into a given photon mode is strongly enhanced whenever that particular mode is already populated by photons provided by a seed beam. Therefore, stimulated Raman process can be used to probe the same properties as spontaneous Raman process, but with potentially much stronger intensity.

In a typical time-resolved stimulated Raman experiment [55, 59–61], such as the one depicted in Figure 3.6, three noncollinear laser pulses are used, all obtained by splitting the fundamental beam from the femtosecond laser. The first pulse hitting the sample is the *pump*, which is resonant with a certain transition of the sample and photoexcites it. To distinguish this pulse from the following two, it is often named *actinic* or *photochemical* pump. After that, a combination of two more pulses is used to probe the vibrational mode pattern of the excited system. One acts as Raman *pump* pulse, long in the time domain, hence narrowband in frequency around ω, as required for spectral resolution; the third pulse, called the Raman *probe* beam, provides the seed responsible of stimulating the Raman emission. The two pulses are overlapping in time and hit the sample together after a certain delay t from the pump. However, the Raman seed is much shorter in time (<100 fs), hence

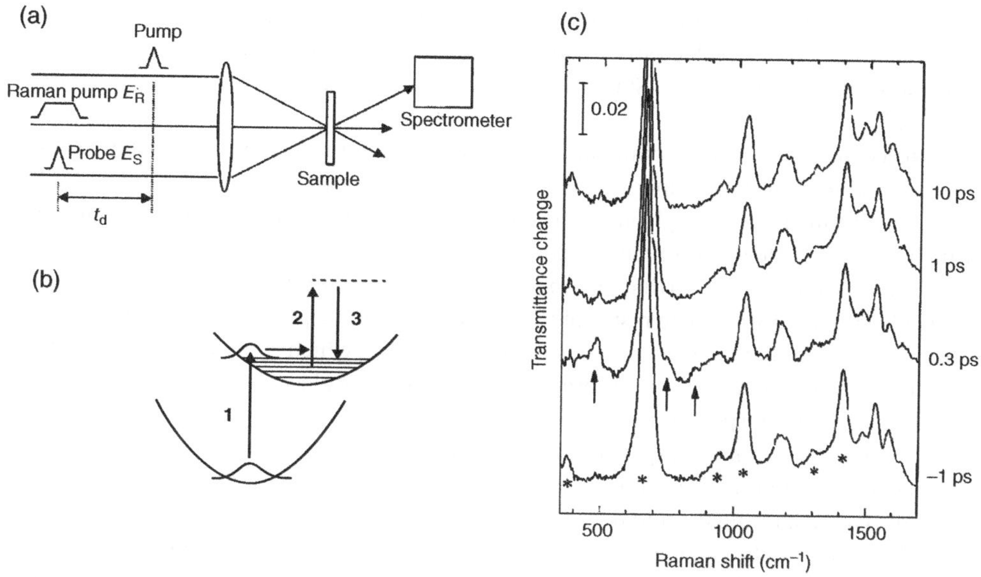

Figure 3.6 Panel (a): Scheme of the three-pulse sequence used in time-resolved stimulated Raman spectroscopy. *Source:* Reprinted with permission from [55]. Copyright (1999) by the American Physical Society. Panel (b): Energy level diagram for time-resolved stimulated Raman. *Source:* Adapted with permission from McCamant et al. [60]. Copyright (2003) American Chemical Society. Panel (c): One of the earliest time-resolved stimulated experiment, probing the trans-cis photoisomerization dynamics of 4-(dicyanomethylene)-2-methyl-6-(4-dimethylaminostyryl)-4H-pyran, dissolved in a 3 mM concentration in DMSO and photoexcited at 397 nm. The Raman pump pulse was tuned at 795 nm (bandwidth 1.5 nm), while the Raman probe pulse, obtained by supercontinuum generation, covered the region 600–1000 nm. Arrows indicate the Raman signals appearing due to the photogenerated transient states, while asterisks indicate the Raman lines of the solvent. *Source:* Reprinted figure with permission from Yoshizawa and Kurosawa [55]. Copyright (1999) by the American Physical Society.

spectrally broadband and contains nonzero spectral density across a wide interval of ω'. In these conditions, one observes a strong enhancement of the Raman scattering at ω', which can amount to several orders of magnitude. If the Raman probe spectrum is dispersed and measured with and without the Raman pump, one can directly obtain the stimulated Raman spectrum as the gain in the probe, that is the natural logarithm of the ratio: $(\text{Probe})_{\text{Pump on}}/(\text{Probe})_{\text{Pump off}}$. Besides, the detection is greatly facilitated by the fact that the stimulated Raman signal is emitted in a defined direction. In fact, femtosecond stimulated Raman spectroscopy allows to acquire high-quality Raman spectra even in the presence of a strong fluorescence [52, 61].

What is more important, however, is that the use of this detection approach provides the trick to achieve very high temporal resolution [61]. In fact, despite Raman scattering occurring during the whole duration of the Raman pump pulse (~1 ps or more), it is only within the duration of the short-lived Raman probe beam that stimulated Raman process occurs. Therefore, the time resolution of the measurement is now controlled by the femtosecond duration of the probe pulse. For example, a narrowband Raman pump pulse at 560 nm, with duration of several picoseconds, can be combined with a probe pulse centered, say, at 580 nm with a FWHM of 20 nm, which is broad enough to stimulate Raman emission over a relatively wide spectral range (~1000 cm^{-1}), and can be compressed down to <25 fs.

Summarizing, high spectral resolution and short temporal resolution are simultaneously achieved by decoupling the two problems, so overcoming the Heisenberg limit: the former is controlled by the FWHM of the Raman pump pulse, while the latter depends on the temporal duration of the probe. Often, time-resolved stimulated Raman is described by the quantity of $\delta t \delta \nu$ used as a figure of merit of the experiment. State-of-the art setups have achieved $\delta t \delta \nu$ of the order of 0.5 ps·cm^{-1} [52].

While femtosecond stimulated Raman can be used as a way to record a ground state Raman spectrum, the final goal of these experiments is usually following the evolution initiated by the actinic pump, with time resolution given by the cross-correlation of the actinic and Raman probe pulses, which can reach a few tens of femtoseconds. To do so, the stimulated Raman spectrum for the unexcited sample is compared to that obtained at variable delays from photoexcitation, in order to follow the effect of relaxation. The result of the experiment can be plotted either in terms of an absolute, time-dependent Raman spectrum (e.g. as in Figure 3.6c), which contains both the features of the unexcited ground molecules and those of the excited molecules, or as a *difference* Raman spectrum, calculated with respect to the unexcited sample.

Figure 3.7 represents a typical experimental setup [60]. The actinic pump is obtained by a NOPA pumped by an amplified Ti:sapphire laser, allowing for tunable excitation as in a typical TA experiment. The remaining portion of the amplifier is used to generate both the Raman pump and probe pulses. In order to narrow down the bandwidth of the Raman pump pulse, the amplifier output traverses narrow band-pass interference filters (BPF in Figure 3.7). Alternatively, one can use more complex prism- or grating-based filters, which allow fine control of both the central

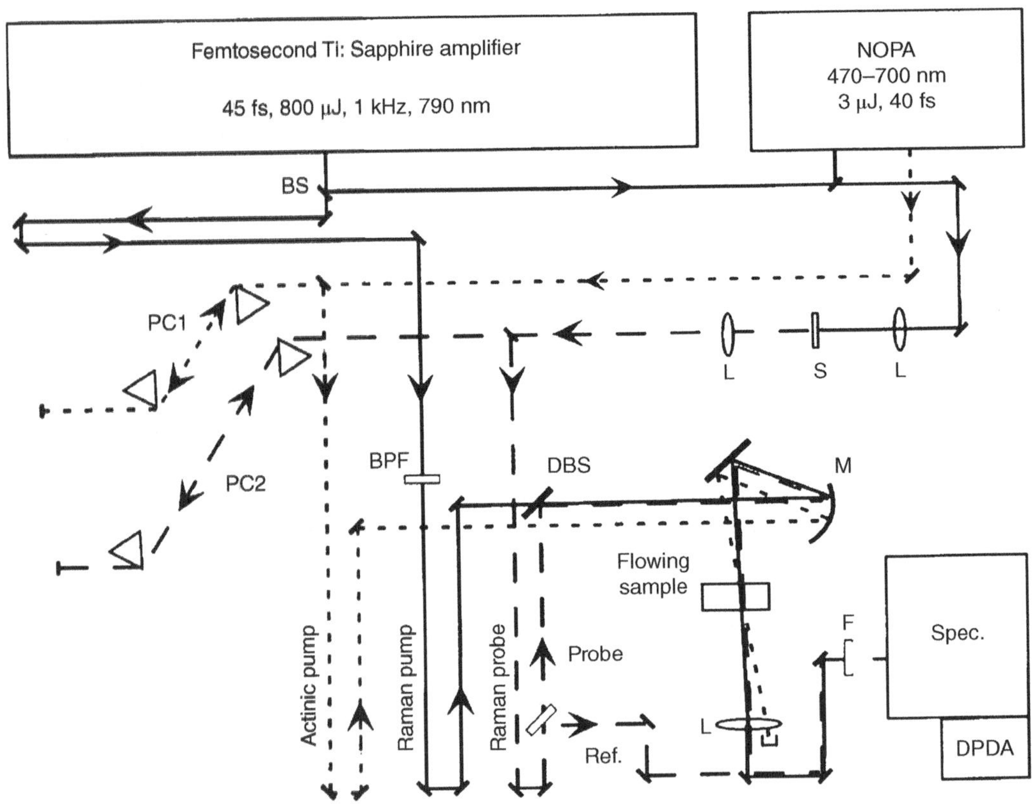

Figure 3.7 A possible configuration for a time-resolved stimulated Raman experiment. *Source:* Adapted with permission from McCamant et al. [60]. Copyright (2003) American Chemical Society.

wavelength and bandwidth. In this particular case, the picosecond Raman pump pulse is centered at 793 nm with bandwidth of 1.1 nm and a duration of several picoseconds. Its intensity is kept relatively high (~1 μJ pulse^{-1}) in order to optimize the amount of Raman scattering produced. The Raman probe pulse is obtained through supercontinuum generation (400–1000 nm) in a sapphire plate, followed by compression in a prism pair wherein only the near-IR portion of the continuum is retained. In this way, a 80 fs pulse is obtained covering the 870–950 nm region, that corresponds to 1000–2000 cm^{-1} Raman shifts from the Raman pump. The intensity of the Raman probe is much lower than that of the Raman pump, typically in the range of tens of nJ pulse^{-1}. Finally, the three pulses are made collinear and focused in the same spot (sized a few tens of μm) of the flow cell containing the sample. After traversing the sample, the Raman probe beam is spatially selected, goes through a spectrograph, and analyzed in an array detector.

Several variants of this general scheme are possible. In some setups, for example, tunable Raman pump and probe pulses are both obtained by NOPAs [62]. In other cases, the fundamental beam from the amplifier was directly used as the Raman probe [63]. In regard to detection, some setups make use of a reference beam [60]: the probe is split in two in order to generate a reference beam, allowing for efficient shot-to-shot normalization of white light fluctuations. Experiments have been conducted over a wide range of Raman pump wavelengths, from the near IR to the UV [64]. This wide flexibility also allows to tune the Raman pump for pre-resonant [63] and resonant [53] Raman process, which provides a convenient route to further enhance the Raman signal, and to single out the Raman contribution from the chromophore of interest.

In regard to data acquisition, several pulse chopping sequences can be used in this type of experiments [52]. By a simple chopper in the Raman pump path, one can acquire the stimulated Raman spectrum from the ratio between Raman-pumped and Raman-unpumped spectra cyclically recorded by the array detector. Then, the Raman spectra of the ground state and photoexcited sample can be obtained sequentially, and compared to each other, by shuttering the actinic pump. If the chopper is positioned in the path of the actinic pump, the ratio between chopped and unchopped signals directly records the difference Raman spectrum induced by photoexcitation. More complex schemes where both the actinic and Raman pump are chopped are sometimes used as well [64].

3.5.3 Data Analysis and Interpretation

Data analysis of TRR data involves a series of additional steps which do not have a counterpart in TA, and the details of which depend on the specific pulse sequence used in the acquisition. A typical analysis workflow is illustrated in [65]. If chopping the actinic pump only, one can directly acquire the difference Raman spectrum induced by photoexcitation. The latter will always "sit" over a broad background due to transient absorption changes of the Raman probe beam. Therefore, one always needs to separately measure the TA and subtract it from the recorded signal. The signal after TA subtraction contains the excited state Raman spectrum, ground state Raman bleach contributions, but also some residual contributions of the solvent, due to attenuation of the Raman pump by photoexcited molecules. Therefore, one needs to separately measure and subtract the pure solvent Raman signal, which is rescaled and subtracted in order to get the final difference spectrum. Some specific additional problems which are not encountered in TA may arise both at the time of setup alignment or data processing. For example, the data may contain spurious contributions due to the cross-phase modulation between Raman pump and probe beams. Several strategies have been proposed to minimize these effects, such as vibrating the Raman pump retroreflector to average out CPM effect [61].

TRR experiments exploiting stimulated Raman scattering are now very well-established in the toolbox of ultrafast spectroscopies. They are powerful tools capable of following the ultrafast dynamics of photoexcited systems by revealing structurally precise details of the undergoing events through the changes of their vibrational mode pattern. The interpretation of time-resolved vibrational data follows basically the same principles as ordinary steady state vibrational spectroscopy, where characteristic fingerprint vibrations are highly structure-specific and, in some frequency regions, can be traced back to specific chemical moieties. The main difference with a steady state Raman experiment is that the vibrational properties change as a result of the electronic redistribution induced by photoexcitation, and of subsequent system dynamics. For example, molecular internal conversion can be pinpointed unambiguously through the appearance and disappearance of the unique fingerprint vibration patterns associated to each of the involved states [66]. On these grounds, TRR methods have been extensively used to interrogate ultrafast photochemical processes such as charge transfer, proton transfer, isomerization, or internal conversion, as demonstrated by many examples in the literature of the last two decades [50, 53, 66, 67].

3.6 Case Studies

3.6.1 Ultrafast Relaxation Dynamics of Molecules in Solution Phase

Transition metal complexes, such as metal–polypyridine complexes, are very interesting for several photochemical and solar energy harvesting applications, such as dye-sensitized solar cells (DSSCs). The excited state dynamics of these molecules display several phenomena which make them a particularly interesting topic in the "ultrafast" community. One of the typical electronic absorption bands of many metal complexes, named metal-to-ligand charge transfer (MLCT), involves an optical charge transfer from the metal to the ligands around it. Excitation within MLCT electronic transitions is the first step toward electron injection in DSSCs, and gives rise to peculiar dynamical behavior which can only be addressed by ultrafast spectroscopic techniques.

It is known [68] that photoexcitation within the singlet ^{1}MLCT state is typically followed by relaxation into a long-lived triplet ^{3}MLCT state, through an intersystem crossing (ISC) process, which can be surprisingly fast, well in the sub-picosecond range. Because of such an extremely short ^{1}MLCT fluorescence lifetime, its quantum yield is so small to make it practically invisible in steady state measurements. For the same reason, a direct observation of ISC in real time is very challenging, but can be achieved by FLUC, which allows to directly detect ^{1}MLCT fluorescence and follow its dynamics as a function of time.

An example of this is reported by Cannizzo et al. [34]. The authors of this paper investigated the dynamics of an archetypal metal complex ($[Ru(bpy)_3]^{2+}$) by ultrafast fluorescence upconversion. In particular, they reported for the first time a FLUC experiment where the detection covered a broad wavelength range (440–690 nm) and, simultaneously, a time resolution of ≈110 fs. Photoexciting the metal complex with a very short excitation pulse at 400 nm allows them to follow the dynamics from the earliest stage, thus capturing the whole relaxation of the system.

The signal observed by FLUC [34] can be described as an emission band, peaking at 19 230 cm^{-1}, which is attributed to ^{1}MLCT emission and grows together with the excitation pulse. This band then decays within 200 fs while a weaker band appears at 16 500 cm^{-1}. The latter corresponds to the steady state emission from the triplet state ^{3}MLCT.

The detailed analysis of the temporal kinetics demonstrates that the bluest ^{1}MLCT band rises and decays during the duration of the excitation pulse, and that also the rise of the redder band occurs during the pulse duration, suggesting that the population of the ^{3}MLCT state is driven by an extremely fast ISC which lasts much less than the experimental time resolution. Deconvolution of the instrumental response function suggests an ISC timescale smaller than 30 fs, which implies in turn an extremely efficient spin–orbit coupling for this complex.

The shape of ^{1}MLCT emission at zero time involves a large excitation–emission Stokes shift immediately after excitation. Analysis of the bandshape initially led to hypothesize that this transient ^{1}MLCT emission is vibrationally "hot," because 400 nm pump pulses selectively excite the $\nu = 2$ level of the Franck–Condon progression in absorption [34]. However, this issue was further addressed by later studies by the same group, which revealed instead that the transient ^{1}MLCT fluorescence is vibrationally relaxed on ultrashort timescales [69], possibly even faster than the ISC process.

3.6.2 Relaxation of Excited Charge Carriers and Excitons in Semiconductor Nanoparticles

Ultrafast transient absorption experiments help to investigate, as already said, the sub-picosecond and picosecond dynamics occurring in different materials after photoexcitation. As a further example, here we report some ultrafast TA studies conducted by Klimov's group ([27]) on CdSe nanocrystals aiming to investigate the fast carrier dynamics of electron–hole pairs and the effects produced by very intense pump excitation, capable of generating multiple excitations in the same nanocrystal.

The experiment consisted in the photoexcitation of CdSe colloidal NCs excited with 100 fs pulses peaking at 400 nm (3.1 eV, that is higher than CdSe bandgap) and in the consequent probing of the absorption changes in the visible-IR range. The size of the studied nanocrystals is small enough to fall in the *strong confinement* regime (diameter smaller than Bohr radius), characterized by an energy spacing between the lowest electron states larger than both the exciton binding energy and the longitudinal optical phonon. The results obtained by TA provided a comprehensive picture of the dynamics of photoexcited charge carriers from the very first moment after photoexcitation.

Figure 3.8a reports typical TA spectra observed in the system at different delays from photoexcitation. The features appearing in the TA spectra can be given a definite identity by a direct comparison with features present in the steady state spectrum. For example, the lowest-energy feature B1 clearly visible in both the TA and steady state spectra is associated to e–h pairs formed by an electron in the 1S state and a hole in the $1S_{3/2}$ state. The spectra at relatively long times (ps) are mostly explained by the bleach of several features of the absorption spectrum, due to the reduced absorption after one (at low pump intensities) or more e–h couples have been generated by photoexcitation. These bleaching signals, also named *state filling* effects, appear as positive signals in Figure 3.8a because the TA is reported with inverted sign ($-\Delta\alpha$). Besides state filling effects, a peculiar derivative-like shape, composed by features B1 and A1, shows up in the TA at short times. This feature is attributed to bi-exciton interactions, that are the Coulombic interactions between the e–h pairs generated by the pump and the probe. On increasing pump–probe time delay, carriers relax from initially high excited to low-energy states, the coulombic contributions diminish and the state filling contributions dominate the spectra. Therefore, the spectral evolution in Figure 3.8a provides direct information on the timescales of the carrier energy cascade in semiconductor NCs.

Beyond the timescales of carrier dynamics, TA signals collected at different pump powers (Figure 3.8b) provide some useful additional information. When the pumping intensity is low

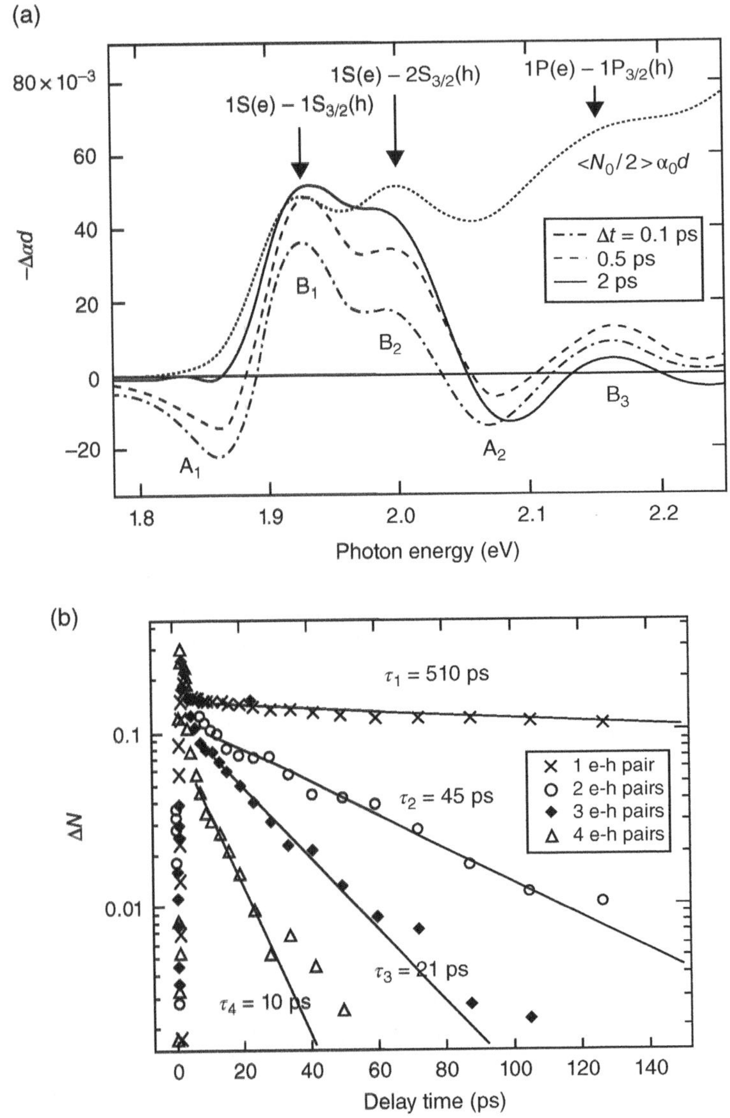

Figure 3.8 (a) TA spectra of CdSe NCs recorded at 0.1, 0.5, and 2 ps compared with the steady state absorption. The features A_i are associated to bi-exciton Coulombic interactions decaying in time, while features B_i are due to depopulation of the ground state. (b) TA kinetics of the 1-, 2-, 3-, and 4- e–h couples fit to a single-exponential decay. *Source:* Adapted with permission from Klimov [27]. Copyright (2000) American Chemical Society.

enough, the photoexcited average NC population can be estimated to be lower than 1, that is less than one e–h pair generated on average per NC. However, if the power is increased in order to have average NC population greater than 1, new decay components show up in the temporal kinetics of the TA signals, as clearly visible in Figure 3.8b. These components are due to the activation of new non-radiative depopulation channels in the sub-nanosecond regime, due to the Auger recombination events which become possible whenever multiple e–h pairs coexist in the same quantum dot.

Interestingly, the Auger lifetimes turn out to be quantized, because the number of generated e–h pairs can only be discrete, yielding a progression of decay lifetimes which depend on pump power (Figure 3.8b). Therefore, the carrier dynamics result to be faster if the power and, consequently, the number of e–h pairs per NC increase.

3.6.3 Ultrafast Relaxation Dynamics of Carbon-based Nanomaterials

Besides the investigation of ultrafast relaxations within a given physical system, such as a molecule, solid, or nanomaterial, femtosecond techniques allow to investigate the interactions of materials with the external environment and with external agents, often characterized by very fast photoreactions. As an example, we report an ultrafast study of the interaction of metal ions with carbon nanodots (CDs), a recently discovered family of fluorescent carbon-based nanoparticles [29]. In particular, the aim of the work was to investigate the changes of the optical properties of CDs induced by metal ions in solution, demonstrating the occurrence of an emission quenching and explaining the underlying mechanism.

The typical photocycle of CDs begins with the instantaneous photogeneration of an electron–hole couple where the electron is exposed to the environment on the surface, while the hole remains inside the core of the nanoparticle [70]. The surface electron first undergoes strong solvent-induced relaxation on femtosecond and picosecond timescales and, at longer times (in the nanosecond range) the exciton recombines radiatively producing the fluorescence. In the presence of even small amounts (μM) of transition metal ions, CDs and ions form stable complexes wherein the fluorescence is strongly quenched. Under the static quenching regime (i.e. quenching of the emission without a variation of the nanosecond lifetime), TA experiments at different concentrations of copper ions were performed in order to clarify the ultimate mechanism responsible for fluorescence deactivation [29].

The TA signal of pure CDs (Figure 3.9a) is characterized at all delays by three contributions: a negative contribution around the pump wavelength, mostly due to ground state bleaching (GSB) associated with the depopulation of the ground state via photoexcitation; a strong stimulated emission (SE) at 520–550 nm, spectrally matching the ordinary fluorescence signal except for its negative sign; and several positive excited state absorption (ESA) signals due to electronic transitions from the excited state toward higher excited states.

The spectral evolution of the signal entirely occurs in the first few picoseconds. The analysis of the observed dynamics shows that the kinetics are dominated by a progressive Stokes shift with negligible signal decay, as can be seen when comparing the peak position of the SE signal at two different times (Figure 3.9a). This behavior can be attributed to excited state solvation, as well-known for other systems [3, 4], and a detailed analysis of data in Figure 3.9a shows that the solvation timescales for CDs are 0.19 ps and 2.1 ps. Adding the external agent in solution (i.e. copper ions, Figure 3.9b) does not modify the shape of the signal. However, the entire signal now undergoes a progressive decay (compare the amplitude of the two signals in Figure 3.9b). Notably, a detailed analysis shows that the two timescales of this decay are found identical to those of solvation.

This result is consistent with a model in which fluorescence quenching is caused by an electron transfer from the CD to the ions: the consequent loss of correlation between electron and hole hinders radiative recombination causing the deactivation of the fluorescence, which we see in real time as a progressive decay of the SE signal in Figure 3.9b. Interestingly, the result that this electron transfer occurs bi-exponentially on the same timescales of aqueous solvation, 0.19 and 2.1 ps, suggests that the photochemical electron transfer reaction is essentially driven by solvation. In this picture, while solvent molecules rearrange around the surface of the photoexcited CD–ion complex,

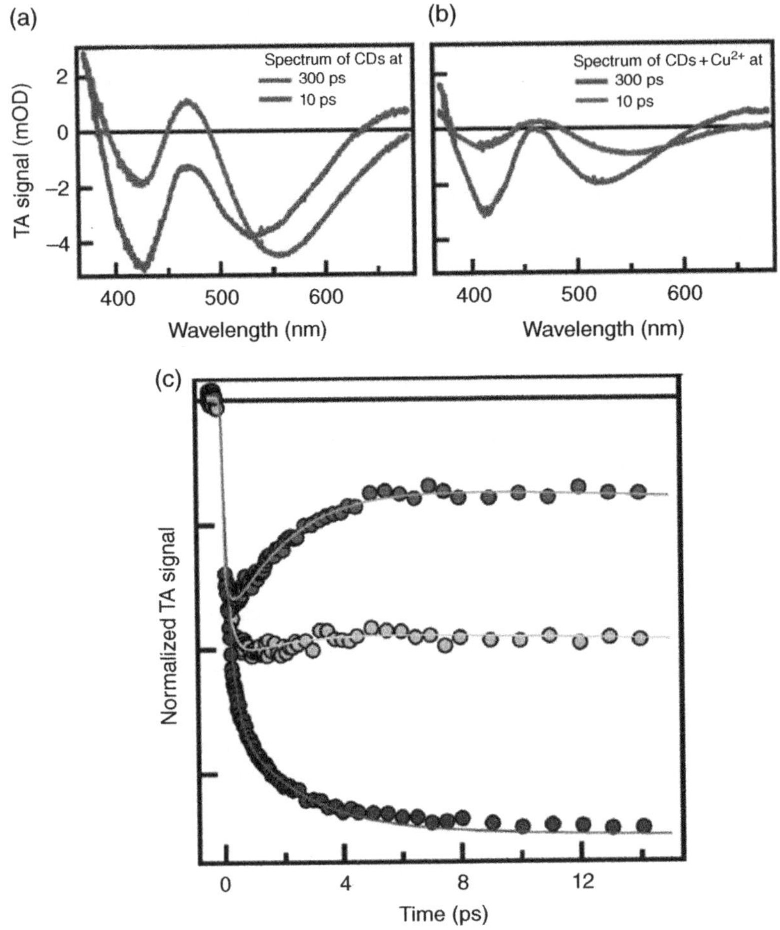

Figure 3.9 TA spectra of bare CDs (a) and with 100 mM of Cu^{2+} (b) recorded at 300 fs and 10 ps after the photoexcitation. (c) Normalized kinetics traces at the SE signal with different amounts of copper ions (0, 20 mM, 100 mM). *Source:* [29]. Adapted by permission of the Royal Society of Chemistry.

on picosecond and sub-picosecond timescales, they are responsible for driving the photoexcited system from the initial state to a final one in which the surface electron has been entirely transferred to the interacting metal ion, so destroying the e–h pair and fully quenching the emission.

References

1 Rosspeintner, A., Lang, B., and Vauthey, E. (2013). Ultrafast photochemistry in liquids. *Annu. Rev. Phys. Chem.* 64: 247–271.

2 Cannizzo, A., Bräm, O., Zgrablic, G. et al. (2007). Femtosecond fluorescence upconversion setup with broadband detection in the ultraviolet. *Opt. Lett.* 32 (24): 3555–3557.

3 Jimenez, R., Fleming, G.R., Kumar, P.V., and Maroncelli, M. (1994). Femtosecond solvation dynamics of water. *Nature* 369: 471–473.

4 Pal, S.K., Peon, J., and Zewail, A.H. (2002). Biological water at the protein surface: dynamical solvation probed directly with femtosecond resolution. *Proc. Natl. Acad. Sci.* 99: 1763–1768.

5 Iwai, S., Murata, S., and Tachiya, M. (1998). Ultrafast fluorescence quenching by electron transfer and fluorescence from the second excited state of a charge transfer complex as studied by femtosecond up-conversion spectroscopy. *J. Chem. Phys.* 109 (14): 5963–5970.

6 Marie, X. and Urbaszek, B. (2015). Ultrafast exciton dynamics. *Nat. Mater.* 14: 860–861.

7 Huber, R., Sp'orlein, S., Moser, J.E. et al. (2000). The role of surface states in the ultrafast photoinduced electron transfer from sensitizing dye molecules to semiconductor colloids. *J. Phys. Chem. B* 104: 8995–9003.

8 Klimov, V.I. (2007). Spectral and dynamical properties of multiexcitons in semiconductor nanocrystals. *Annu. Rev. Phys. Chem.* 58: 635–673.

9 Betz, M., Göger, G., Leitenstorfer, A. et al. (1999). Ultrafast electron-phonon scattering in semiconductors studied by nondegenerate four-wave mixing. *Phys. Rev. B* 60: R11265–R11268.

10 Saule, T., Heinrich, S., Schötz, J. et al. (2019). High-flux ultrafast extreme-ultraviolet photoemission spectroscopy at 18.4 mhz pulse repetition rate. *Nat. Commun.* 10: 458.

11 Mitrano, M., Lee, S., Husain, A.A. et al. (2019). Ultrafast time-resolved x-ray scattering reveals diffusive charge order dynamics in la2–xbaxcuo4. *Sci. Adv.* 5 (8).

12 Neppl, S., Mahl, J., Tremsin, A.S. et al. (2016). Towards efficient time-resolved x-ray absorption studies of electron dynamics at photocatalytic interfaces. *Faraday Discuss.* 194: 659–682.

13 Zhu, Y. and Cheng, J.-X. (2020). Transient absorption microscopy: technological innovations and applications in materials science and life science. *J. Chem. Phys.* 152 (2): 020901.

14 Leone, S.R. and Neumark, D.M. (2016). Attosecond science in atomic, molecular, and condensed matter physics. *Faraday Discuss.* 194: 15–39.

15 Verma, S., Ghosh, A., Das, A., and Ghosh, H.N. (2010). Ultrafast exciton dynamics of j- and h-aggregates of the porphyrin-catechol in aqueous solution. *J. Phys. Chem. B* 114 (25): 8327–8334.

16 Strauss, V., Kahnt, A., Zolnhofer, E.M. et al. (2016). Assigning electronic states in carbon nanodots. *Adv. Funct. Mater.* 26 (44): 7975–7985.

17 Sciortino, A., Gazzetto, M., Buscarino, G. et al. (2018). Disentangling size effects and spectral inhomogeneity in carbon nanodots by ultrafast dynamical hole-burning. *Nanoscale* 10: 15317–15323.

18 Fischer, A.J., Shan, W., Park, G.H. et al. (1997). Femtosecond four-wave-mixing studies of nearly homogeneously broadened excitons in Gan. *Phys. Rev. B* 56: 1077–1080.

19 Rullière, C. (2003). *Femtosecond Laser Pulses Principles and Experiments*. Springer.

20 Diels, J.-C. and Rudolph, W. (2006). *Ultrashort Laser Pulse Phenomena*. Elsevier.

21 Abramczyk, H. (2005). *Introduction to Laser Spectroscopy*. Elsevier.

22 Weiner, A.M. (2009). *Ultrafast Optics*. Wiley.

23 Shah, J. (1988). Ultrafast luminescence spectroscopy using sum frequency generation. *IEEE J. Quantum Electron* 24 (2).

24 Di Carlo, S. and Messina, F. (2017). A collision timing monitor for superKEKB. *Nucl. Inst. Methods Phys. Res.* 869: 95–106.

25 Cerullo, G. and De Silvestri, S. (2002). Ultrafast optical parametric amplifiers. *Rev. Sci. Instrum.*: 74.

26 Alfano, R.R. (2006). *The Supercontinuum Laser Source*. Springer.

27 Klimov, I.V. (2000). Optical nonlinearities and ultrafast carrier dynamics in semiconductor nanocrystals. *J. Phys. Chem. B* 104: 6112–6123.

28 Consani, C., Bräm, O., van Mourik, F. et al. (2012). Energy transfer and relaxation mechanisms in cytochrome c. *Chem. Phys.* 396: 108–115.

29 Sciortino, A., Madonia, A., Gazzetto, M. et al. (2017). The interaction of photoexcited carbon nanodots with metal ions disclosed down to the femtosecond scale. *Nanoscale* 9: 11902–11911.

30 Bragg, A.E., Cavanaugh, M.C., and Schwartz, B.J. (2008). Linear response breakdown in solvation dynamics induced by atomic electron-transfer reactions. *Science* 321: 1817.

31 Marchioro, A., Teuscher, J., Friedrich, D. et al. (2014). Unravelling the mechanism of photoinduced charge transfer processes in lead iodide perovskite solar cells. *Nat. Phot.* 8: 250.

32 Beard, M.C., Knutsen, K.P., Yu, P. et al. (2007). Multiple exciton generation in colloidal silicon nanocrystals. *Nano Lett.* 7: 2506–2512.

33 Ruckebusch, C., Sliwa, M., Pernot, P. et al. (2012). Comprehensive data analysis of femtosecond transient absorption spectra: a review. *J. Photochem. Photobiol.C: Photochem. Rev.* 13: 1–27.

34 Cannizzo, A., van Mourik, F., Gawelda, W. et al. (2006). Broadband femtosecond fluorescence spectroscopy of $[Ru(bpy)_3]^{2+}$. *Angew. Chem* 118: 3246–3248.

35 Ekvall, K., van der Meulen, P., Dhollande, C. et al. (2000). Cross phase modulation artifact in liquid phase transient absorption spectroscopy. *J. Appl.Phys.* 87 (5): 2340–2352.

36 Brixner, T., Stenger, J., Vaswani, H.M. et al. (2005). Twodimensional spectroscopy of electronic couplings in photosynthesis. *Nature* 434: 625–628.

37 Polli, D., Altoè, P., Weingart, O. et al. (2010). Conical intersection dynamics of the primary photoisomerization event in vision. *Nature* 467: 440–443.

38 Gadermaier, C., Kabanov, V.V., Alexandrov, A.S. et al. (2014). Strain-induced enhancement of the electron energy relaxation in strongly correlated superconductors. *Phys. Rev. X* 4: 011056.

39 van Stokkum, I.H.M., Larsen, D.S., and van Grondelle, R. (2004). Global and target analysis of time-resolved spectra. *Biochim. Biophys. Acta* 1657: 82–104.

40 Marciniak, H. and Lochbrunner, S. (2014). On the interpretation of decay associated spectra in the presence of time dependent spectral shifts. *Chem. Phys. Lett.* 609: 184–188.

41 Gerecke, M., Bierhance, G., Gutmann, M. et al. (2016). Femtosecond broadband fluorescence upconversion spectroscopy: spectral coverage versus efficiency. *Rev. Sci. Instrum.* 87 (5): 053115.

42 Messina, F., Bräm, O., Cannizzo, A., and Chergui, M. (2013). Real-time observation of the charge transfer to solvent dynamics. *Nat. Comm.* 4: 2119.

43 Underwood, D.F., Kippeny, T., and Rosenthal, S.J. (2001). Ultrafast carrier dynamics in cdse nanocrystals determined by femtosecond fluorescence upconversion spectroscopy. *J. Phys. Chem. B* 105: 436–443.

44 Zhang, L., Wang, L., Kao, Y.-T. et al. (2007). Mapping hydration dynamics around a protein surface. *Proc. Natl. Acad. Sci.* 104: 18461–18466.

45 Chosrowjan, H. (2017). Fluorescence up-conversion methods and applications. In: *Encyclopedia of Spectroscopy and Spectrometry*, 3ee (eds. J.C. Lindon, G.E. Tranter and D.W. Koppenaal), 654–660. Oxford: Academic Press.

46 Lakowicz, J.R. (2006). *Principles of Fluorescence Spectroscopy*, 3ee. Springer.

47 Xu, J. and Knutson, J.R. (2008). Ultrafast fluorescence spectroscopy via upconversion: applications to biophysics. *Meth. Enzymol.* 450: 159–183.

48 Righini, R. (1993). Ultrafast optical kerr effect in liquids and solids. *Science* 262: 1386–1390.

49 Schmidt, B., Laimgruber, S., Zinth, W., and Gilch, P. (2003). A broadband kerr shutter for femtosecond fluorescence spectroscopy. *Appl. Phys B* 76: 809–814.

50 Kukura, P., McCamant, D.W., Yoon, S. et al. (2005). Structural observation of the primary isomerization in vision with femtosecond-stimulated raman. *Science* 310: 1006–1009.

51 Kukura, P., McCamant, D.W., and Mathies, R.A. (2007). Femtosecond stimulated raman spectroscopy. *Annu. Rev. Phys. Chem.* 58: 461–488.

52 Dietze, D.R. and Mathies, R.A. (2016). Femtosecond stimulated Raman spectroscopy. *Chem. Phys. Chem.* 17: 1224–1251.

53 Kuramochi, H., Takeuchi, S., and Tahara, T. (2012). Ultrafast structural evolution of photoactive yellow protein chromophore revealed by ultraviolet resonance femtosecond stimulated Raman spectroscopy. *J. Phys. Chem. Lett.* 3: 2025–2029.

54 Reid, P.J., Wickham, S.D., and Mathies, R.A. (1992). Picosecond UV resonance Raman spectroscopy of the photochemical hydrogen migration in 1,3,5-cycloheptatriene. *J. Phys. Chem.* 96: 5720–5724.

55 Yoshizawa, M. and Kurosawa, M. (1999). Femtosecond time-resolved raman spectroscopy using stimulated Raman scattering. *Phys. Rev. A* 61: 013808.

56 Adamczyk, K., Prémont-Schwarz, M., Pines, D. et al. (2009). Real-time observation of carbonic acid formation in aqueous solution. *Science* 326: 1690–1694.

57 Schreier, W.J., Schrade, T.E., Koller, F.O. et al. (2007). Thymine dimerization in DNA is an ultrafast photoreaction. *Science* 315: 625–629.

58 Kuramochi, H., Takeuchi, S., and Tahara, T. (2016). Femtosecond time-resolved impulsive stimulated raman spectroscopy using sub-7-fs pulses: apparatus and applications. *Rev. Sci. Instrum.* 87: 043107.

59 Rondonuwu, F.S., Watanabe, Y., Zhang, J.-P. et al. (2002). Internal-conversion and radiative-transition processes among the $1b_u^+$, $1b_u^-$ and $2a_g^+$ states of all-trans-neurosporene as revealed by subpicosecond time-resolved Raman spectroscopy. *Chem. Phys. Lett.* 357: 376–384.

60 McCamant, D.W., Kukura, P., and Mathies, R.A. (2003). Femtosecond time-resolved stimulated raman spectroscopy: application to the ultrafast internal conversion in *β*-carotene. *J. Phys. Chem. A* 107: 8208–8214.

61 McCamant, D.W., Kukura, P., Yoon, S., and Mathies, R.A. (2004). Femtosecond broadband stimulated raman spectroscopy: apparatus and methods. *Rev. Sci. Instrum.* 75: 4971–4980.

62 Dobryakov, A.L., Quick, M., Ioffe, I.N. et al. (2014). Excited-state raman spectroscopy with and without actinic excitation: S 1 Raman spectra of trans-azobenzene. *J. Chem. Phys.* 140: 184310.

63 Kim, H.M., Kim, H., Yang, I. et al. (2014). Time-gated pre-resonant femtosecond stimulated Raman spectroscopy of diethylthiatricarbocyanine iodide. *Phys. Chem. Chem. Phys.* 16: 5312–5318.

64 Laimgruber, S., Schachenmayr, H., Schmidt, B. et al. (2006). A femtosecond stimulated raman spectrograph for the near ultraviolet. *Appl. Phys. B* 85: 557–564.

65 Rhinehart, J.M., Challa, R., and McCamant, D. (2012). Multimode charge-transfer dynamics of 4-(dimethylamino)benzonitrile probed with ultraviolet femtosecond stimulated raman spectroscopy. *J. Phys. Chem. B* 116: 10522–10534.

66 Kardas, T.M., Ratajska-Gadomska, B., Lapini, A. et al. (2014). Dynamics of the time-resolved stimulated raman scattering spectrum in presence of transient vibronic inversion of population on the example of optically excited trans-*β*-apo-8′-carotenal. *J. Chem. Phys.* 140: 204312.

67 Fang, C., Frontiera, R.R., Tran, R., and Mathies, R.A. (2009). Mapping gfp structure evolution during proton transfer with femtosecond raman spectroscopy. *Nature* 462: 200–205.

68 Damrauer, N.H., Cerullo, G., Yeh, t.R. et al. (1997). Femtosecond dynamics of excited-state evolution in $[Ru(bpy)_3]^{2+}$. *Science* 275: 54.

69 Bräm, O., Messina, F., El-Zohry, A.M. et al. (2012). Polychromatic femtosecond fluorescence studies of metal–polypyridine complexes in solution. *Chem. Phys.* 393: 51–57.

70 Sciortino, A., Marino, E., van Dam, B. et al. (2016). Solvatochromism unravels the emission mechanism of carbon nanodots. *J. Phys. Chem. Lett.* 7 (17): 3419–3423.

4

Confocal and Two-Photon Spectroscopy

Giuseppe Sancataldo and Valeria Vetri

Department of Physics and Chemistry – Emilio Segrè, University of Palermo, Palermo, Italy

4.1 Introduction and Historical Perspectives

Over the last centuries, radiation–matter interaction has brought to light plentiful fundamental aspects concerning the composition and the structure of matter itself. The spectroscopic investigation of atoms, molecules and their organization in complex architectures at micro-(nano)-meter scale has been a fundamental tool to investigate condensed and soft matter and it is of outmost importance in many research fields ranging from solid state physics to chemistry and life sciences. The spectroscopic analysis of bulk samples provides useful information at the mesoscopic scale that can be fostered when coupled with microscopy, which allows to integrate the spatial information in the analysis, this being of utmost importance for non-homogeneous samples. In recent years, enormous progresses have been achieved in microscopy methods and instrumentation and in particular in light microscopy methods based on fluorescence spectroscopy.

Fluorescence microscopy can be considered one of the most powerful and adaptable optical imaging methods. It allows to map and to quantify fluorescence signal with high resolution, exploiting hidden properties of matter based on the exquisite sensitivity and specificity of *fluorescence* measurements [1, 2].

At the end of the past century, several technological advances had been introduced, providing increased image quality and spatiotemporal resolution [3]. Technologies such as confocal and two-photon (or multiphoton) microscopy quickly became mainstream methods as they provide the possibility of noninvasive real-time analysis of specimens of different nature, eliciting specific features even in samples with non-ergodic behavior [4, 5]. Fluorescence observables (intensity, spectral features, lifetime) can be measured and localized at diffraction-limited resolution (few hundreds of nanometer) in sharp images sliced out as optical sections from thick samples. Both confocal and two-photon microscopy, indeed, based on different physical principles, which will be described in the following, provide unique possibilities of three-dimensional (3D) optical microscopy that, coupled with digital acquisition and readily available tools of image processing [6], allow researchers quantitative analysis of fluorescence signals as a function of space, time and wavelength [7].

Spectroscopy for Materials Characterization, First Edition. Edited by Simonpietro Agnello.

The idea of confocal microscope was first patented in 1957 by Marvin Minsky, a young postdoctoral fellow at Harvard University, who introduced the idea of confining illumination and detection areas, suitably positioning an opaque screen with a small aperture (pinhole) along the optical pathway, thus reducing image blurring due to scattering and to spurious contribution [8]. Later on in the 1970s, this concept was developed by several groups in Europe and United States. David Egger and Paul Davidovits in Yale University and Colin Sheppard and Tony Wilson in Oxford brought improvements to confocal concepts by introducing laser scanning, photomultiplier tubes as detectors and digital processing till the first commercial laser scanning microscope was produced, in 1983 [9].

In the early 1990s, also the first two-photon microscope was developed in which light confinement, which allows optical sectioning, is obtained by nonlinear light–matter interaction [10]. In two-photon microscopy, fluorescence excitation is obtained by focusing a high-energy infrared laser beam with half the energy with respect to the separation between the electronic levels of the fluorophore within a period shorter than the fluorescence decay time [11]. The excitation is due to a nonlinear effect that was first theoretically described in the doctoral dissertation of Maria Göeppert Mayer in 1931 [12] and observed 30 years later in 1961 in a CaF_2:Eu_2^+ crystal using laser excitation by Wolfgang Kaiser [13].

Confocal and two-photon microscopies share high sensitivity, selectivity, and the ease of use of fluorescence detection. They allow to achieve the theoretical spatial resolution limit imposed by light diffraction thought the apparently simple idea of confining sampling volume to submicrometer size, obtaining images of wide areas through laser scanning. These methods provide noninvasive optical sectioning on samples of various nature, ranging from biological systems (as living cells) to polymeric nanostructured films, crystals, and semiconductor devices, and they are now extensively used in nanotechnology, provided that target molecules are fluorescent [14], [15]. Moreover, both methods are now finding applications as irreplaceable analysis tools for lab-on-chip devices development, giving information on the events dynamics and chemical interactions at resolution down to about 200 nm [16, 17].

Too often, these powerful machines with extremely high potential are used as cameras to take "nice pictures," which, of course, contain a large amount of morphological and spectroscopic information that will be otherwise inaccessible. In this scenario, the knowledge of the basic principles of operations may allow to entirely exploit the complete potential of these measurements, which couple the power of visualizing sample details with the high content of spectroscopic information obtained at single pixel resolution.

4.1.1 Point Spread Function and Optical Resolution

The basic aim of any optical microscope is to make available a magnified image of a sample able to provide information that cannot be seen by naked eye. The capability of an imaging system to resolve details in a specimen defines its spatial resolution, which can be quantified as the minimum distance that can be measured between two objects in an image still allowing to distinguish them as separate.

The spatial resolution of optical microscopy has its roots in the diffraction of light on the objective lens of the imaging system. In 1835, G. Airy observed that when light from a point source passes through a small circular aperture, it does not produce a bright spot, but rather a pattern of circular concentric rings whose intensities decline with their distance from the center. E. Abbe recognized the connection between Airy's observation and the resolution power of optical microscopes that form images by the interference of light coming from each point of a specimen. These light rays

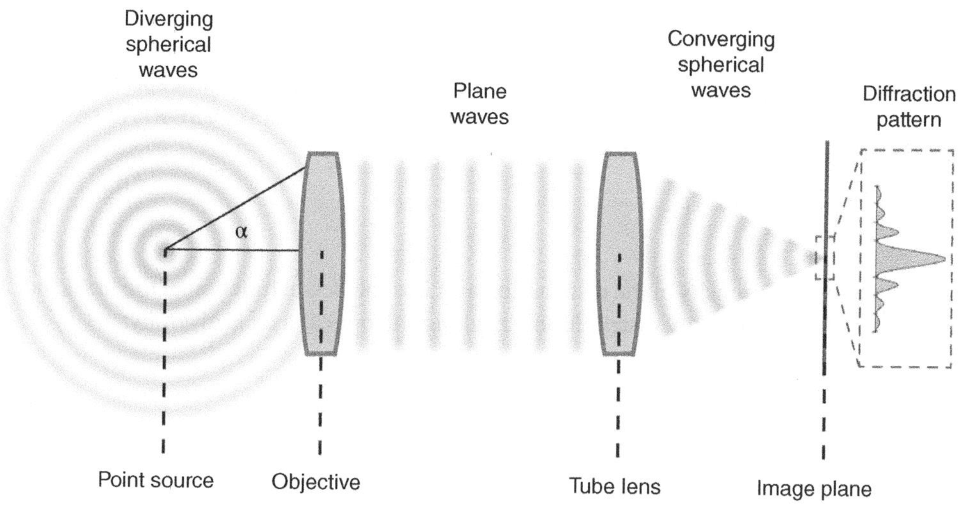

Figure 4.1 Light emitted by a point source is imaged as a diffraction pattern by the optical system. The collected signal (defined by the numerical aperture) is only a fraction of the one emitted by the sample and information is lost.

are collected by the objective lens, which causes the propagating light to interfere constructively at a certain point in space, called the focal point (Figure 4.1). The collected signal is then only a fraction of the one emitted by the sample and information is lost. The objective, indeed, can only collect light within a fixed solid angle, which is determined by the numerical aperture (NA) defined as:

$$\text{NA} = n \cdot \sin\alpha \tag{4.1}$$

with n being the refractive index of the surrounding medium and α half of the angle of the aperture of the objective.

The concept of Abbe's theory was summarized in a simple sentence stating that "The microscope image is the interference effect of a diffraction phenomenon" [18], so the diffraction of light by the specimen and by the objective lens is the determining factor of the image resolution. For an ideal optical instrument, the resolution limit is described by the following law:

$$\Delta x = \frac{\lambda}{2\text{NA}} \tag{4.2}$$

where Δx is the minimum distance between two points in the image that can be seen as separate entities, λ is the wavelength of the collected light, and NA is the numerical aperture of the imaging lens. Under ideal conditions, independently on the magnification of the objective, it is easy to convince yourself that for optical microscopy, the lateral resolution approaches 200 nm. Indeed, the maximum values of sin α approaches 1 (that of course is never reached), the refraction index may assume the value of 1 for a dry objective or 1.5 for immersion objectives and visible light used to image the specimen may range from 400 to 750 nm. It has to be noted that Abbe's law is valid for other optical systems such as telescopes or cameras and applies for all instruments involving the wave–sample interactions. For example, for electron microscopes, where high-energy electron beams ($E \approx 100$ keV) are focused through the sample, spatial resolution reaches picometer ranges. In this case, indeed, wave–particle duality is taken into account and the wavelength of the particle is given by De Broglie law; on the contrary, infrared microscopy is characterized by lateral resolution in the micron scale.

Diffraction-limited resolution can be understood by thinking that the image of an infinitely small point source in the specimen is not an infinitely small point on the focal image plane but is translated by the optical system in a three-dimensional diffraction pattern called point spread function (PSF). In the case of an ideal imaging system, the intensity distribution of the PSF $I(r)$ at position r on a plane perpendicular to the optical axis is given by [19]:

$$I(r) = I_0 \cdot \left[\frac{2J_1(r2\pi\mathrm{NA}/\lambda)}{r2\pi\mathrm{NA}/\lambda}\right]^2 \tag{4.3}$$

where I_0 is the maximum intensity of the distribution and J_1 is the Bessel function of the first kind of order 1. Analogously, the intensity distribution $I(z)$ along the optical axis is given by:

$$I(z) = I_0 \cdot \left[\frac{\sin\left[\left(\frac{z\pi\mathrm{NA}^2}{2\lambda n}\right)\right]}{(z\pi\mathrm{NA}^2/2\lambda n)}\right]^2 \tag{4.4}$$

In a simple description, light coming from a point in the specimen diffracts through the lens aperture and forms a diffraction pattern, described by a well-defined spatial energy distribution, which in ideal systems depends on light wavelength and NA of the lens. In systems without any aberration, the diffraction pattern at the focal point is symmetrical in the lateral (x, y) and axial (z) direction, roughly describing a 3D ellipsoid. The typical shape of the PSF is represented in Figure 4.2 as a function of objective NA.

In the x, y plane, the diffraction pattern, called Airy disk, consists of a central intensity peak surrounded by a series of diffraction rings. As shown in Figure 4.2, the diffraction pattern is

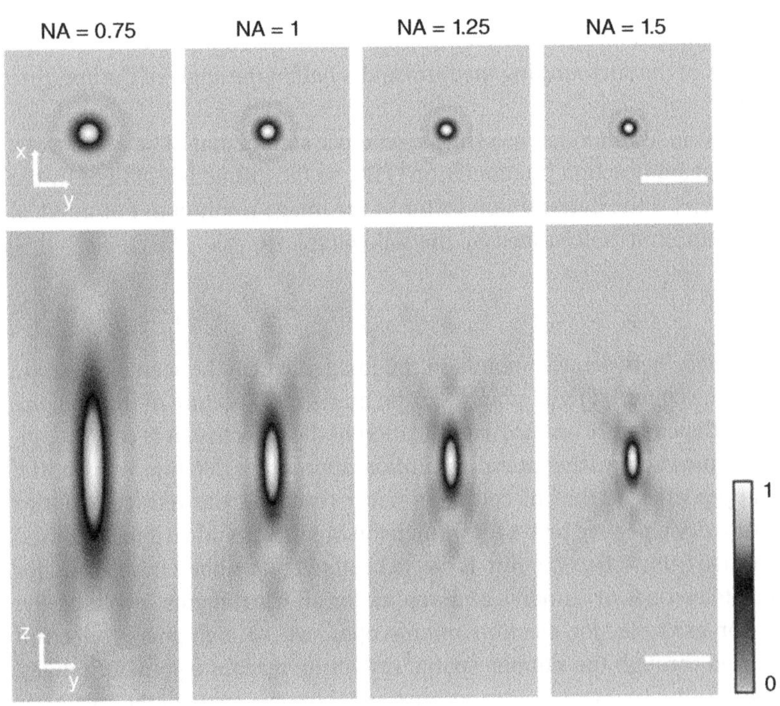

Figure 4.2 PSF as a function of the numerical aperture (NA). Top: intensity distribution in the focal plane. Bottom: intensity distribution along the optical axis. λ = 500 nm, n = 1.33, scale bar = 1 μm.

determined by the NA and its size decreases on increasing numerical aperture; in particular, as NA increases, the radius of the central maximum as well as the distance between the secondary lobes decrease. The shape of the PSF along the optical axis is symmetrical and the light spreading is more pronounced than in *x*, *y* plane, its size being about three times larger than in the *x*, *y* plane.

There are many ways to quantify the resolution of a conventional optical system, all of them based on establishing a limit distance between the PSF central point [3]. The most widely used approach is the Rayleigh criterion, which establishes that two points in an image can be resolved if the distance of the central maxima of the Airy disks representing them is larger than the distance between the first minima. On the *x*, *y* plane, as an image can be considered as the superimposition of a number of Airy disks that are the images of point in the specimen, two points can be resolved if the central maxima of the Airy disks do not overlap. The same concept applies along the optical axis where the resolution is about three times lower. The PSF can be experimentally measured as the image of an object whose size is below the resolution of the microscope in use. This means that in a diffraction-limited instrument, the shape of the PSF determines the resolution of the instrument.

4.1.2 Optical Sectioning and Imaging of 3D Samples

Optical microscopy approaches that provide information from volumetric objects are attractive for a number of applications, ranging from biology to solid state matter [20, 21]. However, in conventional optical microscopy measurements, when collecting images from a thick sample (larger than about 30 μm), not suitable images are obtained since in-focus features are hidden by blur coming from out-of-focus regions. This is because, the objective collects signal (scattered, absorbed, or emitted) coming from different illuminated regions within the sample. Real samples are 3D objects, but standard optical microscopes are designed to produce 2D images; this is because lenses do not discriminate contributions coming from in-focus and out-of-focus points. To avoid this problem, often, the sample is sectioned into thin slices before imaging, but this procedure is highly invasive, time consuming, and, of course, leads to perturbation of structural organization of the sample, possibly inducing artifacts. In this framework, confocal and two-photon microscopes provide the possibility of producing optical sections of the specimen, preserving its structure and functionality [22–24]. An optical section consists in a sharp 2D image of a thin layer of the samples obtained by collecting contributes only from the focal plane (that can be moved along a vertical axis deep within a thick sample) and excluding contributes from other regions. An ideal microscope would examine each point of the specimen and measure the amount of light scattered, absorbed, or emitted by that point, excluding contributions from other parts of the sample. In confocal and two-photon microscopy, 3D measurements are reconstructed by acquiring 2D images at defined planes, moving the focus along a defined axis using a fine stepping as shown in Figure 4.3.

This noninvasive procedure in principle allows to recover the 3D sample morphology; however, the diffraction limit of optical instruments have to be taken into account, slicing performances are limited by the axial dimension of the PSF, then again thinner slices are obtained at shorter wavelengths or using high NA lenses. However, while doing 3D imaging, another factor has to be taken into account, the "free working distance," which defines the linear distance between the objective and the focal plane and in turn the penetration depth in the sample. This parameter usually decreases on increasing NA and measurements are not possible outside the free working distance.

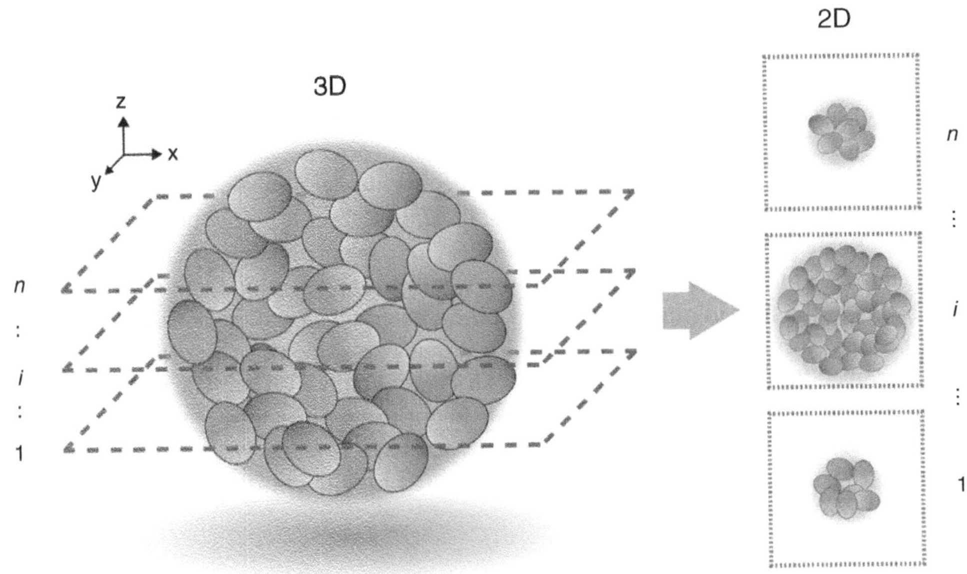

Figure 4.3 Optical sectioning of 3D sample. 3D measurements are reconstructed acquiring 2D images at defined planes, moving the focus along a defined axis using a fine stepping.

4.2 Fluorescence Imaging

Fluorescence microscopy is an optical microscopy technique in which the detected signal arises from the fluorescent molecules of the analyzed materials [25]. Only fluorescent species within the sample are visualized, providing high contrast to images and conferring this method high selectivity and sensitivity. This method has been developed in the past century and it was an essential source of major discoveries in different science fields [3]. Initially, endogenous fluorescence of sample's components was exploited but afterward, the use of exogenous markers brought fundamental advancements. In particular, over the past years, many fluorescent probes have been developed that can be used as spectroscopic molecular reporters of multiple properties of the sample [7, 26]. Fluorescent molecules provide both spatial and functional information: not only their localization but also their molecular features such as their steady state spectrum, lifetime, anisotropy diffusion, fluorescence lifetime, and so forth can be measured. This may provide useful insights on the photophysics of organic or inorganic samples and on the dynamical events with high spatiotemporal resolution down to the single molecule level, dramatically increasing the number of available information on sample structure and dynamics at the micron and nanoscale.

Fluorescence is an exquisitely quantum phenomenon where the emission of light occurs within nanoseconds after the absorption of light that is typically of shorter wavelength (see Chapter 1 for details). Fluorescence excitation is commonly induced by ultraviolet or visible radiation and occurs when the impinging radiation has sufficient energy to induce a molecular transition to an excited electronic level (matching the gap between the ground and the excited state). When the fluorescent molecule returns to the ground state, light is emitted at lower energy with respect to the exciting light and it is seen as fluorescence. The difference between the exciting and emitted wavelengths is known as Stokes shift.

In a conventional wide-field fluorescence microscope (epi-fluorescence), the sample is illuminated through an excitation filter, which selects suitable wavelengths that can be absorbed by the sample, and light is directed on the sample via an objective. The fluorescence emitted by the sample is collected by the same objective and travels through an emission filter to microscope oculars or to a detector. In this way, a 2D image of specimens can be captured at once. The separation between excitation and emission light in the same light path is possible due to Stokes shift and occurs via a dichroic mirror, which selectively transmits (excitation/shorter wavelength) or reflects (emission/longer wavelength) light. Importantly, because fluorescence is observed as light shining on a dark background, fluorescent molecules can be revealed even in extremely small amounts. However, wide-field microscopy shows several drawbacks when imaging is performed in thick samples, since images become blurred by fluorescence generated above and below the objective focal plane. Technical and methodological advances have extended the possibility to investigate matter using advanced fluorescence microscopy. Many methods such as confocal, multiphoton microscopy and other emerging techniques have been developed to image and acquire spectroscopic features of three-dimensional structures.

4.2.1 Laser Scanning Confocal Fluorescence Microscope

Confocal microscopy (or more properly laser scanning confocal fluorescence microscopy, LSCM) is one of the most diffused fluorescence microscopy techniques and, in the last decades, has become an invaluable tool enabling rapid, high-sensitivity, and high-resolution optical imaging [27]. Confocal microscopy is based on the principle that all fluorescence signals emitted by the sample, before reaching the detector, have to pass through a spatial mask, namely the pinhole (a small aperture in an opaque screen). The key role of the pinhole, placed in the detection path, is to select fluorescence from a well-defined detection volume. This allows to reduce the out-of-focus fluorescence light and to increase the signal-to-noise ratio. The optical configuration of a modern confocal scanning laser microscope is reported in Figure 4.4.

The source of illumination in LSCM are, nowadays, lasers. The high energy density, the monochromaticity, and the capability to be focused through lenses make lasers the ideal light sources for confocal microscopes. This coherent light, after passing through an intensity modulator system, passes through the dichroic mirror that splits excitation light (blue, confined, lines) and fluorescence (green, dispersed, lines). The laser light is then reflected by two galvo mirrors that can be rotated perpendicular to each other. In this way, the beam can be scanned in two dimensions. Two lenses, coupled in a telescope, allow translating the rotation of galvo mirrors in a pivoting beam at the objective's back aperture. This results in a rastering of the focus point in the plane x, y within the sample. The excitation beam passes through the objective and excites a spot of the specimen depending on the angle of the two mirrors. The objective focuses the light in the sample and, at the same time, collects fluorescence signal. In is important to note that fluorescence is generated within the double cone of illumination that includes (but is not limited to) the focused spot. Fluorescence, emitted in all the directions (in spherical symmetry) by fluorophores, is collected by the same objective and passes through a beam splitter, which discriminates fluorescence from excitation light. Fluorescence light is then directed to the pinhole. The pinhole aperture is aligned in front of the detection system in a way that only light emitted at the focal point may pass and contribute to image formation. Out-of-focus contributions are stopped by the screen. Fluorescence generated away from the confocal volume (that is not confocal with the pinhole) is stopped as depicted in the box of Figure 4.4.

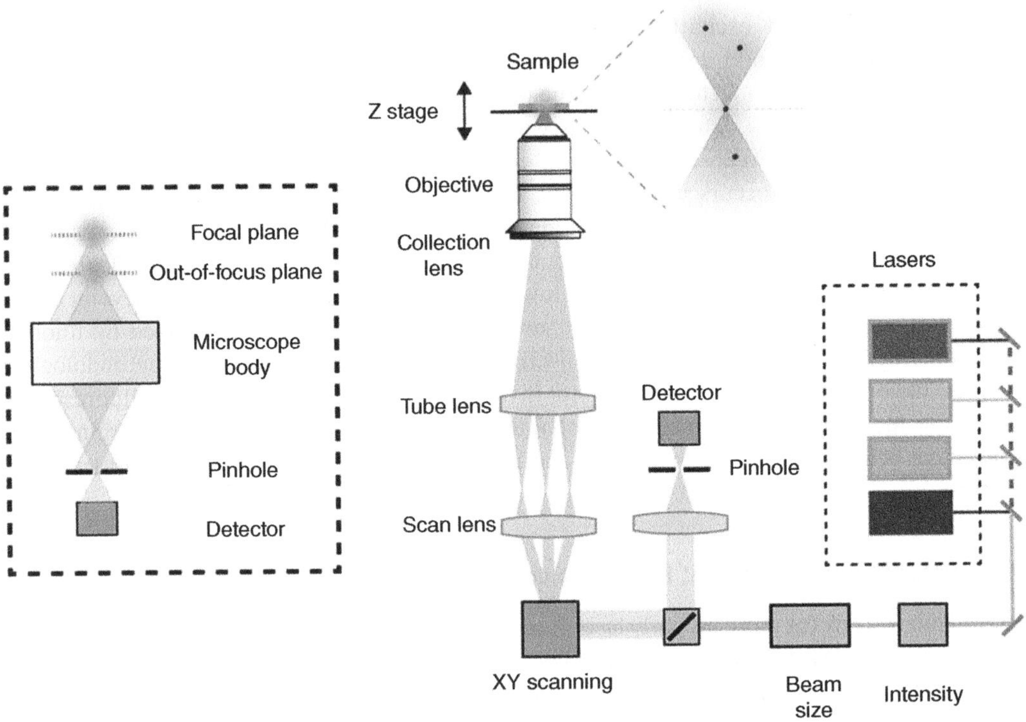

Figure 4.4 Schematic of a laser scanning confocal fluorescence microscope.

This simple idea leads to a very high contrast and a high signal-to-noise ratio. Confocal configuration allows the collection of fluorescence signal only from region at the focus point; thus, to obtain a complete image, excitation light has to be delivered and collected to/from each point in the specimen. In the laser scanning confocal microscope, light is scanned across the sample in sequential lines (raster scan) and the 2D image is digitally reconstructed in a sequential manner point by point in 2D pixel matrices.

Detection system is dependent on the applications: photomultipliers (PMT) or avalanche photodiodes (APDs) are widely used and, nowadays, technological advances are aimed at improving their responsivity and temporal resolution. In the past, the spectral range of fluorescence emission has been discriminated by barrier filters (wavelength band pass). Currently, the detection system can be implemented at higher level of complexity, allowing to modify the spectral range of acquisition, selecting narrow regions or the whole emission spectral range. This allows to use the microscope as a spectrometer and to obtain spectral measurements at each pixel of the image with high temporal resolution and then to separate multiple emissions by their color, or the specific spectrum of the molecule. In combination with appropriate sensors and laser sources, confocal microscope allows to access fluorescence lifetime measurement (FLIM), which provides the measurement of nanosecond fluorescence decays at pixel resolution [28]. What makes this instrument functional to multiple applications is the possibility of obtaining quantitative spectroscopic measurements at diffraction-limited spatial resolution. There is more than just sample morphology: fluorescence intensity in selected spectral range, interactions and dynamics in different spatiotemporal scales can be obtained applying analysis methods like fluorescence recovery after photobleaching

(FRAP), fluorescence correlation spectroscopy (FCS), Förster resonance energy transfer (FRET), which exploit the extraordinary properties of fluorescence and the physicochemical properties of the specimen [29–31].

4.2.2 Two-Photon Microscope

An atom or a molecule could simultaneously interact with two photons, getting them absorbed during the very same quantistic event. The theoretical basis of two-photon absorption was predicted by Maria Göeppert Mayer in 1931 [12] but the experimental observation of this phenomenon in a crystal of CaF_2:Eu_2C was made possible only after the invention of laser source [13]. This quantistic process is described by a time-dependent Schrödinger equation, in which light–matter interaction is represented, in the Hamiltonian, by an electric dipole interaction term: $\overline{E}_\gamma \cdot \overline{r}$, where $\overline{E}_\gamma$ is the electric field vector of the photons and $\overline{r}$ is the position operator [32]. This equation can be solved by perturbation theory and the first-order solution corresponds to the one-photon excitation while the second-order solution describes two-photon excitation. The transition probability between the molecular initial state $|i\rangle$ and the final state $|f\rangle$ is given by:

$$P \sim \left| \sum_m \frac{\langle f \mid \overline{E}_\gamma \cdot \overline{r} \mid m\rangle\langle m \mid \overline{E}_\gamma \cdot \overline{r} \mid i\rangle}{\varepsilon_\gamma - \varepsilon_m} \right| \tag{4.5}$$

where ε_γ is the photon energy associated with the electric field vector $\overline{E}_\gamma$, the summation is over all intermediate states m, and ε_m is the energy difference between the state m and the ground state.

In standard confocal fluorescence microscopy, the fluorescent molecule passes to the excited state, from which fluorescence occurs, by the absorption of a single photon. In two-photon excitation, the excitation of the molecule occurs by the simultaneous absorption of two photons (within a temporal interval of 10^{-16} second as consequence of the Heisenberg uncertainty principle) at lower energy that are individually nonresonant with the electronic transitions [7]. More generally, this phenomenon may occur with multiple-photon absorption (e.g. three-photon excitation needs the absorption of three photons and so on). A Jablonski diagram, as shown in Figure 4.5, can be used to further explain the processes comparing single and multiphoton excitation.

The excited state is reached when a photon (arrow, left) with appropriate energy ($h\nu$) corresponding to the energy difference (ΔE) between the ground and the excited state interacts with the

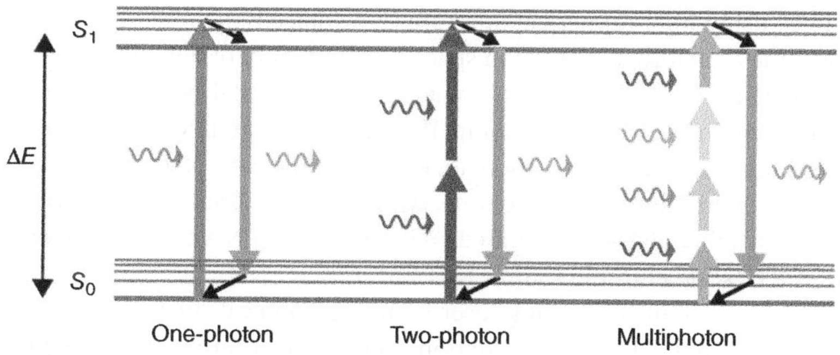

Figure 4.5 Jablonski diagram. Single (blue, left), two-photon (red, center), and multiphoton (pink, right) absorption. The fluorescent molecule passes to the excited state, from which fluorescence occurs, by the absorption of a single photon or of multiple photons.

fluorophore. The excitation to the same energy level can be also obtained when two (arrows, center) or more photons (arrows, right) combine their energy. As reported in previous chapter (Chapter 1), the energy of a photon is inversely proportional to its wavelength. This means that photons (i.e. excitation light) at higher wavelengths (lower energy) can be used to observe fluorescence from the same fluorophore; indeed, independently of the way the molecule is excited, the fluorescence signal has the same properties in terms of spectral shape and lifetime as the molecule emits from the same excited state.

Even if, for practical reasons due to laser excitation, the two photons are usually at the same energy, this is not necessary; to induce electronic transition, their wavelengths should follow the following relation:

$$\lambda_{1P} = \left(\frac{1}{\lambda_1} + \frac{1}{\lambda_2}\right)^{-1} \tag{4.6}$$

where λ_{1P} is the wavelength corresponding to the energy for one-photon excitation and λ_1 and λ_2 are the wavelengths of the two-photon excitation, respectively. In general, one-photon and two-photon selection rules are different [33, 34]; thus, it is possible that forbidden electronic transitions for one of them is allowed for the other [23]. However, experiments usually show that if one-photon excitation occurs at a specific excitation wavelength, two-photon excitation will occur for excitation light at about double wavelength: in general, wider two-photon absorption spectra are found.

Importantly, two-photon excitation is a nonlinear process where the number of photons absorbed per second is proportional to the squared intensity of the incident light. For the same molecule, two-photon cross section is very small compared to that for single photons due to the low probability of two photons interacting simultaneously with the molecule. To qualitatively understand the difference in the occurrence probability between one-photon and two-photon events, it is possible to think that a fluorophore like Rhodamine B in the daylight absorbs a photon about once a second, but a two-photon event may occur once every 10 million years [35]. For this reason, two-photon absorption requires a high concentration of photons in time. This can be obtained with high-power continuous wave lasers or using pulsed laser sources with short pulses. Two-photon excitation in optical microscopy is usually made possible by laser scanning and titanium–sapphire lasers; these are tunable lasers, which generate ultrashort pulses in the near infrared region (usually 670–1050 nm range). By using an ultrashort pulsed laser source, higher peak powers are reached, significantly increasing the probability of two-photon excitation, maintaining low incident power. The use of short pulses at high frequency allows image acquisition without irradiating the sample at high power levels. Furthermore, to increase the probability of the event, photons should be also highly focused in small spatial regions, using high-numerical aperture (NA) objectives.

In a two-photon microscope, the excitation is limited in space as the amount of light absorbed is not constant across the illuminated volume, but shows a maximum at the focal point where the incident intensity is at its maximum. In Figure 4.6, the comparison between one-photon excitation and two-photon excitation of a fluorescent sample is schematized. In the first case, an entire cone of fluorescence is generated, i.e. the entire illumination volume, while in the second, fluorescence is generated only at the focal point in a tiny diffraction-limited volume. For high-NA objective (about 1.4), the fluorescent volume is a fL scale ellipsoid [36].

Often, two photon microscopes are built integrating laser scanning confocal microscopes and tunable mode-locked lasers source that allow the excitation using near-infrared pulsed light. Two-photon excitation in optical microscopy is usually made possible by titanium–sapphire lasers, tunable lasers which generate ultrashort pulses in near-infrared range. The pulsed laser is focused

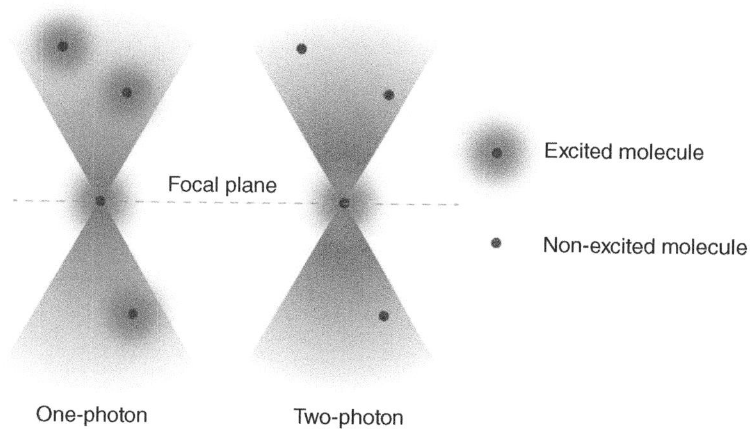

Figure 4.6 A schematic representation of the localization of one-photon and two-photon excitation. In the first case, all the molecules in the illuminated volume are excited by incoming radiation; in the second, only molecules at the focal point emit fluorescence.

to a diffraction-limited spot and scanned in a raster over the sample [37]. When the tiny spot overlaps with fluorophores of the sample, fluorescence signal is generated selectively within the small spot. An image is formed as a well-ordered 2D sequence of points by collecting the fluorescence by a single detector. The temporal signal of the detector is mapped to the corresponding raster-scanned point by the data acquisition computer. Figure 4.7 shows the optical scheme of a typical two-photon microscope.

The intrinsic localization of the excitation enables point-by-point scanning capability, and as the fluorescence signal originates only from the focal point the screening of out of focus contributes (using a pinhole in front of the detector as in the confocal setup) is not necessary [38]. Interestingly, since there are not absorption phenomena in the out-of-focus region, no photobleaching and phototoxicity along the beam path occur, constituting a great advantage for the analysis of degradable samples. The use of two-photon excitation has, then, the advantage of reducing heating and photodamages of samples along the excitation beam trajectory, also delivering less energy to the sample due to longer excitation wavelengths. Two-photon and multiphoton microscopy can be then used to perform minimally invasive experimental measurements also for long acquisition time. The near-infrared excitation enhances the ability to image deeper into a sample as the contribution of light scattering effects is reduced. This is because scattering intensity can be considered, in first approximation, inversely proportional to the fourth power of wavelength of incident light on the sample; then, it is critically reduced if excitation wavelengths are doubled. In this framework, two-photon measurements (with suitable objectives) may allow depth penetration in the sample up to millimeter scale [39]. Importantly, the use of infrared laser and two-photon excitation allows the excitation of fluorescent molecule in the near UV, which can be restricted by the opacity with respect to UV light of some conventional optical components and that would be quickly photodamaged by one-photon excitation. This together with larger spectral separation between the excitation and the emitted light facilitates multicolor experiments, increasing the number of accessible fluorophores that can be measured simultaneously. Multiphoton microscopes allow multiple spectroscopy applications as they provide efficient detection and mapping of fluorescence in 3D, reducing background signal and, due to the spectral distance between excitation and detected light, allow the separation of Raman scattering.

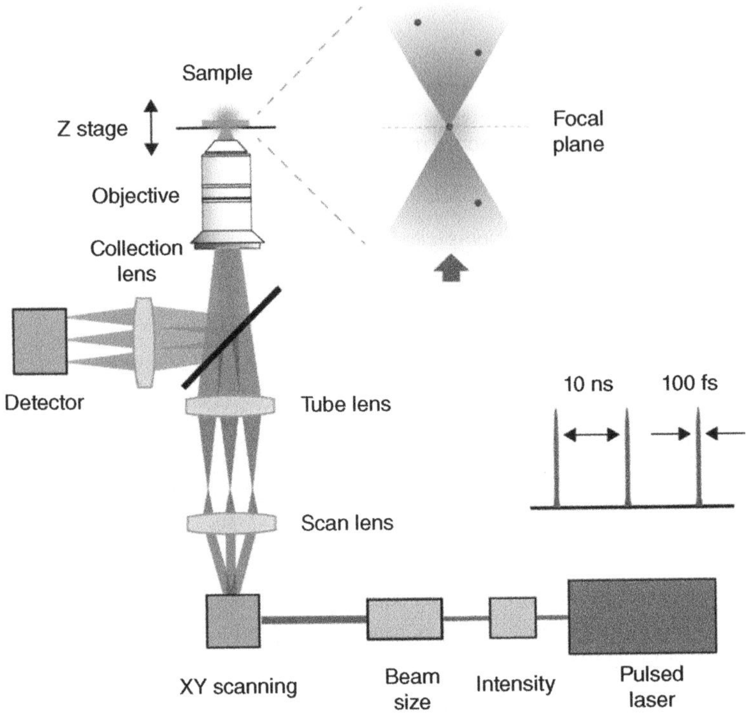

Figure 4.7 A schematic of typical components in a two-photon scanning microscope. The system contains a high-peak-power pulsed laser, a scanning microscope and high-sensitivity detector. Fluorescence is generated only in the focal plane.

4.2.3 The Importance of Sample Preparation from Solid State to Dynamic Specimens

Much of the modern understanding of materials is based on spectroscopy methods; however, in recent years, optical microscopy has brought important advancements, especially because it allows visualizing intrinsically heterogeneous structures (first at microscale level and successively at the nanoscale) with the possibility of mapping spectral properties in a kinetic fashion [40, 41]. For many materials, especially those of soft nature, such as polymers, colloids, and liquid crystals, fluorescence revealed itself as a fundamental contrast method allowing to obtain suitable insights into the structural details. The contrast is achieved from suitable excitation of intrinsic fluorescence of the sample or using specific fluorescent dyes [26, 42]. These are chromophores whose spectral properties are well established that may act as molecular reporters of their environment or occurring interactions. The fluorescent label can be attached covalently or interact via weaker but specific interactions (e.g. electrostatic, hydrophobic, chemical affinity) with the sample that can be exploited and analyzed depending on the specific target [43]. This may characterize topographic structures and evaluate surface chemical properties and establish layer thickness of fluorescent materials.

In designing sample preparation, it is important to be aware of fluorophores' spectral properties and physicochemical specificity, taking into account, according to the application, both their photostability and potential unwanted effects of environmental conditions. Although imaging appears as the easy solution to answer scientific questions, inappropriate preparation of the sample may lead to misleading results or artifacts due to modification of sample integrity.

Fluorescent dyes for multicolor experiments, aimed at localizing molecules, should be selected in such a way that the overlap between their excitation and emission profile is minimized. Compromises have to be made, in many experiments, as the excitation profile should match the available excitation sources. In general, fluorophore size, spectral and lifetime properties play a crucial role in setting the image resolution and, when using quantitative methods like Förster Resonance Energy Transfer (FRET) or fluorescence correlation spectroscopy (FCS), the choice of fluorophores and their concentration in the sample is critical for the success of the experiment since a wrong chose may seriously hinder obtained results.

Every experimental scientist knows that the most important thing to take care of, while performing the experiments, is the sample. Once the right sample is produced and well "matched" with the instrumentation, reaching a meaningful goal is easier.

Confocal and two-photon microscopies excel with bright samples (high concentration of highly efficient fluorophores) indeed, their operation principles intrinsically require "waste of photons." From a practical point of view, these methods undersample the fluorescence in a thick specimen as compared to the conventional microscopies. Sample thickness that can be investigated depends on the free working distance of the selected objective and it is around 100 μm for confocal microscopes and up to millimeter scale for two-photon microscopes. Thicker sample can be visualized but z-maximum penetration depth for images should not exceed the free working distance of the objective. Of course, intrinsic properties of the samples hindering the penetration of laser beams, such as high absorption or turbidity, can greatly affect the portion of sample that can be imaged. In this sense, two-photon excitation, as reported in Section 4.2.2, presents several advantages. Sometimes, it is necessary to look for compromises and select objectives with lower numerical aperture and thus resolution to explore specimens in depth without physical interference. It has also to be taken into account that NA depends on the refraction index of the immersion medium of the objective, thus, oil immersion objectives usually have higher NA with respect to water and dry objectives, and, depending on the sample, one immersion medium can be more advantageous than others. The refraction index should match at best the optics and the immersion medium of the objectives. This is fundamental as mismatched refractive index produces (according to Snell's law) deviation of light and consequently the reduction of collected signal; moreover, aberration in the acquired images may occur. For example, for samples in aqueous solution, the choice of aqueous objectives is preferable especially for sophisticated applications like fluorescence correlation spectroscopy where diffraction index mismatch issues may hinder data acquisition.

4.2.4 Setting Up a Measurement

Setting up the microscope: Thanks to the development of commercial microscopes, both confocal and multiphoton microscopy are now user friendly and make basic experiments much easier than the ones with first custom-made instrumentations. First, you need to turn on the laser and the microscopy hardware according to the manufacturer's instruction and wait for the setup to warm up. Depending on the application, you may need that the light sources are well stabilized and the optics thermalized. Couple of hours will be enough to avoid fluctuation of the measured light intensity.

Selecting an objective lens: The choice of the suitable objective for fluorescence imaging of the sample is determined by three main considerations: numerical aperture (NA), working distance, and magnification. The NA must be large enough to provide enough spatial resolution and to adequately collect fluorescence from the specimen. The working distance of the objective must be great enough to reach the desired plane of focus in depth within the specimen. Magnification must be chosen at first in order to image statistically significant portion of the sample. Choice of immersion medium of the objective is a compromise between accessible resolution and the working distance; depending on the sample, one immersion medium can be more advantageous than others as mismatch between specimen and immersion media refraction index may induce the reduction of collected signal and image aberrations. Usually, the ideal objective for typical two-photon imaging is the lowest magnification lens with the highest available NA, this being a good compromise between resolution and penetration depth.

Configuring and optimizing fluorescence channels: The photophysical properties of fluorescent probes are critical for confocal and two-photon experimental design and their choice also depends on the available excitation source and detection range. For confocal microscopy, a convenient choice is to set laser excitation close to the absorption spectrum maximum. For two-photon microscopy, it is not so simple to predict excitation spectra from the one-photon spectra since they follow fundamentally different quantum mechanical processes and obey very different selection rules. Compared to one-photon spectra, two-photon spectra tend to be broader and shifted toward the blue. The choice of excitation wavelength is generally determined by the fluorophore(s) to be excited. As a rule, select a wavelength that is about two times longer than the wavelength required by a single-photon excitation. Select the appropriate emission range for the fluorophore(s). Commercial microscopes usually allow to select band pass filter or a specific spectral range. In the case of multicolor detection, avoid overlaps among bandwidths to reduce cross talk among different channels. Spend some times in verifying your conditions, trying to maximize the signal avoid bleed-through.

Setting up the specimen: Mount the sample on the microscope stage and focus on it using eyepiece in bright-field or epifluorescence mode. A motor in steps of hundreds of nanometers usually controls the microscope focus. To avoid photobleaching or photodamage, keep the laser power to its minimum intensity. Once the region of interest has been found, optimize laser intensity until a suitable signal-to-noise ratio.

Optimizing spatiotemporal parameters: Commercial confocal and two-photon microscopes offer many different acquisition modes. Volumetric imaging and time-lapse modes are most common modalities and allow to image the specimen over time. Set the frame rate by independently adjusting frame size, scan speed, and number of sections. In the case of advanced fluorescence microscopy technique (i.e. spectral detection and FLIM), set the appropriate parameters. Be prepared to start it over several times; the best practice, indeed, requires that comparison between samples is made using the same experimental parameters.

Data acquisition and visualization: It is important to look for compromise between acquisition parameters that allow the faster possible measurements with higher signal-to-noise ratio. Avoid (if possible) pixel in the image where intensity is out of scale (saturation); in these conditions, light arriving to the detector above this limit will not be taken into account, certainly affecting image quantification. Images with saturated pixels omit a number of spatial features and they are not pretty! Imaging of samples could span a large range of timescales. In the case of long acquisition, check image quality, sample defocusing, and photobleaching. Once data are collected, use appropriate software for visualizing and analysis.

4.3 Spectroscopy Using a Microscope

Modern confocal and multiphoton microscopes allow direct investigation of fluorescent molecules/structures in a spatially resolved manner. Besides the large number of physiological reactions in living system, the possibility to analyze spectroscopic properties of the molecules may allow the analysis of polymeric thin films largely used for organic photovoltaic applications [44], also revealing the active layers, plasmonic effects [45]. Moreover, these methods can be applied to selected (fluorescent) crystal structures [46], liquid crystals allowing revealing specific topological effects, chemical treatments, and catalytic performances in a spatially resolved manner [46, 47]. The possibility of measuring fluorescence and their variations in a time-resolved fashion can be applied to monitor occurring reactions, in real time with low impact on the sample [48].

In recent years, properties of nanomaterials have been exploited to directly generate fluorescent emissions with high quantum yield, excellent photostability, and long fluorescence lifetime; their spectral properties make them suitable probes for many applications. Moreover, some metal nanoparticles are often used to enhance fluorescence of fluorophores located near their surface; these effects can be better analyzed using high-resolution fluorescence microscopy methods.

4.3.1 Observables in Fluorescence Microscopy

In general, fluorescence microscopy can be used to target, track, and characterize fluorescent structures down to the detection of single molecules, revealing information not accessible to ensemble/bulk measurements. Importantly, in the last decades, technological developments in fluorescent microscopy instrumentation made readily accessible the possibility of acquiring spectroscopic measurement mapping the signal at pixel resolution. This means that excitation and emission spectra can be acquired and mapped in 4D (x, y, z, t) measurements and also that fluorescence decays can be quantified at high spatiotemporal resolution. The analysis of excitation and emission spectra of fluorescent molecules can be related to the properties of their electronic levels and in turn to structure and conformation of the molecule and/or to local microenvironment (Figure 4.8a). Fluorescence emission of a molecule depends on the electronic energy levels available for absorbing incident radiation, the efficiency of rearrangements in vibrational levels, and the relaxation processes occurring in the excited states, the population of vibrational levels within this ground state, and the spatial orientation of the molecule with respect to the incident radiation. Fluorescence allows to monitor dynamic events that occur during the excited state (i.e. in the nanosecond timescale), widening the amount of information on fluorophores' fate and environment.

In this context, spectroscopic observables that are of great interest and possess a wide range of applications in microscopy are: (i) fluorescence intensity; (ii) spectral position and shape; (iii) fluorescence lifetime (Figure 4.8b); and also, but less frequently, (iv) fluorescence anisotropy. This provides multiple opportunities, also using combined methods to probe hidden properties of matter at the molecular level. The most commonly used observable in microscopy is the average fluorescence intensity, of the sample measured in a specific spectral range. This measure is often used to localize fluorescent molecules, revealing micron-scale sample topology and existence or coexistence of molecules in the same region. Moreover, fluorescence intensity is related to fluorophores' concentration, which can be suitably calibrated and to their quantum yield. This last property means that experiments can be designed using fluorescence intensity to probe environmental properties such as its polarity, proticity, microviscosity, temperature, charge transfer mechanisms, and accessibility to solvent. Relative changes between radiative and non-radiative rates can be measured in kinetics experiments and related to probe–environment interactions in different conditions.

Furthermore, analysis of spectral shape and position may give similar information but with higher level of details. The Stokes shift, which is the difference between the spectral position of the absorption peak and fluorescence peak maxima, depends on the particular fluorophore and its coupling with the environment. Changes in spectral position are often related to solvation effects

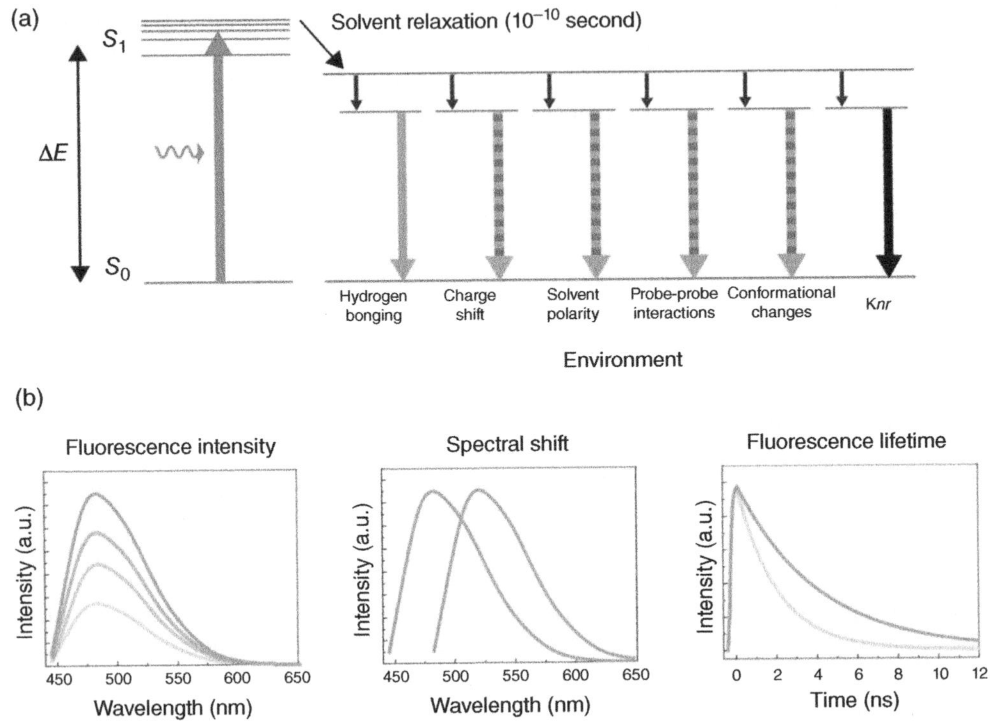

Figure 4.8 Fluorescence emission of a molecule: (a) emission depends on the electronic energy levels available for absorbing incident radiation and on the environment. The excited state is modified by the environment (solvent relaxation effects) and this is reflected on the spectral properties; (b) spectroscopic observables of great interest and with a wide range of microscopy application are: fluorescence intensity, spectral position and shape, fluorescence lifetime.

(solvent relaxation effects). Emission spectral shift can be taken into account using ratiometric methods, i.e. measuring/weighting the ratio of fluorescence intensity at two specific spectral positions, allowing quantitative interpretation of fluorescence data and disentangling the effect of fluorophore concentration.

This is better achieved using fluorescence lifetime, which is a fundamental characteristic of the fluorophore and does not depend on its concentration, this being particularly important in the analysis of intrinsically heterogeneous samples. Fluorescence lifetime imaging microscopy measures and maps fluorescence decays at each pixel of the image, thus giving back information on fluorescence lifetime: the time a fluorophore remains in an excited state before emitting a photon. This measurement detects molecular variations of fluorophores that can be related to a tremendous number of information not accessible to spectral measurements and it is almost totally free from artifacts due to unsought scattering contributions. Eventually, also anisotropy measurements can be performed in imaging experiments whose variations are related to variations on the rotational mobility of the fluorophore, this often being useful in the analysis of the interaction between couple of molecules.

It is possible to state that fluorescence microscopy allows visualization of molecules within hundreds nanometer scale (if super-resolution methods are not taken into account see [41, 49] for further reading) but the analysis of fluorescence spectroscopy observables make accessible under-resolution details at molecular level. For example, molecular interactions and ångström-scale distances can be quantified measuring Förster resonance energy transfer (FRET) [31]. FRET is a non-radiative phenomenon where energy transfers from a fluorescent excited molecule (donor) to non-excited molecule (acceptor) and it occurs when the emission spectrum of the first one (donor) overlaps with the excitation spectrum of the fluorescent acceptor only if the molecules are within a certain distance (Förster radius). Depending on the distance and relative orientation of the two groups, the efficiency of the energy transfer from the excited "donor" molecule to the "acceptor" changes. FRET experiments with high accuracy can be nowadays performed using fluorescence microscopes both using intensity-based or lifetime-based methods, extending experimental resolution to molecular level.

4.3.2 Measuring Dynamics: Gaining Information Below Resolution

Fluorescence microscopy allows a plethora of spectroscopic methods, which give the opportunity to characterize molecules and interactions within the nanometer scale [7, 50]. Methods based on molecular diffusion are the most common and less invasive. In particular, using fluorescence microscopy, both measurements, which quantify the average microscopic motion of particles in an ensemble and direct measurements on the diffusion properties of single fluorescent particle, can be performed. Due to thermal energy, molecules in fluids move in a series of random uncoordinated steps between collisions; this describes the-so called "Brownian motion" whose analysis may provide information on particle size, shape, and the properties of the fluid. For macromolecules moving in "random walks," the mean square displacement $<r^2>$ (which quantifies the actual distance traveled by the particle in a certain time t) is given by:

$$\langle r^2 \rangle = 6Dt \tag{4.7}$$

where D is the diffusion coefficient, which depends on molecular size, shape, and flexibility, and on the viscosity and temperature of the surrounding fluid. In a simplified scheme, the larger diffusion coefficient D is, the larger is the area that the particle covers in a certain time, usually revealing smaller size of the diffusing particles. For a spherical particle, the diffusion coefficient is given by the Stokes–Einstein relation:

$$D = {}^{RT}\!/_{6\pi N_A \eta R_s} \tag{4.8}$$

where R is the ideal gas constant, N_A is the Avogadro's number, T the temperature measured in Kelvin, η is the dynamic viscosity, and R_S the radius of the particle. For example, a 1 nm nanoparticle diffusing in water at room temperature will have a diffusion coefficient of about 200 $\mu m^2\, s^{-1}$.

In this chapter, the two classical methods to measure diffusion of fluorescent molecules will be briefly described: fluorescence recovery after photobleaching (FRAP) and fluorescence correlation spectroscopy (FCS) [29, 51].

Fluorescence recovery after photobleaching (FRAP) is a method based on fluorescence microscopy to measure average (ensemble) diffusion coefficients of fluorescence molecule in a certain region. In FRAP experiments, fluorescence signal of molecules in a small region of the sample is bleached by exposure to high-power laser illumination. After bleaching, the laser power is reduced and the dynamics of fluorescence recovery in the bleached region is measured. The kinetics of fluorescence recovery quantify the mobility of the fluorescent molecules, which is related to the rates of diffusion of fluorescent molecules in the bleached area. In Figure 4.9a, the scheme of typical measurements is shown, together with representations of (b) fluorescence recovery kinetics for full fluorescence recovery, (c) fluorescence recovery kinetics in an area containing immobilized/bound molecules.

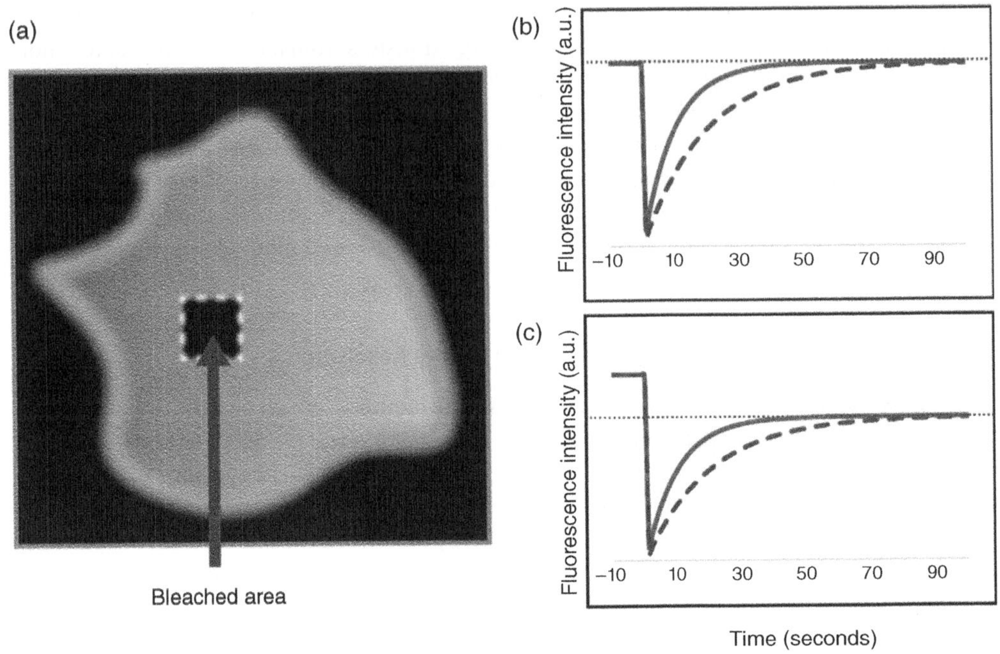

Figure 4.9 Fluorescence recovery after photobleaching (FRAP). Fluorescence signal of molecules in a small region of the sample is bleached by exposure to high-power laser illumination (a). After bleaching, fluorescent molecules diffuse from unbleached regions and a recovery of signal is measured in the bleached area. The kinetics of fluorescence recovery within the bleached area quantify the mobility of the fluorescent molecules: (b) fluorescence recovery kinetics for full fluorescence recovery for fast diffusing molecules (pink, solid line) and slow diffusing molecules (violet, dashed line), (c) fluorescence recovery kinetics in an area containing immobilized/bound molecules, the plateau value sets to a lower value with respect to the one prior to bleaching as the quenched molecules bound at fixed locations cannot be replaced.

Of course, depending on the complexity of sample analysis of fluorescence recovery, the kinetics may be also influenced by transport phenomena and binding interaction, resulting in complex multicomponent kinetics. FRAP experiments may then be set up to understand how fast do molecules move in a certain region of the sample. In homogeneous environment of known viscosity, this can give information on molecular size down to fraction of nanometer radii. More complex experiments and analysis may lead to the understanding of molecular interactions, complex formation, and the presence of compartments in the analyzed spatial region and their connections.

Another way to approach the measure of diffusion in solution is given by FCS. This method, first described in 1970, is based on the measure of temporal fluctuations of fluorescence intensity in a small illuminated volume [52, 53]. Ideally, this method would be performed on a single molecule while in real conditions, few molecules in a femtoliter scale volume are investigated. The observables of interest are the fluorescence intensity fluctuations arising from the diffusion of single fluorescent molecule in and out of the excitation volumes. Small illumination volumes suitable for FCS methods can be obtained using confocal optics or multiphoton excitation at the objective focus, allowing point measurements at specific locations in a 3D sample (Figure 4.10a) [54]. With highly stable fluorophores, the method is highly sensitive as only few molecules are observed at one time. Brownian motion regulates fluorescent molecules getting in and out the excitation volume, thus the rate of fluorescence intensity fluctuation depends on molecular rate of diffusion. Moreover, it is

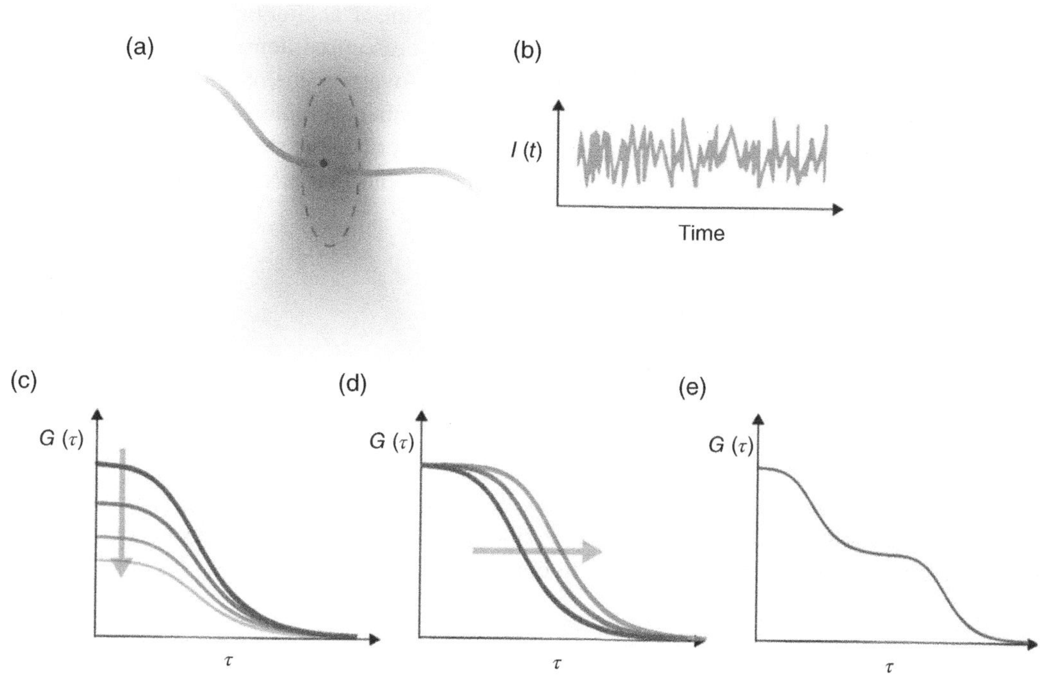

Figure 4.10 Fluorescence correlation spectroscopy (FCS). (a) Fluorescence arises from molecules diffusing in and out from small illumination volumes provided by confocal or multiphoton microscope. *Source:* Based on Berland et al. [54]. (b) Fluorescence signal measured over time, intensity values fluctuate around an average value. Examples of correlation function as a function of molecular concentration (c), diffusion coefficient (d), and presence of multiple (two) species with different diffusion coefficients (e).

easy to visualize that as fluctuations are measured with respect to the average measured intensity, they appear larger at lower molecular concentration. The relative fluctuation amplitude of a signal is inversely proportional to the number of molecules simultaneously measured. The reader will appreciate that, for a femtoliter volume and at a concentration of fluorescent molecules in the nanomolar range, it is possible to observe fluorescence from about tens of molecules or less, down to resolve single molecule's signal. It is important to note that this technique requires a low concentration of diffusing molecules (nM range), as a high number of molecules in the observation volume would bring down the measured fluctuations.

Intensity fluctuations arise from changes in the local concentration of fluorescent molecules within the illuminated volume (Figure 4.10b). FCS data analysis is performed, quantifying these fluctuations by means of temporal autocorrelation of intensity signal. The autocorrelation function describes the time-dependent decay of fluctuations to their equilibrium value (Figure 4.10c–e).

Fluctuations $dF(t)$ as a function of time are estimated as the difference between the fluorescence intensity at a given time $F(t)$ and the average measured fluorescence intensity $<F>$

$$dF(t) = F(t) - \langle F \rangle \tag{4.9}$$

and the autocorrelation function is defined as

$$G(\tau) = \frac{\langle \delta F(t) \cdot \delta F(t + \tau) \rangle}{\langle F \rangle^2} \tag{4.10}$$

where the intensity of the δF at some time t is multiplied with the δF at later time, $t + \tau$. The average of this product is then divided by the square of the average fluorescence intensity for each value of τ. The time 0 value of the autocorrelation function $G(0)$ is the inverse value of the number of molecules. This can be rationalized by considering that the number of molecules contained in the illuminated volume follows a Poissonian distribution as it describes the statistics of particle occupancy fluctuations. For this reason, the average number of molecules $<N>$ and the variance are equal. Thus, $G(0)$ can be written:

$$G(0) = \frac{\langle \delta F(t)^2 \rangle}{\langle F(t) \rangle^2} = \frac{\langle (F(t) - \langle F(t) \rangle)^2 \rangle}{\langle F(t) \rangle^2} \tag{4.11}$$

and for a Poisson distribution, this can be put equal to

$$G(0) = \frac{\text{Variance}}{\langle N \rangle^2} = \frac{1}{\langle N \rangle} \tag{4.12}$$

The shape of $G(t)$, of course, depends on the excitation volume and shape and on the diffusion properties of the molecules. For a 3D Gaussian excitation volume (characterized by a radial dimension ω and axial dimension z), it is possible to obtain:

$$G(\tau) = \frac{\gamma}{< N >} \left(1 + \frac{\tau}{\tau_D}\right)^{-1} \left(1 + \frac{\omega^2}{z^2} \frac{\tau}{\tau_D}\right)^{-\frac{1}{2}} \tag{4.13}$$

where τ_D is the characteristic diffusion time of the particle and it is related to the diffusion coefficient by the relation $D = \frac{\omega^2}{4\tau_D}$ and γ is a geometric scaling factor, which depends on excitation source and it is $\gamma = 1/\sqrt{8}$ for one-photon 3D Gaussian illumination and $\gamma = 3/4\pi^2$ for two-photon Gaussian excitation.

This means that the autocorrelation function analysis provides information on the diffusion properties of the fluorophores and on their absolute concentration. Many experiments can be designed using FCS, which is an established powerful method to study the motions of fluorescent molecules. Importantly, single molecules in the volume of excitation are sufficient to produce a useful signal and the selectivity of fluorescence allows only the molecule of interest to be detected in a complex environment. However, for heterogeneous samples, the fact that information that can be obtained by FCS is localized in a small volume may constitute a limitation as many processes involve large regions of the sample in a inhomogeneous way. In a microscopy experiment, this would mean that measurements are acquired at the focus of the objective, i.e. in a single point of the image, thus making difficult to map diffusion processes in an heterogeneous environment or phenomena occurring in extended spatial scales. For this reason, in recent years, a number of methods that represent an extension of FCS have been introduced. These methods, based on the spatiotemporal image correlation of fluorescence intensity fluctuations, allow mapping fluorescent molecules' stoichiometry and mobility at pixel resolution in a microscopy image [55, 56].

The main idea is that measurements obtained with a laser scanning microscope contain hidden time information structure. In fact, during laser scanning, the position of the illuminated volume changes with time according to a specific relationship. This time information can be exploited to spatially correlate adjacent pixels. As an example, raster scanning image correlation spectroscopy, which can be suited for the measurement of particles with diffusion rate varying in space and time, is briefly described in Section 4.4.1.

4.4 Case Studies

4.4.1 Understanding Microstructures and Mechanistic Aspects in Materials

Both confocal and two-photon excitation microscopy represent user-friendly and noninvasive tools for characterizing sample topology [57, 58]. These methods provide rapid acquisition of longitudinal and cross-sectional images at any position of the sample, quantifying fluorescence intensity. Although these methods are largely applied for quantitative imaging of cell and biological questions, the accessible possibilities for analyzing multiple samples are uncountable, provided that the sample exhibits fluorescence. In Figure 4.11, measurements on samples of different nature are reported assessing microscale topology.

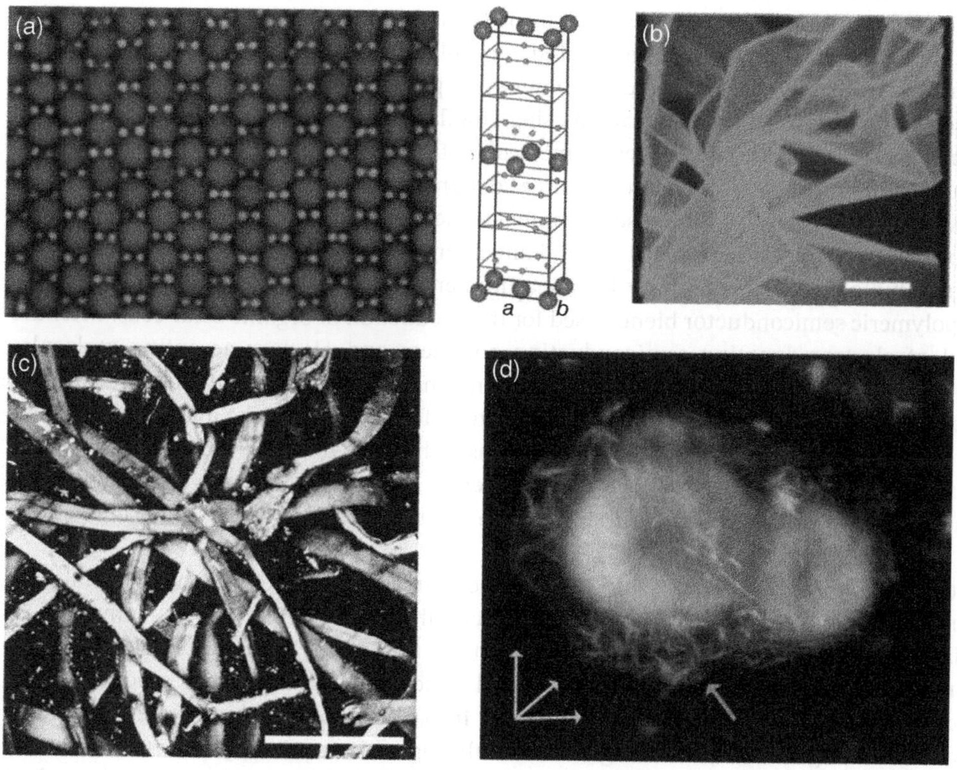

Figure 4.11 (a) Crystal plane consisting of PMMA particles with different charge; positive spheres are stained in green (smaller spheres) and negative spheres are stained in red (larger, spheres). *Source:* Adapted by permission from Nature, Leunissen et al. Copyright (2005) [59]. (b) 3D confocal fluorescence microscopy image of cross-linked CdSe nanoparticles sheets (scale bar is 50 μm); measured signal arises from intrinsic fluorescence of CdSe nanoparticles. *Source:* Adapted with permission from Lin et al. Copyright (2003) American Chemical Society [60]. (c) Fibers organization in a Kinkarakawa wallpaper; the intrinsic fluorescence signal comes from organic pigments (scale bar is 100 μm). *Source:* Adapted under Creative Commons Attribution 4.0 International License from [61]. (d) 3D reconstruction of an insulin spherulite; sample is stained using Thioflavin T dye whose specific properties and fluorescence mechanisms are used to reveal molecular structure, the arrow marks small fibrils surrounding the spherulite; average spherulite linear dimension is 50 μm. *Source:* Adapted from [62], with permission from Elsevier.

In Figure 4.11a, a crystal plane consisting of PMMA (Poly(methyl methacrylate)) particles is shown [59]. Positive spheres are stained using a green fluorescent dye and have a radius 0.36 μm (smaller spheres) while negative ones are stained in red and have a diameter of 1.16 μm (larger spheres). These measurements, in a context of multiple experiments, are aimed at probing how interactions between oppositely charged particles can be tuned to modulate the formation of large ionic colloidal crystals. Figure 4.11b shows a three-dimensional confocal fluorescence microscope image of cross-linked CdSe nanoparticles formed at the toluene/water interface (scale bar is 50 μm), nanoparticles' size is below instrumental resolution (about 3 nm) but their intrinsic fluorescence signal (excitation: 488 nm, detection 540 nm) is used to describe their spatial organization [60]. These measurements in the context of a study of self-assembly of chemically functionalized nanoparticles have shown the possibility of producing ultrathin, composite organic–inorganic membranes. These methods can be also applied to works of arts; in particular, in Figure 4.11c, the analysis of Kinkarakawa wallpapers is reported [61]. Here, the intrinsic fluorescence signal from organic pigments reveals the topology of fibers (scale bar is 100 μm). The 3D reconstruction of an insulin spherulite is shown in Figure 4.11d, a peculiar protein aggregate characterized by a fascinating core–shell morphology, which is common for proteins, synthetic polymers, and minerals, is observed. In this case, sample is stained using Thioflavin T dye whose specific properties and fluorescence mechanisms are used to reveal molecular structure [62].

In parallel with morphological characterization, dynamic events can be also analyzed measuring both steady state spectra and fluorescence intensity decays at selected locations. These studies allow exploring the relation between structural and chemico-physical properties of new materials. As an example, we report in Figure 4.12 measurements on model bulk heterojunction (BHJ) made of polymeric semiconductor blends used for the production of organic solar cells. BHJs are composed of electron-donating semiconducting polymers and electron-accepting molecules blend; morphology is reported to affect recombination dynamics. These devices absorb solar light and generate free charge carriers, which have to be transferred between the donor and acceptor [64]. In these systems, energy transfer mechanisms, which involve exciton diffusion and/or Förster resonance energy transfer at the donor–acceptor interface, can be revealed by fluorescence quenching experiments using both steady state and time-resolved fluorescence microscopy [44, 45].

In Figure 4.12a, a representative confocal microscopy image is reported for prototypical bulk heterojunction, composed of poly((2,7-(9,9-(di-*n*-octyl)fluorene)-alt-5,5-(4′,7′-di-thienyl-2′,1′,3′-benzothiadiazole))) (APFO3) and [6]-phenyl-C61-butyric acid methyl ester (PCBM) [63]. These measurements are aimed at using spatially resolved steady state luminescence spectroscopy to address charge transfer states (CTSs) recombination. In panel a, confocal luminescence signal is color coded according to the spectral region where the signal is acquired, yellow (white) corresponds to signal acquired in the range 495–535 nm and the red (dark gray) color to signal acquired in the spectral region above 650 nm. Phase separation is clearly observed in the sample and it is possible to distinguish roundish red (dark gray) domains (which are identified by the authors as PCBM-rich regions), homogenous region around them, and region at the interface. In panel a, different regions are highlighted using white circles marked with letters where fluorescence spectra are acquired. Correspondent spectra are used to determine the content of different components in selected regions and the charge transfer mechanisms, allowing the authors to state that CTS signal is mostly localized in areas of the sample in which the polymer and the fullerene derivative are more intimately mixed. In particular, the spectrum measured in "A" resembles the one typical of PCBM, spectrum in "B" is the combination of both PCBM and APFO3 components with a major contribution from PCBM. In "C," the spectral shape indicates the dominant contribution of APFO3

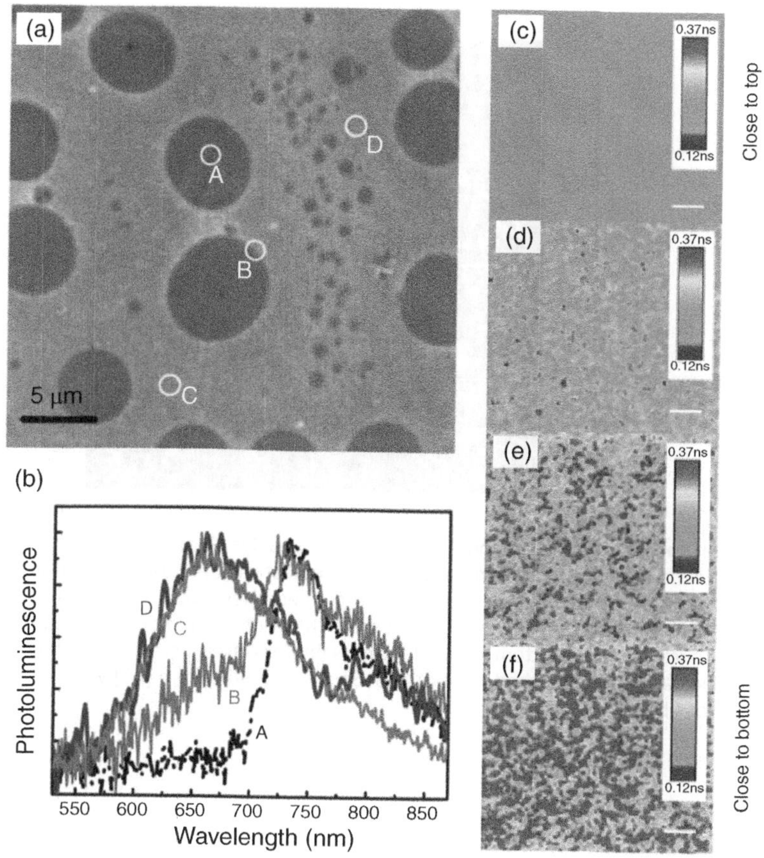

Figure 4.12 (a) Confocal image of a model BHJ made of APFO3/PCBM; yellow (white) signal is acquired in the range 495–535 nm and red (dark gray) signal in the spectral region above 650 nm, scale bar is 5 μm. (b) Luminescence spectra acquired under excitation at 380 nm and recorded in specific regions indicated by white circles marked by letters in panel (a) A – PCBM-rich region, B – interfacial region, C – polymer-rich region, D – region where high intermixing between the two materials is localized. *Source:* Adapted with permission by Royal Society of Chemistry from [63]. (c–f) FLIM images acquired at different heights (500 nm steps) from the bottom layer to the top in 2 μm thick BHJ made of poly(3-hexylthiophene) (P3HT) and PCBM; measures are acquired under excitation by a picosecond laser at 470 nm using a picosecond laser microscope, scale bars are 2 μm. *Source:* Adapted with permission from Huang et al. Copyright (2010) American Chemical Society [44].

to the emission. While in region "D," a new peak centered at about 800 nm appears, which the authors attributed to CTS signal.

BHJ topology and related charge transfer phenomena can be simultaneously studied mapping fluorescence/photoluminescence. In Figure 4.12, FLIM is used to measure exciton lifetime in a poly(3-hexylthiophene) (P3HT) and PCBM BHJ [44]. Representative sections of a z-stack are reported from the top to the bottom, acquired from 2 μm thick samples. Variations in measured lifetime are due to the fact that the electron acceptor PCBM may induce the quenching of P3HT fluorescence due to electron transfer. In the presented measurement, the contrast, given by lifetime, can be attributed to the phase segregation of P3HT and PCBM: long (red, dark gray regions) and short (green, white regions) lifetime domains are assigned to P3HT- and PCBM-rich domains, respectively.

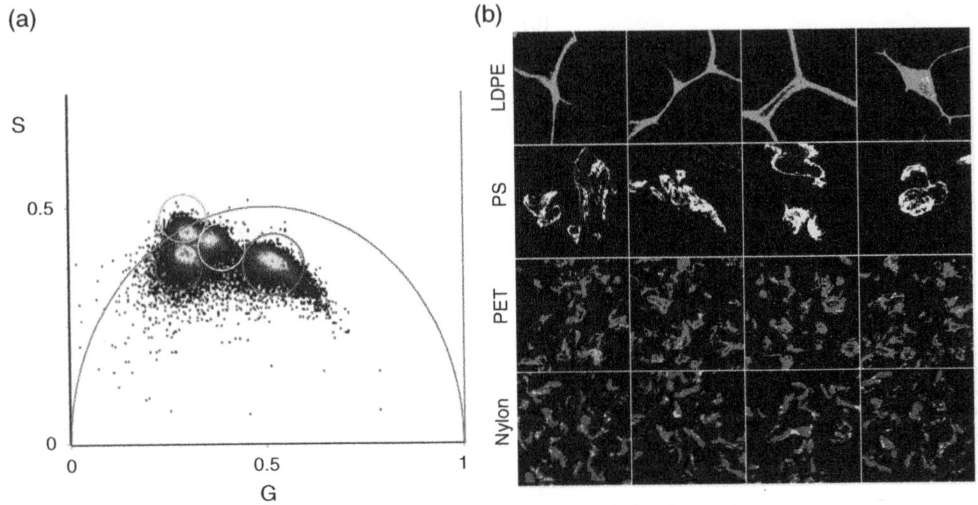

Figure 4.13 Microplastics detection and identification using fluorescent Nile Red dye. Spectral and FLIM data acquired in confocal microscopy allow distinguishing among different microplastics. (a) Phasor FLIM plot, where fluorescence lifetime distributions are represented. It is possible to select them with a colored circle. (b) Microscopy images where pixels are colored according to selection made in the phasor plot: LDPE – top left circle, PS – center circle, PET – right circle, Nylon – bottom left circle. Each sample is characterized by a specific lifetime distribution and then can be easily distinguished from others. *Source:* Adapted from [65] with permission from the Royal Society of Chemistry.

The spectroscopic analysis of the signal of commercial fluorescence dyes can find many practical applications also in the development of new tools, which find applications in many fields. This is the case, for example, of Nile Red, a well-known hydrophobic dye used to stain protein and membranes, whose use is now becoming widespread for the analysis of microplastics, which are micro- and nano-sized plastic fragments recognized as pollutants of environmental concern [65].

In Figure 4.13, we report, as an example, results of a FLIM experiment on Nile Red-stained samples of different natures, which allowed to readily distinguish the chemical composition of microplastics. In particular, the phasor analysis of FLIM through the Fourier transform of the fluorescence decay measured at each pixel of the images provides a graphical description of FLIM data, recovering qualitative and quantitative information on fluorescence lifetime distribution [66–68]. The real and the imaginary components of the Fourier transform are plotted on a polar plot as x axis and y coordinates so that each point in the phasor plot corresponds to a specific fluorescence decay. It is possible to select pixels in the phasor plot by a colored circle (circular region) and the corresponding pixels characterized by selected lifetime are mapped back in the image using the same color code (points with the same lifetime). In Figure 4.13b, the selection map color coded according to different lifetime distributions for four different Nile Red-stained microplastics with different composition, namely low-density polyethylene (LDPE), polystyrene (PS), polyethylene terephthalate (PET), and polyamide (Nylon) dispersed in water, is reported. As can be seen, each plastic displays a distinguishable lifetime distribution and can be thus identified accordingly.

4.4.2 Fluctuation Methods for the Analysis of Nanosystems

As reported in Section 4.3, fluorescence intensity fluctuation analysis reveals the dynamics of fluorescent molecules down to single molecule's analysis. Fluorescence image correlation spectroscopy increases the potential of single point FCS, providing enlarged spatiotemporal coverage. These methods can be used for observing clustering of slowly diffusing components, or processes extending in different temporal scales from nanosecond to millisecond range [55, 69, 70]. Among them, the application of RICS (raster scan image correlation spectroscopy) is spreading very quickly in laboratories as it is accessible to the most existing fluorescent laser scanning microscopes like confocal and two-photon [71]. RICS measures the diffusion of fluorescently labeled molecules from stacks of confocal microscopy images by analyzing correlations within the image. RICS enables the observation of wide region areas of the sample, providing information in the microseconds (time to scan a pixel), milliseconds (time to scan a line), and seconds (time to scan images) timescales.

Due to the nature of the scanning mechanism, fluorescence images inherently capture a wide range of dynamic information, potentially providing details about underlying spatially heterogeneous processes. The laser performs a raster scan from the top left pixel for an entire line and then goes back measuring the following row of pixels from left to right and creates a space–time matrix of pixels within the image in step of defined size and at fixed speed, so that pixel size and pixel dwell time are suitable parameters. (Figure 4.14a and b).

Using the RICS correlation technique, it is possible to generate spatial-temporal maps of dynamics. Fluorescence intensity is measured at each illuminated pixel that can be considered to correspond to a region of the sample with defined width (pixel size) independent of the size of the focal volume of the laser beam [72]. The RICS idea consists in considering that: during a raster scan, a fast-moving fluorescent nanoparticle will be observed across more pixels compared to a slower particle. This can be quantified, calculating from the image the spatiotemporal correlation of fluorescence intensity fluctuations (Figure 4.14c and d).

Image autocorrelation is calculated on each image using the following equation:

$$G(\xi, \psi) = \frac{\langle \delta i(x,y)\delta i(x+\xi, y+\psi)\rangle_{x,y}}{\langle i(x,y)\rangle_{x,y}\langle i(x,y)\rangle_{x,y}} \tag{4.14}$$

where $i(x, y)$ is the intensity at each pixel of the image, ξ and ψ are the x and y spatial correlation shifts, and $\delta(i) = i - \langle i\rangle$ and $\langle i(x,y)\rangle_{x,y}$ is the spatial average of the image. The image correlations of all the frames are averaged. The results are fitted with a relation that correlates to the diffusion coefficient and particles' concentration. Briefly, for raster scanning, the spatial component of the correlation function decays due to movement of the laser beam scanning ($S(\xi, \psi)$) expressed as:

$$S(\xi, \psi) = \exp\left(-\frac{\frac{1}{2}\left[\left(\frac{2\xi\delta_r}{\omega_0}\right)^2 + \left(\frac{2\psi\delta_r}{\omega_0}\right)^2\right]}{\left(1 + \frac{4D(\tau_p\xi + \tau_l\psi)}{\omega_0^2}\right)}\right) \tag{4.15}$$

where ξ and ψ are the x and y spatial lags in pixels, δr (δx and δy) is the pixel size, τ_p (typically in the microsecond range) is the pixel dwell time in x, τ_l is the time between lines in y (typically in the millisecond range), and ω_0 quantifies the radius of the point spread function (the beam waist). The component of the temporal autocorrelation function (ACF) due to diffusion is given by the correlation function:

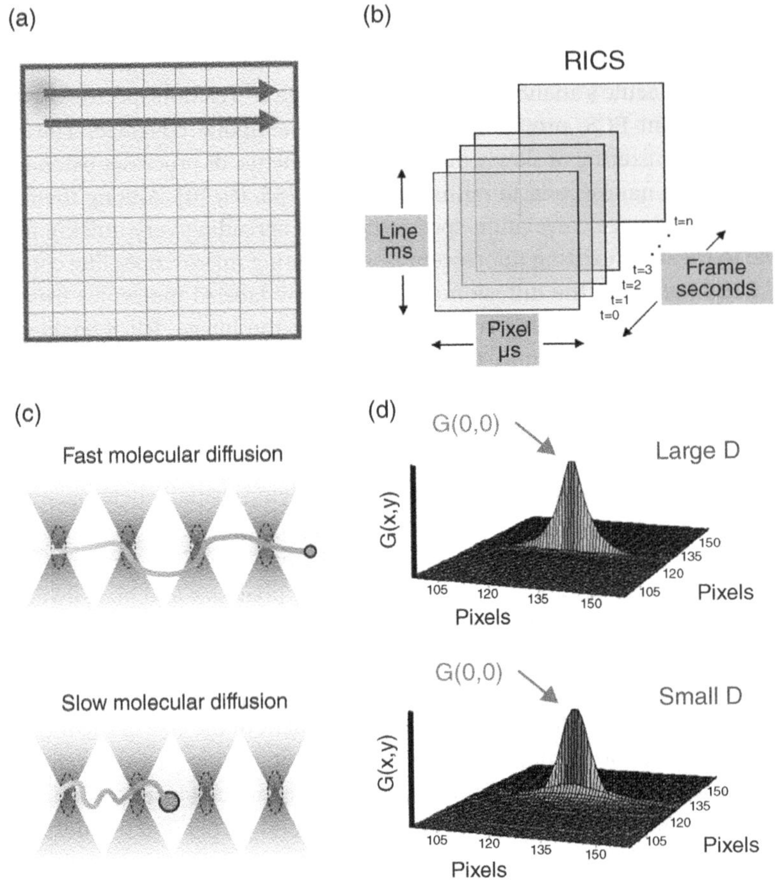

Figure 4.14 RICS analysis. (a) Raster scanning imaging. (b) Spatiotemporal correlation of images. (c) Diffusion of two different particles during laser scanning (fast and slow). During a raster scan, a faster fluorescent particle will be observed across more pixels compared to a slower one. (d) Autocorrelation functions from fast (top) and slow (bottom) diffusion. The diffusion properties affect the shape of the correlation function, the value of G(0,0) is inversely proportional to the diffusing particle concentration.

$$G(\xi,\psi) = \frac{\gamma}{N}\left(1 + \frac{4D(\tau_p\xi + \tau_l\psi)}{\omega_0^2}\right)^{-1}\left(1 + \frac{4D(\tau_p\xi + \tau_l\psi)}{\omega_z^2}\right)^{-1/2} \tag{4.16}$$

where N is the number of particles in the focal volume, D the diffusion coefficient, γ is a shape factor due to uneven illumination across the focal volume, and ω_z is the z-axis beam size. Thus, $G_{\mathrm{RICS}}(\xi, \psi)$ is made of two main terms: $G(\xi, \psi)$, which is related to molecular dynamics, and $S(\xi, \psi)$, which is related to scanning system and optics:

$$G_{\mathrm{RICS}}(\xi,\psi) = S(\xi,\psi)G(\xi,\psi) \tag{4.17}$$

$S(\xi, \psi)$ is described by known instrumental parameters so that G_{RICS} analysis gives back the diffusion coefficient and the concentration of molecules. This last term, following the same rationale described for FCS, is given by the inverse of $G(0, 0)$.

This method is particularly useful when measuring molecular diffusion in heterogeneous systems or when multiple species with critically different sizes are present, or to measure reactions on lab-on-chip devices [17, 70, 71]. The ability to spatially separate the information over different time-scales may allow resolving smaller species also in the presence of large-sized ones that usually hinder classical diffusion measurements like light scattering and FCS.

Analyzing particles' mobility and diffusion properties may also have applications in characterizing the environment/matrices where fluorescent species with known size are diffusing; in a simpler case, not only the viscosity of a homogeneous solution can be recovered but also, for example, the presence of compartments, potential binding sites, or local gradients of crowding agents. A paradigmatic example of a diffusing nanoparticle in a crowded and complex environment is a protein molecule diffusing in a cellular environment: it may encounter barriers, experience crowding due to high concentration of other molecules, or undergo transient binding events and so on. In Figure 4.15, we

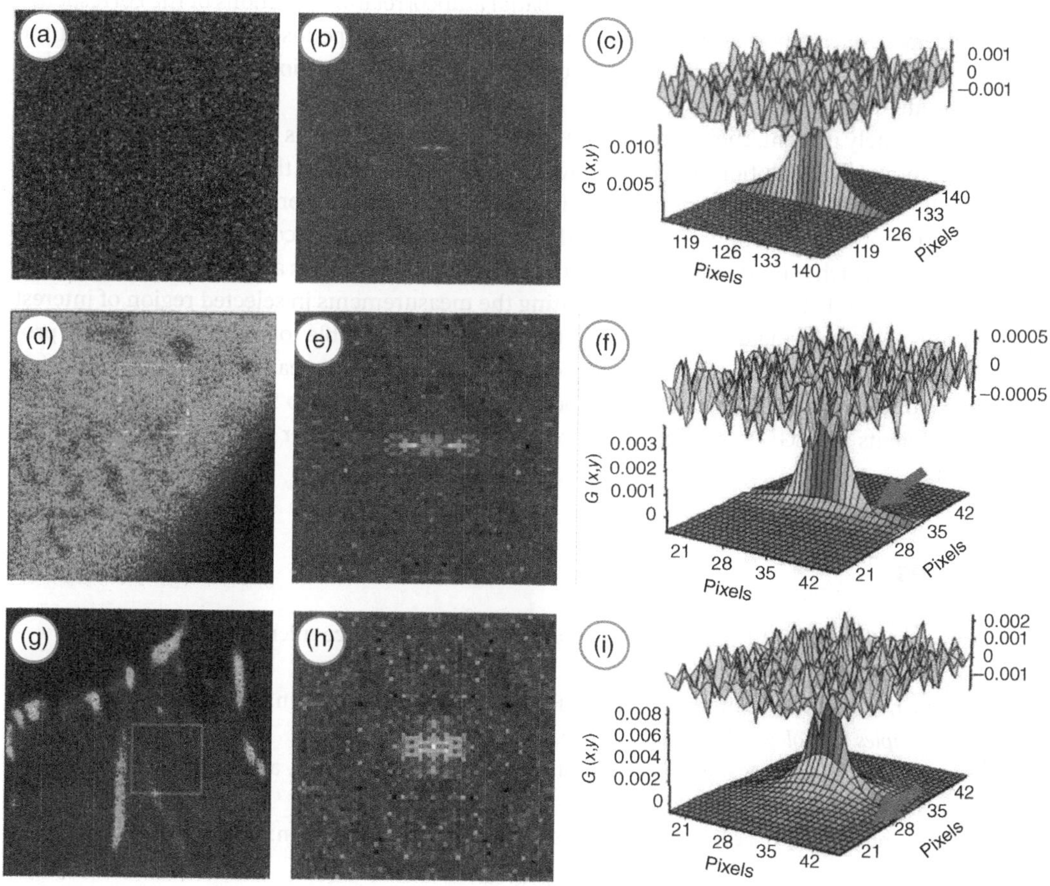

Figure 4.15 Examples of EGFP diffusion in different environments. (a) Intensity image, (b) spatial autocorrelation function, and (c) 3D fit of the autocorrelation function (residuals on the top) for a EGFP solution in water at nanomolar concentration. (d) Intensity image, (e) spatial autocorrelation function, and (f) 3D fit of the autocorrelation function (residuals on the top) for a EGFP diffusing in the cell cytosol. (g) Intensity image, (h) spatial autocorrelation function, and (i) 3D fit of the autocorrelation function (residuals on the top) for a EGFP in a region where focal adhesions are present. Arrows highlight the extension, when present, of the autocorrelation function along the vertical axis. *Source:* Adapted by permission from Nature, Rossow et al. Copyright (2010), [70].

report RICS analysis of a fluorescent protein EGFP (enhanced green fluorescent protein), which can be schematized as a nanoparticle with diameter below 5 nm in water solution, in the cytosol and in complex with paxillin [70].

Paxillin is a protein, which is known to interact with large macromolecular assemblies (focal adhesions) through which cells transmit mechanical force and regulatory signals to the external regions. Paxillin is uniformly distributed and freely diffusing as a monomer in the cytoplasm and its dynamic behavior is critically slowed down at the focal adhesions. In panels a, d, and g of Figure 4.15, representative intensity images of the three different samples are reported. As expected, fluorescence intensity for EGFP diffusing in water is uniformly distributed in the image of panel a. In panel d, an inhomogeneous distribution of the intensity corresponding to slight variations in protein concentration through the cell is evident and finally in panel g, the intensity distribution reveals the morphology of the focal adhesions (bright elongated structures) in which proximity EGFP–paxillin complex is located at higher concentration. Each image in this experiment is 12.8 μm wide. In panels b, e, and h, the spatial autocorrelation functions of the corresponding measurements are reported, the color code (brightness) corresponds to the magnitude of the function. Moreover, in panel c, f, and i, the result of the fit of these functions plotted in 3D together with the residual is reported.

As is immediately evident, the shape of the autocorrelation functions is related to diffusion properties of the protein. In solution, EGFP is moving rapidly compared to the line-to-line scan time, and it results in a 2D autocorrelation function, which resembles an horizontal strip. The slower diffusion in the other two samples results in the elongation of the autocorrelation function along the vertical axis, which is more pronounced at the focal adhesions. This is also highlighted in the 3D fits in images c, f, and i by the arrows. By repeating the measurements in selected region of interest (marked squares), exploring different areas of the sample, it is possible to reveal and compare protein dynamics within cellular environment; for example, for these measurements, the diffusion coefficients for cytosolic EGFP from these experiments range from 20 to 50 $\mu m^2 s^{-1}$, while the diffusion coefficients for EGFP–paxillin range from 1 to 20 $\mu m^2 s^{-1}$, lower D may stem for binding mechanisms.

References

1 Diaspro, A., Bianchini, P., Zanacchi, F.C. et al. (2019). Fluorescence microscopy. In: *Springer Handbooks*, 1039–1088.

2 Dobrucki, J.W. and Kubitscheck, U. (2017). Fluorescence microscopy. In: *Fluorescence Microscopy: From Principles to Biological Applications*, 2e, 85–132. Wiley.

3 Pawley, J.B. (2002). Confocal and two-photon microscopy: foundations, applications and advances. *Microsc. Res. Tech.* 59 (2): 148–149.

4 Olivier, T. and Moine, B. (2013). Confocal laser scanning microscopy. In: *Optics in Instruments*, 1–77. Wiley.

5 So, P.T. (2001). Two-photon fluorescence light microscopy. In: *eLS*. Wiley.

6 Cardullo, R.A. (2003). Fundamentals of image processing in light microscopy. *Methods Cell Biol.* 72 (72): 217–242.

7 Lakowicz, J.R. (2006). *Principles of Fluorescence Spectroscopy*. Springer.

8 Minsky, M. (1957). Microscopy apparatus. US Patent 3013467, vol. 3013467, no. 3013467, p. 5.

9 Sheppard, C.J.R. and Wilson, T. (1981). The theory of the direct-view confocal microscope. *J. Microsc.* 124 (2): 107–117.

10 Denk, W., Strickler, J.H., and Webb, W.W. (1990). Two-photon laser scanning fluorescence microscopy. *Science* 248 (4951): 73–76.
11 Ustione, A. and Piston, D.W. (2011). A simple introduction to multiphoton microscopy. *J. Microsc.* 243 (3): 221–226.
12 Göppert-Mayer, M. (1931). Über Elementarakte mit zwei Quantensprüngen. *Ann. Phys.* 401 (3): 273–294.
13 Kaiser, W. and Garrett, C.G.B. (1961). Two-photon excitation in CaF2: Eu2+. *Phys. Rev. Lett.* 7 (6): 229–231.
14 Konig, K. (2000). Multiphoton microscopy in life sciences. *J. Microsc.* 200 (2): 83–104.
15 Barnard, E.S., Hoke, E.T., Connor, S.T. et al. (2013). Probing carrier lifetimes in photovoltaic materials using subsurface two-photon microscopy. *Sci. Rep.* 3: 1–9.
16 Wu, J., Kwok, T.Y., Li, X. et al. (2013). Mapping three-dimensional temperature in microfluidic chip. *Sci. Rep.* 3: 1–7.
17 Arrabito, G., Cavaleri, F., Montalbano, V. et al. (2016). Monitoring few molecular binding events in scalable confined aqueous compartments by raster image correlation spectroscopy (CADRICS). *Lab Chip* 16 (24): 4666–4676.
18 Zernike, F. (1955). How I discovered phase contrast. *Science* 121 (3141): 345–349.
19 Steen, W. (2000). *Principles of Optics M. Born and E. Wolf*, 7th (expanded)e, vol. 32, no. 5. Cambridge, *1999*: Cambridge University Press.
20 Fischer, R.S., Wu, Y., Kanchanawong, P. et al. (2011). Microscopy in 3D: a biologist's toolbox. *Trends Cell Biol.* 21 (12): 682–691.
21 Lee, T., Senyuk, B., Trivedi, R.P., and Smalyukh, I.I. (2018). Optical microscopy of soft matter systems. In: *Fluids, Colloids and Soft Materials: An Introduction to Soft Matter Physics*, 165–186. Wiley.
22 Conchello, J.A. and Lichtman, J.W. (2005). Optical sectioning microscopy. *Nat. Methods* 2 (12): 920–931.
23 Diaspro, A., Chirico, G., and Collini, M. (2005). Two-photon fluorescence excitation and related techniques in biological microscopy. *Q. Rev. Biophys.* 38 (2): 97–166.
24 Wilson, T. (2011). Resolution and optical sectioning in the confocal microscope. *J. Microsc.* 244 (2): 113–121.
25 Sanderson, M.J., Smith, I., Parker, I., and Bootman, M.D. (2014). Fluorescence microscopy. *Cold Spring Harb. Protoc.* 2014 (10): 1042–1065.
26 Tsien, R.Y. and Waggoner, A. (1995). Fluorophores for confocal microscopy. In: *Handbook of Biological Confocal Microscopy*, 267–279. Springer.
27 Webb, R.H. (1996). Confocal optical microscopy. *Rep. Prog. Phys.* 59 (3): 427–471.
28 Becker, W. (2012). Fluorescence lifetime imaging – techniques and applications. *J. Microsc.* 247 (2): 119–136.
29 Carisey, A., Stroud, M., Tsang, R., and Ballestrem, C. (2011). Fluorescence recovery after photobleaching. *Methods Mol. Biol.* 769: 387–402.
30 Haustein, E. and Schwille, P. (2007). Fluorescence correlation spectroscopy: novel variations of an established technique. *Annu. Rev. Biophys. Biomol. Struct.* 36: 151–169.
31 Lemke, E.A. and Deniz, A.A. (2011). Förster resonance energy transfer. *Compr. Nanosci. Technol.* 1–5: 127–151.
32 So, P.T.C., Dong, C.Y., Masters, B.R., and Berland, K.M. (2000). Two-photon excitation fluorescence microscopy. *Annu. Rev. Biomed. Eng.* 2 (2000): 399–429.
33 Birge, R.R. and Pierce, B.M. (1979). A theoretical analysis of the two-photon properties of linear polyenes and the visual chromophores. *J. Chem. Phys.* 70 (1): 165–178.

34 Friedrich, D.M. and McClain, W.M. (1980). Two-photon molecular electronic spectroscopy. *Annu. Rev. Phys. Chem.* 31 (1): 559–577.

35 Denk, W. and Svoboda, K. (1997). Photon upmanship: why multiphoton imaging is more than a gimmick. *Neuron* 18 (3): 351–357.

36 Sandison, D.R. and Webb, W.W. (1994). Background rejection and signal-to-noise optimization in confocal and alternative fluorescence microscopes. *Appl. Optics* 33 (4): 603.

37 Hoover, E.E. and Squier, J.A. (2013). Advances in multiphoton microscopy technology. *Nat. Photonics* 7 (2): 93–101.

38 Denk, W., Piston, D.W., and Webb, W.W. (2006). Multi-photon molecular excitation in laser-scanning microscopy. In: *Handbook of Biological Confocal Microscopy*, 3e, 535–549. Springer.

39 Sancataldo, G., Silvestri, L., Allegra Mascaro, A.L. et al. (2019). Advanced fluorescence microscopy for in vivo imaging of neuronal activity. *Optica* 6 (6): 758.

40 Diaspro, A. (2011). *Optical Fluorescence Microscopy: From the Spectral to the Nano Dimension*. Springer.

41 Hell, S.W. and Wichmann, J. (1994). Breaking the diffraction resolution limit by stimulated emission: stimulated-emission-depletion fluorescence microscopy. *Opt. Lett.* 19 (11): 780.

42 Aubin, J.E. (1979). Autofluorescence of viable cultured mammalian cells. *J. Histochem. Cytochem.* 27 (1): 36–43.

43 Chalfie, M., Tu, Y., Euskirchen, G. et al. (1994). Green fluorescent protein as a marker for gene expression. *Science* 263 (5148): 802–805.

44 Huang, J.H., Chien, F.C., Chen, P. et al. (2010). Monitoring the 3D nanostructures of bulk heterojunction polymer solar cells using confocal lifetime imaging. *Anal. Chem.* 82 (5): 1669–1673.

45 Sartorio, C., Giuliano, G., Scopelliti, M. et al. (2020). Synergies and compromises between charge and energy transfers in three-component organic solar cells. *Phys. Chem. Chem. Phys.* 22 (16): 8344–8352.

46 Kox, M.H.F., Stavitski, E., Groen, J.C. et al. (2008). Visualizing the crystal structure and locating the catalytic activity of micro- and mesoporous ZSM-5 zeolite crystals by using in situ optical and fluorescence microscopy. *Chem. A Eur. J.* 14 (6): 1718–1725.

47 Ohzono, T., Katoh, K., and Fukuda, J.I. (2016). Fluorescence microscopy reveals molecular localisation at line defects in nematic liquid crystals. *Sci. Rep.* 6.

48 Gentili, D., Durso, M., Bettini, C. et al. (2013). A time-temperature integrator based on fluorescent and polymorphic compounds. *Sci. Rep.* 3: 1–7.

49 Diaspro, A. (2016). *Super-Resolution Imaging in Biomedicine*. Taylor & Francis.

50 Ranjit, S., Lanzanò, L., Libby, A.E. et al. (2021). Advances in fluorescence microscopy techniques to study kidney function. *Nat. Rev. Nephrol.* 17: 128–144.

51 Ries, J. and Schwille, P. (2012). Fluorescence correlation spectroscopy. *Bioessays* 34 (5): 361–368.

52 Magde, D., Elson, E., and Webb, W.W. (1972). Thermodynamic fluctuations in a reacting system measurement by fluorescence correlation spectroscopy. *Phys. Rev. Lett.* 29 (11): 705–708.

53 Elson, E.L. and Magde, D. (1974). Fluorescence correlation spectroscopy. I. Conceptual basis and theory. *Biopolymers* 13 (1): 1–27.

54 Berland, K.M., So, P.T., and Gratton, E. (1995). Two-photon fluorescence correlation spectroscopy: method and application to the intracellular environment. *Biophys. J.* 68 (2): 694–701.

55 Digman, M.A. and Gratton, E. (2011). Lessons in fluctuation correlation spectroscopy. *Annu. Rev. Phys. Chem.* 62: 645–668.

56 Jameson, D.M., Ross, J.A., and Albanesi, J.P. (2009). Fluorescence fluctuation spectroscopy: ushering in a new age of enlightenment for cellular dynamics. *Biophys. Rev.* 1 (3): 105–118.

57 Paddock, S.W. (2000). Principles and practices of laser scanning confocal microscopy. *Appl. Biochem. Biotechnol. B Mol. Biotechnol.* 16 (2): 127–149.

58 Wu, Z. and van Zandvoort, M. (2014). Two-photon microscopy. *Compr. Biomed. Phys.* 4: 165–174.

59 Leunissen, M.E., Christova, C., Hynninen, A.-P. et al. (2005). Ionic colloidal crystals of oppositely charged particles. *Nature* 437 (7056): 235–240.

60 Lin, Y., Skaff, H., Böker, A. et al. (2003). Ultrathin cross-linked nanoparticle membranes. *J. Am. Chem. Soc.* 125 (42): 12690–12691.

61 Piacenza, E., Presentato, A., Di Salvo, F. et al. (2020). A combined physical–chemical and microbiological approach to unveil the fabrication, provenance, and state of conservation of the Kinkarakawa-gami art. *Sci. Rep.* 10 (1): 1–13.

62 De Luca, G., Fennema Galparsoro, D., Sancataldo, G. et al. (2020). Probing ensemble polymorphism and single aggregate structural heterogeneity in insulin amyloid self-assembly. *J. Colloid Interface Sci.* 574: 229–240.

63 Manca, M., Piliego, C., Wang, E. et al. (2013). Tracing charge transfer states in polymer: fullerene bulk-heterojunctions. *J. Mater. Chem. A* 1 (25): 7321–7325.

64 Hedley, G.J., Ruseckas, A., and Samuel, I.D.W. (2017). Light harvesting for organic photovoltaics. *Chem. Rev.* 117 (2): 796–837.

65 Sancataldo, G., Avellone, G., and Vetri, V. (2020). Nile Red lifetime reveals microplastics identity. *Environ. Sci. Process Impacts* 22: 2266–2275.

66 Stringari, C., Cinquin, A., Cinquin, O. et al. (2011). Phasor approach to fluorescence lifetime microscopy distinguishes different metabolic states of germ cells in a live tissue. *Proc. Natl. Acad. Sci. U. S. A.* 108 (33): 13582–13587.

67 Ranjit, S., Malacrida, L., Jameson, D.M., and Gratton, E. (2018). Fit-free analysis of fluorescence lifetime imaging data using the phasor approach. *Nat. Protoc.* 13 (9): 1979–2004.

68 Sancataldo, G., Anselmo, S., and Vetri, V. (2020). Phasor-FLIM analysis of Thioflavin T self-quenching in Concanavalin amyloid fibrils. *Microsc. Res. Tech.* 83 (7): 811–816.

69 Kolin, D.L. and Wiseman, P.W. (2007). Advances in image correlation spectroscopy: measuring number densities, aggregation states, and dynamics of fluorescently labeled macromolecules in cells. *Cell Biochem. Biophys.* 49 (3): 141–164.

70 Rossow, M.J., Sasaki, J.M., Digman, M.A., and Gratton, E. (2010). Raster image correlation spectroscopy in live cells. *Nat. Protoc.* 5 (11): 1761–1774.

71 Digman, M.A., Brown, C.M., Sengupta, P. et al. (2005). Measuring fast dynamics in solutions and cells with a laser scanning microscope. *Biophys. J.* 89 (2): 1317–1327.

72 Brown, C.M., Dalal, R.B., Hebert, B. et al. (2008). Raster image correlation spectroscopy (RICS) for measuring fast protein dynamics and concentrations with a commercial laser scanning confocal microscope. *J. Microsc.* 229 (1): 78–91.

5

Infrared Absorption Spectroscopy

Tiziana Fiore and Claudia Pellerito

Department of Physics and Chemistry – Emilio Segrè, University of Palermo, Palermo, Italy

5.1 Fundamentals

Infrared (IR) absorption spectroscopy is one of the most important analytical techniques available to study different kinds of samples including solids, semisolids, powders, polymers, biological materials, films, liquids, solutions, and gases. One of the great advantages is that IR spectroscopy is a nondestructive technique commonly applied to characterize the chemical structure and identify the chemical bonds in unknown materials. Qualitative identification of organic and inorganic compounds is a primary purpose of IR spectroscopy, even if IR absorption can also be used to provide quantitative measurements of compounds.

The infrared range of the electromagnetic spectrum covers the wavelength region from 0.7 μm to 1 mm, which conforms to the wavenumber range 14,285–10 cm^{-1}. The entire IR region can be divided into the *near IR* region (0.7–2.5 μm, 14,285–4000 cm^{-1}), the *mid-IR* region (2.5–50 μm, 4000–200 cm^{-1}), and the *far-IR* region (50–1000 μm, 200–10 cm^{-1}). Chemists most often work in the mid-IR region where many molecules have strong absorbances and specific vibrational frequencies corresponding to discrete vibrational energy levels, which characterize the infrared spectrum of each compound.

IR spectroscopy became widely used after the development of dispersive infrared spectrometers in the 1940s [1]. The basic components of a dispersive IR spectrometer included a radiation source, monochromator, and detector. This type of instrument helped to spread the use of IR spectroscopy as a common analytical technique for organic compounds characterization in basic and industrial research. However, the development of the Fourier transform infrared (FTIR) spectrometer, and technology advancements in electronics and instruments, marked the abdication of monochromator and the prosperity of interferometer. FTIR instruments employ an interferometer and exploit the well-established computation of Fourier transformation. FTIR spectroscopic analysis has dramatically improved the quality of the spectra and decreased the time needed to obtain information.

IR spectroscopy not only uses the so-called infrared absorption, but also other techniques such as the attenuated total reflection (ATR) method, diffuse reflectance (DR) method, reflection-absorption (RA) method, photoacoustic spectroscopy (PAS), and emission spectroscopy. Furthermore, FTIR can

Spectroscopy for Materials Characterization, First Edition. Edited by Simonpietro Agnello.

be coupled with other analytical techniques (for instance microscopy, thermogravimetric analysis, chromatography, Raman spectroscopy) to retrieve more information about a sample, the so-called hyphenated techniques.

5.1.1 Introduction

IR spectroscopy is a technique based on the vibrations of the atoms of a molecule. Sample molecules with covalent bonds absorb radiation of specific wavelengths when a bond in the molecule vibrates at the same frequency as the incident radiant energy. Consequently, the vibrational energy levels of sample molecules transfer from ground state to excited state. The frequency absorbed depends on many factors such as the strength of the bond, the masses of the bond atoms, the chemical bonds within the molecule, and the geometry of the molecule. Molecular vibrations therefore readily reflect chemical features of a molecule and, as a result, IR spectroscopy contributes to the study of the molecule structure. This chapter covers both the basic theory of FTIR and how FTIR spectrometers work as well as discussing some of the practical aspects of FTIR use.

5.1.2 Basic Principles

When radiation is absorbed, the energy associated with IR radiation is sufficient to induce molecules to rotate (when possible) and to vibrate. The absorption of infrared radiation of frequencies less than ca. 100 cm^{-1} will induce excitation to higher rotational states. This absorption is quantized; thus, a molecular rotation spectrum consists of discrete lines. If the absorption of infrared radiation is in the range 10,000–100 cm^{-1}, the molecule will be excited to higher vibrational states. This absorption is also quantized, but vibrational spectra appear as a sequence of bands rather than lines because a single vibrational energy change is accompanied by a number of rotational energy changes. Basically, infrared light absorption occurs principally based on a transition between vibrational energy levels. For a molecule to show infrared absorptions, it must possess specific features. In general, any transitions with a change in the vibrational quantum number by ± 1 is allowed by the selection rule (*allowed transitions*), while the other transitions are not allowed by the same rule (*forbidden transitions*). Under ordinary conditions, the most likely transition occurs from ground vibrational state, $\nu = 0$, to the first excited vibrational state, $\nu = 1$. This is the so-called *fundamental transition*. Although most bands that are observed in infrared absorption spectra arise from the fundamental transition, in some cases, we can find bands resulting from the excitation from the ground state, $\nu = 0$, to higher vibrational states $\nu = 2, 3$, etc. (the *overtone bands*). These absorptions are multiples of the fundamental absorption frequency. If the fundamental absorption occurs at frequency ν, the overtones will appear at about 2ν (first overtone), 3ν (second overtone), etc. (see Figure 5.1). However, since overtones are forbidden, overtones bands are very weak and may not be observed under real experimental conditions [2–6].

Vibrating atoms may also interact with each other, and this occurs when two fundamental bands absorbing at $\overline{\nu}_1$ and $\overline{\nu}_2$ may couple to produce a new band $\overline{\nu}_3 = \overline{\nu}_1 + \overline{\nu}_2$. This resulting band is called *combination band*. Two frequencies may couple such that $\overline{\nu}_4 = \overline{\nu}_1 - \overline{\nu}_2$, the produced band is called *difference band*. It is also very common that vibrating atoms in the adjacent bonds with similar frequencies may interact with each other, and this interaction is called *coupling interactions*. Coupling interactions are not limited to a singular bond, but may involve a large portion of for example, the carbon backbone, including oxygen or nitrogen atoms if present. In some cases, overtones and combination bands are as strong as fundamental vibrations. Such interaction, known as *Fermi resonance*, occurs when the frequency of an overtone or a combination band is approximately the same as that of a fundamental (or when an overtone and a combination mode have very close

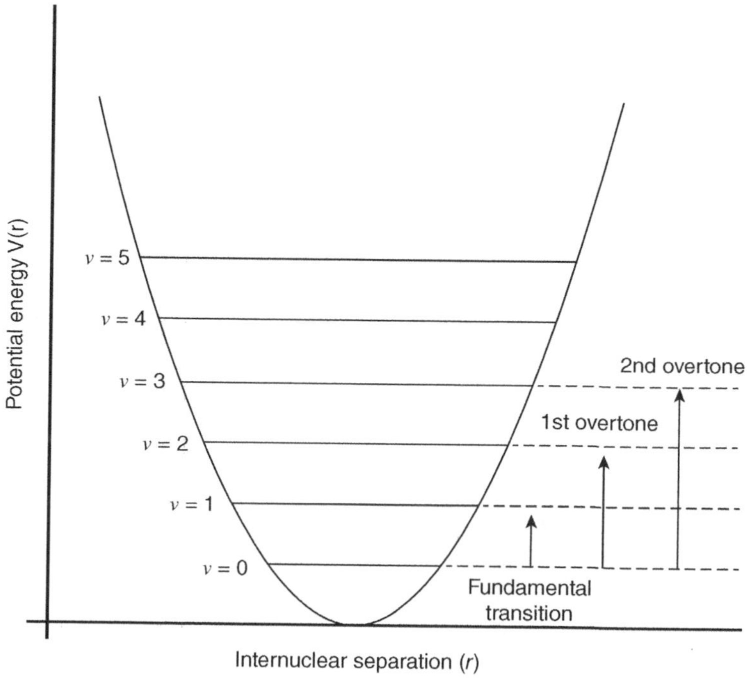

Figure 5.1 Schematic representation of vibrational energy levels for fundamental and overtone infrared transitions for a harmonic oscillator model.

frequencies with each other). The Fermi resonance effect leads to two relatively intense bands in a region where it is expected to observe only one strong fundamental vibration, with frequencies higher and lower than that observed when interaction is absent [2–6].

Another selection rule for infrared absorption is that the molecule must have a change in electric dipole moment during the vibration, in order to absorb infrared radiation. When a molecule vibrates, its charge distribution with respect to that origin may change or not change. This depends upon the symmetry of the molecule. Therefore, only vibrations with a particular molecular structure will cause a change of the electric dipole moment necessary to absorb infrared radiation. A very simple example is the case of *heteronuclear diatomic molecules*. Such molecules containing two atoms with different electronegativities will have the electron density in the bond not equally distributed. This results in a charge separation between the atoms and the bond is said to be polar when such charge separation exists. The molecule is therefore polar and has an electric dipole moment. The electric dipole moment $\overline{\mu}$ is a vector quantity, and is the product of the electric charge, Q, and the vector distance of that charge from some defined origin of coordinates for the molecule, $\overline{r}$

$$\overline{\mu} = Q \cdot \overline{r}. \tag{5.1}$$

Being always a linear molecule, a diatomic molecule vibrates by having the bond stretch and contract, just as a spring. As a molecule vibrates, the dipole moment changes as the bond expands and contracts because the distance between the charges changes. Because the dipole moment changes in a repetitive manner as the molecule vibrates, it is possible for heteronuclear diatomic molecule to absorb IR radiation (infrared-active molecule). It follows that *homonuclear diatomic molecules* (made up of atoms of the same element) have no charge separation because the electronegativity of each atom in the bond is the same. Therefore, its dipole moment remains zero no matter how long the bond is; hence, symmetrical diatomic molecules do not absorb IR radiation (infrared-inactive molecule) [2–6]. Knowledge about vibrations of a molecule is crucially important for understanding IR spectroscopy. The common molecular vibrations, that are excited by IR radiation, are of two types: *stretching* vibrations and *bending* vibrations. Stretching vibration is a rhythmical movement along the bond axis such that the interatomic distance is increasing or decreasing. A bending vibration may consist of a change in bond angle or a change in the position of a group of atoms with respect to the rest of the molecule. The various stretching and bending modes for an AX_2 group (e.g. CH_2, NH_2) appearing as a portion of a molecule are shown in Figure 5.2.

Of the two stretching vibration modes, one can stretch in-phase (symmetric stretching vibrations, shown in Figure 5.2a) and the other can stretch out-of-phase (antisymmetric stretching vibrations,

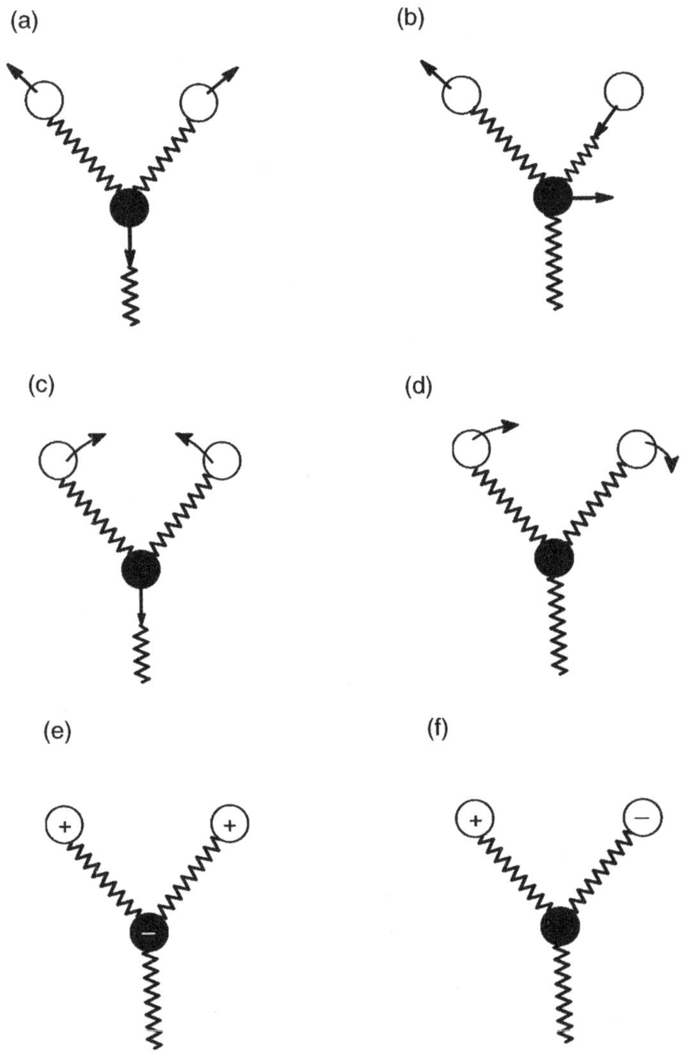

Figure 5.2 Schematic representation of six vibrational modes of an AX_2 group: (a) symmetric stretching vibration, (b) antisymmetric stretching vibration, (c) scissoring bending vibration, (d) rocking bending vibration, (e) wagging bending vibration, and (f) twisting bending vibration. The (+) and (−) signs indicate movement above and below the plane, respectively.

shown in Figure 5.2b). The bending vibrations are of two types: *in-plane* (ip) bending vibrations and *out-of-plane* (oop) bending vibrations. The (ip) bending vibrations are thus named because they happen in the plane of the molecule and are known as *scissoring* and *rocking* vibrations, see Figure 5.2c and d. The (oop) bending vibrations are a change in bond angles across the plane of the molecule and these vibrations are known as *wagging* and *twisting* bending vibrations, see Figure 5.2e and f [2–6].

Molecular vibrations can range from the simple coupled motion of the two atoms of a diatomic molecule to the much more complex motion of each atom in a large polyatomic molecule. To describe the motion of such a molecule, translational, rotational, and vibrational motions must be considered. Molecules with N atoms have $3N$ degrees of freedom, three of which represent translational motion and three represent rotational motion corresponding to the Cartesian coordinates (x, y, z) necessary to describe its position relative to other atoms in the molecule. The remaining $3N-6$ degrees of freedom are vibrational degrees of freedom or *normal vibrations*. A clarification must be made for linear molecules that have $3N-5$ vibrational degrees of freedom. Since rotation of a linear molecule about the axis of the bond does not involve the displacement of any atom, one of the rotational degrees of freedom is lost and linear molecules have an additional vibrational mode [2–6].

For diatomic and triatomic molecules, it is possible to work out the number and kind of vibrations that occur. Regarding **diatomic molecule**, being always a linear molecule, this molecule has only one vibration that corresponds to the stretching and compression of the bond. This accounts for one degree of vibrational freedom ($3 \times 2 - 5 = 1$ normal vibration).

As example of **triatomic molecules**, the normal vibrations of linear molecules, i.e. carbon dioxide (CO_2), and nonlinear molecules, i.e. water (H_2O), will be considered.

For CO_2 molecule, being linear triatomic molecule, it would be expected to have four absorption bands ($3 \times 3 - 5 = 4$ normal vibrations) in its spectrum. In reality, not all possible vibrational modes are IR-active. CO_2 has two equal C=O bond dipoles, but since the molecule has a linear geometry in which the two equal C=O bonds are oriented in the opposite directions at an angle of 180°, the bond dipoles of two C=O bonds counterbalance each other (Figure 5.3a). Therefore, the symmetrical

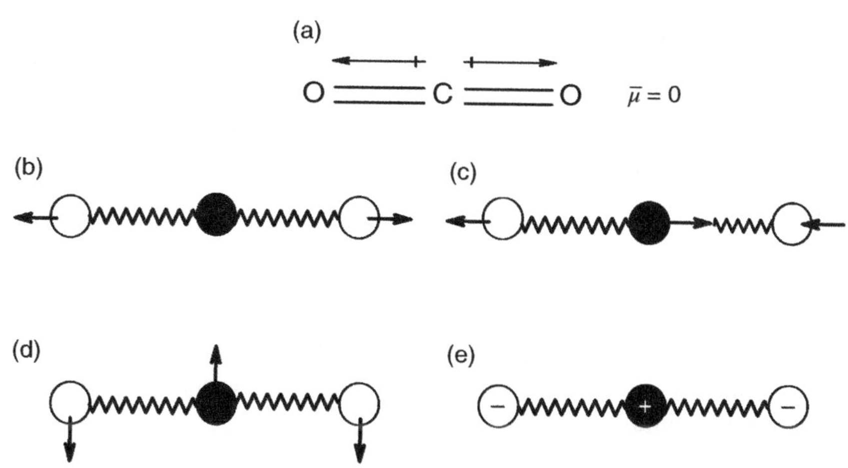

Figure 5.3 (a) Dipole moment of carbon dioxide molecule, and schematic representation of its molecular vibrations: (b) symmetric stretching vibration, (c) antisymmetric stretching vibration, (d) in-plane bending vibration, (e) out-of-plane bending vibration. The (+) and (−) signs indicate movement above and below the plane, respectively.

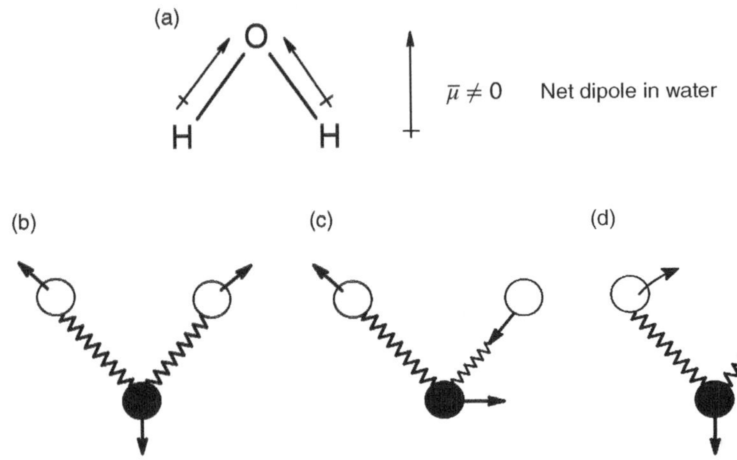

Figure 5.4 (a) Dipole moment of water molecule, and schematic representation of its molecular vibrations: (b) symmetric stretching vibration, (c) antisymmetric stretching vibration, and (d) scissoring bending vibrations.

stretching vibration (Figure 5.3b) is IR-inactive because it produces no change in the dipole moment of the molecule. Furthermore, the two bending modes shown in Figure 5.3d and e are equivalent. They have the same frequency and are said to be *degenerate*. IR spectrum of CO_2 molecule shows therefore only two absorption bands.

H_2O molecule, being a nonlinear triatomic molecule, has three normal vibrations ($3 \times 3 - 6 = 3$, see Figure 5.4). It has two equal O–H bond dipoles, and a bent geometry in which two O–H bonds are oriented at an angle of 104.5°. In this case, the vector sum of the bond dipoles does not cancel and the water molecule has a permanent net dipole moment (Figure 5.4a) and its IR spectrum shows three different absorption bands.

Generally speaking, the analysis of normal vibrations of **polyatomic molecules** is much more complex. Because the molecule consists of a number of atoms, large molecules cannot be easily evaluated, because of their large number of vibrations and interactions between vibrating atoms. As mentioned, the vibrational frequency of a bond depends on the bond strength and the masses of the atoms involved. This frequency is often unchanged (or nearly unchanged) by other atoms connected to the original group. Assignments for frequency of vibrational modes can be approximated by the application of Hooke's law. Two atoms and their connecting bond are treated as a simple harmonic oscillator composed of two masses (M_1 and M_2) linked by a spring.

The following equation derives from Hooke's law:

$$\bar{\nu} = \frac{1}{2\pi c}\sqrt{\frac{f}{\mu}} \tag{5.2}$$

where $\bar{\nu}$ is the *vibrational frequency* (cm^{-1}), c is the *speed of light* ($cm\ s^{-1}$), f is the *force constant* of the considered bond ($dyn\ cm^{-1}$), and μ is the *reduced mass* of the vibrating atoms, expressed combining the individual atomic masses as follows:

$$\mu = \frac{M_1 M_2}{M_1 + M_2} \tag{5.3}$$

Since the force constant of bond measures the stiffness of the chemical bond, and μ is the reduced mass of the vibrating atoms, from Eq. (5.2) we can deduce in a first approximation that the stronger a chemical bond is and the smaller the masses of atoms are, the higher the stretching frequency of a molecule becomes.

The near-constant absorption frequency of a specific bond or chemical group is used to check for the presence or absence of certain bonds in a new compound or for example, to follow the progress of a reaction [2–6].

5.1.3 Infrared Spectra

An infrared spectrum is obtained by passing infrared radiation through a sample and determining what fraction of the incident radiation is absorbed at a particular energy. It is commonly recognizable with absorption intensity (A) or % transmittance ($\%T$) plotted against wavenumber ($\overline{\nu}$). When the spectrum is presented in terms of transmission, the y axis ranges from 0 to 100% transmission while in terms of absorption, the percentage transmission ($\%T$) is converted to absorbance (A) by the following equation

$$A = \log_{10} \frac{100}{\%T}. \tag{5.4}$$

Since absorbance has the advantage of being linearly proportional to the analyte concentration, which can be calculated by the Lambert–Beer law in Eq. (1.8), absorbance units can be used for quantitative analysis. By convention, the x-axis of an infrared spectrum is plotted with high wavenumber to the left and low wavenumber to the right.

Since different molecules have different IR spectra that depend on the structure and mass of the component atoms, it is possible to identify the unknown molecule by matching the absorption spectra of unknown samples with the IR spectra of known compounds. Moreover, since the profile of the spectrum may change with the method of collecting the spectrum, the analyst should know about the sample handling (mull, thin film, KBr pellet, solution, and the solvent) before trying to interpret an IR spectrum.

The interpretation of infrared spectra requires practice and experience; in fact, even simple compounds have relatively complex IR spectra showing many absorption bands and they are more complex than might be expected (an example is shown in Figure 5.5). Additional absorption bands are often present in the spectrum due to overtones and combination bands, coupling, and Fermi resonance. Another factor, which may also contribute to notable changes in infrared spectra, is the hydrogen bonding. The presence of this type of bonding in molecules can introduce additional information into the spectra, and sometimes misinterpretation of bands.

Although the IR spectrum is characteristic of the entire molecule, it is seldom used alone to infer the total structure information of a molecule. Certain groups of atoms give rise to bands at or near the same frequency regardless of the structure of the rest of the molecule. The persistence of these characteristic bands allows chemists to obtain useful structural information by simple inspection and reference to generalized charts of characteristic group frequencies. This permits to identify many of the functional groups that are important in organic chemistry, and provides the basis for qualitative structural identification by IR spectroscopy. These characteristic group vibrational frequencies are tabulated in correlation tables or *correlation charts* [7–9]. Colthup published one of the earliest correlation charts in 1950 [10].

These tables may be found in many textbooks on the interpretation of infrared spectra [11, 12]. The best-known spectral library, consisting of more than 220,000 spectra, is the Sadtler Infrared Spectral Library marketed by the Sadtler Division of Bio-Rad (Philadelphia, PA) [13]. Other important infrared spectral collections are the Perkin-Elmer Library [14], the Sigma-Aldrich Library [15],

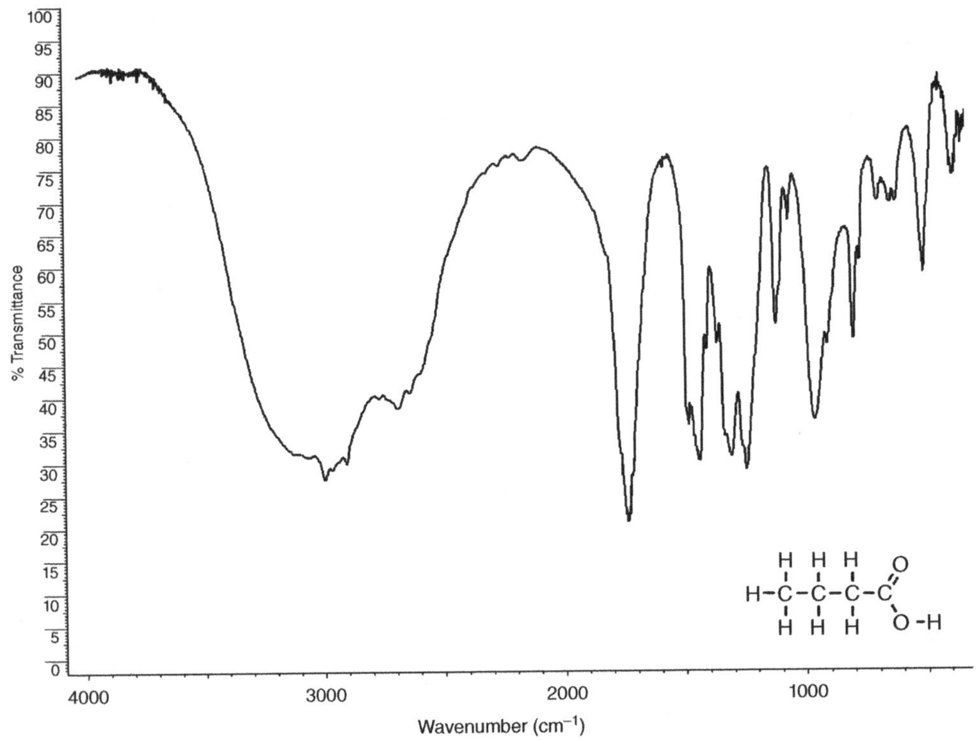

Figure 5.5 Infrared spectrum of butyric acid.

and the web spectral database provided by the National Institute of Advanced Industrial Science and Technology (AIST) from Japan [16].

The energy needed to excite most of the vibrational modes in organic molecules falls in the infrared spectral frequency region 4000–250 cm^{-1}. The two important areas for a preliminary examination of a spectrum are the regions 4000–1300 cm^{-1} and 900–650 cm^{-1}. The region 4000–1300 cm^{-1} contains group frequencies, and is the *functional group region* [11, 12]. Some characteristic stretching frequencies for important functional groups are listed below:

- Broad bands due to O–H and N–H stretching occur in the 4000–2500 cm^{-1} region.
- Weak absorptions from triple bonds due to C≡C, C≡N, and N≡N stretching occur in the 2500–2000 cm^{-1} region.
- Very strong band due to C=O stretching, much weaker absorptions due to C=C stretching and C=N stretching occur in the 2000–1500 cm^{-1} region.
- Set of weak overtone/combination bands if an aromatic ring is present occur in the 2000–1660 cm^{-1} region.

The low-frequency end of the spectrum, 900–650 cm^{-1}, is sometimes called the aromatic region. If there are no strong absorptions in this region, the structure is probably not aromatic. Strong absorptions in this region are due to out-of-plane bending of aromatic ring C–H bonds. Broad absorption bands in this region suggest the presence of nonaromatic amines and amides or carboxylic acid dimers.

The intermediate portion of the spectrum, 1300–900 cm^{-1}, contains low-energy vibrations and is called the *fingerprint region*. The complex absorption patterns are really what make the IR spectrum unique for every molecular species; there are some very important bands in this region, especially the

C–C–O band of alcohols and the C–O–C band of esters. These should be confirmed in conjunction with the appropriate bands (O–H or C=O) in the functional group region of the spectrum [11, 12].

The absorptions observed in the near-infrared region are overtones or combinations of the fundamental stretching bands that occur in the 3000–1700 cm^{-1} region [17]. The bands involved are usually due to C–H, N–H, or O–H stretching. However, there are important differences between the near-infrared positions of different functional groups and these differences can often be exploited for quantitative analysis.

The far-infrared region is more limited than the mid-IR for spectra–structure correlations, but does provide information regarding the vibrations of molecules containing heavy atoms (especially for organometallic compounds, inorganic compounds, and compounds containing halogen atoms); molecular skeleton vibrations; molecular torsions; and crystal lattice vibrations [18].

5.1.4 Fourier Transform Infrared Spectrometers (Interferometers)

Fourier transform infrared (FTIR) spectrometry is based on the use of an interferometer. The most common interferometer used in FTIR spectrometry is a Michelson interferometer, an optical device constructed to cause interference at infrared frequencies [19, 20]. As illustrated in Figure 5.6, a Michelson interferometer consists of two perpendicularly flat mirrors, one fixed and one free to move, and a *beam splitter* at their point of intersection.

The light from an infrared source is separated by a beam splitter into two perpendicular beams. One beam is of fixed length, the other of variable length (movable mirror). The two beams reflect off of their respective mirrors and are recombined when they meet back at the beam splitter. Since the path that one beam travels has a fixed length and the other is constantly changing as its mirror moves, the varying distances between two pathlengths result in a sequence of constructive and

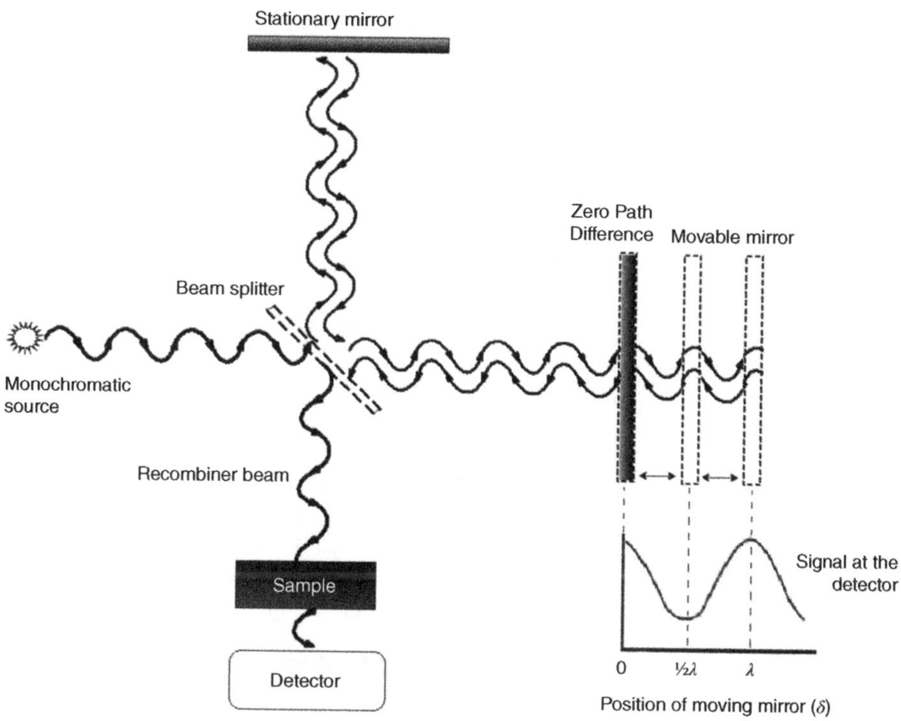

Figure 5.6 Schematic representation of a Michelson interferometer.

destructive interferences and hence variations in intensities. The resulting signal is called *interferogram*. To better understand the processes occurring in a Michelson interferometer, we assume that the source is monochromatic, emitting only a single wavelength of light.

The path difference between the beams traveling to the fixed and movable mirrors and back to the beam splitter corresponds to the *optical path difference* (OPD) of an interferometer, and it is defined as the *retardation* (δ). If there is no difference in pathlength, the retardation is equal to zero ($\delta = 0$) and this condition is called *zero path difference* (ZPD). If the movable mirror moves away from the beam splitter, the displaced distance is defined as the *mirror displacement* and is represented by Δ. The extra distance traveled by the light, which strikes the movable mirror, gives an optical path difference of 2Δ because the light traverses the displaced distance twice on the way to and from the moving mirror. Therefore,

$$\delta = 2\Delta. \tag{5.5}$$

When the two beams are perfectly in-phase on recombination at the beam splitter, the maximum signal will be obtained by the detector, the beams *interfere constructively*. The constructive

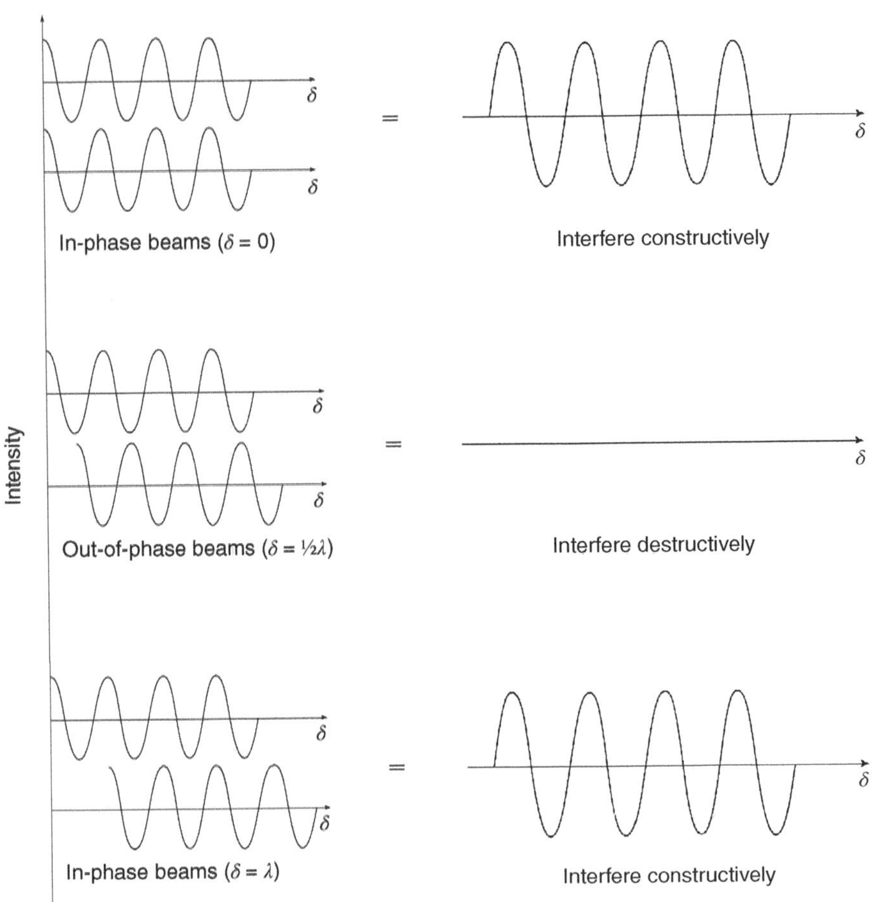

Figure 5.7 Wave interaction: interference.

interference illustrated in Figure 5.7 occurs when the optical path difference (δ) of the two beams is a multiple of their wavelength (λ):

$$\delta = n\lambda \tag{5.6}$$

where $n = 0, 1, 2$, and so on. Conversely, when two out-of-phase beams interfere with each other, their amplitudes will cancel. On recombination at the beam splitter, the beams *interfere destructively*, such that a minimum signal reaches the detector. The destructive interference illustrated in Figure 5.7 can be described by the following equation:

$$\delta = \left(n + \frac{1}{2}\right)\lambda \tag{5.7}$$

where $n = 0, 1, 2$, and so on.

These two situations are two opposite situations. Since the mirror moves back and forth, the interference should be an average between constructive and destructive. Therefore, the intensity of the signal should be between maximum and minimum, and the net signal falling on the detector is a cosine wave. A plot of measured light intensity versus optical path difference is defined as *interferogram*. The frequency of the cosine signal is equal to

$$F = \frac{2}{\lambda}(\nu) = 2\overline{\nu}\nu \tag{5.8}$$

where F is the *Fourier frequency* of the interferogram (Hertz); ν, the velocity of the moving mirror ($cm\ s^{-1}$); λ, the wavelength of radiation (cm); and $\overline{\nu}$, the wavenumber of radiation (cm^{-1}). The frequency (F) gives therefore a direct relationship between the velocity of the mirror and the wavenumber of the incident radiation.

Real IR sources are polychromatic. All the wavenumbers of light emitted by the source pass through the interferometer together. For each of these wavenumbers will be generated a unique cosine wave and the signal at the detector is a result of the sum of these individual cosine waves. Figure 5.8 shows the interferogram of a polychromatic radiation. In the center of the interferogram (δ = ZPD), there is a large burst of infrared intensity (*centerburst*) because all wavelengths of light are in-phase and constructively interfere at ZPD. Because modern FTIR devices scan the moving mirror symmetrically around ZPD, as the OPD changes the infrared intensity falls off quickly on both sides of the centerburst, and these regions are called *wings*. The wings have low intensity because the various Fourier frequencies destructively interfere since they fall increasingly out-of-phase with each other [19, 20].

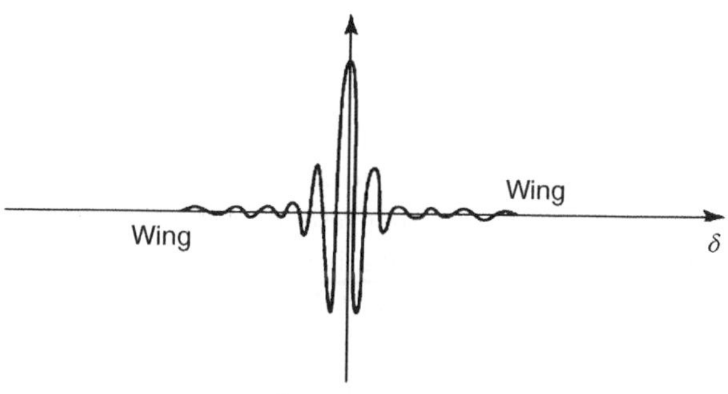

Figure 5.8 Interferogram obtained for polychromatic radiation.

The interferogram holds the spectral information from the source (or sample), and it is a function of time. Therefore, the interferogram is a record of intensity vs. time based on the speed of the moving mirror. The time domain is Fourier-transformed to get a frequency domain, which is deconvoluted to produce a spectrum. A detailed discussion of the mathematics behind the Fourier transform and its use in spectroscopy is available in Ref. [21].

5.2 Sources and Detectors

Before FTIR devices development, the original infrared instruments were of the dispersive type. IR spectra were measured employing a monochromator, which was a scanning dispersive instrument equipped with an optical component of prisms or gratings to separate the wavelengths of light in the spectral range and sequentially present it to the exit slits to isolate the required frequency domain. The essential drawback of the dispersive spectrometer lies with its monochromator. It contains narrow slits at the entrance and exit, limiting the wavenumber range of the radiation reaching the detector to one resolution width. Nowadays, these instruments are used only in selected applications, when the response at one wavelength or a short range of wavelengths is sought. These instruments have been replaced nearly entirely by FTIR spectrometers for several reasons, mainly because of the advantages in rapidity, signal-to-noise ratio (SNR), and precision in determining a spectral frequency that can be obtained from a modern instrument.

A common FTIR spectrometer consists of a source, an interferometer, a sample holder, and a detector. The source generates radiation that passes through an interferometer to the sample before reaching the detector. The signal is amplified and refined from high-frequency contributions by a filter, then the data are converted to a digital signal by an analog-to-digital converter and transferred to the computer for Fourier transformation.

The *infrared source* in an infrared instrument does not depend on the type of instrumentation used to detect the radiation; both dispersive instruments and Fourier-transformed instruments can use the same kind of infrared sources. In the infrared region, glowing blackbody radiation sources are used. The requirements of an optimal radiation source for IR spectroscopy should be continuous emission over the wavenumber range used, constant over long periods of time, and cover a wide wavenumber range.

The ideal source of continuous mid-IR radiation is a high-temperature blackbody. The temperature of an incandescent mid-IR source should be as high as possible to maximize the signal-to-noise ratio (SNR) of the spectrum. Besides temperature, one must also take into account the lifetime and stability of the source, the possibility of saturating the detector or amplifier, and the potential overload of the analog-to-digital converter (ADC).

The more typical sources of IR radiation for the mid-IR region are *Nernst glowers, Globars,* and *heated wires.*

The Nernst glower is a cylindrical bar composed of a mixture of refractory oxides (zirconium oxide, cerium oxide, and thorium oxide) that is electrically heated to a temperature between 1500 and 2000 K. The Nernst glower shows an emission spectrum that resembles that of a blackbody at 1800 K. The source is an insulator at room temperature, but after being preheated becomes a conductor. The biggest disadvantages of the Nernst glower are the short lifetime and mechanical instability. The Nernst glower can easily overheat and burn out because its resistance decreases as the temperature increases. In addition, the emissivity of the oxides from which it is composed is quite low above 2000 cm^{-1}; therefore, the Nernst glower is a better source than a Globar if only the region below 2000 cm^{-1} is of interest. On the other hand, if the entire mid-IR spectrum is to be measured with a Nernst glower, the SNR at high wavenumber will be low [22].

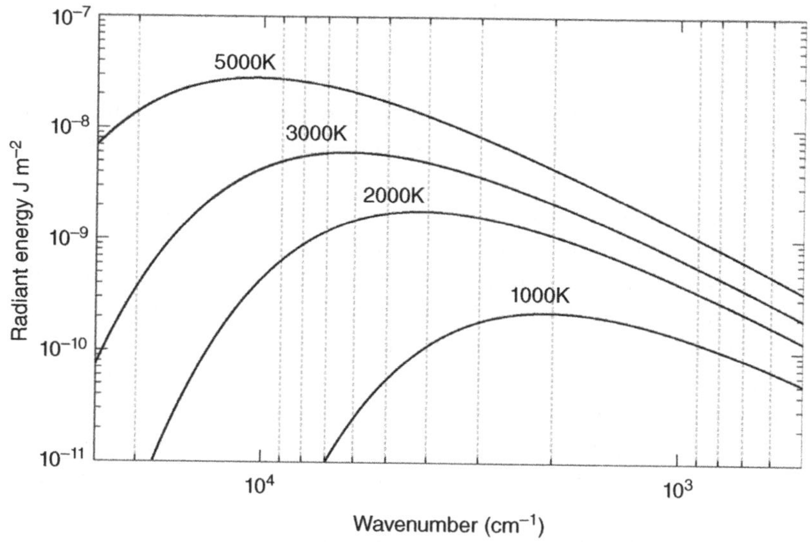

Figure 5.9 Radiant energy received by a surface per unit area (radiant exposure, J m^{-2}), distribution curves for a blackbody source operated at various temperatures.

The Globar source is a rod of silicon carbide (SiC), which is heated electrically to emit continuous IR radiation. The typical temperature at which a Globar is operated is ~1300 K; the temperature is lower than those used by the Nernst glower to avoid air oxidation of the silicon carbide [22].

A third infrared source is made of resistively heated wire coils, such as *nichrome*. This source is simple and rugged, but less intense than the other two. Nichrome wire raises to incandescence by electrical heating in the air at ~900 K. These sources had a lower working temperature and poorer emissivity than a Globar, and are seldom used today [22].

Thanks to the variety of technologically vital applications of the mid-IR spectroscopy (see Section 5.4), significant interest in the development of mid-IR optical systems has been encouraged by the recent development and rapid commercialization of new coherent mid-IR sources. Jung et al. have reported the current state of mid-IR source technology and reviewed potential next-generation optical and optoelectronic materials systems for mid-IR sources [23].

If we take into account the spectral curves for a blackbody source operated at various temperatures (Figure 5.9), it can be seen that operating a mid-IR source at higher temperatures increases the intensity of *near-infrared light* from the source. Since glass and quartz are transparent through the whole NIR region, the simple and low-cost tungsten-filament light bulb or *quartz–tungsten–halogen (QTH) lamp* can be used as the source. The intensity of QTH lamps is very high compared to tungsten light bulbs, but this intensity may sometimes lead to problems of digitization noise and response nonlinearity. The range of light produced by this source is from 25,000 to 2000 cm^{-1} [20].

For the *far-infrared region*, the most common source is the *high-pressure mercury lamp*. This lamp is made of a quartz bulb containing elemental Hg and a small amount of inert gas and produces light as a result of an electrical discharge, generated between two electrodes. When current passes through the lamp, mercury is vaporized, excited, and ionized, forming a plasma discharge at high pressure (>1 atm). The intensity of the radiant energy emitted from the plasma is significantly higher than that emitted by a standard Globar in the far-IR region [20].

The *detector* is one of the most important components of an infrared spectrometer, which converts the analog spectral output into an electrical signal, typically a voltage. Important properties of a detector include its usable wavenumber range, its speed of response, and its noise level. The detectors used in IR spectrometers can be classified as either *thermal detectors* or *photon/quantum detectors* [24, 25].

In the thermal detectors, when absorbed infrared radiation heats the detectors, their temperature increases and fluctuations in the temperature are measured by any temperature-dependent mechanism, such as thermoelectric voltage (e.g. *thermocouples*), electrical resistance change of a conductor (e.g. *bolometers*) or semiconductor material (e.g. *thermistor bolometers*), thermal expansion (e.g. *pneumatic detectors*), or pyroelectric voltage (e.g. *pyroelectric detectors*).

Photon or quantum detectors are based on detection of photons of infrared light through the photoconductive effect that occurs in certain semiconductor materials. Usually, quantum detectors have a much higher sensitivity and a faster response to infrared radiation than thermal detectors; however, since a restricted range of photon energies excites electrons in bound states to the conduction band of a semiconductor, the quantum detectors can operate only over a very limited range of wavelengths. Thermal detectors, on the other hand, can be used on a very wide wavelength range; no specific electronic transitions have to occur, all that is necessary is that the detector absorbs a photon.

The most common types of detectors used in dispersive IR spectroscopy are thermocouples, bolometers, and thermistors, while pyroelectric and photon-sensitive semiconducting detectors are used for FTIR where an appropriate combination of high speed and reasonably good sensitivity is required.

The thermocouple is composed of two wires made of dissimilar metals, chosen to produce a large thermoelectric electromotive force change on heating and cooling, joined to form the thermoelectric junction. When attached directly to the junction, a blackened receiver (usually a thin gold foil) defines the sensitive area. In order to realize a higher temperature sensitivity, several thermocouples are connected in series to form the so-called thermopiles [25]. For a long time, the thermopiles were bulky, expensive, not very sensitive, and their response time was slow but, thanks to innovations in the semiconductor technology, today thermopiles can be optimized for specific applications. In particular, the conventional complementary metal-oxide-semiconductor (CMOS) processes have allowed the integration of the thermopile infrared detectors on standard CMOS chips in several architectures by micromachining process [26–28]. Although thermopiles are not as sensitive as bolometers and pyroelectric detectors, they have relevant advantages such as reliability and a very good compromise between cost-efficiency and performance.

A bolometer is a very sensitive electrical resistance thermometer constructed from a material with a very small thermal capacity and large temperature coefficient so that the absorbed weak thermal radiation produces a large change in the electrical resistance. The most common bolometric detectors are *metal bolometers*, *thermistors*, and *semiconductor bolometers*.

Materials used for metal bolometers are metals with high and long-term stability, and these metals are typically nickel, bismuth, platinum, or antimony. These bolometers based on metals have to be small, so that the thermal capacity is low enough to provide reasonable sensitivity. The resistance of these metals increases about 0.3% per °C. The temperature changes depending on the intensity of incident radiation and on the thermal capacity of the detector. The rate at which the detector heats up or cools down determines how fast the detector is able to respond to a change in radiation intensity. This represents the *response time* of the detector, which is in the range of seconds for the older metal bolometers [29, 30].

Thermistor bolometers are made of a sintered mixture of semiconducting oxides (manganese, cobalt, and nickel) and mounted on electrically insulating but thermally conducting material. They have a higher resistance than metals (2–4% per °C), and are generally much more rugged. Their response time is also slow [31, 32].

Among the thermal detectors, the semiconductor bolometers are the most highly developed. They are the detectors of choice for many applications, especially in the infrared and submillimeter spectral range. The performance of these detectors improves when the devices are cooled because the resistance changes much greater than at room temperature. Moreover, thicker devices can be fabricated to improve IR absorption without increasing the thermal capacity due to the reduced specific heat in the cooled material. Modern well-developed semiconductor bolometer technologies exist to produce micromachined bolometer from silicon. This type of bolometer is only a few micrometers in diameter and has a fast response time. The modern micro-bolometer is particularly useful for detecting far-IR radiation. Detailed information about the fabrication and performance of modern bolometers is given in some review papers [33, 34].

A very common thermopneumatic detector is the Golay detector. This detector consists of a small hollow cell filled with gas (usually xenon for its low thermal conductivity). When a blackened film, placed in the center of the cell, absorbs a photon signal, it produces an increase in temperature. In turn, the film heats the enclosed gas and the consequent expansion of the gas distorts a flexible membrane on which a mirror is mounted. A change in radiation intensity falling on the Golay detector produces a change in the readout from the detector [35]. An important advantage of the Golay cell is that its useful wavelength range is very wide, while the performance of this detector is limited by the temperature noise associated with the thermal exchange between the absorbing film and the detector gas. Moreover, the Golay detector's response time is much faster than that of bolometers, thermistors, or thermocouple detectors.

Pyroelectric materials change their electric polarization spontaneously when their temperature changes, and lose their polarization at some temperature known as the *Curie temperature*, T_C.

Many pyroelectric materials have been investigated for detector applications; the choice depends on many factors such as the required size of the detector, the operating temperature, and the frequency of operation. Characteristics of pyroelectric detector materials have been reported in some papers [36, 37]. These materials can be classified into three categories: *single crystals*, *ceramics (polycrystalline)*, and *polymers*.

Among the single crystals, the most common pyroelectric material in use as an IR detector is *deuterated triglycine sulfate* (DTGS). It possesses appealing properties such as a high pyroelectric coefficient, a reasonably low dielectric constant and thermal conductivity. The major disadvantage is its low T_C. To prevent loss of polarization, DTGS detectors have to be cooled between 20 and 30 °C [38]. *Lithium tantalate* ($LiTaO_3$) is another type of single crystal. It has the advantage of a much higher T_C and high chemical stability, and it does not require cooling; nevertheless, it gives inferior performance to DTGS detectors, due to its lower pyroelectric coefficient [39].

Polycrystalline ferroelectric ceramics is another class of materials consisting of solid solutions of *lead zirconate* ($PbZrO_3$) and *lead titanate* ($PbTiO_3$), and very similar oxides. These materials possess a number of advantages, including low cost to manufacture in large areas, good mechanical properties (both mechanically and chemically robust), and high Curie temperature [40].

Ferroelectric polymers are based on *polyvinylidene fluoride* (PVDF) and on *copolymers with trifluoroethylene* (PVDF-TrFE). This class of materials possesses relatively low pyroelectric coefficient and low dielectric constants, and is attractive for use in pyroelectric vidicon [41, 42].

In *photon detectors*, when semiconductors are exposed to infrared light, the interaction of radiation with the electrons in a solid causes the electrons to be excited to a higher energy state (e.g. from a valence band across a bandgap to a conduction band of a semiconductor), thereby changing the electrical conductivity of the semiconductor. If no radiation falls on them, semiconductor materials are insulators. The response time of a semiconductor detector is represented by the time required to change the semiconductor from an insulator to a conductor (frequently only 1 ns). Since the bandgap of the material used as detector determines wavelength range that can be detected, the semiconductors chosen as IR detectors have to possess bandgaps with appropriate energy. These detectors are generally *p-n* junction semiconductors, with a relatively small bandgap between the valence and conduction bands of the material. Typically, these detectors are comprised of two doped semiconductors. The dopants act as electron donors allowing to control the conductivity level and the response range of the detector. Doping is the addition of small quantities of impurity to a pure semiconductor to modify its electrical conductivity. Some examples of this kind of detectors include lead selenide (PbSe), indium antimonide (InSb), indium gallium arsenide (InGaAs), germanium doped with copper or mercury and mercury cadmium telluride (HgCdTe, also called MCT)

MCT detectors; these are by far the most commonly used photodetectors for mid-IR spectrometry. The MCT material is a mixture of HgTe and CdTe and it can be represented as $Hg_{1-x}Cd_xTe$. MCT needs to operate at liquid nitrogen temperatures of 77 K. The actual spectral range of an MCT detector can be varied by varying the ratio of the blend (HgTe/CdTe ratio) [43–45].

Other detectors such as InGaAs can operate without cooling, but show improved SNR if cooled.

For the far-IR region, doped germanium detectors cooled in liquid helium are sensitive to radiation up to about 120 μm in wavelength [46].

5.3 Techniques and Sampling Methods

The transmission methods are most often used for measurements of infrared spectra of general samples, although reflectance methods, microsampling methods, and the combination of different analytical techniques with IR spectroscopy are also often employed.

5.3.1 Transmission Methods

Transmission spectroscopy is based upon the absorption of infrared radiation at specific wavelengths as it passes through a sample. This technique is the oldest and most basic sampling technique for infrared spectroscopy. It is relatively easy to obtain spectra from samples in solution or in the liquid, solid, or gaseous states. The Lambert–Beer law gives practical measures for the intensities of infrared bands in a transmission spectrum, and transmission analysis can provide both qualitative and quantitative measurements. The sample holder or the material used to contain the sample must be transparent to IR radiation to collect the transmission (absorption) spectra, so the windows of a cell have to pass the desired range of infrared wavelengths. Some representative optical materials used in transmission infrared spectroscopy are summarized in Table 5.1.

5.3.1.1 Solid Samples

Solid samples are usually examined according to three common methods, i.e. alkali metal halide pressed disks, mulls, and films. In the alkali metal halide pressed disks method, the most commonly used alkali metal halide is potassium bromide (KBr) because it is completely transparent in the 4000–400 cm^{-1} region [19], even if some analyses may require spectra extending up to 200 cm^{-1}: in that case, cesium iodide (CsI) can be selected. This is an acceptable method of sampling solids if there is no possibility of cations or anions contained in the sample and in the alkali halide to exchange with each other.

Table 5.1 Selected infrared-transparent materials.

Material	Wavenumber range (cm^{-1})	Refractive index[a]	Water solubility (g/100g, 25°C)
Alkali metal halides			
Sodium chloride (NaCl)	40,000–600	1.49 (10 μm)	36
Potassium chloride (KCl)	33,000–400	1.46 (10 μm)	35
Potassium bromide (KBr)	43,000–400	1.52 (10 μm)	54
Cesium iodide (CsI)	42,000–200	1.74 (10 μm)	86
Cesium bromide (CsBr)	42,000–250	1.66 (10 μm)	123
Fluorides			
Lithium fluoride (LiF)	80,000–1,500	1.39 (5 μm)	0.13
Magnesium fluoride (MgF_2)	1,400–770	1.34 (5 μm)	$7.6 \cdot 10^{-3}$
Calcium fluoride (CaF_2)	75,000–1,000	1.40 (5 μm)	$1.3 \cdot 10^{-3}$
Barium fluoride (BaF_2)	67,000–800	1.45 (5 μm)	0.17
Heavy metal halides			
Silver chloride (AgCl)	25,000–360	1.90 (10 μm)	$1.5 \cdot 10^{-4}$
Thallium bromide/thallium iodide (KRS-5)	12,000–350	2.37 (10 μm)	0.05
Chalcogenides			
Zinc sulfide (ZnS)	27,000–700	2.25 (10 μm)	Insoluble
Zinc selenide (ZnSe)	15,000–650	2.40 (10 μm)	Insoluble
Cadmium telluride (CdTe)	5000–400	2.67 (10 μm)	Insoluble

[a] The refractive index was measured at the wavelength written in parentheses.

In general, a small quantity (1–2 mg) of a very finely ground sample is intimately mixed with approximately 100 mg of a dry potassium bromide (or other alkali halides) powder before compression. The mixture is usually ground with a smooth agate mortar and pestle or, more efficiently, with a small vibrating ball mill, or by lyophilization. Then, it is compacted with special disks under a pressure of 10–15 tons (10,000–15,000 kg) to form a clear transparent disk. The thickness of the disk is important. The disk must not be too thin (thin disks are fragile and difficult to handle) or too thick (thick disks transmit too little radiation). Moreover, the quality of the spectrum also depends on the particle diameter and that should be 2 μm or less to avoid radiation scattering. The complete removal of moisture from the window materials is also very important; if the sample mixed with the alkali metal halide contains moisture, the resultant tablets become opaque. So, the alkali metal halides should be kept desiccated and warm prior to use in order to minimize this effect [6, 20].

The mull method for solid samples involves grinding about 2–5 mg of sample with one or two drops of mineral oil (i.e. a high-boiling petroleum oil, Nujol®) to create a smooth paste of the finely ground sample dispersed in the mineral oil. The mull is placed as a thin film between flat salt plates. The main disadvantage of this method is that the Nujol® bands appear at ca. 2900, 1462, and 1377 cm^{-1}, and they can obscure bands that may be present in the analyzed sample. Hexachlorobutadiene may be used when these bands interfere with the spectrum. The use of both mulling agents makes possible a scan in the 4000–250 cm^{-1} region, essentially free of interfering bands (Figure 5.10) [19].

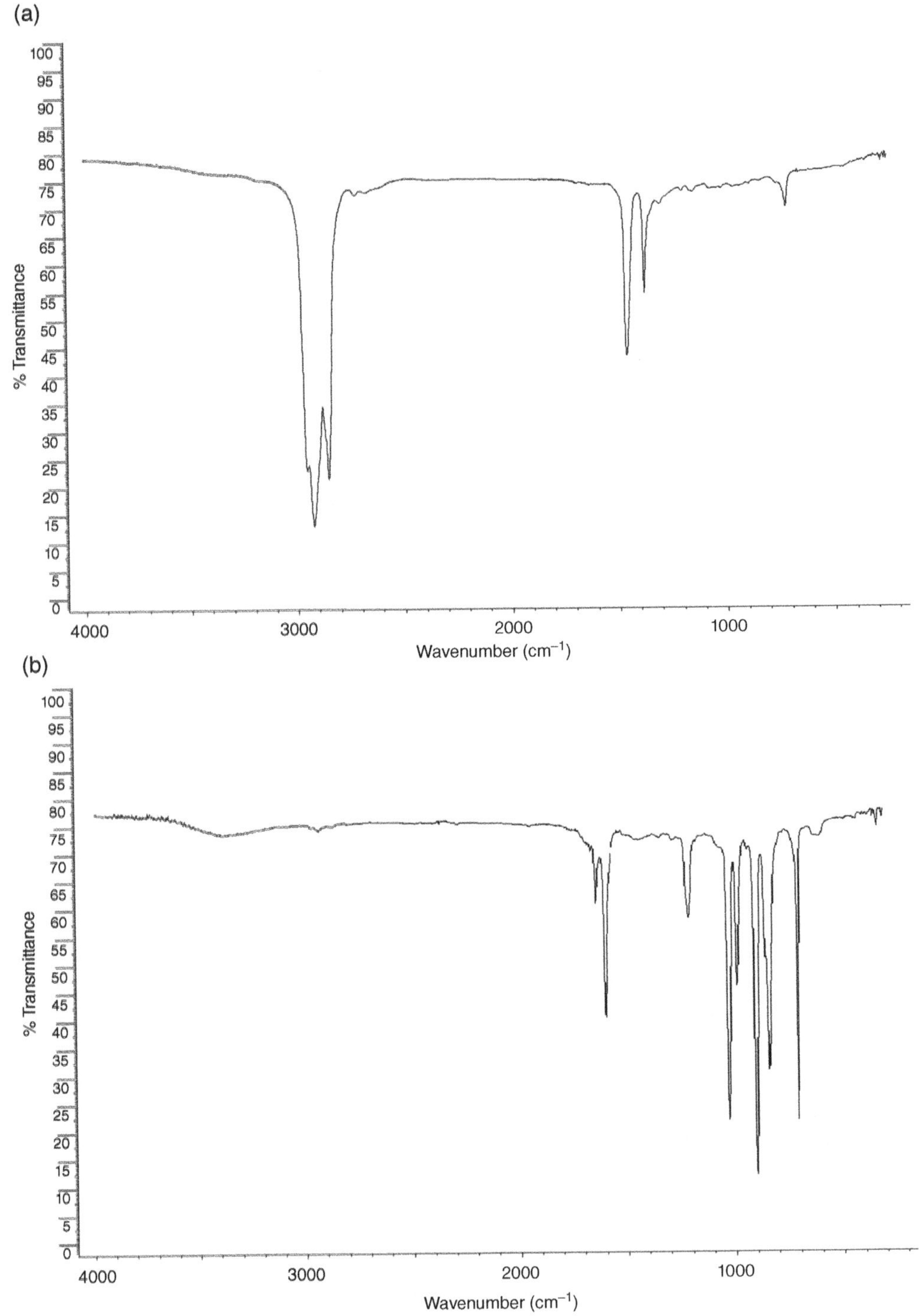

Figure 5.10 Infrared spectrum of Nujol® (a) and hexachlorobutadiene (b) on CsI windows.

Films can be deposited on the surface of a KBr or NaCl plate or onto a disposable "card" by either solvent casting or by melt casting. When the material is deposited from solution, the sample has to be dissolved in a suitable solvent, both, to disperse the sample and produce a uniform film. The concentration depends on the required film thickness.

The thin film approach is particularly useful for obtaining spectra of polymers; the most important issue is to remove all traces of solvent by vacuum treatment or gentle heating before acquiring the spectrum.

Alternatively, solid samples that melt at relatively low temperatures without decomposition can be deposited by casting them straight onto the infrared window [6].

5.3.1.2 Liquid and Solution Samples

Liquids and solutions can be easily measured by FTIR spectroscopy using liquid cells with fixed pathlength (*fixed-path cells*) or with variable pathlength (*demountable cells*). Demountable cells use spacers of different thickness usually made of polytetrafluoroethylene (PTFE, which is known with commercial name "Teflon") that can be selected to establish a suitable pathlength and then the thickness of the liquid film.

Neat liquids with high viscosity and a low vapor pressure may be analyzed with no sample preparation. They can be easily prepared as a capillary film between two flat plates to form a film that is between 10 and 20 μm in thickness. Dilution of a liquid sample may be required to eliminate the problem of total absorption by the strong bands in a neat sample in order to observe the band shape, or in order to study solvent effects. Solutions also give more reproducible spectra. A suitable solvent must be chosen considering some factors. In general, it has to be transparent in the important absorption region of the spectrum, and it should be as nonpolar as possible to minimize solute–solvent interactions [3, 20].

Solvents commonly used for diluting liquid samples include carbon tetrachloride, carbon disulfide, chloroform, benzene, acetone, and some alkanes such as hexane. Unfortunately, no single solvent is transparent over the entire mid-IR region. Since carbon tetrachloride and carbon disulfide are transparent in the region 4000–1350 cm^{-1} and 1330–625 cm^{-1}, respectively, they are often used as a solvent pair. Both solvents must be handled carefully, only under well-ventilated conditions, because they are quite toxic. Carbon tetrachloride may be replaced with the less-toxic tetrachloroethylene or methylene chloride and substitute carbon disulfide with *n*-hexane or *n*-heptane.

Solutions can be measured using a pair of salt plates with a thin spacer in a sealed cell. Typically, solutions of concentration 0.05–10% are handled in IR cells with a pathlength of 0.1–1 mm, while concentration of 10% and cell pathlength of 0.1 mm represent one practical combination.

Fixed pathlength sealed cells are also suitable for volatile liquids. This type of cell is advantageous because it provides a very accurate pathlength and then clearly defined thickness, but cannot be taken apart for cleaning. The filling and cleaning of the cell are done through the use of a pair of hypodermic syringe fittings [19].

Most IR cells must be protected from water, because the salt plates are soluble in water and hygroscopic. If the organic liquid samples are not dried over an appropriate drying agent before their use, the cell surfaces can become opaque and uneven. Absorption bands due to the water are very intense and may obscure some of the sample absorption bands of interest. This problem may be overcome by using heavy water (D_2O) instead of light water (H_2O). Characteristic water bands at 3430 and 1630 cm^{-1} show a downward shift because of the mass dependence of the vibrational wavenumber [20].

Acquiring acceptable IR spectra of samples containing water requires the use of special cells with windows of barium fluoride (BaF_2), silver chloride (AgCl), or KRS-5 (a mixed thallium bromide–thallium iodide) [6].

If quantitative analysis of a sample is required, it is necessary to measure the pathlength in order to use calibration curves obtained with the same cell but at different times. If the internal window surfaces become badly etched, the cell windows have to be removed and polished. This is a serious problem for the quantitative analyses. Light scattering will occur, the pathlength within the cell becomes uneven, and erroneous quantitative results may be obtained [19].

5.3.1.3 Gas Samples

Gas cells can be used to examine gases or low-boiling liquids. These cells are often constructed with a cylindrical glass (or metal) body, two IR-transparent end windows, and a vacuum stopcock for filling from a vacuum-transfer system.

Gases have densities that are several orders of magnitude less than those of liquids or solids; therefore, a gas has many fewer molecules per unit volume than does a condensed phase. To make up for the small concentration of sample molecules in a gas (the C term in Lambert-Beer law, Eq. 1.8), the pathlengths of the gas cells must be correspondingly greater (L is increased). The standard length of the sample cavity for the measurement of the spectra at relatively high partial pressures (0.1–0.01 atm) is about 10 cm long. Gas cells with pathlengths between 1 and 10 cm easily fit in the sample compartments of all benchtop FTIR spectrometers. When longer pathlengths are necessary, the sample compartment size in the instrument is limited, so multi-reflection gas cell is necessary to produce an extended pathlength. In such cells, longer pathlengths are obtained by reflecting the IR beam back and forth many times through the sample using internal mirrors located at the ends of the cell; the IR beam is deflected until it exits the cell after having traveled the required equivalent pathlength. This type of cell provides a pathlength from a few centimeters to 120 m [3, 19, 20].

Sample gas pressure required to produce reasonable spectra depends on the sample absorbance and the cell's pathlength. A single-pass 10 cm cell requires a partial pressure of about 0.0658 atm (50 torr) to obtain a good IR spectrum. However, analysis of gases at concentrations as low as 10 ppb can be successfully performed with multiple reflection cells with long pathlengths. Temperature and pressure must also be controlled if quantitative measurements are being made [3, 19, 20].

5.3.2 Attenuated Total Reflectance (ATR) Method

One of the most commonly used sampling techniques in recent times is the *attenuated total reflectance* (ATR) method based upon internal reflectance phenomenon. In general, reflectance techniques are especially useful for obtaining IR spectra of difficult samples that cannot be readily analyzed by the normal transmission method. They may be used for studying thick or highly absorbing solid and liquid materials, including powders, films, coatings, polymers, and aqueous samples. In reflectance methods, the light is reflected from the surface of the sample. There are several different types of reflection that can take place, depending upon the nature of the reflecting surface [3, 19, 20].

ATR is a contact sampling method that involves a crystal with a high refractive index and excellent IR transmitting properties. It requires little or no sample preparation for most samples. A beam of radiation traveling in a crystalline material with high refractive index (n_c) is reflected when it encounters an interface with a sample of material of a lower refractive index (n_s). When the angle of incidence, between the normal to the surface and the direction of travel of light, (θ_i) at the interface between the crystal and sample is large enough, complete reflection of light occurs at the interface (i.e. total internal reflection). The minimum angle of incidence at which total internal reflectance occurs in a material is called critical angle, θ_c, and depends on the ratio of the two refractive indices of the two materials as follows:

$$\theta_c = \sin^{-1}\left(\frac{n_s}{n_c}\right). \tag{5.9}$$

If a sample of material is placed in close contact with the reflecting surface (called internal reflection element, IRE), an interface is formed. At the interface of the highly refractive IRE and the low refractive sample, a light beam traveling in the IRE will be completely reflected if the critical angle condition is met. The light beam penetrates a fraction of a wavelength (called an evanescent wave) into the low refractive sample, and when the sample can selectively absorb part of the radiation, the light beam loses energy at the wavelength where the sample absorbs, that is the light beam is attenuated. Hence, if the resultant attenuated radiation is measured and plotted as a function of wavelength by the spectrometer, it is possible to obtain a spectrum that resembles a transmission spectrum, but with some differences: the absorption band positions are identical in the two spectra, but the relative intensities of corresponding bands are different. In the ATR method, it is important to know how far the infrared beam penetrates into the sample, i.e. depth of penetration (d_p). The intensity of the evanescent wave decreases exponentially with distance above the surface of the ATR crystal; therefore, the contact between a sample and a crystal must be very smooth. The d_p of the beam depends on a number of factors, including the wavelength of the light within an ATR crystal, λ; the refractive index of the crystal, n_c; and the angle of incident radiation, θ_i, and it is given by:

$$d_p = \frac{\lambda}{2\pi\sqrt{\sin^2\theta_i - \left(\frac{n_s}{n_c}\right)^2}} \tag{5.10}$$

where n_s is the refractive index of the sample. Since the depth d_p is at most a few micrometers in the infrared region, an ATR spectrum provides information regarding a surface and the region close to it [3, 20].

Typical ATR crystal materials are reported in Table 5.2. Such materials have low solubility in water and a very high refractive index.

ATR and transmission approaches are often complementary. It has been reported that ATR and transmission modes ATR-FTIR are suitable for the study of processes at surfaces, like modification, degradation, enzymatic degradation of a substrate film attached to a solid surface, for research applications in food and agricultural fields, for the detection of microbial metabolic products on

Table 5.2 Common ATR IRE materials.

Material	Wavenumber range (cm^{-1})	Refractive index[a]	Water solubility (g/100g, 25 °C)
Germanium (Ge)	5,500–675	4.00 (10 μm)	Insoluble
Silicon (Si)	8,900–1,500	2.41 (0.6 μm)	Insoluble
Zinc selenide (ZnSe)	15,000–650	2.40 (10 μm)	Insoluble
Diamond	30,000–200	2.41 (0.6 μm)	Insoluble

[a] The refractive index was measured at the wavelength written in parentheses.

carbonate mineral surfaces, for studying self-assembled thin films, grafted polymer layers, and adsorption processes of biological and synthetic materials (see Section 5.4) [47–55].

5.3.3 FTIR Microspectroscopy

FTIR instruments with sensitive liquid-nitrogen-cooled MCT detector have permitted the development of the *IR microspectroscopy*, one of the most powerful techniques for chemical characterization of materials at spatial resolutions down to 5–10 μm.

In today's world, where nanomaterials are gaining considerable relevance, there is an ever-increasing need to chemically characterize nanoparticles and domains.

The main limitations of conventional IR microspectroscopy consisted in performance limits due to low brilliance of thermal IR sources and spatial resolution limit set by optical diffraction. The optical diffraction limit along with other practical limitations has limited spatial resolution to the range of 3–30 μm, depending on the wavelength of light and instrumentation used. In fact, the signal-to-noise-ratio drops rapidly when the sample thickness gets much smaller than penetration depth, which is proportional to wavelength.

To deal with those problems, advanced methods and instrumentations were developed. FTIR microscopes coupled to high-brilliance synchrotron IR beamlines can improve the limited performance. *Synchrotron radiation-based Fourier transform infrared* (SR-FTIR) spectromicroscopy is a useful bioanalytical and imaging tool. The technique gives chemical information, with high signal-to-noise ratio at spatial resolutions as fine as 3–10 μm. Thus, it allows researchers to identify, map, and track specific chemical reactions and interactions even in individual living cell or to study ancient paintings and degradation and corrosion of microlayered materials [56].

5.3.4 AFM-IR Spectroscopy

The recent coupling of *atomic force microscopy* (AFM) with pulsed tunable IR laser sources allows to measure and map chemical content in nanoscaled polymer domains, or to characterize films and thin layers in laminated polymer multilayer films, or to perform quality control in biomedical materials, tracking the degradation, thanks to at spatial resolutions down to 100 × 100 nm. In AFM-IR instrument, a tunable infrared laser irradiates a region of a sample in the proximity of a probe tip from an atomic force microscope. If the IR laser is set to a wavelength that corresponds to an absorbing wavelength of the sample, the absorbed light induces a photothermal response in the sample that is then detected with the tip of an atomic force microscope. Thus, the AFM tip itself acts as the IR detector. The most common technique to detect the IR absorption is by a direct measurement of the thermal expansion resulting from optical absorption. Because the thermal expansion signal detected in AFM-IR spectroscopy is directly proportional to the absorption coefficient of the sample, it is possible to readily create an IR absorption spectrum of nanoscale regions of a sample that correlates very well to conventional FTIR absorption spectra [57, 58].

5.3.5 Hyphenated Techniques

IR spectroscopy has also been coupled to other well-established analytical techniques such as *thermal analysis* to provide information about the temperature-dependent physical properties of materials, and *chromatographic techniques* to identify if an unknown sample is either a pure substance or a mixture [59]. For example, coupling a thermal method such as thermogravimetric analysis (TGA) and FTIR spectroscopy (TGA–FTIR) can provide a quantitative assessment of the decomposition process through the TGA curve, plotted as mass loss of sample as a function of temperature, and an identification of the decomposition products from the infrared signatures of the evolved gases [60].

IR spectroscopy may also be coupled with each of a number of possible chromatographic techniques, with gas chromatography–infrared spectroscopy (GC–FTIR) being the most widely used [61].

5.4 Applications and Case Studies

For the advantage of the FTIR technique over other spectroscopic methods discussed in Section 5.1, and for the advantages of the computerized data processing and data storage and the usage of spectral databases, the technique finds applications in many areas of research and industry:

- Chemical and kinetics (identification of an organic compound, structure determination, conformational analysis, study of chemical reaction and kinetics).
- Surfaces (polymers, films, semiconductors, hybrid systems).
- Medical and life science (pharmaceutical, medical, biological, biotechnological).
- Cultural heritage and forensic.
- Environmental and geological.
- Food industry (identification of inorganic and organic materials, purity and control of materials).

Additionally, nowadays portable and handheld FTIR spectrometers make easier the application and use for field analysis.

5.4.1 Chemical Characterization and Kinetics

FTIR spectroscopy equipment is used in industry and academic laboratories to better understand reaction kinetics, mechanism, and pathways as well as catalytic cycles, due to their ability to probe the surface mechanisms at the molecular level. FTIR spectroscopy is used to help scale-up chemical reactions, optimize reaction yield, and minimize impurities and it is a valid tool to ensure that processes are stable, in control, and meet end-product specifications and impurity profiles [62].

Scientists use the fingerprint region of the mid-IR region for measuring reaction trends and profiles in real time, obtaining information about mechanism, kinetics, pathways, and the influence of reaction variables. The Lambert–Beer law gives the connection between the measured absorbance of the reaction species and its concentration.

Monitoring of the reaction process could be useful in organic synthesis, enzymatic reactions, and electrochemistry, for studying keto-enol tautomerism, hydrogenation reactions, polymerization reactions, hydroformylation or oxidations, flow chemistry, Grignard reaction mechanisms, halogenation, catalyzed reactions, biocatalysis, or enzymatic catalysis.

Papers available in the specialist literature report application of IR spectroscopy for the analysis of structural changes [63] as well as for monitoring production (bioproduction) processes [64–66].

A variety of material modifications related to the use of chemical (e.g. enzymatic oxidation, ozone oxidation, surface graft polymerization, site-specific reactions), physical (microwave or UV radiation plasma, electron beam irradiation), or physicochemical treatment were studied using IR spectroscopy [67–73].

Recently, for instance, a kinetics study of two epoxy systems coupling FTIR and differential scanning calorimetry (DSC) [74], or using FTIR microspectroscopy that allowed a full chemical analysis of the surfaces, has been reported.

It is possible also to monitor photoaging of poly(vinyl chloride) (PVC) and poly(methyl methacrylate) (PMMA) and to perform quality control analysis [75].

Micro-FTIR has been used also for characterizing the degradation profiles produced by oxidation in the study "Aerobic oxidation of alcohols with air catalyzed by decacarbonyldimanganese" [76].

In situ real-time IR spectroscopy allows to study reaction processes during the synthesis of a compound. For example, a study about the synthesis of a tetrasubstituted acyclic olefin via ketone enolization and tosylation has been reported [77]. ReactIR, an inline Fourier transform IR immersion probe, was used to monitor ketone enolate formation and tosylation changing the reaction temperature, stoichiometry, and nature of the base for evaluating the impact on the reaction conversion, rate, and ratio of stereoisomers.

5.4.2 Surfaces

The composition of complex materials may be also readily determined by measuring their infrared spectra using a FTIR spectrometer and then comparing the results with a commercially available or specifically prepared spectral database.

In attempt to achieve superior performance regarding toughness, specific electrical or thermal properties, strength, weight, and resistance to degradation, engineering polymers are often a mixture of polymeric components (additives, fillers, compatibilizers, etc.), often with material heterogeneity at the micro- and nanoscales.

For example, today's plastics are complex multicomponent systems made from different polymers, additives, and fillers. Mixing of polymers is an important process in the polymer industry; by combining the strength of different polymers through blending, new products with desirable physical properties can be produced, therefore an essential part of the quality control in the industries is to define the correct composition.

Polymer materials and multilayer laminated films are utilized for many different applications including food packaging and consumer products, in stretch films, electronics, and displays, or in medical devices and in aerospace technologies [78, 79].

Depending on the nature of the blending, different polymer blends exhibit different types of interactions and configurations that affect their mechanical and physical properties. In fact, the physical properties of polymer blends are conditioned by the structure of the molecular chains. FTIR has been used for studying the compatibility of polymer blends via intermolecular hydrogen bonding, and also for investigating their degradation processes [80].

Concerning polymeric materials, three problems will be considered: the identification of a heterogeneous mixture of polymeric materials, the identification of various layers in a multilayer or laminated polymer structure, and the identification of the cause of imperfections in the surface of a polymer.

The process of polymer composite formation and the stability of the composite is a function of temperature, presence of other species, etc. High-spatial resolution techniques such as micro-FTIR allow to analyze the structure of polymer laminates and to find defects, or to map spatial distribution of functional groups [80].

ATR-FTIR spectroscopic imaging was reported as a powerful tool in the study of heterogeneous polymer oxidation and copolymer hydration [81, 82], in the study of depth profile of SiO_2 ultrathin films [83, 84].

In addition, interest in thin films has dramatically grown because of applications in catalysis, microelectronics, sensors, optics, bio- and nanotechnology.

For example, ferroelectric polymers are considered one of the most important electronic polymer groups that have many applications such as transducers, ultrasonic sensors, memory, and storage.

FePt nanoparticles coated with 2 nm thick films of surfactant have been studied as candidates for magnetic recording media [85]. The self-assembly of these nanoparticles is influenced by the properties of their surfaces, which are coated with a layer of mixed oleic acid and oleyl amine. These surfactant-coated FePt nanoparticles were characterized using FTIR spectroscopy. The observation of both ν(COO) and ν(C=O) vibrational modes indicates that oleic acid bonds to the FePt nanoparticles in both monodentate and bidentate forms. The oleylamine bonds to the FePt nanoparticles through electron donation from the nitrogen atom of the NH_2 group [85].

It is very important to characterize the depth change in chemical bonding structures of ultra-thin films.

The AFM-IR imaging and spectroscopy was used to characterize a variety of conducting polymeric materials, for example PEDOT polymer nanostructures [86, 87], as well as to study degradable polymer materials, including observing local variations in crystallinity on the nanoscale associated with annealing [88].

AFM-IR spectroscopy has recently been used also to study the molecular conformation and molecular orientation of electrospun nanofibers [89]. Electrospun fibers are promising systems for drug delivery, tissue engineering, wound dressing and cosmetics, functional materials and devices such as composite reinforcement, filters, protective clothing, and smart textiles, electronics, sensors, and catalysts [90].

By AFM-IR measurements, it is possible to obtain information about molecular orientation of poly(vinylidene fluoride) PVDF fibers under variable polarization [91].

IR spectroscopy was also combined with size exclusion chromatography (SEC) for studying the deformulation of samples of unknown origin, compositional distribution in copolymers, impurity profiling in polymers, and branching in polyolefins [92].

5.4.3 Medical and Life Science (Pharmaceutical, Medical, Biological, Biotechnological)

The number of applications in different domains of science, such as chemistry, biology, and medicine, underline the usefulness of infrared analysis.

Papers have been published about drug characterization, subcellular imaging and spectroscopy, including measurements of bacterial and mammalian cells, metabolites harvested in plants, tissue imaging, protein aggregation/secondary structure, including links between amyloid aggregation and neurodegenerative diseases. Some examples are reported as follows.

FTIR spectroscopy is an excellent technique for pharmaceutical analysis, which offers many advantages since it is easy to use, sensitive, selective, green, and fast (the total analysis time including making the pellets, measurement, identification, and report generation is lower than 10 minutes). Applications include: basic drug research and structural elucidation, formulation development and validation, quality control processes for incoming and outgoing materials, and packaging testing.

FTIR was originally a spectroscopic technique used to identify the functional groups of the chemical constituents, but it has been widely used for the identification of polymorphs, hydrates, solvates, and salts; for the quantitative analysis of material compositions; for studying phase transformation at controlled temperature and humidity; and for performing process analysis for crystallization, blending, granulation, coating, and lyophilization.

In many of these instances, the ability of spectroscopy to differentiate between forms requires differences in chemical structure (e.g. proton transfer) or in crystallographic properties (e.g. polymorphism).

The differentiation of polymorphs is aptly illustrated for atorvastatin calcium, a synthetic lipid-lowering agent acting as inhibitor of 3-hydroxy-3-methylglutaryl-coenzyme A (HMG-CoA) reductase. This enzyme catalyzes the conversion of HMG-CoA to mevalonate, an early and rate-limiting step in cholesterol biosynthesis. Atorvastatin is known to exist in a multitude of polymorphic forms [93].

FTIR spectroscopy can be used to differentiate between drug substances and their potential degradants, as well as to understand some of the more interesting details of crystal packing.

For example, the decomposition of acetylsalicylic acid leads to the formation of salicylic acid, an undesirable reaction. The carboxylic acid group present in acetylsalicylic acid is observed at 1749 cm^{-1}, while the same vibrational mode is shifted to 1653 cm^{-1} for salicylic acid. This shift is due to the fact that in its crystalline state, salicylic acid exists in hydrogen-bonded dimers, but this dimerization is not possible when its hydroxyl group is acetylated. With the use of appropriate standard mixtures, a quantitative method for the amount of salicylic acid in bulk samples of acetylsalicylic acid can readily be developed [93].

The technique is useful in the study of free acids or free bases, especially in the characterization of salt products derived from them: the functional group present in either the free acid or free base will have a characteristic absorption band that disappears upon formation of a salt, and which is replaced by a shifted band corresponding to the transfer of a proton in the salt.

When coupled with FTIR technology, the non-perturbing sampling mode of ATR is probably the most suitable technique for the establishment of the phase identity of a potentially polymorphic substance.

Polymorphs have identical chemical structure but are molecularly packed differently in the crystal lattice, due to possible different hydrogen bond networks and molecular conformation in the lattice. If this occurs, a chemical moiety of one polymorph will be in a different environment than the other forms, which should give rise to distinctly different vibrational frequencies.

The distinctive FTIR spectra make it possible to distinguish the two polymorphs and to determine their polymorphic purities.

Spychala et al. described a procedure to determine the content of individual polymorphs of indomethacin in their mixtures by FTIR spectroscopy based on the characteristic bands of the two forms in the range of 1600–1800 cm^{-1} [94].

AFM-IR spectroscopy has been used to study various drug formulations, especially drug polymer blends. In pharmaceutical sciences, composition, miscibility, dispersion, phase segregation, and crystallinity can influence the stability, efficacy, and bioavailability of a drug formulation. It is possible to apply ATR-FTIR method for studying the distribution of different components in compacted pharmaceutical tablets [95–104].

Calorimetric studies and FTIR spectroscopy were coupled to study the changes in crystalline content of polymer during thermal treatments. FTIR spectra of two major phases (a and b phases) of PVDF have been investigated. Vibrational bands at 530 cm^{-1} (CF_2 bending), 615 cm^{-1} and 765 cm^{-1} (CF_2 bending and skeletal bending), and 795 cm^{-1} (CH_2 rocking) refer to a-phase. Vibrational bands at 510 cm^{-1} (CF_2 bending) and 840 cm^{-1} (CH_2 rocking) correspond to b-phase [105].

It is possible to use FTIR spectroscopy for high-throughput analysis of pharmaceutical formulations, using a drop-on-demand device by preparing arrays of microdrop samples directly on the surface of the ATR crystal.

The development of the macro-ATR FTIR imaging approach allowed to study simultaneously copious samples deposited on the surface of the ATR crystal under a controlled environment.

It is possible to obtain "chemical snapshots" from a spatially defined array of many different polymer/drug formulations under identical conditions [100].

FTIR is also becoming a suitable technique for analysis of herbal medicine.

For instance, Ginseng quality assurance could be studied by FTIR spectroscopy, or phenolic compounds herbal extracts could be compared, and the discrimination of herbal medicines could be performed by molecular spectroscopy [106–110].

FTIR associated with second derivative infrared spectroscopy and 2D-IR was used to study the main constituents in traditional Chinese medicine Angelica and its different extracts [111].

FTIR imaging has not been a sufficient technique for imaging living cells because of sample heterogeneity and scattering, low spatial resolution, lack of depth resolving power, and strong water absorption; but, in medical and life science field, AFM-IR spectroscopy has been used on various tissue samples, including bone, skin, and hair, revealing the distribution of lipids, proteins, minerals, and chemical additives.

For instance, the AFM-IR technique has been used to map the relative distribution of biopolymer (via amide I band associated with protein absorption at 1660 cm^{-1}) versus mineral (via the phosphate band at 1044 cm^{-1}) in the complex structure of bone [112].

Vibrational spectroscopic techniques have become potential tools for noninvasive optical tissue diagnosis and have been applied to study a wide variety of pathologic states [113, 114].

Studies have reported the use of ATR-FTIR spectroscopy for the acquisition of spectral signatures from saliva, blood, and urine for disease diagnostics [115].

AFM-IR spectroscopy can achieve spatial resolutions better than 50 nm, leading to much finer IR mapping of organelles and more highly resolved spectra. The AFM-IR analysis of MCF-7 breast cancer cell has been reported. The IR map of a fixed MCF-7 cancer cell for amide I band (proteins) at 1650 cm^{-1} and phosphate band (DNA, RNA) at 1080 cm^{-1} (symmetrical $P{=}O_2$ stretching vibration), allowed to evaluate the distribution and concentration of biomolecules [116–118].

Thanks to the integration of synchrotron IR sources with IR microspectroscopy and due to improvements in IR detectors, it is possible to use IR spectroscopy in biomedical applications [119]: for studying the cell type, the physiological or disease status, and even environment reactions of the cell or a specific organelle [120].

The analysis of living cells is important to understand their behavior under different conditions and their response when exposed to varying environmental factors. Transmission mode measurements of live cells has been achieved by the use of a powerful synchrotron IR source, thereby minimizing the problematic effect of intense water absorption in the aqueous environment [121–123].

In biological research, FTIR spectroscopy can be used to investigate proteins in hydrophobic membrane environments. Time-resolved (tr)-FTIR spectroscopy allows to monitor reactions of the amino acids, ligands, and specific water molecules in the active center of a protein.

Many correlations have been established between the characteristic infrared bands of peptides and their conformations b-sheets, a-helices, and random coil structures that can exist in one peptide. When the conformation changes during the peptide synthesis, IR absorptions can shift to one direction or the other [124, 125].

Immobilization of bioreceptors on materials can be studied by various IR surface techniques for the elaboration of biosensors and recognition of target molecules [126].

A large variety of high-affinity association processes between two biological partners can be efficiently transduced: the IR probes operate as molecular sensors because often the signal is affected by the association. FTIR ATR-based sensor was set up and utilized by Reiter et al. to monitor the specific recognition of a monoclonal antibody by human tumor necrosis factor-α (TNFα) [127].

IR spectroscopy is suitable not only to study protein conformation and dynamics but also to characterize the structural properties of intrinsically disordered proteins (IDPs) and their induced folding and aggregation through the analysis of the protein absorption spectrum [128].

Recently, ATR FT-IR imaging was used also for studying the fabrication of a chitosan/poly (η-caprolactone) (PCL) hydrogel for tissue engineering purposes and biotechnology applications [129] and the AFM-IR technique was used for monitoring the nanoscale production of biofuel precursors via microorganisms. AFM-IR spectroscopy is specifically able to look inside cells and measure and map the distribution of triglyceride vesicles with nanoscale spatial resolution [130].

5.4.4 Cultural Heritage and Forensic

Analysis of cultural heritage artifacts represents a very important first step in any restoration/conservation attempt. Today, the most widely used method in most museum laboratories is infrared spectroscopy using reference standards and comparing unknowns to these referenced spectra. One of the first freely available web-based databases of only FTIR spectra was published by Castro et al. in 2003 specifically to aid art conservators and professionals [131].

In the analysis of complex materials, such as paints, several analytical methods such as X-ray fluorescence (XRF), scanning electron microscopy (SEM), optical microscopy, and X-ray diffraction (XRD) are usually needed to produce a complete characterization of a sample's components.

When the surface of an artifact, such as a painting, has a complex structure with multiple layers, a cross section sample is taken that incorporates this stratigraphy to facilitate the study of the sample. FTIR spectroscopy, in particular, micro-FTIR provides a distribution map of identified compounds and it permits to study degradation forms of cultural heritage.

Several papers describe the application of FTIR and FTIR-ATR spectroscopy for the analysis of historical metals, paper, ceramic, and fabrics [132–137].

Recently the occasion of the restoration of precious icon "Virgin Hodegetria" was therefore more than a simple opportunity to recover an artifact of unquestionable cultural value, as it represented the possibility of laying the foundations for a study aimed at collecting information on southern icons dating from the end of the XII and the first half of the XIII century and united by a clear Byzantine matrix [138].

Frequently, the aim of scientific researches in cultural heritage is not only to carry out a preliminary study, essential for the restoration, by the characterization of the constitutive materials and the executive techniques, the state of conservation, and the previous treatments but also to answer to some technical inconsistencies emerging at first sight of an artifact.

The problem of attribution, authentication, and/or counterfeit detection of artifacts of artistic, archaeological, or criminalistic interest is of great importance also in forensic analysis.

For instance, multispectral and noninvasive investigation techniques were performed as preliminary recognition and study of the portrait "Ritratto muliebre," attributed to unknown artist and probably dated to the second half of XVI century but inserted in a gilded not original frame [139].

Many other forensic applications have been published about the identification of fibers, hairs, and automobile paints [140].

International drug enforcement agencies, police departments, and customs laboratories rely on spectroscopy to quickly identify illegal drugs, crime scene evidence, banned materials, and counterfeit goods.

FTIR, FTIR-Raman, GC-IR, and IR microscopy techniques can provide fast, easy, and consistent analysis for clandestine labs (chemical evaluation), seized drugs (controlled substances and cutting agents), and textile identification (fibers, coatings, and residues); also, FTIR spectroscopy is a non-destructive analysis of the chemical composition of unknown biological stains.

FTIR spectroscopy allows to distinguish between body fluids and various substances that could be mistaken as body fluid. Blood is one of the most valuable and frequently found biological fluids on a crime scene [141].

5.4.5 Environmental and Geological

In the environmental and geological fields, FTIR spectroscopy can be used to analyze soil samples and to monitor air and water quality and it can address environmental and health concerns that have been caused by the increasing pollution levels. The analysis can be performed to determine not only the nature of pollutants, but also the chemical interactions in case of pollutants removal by sorption processes.

Applications of FTIR spectroscopy to agricultural soil analysis were presented and discussed by Raphael [142]. Soil analysis could have also implications in forensic cases: soil can help link suspects to a crime scene or provide geological information that could narrow down the geographical location where the crime was committed.

The minerals present in soil are predominantly silicates or carbonates but can also include other types of minerals such as sulfates. Clay minerals (aluminosilicates) play an important part in the properties of a soil and have OH groups present with free OH groups (no OH bonding) and give very sharp well-defined absorption bands in the OH stretching region, often at higher frequencies than the organic OH (approximately 3700–3500 cm^{-1}). These bands are highly diagnostic [143].

In geosciences research, FTIR spectroscopy is applied extensively in the analysis of the trace amount of water content in nominally anhydrous minerals (NAMs), in estimating the explosion potential in volcanic setting, in measuring volatile inclusions in glass and minerals, in linking biological affinities of both microfossils and macrofossils, or analyzing chemotaxonomy of early life on earth [144].

FTIR microscopy applied to soil micromorphology is particularly effective for the study of syn- and postdepositional diagenetic processes. Moreover, FTIR microscopy provides molecular information suitable for characterizing the state of preservation of bone, ivory, tooth enamel, and charcoal [145]. Furthermore, FTIR spectroscopy is suitable for estimating temperature transformations of clay minerals, carbonates, phosphates, and sulfates [146].

FTIR spectrometers can detect over a hundred volatile organic compounds (VOCs) emitted from industrial and biogenic sources. Gas concentrations in stratosphere and troposphere were determined using FTIR spectrometers. The basic principle and methods of FTIR spectroscopy of the atmosphere were presented by Bacsik and coworkers in 2004 [147].

FTIR spectroscopy can be used not only to study biomaterials used in depolluting processes, but also to characterize materials obtained after chemical modification of those biomaterials. Studies on heavy metals and organic compounds removal from wastewaters using different natural and synthetic materials are reported. The adsorption of inorganic ions on metal oxides and hydroxides was resolved using in situ ATR-FTIR spectroscopy [148]. In a review, Lefèvre describes and discusses in situ ATR-FTIR spectroscopy used in order to obtain information on the sorption mechanism of sulfate, carbonate, phosphate, perchlorate on hematite, goethite, alumina, silica, and titanium oxide [149].

Volcanoes are considered important natural sources of air pollution. Mori and coworkers used an FTIR spectrometer during a stage of dome lava extrusion of the Unzen volcano [150]. The basic principle of FTIR spectroscopy used in air pollutants detection and measuring is that every gas has its own "fingerprint" or absorption spectrum. The entire infrared spectrum can be monitored and FTIR sensor can read the different fingerprints of the gases present in the air sample [150].

The use of FTIR spectroscopy is also a valuable tool in monitoring emissions from aircraft in airports [151].

5.4.6 Food Industry

FTIR spectroscopy represents a rapid, less destructive, and high-throughput method for the analysis of food products. It is possible to perform the characterization of minor food components down to parts per billion (ppb) levels especially in conjunction with attenuated total reflectance (ATR) technology, dramatically simplifying sample handling. Analysis methods concern foods research, quality assurance and control, environmental and water quality. The determination of product quality and authenticity and the detection of adulteration are major issues in the food industry.

The fingerprint region, located within the MIR region between 1200 and 700 cm^{-1}, contains bands from polysaccharides, proteins, lipids, and carotenoids [152].

The food industry uses FTIR spectroscopy to make sure that they are complying to the required government standards about physical, chemical, and rheological properties of food. Food applications in industry and research include determination of lard content in cakes, chocolates and mixture of fats, adulterations in jam, olive oil, meats and beverages, aflatoxin detection in cereals, coffee quality and adulteration [153, 154].

For instance, FTIR spectra reflect different substitution patterns of triglycerides, differences in chain length of acyl moieties, and differing degrees of unsaturation of fats, or the band shift observed at 3009 cm^{-1} assigned to C–H stretching vibration of cis-double bond allow determination of extra virgin oil adulteration [155].

Franca and Olivera gave a critical overview of the application of FTIR spectroscopy in food processing and engineering [156].

FTIR spectroscopy has gained growing interest and has undergone remarkable innovations that allow many new applications in research and industry: highly sophisticated spectrometers have been developed, for instance, enabling a high sample throughput and FTIR spectroscopy, in conjunction with multidimensional statistical analysis (Chemometrics), has been applied for quantitative determinations.

The intention of this chapter has been to give an overview of current knowledge and strategies of using FTIR spectroscopy in many areas of research and industry, showing its significant potential, limits, and perspectives.

References

1 Grim, W.M., Fateley, W.G., and Grasselli, J.G. (1984). Introduction to dispersive and interferometric infrared spectroscopy. In: *Fourier Transform Infrared Spectroscopy* (ed. T. Theophanides), 25–42. Dordrecht: Springer https://doi.org/10.1007/978-94-009-6418-1_3.

2 Banwell, C.N. and McCash, E.M. (1994). Chapter 3. Infrared spectroscopy. In: Banwell, C.N. and McCash, E.M., editors. *Fundamentals of Molecular Spectroscopy*, 4e. London: McGraw-Hill.

3 Stuart, B.H. (2004). Chapter 1. Introduction. In: Stuart B.H., editor. *Infrared Spectroscopy: Fundamentals and Applications*. Chichester, UK: Wiley.

4 Christy, A.A., Ozaki, Y., and Gregoriou, V.G. (2001). Chapter 3. Theory of infrared spectroscopy. In: Christy, A.A., Ozaki, Y., and Gregoriou, V.G., editors. *Comprehensive Analytical Chemistry. Vol. 35. Modern Fourier: Transform Infrared Spectroscopy*. Elsevier Science.

5 EB Bradley. Infrared and Raman spectroscopy. In: Bauer HH, Christian GD, O'Reilly JE. *Instrumental Analysis*. Allyn and Bacon, Inc; 1978.

6 Robinson, J.W., Frame, E.S., and Frame, I.I.G.M. (2014). Chapter 4. Infrared spectroscopy. In: Robinson, J.W., Frame, E.S., and Frame, I.I.G.M., editors. *Undergraduate Instrumental Analysis*, 7e. Boca Raton, FL: CRC Press.

7 Colthup, N.B., Daly, L.H., and Wiberley, S.E. (1990). *Introduction to Infrared and Raman Spectroscopy*, 3e. New York and London: Academic Press.

8 Seshadri, K.S. and Jones, R.N. (1963). The shapes and intensities of infrared absorption bands – a review. *Spectrochim. Acta* 19 (6): 1013–1085. https://doi.org/10.1016/0371-1951(63)80187-3.

9 Bellamy, L.J. (1980). *The Infrared Spectra and Complexes Molecules: Vol. 2. Advances in Infrared Group Frequencies*, 2e. The Netherlands: Springer.

10 Colthup, N.B. (1950). Spectra-structure correlations in the infrared region. *J. Opt. Soc. Am.* 40 (6): 397–400.

11 Silverstein, R.M., Webster, F.X., Kiemle, D.J., and Bryce, D.L. (2014). Chapter 2. Infrared spectrometry. In: Silverstein, R.M., Webster, F.X., Kiemle, D.J., and Bryce, D.L., editors. *Spectrometric Identification of Organic Compounds*, 8e. Wiley.

12 Pavia, D.L., Lampman, G.M., Kriz, G.S., and Vyvyan, J.R. (2009). *Introduction to Spectroscopy*, 4e. New York: Harcourt College Publishers.

13 Sadtler FTIR Databases Infrared Spectral Libraries. http://www.ir-spectra.com/sadtler/sadtler.htm (accesssed September 2020).

14 Perkin-Elmer Library. https://www.perkinelmer.com/it/product/spectrum-search-plus-l1108755 (accesssed September 2020).

15 Sigma-Aldrich Library. https://www.sigmaaldrich.com/labware/learning-center/spectral-viewer/ft-ir-library.html (accesssed September 2020).

16 National Institute of Advanced Industrial Science and Technology (AIST). Spectral database for organic compounds provided by the National Institute of Advanced Industrial Science and Technology (AIST) from Japan. https://sdbs.db.aist.go.jp/sdbs/cgi-bin/direct_frame_top.cgi (accesssed September 2020).

17 Weyer, L.G. and Lo, S.C. (2002). Spectra–structure correlations in the near-infrared. In: *Handbook of Vibrational Spectroscopy*, vol. 3 (eds. J.M. Chalmers and P.R. Griffiths), 1817–1837. Chichester, UK: Wiley.

18 Griffiths, P.R. (2002). Far-infrared spectroscopy. In: *Handbook of Vibrational Spectroscopy*, vol. 1 (eds. J.M. Chalmers and P.R. Griffiths), 229–239. Chichester, UK: Wiley.

19 Smith, B.C. (2011). *Fundamentals of Fourier Transform Infrared Spectroscopy*, 2e. Boca Raton, FL: CRC Press Chapter 2.

20 Griffiths, P.R. and de Haseth, J.A. (2007). Chapter 2. Theoretical background. In: Griffiths, P.R. and de Haseth, J.A., editors. *Fourier Transform Infrared*, 2e. Wiley.

21 Kauppinen, J. and Partanen, J. (2001). *Fourier Transforms in Spectroscopy*. New York: Wiley.

22 Ramsey, W.Y. and Alishouse, J.C. (1968). A comparison of infrared sources. *Infrared Phys.* 8 (2): 143–152.

23 Jung, D., Bank, S., Lee, M.L., and Wasserman, D. (2017). Next-generation mid-infrared sources. *J. Opt.* 19 (12): 123001. https://iopscience.iop.org/article/10.1088/2040-8986/aa939b.

24 Rogalski, A. (2012). History of infrared detectors. *Opto-Electron. Rev.* 20: 279–308. https://doi.org/10.2478/s11772-012-0037-7.

25 Rogalski, A. (2010). *Infrared Detectors*, 2e. Boca Raton, FL: CRC Press.

26 Graf, A., Arndt, M., Suaer, M., and Gerlach, G. (2007). Review of micromachined thermopiles for infrared detection. *Meas. Sci. Technol.* 18 (7): R59–R75.

27 Ristic, L. (1994). *Sensor Technology and Devices*. Boston, London: Artech House.

28 Baltes, H., Paul, O., and Brand, O. (1998). Micromachined thermally based CMOS microsensors. *Proc. IEEE* 86 (8): 1660–1678.

29 Chasmar, R.P., Mitchell, W.H., and Rennie, A. (1956). Theory and performance of metal bolometers. *J. Opt. Soc. Am.* 46 (7): 469–477. https://doi.org/10.1364/JOSA.46.000469.

30 Novák, R. (1957). The theory of metallic bolometers. *Czech. J. Phys.* 7: 284–291. https://doi.org/10.1007/BF01688034.

31 Wormser, E.M. (1953). Properties of thermistor infrared detectors. *J. Opt. Soc. Am.* 43 (1): 15–21. https://doi.org/10.1364/JOSA.43.000015.

32 De Waars, R. and Wormser, E.M. (1959). Description and properties of various thermal detectors. *Proc. IRE.* 47 (9): 1508–1513.

33 Richards, P.L. (1994). Bolometers for infrared and millimeter waves. *J. Appl. Phys.* 76 (1): 1–24. https://doi.org/10.1063/1.357128.

34 Rogalski, A. (2013). Semiconductor detectors and focal plane arrays for far-infrared imaging. *Opto-Electron. Rev.* 21 (4): 406–426. https://doi.org/10.2478/s11772-013-0110-x.

35 Golay, M.J.E. (1949). The theoretical and practical sensitivity of the pneumatic infra-red detector. *Rev. Sci. Instrum.* 20 (11): 816–820. https://doi.org/10.1063/1.1741396.

36 Whatmore, R.W. (2004). Pyroelectric arrays: ceramics and thin films. *J. Electroceram.* 13 (1–3): 139–147.

37 Muralt, P. (2001). Micromachined infrared detectors based on pyroelectric thin films. *Rep. Prog. Phys.* 64 (10): 1339–1388.

38 Neumann, N. (1993). Modified triglycine sulphate for pyroelectric infrared detectors. *Ferroelectrics* 142 (1): 83–92. https://doi.org/10.1080/00150199308237885.

39 Norkus, V., Gerlach, G., and Hofmann, G. (1999). Process technologies for high-resolution infrared detectors based on $LiTaO_3$. *Proc. SPIE.* 3892: 233–240. https://doi.org/10.1117/12.364489.

40 Jiménez, R. and Jiménez, B. (2011). Pyroelectricity in polycrystalline ferroelectrics. In: *Multifunctional Polycrystalline Ferroelectric Materials* (eds. L. Pardo and J. Ricote), 573–616. Dordrecht: Springer https://doi.org/10.1007/978-90-481-2875-4_12.

41 Lang, S.B. and Muensit, S. (2006). Review of some lesser-known applications of piezoelectric and pyroelectric polymers. *Appl. Phys. A.* 85: 125–134.

42 Hammes, P.C.A. and Regtien, P.P.L. (1992). An integrated infrared sensor using the pyroelectric polymer PVDF. *Sens. Actuators A Phys.* 32 (1–3): 396–402. https://doi.org/10.1016/0924-4247(92)80019-Y.

43 Kinch, M.A., Borrello, S.R., and Simmons, A. (1977). 0.1 eV HgCdTe photoconductive detector performance. *Infrared Phys.* 17: 127–135.

44 Gertner, E.R., Tennant, W.E., Blackwell, J.D., and Rode, J.P. (1985). HgCdTe on sapphire. A new approach to infrared detector arrays. *J. Cryst. Growth* 72 (1–2): 462–467.

45 Vydyanath, H.R., Nathan, V., Becker, L.S., and Chambers, G. (1999). Materials and process issues in the fabrication of high-performance HgCdTe infrared detectors. *Proc. SPIE.* 3629: 81–87.

46 Fujiwara, M., Hirao, T., Kawada, M. et al. (2003). Development of a gallium-doped germanium far-infrared photoconductor direct hybrid two-dimensional array. *Appl. Optics* 42 (12): 2166–2173. https://doi.org/10.1364/ao.42.002166.

47 Freger, V. and Ben-David, A. (2005). Use of attenuated total reflection infrared spectroscopy for analysis of partitioning of solutes between thin films and solution. *Anal. Chem.* 77 (18): 6019–6025. https://www.researchgate.net/publication/7604036. https://doi.org/10.1021/ac50689w.

48 Janorkar, A.V., Luo, N., and Hirt, D.E. (2004). Surface modification of an ethylene-acrylic acid copolymer film: grafting amine-terminated linear and branched architectures. *Langmuir* 20 (17): 7151–7158. https://pubmed.ncbi.nlm.nih.gov/15301499/. https://doi.org/10.1021/la049715w.

49 Bokria, J.G. and Schlick, S. (2002). Spatial effects in the photodegradation of poly(acrylonitrile-butadiene-styrene): a study by ATR-FTIR. *Polymer* 43 (11): 3239–3246.

50 Snabe, T. and Petersen, S.B. (2002). Application of infrared spectroscopy (attenuated total reflection) for monitoring enzymatic activity on substrate films. *J. Biotechnol.* 95 (2): 145–155.

51 Sethuraman, A. and Belfort, G. (2005). Protein structural perturbation and aggregation on homogeneous surfaces. *Biophys. J.* 88 (2): 1322–1333. https://www.ncbi.nlm.nih.gov/pmc/articles/PMC1305134/.

52 Gershevitz, O. and Sukenik, C.N. (2004). In situ FTIR-ATR analysis and titration of carboxylic acid-terminated SAMs. *J. Am. Chem. Soc.* 126 (2): 482–483. https://doi.org/10.1021/ja037610u.

53 Al-Hosneya, H.A. and Grassian, V.H. (2005). Water, sulfur dioxide and nitric acid adsorption on calcium carbonate: a transmission and ATR-FTIR study. *Phys. Chem. Chem. Phys.* 7 (6): 1266–1276.

54 Mangoni, M.L., Papo, N., Mignogna, G. et al. (2003). Ranacyclins, a new family of short cyclic antimicrobial peptides: biological function, mode of action, and parameters involved in target specificity. *Biochemistry* 42 (47): 14023–14035. http://www.ncbi.nlm.nih.gov/pubmed/23988198. https://doi.org/10.1021/bi034521l.

55 Freger, V., Gilron, J., and Belfer, S. (2002). TFC polyamide membranes modified by grafting of hydrophilic polymers: an FT-IR/AFM/TEM study. *J. Membr. Sci.* 209 (1): 283–292. https://www.sciencedirect.com/science/article/pii/S0376738802003563. https://doi.org/10.1016/S0376-7388(02)00356-3.

56 Nasse, M.J., Walsh, M.J., Mattson, E.C. et al. (2011). High-resolution Fourier-transform infrared chemical imaging with multiple synchrotron beams. *Nat. Methods* 8 (5): 413–416.

57 Dazzi, A. and Prater, C.B. (2017). AFM-IR: technology and applications in nanoscale infrared spectroscopy and chemical imaging. *Chem. Rev.* 117 (7): 5146–5173.

58 Dazzi, A., Prater, C.B., Hu, Q. et al. (2012). AFM-IR: combining atomic force microscopy and infrared spectroscopy for nanoscale chemical characterization. *Appl. Spectrosc.* 66 (12): 1365–1384.

59 Lin, S.-Y. (2016). An overview of advanced hyphenated techniques for simultaneous analysis and characterization of polymeric materials. *Crit. Rev. Solid State Mater. Sci.* 41 (6): 482–530. https://doi.org/10.1080/10408436.2016.1186598.

60 Wilkie, C.A. (1999). TGA/FTIR: an extremely useful technique for studying polymer degradation. *Polym. Degrad. Stab.* 66 (3): 301–306. https://doi.org/10.1016/S0141-3910(99)00054-3.

61 Kalasinsky, K.S., Levine, B., Smith, M.L. et al. (1993). Detection of amphetamine and methamphetamine in urine by gas chromatography/Fourier transform infrared (GC/FTIR) spectroscopy. *J. Anal. Toxicol.* 17 (6): 359–364. https://doi.org/10.1093/jat/17.6.359.

62 Fath, V., Szmais, S., Lau, P. et al. (2019). Model-based scale-up predictions: from micro- to millireactors using inline Fourier transform infrared spectroscopy. *Org. Process Res. Dev.* 23 (9): 2020–2030. https://doi.org/10.1021/acs.oprd.9b00265.

63 Spiridon, I., Teaca, C.A., and Bodirlau, R. (2011). Structural changes evidenced by FTIR spectroscopy in cellulosic materials after pretreatment with ionic liquid and enzymatic hydrolysis. *Bioresources* 6 (1): 400–413.

64 Roychoudhury, P., Harvey, L.M., and McNeil, B. (2006). The potential of mid infrared spectroscopy (MIRS) for real time bioprocess monitoring. *Anal. Chim. Acta* 571 (2): 159–166. https://pubmed.ncbi.nlm.nih.gov/17723434/. https://doi.org/10.1016/j.aca.2006.04.086.

65 Scholz, T., Lopes, V.V., and Calado, C.R.C. (2012). High-throughput analysis of the plasmid bioproduction process in *Escherichia coli* by FTIR spectroscopy. *Biotechnol. Bioeng.* 109: 2279–2285. https://doi.org/10.1002/bit.24502.

66 Villar, A., Gorritxategi, E., Aranzabe, E. et al. (2012). Low-cost visible–near infrared sensor for on-line monitoring of fat and fatty acids content during the manufacturing process of the milk. *Food Chem.* 135: 2756–2760. https://doi.org/10.1016/j.foodchem.2012.07.074.

67 Asadinezhad, A., Novak, I., Lehocky, M. et al. (2010). Physicochemical approach to render antibacterial surfaces on plasma-treated medical-grade PVC: irgasan coating. *Plasma Processes Polym.* 7: 504–514. https://doi.org/10.1002/ppap.200900132.

68 He, C., Wang, M., Cai, X. et al. (2011). Chemically induced graft copolymerization of 2-hydroxyethyl methacrylate onto polyurethane surface for improving blood compatibility. *Appl. Surf. Sci.* 258 (2): 755–760. https://doi.org/10.1016/j.apsusc.2011.08.074.

69 Otto, D.P., Vosloo, H.C.M., Liebenberg, W., and de Villiers, M.M. (2008). Development of microporous drug-releasing films cast from artificial nanosized latexes of poly(styrene-co-methyl methacrylate) or poly(styrene-co-ethyl methacrylate). *Eur. J. Pharm. Biopharm.* 69 (3): 1121–1134. https://doi.org/10.1016/j.ejpb.2008.02.004.

70 Abenojar, J., Martınez, M.A., Encinas, N., and Velasco, F. (2013). Modification of glass surfaces adhesion properties by atmospheric pressure plasma torch. *Int. J. Adhes. Adhes.* 44: 1–8. https://doi.org/10.1016/j.ijadhadh.2013.02.002.

71 Aksoy, A.E., Hasirci, V., and Hasirci, N. (2008). Surface modification of polyurethanes with covalent immobilization of heparin. *Macromol. Symp.* 269 (1): 145–153. https://doi.org/10.1002/masy.200850918.

72 Zhou, X., Zhang, T., Jiang, X., and Gu, N. (2010). The surface modification of medical polyurethane to improve the hydrophilicity and lubricity: the effect of pretreatment. *J. Appl. Polym. Sci.* 116 (3): 1284–1290. https://doi.org/10.1002/app.31499.

73 You, D., Liang, H., Mai, W. et al. (2013). Microwave-assisted functionalization of polyurethane surface for improving blood compatibility. *J. Ind. Eng. Chem.* 19 (5): 1587–1592. https://doi.org/10.1016/j.jiec.2013.01.027.

74 Achilias, D.S., Karabela, M.M., Varkopoulou, E.A., and Sideridou, I.D. (2012). Cure kinetics study of two epoxy systems with Fourier tranform infrared spectroscopy (FTIR) and differential scanning calorimetry (DSC). *J. Macromol. Sci. A.* 49 (8): 630–638. https://www.tandfonline.com/doi/abs/10.1080/10601325.2012.696995.

75 Lacoste, J., Arnaud, R., Singh, R.P., and Lemaire, J. (1988). Modelization of the photocatalyzed oxidation of polyolefins. Oxidation of model hydrocarbons with ZnO. *Macromol. Chem. Phys.* 189 (3): 651–666. https://onlinelibrary.wiley.com/doi/abs/10.1002/macp.1988.021890317.

76 Meng, S.S., Rong, L.L., Luo, X. et al. (2019). Aerobic oxidation of alcohols with air catalyzed by decacarbonyldimanganese. *Green Chem.* 21: 6187–6193. https://doi.org/10.1039/C9GC02446H.

77 Rao Kallakuri, S., St-Jean, F., and Archana, A. (2019). Quantitation of a ketone enolization and a vinyl Sulfonate stereoisomer formation using inline IR spectroscopy and modeling. *Org. Process Res. Dev.* 23 (5): 945–951. https://pubs.acs.org/doi/10.1021/acs.oprd.9b00042.

78 Riaz, U. and Ashraf, S.M. (2014). Characterization of polymer blends with FTIR spectroscopy. In: *Characterization of Polymer Blends* (eds. S. Thomas, Y. Grohens and P. Jyotishkumar). Wiley Online Chapter 20. https://doi.org/10.1002/9783527645602.ch20.

79 Bhargava, R., Wang, S.Q., and Koenig, J.L.F.T.I.R. (1970). Microspectroscopy of polymeric systems. *Adv. Polym. Sci.* 163: 137–191. https://doi.org/10.1007/b11052.

80 Tolstoy, V.P., Chernyshova, I., and Skryshevsky, V.A. (2003). *Handbook of Infrared Spectroscopy of Ultrathin Films*, 352–355. Hoboken, NJ: Wiley.

81 Nagai, N., Matsunobe, T., and Imai, T. (2005). Infrared analysis of depth profiles in UV-photochemical degradation of polymers. *Polym. Degrad. Stab.* 88 (2): 224–233. https://doi.org/10.1016/j.polymdegradstab.2004.11.001.

82 Nagle, D.J., George, G.A., Rintoul, L., and Fredericks, P.M. (2010). Use of micro-ATR FT-IR imaging to study heterogeneous polymer oxidation by direct solvent casting onto the ATR IRE. *Vib Spectrosc.* 53 (1): 24. https://doi.org/10.1016/j.vibspec.2010.01.018.

83 Nagai, N. and Hashimoto, H. (2001). FT-IR-ATR study of depth profile of SiO_2 ultra-thin films. *Appl. Surf. Sci.* 172 (3–4): 307–311. https://doi.org/10.1016/S0169-4332(00)00867-9.

84 Mastel, S., Govyadinov, A.A., Thales, V.A. et al. (2015). Nanoscale-resolved chemical identification of thin organic films using infrared near-field spectroscopy and standard Fourier transform infrared references. *Appl. Phys. Lett.* 106: 023113. https://doi.org/10.1063/1.4905507.

85 Shukla, N., Liu, C., Jones, P.M., and Weller, D. (2003). FTIR study of surfactant bonding to FePt nanoparticles. *J. Magn. Magn. Mater.* 266 (1–2): 178–184. https://doi.org/10.1016/S0304-8853(03)00469-4.

86 Castagnola, V., Bayon, C., Descamps, E., and Bergaud, C. (2014). Morphology and conductivity of PEDOT layers produced by different electrochemical routes. *Synth. Met.* 189: 7–16. https://doi.org/10.1016/j.synthmet.2013.12.013.

87 Massonet, N., Carella, A., De Geyer, A., and Faure Vincent, J. (2015). Metallic behaviour of acid doped highly conductive polymers. *Chem. Sci.* 6 (1): 412–417. https://doi.org/10.1039/C4SC02463J.

88 Marcott, C., Lo, M., Kjoller, K. et al. (2011). Spatial differentiation of sub-micrometer domains in a poly(hydroxyalkanoate) copolymer using instrumentation that combines atomic force microscopy (AFM) and infrared (IR) spectroscopy. *Appl. Spectrosc.* 65 (10): 1145–1150. https://doi.org/10.1366/11-06341.

89 Gong, L., Chase, D.B., Noda, I. et al. (2015). Discovery of β-form crystal structure in electrospun poly[(*R*)-3-hydroxybutyrate-co-(*R*)-3-hydroxyhexanoate] (PHBHx) nanofibers: from fiber mats to single fibers. *Macromolecules* 48 (17): 6197–6205. https://doi.org/10.1021/acs.macromol.5b00638.

90 Lu, P. and Ding, B. (2008). Applications of electrospun fibers. *Recent Pat. Nanotechnol.* 2 (3): 169–182. https://doi.org/10.2174/187221008786369688.

91 Ma, X., Liu, J., Ni, C. et al. (2012). Molecular orientation in electrospun poly(vinylidene fluoride) fibers. *Macro. Lett.* 1 (3): 428–431. https://doi.org/10.1021/mz3000122.

92 Willis, J.N., Dwyer, J.L., and Liu, M.X. (1997). Polymer characterization using SEC-FTIR. *Int. J. Polym. Anal. Charact.* 4 (1): 21–29. https://doi.org/10.1080/10236669708033934.

93 Brittain, H.G. (2018). Mid-infrared spectroscopy of pharmaceutical solids. In: *Profiles of Drug Substances, Excipients and Related Methodology*, vol. 43 (ed. H.G. Brittain), 321–358. Elsevier https://doi.org/10.1016/bs.podrm.2017.12.002.

94 Spychala, S., Butkiewicz, K., and Pakula, R. (1976). Polymorphism of indomethacin. Part II. Identification and rapid determination of polymorphic forms of indomethacin by IR spectrometry. *Pol. J. Pharmacol. Pharm.* 29 (2): 157–160.

95 Van der Weerd, J. and Kazarian, S.G. (2004). Combined approach of FTIR imaging and conventional dissolution tests applied to drug release. *J. Control. Release* 98 (2): 295–305. https://doi.org/10.1016/j.jconrel.2004.05.007.

96 Van der Weerd, J., Chan, K.L.A., and Kazarian, S.G. (2004). An innovative design of compaction cell for in situ FTIR imaging of tablet dissolution. *Vib. Spectrosc.* 35: 9–13. https://doi.org/10.1016/j.vibspec.2003.11.004.

97 Chan, K.L.A., Elkhider, N., and Kazarian, S.G. (2005). Spectroscopic imaging of compacted pharmaceutical tablets. *Chem. Eng. Res. Des.* 83 (11): 1303–1310. https://doi.org/10.1205/cherd.05088.

98 Elkhider, N., Chan, K.L.A., and Kazarian, S.G. (2006). Effect of moisture and pressure on tablet compaction studied with FTIR spectroscopic imaging. *J. Pharm. Sci.* 95 (2): 351–360. https://doi.org/10.1002/jps.20805.

99 Kazarian, S.G. and Chan, K.L.A. (2003). "Chemical fhotography" of drug release. *Macromolecules* 36 (26): 9866–9872. https://doi.org/10.1021/ma035210l.

100 Chan, K.L.A. and Kazarian, S.G. (2005 Mar). Fourier transform infrared imaging for high-throughput analysis of pharmaceutical formulations. *J. Comb. Chem.* 7 (2): 185–189. https://doi.org/10.1021/cc049840q.

101 Chan, K.L.A. and Kazarian, S.G. (2004 Jul). FTIR spectroscopic imaging of dissolution of a solid dispersion of nifedipine in poly(ethylene glycol). *Mol. Pharm.* 1 (4): 301–305. https://doi.org/10.1021/mp049973m.

102 Van der Weerd, J. and Kazarian, S.G. (2005). Release of poorly soluble drugs from HPMC tablets studied by FTIR imaging and flow-through dissolution tests. *J. Pharm. Sci.* 94 (9): 2096–2109. https://doi.org/10.1002/jps.20428.

103 Wray, P.S., Clarke, G.S., and Kazarian, S.G. (2011 Nov). Application of FTIR spectroscopic imaging to study the effects of modifying the pH microenvironment on the dissolution of ibuprofen from HPMC matrices. *J. Pharm. Sci.* 100 (11): 4745–4755. https://doi.org/10.1002/jps.22667.

104 Kazarian, S.G. and Ewing, A.V. (2013). Applications of Fourier transform infrared spectroscopic imaging to tablet dissolution and drug release. *Expert Opin. Drug Discovery* 10: 1207–1221. https://doi.org/10.1517/17425247.2013.801452.

105 Salimi, A. and Yousefi, A.A. (2003). Analysis Method: FTIR studies of β-phase crystal formation in stretched PVDF films. *Polym. Test.* 22 (6): 699–704. https://doi.org/10.1016/S0142-9418(03)00003-5.

106 Angelova, N., Kong, H.W., van der Heijden, R. et al. (2008). Recent methodology in the phytochemical analysis of ginseng. *Phytochem. Anal.* 19 (1): 2–16.

107 Yap, K.Y.L., Chan, S.Y., and Lim, C.S. (2007). Authentication of traditional Chinese medicine using infrared spectroscopy: distinguishing between ginseng and its morphological fakes. *J. Biomed. Sci.* 14: 265–273. https://doi.org/10.1007/s11373-006-9133-3.

108 Fontenla, V.E., Santos, J., Freire, M.S. et al. (2008). Antioxidant activity and phenolic content of chestnut (*Castanea sativa*) shell and eucalyptus (*Eucalyptus globulus*) bark extracts. *Ind. Crop Prod.* 28 (3): 279–285. https://doi.org/10.1016/j.indcrop.2008.03.003.

109 Liu, H.X., Zhou, Q., Sun, S.Q., and Bao, H.J. (2008). Discrimination of different Chrysanthemums with fourier transform infrared spectroscopy. *J. Mol. Struct.* 883–884: 38–47. https://doi.org/10.1016/j.molstruc.2007.12.010.

110 El-Bahy, G.M.S. (2005). FTIR and Raman spectroscopic study of Fenugreek (*Trigonella foenum graecum* L.) seeds. *J. Appl. Spectrosc.* 72: 111–116. https://link.springer.com/content/pdf/10.1007/s10812-005-0040-6.pdf.

111 Liu, H.X., Sun, S.Q., Lu, G.H., and Chan, K.K.C. (2006). Study on Angelica and its different extracts by Fourier transform infrared spectroscopy and two-dimensional correlation IR spectroscopy. *Spectrochim. Acta A Mol. Biomol. Spectrosc.* 64 (2): 321–326. https://doi.org/10.1016/j.saa.2005.07.026.

112 Gourion-Arsiquaud, S., Marcott, C., Hu, Q., and Boskey, A. (2014). Studying variations in bone composition at nano-scale resolution: a preliminary report. *Calcif. Tissue Int.* 95 (5): 413–418. https://doi.org/10.1007/s00223-014-9909-9, https://doi.org/10.1007/s00223-014-9909-9.

113 Movasaghi, Z., Rehman, S., and ur Rehman, I. (2007). Raman spectroscopy of biological tissues. *Appl. Spectrosc. Rev.* 42 (5): 493–541. https://doi.org/10.1080/05704920701551530.

114 Movasaghi, Z., Rehman, S., and ur Rehman, I. (2008). Fourier transform infrared (FTIR) spectroscopy of biological tissues. *Appl Spectrosc Rev.* 43 (2): 134–179. https://doi.org/10.1080/05704920701829043.

115 Khulla, N., Salmann, A., and Javaria, Q. (2020). ATR-FTIR spectroscopy as the future of diagnostics: a systematic review of the approach using bio-fluids. *Appl. Spectrosc. Rev.* 8453 https://doi.org/10.1080/05704928.2020.173.

116 Salzer, R. and Siesler, H.W. (2014). *Infrared and Raman Spectroscopic Imaging: Second, Completely Revised and Updated Edition*, 111–218. Wiley-VCH Verlag GmbH & Co. KGaA; 479–500.

117 Jamin, N., Miller, L., Moncuit, J. et al. (2003). Chemical heterogeneity in cell death: combined synchrotron IR and fluorescence microscopy studies of single apoptotic and necrotic cells. *Biopolymers (Biospectroscopy)* 72: 366–373. https://doi.org/10.1002/bip.10435.

118 Lee, S.H., Kim, O.K., Lee, S., and Kim, J.K. (2018). Local-dependency of morphological and optical properties between breast cancer cell lines. *Spectrochim. Acta A* 205: 132–138. https://doi.org/10.1016/j.saa.2018.07.024.

119 Clède, S., Lambert, F., Sandt, C. et al. (2013). Detection of an estrogen derivative in two breast cancer cell lines using a single core multimodal probe for imaging (SCoMPI) imaged by a panel of luminescent and vibrational techniques. *Analyst* 138 (19): 562738. https://doi.org/10.1039/c3an00807j.

120 Policar, C., Waern, J.B., Plamont, M.A. et al. (2011). Subcellular IR imaging of a metal–carbonyl moiety using fhotothermally induced resonance. *Angew. Chem. Int. Ed.* 50 (4): 860–864. https://doi.org/10.1002/anie.201003161.

121 Holman, H.Y.N., Bechtel, H.A., Hao, Z., and Martin, M.C. (2010). Synchrotron IR spectromicroscopy: chemistry of living cells. *Anal. Chem.* 82 (21): 8757–8765. https://doi.org/10.1021/ac100991d.

122 Chen, L., Holman, H.Y.N., Hao, Z. et al. (2012). Synchrotron infrared measurements of protein phosphorylationin living single PC12 cells during neuronal differentiation. *Anal. Chem.* 84 (9): 4118–4125. https://doi.org/10.1021/ac300308x.

123 Munro, K.L., Bambery, K., Carter, E. et al. (2010). Synchrotron radiation infrared microspectroscopy of arsenic-induced changes to intracellular biomolecules in live leukemia cells. *Vib Spectrosc.* 53 (1): 39–44. http://dx.doi.org/10.1016/j.vibspec.2010.02.004.

124 Henkel, B. and Baye, E. (1998). Monitoring of solid phase peptide synthesis by FTIR spectroscopy. *J. Pept. Sci.* 4 (8): 461–470. https://doi.org/10.1002/(sici)1099-1387(199812)4:8%3c461::aid-psc165%3e3.0.co;2-c.

125 Pradier, M., Salmain, M., and Boujday, S. (2011). IR spectroscopy for biorecognition and molecular sensing. In: *Biointerface Characterization by Advanced IR Spectroscopy* (eds. C.M. Pradier and Y.J. Chabal), 167–216. Elsevier https://doi.org/10.1016/C2009-0-61094-1.

126 Hamers, R.J., Wang, X., Franking, R. et al. (2011). Infrared spectroscopy for characterization of biomolecular interfaces. In: *Biointerface Characterization by Advanced IR Spectroscopy* (eds. C.M. Pradier and Y.J. Chabal), 57–80. Elsevier https://doi.org/10.1016/C2009-0-61094-1.

127 Reiter, G., Hassler, N., Weber, V. et al. (2004). In situ FTIR ATR spectroscopic study of the interaction of immobilized human tumor necrosis factor-α with a monoclonal antibody in aqueous environment. *Biochim. Biophys. Acta* 1699 (1-2): 253–261. https://doi.org/10.1016/j.bbapap.2004.03.005.

128 Natalello, A. and Doglia, S.M. (2010). Instrumental analysis of intrinsically disordered proteins: assessing structure and conformation. In: *Intrinsically Disordered Proteins and Induced Folding Studied by FTIR Spectroscopy* (eds. V. Uversky and S. Longhi), 225–250. Wiley.

129 Zhong, X., Ji, C., Chan, A.K. et al. (2011). Fabrication of chitosan/poly(ε-caprolactone) composite hydrogels for tissue engineering applications. *J. Mater. Sci. Mater. Med.* 22 (2): 279–288. https://doi.org/10.1007/s10856-010-4194-2.

130 Deniset-Besseau, A., Prater, C.B., Virolle, M.J., and Dazzi, A. (2014). Monitoring triacylglycerols accumulation by atomic force microscopy based infrared spectroscopy in *Streptomyces* species for biodiesel applications. *J. Phys. Chem. Lett.* 5 (4): 654–658. https://doi.org/10.1021/jz402393a.

131 Castro, K., Perez, M., Rodriguez-Laso, M.D., and Madariaga, J.M. (2003). FTIR spectra database of inorganic art materials. *Anal. Chem.* 75 (9): 214A–221A. https://doi.org/10.1021/ac031320e.

132 Ormsby, B., Kampasakali, E., Miliani, C., and Learner, T. (2009). An FTIR-based exploration of the effects of wet cleaning treatments on artists' acrylic emulsion paint films. E-Preservation. *Science* 6: 186–195. https://www.researchgate.net/publication/49584109_An_FTIR-based_exploration_of_the_effects_of_wet_cleaning_treatments_on_artists'_acrylic_emulsion_paint_films/link/53e4cb3d0cf21cc29fc92339/download.

133 Vagnini, M., Miliani, C., Cartechini, L. et al. (2009). FT-NIR spectroscopy for non-invasive identification of natural polymers and resins in easel paintings. *Anal. Bioanal. Chem.* 395: 2107–2118. https://doi.org/10.1007/s00216-009-3145-6.

134 Bitossi, G., Giorgi, R., Mauro, M. et al. (2005). Raman spectroscopy in forensic science. *Appl Spectrosc Rev.* 40: 187–228. https://doi.org/10.1081/ASR-200054370.

135 Derrick, M.R., Stulik, D., and Landry, J.M. (1999). *Infrared Spectroscopy in Conservation Science. Scientific Tools for Conservation.* Los Angeles, CA: Getty Conservation Institute http://hdl.handle.net/10020/gci_pubs/infrared_spectroscopy.

136 Prati, S., Joseph, E., Sciutto, G., and Mazzeo, R. (2010). New advances in the application of FTIR microscopy and spectroscopy for the characterization of artistic materials. *Acc. Chem. Res.* 43 (6): 792–801. https://doi.org/10.1021/ar900274f.

137 Platania, E., Streeton, N.L.W., Lluveras-Tenorio, A. et al. (2020). Identification of green pigments and binders in late medieval painted wings from Norwegian churches. *Microchem. J.* 156: 104811. https://doi.org/10.1016/j.microc.2020.104811.

138 Pellerito, C., Amadori, M.L., Raspugli, V. et al. (2019). *Le indagini scientifiche per lo studio e la conservazione dei beni culturali: un approccio analitico integrato, la conoscenza dei materiali costitutivi e della tecnica esecutiva della Madonna Odigitria di Monreale* (eds. M. Sebastianelli and G. Travagliato), 131. Collana Artes: Palermo University Press.

139 Pellerito, C., Di Marco, A.E., Di Natale, M.C. et al. (2016). Scientific studies for the restoration of a wood painting of the Galleria Interdisciplinare Regionale della Sicilia-Palazzo Mirto di Palermo. *Microchem. J.* 124: 682–692. https://doi.org/10.1016/j.microc.2015.10.033.

140 Bartick, E.G., Tungol, M.W., and Reffner, J.A. (1994). A new approach to forensic analysis with infrared microscopy: internal reflection spectroscopy. *Anal. Chim. Acta* 288: 35–42. https://doi.org/10.1016/0003-2670(94)85114-X.

141 Mistek, E. and Lednev, I.K. (2018). FT-IR spectroscopy for identification of biological stains for forensic purposes. *Spectroscopy.* https://www.spectroscopyonline.com/view/ft-ir-spectroscopy-identification-biological-stains-forensic-purposes.

142 Raphael, L. (2011). Application of FTIR spectroscopy to agricultural soils analysis. In: *Fourier Transforms – New Analytical Approaches and FTIR Strategies* (ed. G. Nikolic), 385–401. Intech Open.

143 Robertson, A.H.J., Main, A.M., Robinson, L.J., and Dawson, L.A. (2015). In situ FTIR analysis of soils for forensic applications. *Spectroscopy* 30 (8): 22–30. https://www.spectroscopyonline.com/view/situ-ftir-analysis-soils-forensic-applications.

144 Chen, Y., Zou, C., Mastalerz, M. et al. (2015). Applications of micro-Fourier transform infrared spectroscopy (FTIR) in the geological sciences – a review. *Int. J. Mol. Sci.* 16: 30223–30250. https://dx.doi.org/10.3390%2Fijms161226227.

145 Goldberg, P. and Berna, F. (2010). Micromorphology and context. *Quat. Int.* 214 (1–2): 56–62. http://dx.doi.org/10.1016/j.quaint.2009.10.023.

146 Berna, F., Behar, A., Shahack-Gross, R. et al. (2007). Sediments exposed to high temperatures: reconstructing pyrotechnological processes in Late Bronze and Iron Age Strata at Tel Dor (Israel). *J. Archaeol. Sci.* 34 (3): 358–373. http://dor.huji.ac.il/Download/Article/Berna2007.pdf.

147 Bacsik, Z., Mink, J., and Keresztuny, G. (2004). FTIR spectroscopy of the atmosphere. I. Principles and methods. *Appl Spectrosc Rev.* 39 (3): 295–363. https://doi.org/10.1081/ASR-200030192.

148 Simonescu, M.C. (2012). Application of FTIR spectroscopy in environmental studies combination with other techniques, advanced aspects of spectroscopy. In: *Advanced Aspects of Spectroscopy* (ed. M.A. Farrukh), 49–84. Intech Open https://doi.org/10.5772/48331.

149 Lefèvre, G. (2004). In situ Fourier-transform infrared spectroscopy studies of inorganic ions adsorption on metal oxides and hydroxides. *Adv. Colloid Interface Sci.* 107: 109–123. https://doi.org/10.1016/j.cis.2003.11.002.

150 Mori, T., Notsu, K., Tohjima, Y., and Wakita, H. (1993). Remote detection of HCl and SO_2 in volcanic gas from Unzen volcano, Japan. *Geophys. Res. Lett.* 20: 1355–1358. https://doi.org/10.1029/93GL01065.

151 Arrigone, G.M. and Hilton, M. (2005). Theory and practice in using Fourier transform infrared spectroscopy to detect hydrocarbons in emissions from gas turbine engines. *Fuel* 84: 1052–1058. https://doi.org/10.1016/J.FUEL.2005.01.018.

152 Ismail, A., van de Voort, F.R., and Sedman, J. (1997). Fourier transform infrared spectroscopy: principles and applications. In: *Instrumental Methods in Food Analysis* (eds. J.R.J. Paré and J.M.R. Belanger), 93–136. Elsevier.

153 van de Voort, F.R. (1992). Fourier transform infrared spectroscopy applied to food analysis. *Food Res. Int.* 25 (5): 397–403. https://doi.org/10.1016/0963-9969(92)90115-L.

154 Rodriguez-Saona, L.E. and Allendorf, M.E. (2011). Use of FTIR for rapid authentication and detection of adulteration of food. *Annu. Rev. Food Sci. Technol.* 2: 467–483. https://www.researchgate.net/deref/http%3A%2F%2Fdx.doi.org%2F10.1146%2Fannurev-food-022510-133750.

155 Guillen, M.D. and Cabo, N. (1997). Infrared spectroscopy in the study of edible oils and fats. *J. Sci. Food Agric.* 75 (1): 1–11. https://doi.org/10.1002/(sici)1097-0010(199709)75:1%3C1::aid-jsfa842%3E3.0.co;2-r.

156 Franca, A. and Oliveira, L. (2011). Potential uses of fourier transform infrared spectroscopy (FTIR) in food processing and engineering. In: *Food Engineering* (ed. C. Siegler Brendan), 211–258.

6

Raman and Micro-Raman Spectroscopy

Giuliana Faggio, Rossella Grillo, and Giacomo Messina

Department of Information Engineering, Infrastructure and Sustainable Energy, University Mediterranea of Reggio Calabria, Reggio Calabria, Italy

6.1 Basic Theory

6.1.1 Introduction

In 1928 in India, C.V. Raman observed an effect of inelastic scattering of the light called by Raman himself in his original paper "A new radiation" [1] only using the sun as light source, a telescope to collect the sunlight, and his eyes as detector. The inelastic scattering of light by molecules had been theoretically predicted by the Austrian theoretical physicist A. Smekal in 1923 as analogous to the Compton effect, the change in the wavelength of X-rays scattered by electrons. Raman was awarded the Nobel prize in Physics in 1930 for his discovery.

The Raman effect consists in the change in frequency of a very small fraction of monochromatic incident light due to the interaction with a molecule. Most of incident photons are elastically scattered; a small percentage of the incident light (about 1 in 10^7 photons) is diffused at energies both lower and higher than the energy of incident light. The shift from the energy of the incident beam involves transitions between the vibrational levels of the molecule. Thus, the Raman spectrum can be regarded as the fingerprint of the molecule.

There are tens of excellent books on Raman spectroscopy in the literature, with different degrees of depth. Here, we just try to present an introduction to Raman spectroscopy that is a compromise between clarity and depth. Readers interested in more complete descriptions of Raman spectroscopy and its applications are encouraged to consult, among the others, the contents of refs [2–6]. The authors of this chapter themselves have studied these texts.

6.1.2 Spectroscopic Units

The interaction of electromagnetic (e.m.) radiation with matter can be studied by measuring the absorption of infrared (IR) radiation or the scattering of light. These two phenomena are governed by completely different interaction mechanisms, but both are related to the same type of transitions between vibrational states.

Spectroscopy for Materials Characterization, First Edition. Edited by Simonpietro Agnello.

Vibrational spectroscopies (both IR and Raman spectroscopy) deal with all radiation–matter interaction processes that determine changes in the vibrational motions of molecules. A great deal of information is contained in vibrational spectra, strictly related to the molecular structure. Generally speaking, the vibrational frequencies depend on the masses of the atoms constituting the molecule, on the binding forces, and the molecular geometry. Therefore, if masses and geometry are known, the binding forces can be deduced from the measured vibration frequencies.

The frequency of the e.m. radiation is rarely used in vibrational spectroscopy. The unit of measurement commonly used in vibrational spectroscopy is the so-called "wavenumber" defined as:

$$\tilde{\nu} = \frac{1}{\lambda} \tag{6.1}$$

representing the number of wavelengths per centimeter. It has the dimension of cm^{-1}.

The frequency ν of the wave (in *Hertz*, representing the number of wave periods per second) is related to the wavenumber $\tilde{\nu}$ by the relation:

$$\nu = \frac{1}{T} = \frac{c}{\lambda} = c\tilde{\nu} \tag{6.2}$$

where c is the velocity of light. Therefore, the photon energy $E = h\nu = hc\tilde{\nu}$ is proportional to the wavenumber $\tilde{\nu}$.

Although ν and $\tilde{\nu}$ differ by the factor c (and therefore have different dimensions, *Hertz* and cm^{-1} respectively), in vibrational spectroscopy the word *frequency* is commonly used for both quantities.

An infrared (IR) absorption spectrum is obtained by passing IR radiation through a sample and measuring what fraction of the incident radiation is absorbed at a given frequency. The frequency at which an absorption peak is observed in the IR absorption spectrum corresponds to a specific vibration frequency of the molecule.

Although Raman scattering originates from molecular vibrations in the IR region, the spectra using visible laser light are obtained in the visible range as a *shift* from the excitation frequency. However, if the intensity of the Raman scattered radiation is plotted versus the *frequency shift*, that is the difference between laser frequency and scattered frequency, Raman spectra can be directly compared with the IR counterpart.

6.1.3 Molecular Vibrations

Molecular vibrations can be quantitatively treated by considering the molecule as a system of coupled oscillators. The simplest case is that of a diatomic molecule, where two atoms of masses m_1 and m_2 are connected by a chemical bond.

The molecule potential energy around the minimum point (equilibrium point) can be approximated by a parabola; thus, the restoring force is linear, which implies that the molecule will undergo simple harmonic motion.

The mechanical model of the two-body oscillator is used to predict the classical vibrational frequency of the diatomic molecule. The chemical bond is described by a spring that follows Hooke's law,

$$F = -kx \tag{6.3}$$

where x is the displacement from the equilibrium position and k the force constant (equal to the second derivative of potential energy at the equilibrium position), related to the molecular stiffness.

The two-body oscillator is equivalent to a simple harmonic oscillator with a mass μ equal to the reduced mass $\mu = m_1 m_2 /(m_1 + m_2)$ and attached to a spring of the same force constant k.

In a diatomic molecule, the vibration occurs along the axis of the chemical bond at a frequency given by:

$$\nu = \frac{\omega}{2\pi} = \frac{1}{2\pi}\sqrt{\frac{k}{\mu}}. \tag{6.4}$$

In a polyatomic molecule, instead of a simple two-body oscillator, a system of coupled oscillators must be considered. It is possible to show that the most general motion of the molecule can be expressed as a superposition of *normal modes of vibration*. The normal modes of vibration (or simply the *normal vibrations*) are nothing more than global oscillations of the whole molecule in which all atoms oscillate at the same frequency. Normal modes are completely independent of each other. The frequency associated with a normal mode of vibration is called *normal frequency* of vibration.

The number of possible vibrations in a polyatomic molecule can be deduced in the following way. In order to localize a single particle in space, three coordinates are needed. For a molecule with N atoms, $3N$ coordinates are needed ($3N$ degrees of freedom). Three of the $3N$ coordinates are used to localize the position of the center of mass (*CM*) of the molecule (and therefore to describe the translation of the whole molecule). Three more coordinates are used to define the orientation of the molecule with respect the three principal axes of rotation (and therefore to describe the rotations of the whole molecule considered as a rigid structure).

The remaining $3N$–6 degrees of freedom correspond to the internal motions of the atoms of the molecule around their equilibrium positions. Thus, the number of normal vibrations of a polyatomic molecule with N atoms is $3N$–6.

A linear molecule is a special case. Only two degrees of freedom are needed to describe the rotational motions, since the rotation around the molecule axis does not occur. Thus, the number of possible modes of vibration of a linear molecule with N atoms is $3N$–5.

6.1.4 Classical Theory of the Raman Scattering

The classical theory of the Raman effect relies on the hypothesis that the electric field of the electromagnetic wave incident on a sample induces in the molecules a dipole moment oscillating at the same frequency of incident radiation, which therefore acts as a source of e.m. waves propagating in all directions.

To be concrete and for sake of simplicity, we consider a diatomic molecule constituted by two identical atoms with their nuclei in a fixed equilibrium position. The electric dipole moment of this symmetric (nonpolar) molecule is obviously null. If a static electric field is applied to the molecule, the negative electron cloud and the positively charged nuclei move in opposite direction and the molecule suffers a distortion.

The charge displacement causes the molecule to acquire an electric dipole moment proportional to the external electric field:

$$p = \alpha E \tag{6.5}$$

where the proportionality constant α is called *polarizability* of the molecule, which qualitatively describes how easily the electron cloud of the molecule can be distorted by the external electric field.

In the general case, the mathematical relationship between induced dipole moment vector and electric field vector is not of simple proportionality. In fact, in a real molecule, the direction of the chemical bonds affects the polarization of the molecule itself; therefore, the response of the electron cloud to an external electric field is generally different in the different directions of space.

The relationship between $\boldsymbol{p}$ and $\boldsymbol{E}$ is a matrix equation:

$$\begin{pmatrix} p_x \\ p_y \\ p_z \end{pmatrix} = \begin{pmatrix} \alpha_{xx} & \alpha_{xy} & \alpha_{xz} \\ \alpha_{yx} & \alpha_{yy} & \alpha_{yz} \\ \alpha_{zx} & \alpha_{zy} & \alpha_{zz} \end{pmatrix} \begin{pmatrix} E_x \\ E_y \\ E_z \end{pmatrix} \tag{6.6}$$

where α is the so-called "polarizability tensor." In almost every case, α is a symmetric tensor ($\alpha_{xy} = \alpha_{yx}$, $\alpha_{xz} = \alpha_{zx}$, $\alpha_{yz} = \alpha_{zy}$).

Let us suppose that the molecule is subjected to the electric field E of an electromagnetic wave oscillating at the frequency ν_0 with amplitude E_0

$$E = E_0 \cos 2\pi\nu_0 t. \tag{6.7}$$

The dipole moment induced in the molecule is now:

$$p = \alpha E = \alpha E_0 \cos 2\pi\nu_0 t. \tag{6.8}$$

If the molecule polarizability were a constant quantity, the dipole moment would oscillate at the same frequency ν_0 of the driving field and would emit e.m. waves at the frequency ν_0. According to the classical e.m. theory, the power P irradiated by an oscillating dipole is given by:

$$P = \frac{4\pi^3\nu_0^4}{3\varepsilon_0 c^3}|p|^2 = \frac{4\pi^3\nu_0^4}{3\varepsilon_0 c^3}\alpha^2 E_0^2 \cos^2(2\pi\nu_0 t). \tag{6.9}$$

Thus, in the case of a molecule with fixed nuclei (α constant), the scattered radiation has the same frequency as the incident wave.

Now let us remove the constraint that the molecule has fixed nuclei and suppose that it is vibrating with a frequency ν_{vib} (vibration normal mode). Let us indicate with q the variation of the nuclear distance during a vibration, the so-called "nuclear displacement" or "nuclear distortion." Such a distortion is given by:

$$q = q_o \cos(2\pi\nu_{\text{vib}} t + \varphi) \tag{6.10}$$

where q_0 is the maximum distortion and φ is a phase factor different from molecule to molecule.

The molecular vibration can affect the molecular polarizability α, because the displacement of the nuclei during a normal vibration modifies the conformation of the molecule. The variation of the polarizability with the nuclear displacement q can be written (at the first order) by:

$$\alpha = \alpha_o + \left(\frac{\partial\alpha}{\partial q}\right)_{q=0} q + \cdots \tag{6.11}$$

where α_o is the polarizability of the molecule at fixed nuclei and $(\partial\alpha/\partial q)_{q=o}\cdot q$ represents the variation $\Delta\alpha$ of the polarizability due to the molecular oscillation, evaluated at the equilibrium position $q = 0$.

The polarizability α varies with time as

$$\alpha(t) = \alpha_o + \left(\frac{\partial\alpha}{\partial q}\right)_{q=0} q_o \cos(2\pi\nu_{\text{vib}} t + \varphi). \tag{6.12}$$

After substituting this expression of $\alpha(t)$ in the equation of the electric dipole moment, the expression for the induced dipole moment becomes:

$$p = \alpha_o E_0 \cos(2\pi\nu_0 t) + \left(\frac{\partial\alpha}{\partial q}\right)_{q=0} q_o E_0 \cos(2\pi\nu_{vib} t + \varphi) \cos(2\pi\nu_0 t). \tag{6.13}$$

This can be rewritten as

$$\begin{aligned} p(t) &= \alpha_o E_0 \cos(2\pi\nu_0 t) \\ &+ \left(\frac{\partial\alpha}{\partial q}\right)_{q=0} \frac{q_o E_0}{2} \cos[2\pi(\nu_0 + \nu_{vib})t + \varphi] \\ &+ \left(\frac{\partial\alpha}{\partial q}\right)_{q=0} \frac{q_o E_0}{2} \cos[2\pi(\nu_0 - \nu_{vib})t - \varphi]. \end{aligned} \tag{6.14}$$

According to the classical theory, an e.m. wave with frequency ν_0 incident on a molecule vibrating with frequency ν_{vib} induces a dipole moment that oscillates (and therefore emits) not only at the frequency ν_0 of the incident radiation (elastic scattering or Rayleigh scattering), but also at higher frequency $\nu_0 + \nu_{vib}$ (*anti-Stokes Raman scattering*) and at lower frequency $\nu_0 - \nu_{vib}$ (*Stokes Raman scattering*).

In order to observe Raman scattering, the rate of change $\left(\frac{\partial\alpha}{\partial q}\right)_{q=0}$ of the molecular polarizability α with the normal vibration must not be zero. The vibration is said *Raman-active*. On the contrary, if $\left(\frac{\partial\alpha}{\partial q}\right)_{q=0} = 0$, the vibration is said to be *not Raman-active*.

The elastic scattering (*Rayleigh scattering*), in-phase with the incident light, is *coherent*. The inelastic Stokes and anti-Stokes scattering is incoherent because of the phase factor φ, which is in general different from molecule to molecule.

According to the classical e.m. theory, the intensity of the scattered radiation, proportional to the squared second derivative of the dipole moment, is given by:

$$\begin{aligned} I \propto E_o^2 \{ &\nu_0^4 \alpha_o^2 \cos^2(2\pi\nu_0 t) \\ &+ (\nu_0 + \nu_{vib})^4 \cdot \left[\left(\frac{\partial\alpha}{\partial q}\right)_{q=o} \frac{q_o}{2}\right]^2 \cdot \cos^2[2\pi(\nu_0 + \nu_{vib})t + \varphi] \\ &+ (\nu_0 - \nu_{vib})^4 \cdot \left[\left(\frac{\partial\alpha}{\partial q}\right)_{q=o} \frac{q_o}{2}\right]^2 \cdot \cos^2[2\pi(\nu_0 - \nu_{vib})t - \varphi] \} \\ &+ \text{cross-terms} \end{aligned} \tag{6.15}$$

where the cross-terms can be neglected since the power they irradiate averages to zero over sufficiently long time intervals [6].

Thus, in a simple experiment of light scattering, the classical approach correctly predicts the existence of Stokes and anti-Stokes lines and their dependence from the fourth power of the frequency. However, classical theory fails to predict the intensity ratio between Stokes and anti-Stokes lines and its dependence on temperature. The relative intensity of the Stokes and anti-Stokes lines is predicted to be

$$\frac{I_{\text{Stokes}}}{I_{\text{anti-Stokes}}} = \left(\frac{\nu_0 - \nu_{vib}}{\nu_0 + \nu_{vib}}\right)^4 < 1 \tag{6.16}$$

and temperature independent. This prediction is not in accordance with the experimental results, which show that Stokes lines are always more intense than anti-Stokes lines and strongly dependent on temperature.

This problem is overcome by using a quantum mechanical treatment of Raman scattering.

6.1.5 Simplified Quantum Approach to Raman Scattering

The simplest way of visualizing the mechanisms at the basis of Raman effect is through a simplified quantum description of the interaction between light and molecule, where the radiation field is described classically. The interaction between the molecule and the classical radiation field at frequency ν_0 is treated using the quantum mechanical time-dependent perturbation theory. The classical induced electric dipole moment is replaced by the electric dipole matrix element associated with the transition of the molecule from an initial state to a final state. An excellent discussion of this topic can be found in Chapter 4 of Ref. [2], whereas here we only provide a brief simplified description.

The main features of the radiation–molecule interaction are determined by the energy levels associated to the degrees of freedom of the molecule. The motion of the electrons in the potential generated by the nuclei determines the electronic energy levels of the molecule; the movement of atoms in the molecule determines the vibrational and rotational energy levels.

The molecule's energy can be thought as constituted by three terms associated with the motion of electrons in the molecule, the vibrations of the atoms constituting the molecule, the rotation of the molecule as a whole. The possibility of separating nuclear and electronic motions depends on the large difference between the mass of the electrons and the mass of the nuclei [6].

Figure 6.1 illustrates the energy levels of a diatomic molecule. The spacing (not in scale) of the vibrational levels (typically in the 10^2–10^4 cm^{-1} region, or 10^2–1 μm in wavelengths) is much smaller than that of the electronic levels (typically in the 10^4–10^6 cm^{-1} region, or 1–10^{-2} μm in wavelengths), and usually much larger than those of the rotational levels (typically in the 1–10^2 cm^{-1} region, or 10^4–10^2 μm in wavelengths).

Let us consider a molecule characterized by a ground electronic level and by at least one excited electronic level. The exciting radiation of frequency ν_0 is not able to bring the system from the ground state to the first excited electronic state (normal Raman scattering). The scattering process can be visualized as a two-step process, combination of absorption of the incident photon and simultaneous emission of the scattered photon.

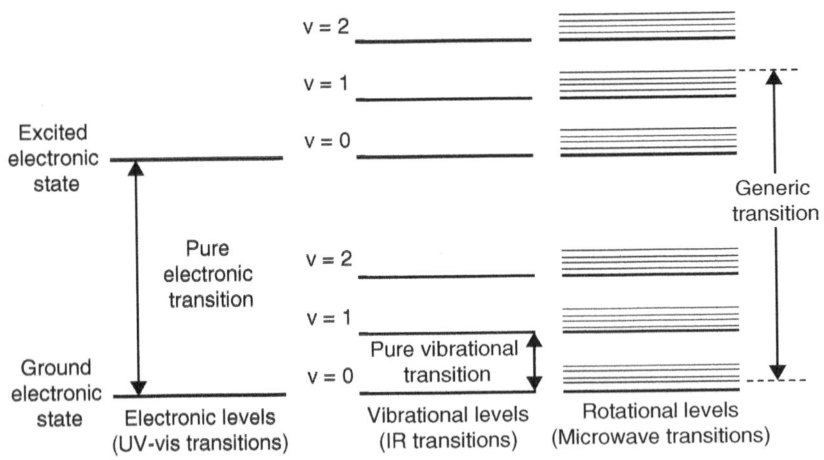

Figure 6.1 Energy levels of a diatomic molecule. Pure electronic and vibrational transitions are represented. A generic transition involving changes in all quantum numbers (electronic, vibrational, and rotational) is also represented.

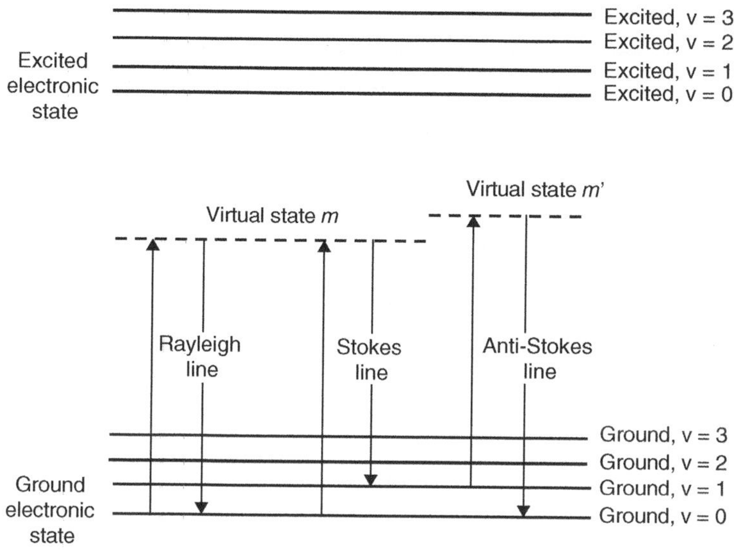

Figure 6.2 Energy levels scheme showing the basic transitions involved in normal Raman scattering.

The fundamental steps of a scattering process of light from a molecule, described in a simplified quantum language derived from the quantum mechanical perturbation theory, are the following (Figure 6.2):

- The system undergoes a transition from the ground electronic state g and ground vibrational state ($v = 0$) to an intermediate virtual level m.
- The system can immediately decay from the virtual level m to the starting state (ground vibrational state ($v = 0$) of the ground electronic level g). The emitted radiation will have the same frequency ν_0 of the incident radiation (Rayleigh line).
- The system can immediately decay from the virtual level m to the first vibrational level ($v = 1$) of the ground electronic level g. The emitted radiation will have frequency $(\nu_0 - \nu_{vib})$ (Stokes line).

If the molecule is in an excited vibrational state of the electronic ground state (in Figure 6.2 corresponding to the vibrational quantum number $v = 1$), the transition is toward another virtual level m'; from the virtual state m', the system can decay to the ground vibrational state ($v = 0$) of the ground electronic level g. The emitted radiation will have frequency $(\nu_0 + \nu_{vib})$ (anti-Stokes line).

In normal Raman scattering, no electronic energy levels are close to the energy of the incident photons. As the energy of the excitation light approaches that of electronic transitions, the intensity of the Raman bands can be enhanced from three to five–six orders of magnitude. This phenomenon is called resonance Raman scattering (RRS) (Figure 6.3).

In a RRS experiment, the frequencies of the vibrational modes depend only on the ground state of the molecule, whereas their intensities depend on the electronic excited states. RRS can be very useful for the analysis of specific molecules at very low concentrations, from three to five–six orders of magnitudes lower than those used in normal Raman scattering experiments.

A detailed treatment of this issue can be found in Refs. [2, 3].

At first glance, Raman scattering may appear quite similar to a fluorescence process, since in both cases the incident photon is replaced by another photon, scattered or emitted by fluorescence. However, the timescales of Raman scattering processes and fluorescence are very different.

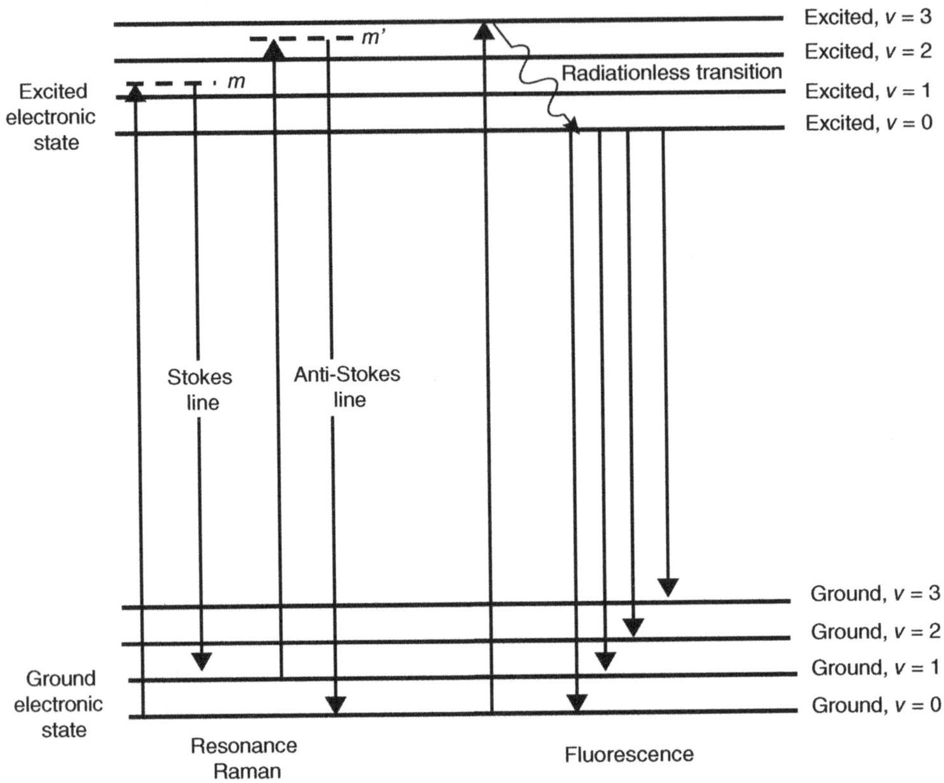

Figure 6.3 Scheme of energy levels for resonance Raman and fluorescence.

Fluorescence consists in the emission of radiation following the absorption of the exciting radiation, generally in the UV-visible range. Firstly, the molecule absorbs a photon and goes in an electronic and vibrational excited state (one-photon process). Then, it decays to the lowest vibrational state of the excited electronic state via radiation-less transitions (represented in Figure 6.3 by a wavy arrow). Finally, the molecule goes back to the ground electronic state, emitting a fluorescence photon (again a one-photon process, temporally well separated from the absorption process). The radiation emitted in the fluorescence process always has lower energy than that of the incident radiation (Stokes shift). Conversely, Raman scattering always shows, in addition to the Stokes line at lower energy, an anti-Stokes line at higher energy. Unlike fluorescence, in resonance Raman scattering, the relaxation to the ground state is not preceded by a prior relaxation to the lowest vibrational level of the excited electronic state.

In the Raman effect, the two acts of "absorption" and "re-emission" used to pictorially describe the scattering process are simultaneous: the scattering processes are nearly instantaneous. Thus, Raman scattering may occur even if no real electronic transition exists at the incident radiation energy. Therefore, the virtual state of Figure 6.2, as its name suggests, does not have in most cases any physical reality and should be considered in some way only a "pictorial aid" for describing the scattering process, whose mathematical description requires the sophisticate approaches of the quantum mechanics.

Raman scattering occurs in a timescale of the order of 10^{-15} second, much shorter than the timescale of fluorescence process (10^{-9}–10^{-12} second). Raman scattering is not a fluorescence phenomenon. Fluorescence is a one-photon phenomenon, while Raman scattering is a two-photon process.

The intensity of the Raman scattering is related to the population of the initial state of the molecule: the Stokes line is related to the population N of the ground state ($v = 0$), while the anti-Stokes line to population N' of the first excited state ($v = 1$).

The population ratio N'/N is given by the Maxwell–Boltzmann distribution law $N'/N = \exp[-\Delta E/kT]$, where $\Delta E = h\nu_{\text{vib}}$ is the energy difference between the two vibrational states, h is Planck's constant, k is Boltzmann's constant, and T is the absolute temperature.

The ratio of anti-Stokes to Stokes intensity is found to be [2–6]:

$$\frac{I_{\text{anti-Stokes}}}{I_{\text{Stokes}}} = \frac{(\nu_0 + \nu_{\text{vib}})^4}{(\nu_0 - \nu_{\text{vib}})^4}\frac{N'}{N} = \frac{(\nu_0 + \nu_{\text{vib}})^4}{(\nu_0 - \nu_{\text{vib}})^4}\exp\left[-\frac{\Delta E}{kT}\right] \tag{6.17}$$

according to the experimental results. The anti-Stokes line of a Raman spectrum is much weaker than the Stokes line and the intensities of the peaks vary exponentially with the energy of the vibration $h\nu_{\text{vib}}$. Therefore, the quantum treatment of the Raman scattering process predicts the correct intensity ratio between Stokes and anti-Stokes lines and its temperature dependence.

Figure 6.4 shows the Raman spectrum of cysteine obtained by using the spectrometer HORIBA Scientific LabRAM HR Evolution with an excitation wavelength of 633 nm.

In the abscissa of Figure 6.4, the wavenumber shift is reported, defined as the difference in wavenumbers between the wavenumber of the excitation light and that of the scattered radiation. The Stokes lines are found at wavenumbers smaller than the laser line, anti-Stokes lines at greater than the wavenumber of the laser excitation. As evidenced in the above discussion, the pattern of the Raman lines on each side of the excitation line is identical. Since the anti-Stokes lines have lower intensity than the corresponding Stokes lines, only the Stokes part of a spectrum is generally used. Despite its lower intensity, anti-Stokes scattering may be very useful in the case of fluorescing samples, since fluorescence may seriously interfere with the detection of Stokes lines (at lower energy than excitation source, as fluorescence), but does not affect the observation of anti-Stokes lines that are at energy higher than fluorescence emission.

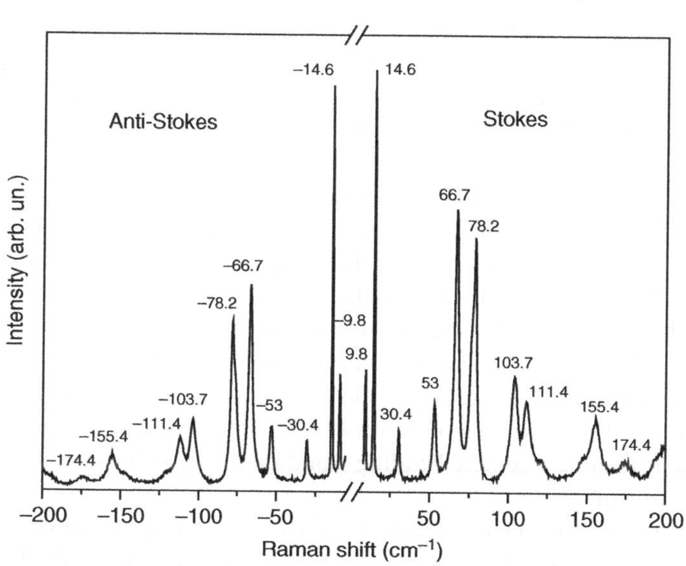

Figure 6.4 Raman spectrum of cysteine obtained by using the spectrometer LabRam HR-evolution (HORIBA scientific) with an integrated Olympus BX41 microscope (excitation wavelength λ = 633 nm).

Finally, it is noteworthy that Raman spectroscopy can be utilized for an optical measurement of the temperature, in cases where a direct measurement is not possible. In fact, by inverting Eq. (6.17), the temperature T of the sample can be obtained through quantities directly obtained by the Raman spectrum:

$$T = \frac{h\nu_{\text{vib}}}{k} \frac{1}{\ln\left[\frac{I_{\text{Stokes}}}{I_{\text{anti-Stokes}}} \frac{(\nu_0 + \nu_{\text{vib}})^4}{(\nu_0 - \nu_{\text{vib}})^4}\right]}. \tag{6.18}$$

6.1.6 Raman and IR Activities

It is possible to deduce the number and types of normal vibrations in simple diatomic or triatomic molecules and if these normal vibrations give rise to IR and Raman activities.

Vibrations can involve either a change of bond length (stretching mode) or bond angle (bending mode).

Symmetric stretching occurs if the bonds stretch in-phase. Asymmetric stretching occurs if the bonds stretch out-of-phase.

The vibration motions may generate changes in the electric dipole moment and/or in the polarizability of the molecule, giving rise to spectroscopic effects.

A normal vibration is *IR-active* if the electric dipole moment of the molecule changes during the vibration: infrared absorption is observed at the frequency of the normal vibration. The larger the change in dipole moment, the more intense the absorption band will be.

A normal vibration is *Raman-active* if the molecular polarizability changes during the vibration: the incident radiation is inelastically scattered at the shifted frequency $\nu_L \pm \nu_{\text{vib}}$.

In the case of small molecules, the IR and Raman activities can be determined quite easily by a careful examination of the normal vibration.

To be concrete, we consider some model cases: symmetric diatomic molecule (i.e. H_2, O_2), asymmetric diatomic (i.e. CO, HCl), and symmetric linear triatomic molecule (i.e. CO_2).

A symmetric diatomic molecule has no net dipole moment. During a stretching vibration, the dipole moment remains zero, so the vibration is not IR-active (see Figure 6.5a).

During the vibration, when the bond is elongated ($q > 0$), the electrons involved in the bond are less firmly held by the nuclei, so the molecule is more easily polarizable (larger α). On the contrary, when the bond is compressed, the polarizability decreases (smaller α). Since the vibration causes a change in the molecule polarizability, it is Raman-active (see Figure 6.5a).

An asymmetric diatomic molecule has a nonzero dipole moment due to the difference in electronegativity of the two atoms. The vibration causes a change of the dipole moment, so it is IR-active. Furthermore, the variation of the bond length during the vibration causes a change in the molecule polarizability. Thus, the vibration is also Raman-active (see Figure 6.5b).

A symmetric linear triatomic molecule (such as CO_2) has four normal modes ($3N-5$, with $N = 3$), schematically shown in Figure 6.5c: a symmetric stretching mode (ν_1), an asymmetric stretching mode (ν_3), and two bending modes (ν_2). The bending modes constitute a doubly degenerate pair; the vibrations occur on perpendicular planes. A generic bending vibration in any other direction can be expressed through their linear combination.

For symmetry reasons, the linear triatomic molecule in the ground state has no dipole momentum (see Figure 6.5c). If the two bonds are stretched symmetrically, there is no change in the dipole moment; so the vibration is not IR-active. The variation of the bond length during the vibration modifies the molecule's polarizability, which becomes greater than the equilibrium value α_0 when

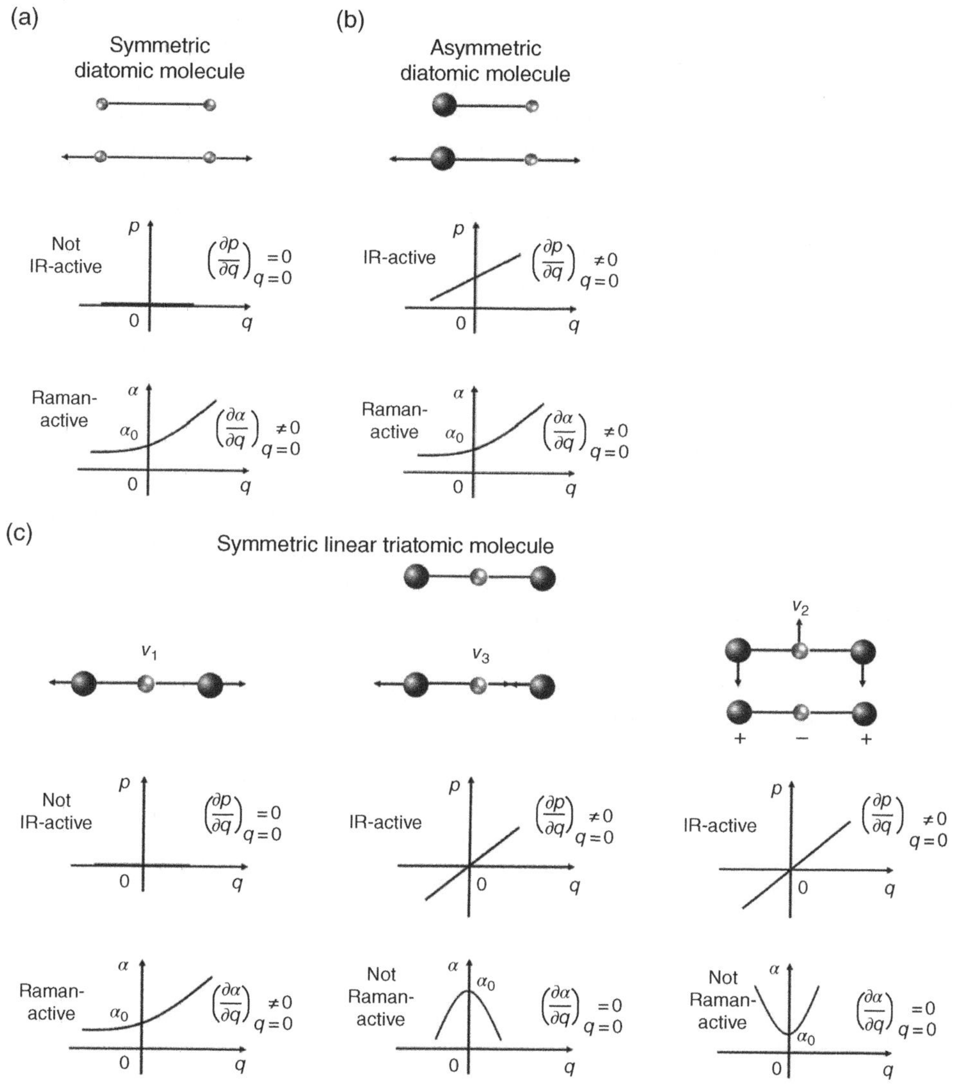

Figure 6.5 Selection rules for IR and Raman activity of (a) symmetric, (b) asymmetric diatomic molecules, and (c) symmetric linear triatomic molecule. Note that + and − indicate upward and downward vibrations in direction perpendicular to the page plane.

the bonds stretch ($q > 0$), and smaller when the bonds contract ($q < 0$). Thus, the symmetric stretching mode is Raman-active.

In an asymmetric stretching vibration, the two bonds have different length and hence a net dipole moment arises. Therefore, the asymmetric stretching mode is IR-active. As concerns the Raman activity, the polarizability decreases equally for positive and negative q. The molecule is chemically identical at the two extremes of the vibration. The polarizability changes induced by one part of the molecule are balanced by the other part. The derivative of the polarizability $\left(\frac{\partial \alpha}{\partial q}\right)_{q=0}$ with respect to the normal coordinate q is zero at $q = 0$. The asymmetric stretching mode is not Raman-active.

In the bending vibrations, a net dipole moment arises in the direction perpendicular to molecule axis. Hence, the bending mode is IR-active. As concerns the Raman activity, bending vibrations cause an increase in polarizability, but the change is the same for both positive and negative q. Hence, the bending mode is not Raman-active for the same reason as that of the asymmetric stretching mode.

The preceding examples show a quite general conclusion. If a vibration mode is symmetric with respect to the center of symmetry of the molecule, then that vibration is Raman-active, but not IR-active. On the contrary, if the vibration mode is antisymmetric with respect to the center of symmetry, then the vibration is IR-active, but not Raman-active (antisymmetric stretching and bending in Figure 6.5b and c). This statement, valid for molecules with a center of symmetry, is known as *rule of mutual exclusion*.

If a polyatomic molecule does not have a center of symmetry, some vibrations (but not necessarily all) can be both IR-active and Raman-active. Obviously, it is by no means certain that the lines of the IR spectrum are comparable in intensity with the corresponding lines of the Raman spectrum. The presence of a line at a given vibrational frequency in both Raman and IR spectrum generally indicates that the molecule does not have a center of symmetry.

Clearly, the approach based on the inspection of the vibration normal mode is hardly applicable to complex molecules. Group theory provides powerful tools to determine IR and Raman activities of such molecules, but its treatment is beyond the scope of this chapter.

Both IR and Raman spectroscopy provide information on the vibrational properties of materials. There are limitations typical of each technique, which make them complementary rather than competing. For instance, water is a weak Raman scatterer, whereas it strongly absorbs in IR region. Thus, Raman spectroscopy is extremely suitable for the study of compounds in aqueous solutions, such as biologic samples.

The selection rules for each process, absorption and scattering, are different. Raman-active vibrations may not be active in IR and vice versa, as occurs in molecules having a center of symmetry (rule of mutual exclusion). Furthermore, the strongly active vibrations in one technique are generally weak in the other. As thumb rule, covalent bonds respond best in Raman, while ionic bonds respond best in IR. In Raman spectra, stretching vibrations are generally more intense than bending vibrations.

6.1.7 Crystal Vibrations

In order to discuss the selection rules of Raman scattering in solids, it is useful to briefly reconsider the theory of crystal vibrations [7, 8].

Let us consider a diatomic linear chain with two atoms per primitive cell as shown in Figure 6.6.

The atoms have masses M_1 and M_2, connected by springs of spring constant C. The parameter a denotes the repeat distance of the lattice. This simple model provides a good description of the main features of the elastic vibrations in a real three-dimensional lattice.

Let us assume that the displacements of M_1 and M_2 from their equilibrium positions in the sth primitive cell are u_s and v_s, respectively. The equations of motion for the atoms M_1 and M_2 in the primitive cell are [5, 7, 8]:

$$M_1 \frac{d^2}{dt^2} u_s = C(v_s + v_{s-1} - 2u_s) \tag{6.19a}$$

$$M_2 \frac{d^2}{dt^2} v_s = C(u_{s+1} + u_s - 2v_s). \tag{6.19b}$$

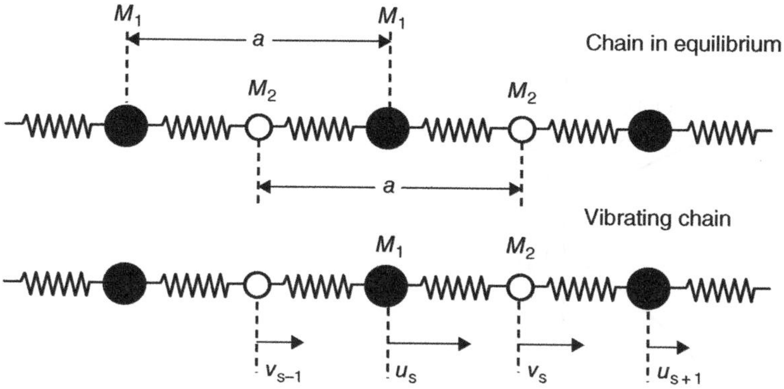

Figure 6.6 Diatomic chain with masses M_1 and M_2 connected by springs of force constant C. The displacements from the equilibrium positions of atoms M_1 are denoted by u_s; those of atoms M_2 by v_s. The dimension of the primitive cell is a.

We seek solutions representing traveling waves with different amplitudes u and v for the different atoms:

$$u_s = ue^{i(qsa - \Omega t)} \tag{6.20a}$$

$$v_s = ve^{i(qsa - \Omega t)} \tag{6.20b}$$

where q is the wavevector and Ω is the angular frequency. If Eq. (6.20) is substituted in Eq. (6.19), the following homogeneous linear system of equations is obtained:

$$-\Omega^2 M_1\, u = Cv\left(1 + e^{-iqa}\right) - 2Cu \tag{6.21a}$$

$$-\Omega^2 M_2\, v = Cu\left(1 + e^{iqa}\right) - 2Cv. \tag{6.21b}$$

A nontrivial solution of this pair of homogeneous linear equations exists only if the determinant of the coefficients vanishes:

$$\begin{vmatrix} 2C - M_1\Omega^2 & -C(1 + e^{-iqa}) \\ -C(1 + e^{iqa}) & 2C - M_2\Omega^2 \end{vmatrix} = 0 \tag{6.22}$$

$$M_1 M_2 \Omega^4 - 2C(M_1 + M_2)\Omega^2 + 2C^2(1 - \cos qa) = 0. \tag{6.23}$$

Equation (6.23) can be solved exactly for Ω^2, giving the dispersion relation:

$$\Omega^2 = C\left(\frac{1}{M_1} + \frac{1}{M_2}\right)\left[1 \pm \sqrt{1 - \frac{4M_1M_2}{(M_1 + M_2)^2}\sin^2\left(\frac{qa}{2}\right)}\right]. \tag{6.24}$$

For a given wavevector q, there are two possible values of Ω , one relative to the sign (+) (optical branch) and the other relative to the sign (−) (acoustic branch).

The periodic nature of the function $\sin^2\left(\frac{qa}{2}\right)$ (that has period π) allows to limit the values of the argument $\frac{qa}{2}$ to the interval $\left[-\frac{\pi}{2}; \frac{\pi}{2}\right]$. As a consequence, the range of independent values of the wavevector q is

$$-\frac{\pi}{a} \le q \le \frac{\pi}{a}. \tag{6.25}$$

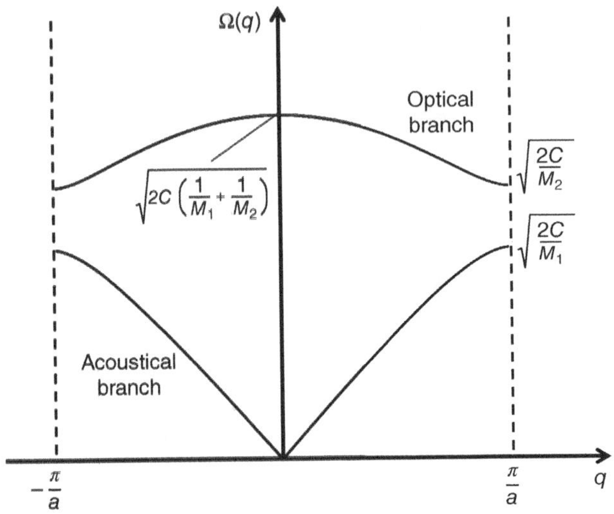

Figure 6.7 Dispersion relation for a diatomic linear chain. The lower branch is the acoustical branch; the upper branch is the optical branch.

This range is called "first Brillouin zone" (BZ), where a is the lattice periodicity. Values of the wavevector q beyond the first BZ simply reproduce lattice motions already described by q values within the interval $\left[-\frac{\pi}{a};\frac{\pi}{a}\right]$. The two Ω vs. q curves, representing the two branches of the dispersion relation, are plotted in Figure 6.7 within the first Brillouin zone.

For small values of qa (that is at the center of the BZ), the two solutions become

$$\Omega \cong \sqrt{2C\left(\frac{1}{M_1}+\frac{1}{M_2}\right)} \tag{6.26}$$

for the optical branch, and

$$\Omega \cong \sqrt{\frac{C}{2(M_1+M_2)}}\,qa \tag{6.27}$$

for the acoustical branch. Note that, for $q \to 0$, Ω is independent of q in the optical branch. In the acoustical branch, Ω varies linearly with q as in acoustical waves in a continuous medium.

At the extremes of the BZ, that is $= \pm\frac{\pi}{a}$, the two solutions are (with $M_1 > M_2$):

$$\Omega = \sqrt{\frac{2C}{M_2}} \tag{6.28}$$

for the optical branch, and

$$\Omega = \sqrt{\frac{2C}{M_1}} \tag{6.29}$$

for the acoustical branch, respectively. For angular frequencies Ω in the range

$$\sqrt{\frac{2C}{M_1}} < \Omega < \sqrt{\frac{2C}{M_2}} \tag{6.30}$$

wavelike solutions do not exist. This forbidden frequency gap is a general feature of elastic waves in polyatomic lattices and it is strictly related to the periodicity of the crystalline lattice.

For the optical branch at $q = 0$, after substituting Eq. (6.26) in the linear system Eq. (6.21), we obtain:

$$\frac{u}{v} = -\frac{M_2}{M_1} \tag{6.31}$$

the two atoms in the cell move against each other (180° out-of-phase). If the two atoms M_1 and M_2 have opposite electric charges, this vibration can be excited by the electric field of a light wave. For this reason, this type of motion has been called "optical branch."

In the same way, for the acoustic branch at $q = 0$ (Eq. (6.27)), we find:

$$u = v \tag{6.32}$$

The two atoms in the cell move together (in-phase) as in sound waves, whence the name "acoustical branch."

In the general case of a three-dimensional lattice with a basis of N atoms, there are $3N$ branches. Three of these are acoustic (two transverse (T) and one longitudinal (L)). The other $3N$–3 branches are optical. These branches can be regarded as the generalization to the crystalline case of the degrees of freedom of a polyatomic molecule (see Section 6.1.3) [7, 8].

6.1.8 Raman Scattering in Crystals

In the process of inelastic scattering of light from a crystal, an incident photon of wavevector $\mathbf{k}_i$ and frequency ω_i is scattered in the direction identified by the wavevector $\mathbf{k}_s$. The frequency of the scattered photon changes to the new value ω_s, while a phonon of wavevector $\mathbf{q}_j$ and frequency Ω_j is created or destroyed. In the scattering process, the laws of conservation of energy and moment must be respected:

$$\hbar\omega_i = \hbar\omega_s \pm \hbar\Omega_j \tag{6.33a}$$

$$\hbar\mathbf{k}_i = \hbar\mathbf{k}_s \pm \hbar\mathbf{q}_j \tag{6.33b}$$

the plus (+) sign is valid for the creation of a phonon (Stokes scattering), the minus (−) sign for the annihilation of a phonon (anti-Stokes scattering). In Figure 6.8a, a schematic of the process is reported.

The frequency of the incident radiation ω_i, usually in the visible (or close to the visible) region, is generally much greater than the frequency of the lattice vibrations Ω_j. Therefore, the conservation of energy $\hbar\omega_i = \hbar\omega_s + \hbar\Omega_j$ implies that the frequency of scattered photon is approximately equal to that of the incident photon: $\omega_s \cong \omega_i$. Therefore, taking into account that $\omega = \frac{2\pi}{T} = \frac{2\pi c}{\lambda} = ck$ where T is the period of the wave and c the speed of light, the modules of the wavevectors $\mathbf{k}_s$ and $\mathbf{k}_i$ will also be approximately equal $k_s \cong k_i$.

Thus, the conservation of momentum (see Figure 6.8b) gives

$$|\underline{\mathbf{k}}_i - \underline{\mathbf{k}}_s| = q_j = 2k_i \sin\frac{\phi}{2}. \tag{6.34}$$

The wavevector $\mathbf{q}_j$ of the phonon involved in the scattering process depends on the scattering angle. In the usual scattering geometries (90° or 180°), it is comparable to that of the incident photon $\mathbf{k}_i$.

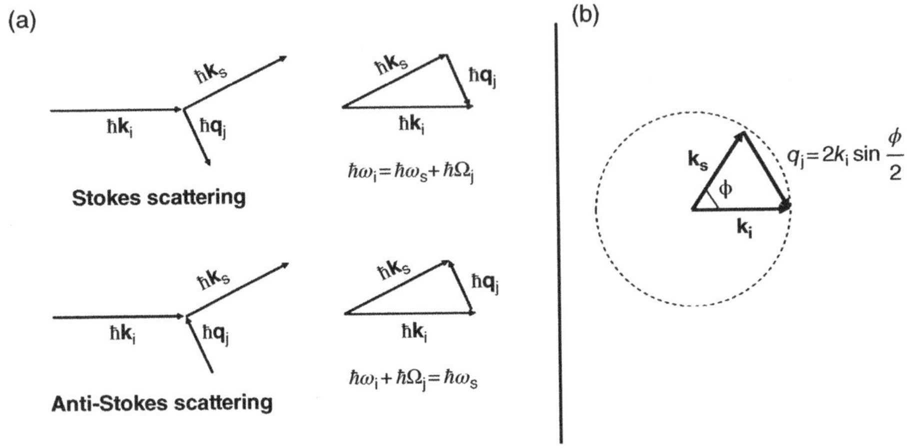

Figure 6.8 (a) Schematics of the inelastic scattering of light from a crystal. (b) Geometric relation among incident photon wavevector $\mathbf{k}_i$, scattered photon wavevector $\mathbf{k}_s$, and phonon wavevector $\mathbf{q}_j$. φ denotes the scattering angle. The triangle of sides $\mathbf{k}_i$, $\mathbf{k}_s$, $\mathbf{q}_j$ is isosceles. The basis q_j is simply $2k_i \sin(\varphi/2)$.

For excitation in the visible range, the wavevector $k_i = 2\pi/\lambda_i$ of the incident light is of the order of $10^5\ \mathrm{cm}^{-1}$. For example, for $\lambda = 514.5$ nm (the green line of an Argon ion laser), the corresponding wavenumber is $\tilde{\nu} = \frac{1}{\lambda} \cong 19\,436\ \mathrm{cm}^{-1}$ and the corresponding wavevector is $k = \frac{2\pi}{\lambda} = 2\pi\tilde{\nu} \cong 1.22{\cdot}10^5\ \mathrm{cm}^{-1}$.

The crystal phonons have wavevectors q_j within the first Brillouin zone, whose extreme $q_{max} = \pi/a$ is of the order of $10^8\ \mathrm{cm}^{-1}$. Therefore, the phonon involved in the first-order scattering process from a crystal must have a wavevector q_j (comparable in magnitude with that of the incident light $k_i \approx 10^5\ \mathrm{cm}^{-1}$) included in the first thousandth of the Brillouin zone, that is near the center of the BZ:

$$q_j \approx 0 \tag{6.35}$$

this represents the "selection rule" for the first-order (one phonon) Raman scattering from a perfect crystal.

If two phonons are involved in the scattering process of the incident light, a second-order Raman scattering occurs. The incident photon can create two phonons, giving a Stokes component in the scattered light; it can absorb two phonons, producing an anti-Stokes component; or it can create one and absorb another, generating a Stokes and anti-Stokes component.

Let us suppose two phonons are created with wavevectors $\mathbf{q}_\sigma$ and $\mathbf{q}'_{\sigma'}$, where σ and σ' are the so-called *branch indices*. The energy and momentum conservation laws will then have the form:

$$\hbar\omega_i = \hbar\omega_s + \hbar\Omega_{\sigma q} + \hbar\Omega_{\sigma' q'} \tag{6.36a}$$

$$\hbar\mathbf{k}_i = \hbar\mathbf{k}_s + \hbar\mathbf{q}_\sigma + \hbar\mathbf{q}'_{\sigma'}. \tag{6.36b}$$

The energy of the incident photon is much greater than the phonon energy $\hbar\omega_i \gg \hbar\Omega_q$. Therefore, as in the case of first-order Raman scattering, we have $\omega_s \cong \omega_i$ and $k_s \cong k_i$.

However, in two-phonon scattering, there is no restriction on the wavevectors $\mathbf{q}_\sigma$ and $\mathbf{q}'_{\sigma'}$ of the phonons involved in the scattering process. The only requirement is that their sum must equal the difference between the wavevectors of the incident photon $\mathbf{k}_i$ and the scattered photon $\mathbf{k}_s$:

$$\mathbf{k}_{\mathrm{i}} - \mathbf{k}_{\mathrm{S}} = \mathbf{q}_{\sigma} + \mathbf{q}'_{\sigma'}. \tag{6.37}$$

Thus, the phonon wavevectors $\mathbf{q}_{\sigma}$ and $\mathbf{q'}_{\sigma'}$ need not be small as in first-order Raman scattering, but they can range over the entire Brillouin zone. As the photon wavevectors $\mathbf{k}_{\mathrm{i}}$ are always much smaller than most of the phonon wavevectors q, the law of conservation of the moment can be written as

$$\mathbf{q}_{\sigma} + \mathbf{q}'_{\sigma'} = 0. \tag{6.38}$$

Thus, the wavevectors $\boldsymbol{q}$ and $\boldsymbol{q}'$ of the created phonons must be equal and opposite. Taking into account that the vibrational frequencies are the same at $+\mathbf{q}$ and $-\mathbf{q}$ within the same branch, the energy conservation equation (Eq. 6.36a) can be written as

$$\hbar\omega_{\mathrm{i}} - \hbar\omega_{\mathrm{S}} = \hbar\Omega_{\sigma q} + \hbar\Omega_{\sigma' q}. \tag{6.39}$$

If the created phonons belong to the same branch, we speak of overtones. If the created phonons belong to different branches, we speak of combinations. The second-order Raman spectrum consists of a more or less wide band containing structures that can be explained in terms of combinations and overtones of the phonons of the crystal. The intensity of the second-order spectrum is obviously much lower than that of the first order, because a two-phonon phenomenon is always less likely than a single-phonon process.

6.1.9 Surface-Enhanced Raman Scattering (SERS)

In this paragraph, we give a brief introduction to the fundamentals of the SERS effect. The Raman signal enhancements observed by Fleischman et al. [9] in 1974 on pyridine on electro-chemically roughened silver electrodes were interpreted as due to the increased number of adsorbed molecules resulting from the larger surface area due to the roughening of the electrodes.

When a molecule is in close proximity to a metallic surface, the optical processes that depend directly on the intensity of the incident electric field can be strongly affected by the presence of an enhanced local field at the metal surface.

As described in Sections 6.1.4 and 6.1.5, the Raman effect consists of an inelastic scattering process of the incident light from a molecule, mediated by the vibrational modes of the molecule. The Raman signal depends on the induced electric dipole momentum $p = \alpha E$ and its intensity roughly scales as the product of the squared incident electric field amplitude E_0^2 and the squared derivative $\left(\frac{\partial \alpha}{\partial q}\right)^2$ of the molecular polarizability (see Eq. 6.15). Thus, it is reasonable to think that the enhancement of the Raman signal can come from an enhancement of either the molecular polarizability α or the electric field E incident on the molecule.

The enhancement of the molecular polarizability can be associated to chemical effects, such as charge transfer or bond formation between metal and adsorbed molecule and is therefore termed *chemical enhancement*. In order to show chemical enhancement, the molecule must be directly adsorbed on the nanostructured metallic surface. The chemical mechanism typically provides an enhancement factor of only one or two orders of magnitude [4, 10, 11].

The enhancement associated with the electric field E is termed the electromagnetic effect. Since the measured SERS enhancement factors are typically in the range 10^3–10^{10}, then the main contribution to the SERS enhancement factor is of electromagnetic nature.

The electromagnetic enhancement occurs because the local electromagnetic field near rough metal surfaces or at the surface of metal nanoparticles is significantly increased with respect to the incident field. The laser field of amplitude E_0 and frequency ν_L incident on the metal surface excites localized surface plasmons of the metal nanostructure, giving rise to an enhanced local field E_{loc} given by:

$$E_{loc}(\nu_L) = \Gamma_{exc}(\nu_L) E_0(\nu_L) \tag{6.40}$$

where $\Gamma_{exc}(\nu_L)$ is the enhancement factor, a simple scalar in case of systems with isotropic plasmonic response or a tensor in the general case.

The molecule near the metal surface will be excited by an enhanced local field E_{loc} that will produce an induced electric dipole momentum $p = \alpha E_{loc}$. Thus, the electric field E_{Raman} associated to the radiation scattered by the molecule will be proportional to the local electric field E_{loc} (see Eq. (6.15) with E_0 replaced by E_{loc}).

The Raman electric field E_{Raman} at the shifted frequency ν_R also benefits from the plasmonic amplification, the so-called re-radiation enhancement [4, 10, 11]. The enhanced re-radiated field E_{rad} can be written as

$$E_{rad}(\nu_R) = \Gamma_{rad}(\nu_R)\, E_{Raman}\,(\nu_R). \tag{6.41}$$

Hence, the total intensity of the Raman process benefits from both excitation enhancement and re-radiation enhancement.

A total SERS enhancement factor EF_{SERS} can be defined as the ratio between the intensity of the Raman signal in presence of SERS enhancement and the Raman intensity acquired under non-SERS conditions for the same sample.

The total SERS enhancement factor EF_{SERS} can be expressed as:

$$EF_{SERS} = \frac{I_{SERS}}{I_{Raman}} = |\Gamma_{exc}(\nu_L)|^2 \cdot \Gamma_{rad}(\nu_R)^2 = \left|\frac{E_{loc}(\nu_L)}{E_0(\nu_L)}\right|^2 \cdot \left|\frac{E_{rad}(\nu_R)}{E_{Raman}(\nu_R)}\right|^2. \tag{6.42}$$

Since in *SERS* experiments, the difference in frequency between excitation and Raman signal is typically small compared to the linewidth of the plasmon resonance where the electromagnetic enhancement is relevant, the enhancement factors $\Gamma_{exc}(\nu_L)$ at the excitation frequency ν_L and $\Gamma_{rad}(\nu_R)$ at the frequency ν_R of the Raman signal can be considered approximately equal:

$$\Gamma_{rad}(\nu_R) \cong \Gamma_{exc}(\nu_L). \tag{6.43}$$

The above equation can be written as:

$$EF_{SERS} = |\Gamma_{exc}(\nu_L)|^2 \cdot \Gamma_{rad}(\nu_R)^2 \cong |\Gamma_{exc}(\nu_L)|^4 = \left|\frac{E_{loc}(\nu_L)}{E_0(\nu_L)}\right|^4. \tag{6.44}$$

The *SERS* enhancement factor EF_{SERS} scales with the fourth power of the local field enhancement. This is the so-called E^4 approximation [4, 10, 11].

Despite many simplifications, it provides a useful estimate for the actual experimental *SERS* enhancements in the case of a single molecule. However, in most SERS experiments, the contribute from different molecules is detected with very different enhancement factors at the different positions on the metal surface (hot spots). In such cases, using a surface-averaged SERS enhancement factor is a more appropriate approach to evaluate the overall performance of a SERS substrate.

The approximation approach briefly presented above neglects some aspects of the problem, such as the influence of the polarization of both incoming and scattered field, or the frequency dependence of the re-mission amplification factor. These features, known as aspects "beyond the E^4 model," have been object of several papers in recent literature [10, 12–16].

6.2 Instrumentation

The main parts of a Raman spectrometer can be summarized as follows: laser source, optical filters, monochromator, and detector. A summary diagram of a Raman system is shown in the left-hand side of Figure 6.9. Briefly, the light scattered by the sample is filtered to separate the weak Raman signal from the stronger Rayleigh scattering. The Raman light is sent into a monochromator, which disperses it into its various components that are then collected by a detector.

Raman scattering can be observed from any angle, but the most used scattering geometries are 90° and 180° (right-hand side of Figure 6.9). In the first case, the Raman signal collection is at 90° with respect to the direction of the excitation laser beam, while in the 180° geometry (back-scattering) the Raman signal is collected in the same direction as the incident laser beam.

The 90° scattering geometry is typically used for the analysis of transparent samples on the macro scale. The back-scattering geometry, instead, is commonly employed in micro-Raman systems, i.e. a Raman spectrometer combined with an optical microscope.

6.2.1 Laser Sources and Optical Filters

Laser is the standard light source used in Raman spectroscopy. Mercury arc lamps were used as a light source before the invention of laser in 1960. Monochromaticity, linear polarization, and high power focusing on a small area are the characteristics that make lasers the ideal excitation source for Raman scattering [3]. However, the high power focused on small areas requires special care to avoid thermal damage of the sample.

Laser wavelengths ranging from ultraviolet (UV) through visible to near infrared (NIR) are typically used for Raman spectroscopy. In the past, gas lasers such as He-Cd (325 and 441.6 nm), He-Ne (632.8 nm), Ar-ion (514.5, 488, 457.9 nm), and Kr-ion (406.7–676.4 nm) have been widely used, but now laser diodes, more compact and cheaper, are preferred. A description of the operating principles of the light sources is given in Chapter 1.

The choice of laser wavelength depends on sample properties as well as on the availability of the equipment in the laboratory. As shown in Section 6.1, Raman scattering intensity is inversely proportional to the fourth power of the excitation wavelength, meaning that shorter wavelength gives stronger Raman signal than longer wavelengths. However, some materials show fluorescence when excited with visible light, which tends to overcome the weaker Raman signal. In this case, red or

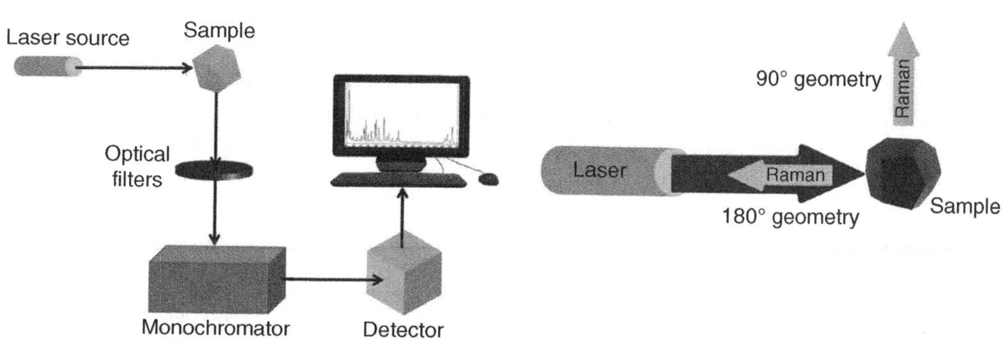

Figure 6.9 On the left: Schematic diagram of the main parts of a Raman spectrometer. On the right: The two most used Raman scattering geometries.

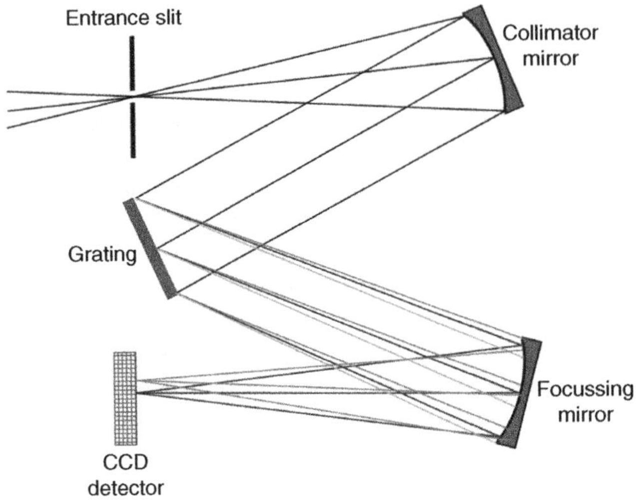

Figure 6.10 Schematic of a Czerny–Turner monochromator. The divergent light that passes through the entrance slit is collimated by a concave mirror and directed onto a diffracting grating. The light is then focused by a second concave mirror and imaged onto the CCD detector.

NIR light is usually good for fluorescence suppression [17]. Visible lasers are most popular for inorganic materials, resonance Raman experiments, and surface-enhanced Raman scattering (SERS).

Optical filters are fundamental components in Raman spectroscopy as they selectively block the laser light, letting only the Raman signal to pass. These optical components are placed in the Raman beam path and for each laser wavelength, an individual filter is needed to remove the Rayleigh scattering, which is usually 10^3–10^5 higher than the Raman signal. The mainly used filters are:

- Notch filter: It is a band filter, which absorbs radiation at the laser wavelength, in a range of a few nanometers, and lets the rest of the radiation through. The advantage of the notch is that it allows simultaneous measurements of Stokes and anti-Stokes spectra, but it degrades over time and has a finite lifetime.
- Edge filter: It is a long pass filter that blocks all light up to a determined wavelength and then transmits it with high efficiency. It passes Raman signal to the long-wavelength side of excitation, allowing the recording of Stokes spectra only, but it has the advantage of blocking very close to the laser line and it is environmentally stable with very long lifetimes.

An alternative way to separate Raman scattering from Rayleigh scattering is to use three monochromators linked together. The advantage of using this configuration is that the laser filter is infinitely variable and can be used to work with any laser source. However, this setup is bulky and more complex to use compared to the more standard filter-based single monochromator systems[1].

6.2.2 Monochromators

Monochromators typically use planar gratings. The most common design is the Czerny–Turner consisting of two concave mirrors and one planar diffraction grating (Figure 6.10). The grating splits

1 www.horiba.com 2020

the incident radiation into its component wavelengths, so that they can be revealed on a charge coupled device (CCD) detector and sent to the computer to be processed.

Monochromator focal length (the distance between focusing mirror and detector) and grating affect spectral resolution, which is the ability of a spectroscopic system to separate adjacent peaks. High spectral resolution is achieved with a long focal length and with a grating with a high number of grooves per millimeter (the more grooves the greater the dispersion). Modern spectrometers are more compact (focal length ranging from 200 to 800 mm) than those used in the past (focal length up to one meter for high resolution), thanks to the use of notch filters, which suppress the intense Rayleigh scattered light and the stray light (any unwanted light that reaches the detector). The choice of grating also depends on the wavelength used. Blazed gratings are optimized to provide high diffraction efficiency (typically ≥80%) at a specific wavelength. The blazed wavelength is the wavelength at which scattering is more efficient.

6.2.3 Detectors

As already discussed, the Raman signal is very weak and therefore its detection is an important topic. In the past, photomultiplier tubes (*single-channel technique*) were used to detect the scattered light, at a given wavelength. The spectrum in a given spectral region is obtained step by step through the rotation of the diffraction grating, resulting in quite long acquisition times. The introduction of solid state detectors allowed to speed up measurement times and to increase the sensitivity of the spectrometer. A charge coupled device (CCD) is a silicon-array detector of UV, visible, and NIR (up to 800 nm) light. It is typically composed of thousands or millions of individual detector elements (also known as pixels). Such devices are greatly sensitive to light and they are able to detect an entire Raman spectrum in a single acquisition (*multichannel technique*). The choice of the correct CCD strongly influences the performance of the instrument and it depends on many factors, such as the spectral range and spectral resolution, among others (See Footnote 1). For example, the spectral range depends on the size of the CCD and the focal length of spectrograph; the pixel size affects the spectral resolution (the smaller the pixels, the higher the spectral resolution). Typically, CCDs are cooled to reduce the dark noise and thus increase their sensitivity. Most Raman systems use Peltier-cooled (down to −90°) detectors, but for certain specialized applications, liquid nitrogen-cooled detectors are preferred.

Finally, it is worth mentioning that at longer wavelength silicon becomes transparent to the incident photons. Therefore, detectors based on other semiconductor materials, such as indium gallium arsenide (InGaAs), are commonly used in the detection range from 800 to 1800 nm [17, 18].

6.2.4 Raman Microscopy and Raman Mapping

As mentioned above, a micro-Raman system combines a Raman spectrometer with an optical microscope. Raman microscopy has extended the field of application of the Raman technique to the characterization of small samples (down to micron scale) and multilayer films as well as the generation of Raman maps in an unprecedent manner.

The exciting laser beam is focused, through the microscope objective lenses, into a microscopic spot on the sample, then the Raman scattered light is collected by the same objective (backscattering geometry) and sent to the spectrometer.

A high spatial resolution is obtained by appropriately choosing the excitation wavelength, λ, and the numerical aperture, NA, of the objective.

The laser spot size is limited by diffraction and its diameter can be calculated using the following equation (Airy disk)

$$\text{Laser spot diameter} = 1.22\lambda/\text{NA}. \tag{6.45}$$

For example, with a 532 nm laser, and a 0.9/100× objective, the spot diameter is 721 nm. From the Eq. (6.45), it can be seen that lower wavelengths lasers offer higher spatial resolution (i.e. a 473 nm laser will have a smaller spot size than a 633 nm laser if the same objective is used), with the same NA of the objective.

The theoretical spatial resolution on a plane can be estimated by the Rayleigh criterion

$$\text{Spatial resolution} = 0.61\lambda/\text{NA}. \tag{6.46}$$

A theoretical spatial resolution of 361 nm is achieved with a 532 nm laser and a 0.9/100× objective. However, Eq. (6.46) is applicable for standard light microscopy. The Raman spatial resolution is limited by more complex optical processes, such as scattering of the laser and Raman photons and interaction with interfaces in the sample. Thus, with standard samples, a typical Raman spatial resolution is in the order of 1 μm.

A higher degree of lateral (*XY*) spatial resolution, approaching the diffraction limit, along with depth (*Z*) spatial resolution can be achieved in semitransparent materials by using a confocal microscope. The principle of confocal design (see left side of Figure 6.11) is to attenuate the radiation scattered by the out-of-focus regions of the sample by means of a fully adjustable confocal pinhole aperture [18]. The analysis of well-defined sample volume is thus possible, with an achievable depth resolution in the order of 1–2 μm that allows to analyze individual layers of a sample. It

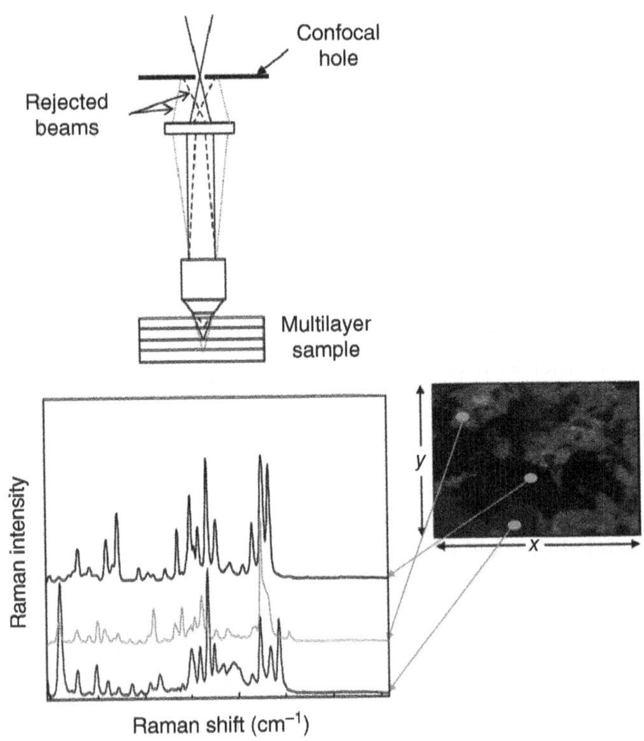

Figure 6.11 On the left: Principle of confocality. Suppression of radiation scattered by the out-of-focus regions of the sample by means of a confocal hole. Note that if the sample is opaque, the laser will not penetrate below the focal plane. On the right: Pictorial description of Raman map. Each pixel of the false color map corresponds to a Raman spectrum.

should be noted that if the sample to be analyzed is opaque, the laser will not penetrate below the focal length. Other advantages of confocal Raman microscopy over conventional Raman microscopy are the rejection of stray light and fluorescence.

A detailed treatment of Raman microscopy is beyond the scope of this chapter. The interested reader may refer to Refs. [17–19].

Raman mapping is an emerging imaging technique that allows to detect variation in heterogeneous samples with submicron spatial resolution. The map is obtained by scanning point-by-point a well-defined region of the sample. This can be obtained by moving the sample with a motorized *XYZ* sample stage or by scanning the laser in the sample plane. Hundreds of Raman spectra are collected, one for each pixel, resulting in a false color map based on material composition, phase, crystallinity, and strain (see right-hand side of Figure 6.11). The image is then constructed depending on the Raman feature under investigation (band frequency, bandwidth, band intensity, band ratio, etc.). Spatial resolution in Raman mapping is governed by the laser spot size and the distance between two acquisition points on the sample.

6.3 Case Studies

Raman spectroscopy has become a crucial technique for material characterization. As described in Section 6.1.3, the set of normal vibrations of a molecule is determined by the structure of the molecule itself. This makes vibrational spectroscopy, and consequently Raman spectroscopy, a suitable tool for substance identification. This section is aimed to offer a practical guide for interpreting a Raman spectrum. Concrete examples will be also given.

6.3.1 Raman Indicators

A Raman spectrum is commonly plotted as intensity in function of the Raman shift. The Raman shift, which is proportional to the energy difference between the incident and the scattered light, is usually expressed in wavenumbers (number of wavelengths per centimeter). The analysis of the Raman spectra provides exhaustive structural and physical information on the different materials, for example the presence of stress, defects, doping, and the degree of disorder in their structure.

The different features of a Raman spectrum are related to different properties of the analyzed material. The first band features to examine when looking at a Raman spectrum, in order to get preliminary information of the substance under study, are: frequency position, intensity, width, and shift in the frequency position. The band frequency position is observed to identify the material. As mentioned in Section 6.1.1, the vibrational spectrum is unique for each molecule, allowing to obtain the distinctive "fingerprint" of the material. The intensity of the Raman band is related to the substance concentration. Moreover, if the vibrations of the molecule produce a change in the polarizability of the electron cloud around the molecule, intense Raman bands appear. Finally, the bandwidth provides information on the crystallinity of the structure and the presence of defects and, together with a shift in the frequency position, it can evidence stress and/or doping in the sample.

6.3.2 Identification of Materials and Crystalline Quality

A first example of material identification can be seen in Figure 6.12a. The figure compares the spectra of two semiconductors: crystalline silicon and single crystal diamond. Diamond, which is composed of sp^3 hybridized carbon atoms, shows a well-defined Raman spectrum.

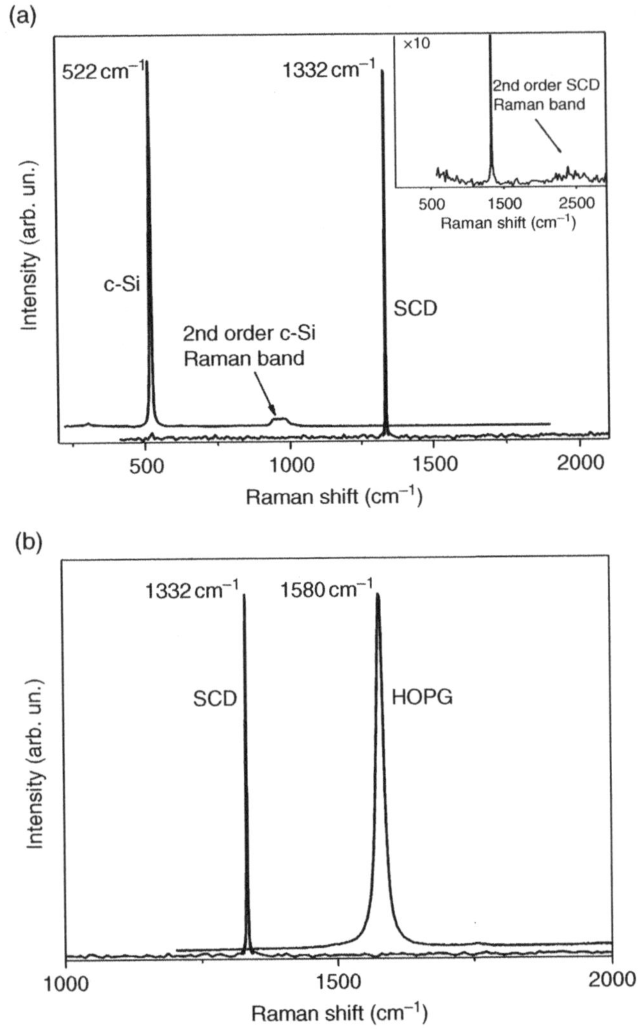

Figure 6.12 (a) Comparison between the Raman spectra of crystalline silicon (c-Si) and single crystal diamond (SCD). The arrow indicates the second-order Raman band of crystalline silicon, while the second-order Raman band of single crystal diamond is shown in the inset. (b) Comparison between the Raman spectra of single crystal diamond and highly oriented pyrolytic graphite (HOPG). Raman measurements were performed at room temperature on an Instrument S.A. Ramanor U1000 double monochromator (HORIBA Jobin Yvon), equipped with a microscope Olympus BX40 for micro-Raman sampling. The 514.5 nm line of a Ar ion laser was utilized as exciting source.

The first-order Raman band of single crystal diamond is a single sharp and intense line occurring at 1332 cm^{-1} with a full width at half maximum (FWHM) of about 2 cm^{-1}, which results from the inelastic scattering of the laser light by the triply degenerate optical phonon at the Brillouin zone center [20–22]. However, the Raman spectrum of diamond includes more bands that manifest themselves as second-order Raman lines, although their intensities are negligible compared to that of the first-order band (see the inset of Figure 6.12a).

Crystalline silicon is also composed of sp^3 hybridized atoms. The Raman spectrum shows a narrow and symmetric peak centered at 522 cm^{-1} with a full width at half maximum (FWHM) of about

4 cm^{-1}. This is the first-order optical phonon mode, which is the most intense peak in the Raman spectrum of crystalline silicon. Moreover, another broad peak appears around 960 cm^{-1}, resulting from a second-order phonon mode [23].

In short, the two materials possess the same crystal structure, are both sp^3 hybridized and both semiconductors, nevertheless produce easily distinguishable Raman spectra. This demonstrates that with Raman spectroscopy, it is possible to identify and differentiate even materials that possess the same crystal structure.

As explained in Section 6.1.3, the frequency of vibration of a molecule is strongly affected by the stiffness of the bonds between the atoms involved. With the same mass of atoms, weak bonds have lower Raman shifts compared to strong bonds. To give one example, Figure 6.12b shows the comparison between two allotropes of carbon: diamond and graphite. In sp^2 hybridized carbon system, the in-plane sp^2 bonds have higher stiffness compared to the sp^3 ones. That is why the first-order Raman band for diamond occurs at lower Raman shift than the first-order band in graphite, as shown in Figure 6.12b.

To show how Raman spectroscopy allows to detect the crystalline quality of a material, the spectra of single crystal diamond (SCD) and polycrystalline diamond (poly-D) are compared in Figure 6.13.

In the figure, the first-order Raman band of polycrystalline diamond film shows a shift and a significant increase in FWHM of about 7 cm^{-1}. It has been shown that the first-order band in polycrystalline diamond can reach a width ranging from 5 to 15 cm^{-1} [24] and that its shape and position are strongly affected by stress [25, 26].

In presence of stress, the triple degeneracy of the optical phonon mode is lifted, leading to diamond peak splitting.

The magnitude of the splitting and shift of the Raman band is affected by the degree and nature of the stress. Moreover, the inhomogeneity of local stress distribution may cause a broadening of the band [27]. Since the different grains of the polycrystal are under different stress levels, their corresponding Raman lines are consequently centered at different positions. Thus, if the laser probe samples an area enclosing a great deal of grains, a broad and nearly symmetric peak is observed (see Figure 6.13a), arising from the convolution of the responses coming from the individual grains, as shown in the simulation of Figure 6.13b. Hence, the large linewidth observed in the Raman spectrum of the polycrystalline diamond is caused not only by a low crystalline quality of the sample, but also by a large anisotropic stress inhomogeneously distributed in the whole film. The local stress distribution is clearly visible on a smaller scale, when a single grain is investigated. As shown in the inset of Figure 6.13a, the Raman band is split into two components, at 1329 and 1333 cm^{-1}.

6.3.3 Graphene and Graphite Raman Spectra

Over the years, Raman spectroscopy has become a crucial technique for the characterization of carbon-based materials, being widely used to study different carbon allotropes such as diamond, carbon nanotubes, graphene, and highly oriented pyrolytic graphite (HOPG), among others. Graphene, a single flat sheet of carbon atoms arranged in honeycomb 2D structure, is considered the mother of all graphitic materials since it is the basic building block for all sp^2 hybridized materials of every other dimensionality [28]. A common feature of all sp^2 carbon materials is the so-called G band, peaked at 1580 cm^{-1}, which is produced by the in-plane stretching of the C–C bond in both rings and chains. By looking at the G band, one can get information about any perturbation in the structure of graphene. In particular, the frequency position and shape of the G band are strongly affected by strain effects [29]. Besides the G band, another band occurring

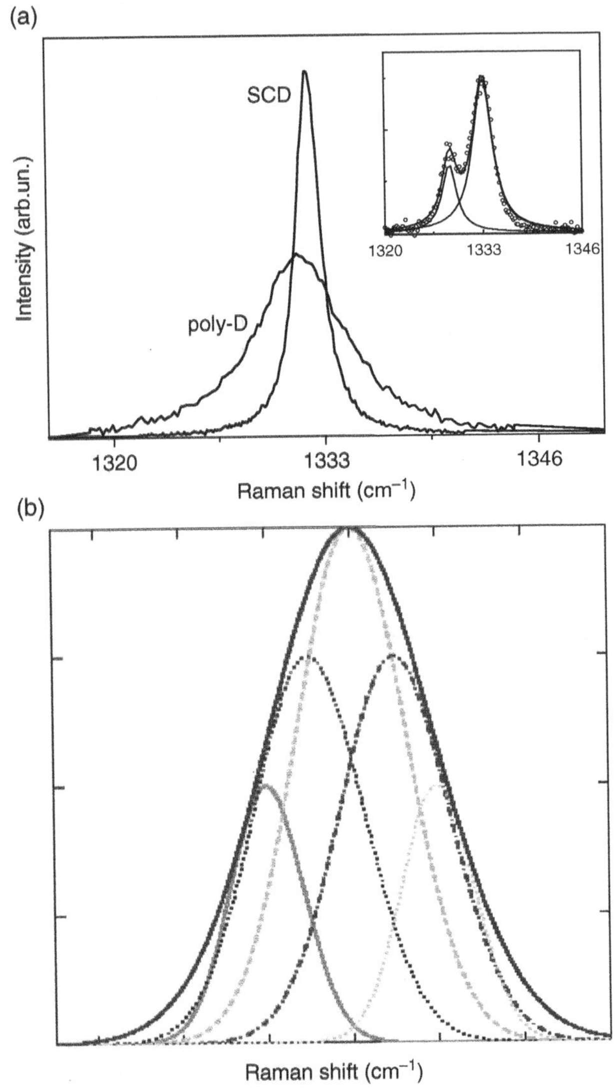

Figure 6.13 (a) Raman spectra of single crystal diamond and polycrystalline diamond. The inset shows the Raman spectrum of a single grain. (b) Simulation showing the convolution of the responses coming from the individual grains of polycrystalline diamond. Raman measurements were performed at room temperature on an Instrument S.A. Ramanor U1000 double monochromator (HORIBA Jobin Yvon), equipped with a microscope Olympus BX40 for micro-Raman sampling. The 514.5 nm line of a Ar ion laser was utilized as exciting source.

in the frequency range of 2500–2800 cm^{-1} is the second common factor of all kinds of sp^2 carbon materials. The so-called 2D band is a second-order Raman band that originates from a double resonance Raman process [30].

This band is highly sensitive to any modification of the electronic structure. The observation of this Raman band allows to distinguish single-layer graphene from crystalline graphite, as shown in Figure 6.14. In particular, shape, width, and frequency position of the 2D peak deliver detailed information on the number and stacking order of graphene layers [30]. Single-layer graphene

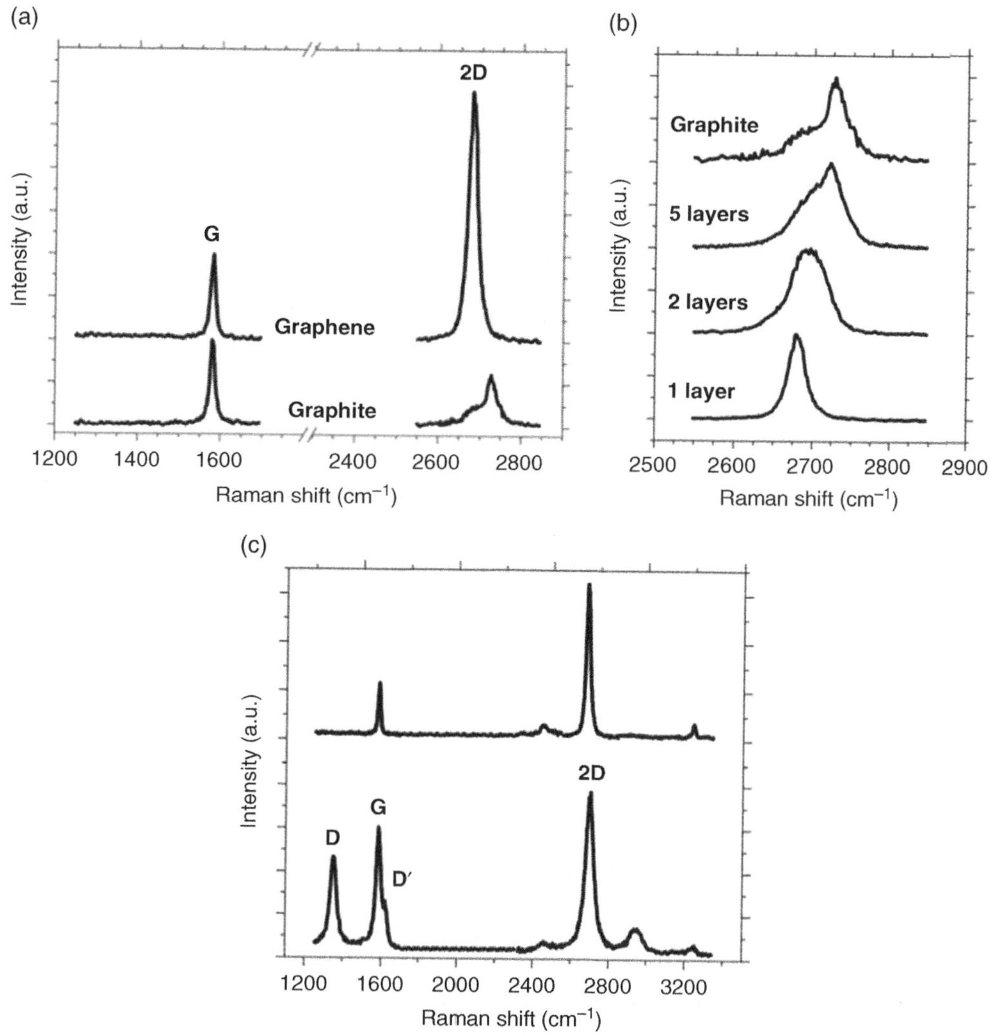

Figure 6.14 (a) Raman spectra of graphite and mechanically exfoliated graphene. (b) Modification of the 2D peak occurring with the increasing of graphene layers. (c) Single (top) and multilayer graphene produced via chemical vapor deposition. The occurrence of the D and D′ bands indicates the presence of defects in the sample. Raman measurements were obtained by using the spectrometer LabRam HR-evolution (HORIBA Scientific) with an integrated Olympus BX41 microscope (excitation wavelength $\lambda = 532$ nm).

produced by mechanical exfoliation of bulk graphite shows a sharp 2D peak, well fitted to a single Lorentzian component with FWHM of about 24 cm^{-1} (see Figure 6.14a). In bilayer graphene, instead, the 2D peak is asymmetric and can be fitted to four Lorentzian components. If the graphene layers are more than five, the Raman spectrum becomes hardly distinguishable from that of bulk graphite, as displayed in Figure 6.14b. In unintentionally doped graphene produced via chemical vapor deposition (see Figure 6.14c), the shape of the 2D peak does not show significant variations from single to multilayer, due to a lower electronic coupling between graphene layers with no stacking order [30]. In fact, a narrow Lorentzian line (FWHM of about 30–40 cm^{-1}) can be used to fit the 2D peak of both single-layer and bilayer CVD graphene, and for regions with three or more layers, the 2D (Lorentzian) peak increases to a FWHM of about 70 cm^{-1}. In this case, the indicator of number of graphene sheets is given by the G-to-2D intensity ratio (I_G/I_{2D}): single layer (1L) or bilayer (2L) is characterized by $I_{2D}/I_G > 1$, while in multilayer graphene, $I_{2D}/I_G < 1$ [31, 32].

Raman spectroscopy is a key tool to identify disorder in the lattice of carbon structures. Concerning sp^2 hybridized carbon systems, the presence of disorder or structural defects is established by the occurrence of the so-called D band at around 1350 cm^{-1} [33]. The D band originates from a double resonance process and requires a defect for its activation. It arises from a breathing mode of six-atom rings [34]. Its intensity is correlated to the amount of defects [35]. In particular, the intensity ratio between the D band and the G band I_D/I_G is the Raman indicator of crystalline quality of carbon-based materials. Besides the D band, another band is also observed in samples with defects, called D′ (1620 cm^{-1}). The D′ band arises from an intravalley double resonance scattering process [36].

6.3.4 Doping Detection

As already mentioned, Raman spectroscopy is able to provide useful information on the effect of dopants, impurities, and defects in materials since their presence breaks the symmetry of the crystal, affecting the lattice vibrational modes. To provide a clear example, intrinsic and doped silicon carbide (SiC) films were both characterized by means of micro-Raman spectroscopy.

Crystalline SiC forms long-range ordered structures and, according to the possible stacking sequence along the crystallographic axis direction, it is characterized by many different polytypes, each one identified by their own featuring electrical and optical properties [37]. Cubic, hexagonal, and rhombohedral are the main SiC structures.

Among all of the polytypes, the hexagonal structure 4H-SiC is viewing a rapid increase of its use as a suitable material for optoelectronic applications [38]. Indeed, it is characterized by visible blindness [39], high sensitivity to UV photons, and radiation-hardness [40]. Such properties combined with controllable n- and p-type doping make it an appropriate material for several potential applications. Figure 6.15 shows typical Raman data for intrinsic and variously doped 4H-SiC, taken at room temperature. In the spectrum of the intrinsic sample, the peaks at 204.3 cm^{-1} and at 609.8 cm^{-1} originate from a transverse acoustic (TA) and a longitudinal acoustic (LA) mode, respectively. The peaks at 776.4 cm^{-1} and at 797.5 cm^{-1} derive from two transverse optical (TO) modes and the peak appearing at 963.6 cm^{-1} originates from a longitudinal optical (LO) mode [41]. Furthermore, two weaker peaks are observed in the spectrum at 195.7 and 265.5 cm^{-1} arising from two transverse acoustic modes.

4H-SiC crystal is regarded such an important material especially in its doped configuration. In Figure 6.15a, p-type doped sample spectrum is compared with the intrinsic one. The strongest modes, quite sensitive to the type and degree of crystallization, are roughly centered as for the intrinsic sample. In particular, all the Raman peaks shifted by about 1 cm^{-1} to higher frequency and an increase of the low frequency baseline, proportional to the doping level, is evidenced too [42]. A new peak, characterized by low intensity and high width, is registered at about 164 cm^{-1}, as shown in the inset of Figure 6.15a. Raman spectrum of n-doped sample is shown in Figure 6.15b. The featuring peaks are still the same as reported for the intrinsic sample, unless for the LO mode at about 967 cm^{-1}. Significant changes are observed in the shape and position of this band as a function of increasing doping concentration [43]. The presence of nitrogen determines a broadening and a shift to higher frequency of the peak, while its intensity decreases sharply.

6.3.5 Basic Examples of SERS

The examples of Raman spectroscopy presented so far refer to materials that are all good Raman scatterers. On the other hand, when the materials under examination do not show a strong Raman

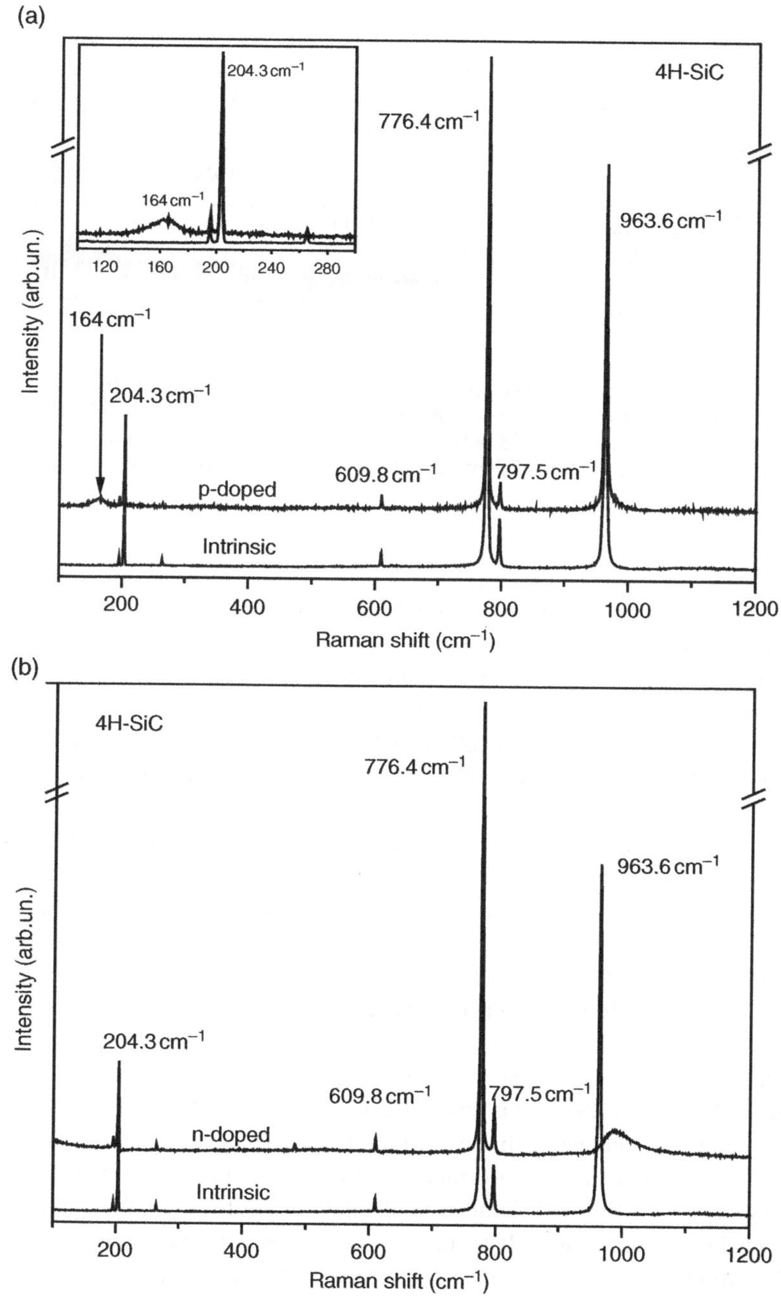

Figure 6.15 Raman spectra of (a) p-doped and (b) n-doped 4H-SiC samples, both compared with the intrinsic 4H-SiC. The inset in panel a is a magnification of the low Raman shift region. Raman measurements were obtained by using the spectrometer LabRam HR-evolution (HORIBA Scientific) with an integrated Olympus BX41 microscope (excitation wavelength λ = 532 nm).

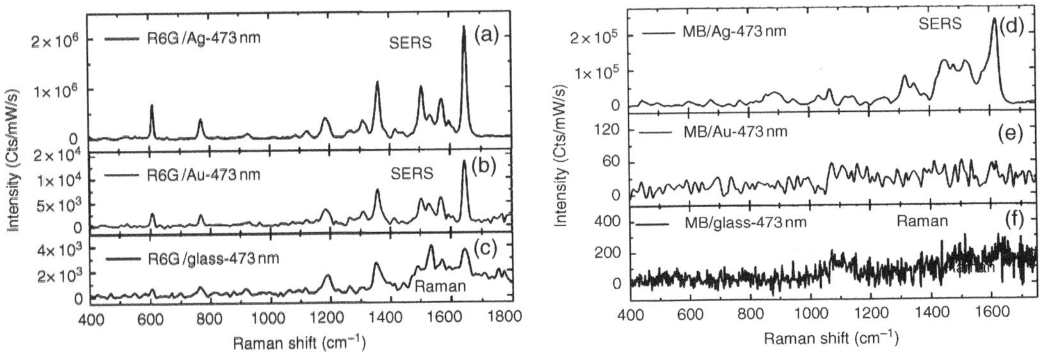

Figure 6.16 On the left: Comparison between the Raman spectra obtained from (a) R6G/Ag, (b) R6G/Au, and (c) R6G on glass (λ_{exc} = 473 nm). On the right: Comparison between the Raman spectra obtained from (d) MB/Ag, (e) MB/Au, and (f) MB on glass (λ_{exc} = 473 nm). Raman measurements were obtained by using the spectrometer LabRam HR-evolution (HORIBA Scientific) with an integrated Olympus BX41 microscope.

signal, SERS technique is used. As described in Section 6.1.9, SERS effect consists in the enhancement of the electromagnetic radiation diffused by molecules adsorbed on plasmonic nanostructures, allowing to detect Raman scattering by even a single molecule.

To give an example of the enhancement effect of SERS, two standard dye molecules, Rhodamine 6G (R6G, $C_{28}H_{31}N_2O_3Cl$) and methylene blue (MB, $C_{16}H_{18}C_lN_3S$) were mixed with gold nanoparticles (from now on referred as R6G/Au and MB/Au) and silver nanoparticles (henceforward indicated as R6G/Ag and MB/Ag).

The mixtures were then deposited on glass microscope slides via drop casting. Silver and gold plasmonic nanostructures are commonly used in SERS experiments because their plasmon resonance frequencies fall within the wavelength range of the radiation applied to excite the Raman modes.

The left-hand side of Figure 6.16 shows the SERS spectra of the R6G/Ag (see panel a) and R6G/Au (see panel b) mixtures compared with the Raman signal of R6G deposited on glass (see panel c). The excitation wavelength was 473 nm.

The results show an enhancement of the Raman signal of about 10^3 in the case of R6G/Ag and by a factor of 5 in the case of R6G/Au. The SERS effect is more intense in the case of R6G/Ag, with the laser excitation wavelength being in resonance with the plasmon frequency of the silver nanoparticles (450 nm). The same measurements were performed for methylene blue with unchanged experimental conditions. As shown on the right-hand side of Figure 6.16, the Raman signal of MB is indistinguishable from instrumental noise both in the case of MB/Au mixture as well as for MB on glass (see panels e and f). On the other hand, in the case of MB/Ag (see panel d) a very intense SERS signal is obtained, especially at high vibrational frequencies.

References

1 Raman, C.V. (1928). *Indian J. Phys.* 2: 387–398.

2 Long, D.A. (2002). *The Raman Effect: A Unified Treatment of the Theory of Raman Scattering by Molecules*. Wiley.

3 Ferraro, J.R., Nakamoto, K., and Brown, C.W. (2003). *Introductory Raman Spectroscopy*. Elsevier.

4 Le Ru, E. and Etchegoin, P. (2009). *Principles of Surface Enhanced Raman Spectroscopy and Related Plasmonic Effects*. Elsevier.

5 Grasselli, J.G. and Bulkin, B.J. (eds.) (1991). *Analytical Raman Spectroscopy*. Wiley.

6 Califano, S. (1976). *Vibrational States*. Wiley.

7 Ashcroft, N.W. and Mermin, N.D. (1976). *Solid State Physics*. Hacourt College Publishers.

8 Kittel, C. (2005). *Introduction to Solid State Physics*, 8e. Wiley.

9 Fleischman, M., Hendra, P.J., and McQuillan, J. (1974). *Chem. Phys. Lett.* 26: 163–166.

10 Gucciardi, P.G., Lami de la Chapelle, M., and Lidgi-Guigui, N. (2005). *Handbook of Enhanced Spectroscopy*. Pan Stanford Publishing.

11 Jones, P., Maragò, O.M., and Volpe, G. (2015). *Optical Tweezers, Principles and Applications*. Cambridge University Press.

12 Fazio, B., D'Andrea, C., Bonaccorso, F. et al. (2009). Re-Radiation Enhancement in Polarized Surface Enhanced Resonant Raman Scattering of Randomly Oriented Molecules on Self-Organized Gold Nanowires. *ACS Nano* 5: 5945–5956.

13 Foti, A., D'Andrea, C., Messina, E. et al. (2015). On the SERS depolarization ratio. *Nanospectroscopy* 1: 26–32.

14 Le Ru, E.C., Grand, J., Félidj, N. et al. (2008). Experimental verification of the SERS electromagnetic model beyond the $|E|^4$ approximation: polarization effects. *J. Phys. Chem. C* 112: 8117–8121.

15 Etchegoin, P.G., Galloway, C., and Le Ru, E.C. (2006). Polarization-dependent effects in surface-enhanced Raman scattering (SERS). *Phys. Chem. Chem. Phys.* 8: 2624–2628.

16 Wei, H., Hao, F., Huang, Y. et al. (2008). Polarization dependence of surface-enhanced Raman scattering in gold nanoparticle–nanowire systems. *Nano Lett.* 8: 2497–2502.

17 Smith, E. and Dent, G. (2019). *Modern Raman Spectroscopy: A Practical Approach*, 2e. Wiley.

18 Lewis, I.R. and Edwards, H.G.M. (2001). *Handbook of Raman Spectroscopy: From the Research Laboratory to the Process Line*. Taylor & Francis Group, LLC.

19 Turrell, G. and Corset, J. (1996). *Raman Microscopy: Developments and Applications*. Elsevier Ltd.

20 Donato, M.G., Messina, G., Verona Rinati, G. et al. (2009). Exciton condensation in homoepitaxial chemical vapor deposition diamond. *J. Appl. Phys.* 106: 053528. (1–5).

21 Pickard, C.D.O., Davis, T.J., Wang, W.N., and Steeds, J.W. (1998). Mapping crystalline quality in diamond films by micro-Raman spectroscopy. *Diamond Relat. Mater.* 7: 238–242.

22 Alfieri, D., Almaviva, S., De Sio, A. et al. (2010). Single-crystal diamond MIS diode for deep UV detection. *Radiat. Eff. Defects Solids* 165: 737–745.

23 Liu, J., Niu, J., Yang, D. et al. (2004). Raman spectrum of array-ordered crystalline silicon nanowires. *Physica E* 23: 221–225.

24 Ager, J.W., Kirk Veirs, D., and Rosenblatt, G.M. (1991). Spatially resolved Raman studies of diamond films grown by chemical vapor deposition. *Phys. Rev. B* 43: 6491–6499.

25 Stuart, S.A., Prawer, S., and Weiser, P.S. (1993). Growth-sector dependence of fine structure in the first-order Raman diamond line from large isolated chemical-vapor-deposited diamond crystals. *Appl. Phys. Lett.* 62: 1227–1229.

26 Fayette, L., Marcus, B., Mermoux, M. et al. (1997). A sequential Raman analysis of the growth of diamond films on silicon substrates in a microwave plasma assisted chemical vapor deposition reactor. *J. Mater. Res.* 12: 2686–2698.

27 Donato, M.G., Faggio, G., Marinelli, M. et al. (2001). *Diamond Relat. Mater.* 10: 1535–1543.

28 Geim, A.K. and Novoselov, K.S. (2007). The rise of graphene. *Nat. Mater.* 6: 183–191.

29 Dresselhaus, M.S., Jorio, A., Hofmann, M. et al. (2010). Perspectives on carbon nanotubes and graphene Raman spectroscopy. *Nano Lett.* 10: 751–758.

30 Malard, L.M., Pimenta, M.A., Dresselhaus, G., and Dresselhaus, M.S. (2009). Raman spectroscopy in graphene. *Phys. Rep.* 473: 51–87.

31 Reina, A., Jia, X., Ho, J. et al. (2008). Large area, few-layer graphene films on arbitrary substrates by chemical vapor deposition. *Nano Lett.* 9: 30–35.

32 Faggio, G., Messina, G., Lofaro, C. et al. (2020). Recent advancements on the CVD of graphene on copper from ethanol vapor. *C* 6: 14–27.

33 De Luca, O., Grillo, R., Castriota, M. et al. (2018). Different spectroscopic behavior of coupled and freestanding monolayer graphene deposited by CVD on Cu foil. *Appl. Surf. Sci.* 458: 580–585.

34 Ferrari, A.C. and Basko, D.M. (2013). Raman spectroscopy as a versatile tool for studying the properties of graphene. *Nat. Nanotechnol.* 8: 235–246.

35 Capasso, A., Rodrigues, J., Moschetta, M. et al. (2020). Interactions between primary neurons and graphene films with different structure and electrical conductivity. *Adv. Funct. Mater.*: 2005300. (1–11) (Early View, https://doi.org/10.1002/adfm.202005300).

36 Basko, D.M., Piscanec, S., and Ferrari, A.C. (2009). Electron–electron interactions and doping dependence of the two-phonon Raman intensity in graphene. *Phys. Rev. B* 80: 165413. (1-10).

37 Neudek, P.G. (1995). Progress in silicon carbide semiconductor electronics technology. *J. Electron. Mater.* 24: 283–288.

38 Zhao, F. et al. (2009). Optically activated 4H-SiC p-i-n diodes for high-power applications. *IEEE Electron Device Lett.* 30: 1182–1184.

39 Liu, X.F., Sun, G.S., Li, J.M. et al. (2007). Vertical PIN ultraviolet photodetectors based on 4H-SiC homoepilayers. *Phys. Stat. Sol. (c)* 4: 1609–1612.

40 Akturk, A., Goldsman, N., Aslam, S. et al. (2008). *J. Appl. Phys.* 104 (026101): 1–3.

41 Burton, J.C., Sun, L., Pophristic, M. et al. (1998). Spatial characterization of doped SiC wafers by Raman spectroscopy. *J. Appl. Phys.* 84: 6268–6273.

42 Sartel, C., Soulière, V., Zielinski, M. et al. (2005). Control of Al-doping in 4H–SiC homo-epitaxial layers grown with a HMDS/TMA/P mixture. *Phys. Stat. Sol. (a)* 202: 550–554.

43 Nakashima, S., Nakatake, Y., Ishida, Y. et al. (2001). Detection of defects in SiC crystalline films by Raman scattering. *Phys. B Condens. Matter* 308: 684–686.

7

Thermally Stimulated Luminescence

Federico Moretti

Lawrence Berkeley National Laboratory, Berkeley, CA, USA

Thermally stimulated luminescence (TSL) – also called thermoluminescence – is a very well-established experimental technique that finds use on a daily basis in several fields, ranging from radiation dosimetry, to archeological and geological dating, and to investigation of defects in insulators and semiconductors. Several books and papers dedicated to TSL have been published over the years, both on fundamental aspects and on its applications. This chapter is only a quick introduction to the technique itself and the fundamental physical phenomena underlying it. More advanced topics can be found in the excellent books by McKeever [1], Chen and McKeever [2], and Sunta [3], which have been consulted several times in the writing of this chapter. Other, shorter, introductory papers that are highly recommended are from Martini and Meinardi [4], and Bos [5, 6].

Few examples on TSL application taken from the literature are also used to highlight the versatility of TSL as a research tool, also in combination with other experimental methods, notably EPR, present in this book.

Before starting to delve into the subject, it is appropriate to give a definition of what thermally stimulated luminescence is:

TSL is the light emitted by an insulator or a semiconductor while it is heated up after a previous absorption of energy from radiation.

This definition encompasses all the fundamental elements necessary for the TSL phenomenon to occur: The investigated material must possess luminescent properties, meaning that the material will emit light in the ultraviolet (UV), visible, or near-infrared region of the electromagnetic spectrum given the right excitation; it must have absorbed energy at some time by irradiation, usually ionizing radiation (X- and γ-rays, α- and β-particles, etc.), but in some cases UV light might be sufficient; and finally, the heating procedure triggers the release of the energy stored in the material leading to the luminescence phenomenon, and it is not by itself the cause of the luminescence (incandescence is not TSL). What is not apparent from the definition, however, is that the TSL is a transient phenomenon. Once heated up, a material cannot be made to emit TSL by simply cooling and reheating it, but it has to be irradiated again.

Spectroscopy for Materials Characterization, First Edition. Edited by Simonpietro Agnello.

7.1 Theory of Thermally Stimulated Luminescence

As useful as the above definition is in the description of the TSL phenomenon, it does not give any information about the physical processes that are responsible for it. To understand TSL, we have to refer to the energy band theory of solids [7]. The solution of the Schrödinger equation for electrons in a periodically varying potential shows that there are bands in energy in which the electrons are allowed to stay. Other possible energy values are "forbidden zones" or energy bandgaps. The highest energy band that contains electrons is known as the valence band, while the next available one in the energy structure is called the conduction band.

In the case of insulators and semiconductors (at 0 K), the valence band is completely filled, which implies that it is not possible to obtain a net transport of charge in the direction of an applied electric field. Electronic conduction is possible only if enough energy is given to the electrons to cross the energy gap, E_g, between the valence and the conduction band. The conductivity of these materials is characterized by a temperature dependence of the form $\exp(-E_g/kT)$, where k is the Boltzmann constant and T is the absolute temperature, and tends to zero as the temperature goes to zero. Additionally, for ideal materials, optical absorption only takes place for photon energies at or above E_g.

This picture is true only for perfect crystals: the presence of lattice imperfections breaks the periodicity of the crystal, giving rise to localized energy states in the bandgap, which can be occupied by electrons or holes given the right conditions, and might also give rise to optical absorption phenomena for photon energies lower than E_g.

There are different kinds of lattice imperfections, or defects, but they can be subdivided into two main categories: intrinsic and extrinsic, according to whether they are inherently present in a material or caused by external means, respectively. Intrinsic defects (examples are missing or misplaced atoms in the structure – vacancies, interstitials, and antisites – or extended defects like stacking faults and grain boundaries) appear for thermodynamic reasons during the material manufacture or natural growth. Extrinsic defects, on the other hand, are caused by the presence of ions that do not "belong" to the pure crystal chemical composition and are incorporated in the crystal itself due to the purity of the raw materials or are purposefully added; in the latter case, they are referred to as dopants. The wide variety of defects can hardly be summarized in this chapter and more information on the subject can be found in books and papers [8–10].

The presence of defects, as mentioned few lines above, in this section, causes the formation of localized levels inside the bandgap. The level position in the bandgap is determined by the type of defect and its effective charge; this, in turn, also determines which charge carrier (electrons or holes) is likely to become trapped on a specific defect. An anion vacancy in alkali halides (a missing X^-, where X = F, Cl, Br, I), for instance, behaves as a Coulomb attractive center for electrons freed from the valence band and moving through the crystal. This attraction causes the electron to become trapped on the vacancy, forming an *F* center. The energy required to release the electron from the trap is, however, less than that needed to release an electron from an X^- ion; thus, the anion vacancy has associated with it an energy level that lies somewhere between the valence and the conduction band. A similar situation occurs also in the case of cationic vacancies, save that in this case holes are trapped on this defect. Although the just presented examples involve a net charge on the defect, this might not be always the case, as exemplified by isovalent antisite defects or germanium impurities in silicon.

All luminescence processes are characterized by a change in occupancy of the various localized energy states. These modifications in the populations result from electronic transitions from one energy state to another. Several kinds of transitions are possible and some are presented in Figure 7.1 for both electrons and holes. Transition 1 is the excitation of a valence electron from

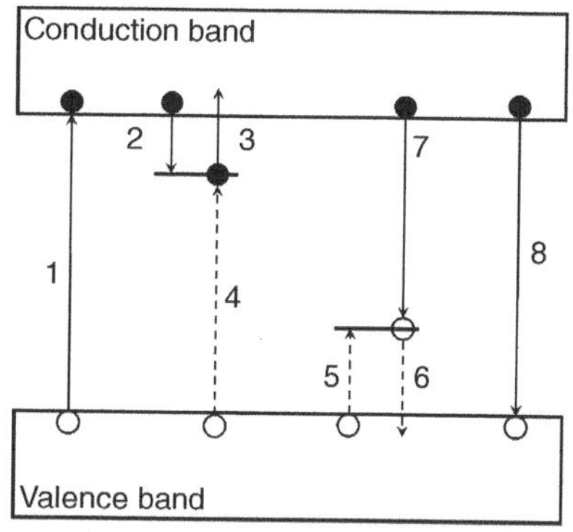

Figure 7.1 Typical electronic transitions in semiconductors and insulators: (1) ionization; (2) and (5) electron and hole trapping, respectively; (3) and (6) electron and hole release; (4) and (7) indirect recombination; (8) direct recombination. Electrons, solid circles; electronic transitions, solid arrows; holes, open circles; hole transition, dashed arrows, energy levels localized inside the bandgap are marked by horizontal lines.

an atom in the material into the conduction band in which the electron possesses enough energy to move freely through the lattice. This transition corresponds to the ionization process and it is the result of energy absorption from an external source like, for example, X-ray or UV radiation. Each electron promoted in the conduction band leaves in the valence band a hole, which is also free to move through the material. Both charges remain delocalized in the respective bands until they arrive at a defect center, becoming localized. This results in the trapping of electrons (transition 2) and holes (transition 5). The localized electrons and holes may be released from their traps by thermal or optical excitation (transitions 3 and 6) back into the delocalized bands where they are again free to move through the crystal. Alternatively, electrons and holes can recombine with charge carriers of opposite sign, either directly (transition 8) or indirectly by recombination with previously trapped carriers (transitions 4 and 7). If either of the recombination mechanism is radiative, then light is emitted. The localized levels inside the bandgap can, thus, behave as traps or recombination centers and it is important to clarify what is the distinction between these two centers.

The classification is usually done on the relative probabilities of recombination and thermal excitation: if transition 3 in Figure 7.1 is more probable than transition 4, then the center can be classified as a trap; if, on the other hand, transition 4 is more probable than transition 3, then the energy level corresponds to a recombination center.

The probability p for unit time for a trapped charge to be thermally promoted in the delocalized band is:

$$p = s \exp\left(-\frac{E}{kT}\right) \tag{7.1}$$

where E is the thermal depth of the trap, or the energy difference between the trap and the edge of the corresponding delocalized band, k is the Boltzmann constant, s is the frequency (or attempt-to-escape) factor, and T is the absolute temperature. It follows that, for a given temperature, centers with small E are more likely to behave as traps than those with larger energy values. It is also easy to see that a trap at a given temperature might behave as a recombination center at lower temperatures, and vice versa.

The above classification also implies that there might be a given temperature at which the probability of carrier recombination and thermal release in the delocalized band is equal for a certain defect level. Such a level, of depth D, represents a demarcation between the traps and the recombination centers, such that a center with energy depth $E < D$ would be a trap, and, conversely, a center with $E > D$ would be a recombination center. Considering that the depth of a center is the energy difference between the localized level and the respective delocalized band (the conduction band for electrons, and the valence band for holes), it follows that there are two demarcation levels, one for electrons and the second for holes. From the definition of the demarcation levels, it is then possible to write

$$s_e \exp\left(-\frac{D_e}{kT}\right) = n_r A_r^e \tag{7.2}$$

$$s_h \exp\left(-\frac{D_h}{kT}\right) = m_r A_r^h \tag{7.3}$$

for electrons and holes, respectively. The left-hand side of these equations represents the probability per unit time of thermal excitation of electrons (Eq. 7.2) and holes (Eq. 7.3) from their respective demarcation levels, thus corresponding to transitions 3 and 6 in Figure 7.1. The right-hand side of the equations is the probability for unit time of a trapped charge carrier recombining with a delocalized one with opposite charge, transitions 4 and 7. A_r^e and A_r^h are the recombination transition coefficients (in $m^3\ s^{-1}$), while n_r and m_r are the concentrations (in m^{-3}) of trapped electrons and holes, respectively. The transition coefficients can be viewed [11] as the products $v_e\sigma_r^e$ and $v_h\sigma_r^h$, where v_e and v_h are the thermal speed of electrons and holes in the delocalized bands, σ_r^e and σ_r^h are the carrier capture cross sections for electron and holes by the corresponding trapped charges. Analogously, it is possible to define transition coefficients for electron and hole trapping on unpopulated centers.

The carrier capture cross section value strongly depends on the potential distribution around the defect and has values of the order of $10^{-17}\ m^2$ for a Coulomb attractive center, to $10^{-19}\ m^2$ for a neutral one and $10^{-26}\ m^2$ for a repulsive one [12]. Since a center that is attractive for charge carriers of one sign is neutral or repulsive for the other one, it follows that each center is characterized by two different capture cross sections, one for electrons and the other for holes, and it is their relative magnitude that determines the behavior of a center in the trapping and the recombination processes. Shockley and Read [13], in their analysis of trapping and recombination process in semiconductors, indeed assume that the charge of a center will have two values, one for electrons and the other for holes, differing by one elementary charge. The trapping of an electron (or the release of a hole) means that the center becomes more negative and less likely to attract an additional electron, and similarly for hole trapping. The cross sections are known only for a few materials, mostly semiconductors, and defect types [14–16].

In the presentation of Eq. (7.1), we introduced the pre-exponential term s (the frequency factor or attempt-to-escape frequency) without giving a justification or a possible physical meaning, and it requires a discussion. The thermal promotion of a bound electron or hole on a trap with energy E to the delocalized band involves two steps. The first is the thermal elevation from the trap ground state to an energy level E above it by interaction with the matrix phonons. The fraction of the trapped electron population that can achieve this energy is given by the Boltzmann function, $\exp(-E/kT)$. These electrons, however, are still not free from the traps and require further assistance to transit into the free state: The frequency factor is the rate of this latter transition [17], and can be written as the product of phonon frequency (ν) with a transition coefficient (κ) and an entropy-related term [17]. According to Williams and Eyring [18], the release rate of trapped electrons can be written as:

$$p_r = \nu\kappa \exp\left(-\frac{\Delta F}{kT}\right) \tag{7.4}$$

where ΔF is the Helmholtz free energy, given by $\Delta F = \Delta U - T\Delta S$, where ΔS is the entropy change during the reaction and ΔU is the activation energy. Equation (7.4) becomes:

$$p_r = \nu\kappa \exp\left(\frac{\Delta S}{k}\right) \exp\left(-\frac{\Delta U}{kT}\right) \tag{7.5}$$

from which it follows that:

$$s = \nu\kappa \exp\left(\frac{\Delta S}{k}\right). \tag{7.6}$$

Considering all these factors, it is easy to see that the values the frequency factor can have are at most equal to the lattice vibration frequency (of the order of 10^{12}–10^{14} s^{-1}), or Debye frequency.

From detailed balance arguments [19], the frequency factor can also be related to the charge carrier capture cross section σ of the trap:

$$s = N_c \sigma v, \tag{7.7}$$

where N_c is the effective density of states in the delocalized band and v is the free carrier thermal velocity. Note that all the factors on the right-hand side of Eq. (7.7) are temperature dependent, suggesting that the frequency factor should also vary with the temperature. However, its variation is small compared to the strong exponential temperature dependence of Eq. (7.1), and it is usually not considered.

7.1.1 Simple Model

Now that we have all the elements, we can introduce the simplest possible model for TSL. This model is based on a one-trap and one-recombination center (OTOR) scheme and, although it is inadequate to describe the typical TSL curves (glow curves) encountered in real materials, it represents the basis of many other more complex models that have been developed over the years.

Figure 7.2 presents a simplified band energy scheme with only two localized levels in the band-gap, one for electrons and behaving as a trap, and the other for holes and behaving as a recombination center. In the case that will be discussed, it is assumed that only electrons can be thermally freed from the trap. The arguments here reported are valid also in the case of holes localized on traps and electrons on recombination centers.

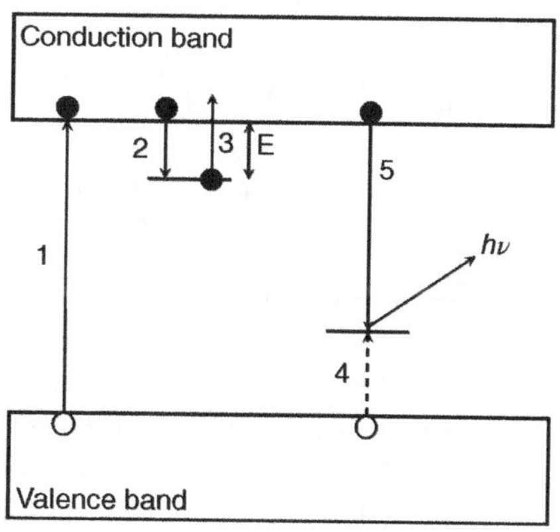

Figure 7.2 Simple two-level model for thermally stimulated luminescence. The allowed transitions are: (1) ionization, (2) and (4) electron and hole trapping, respectively; (3) electron thermal release; (5) radiative recombination and emission of light. Electrons are, in this case, the active carriers, but analogous situation arises for holes. Electron, solid circles; electron transitions solid arrows; holes, open circles; hole transition, dashed arrow.

The absorption of radiation with energy $h\nu$ greater than the bandgap energy E_g results in the ionization of the materials with the production of energetic electrons and holes which, after thermalization, will produce free electrons in the conduction band and free holes in the valence band, transition 1. These free charges can recombine with each other or become trapped. The third option of the charge carriers remaining free in the delocalized bands is not considered since this would imply a greater stable conductivity of the material after irradiation than before. This is in contradiction with experimental results and will not be further discussed.

The direct recombination of electrons and holes, or band to band recombination, is usually not favorable in large-bandgap insulators and semiconductors compared to the indirect, Shockley and Read recombination on a localized state. This implies that the holes must first become trapped on the R centers in order to have recombination, transition 4. The recombination occurs via the annihilation of the trapped holes with the electrons in the conduction band (transition 5). If the recombination is radiative, then light emission from a sample occurs. This phenomenon is usually called radioluminescence, if ionizing radiation is used as excitation source.

The electrons can also become localized on the trap level (transition 2), and in such a case the recombination can occur only if the electron absorbs enough energy E to be promoted back into the valence band (transition 3). The luminescence is therefore delayed by an amount of time τ spent by the electrons in the trap and governed by Eq. (7.1), with $\tau^{-1} = p$.

If the trap depth E is large compared to kT_0, where T_0 is the irradiation temperature, the electrons that are trapped will remain so for a sizeable amount of time, so that even after the removal of the irradiation there will be a substantial population of trapped electrons. Considering also that the free electrons and holes are created and annihilated in pairs, there must be an equal population of holes on the R centers. These populations of trapped holes and electrons represent an out-of-equilibrium state. The return to equilibrium is possible, but since the irradiation was performed at low temperature, with respect to E/k, the relaxation rate is going to be very slow, as easily visible from Eq. (7.1). The return to the equilibrium state can be accelerated by increasing the temperature to a value $T \geq E/k$, where the electron release to the conduction band is much faster. The recombination of the electron with the holes results in the TSL. The intensity of thermally stimulated luminescence is proportional to the recombination rate of holes and electron on the level R.

All the statements above can be easily summarized mathematically with a set of rate equations describing how the population of electrons and holes in both delocalized bands and localized levels changes over time. The equations reported below, however, describe only the population evolution after the irradiation end. Several examples of more general cases, which consider also the irradiation stage, can be found in the literature [20, 21].

The set of rate equations is:

$$\frac{dn_c}{dt} = ns\exp\left(-\frac{E}{kT}\right) - n_c(N-n)A_e - n_c m A_r \tag{7.8}$$

$$\frac{dn}{dt} = -ns\exp\left(-\frac{E}{kT}\right) + n_c(N-n)A_e \tag{7.9}$$

$$\frac{dm}{dt} = -n_c m A_r \tag{7.10}$$

where n_c and n are the concentration of electrons in the conduction band and on the trap, respectively; m is the concentration of holes in the recombination center; and N is the concentration of traps. A_e and A_r are the transition coefficients for electrons on the trap and recombination centers, respectively. The various terms on the right-hand side of these equations are: the electron detrapping rate, $ns\exp(-E/kT)$; the electron retrapping rate, $n_c(N-n)A_e$; and the electron recombination on holes, $n_c m A_r$.

Charge neutrality also dictates that

$$n_c + n = m. \tag{7.11}$$

Finally, the amplitude of the emitted TSL light during the return of the system to equilibrium is given by

$$I(t) = -\eta\frac{dm}{dt} \tag{7.12}$$

where η is the efficiency of the emitted light detection. If all recombination events produce photons and all the photons are detected then $\eta = 1$. In the following, unless specifically stated, this η is to be assumed as equal to 1.

Equations (7.8) to (7.10) are coupled, first-order, nonlinear differential equations and, in general, are not analytically soluble. Note also that all concentrations (n_c, n, and m) are time and temperature dependent, with the t and T dependencies linked through a heating rate function $T = f(t)$.

In order to arrive at an analytical solution, then, more approximations are needed. The assumptions that are usually used are:

(i) The free carrier concentration in the conduction band is always much smaller than that of the trapped carriers,

$$n_c \ll n \tag{7.13}$$

and (ii) the rate of change of the free carrier concentration is always slower than that of trapped charge concentration

$$\left|\frac{dn_c}{dt}\right| \ll \left|\frac{dn}{dt}\right|, \left|\frac{dm}{dt}\right|. \tag{7.14}$$

These two assumptions are generally called the quasi-equilibrium approximation. These two approximations together imply that the free carrier lifetime is much smaller than that of the trapped charges, giving rise to a quasi-stationary free carrier concentration. Applying these inequalities to Eqs. (7.8)–(7.10) and (7.12), it is easy to see that

$$-\frac{dn}{dt} \approx -\frac{dm}{dt} = I(t) \tag{7.15}$$

and, thus, it is possible to write the TSL amplitude as:

$$I(t) = \frac{ns\exp(-E/kT)\, mA_r}{(N-n)A_e + mA_r} = ns\exp\left(-\frac{E}{kT}\right)\left[1 - \frac{(N-n)A_e}{(N-n)A_e + mA_r}\right]. \tag{7.16}$$

The term in the square brackets is the probability that a thermally freed electron will not be retrapped, and $(N-n)A_e/mA_r$ is the ratio of the retrapping rate to the recombination rate. This equation can now be solved by using further approximations on the relative ratio between the retrapping and the recombination rates.

7.1.1.1 First-Order Kinetics

The first solution, given by Randall and Wilkins [22], is obtained by considering the retrapping rate as negligible compared to the recombination one: $(N-n)A_e \ll mA_r$. Equation (7.16), then, reduces to:

$$I(t) = -\frac{dn}{dt} = ns \exp\left(-\frac{E}{kT}\right). \tag{7.17}$$

This differential equation describes the charge transport in the lattice as a first-order process. If the temperature is kept constant, then the electron release rate $p = s\exp(-E/kT)$ remains constant and the luminescence amplitude can be found from Eq. (7.17) by integration

$$I(t) = I_0 \exp(-tp), \tag{7.18}$$

where I_0 is the amplitude at $t = 0$. At a constant temperature, the luminescence amplitude decreases exponentially with time. This phenomenon is usually called phosphorescence[1] or afterglow.

If, instead, the temperature varies with time, p is no longer a constant. Rearranging Eq. (7.17), we have

$$\frac{dn}{n} = -s \exp\left(-\frac{E}{kT(t)}\right) dt \tag{7.19}$$

integrating the equation above, we obtain the value n at any time t during the measurement

$$n = n_0 \exp\left[-s\int_0^t \exp\left(-\frac{E}{kT(t')}\right) dt'\right] \tag{7.20}$$

where n_0 is the initial population of trapped electrons at the beginning of the measurement. Substituting the expression for n back in Eq. (7.17), we obtain that

$$I(t) = -\frac{dn}{dt} = n_0 s \exp\left(-\frac{E}{kT(t)}\right) \exp\left[-s\int_0^t \exp\left(-\frac{E}{kT(t')}\right) dt'\right]. \tag{7.21}$$

As the temperature increases, the probability of charge carrier detrapping increases, leading to an initial increase in the TSL amplitude. This is then followed by a maximum and a decrease in the luminescence amplitude as the trap becomes depleted.

TSL is usually observed as the temperature is increased linearly in time: $T(t) = T_0 + \beta t$, where β is the constant heating rate and T_0 is the temperature at $t = 0$. This allows us to rewrite the TSL amplitude as a function of the temperature as

$$I(T) = -\frac{1}{\beta}\frac{dn}{dt} = n_0 \frac{s}{\beta} \exp\left(-\frac{E}{kT}\right) \exp\left[-\frac{s}{\beta}\int_{T_0}^{T} \exp\left(-\frac{E}{kT'}\right) dT'\right] \tag{7.22}$$

which is the Randall–Wilkins first-order recombination expression of a single glow peak. It has to be noted that in other books and articles, see for instance Chen and McKeever [2], the first β on the right-hand side of Eq. (7.22) is not present. This difference is related to how the experimental apparatus used in obtaining the measurements works or, in other words, whether the points composing the glow curves are obtained per unit of time or per unit of temperature. The equation reported here is for the case of data collected per unit of temperature. The difference is minimal if all the measurements are obtained with the same heating rate, but it has consequences on the amplitude of the glow curves obtained at different heating rates, as we will see later.

The properties of the Randall–Wilkins equation are illustrated in Figure 7.3 by varying one at a time each of the parameters in the equation and solving it numerically. Panel a reports the TSL amplitude for different initial concentrations of trapped charges and with $E = 0.7$ eV, $\beta = 0.1$ K s^{-1}, and $s = 10^{13}$ s^{-1}. Note, in particular, that: (i) all the glow curves are asymmetrical,

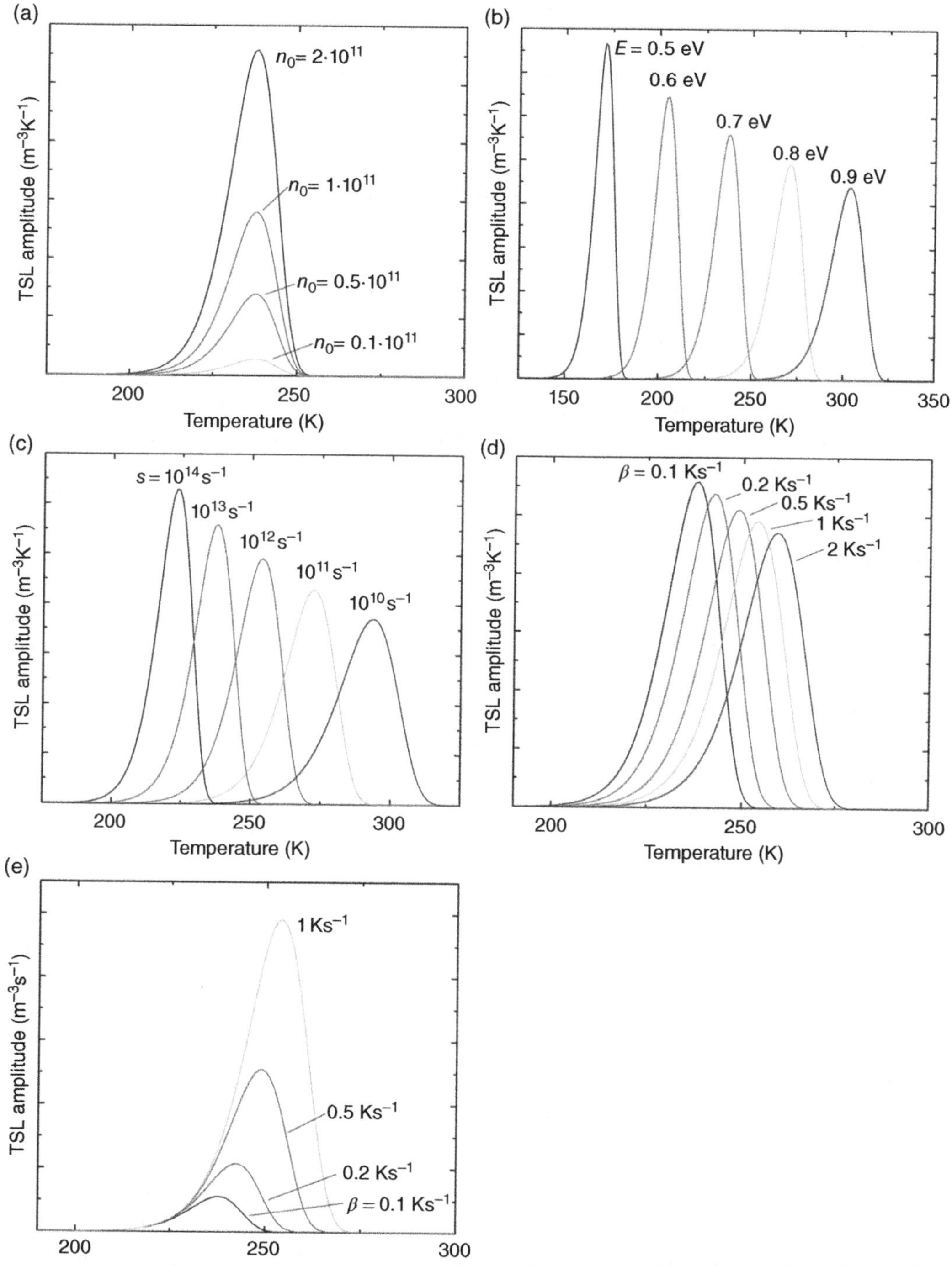

Figure 7.3 Calculated glow curves for first-order Randall and Wilkins recombination obtained by using $E = 0.7$ eV, $s = 10^{13}$ s^{-1}, $n_0 = 2 \cdot 10^{11}$ m^{-3}, and $\beta = 0.1$ K s^{-1} as initial parameters, and modifying each one at a time to study how position, shape, and amplitude are affected by: (a) initial trapped charges n_0 in m^{-3}; (b) trap energy E; (c) frequency factor s; and (d) heating rate β. Plot (e) reports the same data in (d) but using a different ordinate axis.

(ii) n_0 is simply a multiplication factor, and (iii) both the shape and the position of the glow peak are not affected by n_0. Only the amplitude and its integral (the intensity) are increased by increasing the initial concentration of trapped charges. This is of fundamental importance in dosimetry since it is possible to have a strictly linear relation between the TSL amplitude and the dose absorbed by a dosimeter (at least for $n \ll N$).

The temperature of the amplitude maximum (T_m) can be calculated by evaluating the first derivative of Eq. (7.21) and equating it to zero: from this condition, we get

$$\frac{\beta E}{kT_m^2} = s\ \exp\left(-\frac{E}{kT_m}\right), \tag{7.23}$$

which clearly shows that T_m does not depend on n_0. This independence is a specific characteristic of first-order glow curves. It is also exploited to evaluate experimentally if the peaks of a glow curve do follow a first-order recombination kinetics by performing a series of measurements varying the initial population of the traps, which should be one of the first checks to be performed in the characterization of a new material.

In Figure 7.3b, the activation energy has been varied between 0.5 and 0.9 eV, and the other parameters have been kept constant ($s = 10^{13}\ s^{-1}$, $\beta = 0.1\ K\ s^{-1}$, and $n_0 = 2 \cdot 10^{11}\ m^{-3}$). As E increases, the peak shifts toward higher temperatures with a decrease in the amplitude and an increase in the width, the intensity (the area underneath the curve) remains constant. The shift toward higher temperatures can be understood by realizing that for higher E values, or deeper traps, more energy (higher temperature) is needed to release the charge carriers. Similar changes are also obtained as the frequency factor s is varied, Figure 7.3c, although in this case, the peak shifts toward lower temperatures and becomes narrower as s is increased. A charge carrier trapped on a high frequency factor center needs less energy to be released in the delocalized band.

Finally, the glow curve dependence upon the heating rate β is reported in Figure 7.3d and e. Figure 7.3d, in particular, reports the amplitude in counts per *Kelvin* per cubic meter (directly from Eq. 7.22), and as clearly visible, an increase in the heating rate results in the shift to higher temperatures of the glow peak. The amplitude of the glow peak decreases, while its width slightly increases, just as in the case of decreasing s. In any case, the intensity of the peak is maintained. Figure 7.3e, on the other hand, reports the amplitude in counts per *second* per cubic meter, related to Eq. (7.22) by removing the first β. In this case, the TSL amplitude increases as the heating rate is increased, leading to an apparent increase in the intensity. This increase is, however, only apparent, since the integral of the glow curve done in terms of $I(t)\mathrm{d}t$ leads to the same value, irrespective of the heating rate. As mentioned above, the difference between these two visualizations of the glow curves is not important when comparing data obtained with the same heating rate. It can become an issue if the amplitude as a function of the heating rate is to be interpreted and the ($1/\beta$ from time to temperature) conversion factor between the curves is not considered, possibly leading to wrong interpretations. On a side note, since the position in temperature of the glow peak is affected by the heating rate, it is always necessary to report the heating rate when publishing experimental data: saying that a glow peak has a maximum at a certain temperature without also citing the heating rate makes the comparison among other published data impossible.

Of all the four parameters, the activation energy E and the frequency factor s are intrinsically related to the trap under study; they are usually referred to as the trapping parameters and are not under experimental control, in contrast with the heating rate β and the concentration of trapped charges n_0, with the latter proportional to the absorbed dose. The investigation of a new material, or sample, should therefore start by studying the glow peak behavior, varying the absorbed dose and the heating rate.

7.1.1.2 Second-Order Kinetics

Starting from the same model employed by Randal and Wilkins, Garlick and Gibson [23] modified it by considering the possibility that the thermally freed charges can be captured back in the trap instead of recombining on the luminescence center, and that the retrapping is the dominating phenomenon. This can be expressed as: $mA_r \ll (N-n)A_e$. They also considered that the population of the trap is far from saturation ($N \gg n$) and $m = n$.

With this in mind, it is the possible to rewrite Eq. (7.16) as:

$$I(t) = -\frac{dn}{dt} = s\frac{A_r}{A_e}\frac{n^2}{N}\exp\left(-\frac{E}{kT}\right), \tag{7.24}$$

assuming also $A_e = A_r$, the solution for a constant temperature becomes:

$$I(t) = \frac{I_0}{(1 + n_0\alpha t)^2} \tag{7.25}$$

with $\alpha = \frac{s}{n}\exp(-E/kT)$. In this case, the decay becomes hyperbolic instead of the exponential one obtained in the case of first-order recombination.

Integrating Eq. (7.24) by considering, again, a linear heating rate, it is possible to obtain

$$I(T) = \frac{n_0^2}{N}\frac{s}{\beta}\exp\left(-\frac{E}{kT}\right)\left[1 + \frac{n_0 s}{N\beta}\int_{T_0}^{T}\exp\left(-\frac{E}{kT'}\right)dT'\right]^{-2}. \tag{7.26}$$

This is the Garlick–Gibson TSL equation for second-order kinetics.

Although we imposed, for simplicity, $A_e = A_r$, it is still possible to obtain a second-order kinetics without this simplification. The required condition is still $n \ll N$, so that NA_e becomes much larger than nA_r,and in (7.26), n/N is replaced by nA_r/NA_e, and s/N is replaced by sA_r/NA_e. In this case, however, the second-order kinetics is obtained only in the limit for $n \ll N$, while in the case of the Garlick–Gibson relation, it is valid for all the trap populations.

The main feature of the Garlick–Gibson curve is that it is nearly symmetrical, with the high-temperature half of the curve slightly broader than the low-temperature half, Figure 7.4a. This can be understood considering that the released charges become trapped before they can recombine, leading to a delay in the luminescence emission, which becomes spread out toward higher temperatures. The other main characteristic is that the carrier population in the trap n_0 is no longer a pure multiplicative factor, as in the case of the first-order kinetics, but it also changes the shape of the curve itself as different trap populations are considered. As visible in Figure 7.4a, the glow peak shifts toward lower temperature as higher trap population is considered. The behavior of the curve by varying the energy, frequency factor, and the heating rate is, on the contrary, similar to the first-order case.

Analogous to the first-order Randall–Wilkins case, the second-order peaks are characterized on the low-temperature side by an initial growth of the form $s\exp(-E/kT)$.

7.1.1.3 General-Order Kinetics

The first and the second-order TSL equations have been obtained considering several simplifying assumptions on the relative size of the retrapping and the recombination rates. When these approximations are not satisfied, however, the TSL peak will not follow either the first or the second-order

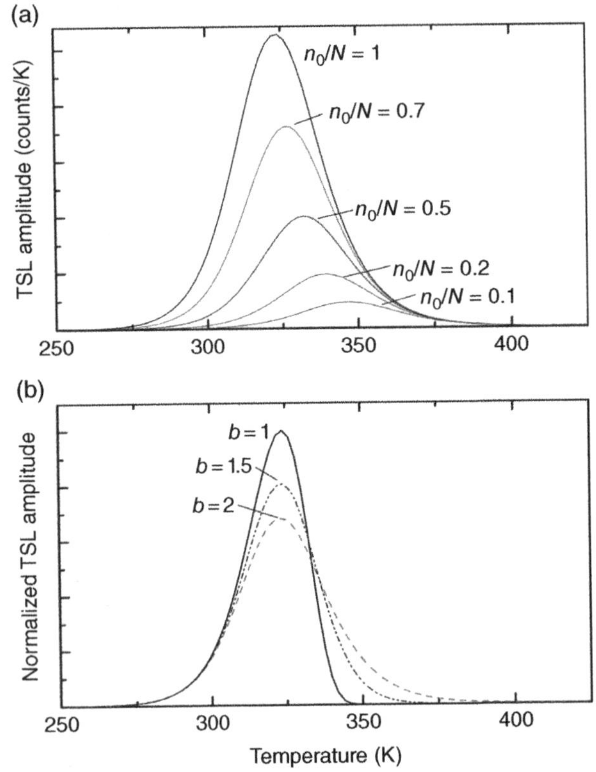

Figure 7.4 (a) Evolution of a calculated second-order Garlick and Gibson glow curve as a function of the initial population of trapped charges (n_0) with respect to the concentration of traps (N). (b) Calculated dependence of a general-order glow curve as a function of the kinetic order parameter b. The values used in the calculation are: $E = 0.9$ eV, $s = 10^{13}$ s^{-1}, and $\beta = 1$ K s^{-1}.

kinetics. May and Partridge [24] wrote an empirical expression, starting from Eq. (7.17) and (7.24), of a general-order TSL kinetics:

$$I(t) = -\frac{\mathrm{d}n}{\mathrm{d}t} = n^b s' \exp\left(-\frac{E}{kT}\right) \tag{7.27}$$

where s' has the dimension of $m^{+3(b-1)}s^{-1}$ and b is defined as the general-order parameter and it is not necessarily equal to 1 or 2. The integration of this equation for $b \neq 1$ yields

$$I(T) = \frac{s''}{\beta} n_0 \exp\left(-\frac{E}{kT}\right)\left[1 + (b-1)\frac{s''}{\beta}\int_{T_0}^{T} \exp\left(-\frac{E}{kT'}\right)\mathrm{d}T'\right]^{-b/(b-1)}, \tag{7.28}$$

where $s'' = s' n_0^{b-1}$ with unit s^{-1}. This equation includes the second-order case ($b = 2$) and reduces to Eq. (7.22) for $b \to 1$.

The complexity of this empirical development of the general-order TSL equation lies in the meaning of the s' parameter, the dimension of which changes with the kinetic order parameter b, making s' of difficult interpretation. Still, this general-order case is useful since intermediate cases can be dealt with and it smoothly transits from first order to second order, see Figure 7.4b. For non-first order kinetics, the temperature of the peak maximum is still dependent on the initial concentration of the charge carrier n_0.

7.1.2 Localized Transitions

In the model discussed so far, the recombination of charge carriers freed from traps with the luminescent centers occurs through the promotion of the carrier into the delocalized band. In this way, a substantial diffusion of the carrier through the lattice can happen connecting distant trap–center pairs. A consequence of the carrier presence in the delocalized band is that a change in the conductivity of the material upon temperature change should be expected: this phenomenon is called thermally stimulated conductivity. There are, however, experimental evidences that the recombination process does not always proceed through the delocalized bands but instead involves only the trap and the recombination center levels [25–30]. Such a phenomenon is usually called "trap-to-center recombination" or "localized transition." Several other terms are also employed, sometimes used as synonyms but implying somewhat different mechanisms, and they are: "tunneling recombination" (athermal or thermally assisted), "anomalous fading" (predominantly used in dosimetry, archeometry, and geochronology), and recombination within a defect cluster. The localized transition can occur if the trap and the recombination centers are sufficiently close to each other to allow the possible tunneling of the localized charge from the trap level(s) to an excited state of the recombination center.

Two different mechanisms can be considered: one in which the trapped charge can readily tunnel through the barrier (athermal tunneling), and the second in which the trapped charge must acquire enough thermal energy to arrive to a higher vibrational level of the trap from which it can tunnel to the recombination center (thermally assisted tunneling).

In both cases, the phenomenon can be described by taking into consideration the presence of a one-dimensional rectangular barrier between the trap and the recombination center: the tunneling rate, P_T, is commonly described as the product of two terms [31–34]

$$\mathrm{P_T} = C{\cdot}W(r) \tag{7.29}$$

where $W(r)$ is the tunneling probability which, for a rectangular barrier, is equal to:

$$W(r) = \exp(-Kr) \tag{7.30}$$

with r the distance between the trap and the recombination center, and $K = (4\pi/h)\sqrt{2mE}$ where m is the electron (effective) mass, h is the Planck's constant, and E is the trap energy depth. The term C has the dimensions of the inverse of time.

In the case of athermal tunneling, it is easy to show that the luminescence decay is

$$I(t) = -\frac{\mathrm{d}n}{\mathrm{d}t} = I_0 \exp(-CWt). \tag{7.31}$$

Such decay is strongly dependent on the barrier configuration, and not on the temperature, hence the "athermal" term used to describe this kind of recombination.

It has to be remarked, however, that the use of a rectangular barrier is quite a simplification, and more accurate descriptions should take into account more complex barrier profiles [35, 36]. This single exponential behavior, moreover, is not commonly observed. This can be easily understood by considering that there are likely several different trap–center distances due to their random distributions in the crystal. The signal is, therefore, the combination of all tunneling contributions from pairs at different distances, resulting, after an initial transient time, in a $1/t$ decay law [33, 34].

Among the consequences of athermal tunneling between traps and recombination centers is that the number of trapped charges diminishes over time after the end of the irradiation even for $E \gg kT$, which leads to the "fading" of the subsequent TSL signal, i.e. the longer the time interval between the irradiation end and the heating, the weaker is the observed TSL. Moreover, the fading rate is unaffected by the temperature at which the sample is stored after the initial irradiation.

Since this phenomenon has a crucial parameter in the distance between traps and recombination centers, a dependence upon the concentration of traps and recombination centers is to be expected, and indeed it was rather clearly evidenced by Dobrowolska et al. in the case of YPO_3:Ce, Ln (Ln = Er, Ho, Nd, Dy) for different Ln concentrations [35].

In the case of thermally assisted tunneling, the trapped charge has to be promoted to a higher vibrational level of the trap from which it can tunnel toward the excited state of the recombination center, where it can undergo radiative recombination with the emission of photons. When this process occurs during a TSL measurement, it gives rise to a glow peak that is apparently identical to that obtained in the case of the first-order, Randall–Wilkins, recombination process (Eq. 7.22), which involves the release of the charge in the delocalized bands [37]. So, it is not possible to distinguish a thermally stimulated tunneling process only by the simple inspection of the glow peak. The analysis of the glow peak with one of the methods reported below in Section 7.2, however, might result in frequency factors s characterized by very small values, by several orders of magnitude (e.g. 10^3–10^4 [38]), compared to the expected Debye frequency. This can be understood considering Eq. (7.6) and the meaning of the κ coefficient as the transmission coefficient of the barrier: combining Eqs. (7.6) and (7.30), it can be shown that the frequency factor becomes

$$s = \exp\left(-Kr\right) \nu \exp\left(\frac{\Delta S}{k}\right) = \exp\left(-r\,(4\pi/h)\sqrt{2mE}\right) \nu \exp\left(\frac{\Delta S}{k}\right) \tag{7.32}$$

which clearly evidences that the frequency factor would strongly depend on the distance between the recombination center and the trap, leading to anomalously lows values. This also implies that, in the right circumstances, a single trap can give rise to multiple (first-order) glow peaks at different temperatures characterized by the same thermal activation energy and markedly different frequency factors. These multiple peaks can be related to the different discrete distances between the trap and the recombination centers. One of such examples can be found in Ref. [39] for Lu_2SiO_5:Ce and $(Lu,Y)_2SiO_5$:Ce oxyorthosilicate crystals.

A more in-depth discussion on tunneling phenomena can be found on a review paper written by Vedda and Fasoli [40].

7.1.3 Beyond the Ideal Behavior

The one-trap/one-recombination (OTOR) model so far presented and discussed is very successful in explaining the TSL features observed in real materials, namely, the shape of the detected glow peaks and their dependence upon trapped charge concentration, heating rate, and trap depth. However, there are no real materials for which it can be claimed that there is only one trap and one recombination center; see Figure 7.5 for an example of a complex glow curve with more than 10 different glow peaks in the 85–650 K region. The OTOR model, also, is not able to give a reasonable explanation of why in most of the materials, the first-order recombination appears as the dominant mechanism.

One logical extension of the above presented model is to consider other, deeper traps (usually called "thermally disconnected traps") that are considered stable during the emptying of the active one and thus are retaining their trapped charge population during the emptying of the shallower trap. The result is that the concentration of charges in the active trap is smaller than that on the recombination centers ($m = n + h$, where h is the amount of charges trapped on the thermally disconnected centers), favoring the recombination process over the retrapping. A further possibility is that also the thermally disconnected traps can capture the charge carriers released from the

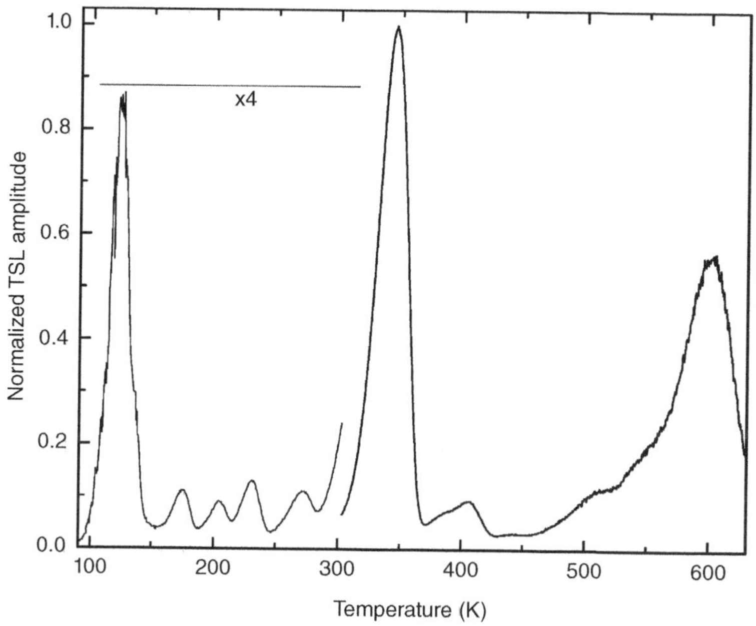

Figure 7.5 Glow curve obtained after irradiation with X-rays at 85 K of a Lu_2SiO_5:Ce single crystal scintillator. More than 10 different glow peaks can be easily seen. The heating rate was set to 0.1 K s^{-1}.

shallower defect in the delocalized band, further decreasing the probability of retrapping. So, the presence of thermally disconnected traps, together with localized transitions, can explain the abundance of the reported first-order kinetic with respect to the other kinetics, which would not be easy to justify in terms of the A_e and A_r transition coefficients [41, 42]. As a consequence of this, it can be stated that, in the case of a system in which recombination dominates over retrapping, glow curves with multiple discrete peaks can be represented accurately by the superposition of individual first-order peaks, each following Eq. (7.22).

Lewandowski and colleagues examined theoretically the possibility that the quasi-equilibrium approximation does not hold during heating and the effect that this has on the shape of the glow peaks [43, 44]. They developed a modified equation that describes a perfectly general TSL curve, though within the limits of the model they used, since neither the quasi-equilibrium nor a particular kinetic order is assumed. Their detailed analysis showed that the shape of the glow peak derived from this general equation differs very little from the Randall–Wilkins, first-order glow peak. On the contrary, similar conclusions were not obtained for non-first order kinetics.

Section 7.1.3.1 and 7.1.3.2 will discuss two other phenomena that can lead to significant deviations from the expected behavior of the glow curves and which should be considered in the analysis of experimental data.

7.1.3.1 Luminescence Quenching

In the discussion so far presented, it was assumed that the η term in Eq. (7.12) was equal to 1 and independent from the temperature. This term accounts for the probability of photon emission and detection from the sample under study, and it is the product of several physical phenomena that influences the number of photons that can actually be detected during a TSL measurement [45]. One of these parameters is the quantum efficiency (Q) of the recombination center or, in other words, the probability that the recombination centers emits light once the free charge carrier has been captured by the recombination center itself. The quantum efficiency strongly depends on the temperature and this can have profound effects on the obtained glow curves. Generally, the quantum efficiency remains almost constant up to a critical temperature above which the efficiency decreases rapidly: this temperature dependence is called thermal quenching and it is usually described with the Mott–Seitz model:

$$Q(T) = \frac{1}{1 + C\ \exp\left(-W/kT\right)} \tag{7.33}$$

where C and W are the quenching parameters. The effect is particularly evident when a set of TSL curves are obtained by varying the heating rate. As the TSL peak shifts toward higher temperatures by increasing the heating rate, the luminescence efficiency decreases and a reduced TSL glow peak area and peak height are obtained. The shape of the glow peak too can be affected, particularly on the high-temperature side, resulting in peak shapes that do not conform to the ones expected from kinetic considerations. If not taken into consideration, this effect can clearly undermine the validity of the results obtained by using any of the glow curve analysis methods that will be presented later. Examples of the effect that thermal quenching has on the shape of the glow curve and the subsequent analysis are Al_2O_3:C [46, 47], Lu_2SiO_5:Ce [39], and quartz [48].

7.1.3.2 Trap Energy Distributions

The discussion so far assumed that the trap depths associated with the localized states are single-valued in energy. Although this might be the case in high-quality single crystalline materials, it can be expected that in highly defective and, especially, in amorphous materials, the energy associated with a defect might be smeared out over a range of values rather than being uniquely defined. The presence of slightly different sites, in both bond angles and bond lengths, hosting the same kind of defect results in the inhomogeneous broadening of the energy depth of the defect itself, and in the formation of distribution in the energy related to a single defect. Although this is clearly the case in glasses and amorphous materials [49, 50], it can also appear in polycrystalline materials, due to the high surface-to-volume ratio of the crystalline domains and the presence of a great variety of bulk and surface defects [51–53], or even in single crystals characterized by compositional disorder [54].

The presence of energy distributions results in broad glow peaks that can be described as the sum of a large number of first-order peaks, whose shape can be reminiscent of second-order, more symmetrical peaks. In the case of amorphous materials, moreover, the glow peak can be so broadened to appear as a wide almost featureless hump [49, 50]. Note that, in the case of distributions, the position in the temperature of the TSL maximum is not affected by the initial concentration of trapped charges, in contrast with what is expected by real second-, and general-order peaks, see Sections 7.1.1.2 and 7.1.1.3. Moreover, the partial cleaning (see Section 7.2.1) of the trap results in shift toward higher temperatures of the glow peak maximum, as in the case of second-order peaks. The presence of energy distribution can as well result in isothermal decays that tend to follow a $1/t$ law but, unlike the case of athermal tunneling, they do show some temperature dependence [49, 55].

7.2 Data Analysis Methods

The glow curves are useful in indicating at which temperature, at a given heating rate, the traps become unstable; but, alone they cannot give the experimenter much information about the trap stability over time at a given temperature, for instance in the case of thermoluminescence dosimetry or dating applications, or its contribution to afterglow or phosphorescence phenomena, in the case of scintillators and persistent phosphors. It is, thus, often necessary to analyze the glow curves to obtain the trap parameters. Of all parameters that characterize a trap, only the energy E, the frequency factor s, and the kinetic order b are accessible from the analysis of a glow curve.

Although not strictly related to glow curve analysis, the energy and frequency factor values can be obtained by studying the afterglow of a sample. Remembering that for first-order recombination processes the decay time of a trap is the inverse of Eq. (7.1), measurements of the decay profile at different temperatures allows to evaluate the temperature dependence of the decay time (τ). Plotting the log $(1/\tau)$ as a function of $1/T$ yields a straight line whose slope is equal to $-E/k$ and whose intercept with the ordinate axis is log(s). Similar evaluations can be done also for second-order and general-order recombination [2]. In principle, such an evaluation should be performed on an as wide temperature range as possible. This is, however, not possible, due to limitations in the used instrument (the decay can become too long, or too short, to be detected) and the presence of several other glow peak contributions that make the evaluation of the decay time more complicated.

In the following, four different glow curve analysis methods will be presented and discussed in view of the practical difficulties that are often encountered in the application of a method to a real glow curve. Many more methods have been developed over time and a broader picture can be found in McKeever's book [1].

7.2.1 Initial Rise

The initial rise method [23] is based on the assumption that, for sufficiently low temperatures, the trap population n is not particularly affected by the emptying of the trap and, from Eqs. (7.22), (7.26), and (7.28), it follows that the TSL glow curve is an exponential-like function and the emitted light is proportional only to the factor $\exp(-E/kT)$, independently from the order of kinetics. When plotted in an Arrhenius plot ($\ln(I)$ vs. $1/T$), this region will appear as a straight line with slope $-E/k$, see Figure 7.6, from which the activation energy of the trap is easily found. The suitable initial portion of the glow curve that can be used for this evaluation has been debated [56, 57], but in the case of isolated peaks it is generally accepted that only the initial regions with amplitude below 15–20% of the maximum should be used [58, 59]. However, as previously mentioned, real glow curves are generally composed by several peaks and it is quite possible that the initial region of a glow peak is obscured by the tail of the preceding one, which makes the use of the initial rise method not feasible, since it is possible that no straight section can be resolved in the Arrhenius plot.

To solve this issue, it is possible to use the "partial cleaning" method. This procedure simply consists in the preheating of the sample at a temperature high enough to completely empty the peak at low temperature but not to strongly affect the following one. After preheating, the sample is cooled down followed by a complete TSL measurement. Applying this method multiple times and at different temperatures, it is possible to remove in stages the peaks at lower temperatures throughout the whole TSL curve, and to use the resulting glow curves to evaluate the energy of each trap. There is, however, the risk for closely spaced peaks that the preceding one has not been perfectly removed, leading to underestimation of the activation energy. So, multiple partial cleaning and initial rise cycles at slightly different temperatures should be considered for each peak.

Once the value of the trap energy is known, it is possible to calculate the frequency factor of a trap by using Eq. (7.23) in the case of a first-order recombination peak. It is to be noted that, considering the possible errors in temperature of the maximum and in the evaluated energy, only the order of magnitude of the frequency factor can be obtained with confidence. Although Eq. (7.23) is true only

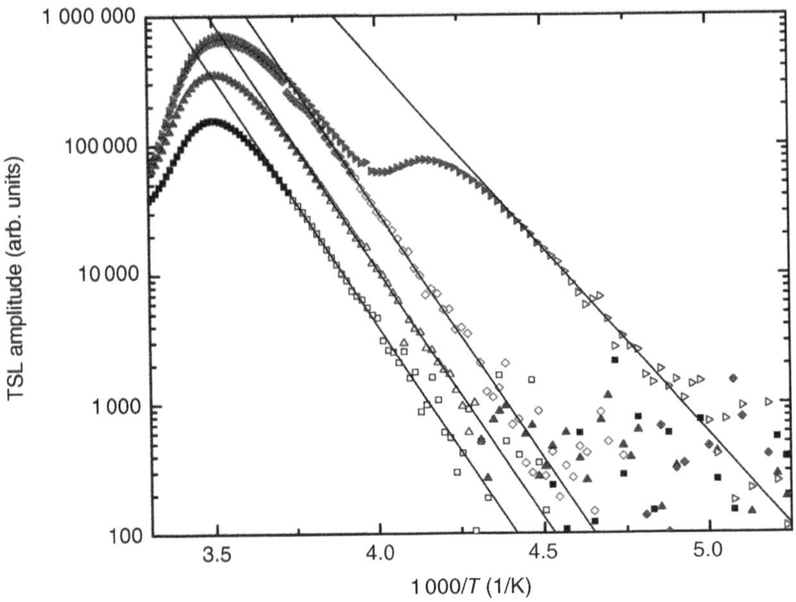

Figure 7.6 Partial cleaning and initial rise techniques applied to glow curves obtained after irradiation at 10 K on $Lu_3Al_5O_{12}$:Ce single crystal. Heating rate: 0.1 K s^{-1}. The regions with open symbols are used for the fits in term of the Arrhenius equation. The lines represent the fit results. Note, in particular, the change in slope between the line on the right and the three on the left, indicating the presence of two traps with different activation energy. Moreover, the three lines on the left are practically parallel, indicating a single-valued energy trap. The curves, from right to left, have been obtained after partial cleaning at 220, 257, 270, and 280 K, respectively.

for first-order kinetic, it can be used also in the case of non-first order ones since the error introduced in s is usually negligible [60].

The initial rise method is strongly suggested since it works for any kinetic order and it can deal also with trap energy distributions [52, 61]. However, it may fail in the case of thermal quenching phenomena. Also, since it takes into consideration only points in the rise of the glow curve, it can also give incorrect results in the case of weak glow peaks where the signal-to-noise ratio is rather low.

7.2.2 Peak Shape

In contrast with the initial rise technique, methods based on the shape of the peak rely only on two or three points of the glow curve: these are usually the temperature of maximum of the peak (T_m) and either, or both, the low- and high-temperature half-heights at T_1 and T_2 (see Figure 7.7a). These methods are variations upon the same idea and different formulas have emerged over time to deal with first-order or general-order kinetics [37, 63–66]. Chen analyzed all these methods to their full extent and only Chen's formula will be reported in this section [62, 67]. The full width at half maximum is given by $\omega = T_2 - T_1$, the low-temperature half width by $\tau = T_m - T_1$, and the high-temperature one by $\delta = T_2 - T_m$. Halperin and Braner also suggested the use of the symmetry factor μ defined as $\mu = \delta/\omega$ [37]. The resulting equation is:

$$E = c_\gamma (kT_m^2/\gamma) - b_\gamma (2kT_m) \tag{7.34}$$

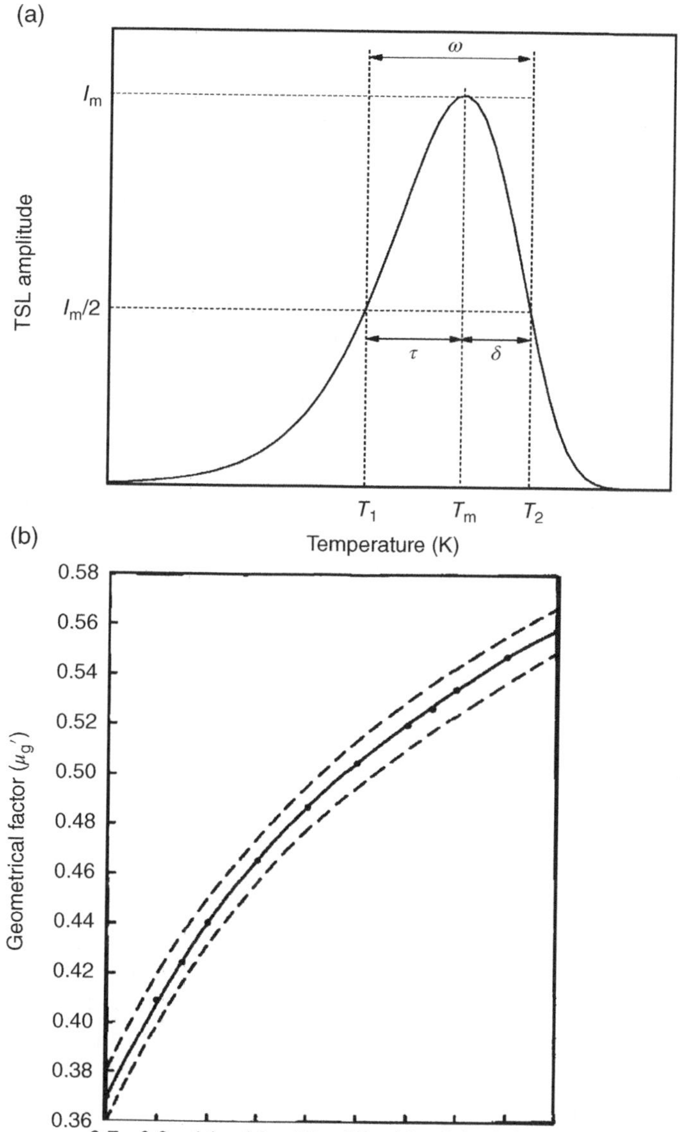

Figure 7.7 (a) Normalized isolated first-order glow peak. The parameter used in the Chen's method to calculate the energy of a glow peak are also reported in the graph. The symmetry parameter $\mu = \delta/\omega$ is equal to 0.42. (b) Calculated symmetry factor μ as a function of the given kinetic order. The central line gives the average values, upper and lower lines give the largest possible variations. Source: Reproduced with permission from [62]. Glow curves with general-order kinetics, by Reuven Chen, *Journal of the Electrochemical Society*, © 1969, IOP Publishing.

where γ is ω, τ, or δ. The constants c_γ and b_γ are listed in Table 7.1, also considering Halperin and Braner's symmetry factor. Concerning μ, Chen numerically solved the general-order expression for TSL and calculated values of μ for different values of the kinetic order b. The resultant values of μ are between 0.42 and 0.52 for first- and second-order kinetics, respectively, and can be evaluated from Figure 7.7b [62].

Table 7.1 values for the constants c_γ and b_γ in Eq. (7.34).

γ	c_γ	b_γ
ω	$1.51 + 3\,(\mu - 0.42)$	$1.58 + 4.2\,(\mu - 0.42)$
τ	$0.976 + 7.3\,(\mu - 0.42)$	0
δ	$2.52 + 10.2\,(\mu - 0.42)$	1

The peak shape is an approximate method in the sense that the values of the peak shape factor μ is not totally independent of the energy and frequency factor values, and it varies up to ±7% from its mean value for different E and s.

The application of this method to an experimental glow curve can be difficult since it relies on the precise determination of only two or three points. Any slight inaccuracy, due for instance to inefficient thermal cleaning of the lower temperature peaks, will inevitably produce erroneous values for the trap depth. Other difficulties arise in the case of glow curves that are affected by thermal quenching. Moreover, this method cannot work in the case of large energy distribution of the trap. Thus, the application of this method must be made with caution.

7.2.3 Heating Rate Method

The heating rate used to collect a glow curve influences the position in temperature of the glow peak, with the latter shifting toward higher temperature as the heating rate is increased. This can also be seen by rearranging Eq. (7.23) in the following way:

$$\beta = \frac{sk}{E} T_{\mathrm{m}}^2 \exp\left(-\frac{E}{kT_{\mathrm{m}}}\right). \tag{7.35}$$

Increasing the heating rate β on the left-hand side must necessarily result in an increase in the right-hand side. Since the term $T_{\mathrm{m}}^2 \exp(-E/kT_{\mathrm{m}})$ is an always increasing function of T_{m}, a growth in the value of the function necessarily implies an increase in the argument T_{m}. This relationship between E, s, β, and T_{m} has generated a series of methods to calculate the energy by noting the change in T_{m} for a certain change in β. Using Eq. (7.23), Hoogenstraaten suggested to employ several different heating rates and recording the value of T_{m} for each measurement [68]. Plotting $\ln(T_{\mathrm{m}}^2/\beta)$ against $1/T_{\mathrm{m}}$ should yield a straight line with slope E/k and intercept $\ln(E/sk)$, from which both the energy and the frequency factor can be calculated.

Although this is strictly true for first-order recombination, it has been demonstrated that it is still a good approximation for higher order recombination, provided that the initial number of trapped charges remains the same [60].

The major advantage of this method is that it only requires data to be taken from a peak maximum, which, in the case of a large peak surrounded by smaller ones, can be reasonably determined from the glow curve. Furthermore, the calculation of E is less influenced by thermal quenching phenomena, as in the case of the initial rise method. If, however, the energy of the satellite peaks has to be determined, or the peak is the superposition of more than one component, the results might not be accurate. From a theoretical point of view, this method should be done by using the widest possible change in the heating rate. This, however, is limited by experimental constraints: at too low β, the resulting glow curve will be very noisy, making the evaluation of T_{m} difficult; on the other hand, a high β might result in thermal lag between the measured temperature and the actual one of the sample, and thus in errors in the value of T_m. For relatively large samples, a high heating rate might result also in thermal gradients in the sample itself.

7.2.4 Glow Curve Fit

All the analysis methods discussed so far consider only a limited amount of data of the glow curve in order to evaluate the energy and the frequency factor of a given glow peak. Curve fitting procedures, in which the experimental data are fit to a previously defined function, have the advantage to use practically all the data in the glow curve and the analysis is effectively done in one go even in the case of complex glow curves. Several functions have been used for this analysis [69], but in most of the cases these are based on approximations of the first or general-order recombination equations presented above, where the analytically unsolvable integral in the equations is approximated by a polynomial or a series expansion [70, 71].

The difficulties arise from the fact that the position and the full width at half maximum of a glow peak are dependent on both the trap energy and the frequency factor. This implies that several pairs of E and s produce almost the same fit [4]: this is, from a numerical point of view, an ill-conditioned problem. Indeed for a single first-order peak, the least squares function is, in the $E - \ln(s)$ plane, a narrow valley whose bottom shows only a very gentle curve, making the search for the actual minimum computationally complicated [72]. As a consequence, in case of less than optimal glow curves, the results obtained from a fit procedure might not reflect accurately the actual trap parameters. One possible way to solve this problem is to use the E and s values obtained with one of the methods presented above as initial guesses for the fit routine, thus reducing the probability that the fit routine arrives at a local minimum of the minimization function.

The second issue with fit procedures is that they imply that the glow curve is a superposition of independent glow peaks: this is only true for first-order recombination kinetic. Non-first order processes imply the presence among trapping centers of interactions which are not taken into consideration by the fit in terms of a sum of non-first order glow peaks.

As mentioned several times, real materials contain several trapping states and it is possible that some of these traps might have rather similar parameters: the resulting glow peaks might be sufficiently close to look like a single first-order peak, but with anomalous low evaluated energy and frequency factor [73]. Also, the choice of how many components to use in the fit might not be always straightforward.

The last comment is not on the method itself, but on the way it is sometimes used and in particular on the choice of the fit function. As mentioned few lines above, the experimenter has usually to choose whether to use a first-order or a general-order function to fit the data: the selection should not be based upon the shape of the glow curve itself, but on additional measurements done as a function of the dose, or irradiation duration, imparted to the sample. All non-first order recombination processes, in fact, show a dose dependence on the temperature of the peak maximum (it decreases by increasing the dose) that first-order recombination does not show. In other words, the fit procedure should not be used to determine the recombination kinetics of a glow curve.

7.3 Instrumentation and Considerations on Samples

A typical TSL recording instrument is composed by a sample chamber with a temperature-controlled heating system, a light detector, and the associated electronics for signal processing. The heating system should provide reproducible and controllable heating profiles. The heating is usually provided, in the case of measurement above room temperature, by passing a high current through a thin, low-impedance metal strip, whose temperature is monitored by a suitable thermocouple. The sample is placed directly in contact with the heating element or, in case of powders, they are glued to a metal disc. The measurements are performed in an inert gas atmosphere at reduced pressure (0.5 atmospheres or so) to ensure that the sample temperature is as close as possible to the heating element one. In the case of measurements performed below room temperature, usually down to 10 K, the system also includes a sample cooling system, e.g. a cryocooler, with the provision of a way to irradiate the sample in situ. For obvious reasons, low-temperature measurements are performed in vacuum and the sample is attached to the cooling-heating element with a thermally conductive glue. The usually employed heating rates strongly depend on the system design but they are usually of the order of few $K\,s^{-1}$, in the case of high-temperature systems, and of few tenths of Kelvin per second in systems working below room temperature.

The second component of a TSL reader is the light detection system, and different methods are employed based on the application for which the system is used. In the case of dosimetry and chronology, where the signals to be measured might be low, the systems use a photomultiplier working in single photon counting, which allows to detect very low light signals and have a relatively large linearity range. Filters might be used to select only a relatively narrow band in the sensitivity region of the photomultiplier, mostly to reduce the amount of light coming from the incandescence of the heating element and of the sample. In materials research, on the other hand, it is also important to have more information on the spectral characteristics of the emitted light since this can allow the identification of the recombination centers and, in some cases, also to have a clearer picture of whether a certain glow peak is related to an electron or hole trap [74]. In this case, the common way to extract the spectral information is to equip the heating system with a polychromator and a CCD camera, allowing the measurement of the TSL amplitude as a function of the temperature and of the wavelength simultaneously.

Sample excitation can be performed in various ways by using radioactive sources, X-ray generators, and UV-Vis lamps. Sometimes, the irradiation source is integrated in the TSL system, which makes it possible to collect automatically a series of measurements. Other important information can be extracted by measuring the so-called TSL excitation spectrum, by illuminating the sample with monochromatic light obtained from a UV-Vis source coupled to a monochromator. These measurements are performed for different excitation wavelengths; plotting the area of a selected glow peak as a function of the excitation wavelength, it is possible to obtain the TSL excitation spectrum for that glow peak [75].

Several TSL readers are available on the market, some of which have high versatility in both excitation and detection [76–79]. Low-temperature systems, particularly at cryogenic temperatures, are usually custom built by the researchers from various already available parts.

Samples can be either powders or relatively small chips. In the latter case, the sample should be flat and relatively thin, 1 mm or less, in order to avoid possible thermal gradients in the sample itself. The presence of thermal gradients usually results in glow curves that are larger than expected and sometime shifted toward higher temperatures.

7.4 Case Studies

The last section of this chapter, Section 7.4, is dedicated to the presentation of a few examples where TSL is used primarily as a research tool to obtain more information about both defects in materials as well as to gain information on the energy position of the ground state levels of dopant ions inside the bandgap of the materials themselves. One of the examples also presents the use of TSL in conjunction with electron paramagnetic resonance (EPR).

7.4.1 Lanthanide Energy Level Position in the Bandgap

The relative position of ground and excited levels of a luminescent dopant ion in the bandgap of inorganic materials determines whether the ion can emit light or not. If, for instance, the excited levels are completely buried in the conduction band, the promotion of an electron of the ion to one of the excited states is likely to result in the electron escaping from the excited state itself and becoming delocalized in the conduction band, instead of the emission of light due to the electron transition back from the excited state to the ground state of the ion. Dorenbos developed a simple phenomenological model able to describe 4f and 5d level positions of both divalent and trivalent lanthanide ions in different matrices using only three parameters [80–83]. This allows to design new luminescent materials not by trial and error but by carefully informed design of the material itself.

Dorenbos' model also allows to predict the behavior of a lanthanide in a given matrix. For instance, when the ground state of divalent lanthanides is below the conduction band, the corresponding trivalent ion might behave as a trap for free electrons. The model suggests, moreover, that in all matrices both divalent and trivalent lanthanide ions ground state levels follow the same zig-zag curve, and thus in a predictable position of the electron trap depth in the case of divalent ions [84]. The model predictions have been confirmed experimentally by TSL studies on YPO_4:Ce,Ln (with Ln = Pr, Nd, Sm, Dy, Ho, Er, Tm, and Yb) [74]. Figure 7.8 reports a sketch of the zig-zag curve for the divalent Ln ions in correlation with the experimentally measured glow curves. In YPO_4, and in several other matrices, during irradiation with ionizing radiation, Ce^{3+} behaves as a trap for holes, becoming temporarily tetravalent and thus able to subsequently capture an electron giving rise to luminescence. The second rare earth in the material, instead, behaves

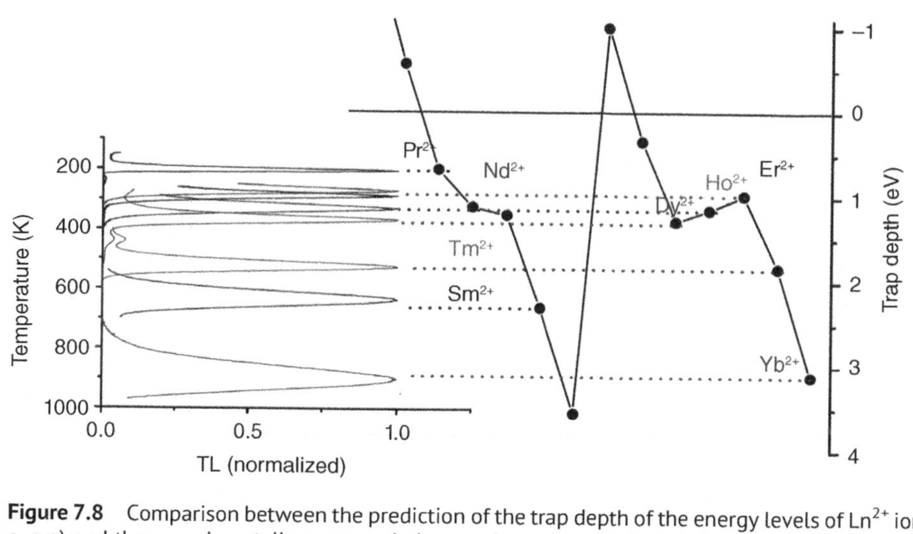

Figure 7.8 Comparison between the prediction of the trap depth of the energy levels of Ln^{2+} ions (the zig-zag curve) and the experimentally measured glow peak maxima of YPO_4:Ce,Ln. Source: Figure reproduced with permission from [74]. Study of TL glow curves of YPO_4 double doped with lanthanide ions, by Adrie J.J. Bos et al., *Radiation Measurements*, December 2011, © 2011, Elsevier.

mostly as an electron trap, effectively impeding the electron recombination on Ce^{4+}. Since the Ln^{3+} ions are substitutional to Y^{3+} and no charge compensation is needed, the trapping site for all the Ln ions is the same and so is the frequency factor s. It follows that, for the same heating rate, the temperature of the glow peak maximum is directly proportional to the trap depth. The substitution of Ce^{3+} with Pr^{3+} (another rare earth that mainly behaves as a hole trap) yields an identical dependence of the glow curve position as a function of the codopant ion [85], but now with the Pr^{3+} luminescence. Similar results have also been obtained for other materials, substantially confirming the systematic trends for the lanthanide energy levels. This property has been used, for instance, to design new dosimetric materials [86] and persistent phosphors [85], as well as in the clarification of the role of defects in luminescence hysteresis phenomena occurring in scintillators [87].

7.4.2 Bandgap Engineering

The presence of point defects acting as traps for delocalized carriers in scintillators is the cause of degradation of the luminescence properties of this class of materials since they can delay significantly the carrier recombination on the luminescence centers, when not stopping it altogether. Several methods have been proposed and exploited in order to reduce the concentration of defects in this class of materials [88], some examples are aliovalent codoping and synthesis methods with a lower process temperature than conventional single crystal growth. In lutetium aluminum garnets ($Lu_3Al_5O_{12}$, LuAG), one of the main defects that causes significant troubles is related to a cation antisite (Lu^{3+} on Al^{3+} site). The presence of this defect leads to a rather shallow trap, which is the cause of relatively long afterglow phenomena [89]. The reduction of the concentration of cation antisites in garnets is not trivial since the formation energy for these defects is low, and they are also charge neutral. Therefore, it cannot be removed by aliovalent cooping. On the other hand, while it is possible to obtain garnets by other methods, for instance in optical ceramic or by liquid phase epitaxy, in order to avoid the formation of antisites, these synthesis methods do not allow to obtain garnets in bulk quantities. The addition of Ga^{3+}, however, proved to be a particularly successful way to improve LuAG scintillation properties, since it removes the TSL peak related to antisites [90]. These improvements are related to the reduction of the material bandgap by Ga addition, with a lowering of the conduction band minimum below the antisite defect states, thus effectively removing this defect from the bandgap [91, 92].

To understand precisely how Ga admixture modifies the electronic structure of the aluminum garnets, Fasoli and colleagues [93] performed TSL measurements on LuAG doped with either Ce^{3+} or Eu^{3+} and containing different amounts of Ga. Both Ce^{3+} and Eu^{3+} in garnets behave as recombination centers; however, due to their different electronic properties, Ce prefers to capture holes as a first step, while Eu captures electrons. This different behavior has been exploited by the authors to monitor the shift of either the conduction band or the valence band, for Ce and Eu respectively, caused by the addition of Ga. In both cases, TSL glow curves show a clear shift toward lower temperatures, see Figure 7.9, indicating a rather substantial reduction in the thermal depth of different traps, and thus in a reduction of the energy difference between the various trap levels and the delocalized band. The authors also compared the TSL results with those obtained by first-principles calculations (Density Functional Theory, DFT), obtaining very similar values in the reduction of the bandgap energy and of the relative shift of the bottom of the conduction band and the top of the valence band. The authors were able, in conclusion, to identify the mechanism by which isovalent substitution of Al with Ga in LuAG improves the scintillation performance and predicted other doping strategies that could provide even better performance.

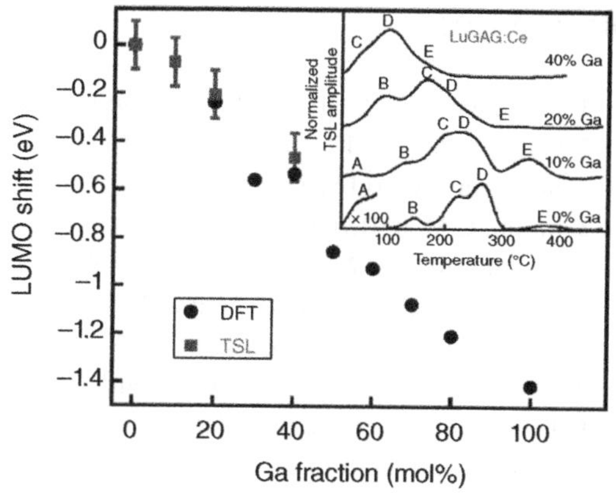

Figure 7.9 Comparison between Density Functional Theory and TSL experimental results of the LUMO (lowest unoccupied molecular orbital) shift in LuAG as a function of Ga concentration. The inset shows the shift of TSL peaks to lower temperatures with Ga substitution. Source: Reproduced with permission from [93]. Band-gap engineering for removing shallow traps in rare-earth $Lu_3Al_5O_{12}$ garnet scintillators using Ga^{3+} doping, by M. Fasoli et al., *Physical Review B*, 26 August 2011, © 2011, American Physical Society.

7.4.3 Correlation of TSL Data with EPR Results

Although TSL is a very powerful technique to gain insight in the defectuality of a material, the information that can be obtained does not allow to identify the defect responsible for a certain glow peak. To obtain more information about the defects themselves, it is necessary to compare TSL results with those obtained by other techniques, notably optical absorption and electron paramagnetic resonance (EPR, also called electron spin resonance, ESR). EPR, in particular, is particularly useful because it can clarify the nature of paramagnetic defects or impurities present in a material. These paramagnetic defects might be involved in the TSL process, either as traps or as recombination centers. Comparing the temperature dependence of the EPR signals with the TSL glow peaks, it is possible to arrive at a clear identification of the defect responsible for the TSL peak. This knowledge can, thus, be used to improve the material under investigation, whatever the characteristic the researcher is interested in or, more generally, to have a better understanding of the investigated material. Several examples can be found in the literature of the direct correlation between EPR temperature dependence and TSL measurements [94–98].

The combination of EPR and TSL results can also be used to obtain more information on possible competitive electronic process occurring in materials. Hu et al. [99, 100] studied the role of a paramagnetic hole trapping center in the scintillation properties of $Lu_3Al_5O_{12}$:Ce,Mg (LuAG) optical ceramics, and compared it to TSL glow curves related to the recombination of electrons on the Ce^{3+} luminescent center. The results of this study put in evidence the presence in LuAG of O^- centers related to holes trapped at oxygen ions bound to perturbing defects. The O^- defect concentration temperature dependence shows four rather clear decrease steps. The authors were able to reconstruct this decrease by using information from TSL, thus suggesting that the O^- centers compete for electron capture with the Ce^{4+} centers also created during the initial irradiation. In the case of electron recombination on the O^- centers, the process is non-radiative, and thus represents a possible mechanism leading to the reduction of the scintillation light yield.

Note

1 In Chapter 1 of this book, and in many other texts, the term phosphorescence indicates the photoluminescence emission from forbidden transitions between energy levels of a single center, in contrast to fluorescence, which applies to allowed transitions. However, in the TSL case, the term phosphorescence is used to indicate the emission mechanism involving a trap and a recombination center as distinct entities.

References

1 McKeever, S.W.S. (1985). *Thermoluminescence of Solids*. Cambridge: Cambridge University Press.
2 Chen, R. and McKeever, S.W.S. (1997). *Theory of Therrmoluminescence and Related Phenomena*. Singapore: World Scientific.
3 Sunta, C.M. (2015). *Unraveling Thermoluminescence*. Mumbai: Springer India.
4 Martini, M. and Meinardi, F. (1997). *Riv. Nuovo Cimento* 20: 1.
5 Bos, A.J.J. (1997). *Radiat. Meas.* 41: 545.
6 Bos, A.J.J. (2017). *Materials* 10: 1357.
7 Kittel, C. (1996). *Introduction to Solid State Physics*, 7e. Wiley.
8 Enderson, B. and Hughes, A.E. (eds.) (1976). *Defects and Their Structures in Nonmetallic Solids*. New York: Plenum Press.
9 Stoneham, A.M. (2001). *Theory of Defects in Solids: Electronic Structure of Defects in Insulators and Semiconductors*. Oxford: Oxford University Press.
10 Klick, C.C. (1965). *Science* 150: 451.
11 Adirovitch, I.E. (1956). *J. Phys. Rad* 17: 705.
12 Rose, A. (1955). *Phys. Rev.* 97: 322.
13 Shockley, W. and Read, W.T. (1952). *Phys. Rev.* 87: 835.
14 Lax, M. (1960). *Phys. Rev.* 119: 1502.
15 Mitonneau, A., Mircea, A., Martin, G.M., and Pons, D. (1979). *Rev. Phys. Appl.* 14: 853.
16 Khanin, V., Vrubel, I.I., Polozkov, R.G. et al. (2019). *J. Phys. Chem. C* 123: 22725.
17 Curie, D. (1960). *Luminescence in Crystals*. London: Methuen.
18 Williams, F.E. and Eyring, H. (1947). *J. Chem. Phys.* 15: 289.
19 Bube, R.H. (1960). *Photoconductivity of Solids*. New York: Wiley.
20 Chen, R., McKeever, S.W.S., and Durrani, S.A. (1981). *Phys. Rev. B* 24: 4931.
21 Maxia, V. (1980). *Phys. Rev. B* 21: 749.
22 Randall, J.T. and Wilkins, M.H.F. (1945). *Proc. R Soc. Lond.* 184: 365.
23 Garlick, G.F.I. and Gibson, A.F. (1948). *Proc. Phys. Soc.* 60: 574.
24 May, C.E. and Partridge, J.A. (1964). *J. Chem. Phys.* 40: 1402.
25 Schulman, J.H., Ginther, R.J., Gorbics, S.G. et al. (1969). *Int. J. Appl. Radiat. Isot.* 20: 523.
26 Burgkhardt, B., Herrera, R., and Piesch, E. (1976). *Nucl. Instrum. Methods* 137: 41.
27 Zou, Z., Wu, J., Xu, H. et al. (2015). *J. Mater. Chem. C* 3: 123518.
28 Matsuzawa, T., Aoki, Y., Takeuchi, N., and Murayama, Y. (1996). *J. Electrochem. Soc.* 143: 2670.
29 Laguta, V.V., Nikl, M., Rosa, J. et al. (2008). *J. Appl. Phys.* 104: 103535.
30 Fasoli, M., Fontana, I., Moretti, F. et al. (2008). *IEEE Trans. Nucl. Sci.* 55: 1114.
31 Delbeq, C.J., Toyozawa, Y., and Yuster, P.H. (1974). *Phys. Rev. B* 9: 4497.
32 Thioulouse, P., Giess, E.A., and Chang, L.F. (1982). *J. Appl. Phys.* 53: 9015.
33 Avouris, P. and Morgan, T.N. (1981). *J. Chem. Phys.* 74: 4347.
34 Huntley, D.J. (2006). *J. Phys. Condens. Matter* 18: 1359.
35 Dobrowolska, A., Bos, A.J.J., and Dorenbos, P. (2014). *J. Phys. D Appl. Phys.* 47: 335301.
36 Dorenbos, P., Bos, A.J.J., and Poolton, N.R.J. (2010). *Phys. Rev. B* 82: 195127.
37 Halperin, A. and Braner, A.A. (1960). *Phys. Rev.* 117: 408.
38 Vedda, A., Moretti, F., Fasoli, M. et al. (2009). *Phys. Rev. B* 80: 045104.
39 Vedda, A., Nikl, M., Fasoli, M., Mihokova, E. et al. (2008). *Phys. Rev. B* 78: 195123.
40 Vedda, A. and Fasoli, M. (2018). *Radiat. Meas.* 118: 86.

41 Sunta, C.M., Ayta, W.E.F., Cubaci, J.F.D., and Watanabe, S. (2001). *J. Phys. D Appl. Phys.* 34: 2690.

42 Pagonis, V. and Kitis, G. (2012). *Phys. Status Solidi B* 249: 1590.

43 Lewandowski, A.C. and McKeever, S.W.S. (1991). *Phys. Rev. B* 43: 8163.

44 Lewandowski, A.C., Marley, B.G., and McKeever, S.W.S. (1994). *Phys. Rev. B* 49: 8029.

45 Bos, A.J.J. (2001). *Radiat. Meas.* 33: 737.

46 Akselrod, M.S., Larsen, N.A., Whitley, V., and McKeever, S.W.S. (1998). *J. Appl. Phys.* 64: 3364.

47 Kitis, G., Papadopulos, J.G., Charalambous, S., and Tuyin, J.W.N. (1994). *Radiat. Prot. Dosimetry* 55: 183.

48 Subedi, B., Oniya, E., Polymeris, G.S. et al. (2011). *Nucl. Instrum. Methods Phys. Res., Sect. B* 269: 572.

49 Martini, M., Meinardi, F., Rosetta, E., Spinolo, G. et al. (1998). *IEEE Trans. Nucl. Sci.* 45: 1396.

50 Vedda, A., Chiodini, N., Di Martino, D. et al. (2005). *J. Non Cryst. Solids* 351: 3699.

51 Kiisk, V., Puust, L., Utt, K. et al. (2016). *J. Lumin.* 174: 49.

52 Van de Eeckhout, K., Bos, A.J.J., Poelman, D., and Smet, P.F. (2013). *Pys. Rev. B* 87: 045126.

53 Shen, Y., Shi, Y., Feng, X. et al. (2012). *J. Am. Ceram. Soc.* 95: 2130.

54 Brylew, K., Drozdowski, W., Wojtowicz, A.J. et al. (2014). *J. Lumin.* 154: 452.

55 Medlin, W.L. (1961). *Phys. Rev. B* 123: 502.

56 Chen, R. and Haber, G.A. (1968). *Chem. Phys. Lett.* 2: 483.

57 Haake, C.H. (1957). *J. Opt. Soc. Am.* 47: 649.

58 Kivits, P. and Hagebeuk, H.J.L. (1977). *J. Lumin.* 15: 1.

59 Pagonis, V., Kitis, G., and Furetta, C. (2006). *Numerical and Practical Exercises in Thermoluminescence*. New York: Springer Verlag.

60 Chen, R. and Winer, S.A. (1970). *J. Appl. Phys.* 41: 5227.

61 Coleman, A.C. and Yukihara, E.G. (2018). *Radiat. Meas.* 117: 70.

62 Chen, R. (1969). *J. Electrochem. Soc.* 116: 1254.

63 Grossweiner, L.I. (1953). *J. Appl. Phys.* 24: 1306.

64 Dussel, G.A. and Bube, R.H. (1967). *Phys. Rev.* 155: 764.

65 Keating, P.N. (1961). *Proc. Phys. Soc.* 78: 1408.

66 Land, P.L. (1681). *J. Phys. Chem. Solid* 30: 1969.

67 Chen, R. (1969). *J. Appl. Phys.* 40: 570.

68 Hoogenstraaten, K. (1958). *Philips Res. Rep.* 13: 515.

69 Bos, A.J.J., Pitiers, T.M., Gomez-Ros, J.M., and Delgado, A. (1993). *Radiat. Prot. Dosimetry* 47: 473.

70 Horowitz, Y.S. and Yossian, D. (1995). *Radiat. Prot. Dosimetry* 60: 3.

71 Kitis, G., Gomez-Ros, J.M., and Tuyn, J.W.N. (1998). *J. Phys. D Appl. Phys.* 31: 2636.

72 van Dijk, J.W.E. (2006). *Radiat. Prot. Dosimetry* 119: 332.

73 Chen, R., Pagonis, V., and Lawless, J.L. (2016). *Radiat. Meas.* 91: 21.

74 Bos, A.J.J., Dorenbos, P., Bessiere, A. et al. (2011). *Radiat. Meas.* 46: 1410.

75 Bos, A.J.J., Van Duijvenvoorde, R.M., Van der Kolk, E. et al. (2011). *J. Lumin.* 131: 1465.

76 Thomsen, K.J., Botter-Jensen, L., Jain, M. et al. (2008). *Radiat. Meas.* 43: 414.

77 Lapp, T., Jain, M., Thomsen, J.J. et al. (2012). *Radiat. Meas.* 47: 803.

78 Richter, D., Richter, A., and Dornich, K. (2013). *Geochronometria* 40: 220.

79 Richter, D., Richter, A., and Dornich, K. (2015). *Geochronometria* 42: 202.

80 Dorenbos, P. (2003). *J. Phys. Condens. Matter* 15: 8417.

81 Dorenbos, P. (2013). *ECS J. Sol. State Sci. Technol.* 2: R3001.

82 Dorenbos, P. (2014). *J. Lumin.* 151: 224.

83 Dorenbos, P. (2017). *Opt. Mater.* 22: 8.

84 Dorenbos, P. (2005). *Chem. Mater.* 17: 6452.

85 Lecointre, A., Bessiere, A., Bos, A.J.J. et al. (2011). *J. Phys. Chem. C* 115: 4217.
86 Yukihara, E.G., Milliken, E.D., Oliveira, L.C. et al. (2013). *J. Lumin.* 133: 203.
87 Moretti, F., Patton, G., Belsky, A. et al. (2014). *J. Phys. Chem. C* 118: 9670.
88 Dujardin, C., Auffray, E., Bourret-Courchesne, E. et al. (2018). *IEEE Trans. Nucl. Sci.* 65: 1977.
89 Nikl, M., Mihokova, E., Pejchal, J. et al. (2005). *Phys. Status Solidi B* 242: R119.
90 Mihokova, E., Nikl, M., Mares, J.A. et al. (2007). *J. Lumin.* 126: 77.
91 Zorenko, Y. (2000). *Opt. Spectrosc.* 88: 551.
92 Nikl, M., Vedda, A., and Laguta, V.V. (2007). *Radiat. Meas.* 42: 509.
93 Fasoli, M., Vedda, A., Nikl, M. et al. (2011). *Phys. Rev. B* 84: 081102(R).
94 Buryi, M., Laguta, V., Fasoli, M. et al. (2017). *J. Lumin.* 192: 767.
95 Vaccaro, G., Panzeri, L., Paleari, S. et al. (2017). *Quat. Geochronol.* 39: 99.
96 Laguta, V.V., Martini, M., Vedda, A. et al. (2003). *Phys. Rev. B* 67: 205102.
97 Born, G.K., Grasser, R.J., and Scharmann, A.O. (1968). *Phys. Status Solidi* 28: 583.
98 Katzitr, A., Suss, J.T., Zunger, A., and Halperin, A. (1975). *Phys. Rev. B* 11: 2370.
99 Hu, C., Liu, S., Fasoli, M. et al. (2015). *Phys. Status Solidi RRL* 9: 245.
100 Hu, C., Liu, S., Fasoli, M. et al. (2015). *Opt. Mater.* 45: 252.

8

Spectroscopic Studies of Radiation Effects on Optical Materials

Sylvain Girard, Vincenzo De Michele, and Adriana Morana

Laboratoire Hubert Curien, Université Jean Monnet, Saint-Etienne, France

8.1 Introduction

Optical materials, bulk glasses or optical fibers, offer many advantages for implementation in radiation environments such as the space, high-energy physics, nuclear industry, or medicine. Exploiting their unique optical properties, these materials are functionalized to design high-speed, high-bandwidth data links, laser or plasma diagnostics, point or distributed sensors, allowing to provide real-time maps of environmental parameters' evolutions. Radiation often alters the macroscopic optical properties of glass, implying first an evaluation the material's vulnerability to a given profile of use before implementing it in harsh environments and sometimes further application of radiation hardening techniques to improve it in order to meet application requirements. The qualification of the materials then implies to develop "ad hoc" experimental spectroscopic techniques able to monitor the levels and kinetics of these radiation induced changes. Furthermore, other tools are needed to identify microscopic point defects at their origins to facilitate the identification of mitigation solutions of radiation effects. Those techniques and tools will be described in this chapter, highlighting the specificities related to the radiation testing and radiation data analysis compared to the approaches developed for in-lab experiments.

8.1.1 Radiation Environments

A radiation environment is characterized by the presence of high-energy photons such as γ-rays or X-rays, and/or particles, such as neutrons, protons, electrons, and even heavy ions. Radiation can be classified as ionizing or nonionizing depending on the way it interacts with matter. The former, such as γ-rays, mainly interacts with matter through ionization process: an electron is excited by the valence band to the conduction band, leaving a hole in the valence band. The generated electron–hole pairs can recombine either radiatively by emitting photons or non-radiatively, through the excitation of phonons or secondary radiolytic processes, inducing point defects. The non-ionizing radiation, such as neutrons, instead, interacts with matter through direct atomic displacements, also known as knock-on processes [1]. Some of the studied harsh environments are natural such

Spectroscopy for Materials Characterization, First Edition. Edited by Simonpietro Agnello.

as space while some are artificial, i.e. nuclear power plants, nuclear waste repositories, and high-energy physics facilities. They are classified according to their main characteristics [2, 3]:

- The nature of radiation.
- The total ionizing dose (TID): TID is the quantity of energy deposited in the material, the unit of measurement is Gray (Gy), sometimes the older unit rad is still used: $1\ Gy = 1\ J\ kg^{-1} = 100$ rad. TID depends on the material; the dose is then usually explicitly related to a given material, e.g. Gy(SiO_2).
- The dose rate is the deposition rate of energy in the material, its unit of measurement is then $Gy\ s^{-1}$.
- The irradiation temperature.
- Other environmental parameters like the pressure, the presence of chemical species such as the molecular hydrogen released by the nuclear wastes, and the presence of liquid water and vapor for the sensors operating at storage fuel pools.

Table 8.1 reports the dose rate, dose, and temperature values characterizing some of the radiation environments in which optical materials play a key role [3]. For example, space missions are mostly associated with very low dose rate, between 10^{-5} and $10^{-3}\ Gy\ h^{-1}$, but as their duration can be quite long (e.g. up to 20 years), the dose levels can be as large as 10 kGy. In a nuclear reactor core, the temperature can increase from room temperature (RT) to 800 °C while very large fast neutron fluence up to $10^{20}\ n\ cm^{-2}$ and large TID levels (>GGy) at a flux of ~$10^{15}\ n\ cm^{-2}\ s^{-1}$ (dose rate up to 10 $kGy\ h^{-1}$) are attained.

8.1.2 Applications for Optical Materials

For safety improvement, there has been a growing interest in innovative systems monitoring the structural health of facilities by measuring a set of environmental parameters such as temperature, pressure, strain, dose, or dose rate. For all these monitoring operations, in the last decades, researchers focused their interest on optical fiber sensors (OFSs). The latter, indeed, benefit from all the nice properties of the optical fibers (OFs), such as light weight, small size, and an excellent immunity to electromagnetic perturbations. Moreover, several sensing points can be achieved

Table 8.1 Dose rate, dose, and temperature ranges for the different radiation environments where optical materials are implemented.

	Dose rate: <1 $Gy\ s^{-1}$ Dose: 0.01 to 50 Gy T range: RT	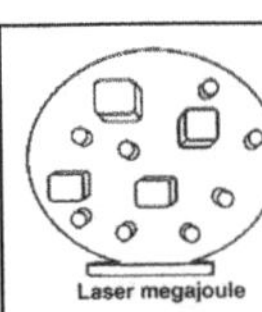	Dose rate: >1 $MGy\ s^{-1}$ Dose: <1 kGy T range: RT
	Dose rate: 10^{-5} to 10^{-3} $Gy\ h^{-1}$ Dose: <10 kGy T range: −200 to 300 °C		Dose rate: <0.1 $Gy\ h^{-1}$ Dose: <100 kGy T range: RT
	Dose rate: <10 $Gy\ s^{-1}$ Dose: <10 MGy T range: RT to 90 °C		Dose rate: $10^{15}\ n\ cm^{-2}\ s^{-1}$, $<10^{15}\ n\ cm^{-2}\ s^{-1}$ Dose: >GGy, $10^{20}\ n\ cm^{-2}$ T range: RT to 800 °C

along a unique fiber. For example, OFSs exploiting the light backscattering phenomena (Rayleigh, Raman, and Brillouin) are known as distributed since the whole fiber length acts here as a continuous series of sensitive probes [4]. Other OFS point technologies such as fiber gratings can be inscribed at different locations along the same fiber to increase the number of measurement points [5] while keeping very high acquisition rates (up to MHz). However, as radiation influences the performances of these sensors, it remains mandatory to estimate their potential for different targeted applications considering the radiation constraints. Furthermore, to obtain the most radiation-resistant OFSs possible, hardening techniques have been developed to improve the hardness of the various classes of sensor either at the system level (e.g. *using a different sensor architecture*) or at the component level (e.g. *by adjusting the OF composition*) [3]. Among the different classes of optical fibers, the most widely used in such harsh environments are the ones with pure or fluorine-doped silica cores, since their behavior is less affected under steady state irradiation [6].

Another crucial need in harsh environments consists in imaging applications, implying the free-space image transport through optical relays as well as radiation-hardened cameras. This is the case for ITER remote handling operations or for the visual inspection of nuclear wastes. If the camera is a quite complex system constituted by a CMOS image sensor with its readout electronics, an illumination system and an optical system [7], the darkening of the lenses remains one of the factors limiting the camera use at high dose level (>MGy) since today the CMOS image sensor can be hardened-by-design [8]. Concerning the glass-made lenses constituting the optical systems, their properties (transmission, refractive index) evolved under irradiation, impacting the image quality [7, 9]. Among the different classes of bulk glasses, the most used in radiation environments are the ones codoped with the Cerium element that was shown to limit their darkening under γ-ray and proton irradiation [10].

8.2 Radiation-Induced Effects on Optical Materials and Optical Fibers

Radiation–matter interaction, through ionization or displacement damage processes, gives rise to three main macroscopic effects: radiation-induced attenuation (or briefly RIA), radiation-induced emission (RIE), and radiation-induced compaction (RIC). Both RIA and RIC can lead to a radiation-induced refractive index change (RIRIC). Although these processes are today well known [1], their consequences are difficult to predict as their amplitudes and kinetics depend on many parameters such as:

- The characteristics of the optical material under test (composition and impurities presence) and its manufacturing process and history: pretreatments (thermal, gas loading, pre-irradiation).
- The irradiation parameters: dose, dose rate, temperature.
- The material's profile of use, e.g. the injected power level and the wavelength of the signal for sensors or data links.

8.2.1 Radiation-Induced Attenuation – RIA

The RIA is the most widely studied phenomenon as it is limits the applicability of most materials. RIA consists in a decrease of the transmission properties of the material over the whole bandgap. RIA is caused by the appearance of absorption bands in the transmission window of the material, these bands being related to microscopic point defects (sometimes called color centers), causing the

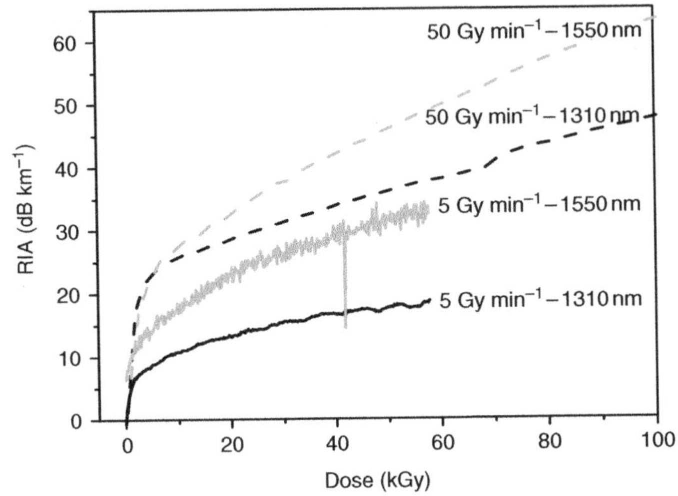

Figure 8.1 Radiation-induced attenuation (RIA) of the SMF28e+ from Corning under γ-rays as a function of the dose, at RT, for two different dose rates (5 Gy min^{-1}; 50 Gy min^{-1}) and two wavelengths: 1310 nm; 1550 nm. Before irradiation, the optical losses of this fiber are below 0.5 dB km^{-1} at both wavelengths.

appearance of energy levels in the gap of amorphous glasses. As an example, Figure 8.1 illustrates the RIA growth as a function of the TID measured at 1310 and 1550 nm in a single-mode germanosilicate (Ge-doped silica) fiber SMF28e+ from Corning, under steady state γ-rays at different dose rates and at room temperature (RT).

First of all, the RIA features depend on the signal wavelength (or energy range), as shown in Figure 8.1. This can be easily explained by the fact that the various generated defects present absorption bands differing in terms of peak positions and widths [11]. Usually, RIA levels increase at shorter wavelength because most of the defects are absorbing in the UV and visible domains. Besides its dependence on the material itself, RIA also depends on the irradiation conditions. As highlighted in Figure 8.1, generally RIA increases with increasing TID. However, the dependence can be as simple as a linear one or very complex. The kinetics again depend on the generation mechanisms of the various defects, on the possible interconversion between them, and on their bleaching mechanisms. As a consequence, the resulting RIA kinetics are governed by the competition between all these mechanisms. For most of the materials, RIA first quickly increases up to kGy dose levels as defects are generated from pre-existing "weak" precursor sites and then slows down or sometimes saturates when the point defects are less efficiently generated from regular bonds. At the irradiation end, RIA can remain stable if the defects are thermally stable or decrease with various kinetics if thermal bleaching takes place: this phase is known as RIA recovery. Understanding the physics at stake during irradiation remains mandatory as for most applications, accelerated qualification tests are done and RIA behaviors should be extrapolated for the targeted mission. For this, several semiempirical models or physical models have been proposed in literature and have been reviewed in [2, 3].

The dose rate and the irradiation temperature, indeed, control the equilibrium between the defect generation and bleaching processes and then the RIA kinetics. Generally, for a same value of TID, the higher the dose rate, the higher the RIA, because the run duration is smaller and a lower proportion of defects will be bleached than during a longer run. On the contrary, a higher irradiation temperature was found to accelerate the bleaching process; this phenomenon is known as thermal bleaching. However, this is not always accompanied by a decrease of RIA as it appears that the irradiation temperature could also enhance the defect generation efficiency, e.g. some defects such as the SiE' will be more efficiently created at 100 °C than at RT. The combined effects of temperature and radiation are still the subject of active research [12, 13].

Another parameter to consider is the injected light power chosen to measure the RIA and its representativity regarding the application needs, which can vary from a few microwatts to tens of watts. Indeed, two opposite phenomena can be activated by the probe light; their impact strongly depends on the material under study. The first one is known as photobleaching, where high power can facilitate the defect recombination or transformation and then decrease the RIA levels and accelerate RIA recovery kinetics. The second one is known as photo-darkening, which is an attenuation increase due to the defects generated by the probe light itself (in opposition to the radio-darkening causing the RIA).

Every defect structure possesses its own generation, conversion, or bleaching mechanisms, and a systematic study of the influence of the fiber intrinsic or extrinsic parameter has to be carried out to fully assess the device's vulnerability to a given harsh environment. As the nature of radiation can influence the material response, the fiber's vulnerability is usually investigated for each of the targeted applications and each of the targeted environments. In the future, multi-scale simulation tools should allow to complement radiation testing through predictive analysis and facilitate the deployment of optical materials in the most challenging environments [11].

8.2.2 Radiation-Induced Emission – RIE

Radiation can also generate light in the exposed material, this is called radiation-induced emission. RIE could consist of two different contributions. The first one is the Cerenkov emission that is produced when a charged particle propagates in a medium with a speed greater than that of light in that material (which is the speed of light in vacuum divided by the material refractive index). This light is emitted on a cone around the particle's propagation direction, consequently only during irradiation, and has a continuous spectrum whose intensity is proportional to the inverse of the squared wavelength and to the particle flux [14]. An example of Cerenkov light is the blue emission observed in a pool containing a nuclear reactor.

The second contribution is known as radiation-induced luminescence (RIL) or briefly radioluminescence (RL). It is the luminescence emitted from pre-existing or radiation-induced point defects excited by the radiation itself. Under irradiation, the electrons are excited from the valence band to the conduction one, but, when they try to come back to the valence band, they can be trapped on pre-existing or radiation-induced centers. A part of these electrons manages to be detrapped and a luminescence appears. The RL increases as soon as the irradiation starts, because there is a competition between the trapping and detrapping of the electrons. When this phenomenon reaches an equilibrium, the RL does too. Initially, the saturation value increases with the dose rate and, at dose rates so high that all the traps are filled, it stabilizes. In conclusion, the RL intensity depends on the dose rate on a large range. In this range of dose rate, the RL of some materials could be exploited to monitor the radiation flux in harsh environments.

When the irradiation stops, the RL signal goes to zero. However, a part of the involved electrons remains trapped and needs a stimulation (either thermal or optical) to be detrapped. Indeed, if after the irradiation, a laser with appropriate wavelength is injected on the sample, a signal known as optically stimulated luminescence (OSL) could appear [15]. In the most interesting case, the OSL signal is proportional to the number of trapped sites and, consequently, on the TID. However, it depends also on the time delay between the irradiation end and the laser injection, because some electrons can be thermally detrapped before the optical stimulation: this phenomenon is called "fading."

It is very important to highlight that the Cerenkov emission originates only when the incident particles (or the secondary particles) are charged; whereas the RL depends mainly on the material, often scintillating elements are added to the matrix (i.e. silica-based or plastic optical fibers) in order to enhance their emission. Typical elements used today are copper, cerium, and gadolinium.

8.2.3 Radiation-Induced Compaction – RIC and Refractive Index Change – RIRIC

The radiation-induced compaction is the variation of density ($\Delta\rho$) under irradiation. Its amplitude depends on the dose (D), for example according to a power law for silica-based materials:

$$\frac{\Delta\rho}{\rho} \propto D^k \tag{8.1}$$

where k depends on the irradiation nature, its value is of ~1 for fast neutrons and ~2/3 for γ-rays [16]. The RIC depends on the material, but Primak observed that it saturates at the same value for both amorphous and crystalline silicon dioxide when irradiated at a high neutron fluence of 10^{19} n cm^{-2} even if these two materials are totally different before irradiation. This suggests that the amorphous and crystal phases of the silicon dioxide tend to the same structure when they are exposed to very high neutron fluence: this phase is known as metamict [16].

Both the RIA and the RIC gives rise to radiation-induced refractive index change, a crucial parameter of optical materials and devices. The latter is linked to the absorption band variation ($\Delta\alpha(\lambda')$) through the Kramers–Kronig relation [17]:

$$\Delta n(\lambda) = \frac{\lambda^2}{2\pi^2} \int_0^\infty \frac{\Delta\alpha(\lambda')}{\lambda^2 - \lambda'^2} \mathrm{d}\lambda' \tag{8.2}$$

whereas the RIRIC dependence on the density change ($\Delta\rho$) and on the molar refractivity (ΔR) is obtained by the Lorentz–Lorenz formula:

$$\Delta n(\lambda) = \frac{(n^2+2)(n^2-1)}{6n} \left(\frac{\Delta\rho}{\rho} + \frac{\Delta R}{R} \right). \tag{8.3}$$

Consequently, for the case of the silicon dioxide under a neutron fluence of 10^{20} n cm^{-2}, Primak observed a refractive index increase for amorphous silica from 1.46 to 1.4670 and a refractive index decrease for quartz from 1.555 to 1.4680 [16]. The values of the pristine samples are completely different but tend to a similar one after a high neutron fluence. It is worth noticing that a refractive index change can affect the properties of optical fibers and fiber sensors, and also of the lenses constituting the optical systems of cameras, resulting in a blurring of the images [18].

8.2.4 Origins of Radiation-Induced Optical Changes

The basic mechanisms of the macroscopic effects of radiation (RIA, RIE, RIRIC) on optical materials have to be understood to precisely estimate the vulnerability of a given application for a future mission. As the physics of the point defects is complex, especially in optical fiber cases, it is not yet possible to predict the fiber response without performing radiation tests. As a consequence, it is crucial to be able to experimentally well characterize the levels and kinetics of RIA, RIE, or RIRIC phenomena in optical fibers or bulk glasses. Depending on the application, one should be able to perform those measurements in conditions representative of the targeted environment to avoid to overestimating or underestimating the damage levels or kinetics due to combined effects such as temperature – radiations or photobleaching – photo-darkening – radio-darkening. Particularly, it is crucial to promote online or in situ radiation measurements to consider both the transient and permanent radiation effects on glasses.

8.3 Radiation-Induced Attenuation Measurements

The attenuation (or absorption) phenomenon is described by the Lambert–Beer law, already reported in Eq. (1.4), with the attenuation coefficient often expressed in cm^{-1} for bulk glasses and in $dB\,m^{-1}$ or $dB\,km^{-1}$ for optical fibers.

8.3.1 Postirradiation RIA Measurements

When the irradiation facility is not easily accessible and cannot easily be instrumented, such as nuclear reactor cores, RIA measurements are very difficult to perform online; so, the only option is to irradiate the samples without monitoring (passive tests) and to get them back after irradiation. In this case, the attenuation measurements have to be performed before and after the irradiation and only the permanent RIA levels will finally be obtained [19]. Indeed, this type of measurement is affected by the recovery processes taking place in the material between the irradiation and the measure. They are then unable to describe the transient radiation response of the material; e.g. for the SMF28 fiber, more than 90% of the RIA is recovered a few seconds after an X-ray pulse [2]. These postirradiation tests give information only on the stable defects and then always underestimate the real fiber degradation during exposure. However, they could provide very useful preliminary information to better prepare complex online experiments, e.g. in nuclear reactor core. As a consequence, active and passive testing configurations are usually combined for bigger research projects.

8.3.1.1 Bulk Glasses

For bulk glasses, postirradiation attenuation measurements can be performed using an absorption spectrometer, whose representation is reported in Figure 1.7. In this case, the permanent RIA spectral dependence (RIA vs. wavelength) is obtained by comparing the spectra acquired before and after irradiation.

8.3.1.2 Optical Fibers

The postirradiation RIA measurements on optical fibers are performed through the *cut-back technique*. The transmission spectra of two different lengths, a longer one (L) and a shorted one (ℓ), of the fiber under test (briefly, FUT) have to be acquired by using the same acquisition system, constituted by a white source, generally a deuterium–halogen lamp, and a spectrometer. In this case, it is very important not to change the injection conditions; for this reason, the shorter length is obtained by cutting a fiber length ($\Delta\ell = L - \ell$) from the longer one, without touching the connections, especially at the source injection. The spectral attenuation (α corresponding to the whole fiber loss) is obtained by the following formula:

$$\alpha(\lambda) = -\frac{10}{\Delta\ell}\cdot\log_{10}\left(\frac{I_L(\lambda) - I_{\text{noise}}(\lambda)}{I_\ell(\lambda) - I_{\text{noise}}(\lambda)}\right) \tag{8.4}$$

where $I_L(\lambda)$ and $I_\ell(\lambda)$ are, respectively, the transmission spectra of the long and short length and $I_{noise}(\lambda)$ is the noise spectrum, acquired by the spectrometer when the source is off. This latter is due to spectrometer electrical noise and decreases with decreasing temperature. As already said, the RIA $\alpha_{RIA}(\lambda)$ is obtained by subtracting the attenuation of the pristine fiber $\alpha_0(\lambda)$ to the one of the irradiated sample:

$$\alpha_{RIA}(\lambda) = \alpha(\lambda) - \alpha_0(\lambda). \tag{8.5}$$

Typical example of RIA measurements is given in [19] on neutron irradiated samples.

8.3.2 In Situ RIA Measurements

Contrary to the postmortem radiation-induced attenuation measurements, which allow to observe only the stable defects, the ones performed in situ provide in real time the growth and decay kinetics of the RIA and, then, of the corresponding transient and stable point defects. In this case, the resulting RIA is a function of the time (or of the dose during the irradiation). To obtain online the RIA spectra, the transmission spectra of a sample with length ℓ have to be acquired as a function of the time before the irradiation starts, while the fiber is exposed to the radiation beam. Afterward, the RIA is defined as:

$$\mathrm{RIA}(\lambda, t) = -\frac{10}{\ell} \cdot \log_{10}\left(\frac{I(\lambda, t) - I_{noise}(\lambda)}{I(\lambda, t_0) - I_{noise}(\lambda)}\right) \tag{8.6}$$

where $I(\lambda, t)$ is a transmission spectrum acquired at the time t, t_0 is the irradiation start, and $I(\lambda, t_0)$ is the reference spectrum recorded just before the irradiation start. As before, $I_{noise}(\lambda)$ is the noise spectrum. Generally, white light sources such as deuterium/halogen are used, they have light power levels of few tens of microwatts. However, for some fiber tests with small-sized fiber cores, most powerful sources such as SuperContinuum sources or laser-driven light source (LDLS) are preferred, allowing to increase the available power by at least three or four orders of magnitude compared to a halogen source. It is worth noticing that such a high power can give rise to both photo-darkening and photobleaching effects and should be used with caution.

8.3.2.1 Bulk Glasses

When studying the RIA of bulk glasses at a radiation facility, it is not possible to install a spectrometer inside the irradiation room. For those tests, the most difficult task is to bring the light from a source located in the instrumentation zone to the sample exposed to radiations and from the latter to the spectrometer, also in the radiation-free zone. To do so, the probe light can be transported by radiation-resistant multimode optical fibers, i.e. whose RIA is orders of magnitude lower than the one of the bulk samples under test. Usually, F-doped fibers with low RIA degradation are selected as transport fibers. Nevertheless, these fibers should still be shielded (e.g. with lead) as despite their low radiation sensitivity, as their length (several meters) is very long compared to the one of the bulk glass (mm), they could prevent the characterization of the small RIA levels at the beginning of the tests or the characterization of radiation hardened bulk glasses.

A possible setup for the online characterization of the RIA in bulk glass samples or fiber preforms is given in Figure 8.2. The glass sample is placed between two fibered parabolic mirrors, the first one is illuminated by a cone of light exiting from the transport fiber connected to the source. The parabolic mirror reflects the light beam and makes it parallel. The beam size depends on the distance from the input fiber end and the mirror and on the fiber numerical aperture; generally, it is about 2 mm. Then, the beam propagates through the sample (its length parallel to the light propagation is

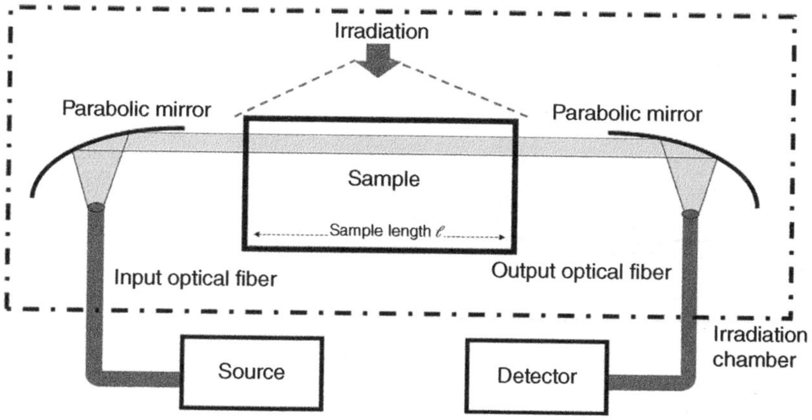

Figure 8.2 Setup for in situ RIA measurements on bulk glasses.

the one to be used in the Eq. (8.6)) and reaches the second mirror that couples it into the fibered detector.

8.3.2.2 Optical Fibers

In the case of optical fibers, *polychromatic* measurements can be performed by placing a length ℓ of the investigated sample under a homogenous radiation beam and simply splicing it to radiation-resistant transport fibers, connected to a white source and a spectrometer. The setup is illustrated in Figure 8.3a. As for bulk samples, if the FUT presents low RIA levels, the transport fibers have to be shielded, above all if the FUT length is very short, as for the tests in the ultraviolet and visible domains.

Often, *monochromatic* measurements are preferred. A way to do so consists in replacing the white source and the spectrometer with a laser and a photodetector. In this case, only the intensity at the laser wavelength is monitored and from the same Eq. (8.5), the RIA at that wavelength can be calculated. This procedure has to be employed, for example, when investigating photobleaching phenomena, since the injected laser power can be easily varied, for instance by changing the laser diode current or by using an attenuator. Another technique to obtain monochromatic RIA measurements consists in using an *optical time domain reflectometer* (OTDR): a pulsed signal at wavelength λ_0 is injected in a fiber end (hereafter indicated as $z = 0$) and the backscattered Rayleigh signal is acquired from the same end, as a function of position (z) along the fiber length. This latter is measured through the time of flight time to reach from the fiber input the position z and to come back to the same fiber end. Since the light propagates twice along the distance z, the inject light intensity will be attenuated as follows:

$$I(\lambda_0, z, t) = I(\lambda_0, z = 0, t)\, e^{-\alpha(\lambda_0)2z}. \tag{8.7}$$

The overall attenuation $\alpha(\lambda_0)$ corresponds to half of the slope of the curve $I(\lambda_0, z, t)$ as a function of z, for each time t. Then, to calculate the RIA, the contribution of the pristine sample has to be subtracted: $\alpha_{RIA}(\lambda_0, t) = \alpha(\lambda_0, t) - \alpha_0(\lambda_0)$. The main limitation of this technique is that the OTDR instrument wavelengths usually fall into telecom windows, as 850, 1310, 1550, and 1625 nm limiting its application in the UV spectral range to visible domain. However, it possesses the main advantage of being single ended, meaning that only the FUT end has to be connected to the instrument, as shown in Figure 8.3b.

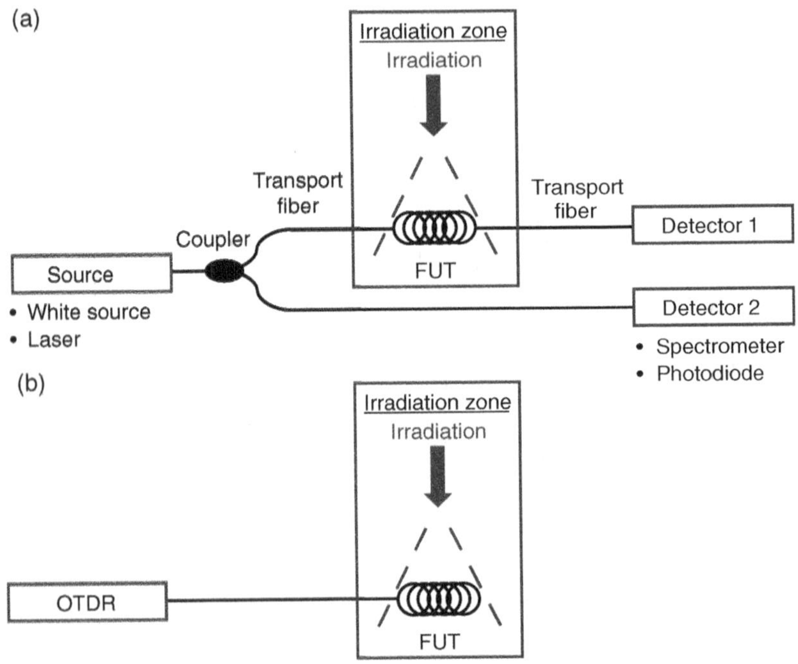

Figure 8.3 Setup for in situ RIA measurements on optical fibers: (a) double-ended scheme and (b) single-ended case with the OTDR instrument.

8.3.2.2.1 *Steady State Irradiation*

The term "steady state" indicates an irradiation whose dose rate does not change during the exposure. An example of steady state irradiation is constituted by a classical ^{60}Co source emitting γ-rays. For this kind of irradiation, all the in situ RIA techniques, both polychromatic (with a white source and a spectrometer) and monochromatic (with a laser and a photodetector, or with an OTDR) can be employed as there is no strong requirement in terms of acquisition time. However, generally, as such tests can last several days or weeks, the source (whatever the selected technology) can fluctuate and degrade the precision of radiation-induced attenuation measurement. To avoid this issue, a possible procedure consists in continuously monitoring the source intensity on a reference channel. This is possible by adding Y junction at the output of the source, as illustrated in Figure 8.3a, and by correcting the RIA calculation using the monitored source fluctuations. In detail, the Y junction divides the input signal between two branches either equally or at a controlled ratio. One of the branches is connected to the transport fiber connected to the FUT, while the second output goes directly to a second detector (a spectrometer or a powermeter). It should be noted that the OTDR technique allows to get rid of source fluctuations, since, as shown in Eq. (8.6), a variation of $I(\lambda_0, z = 0, t)$ will not influence the curve slope that is used to monitor the attenuation evolution with time (dose).

8.3.2.2.2 *Pulsed Irradiation*

As the term itself suggests, in this case, the irradiation is not emitted continuously over time, but by pulse(s) as short as a few nanoseconds. Such environments have to be considered for military applications and also for fusion by inertial confinement experiments (see [2] for more details). Contrary to the steady state irradiations, for which the acquisition time of the setup is not really limiting (except for particular cases), most of the acquisition systems fail to follow the RIA kinetics at short times during or after such radiation pulse. Among the three experimental setups previously illustrated, the acquisition time of an OTDR trace exceeds one second, generally even 30 or 60 seconds; so, it is completely useless to characterize the optical material's transient response ($t < 1$ second). The integration time for a classical spectrometer, instead, can vary from few milliseconds to several seconds, which corresponds to an acquisition rate between 1 kHz and 0.01 Hz. It depends on the source transmission spectrum and a compromise between this rate and the signal-to-noise ratio has to be reached too. With this setup, often no information can be obtained on the defect growth kinetics during the pulse, but the recovery phase at $t > 1$ ms can be investigated. The best solution is constituted by a continuous wave laser and a high-speed photodetector whose acquisition rate can reach tens of GHz. It allows a measurement of the transmitted signal level each nanosecond. If the detector is linked to an oscilloscope, the latter records the time evolution of the transmitted signal before, during, and after the irradiation, opening the way to the characterization of the RIA kinetics at times shorter than 1 μs an up to thousands of seconds after a pulse (see [2] for more details). A possible future improvement of such high-speed measurements could be based on the use of ultrafast camera allowing to record the RIA spectra in the nanosecond to second range and not only the RIA at the wavelength of the probe laser.

8.3.2.2.3 Hints for Particular Environments, Applications

In the last two sections, the radiative environments have been divided into two categories, on the basis of the time scale characterizing them. However, it is not the only important feature to design an experiments. Often harsh environments are not easily accessible, implying that tests be performed at dedicated facilities (far from the lab) during an irradiation campaign of limited duration, from a few hours to weeks of irradiation. For TID tests, two main classes of radiation sources are used: X-rays and γ-rays. For displacement damages, tests with neutrons, protons, or ions are necessary.

X-Ray Testing/Principle An *X-ray tube* is a simple machine, constituted by two electrodes: when a high tension (generally one or few hundreds of kilovolts) is applied between these two, an electrical current goes from the cathode to the anode, also known as target, where the electrons lose energy and produce X-rays and heat. The X-rays are due to two phenomena:

- The Bremsstrahlung, which is a radiation emission induced by the deceleration of a charged particle approaching another charged particles, in this case an electron by atomic nucleus, and gives rise to a wide spectrum from 0 eV to the charge of an electron times the applied tension.
- The transition of an outer electron of an atom to an inner shell with a vacancy (Auger effect): this transition gives rise to characteristic X-rays, very narrow emissions typical of the atoms of the target.

It is worth noticing that the mean energy of the emitted photons depends on the applied tension, whereas the dose rate depends on the current and on the inverse of the squared distance from the tube.

X-Ray Testing/Implementation The X-ray tube is enclosed into a lead cabin to shield the surrounding environment from the radiation, but it remains easily accessible and very safe, since the X-rays are not produced anymore when no tension is supplied.

X-Ray Testing/Advantages and Limitations Depending on the X-rays' energy, often they are not very penetrating in matter. For example, at a tension of 100 kV, a tube in tungsten emits photons having mean energy of ~40 keV. These penetrate homogenously only one silica-based optical fiber layer of 125 μm diameter; if two fibers are overlapped, the one put on the top will shield the second one, which will receive a dose reduced by ~30% compared to that of the unshielded fiber. For this reason, when long samples have to be irradiated, these need to be coiled into a spiral with no overlapping; moreover, the dose rate has to be homogenous over all the surface sample. Despite these limitations, X-rays are a very efficient tool in the case of optical materials; they are also considered for microelectronic tests but in this case, more precautions have to be taken.

γ-Ray Testing/Principle A *γ-ray source* consists of rods of a radioactive element, the isotope Cobalt-60 (^{60}Co), which decays by emitting photons with energy of ~1 MeV. Since these photons are more energetic, they are more penetrating and there is no problem, such as the overlapping encountered for the X-rays. However, this is not a source that can be switched off, so only specialized personnel can operate it.

γ-Ray Testing/Implementation The shielding of the source is thicker than the one for the X-rays (for example, the door cannot be opened by a human, but only with an electrical system). In this case, the transport fibers have to be longer and to pass through small windows often placed on the top of the irradiation room, in order to be as far as possible from the source and not at human level. Consequently, these have to be sufficiently long in order to go from the control room to the irradiation room and they should be packaged as fiber cable in order to be less fragile.

γ-Ray Testing/Advantages and Limitations The main limitation of γ-ray tests is their limited access and high cost. They are usually preferred for qualification tests, e.g. for space application, and are mandatory for those devices that cannot be tested with X-rays.

Other Radiation Facilities Other tests with protons, neutrons or heavy ions are sometimes mandatory to evaluate the vulnerability of glass, especially to displacement damages. For those tests, the implementation of the measurement setups usually implies finely studying the facility constraints in terms of accessibility (restricted access zone) and beam parameters (beam size, ranges of flux, fluences, associated dosimetry). In some installations, such as the Neutron-Irradiation Facility (NIF) at TRIUMF (Vancouver, Canada), it is only possible to have access to only one fiber end, for example because the neutron source is inside a very deep and narrow well. In this case, RIA measurements with the OTDR setup have to be favored and the setting-up has to be as fast as possible. For those installations, specific procedures have to be followed to ensure worker protection against the risks due to ionizing radiation, implying that the researchers qualify for such testing. As an example, in France, to determine the medical follow-up of persons exposed to ionizing radiation, the employer must classify them into category A or B according to the level of exposure likely to be reached under normal working conditions. Another constraint concerns the activation of the samples and sometimes of part of the equipment used in the irradiation zone to carry out the RIA or RIL measurements. In other installations, instead, tests have to be done under vacuum (e.g. within the target chamber of the OMEGA laser at Rochester facility) and the transport fiber should pass through a suitable feedthrough, which could create interferences or degradation of the measurements.

In conclusion, when performing experiments outside your own laboratory, it is very important to:

- Prepare very well the irradiation campaign by fulfilling all the documents needed to access the facilities and to send the equipment (two-month delays can be needed for oversea testing).
- Prepare a detailed Design of Experiments, the schedule of the test.
- List all the material, samples that could be necessary and be prepared to have backup solutions. Indeed, if something goes wrong when you are far from your lab, it will not be possible to come back to overcome the issues.
- Test all the samples with the planned instruments in the exact same conditions (e.g. consider the differences of voltages between countries) to avoid bad surprises.
- Have compact and light instruments, above all if you have to flight overseas.
- Always monitor the temperature fluctuations during the irradiation run. Temperature can largely change due to night and day fluctuations or changes in the weather conditions for long irradiation run.

8.3.3 Exploitation of RIA Spectra: Point Defect Identification

The measured RIA spectrum results from the contribution from the various absorption bands associated with each of the radiation-induced point defects [11]. In the case of optical fibers, in addition to this material radiation response, the light guidance properties also have to be considered, especially for single-mode optical fibers in the infrared region. Until today, the researchers have managed to identify the defect structures associated with a large set of absorption bands by combining the absorption measurements with other complementary spectroscopic techniques such as electron paramagnetic resonance (EPR, see Chapter 9) or photoluminescence (PL, see Chapter 2) while submitting identical samples at various treatments such as isochronal or isothermal treatments, irradiation, and photobleaching experiments in order to change the defect concentration and identify correlated behaviors between optical properties and EPR signatures. An example of such successful results concerns the non-bridging oxygen hole center (NBOHC), which is a common defect of silica-based samples, leading to a photoluminescence signal at 1.9 eV. To associate this defect to its absorption bands, their intensities have been correlated with those of the PL for samples differing in terms of NBOHC concentrations [20]. Another example concerns the twofold coordinated silicon or ODC(II), which is a silicon atom coordinated with two oxygen atoms and a lone pair: samples have been treated at different temperatures in presence of CO_2; during the treatment, the defect interacts with molecular oxygen to transform itself into a four coordinated silicon [21]. Today, there is a great amount of interest in the spectroscopic techniques allowing to localize the defects in complex structures such as optical fibers and compare their distributions with the ones of chemical elements or strain. Confocal microscopy of luminescence (see Chapter 4) and cathodoluminescence have been shown as very promising tools for point defect studies in glasses.

8.3.3.1 Spectral Decomposition

The absorption bands of most of point defects present a Gaussian shape, as already explained in Section 1.2.3. The exceptions are rare, as the absorption peaked at 1.9 eV of the NBOHCs. In order to better highlight the radiation-induced defects, the radiation-induced attenuation can at the first order be decomposed as the sum of several Gaussian bands, defined as:

$$G(x) = \frac{A}{\sqrt{2\pi\sigma^2}} e^{-\frac{(x-x_c)^2}{2\sigma^2}} \tag{8.8}$$

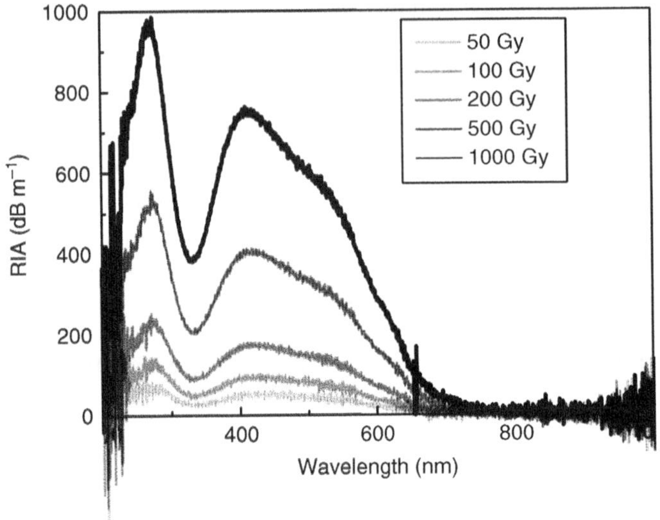

Figure 8.4 In situ RIA spectral measurements at different TID levels from 50 to 1000 Gy of a P-doped multimode optical fiber.

where A is the band area, x_c is the optical absorption band center, and σ (the standard deviation) is linked to the full width at half maximum (FWHM) as:

$$\text{FWHM} = 2\sqrt{2\ln(2)}\,\sigma. \tag{8.9}$$

The fitting routine consists in using a set of Gaussian bands with fixed parameters, such as the center of the Gaussian and its FWHM, taken from literature, as in [11]. The Gaussian area is the free parameter that evolves with the dose, dose rate, or temperature. We illustrate the usual decomposition approach with the case of the phosphorus-doped optical fibers. Many studies have shown a specific radiation response of optical fibers containing even small traces of phosphorus, either in their core or in the cladding. These fibers present high levels of radiation-induced attenuation in the whole spectral domain that are due to P-related defects. This peculiarity prohibits the use of P-doped fibers for data transfer or temperature and strain monitoring under irradiation. On the contrary, the high radiation sensitivity makes these fibers suitable for radiation detection or dosimetry purposes.

In the present case of study, we will analyze the RIA dose and wavelength dependence of a P-doped multimode optical fiber under steady state X-rays at room temperature (RT). Furthermore, in order to better highlight the defects contributing to the fiber darkening in the visible and near-IR domains, a Gaussian deconvolution will be performed on the RIA spectra, relying on a set of P-related defects already identified in [11]. In Figure 8.4 are depicted the in situ RIA spectral measurements at different TIDs.

Considering that the single defect contribution could be described by a Gaussian function versus energy, we reproduce the RIA spectra of Figure 8.4 based on a set of already known absorbing defects whose properties are listed in Table 8.2. The best fits are reported in the upper part of Figure 8.5. The bottom panel of Figure 8.5 gives the percentage of residuals between the experimental data and the global fitting function, showing the quality of the fit within the 10% of error. This means that the used set of defects is able to explain the measured RIA growth kinetics in this class of fibers.

Table 8.2 Gaussian absorption bands peak position, width, and amplitude at different irradiation doses of the set of P-related defects [11] used to reproduce the measured RIA spectra reported in Figure 8.5.

Defect	OA Peak (eV)	FWHM (eV)	50 Gy (dB eV m^{-1})	500 Gy (dB eV m^{-1})	1000 Gy (dB eV m^{-1})
s-POHC	2.2	0.35	3.7	30	55
s-POHC	2.5	0.66	27	190	345
m-POHC	3.1	0.78	37	300	560
P2	4.5	1.2	65	500	910
P4	4.8	0.71	41	210	360
s-POHC	5.3	0.73	16	120	225

8.3.3.2 Point Defect Kinetics

As already reported in Eq. (1.61), the area of each absorption band is proportional to the point defect concentration and its oscillator strength. Consequently, by decomposing all the RIA spectra acquired as a function of the time, it is possible to obtain the band area as a function of the time, which corresponds to the growth and decay kinetics of the point defects. In Figure 8.6, we report the growth kinetics of the P-related defects used to reproduce the spectra shown in Figure 8.4, having the possibility to evaluate the single defect contribution as a function of the total dose.

8.4 Radiation-Induced Luminescence (RIL)

Contrary to the RIA, the RIL (or RL) measurements do not need any source, since the signal to be detected is generated by the radiation itself. The detector can be:

- A spectrometer, which allows to obtain the spectral shape of the luminescence band and consequently to understand the involved luminescent defects.
- A photomultiplier tube (briefly, PMT), which integrates all the light in its spectral window, except if an optical filter is used to isolate a contribution in a narrow spectral region. PMT setup gives information only about the RL intensity value, but it allows measuring very low signal intensities, above all if the signal is amplified by an electrometer.

8.4.1 Architectures of Fiber-Based Sensors: Extrinsic and Intrinsic

As already introduced in Section 8.2.2, the RL signal appears only in presence of radiations and it quickly disappears when the irradiation stops. Moreover, its intensity depends on the dose rate; for this reason radioluminescent materials are promising radiation detectors, and sometimes as dosimeters.

There exist two types of such sensors. The first is an extrinsic one, since the sensitive element is a luminescent bulk material, connected to a fiber for transporting the RL signal to the detector. The sensitive part can be a cavity micromachined at one fiber end and filled with a scintillator powder, as shown in [22], or an element spliced to the transport fiber, such as a glassy doped rod, having ~1 mm diameter and a few centimeters long [23]. The second type of RL based dosimeter is known as intrinsic, since the sensitive element is the fiber itself, as the N-doped fiber, which presents a RL signal around 550 nm [24]. In both cases, it is very important that the fiber transporting the RL

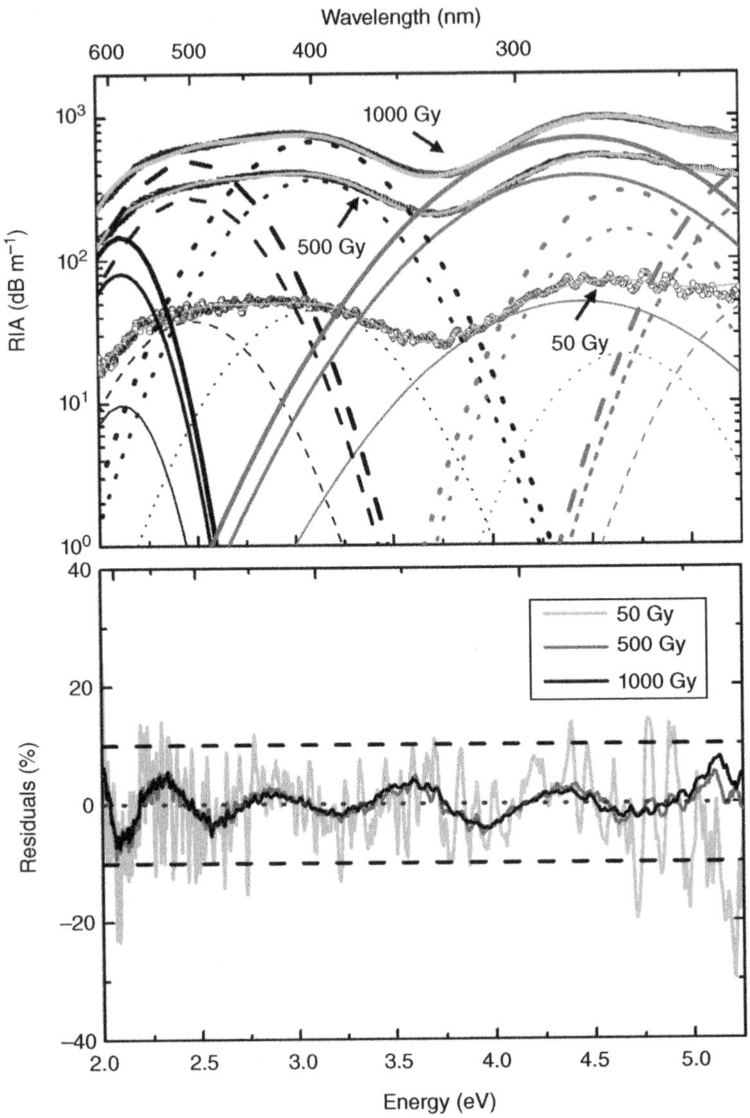

Figure 8.5 Upper panel: experimental RIA measured at 50, 500, and 1000 Gy of a P-doped multimode optical fiber: in light gray, the total fitting function using the defect set of Table 8.2. Bottom panel: the percentage of residual between the experimental function and the global fitting function for TIDs from 50 to 1000 Gy. The dashed black and the dotted gray lines are guide lines for the ±10% and the 0%, respectively.

signal of the detector is radiation resistant in the considered spectral domain. Whereas it is easy to do so for extrinsic sensors, for the intrinsic sensors it has to be demonstrated that the luminescent fiber does not present high RIA level at the RL wavelength. Indeed, by increasing the fiber sample, two phenomena take place: both RL intensity and radiation-induced losses increase. So, the optimized sensing length will be a compromise that maximizes the RL intensity and minimizes the RIA issues. Finally, since the PMT measurements can be perturbated by the ambient light, it is very important to maintain the PMT, probe, and transport fibers in the dark to reduce the background signal and optimize the dose rate detection threshold. This aspect is really important as usually

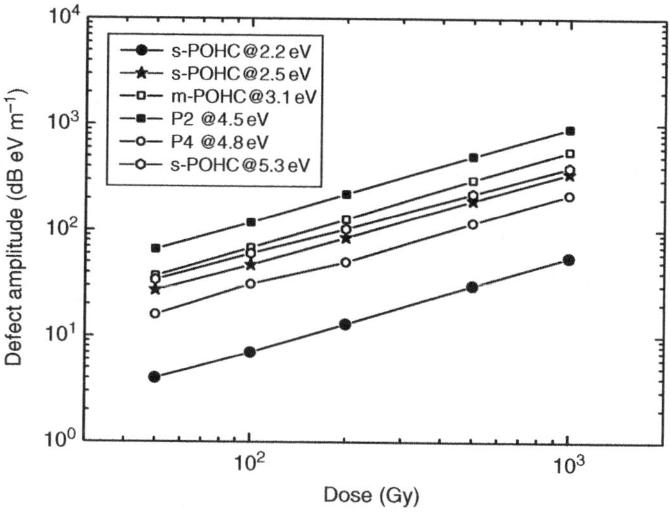

Figure 8.6 Growth kinetics of the defects in P-doped fibers as extracted from the described fitting routine for the data acquired at different total accumulated doses. The symbols are: circles for the s-POHC peaked at 2.2 eV; stars for the s-POHC at 2.5 eV; open squares for the m-POHC at 3.1 eV; squares the P2 at 4.5 eV; open circles for the P4 at 4.8 eV; and open hexagons for the s-POHC at 5.3 eV.

intermittent lights signalizing the beam presence are distributed inside the irradiation chamber and cannot be switched off.

8.4.2 Calibration of the RIL Versus Proton Flux

As already said, the RIL signal depends on the dose rate. Each proton fluence corresponds to an ionizing dose that can be calculated through radiation–matter simulation codes. The conversion coefficients depend on the proton energy and on the considered material.

To obtain a good dosimeter, the RIL intensity should linearly depend on the dose rate and only slightly depend on the proton energy. For this reason, the RIL dependence on proton flux, which is changed by varying the beam current, and consequently on the dose rate, has to be verified as a function of the proton energy. Generally, no big difference is observed depending on the proton energy, but, for example, the RIL of the Ce-doped silica-based rod has been demonstrated to be a proton counter, whose intensity is only dependent on the flux [23]. Moreover, the RIL response has to be independent from the accumulated dose; generally, this is true up to a certain TID, which will be the maximum TID at which the sensor can be used.

8.4.3 Bragg Peak Measurements for Proton-Therapy Applications

Whereas the photon beam energy decreases exponentially with increasing depth in the material, the energy transferred by protons depends inversely on their initial energy [25]: the higher the proton energy, the less the protons interact with the material and higher is their penetration depth. But, the protons progressively lose energy during the interaction and then interact more and more with the material up to the point that the protons give up all their energy and stop at a certain depth. For this reason, the proton energy decreases, giving rise to a typical track of the dose as a function of the depth, known as *Bragg peak* (BP), reported in Figure 8.7a. Protons of different energies will have BPs placed at different depths inside the material.

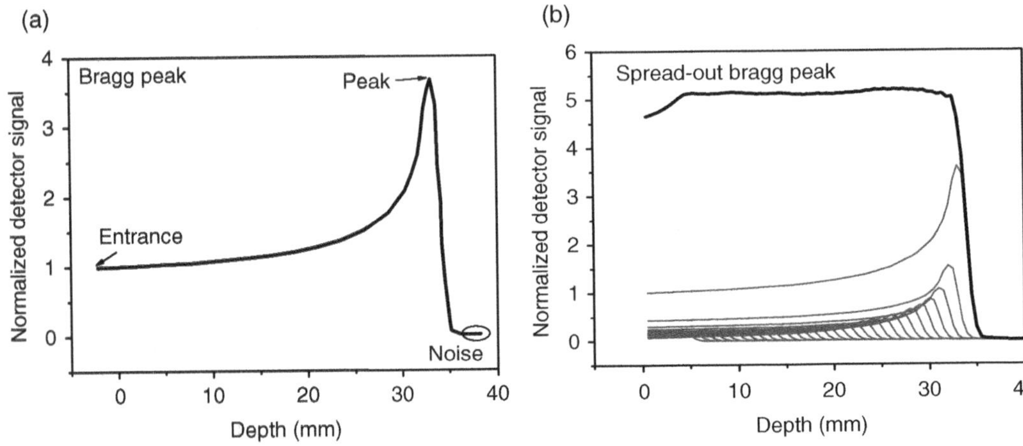

Figure 8.7 (a) Bragg peak detected with a Markus chamber for a 74 MeV proton beam in a water phantom. The detector signal is proportional to the TID. (b) Spread-out Bragg peak (black curve) obtained as the sum of different Bragg peaks (gray curves).

During a proton therapy, the objective is to overlap as much as possible the proton's Bragg peak with the tumor to be treated. The cancer tumors differ in size from one patient to another and can be larger than the small region where all the energy of a monoenergetic proton beam is deposited, having a very sharp Bragg peak; consequently, to cover all the cancer area, proton beams of different energy have to be combined. The proton beam energy is modulated by rotating a modulator wheel, whose thickness changes by steps [22]: at each step, the proton beam energy changes and gives rise to a Bragg peak at a particular depth. The result obtained as the superposition of all these BPs is known as *spread-out Bragg peak* (SOBP) and is illustrated in Figure 8.7b [25].

To calibrate a RL-based dosimeter, its sensitive part has to be mounted on a 3D stage and placed inside a water container, called water phantom and simulating the human body; then, it will be translated along the proton beam direction. The BP and SOBP curves are obtained, respectively, at constant proton energy or by changing it with the modulator wheel. The reference dosimeter is generally a Markus chamber, which presents a BP with peak-to-entrance ratio of 3.7, as shown in Figure 8.7a, once the noise after the Bragg peak has been subtracted, and a flat SOBP [26]. Among the RL-based dosimeters, it has been demonstrated, for example, that a Gd^{3+}-doped silica is a very promising material, since its peak-to-entrance ratio of 3.5 is the closest to the one of the gold standard [26].

8.5 Case Studies

This section is dedicated to two examples concerning the RIA on bulk and on optical fibers, and one about the RIL.

8.5.1 Characterization of Bulk Glasses for Space Optical Systems

To study the radiation resistance of bulk glasses for space applications, different samples were irradiated under 480 MeV protons, at a flux of ~$2.7 \cdot 10^9$ p cm^{-2} s^{-1} up to a fluence of ~$6.6 \cdot 10^{12}$ p cm^{-2}, at the Proton Irradiation Facility (PIF) of TRIUMF (Vancouver, Canada). This corresponds to a TID

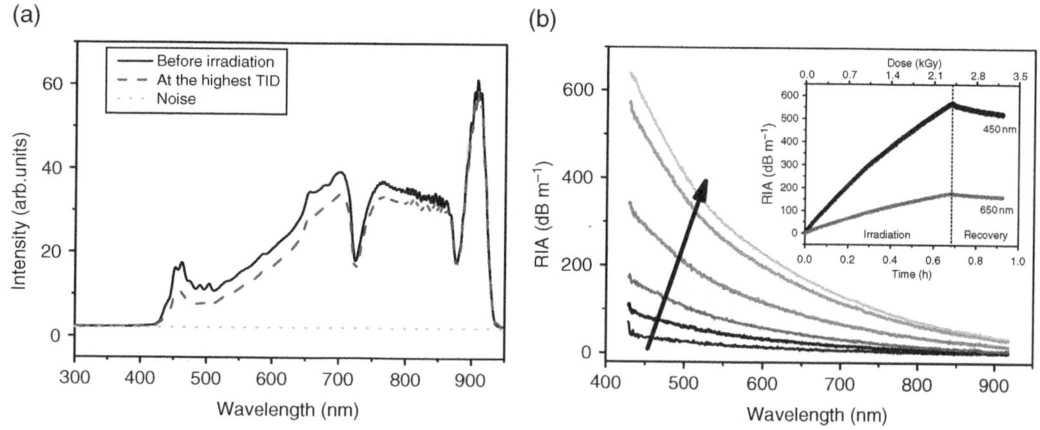

Figure 8.8 (a) Transmission spectra recorded before irradiation (continuous black line) and at the highest TID (dashed gray line) for a BAF4 bulk glass under 480 MeV protons and the noise recorded on the UV-Vis spectrometer (dotted light gray line). (b) Spectral RIA at different equivalent doses (100, 250, 500, 1000, 2000, 2400 Gy: from the darkest to the lightest curves). The arrow indicates the increasing dose. In the inset, kinetics as a function of time (or dose, in the upper axis) for the RIA at two wavelengths: 450 nm (black line) and 650 nm (gray line). The dashed vertical line indicates the irradiation end.

of ~2.4 kGy (SiO_2). In situ RIA measurements were performed with the setup illustrated in Figure 8.2, by acquiring the transmission spectra with a deuterium/halogen source and a UV-Vis spectrometer. Figure 8.8a reports an example of the transmission spectra recorded before and at the highest fluence (or TID) and the noise. The noise subtraction is a very important task, above all for the UV region where the transmitted intensity is not much higher than the noise. Since no signal is transmitted for wavelengths lower than 300 nm, at this wavelength the noise level must coincide with the transmission intensity. From these recorded spectra, through Eq. (8.5), the RIA spectrum is calculated and few examples, at different doses, are reported in Figure 8.8b. By analyzing all the recorded spectra and by considering the losses at specific wavelengths, the RIA growth and decay kinetics can be analyzed. Typical results from the same data set at 450 and 650 nm are also shown in the inset of Figure 8.8b.

8.5.2 Fiber-Based Dosimetry with Phosphorus-Doped Optical Fibers

While most of the research has been devoted to the identification of radiation-hardened optical fibers, some kinds of fibers have also been identified as very radiation sensitive, with very high radiation-induced attenuation levels, opening the way to the design of point or distributed radiation detectors. To act as a dosimeter, the fiber RIA should in addition present very specific features [27] such as a linear dose dependence, a dose rate and temperature independence (*in the respective ranges of interest*), no recovery post-irradiation, and ideally the dosimeter should be reusable meaning that the induced losses are bleachable and that the fiber keep its characteristics after the treatment. P-doped optical fibers possess such properties at least in the IR (near 1550 nm, radiation sensitivity coefficient around 3.5 dB km^{-1} Gy^{-1}) [27] and in the visible domain (around 650 nm; ~100 dB km^{-1} Gy^{-1}) [28]. Researchers at CERN showed that TID real-time mapping of their accelerators can be obtained through a fiber distributed sensor exploiting the radiation response of a single-mode P-doped optical fiber monitored with an OTDR at 1550 nm [27]. Figure 8.9a illustrates the dose dependence of the RIA at 1550 nm of this single-mode optical fiber at four different

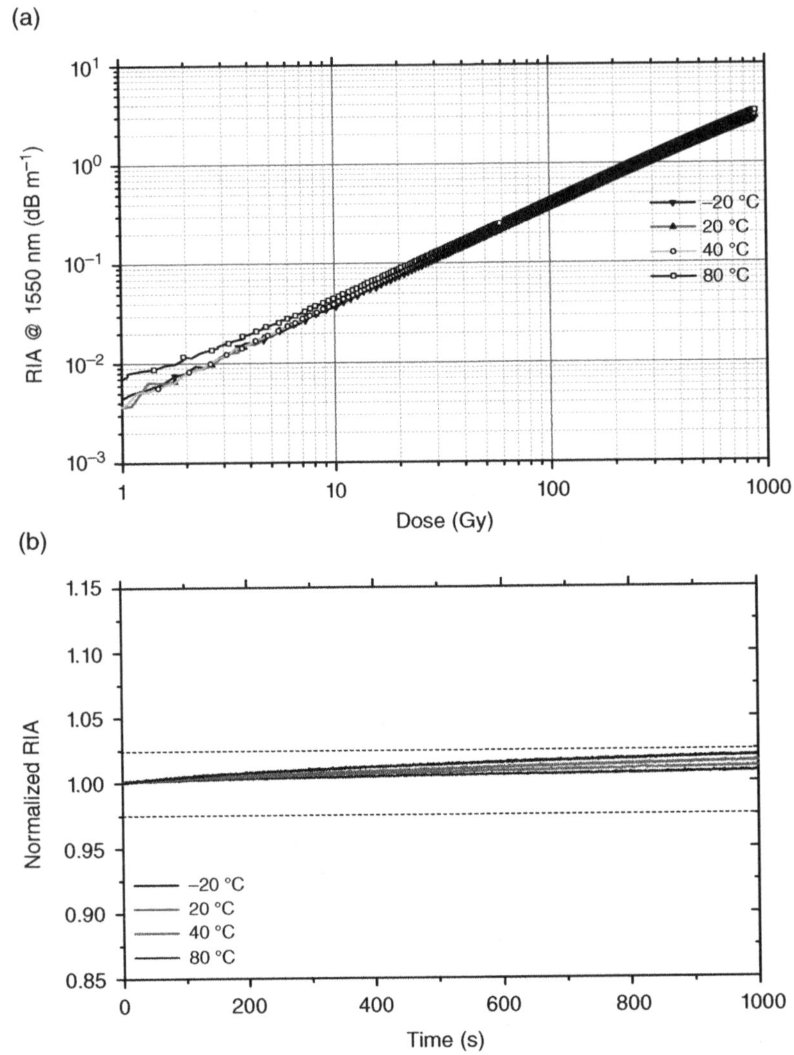

Figure 8.9 (a) RIA of a P-doped single-mode fiber at 1550 nm as a function of the dose, for four different temperatures, dose rate being 1 Gy(SiO_2)/s, up to TID of ~1 kGy(SiO_2). (b) Normalized RIA with respect to the value at the higher TID, as a function of the time after the irradiation stop, for the four temperatures.

temperatures (from −20 to 80 °C) of X-ray irradiation, demonstrating the very slight temperature dependence of the RIA. Such a result could typically serve as the calibration curve of the dosimeter, allowing to convert the measured losses (in dB km^{-1}) into the TID (in Gy(SiO_2)). Figure 8.9b shows the absence of RIA recovery following the irradiation run. It is observed that at higher temperatures, the RIA at this wavelength slightly increases (less than 2.5%); this very specific behavior is explained by a conversion between two of the P-related defects, the POHC and the P1 ones [11].

Another important result regarding this dosimetry tool is that the RIA only depends on the ionization dose and that this TID level can be measured even in mixed environments, combining radiation of various natures [29].

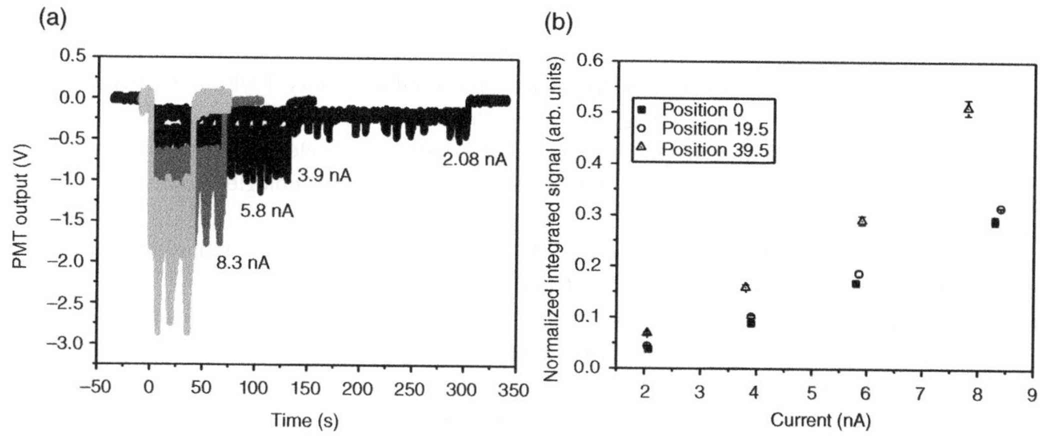

Figure 8.10 (a) PMT output as a function of time for different currents (the darkest curve corresponds to the lowest current), for 63 MeV proton beams. (b) Integrated signal normalized to the irradiation time (to have the average RIL amplitude) and to the deposited dose, since, from one run to another there could be slight variations, as a function of the current (proportional to proton flux) and for three different positions of the modulator wheel (position 0 corresponds to a proton energy of 63 MeV).

8.5.3 Proton Flux Measurements Through the RIL of Optical Fibers

To study the dosimetry properties of a RIL, one has to investigate its dependence on flux, fluence, and energy of protons. In this case study, the RIL of a Cu-doped silica-based rod, whose diameter is ~2 mm and length 1 cm, is investigated under 63 MeV proton beam. As already explained, the proton energy can be modulated by a wheel having different thicknesses in different positions, whereas the flux is varied by changing the source current. Figure 8.10a illustrates typical RIL curves acquired by a PMT and an oscilloscope, for an energy of 63 MeV and different currents, whereas the total fluence is kept constant for each run. It is worth noting that by increasing the current, and consequently the flux, the signal amplitude, that means the difference between the value before the irradiation start, due to the background noise and that has been put to zero in this figure, and the one during the irradiation, increases, whereas the irradiation duration decreases, as the fluence is kept constant. In conclusion, the radiation-induced luminescence amplitude is proportional to the flux, while the information on the fluence is obtained by integrating the curves. Figure 8.10b reports the results obtained by analyzing the previous curves, for example, by integrating them and normalizing with respect to the irradiation time (to have the average RIL amplitude) and to the exact deposited dose (since it can slightly vary from one run to another) for different currents and different modulator wheel positions, which correspond, respectively, to different proton fluxes and energies. These results show that the Cu-doped rod is an efficient radiation detector through the exploitation of the RIL and that, after calibration, such an optical material could serve to monitor in real time the flux of proton beam used for radiation testing or medical applications.

References

1 Griscom, D.L. (1985). Nature of defects and defect generation in optical glasses. *Proc. SPIE* 541: 38–59.

2 Girard, S., Kuhnhenn, J., Gusarov, A. et al. (2013). Radiation effects on silica-based optical fibers: recent advances and future challenges. *IEEE Trans. Nucl. Sci.* 60: 2015–2036.

3 Girard, S., Morana, A., Ladaci, A. et al. (2018). Recent advances in radiation-hardened fiber-based technologies for space applications. *J. Opt.* 20: 093001.

4 Hartog, A.H. (2018). *An Introduction to Distributed Optical Fibre Sensors.* Taylor & Francis Group.

5 Kashyap, R. (2013). *Fiber Bragg Gratings.* Academic Press Inc.

6 Brichard, B. (2008). Systèmes à fibres optiques pour infrastructures nucléaires: du durcissement aux radiations à l'application. Thèse de doctorat en Electronique, nanotechnologies, composants et systèmes, Université de Montpellier 2.

7 Girard, S., Allanche, T., Paillet, P. et al. (2018). Growth and decay kinetics of radiation-induced attenuation in bulk optical materials. *IEEE Trans. Nucl. Sci.* 65: 1612–1618.

8 Goiffon, V., Rizzolo, S., Corbière, F. et al. (2018). Total ionizing dose effects on a radiation hardened CMOS image sensor demonstrator for ITER remote handling. *IEEE Trans. Nucl. Sci.* 65: 101–110.

9 Allanche, T., Paillet, P., Goiffon, V. et al. (2018). Vulnerability and hardening studies of optical and illumination systems at MGy dose levels. *IEEE Trans. Nucl. Sci.* 65: 132–140.

10 Fruit, M., Gussarov, A., Berghmans, F. et al. (2017). Radiation impact on the characteristics of optical glasses test results on a selected set of materials. *Proc. SPIE 10569, International Conference on Space Optics*. 105691R.

11 Girard, S., Alessi, A., Richard, N. et al. (2019). Overview of radiation induced point defects in silica-based optical fibers. *Rev. Phys.* 4: 100032.

12 Girard, S., Marcandella, C., Morana, A. et al. (2013). Combined high dose and temperature radiation effects on multimode silica-based optical fibers. *IEEE Trans. Nucl. Sci.* 60: 4305–4313.

13 Campanella, C., Morana, A., Girard, S. et al. (2020). Combined temperature and radiation effects on radiation sensitive single-mode optical fibers. *IEEE Trans. Nucl. Sci.* 67: 1643–1649.

14 Robin, S. (1950). Effet Tcherenkov. *J. Phys. Radium* 11: 1–6.

15 Yukihara, E.G. and McKeever, S.W.S. (2011). *Optically Stimulated Luminescence: Fundamentals and Applications*. Wiley.

16 Primak, W. (1958). Fast-neutron-induced changes in quartz and vitreous silica. *Phys. Rev.* 110: 1240–1254.

17 Peiponenand, K.-E. and Vartiainen, E.M. (1991). Kramers-Kronig relations in optical data inversion. *Phys. Rev. B.* 44: 8301–8303.

18 Muller, C., Lépine, T., Allanche, T. et al. Vulnerability and hardening studies of optical and illumination systems at MGy dose levels. *17th European Conference on Radiation and Its Effects on Components and Systems (RADECS)* 2017: 1–7.

19 Morana, A., Girard, S., Cannas, M. et al. (2015). Influence of neutron and gamma-ray irradiations on rad-hard optical fiber. *Opt. Mater. Express* 5: 898–911.

20 Cannas, M., Vaccaro, L., and Boizot, B. (2005). Spectroscopic parameters related to non-bridging oxygen hole centers in amorphous-SiO_2. *J. Non Cryst. Solids* 352: 203–208.

21 Radzig, V. (1993). Reactive silica – a new concept of the structure of active sites. *Colloids Surf. A* 74: 91–100.

22 Penner, C., Hoehr, C., Lindsay, C. et al. (2018). Timing resolution for an optical fibre-based detector in a 74 MeV proton therapy beam. *J. Phys. Conf. Ser.* 1067: 062001.

23 Girard, S., Capoen, B., El Hamzaoui, H. et al. (2017). Potential of Copper- and Cerium-doped optical fiber materials for proton beam monitoring. *IEEE Trans. Nucl. Sci.* 64: 567–573.

24 Girard, S., Di Francesca, D., Morana, A. et al. (2019). X-rays, γ-rays, and proton beam monitoring with multimode nitrogen-doped optical fiber. *IEEE Trans. Nucl. Sci.* 66: 306–311.

25 Paganetti, H. (2012). *Proton Therapy Physics.* CRC Press.

26 Hoehr, C., Morana, A., Duhamel, O. et al. (2019). Novel Gd^{3+}-doped silica-based optical fiber material for dosimetry in proton therapy. *Sci. Rep.* 9: 16376.

27 Di Francesca, D., Li Vecchi, G., Girard, S. et al. (2019). Qualification and calibration of single mode phosphosilicate optical fiber for dosimetry at CERN. *J. Lightw. Tech.* 37: 4643–4649.

28 Girard, S., Ouerdane, Y., Marcandella, C. et al. (2011). Feasibility of radiation dosimetry with phosphorus-doped optical fibers in the ultraviolet and visible domain. *J. Non Cryst. Solids* 357: 1871–1874.

29 Di Francesca, D., Infantino, A., Vecchi, G. et al. (2019). Dosimetry mapping of mixed field radiation environment through combined distributed optical fiber sensing and FLUKA simulation. *IEEE Trans. Nucl. Sci.* 66: 299–305.

9

Electron Paramagnetic Resonance Spectroscopy (EPR)

Antonino Alessi[1,2] *and Franco Gelardi*[3]

[1] *Fraunhofer Institute for Technological Trend Analysis, Appelsgarten 2, Euskirchen, Germany*
[2] *Laboratoire des Solides Irradiés, CEA/DRF/IRAMIS, CNRS, Ecole polytechnique, Institut Polytechnique de Paris, Palaiseau, France*
[3] *Department of Physics and Chemistry – Emilio Segrè, University of Palermo, Palermo, Italy*

9.1 Introduction

Electron paramagnetic resonance (EPR) or electron spin resonance (ESR) spectroscopy [1–5] is an experimental technique that allows to obtain information on the quantity, the chemical composition, and the microscopic structure of the paramagnetic centers under study [1–10]. The relation, between the EPR signal and the concentration of the paramagnetic defects, can be used to study the dynamics of generation or conversion of such defects as a function of some physical parameters, while a detailed analysis of the spectra can be applied to deduce the microscopic structures of the investigated system; furthermore, the presence of atoms with nonzero nuclear magnetic moment can be detected by the presence of hyperfine signals, for obtaining additional information on the environment of a given center.

Due to the physic phenomena that are at the base of this technique, EPR can be employed to study systems having nonzero electronic magnetic moment (both electron spin and orbital angular momenta) [1–10]. So, the technique can be applied to investigate: free radicals in solid, liquid, or gaseous phases; isolated atoms and ions; ions of the transition metal and actinide group in solids; localized imperfections in solids (point defects); and systems with conduction electrons [1–10]. As a consequence, the technique has been successfully employed in studying radiation processes [9, 11], in solid state radiation chemistry [9], in biochemistry, biophysics, biology and medical applications [7, 9, 12, 13], in materials science [9, 11], radiation metrology [9], geology [9, 14], physics [9, 11], and other fields.

9.2 Basic Principle of EPR

Paramagnetic systems are studied by introducing the effective spin Hamiltonian or more simply spin Hamiltonian, expressed in terms of the effective spin operator **S** selected in such a way that $2S + 1$ is equal to the states number. Using the spin Hamiltonian, we can study the effects of

Spectroscopy for Materials Characterization, First Edition. Edited by Simonpietro Agnello.

the external magnetic fields on unpaired electrons and the interactions of the latter with the surrounding environment.

To understand the basis of the technique, we start by considering a simple case, in which the magnetic moment is due only to the spin of a free (isolated or not interacting) unpaired electron. In this case, the general expression for the Zeeman interaction between a microscopic magnetic moment ($\boldsymbol{\mu}$) and an external magnetic field (**B**) (Eq. 9.1) [1–5] can be written as in Eq. (9.2).

$$\hat{\mathbf{H}}_{\text{Zeeman}} = -\boldsymbol{\mu}^{\mathbf{T}}\cdot\mathbf{B} \tag{9.1}$$

$$\hat{\mathbf{H}}_{\text{Zeeman}} = g_e\beta\mathbf{B}^{\mathbf{T}}\cdot\mathbf{S} \tag{9.2}$$

from now on, the T apex symbol adjacent to an operator indicates the transpose of the matrix. In Eq. (9.2), $g_e = 2.002\,32$ is the free electronic splitting factor, $\beta = 9.274\,08 \times 10^{-24}\,\text{J}\,\text{T}^{-1}$ ($10^{-28}\,\text{J}\,\text{G}^{-1}$) is the Bohr magnetic moment, and **S** is the spin operator in units of $\hbar$, the Planck constant.

In this case, the corresponding energy values of the system are obtained by substituting the eigenvalues of the **S** component along the **B** direction $m_s = \pm 1/2$:

$$E = g_e\beta B m_s. \tag{9.3}$$

In other words, according to Eq. (9.3), the interaction of the paramagnetic system with the external static magnetic field originates a splitting of the energy levels having different m_s. In the considered case, the separation between the two states is $\Delta E = g_e\beta B$ and it grows by increasing the applied magnetic field. Figure 9.1 illustrates the splitting induced by a static magnetic field on the energy levels of a free (isolated or not interacting) unpaired electron as a function of the magnetic field. It is worth noting that according to the selection rule $\Delta m_s = \pm 1$ [1–5], it is possible to induce a transition between the two states.

In continuous wave (cw) EPR experiments, the sample is subjected to a sweeping magnetic field **B**, which induces energy splitting, and to a second field B_1 ($|B_1| \ll |B|$, B_1 is orthogonal to B) oscillating at one fixed microwave frequency, ν, which induces the transition between the two states [1–5], when the resonance condition $h\nu = \Delta E = g_e\beta B_r$ is fulfilled, where B_r, is the resonance value of the static field. In other words, B is varied crossing the condition in which it induces an energy splitting equal to the microwave photons energy. Different from other techniques like optical absorption, in typical EPR experiments, the energy of the photons used to probe the systems is kept constant. Furthermore, EPR is based on the interaction of the electromagnetic incident radiation with magnetic moments instead of electric dipole moments.

In the most common case, for reasons that will be described in Section 9.7, the acquired cw EPR signal represents the first derivative of the energy absorption. The latter, under the most commonly

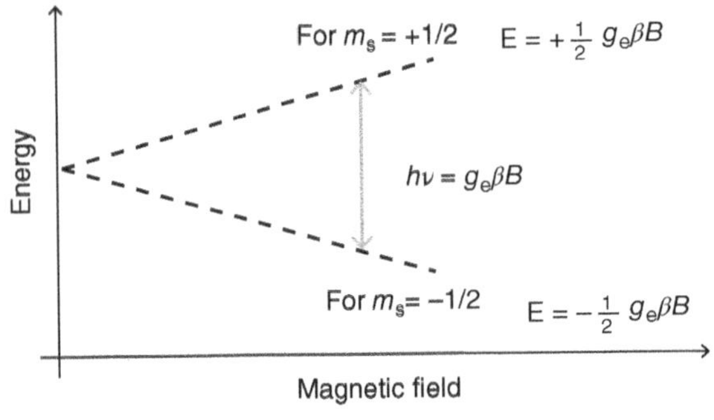

Figure 9.1 Splitting of the energy levels for a system with $S = 1/2$ in a magnetic field B.

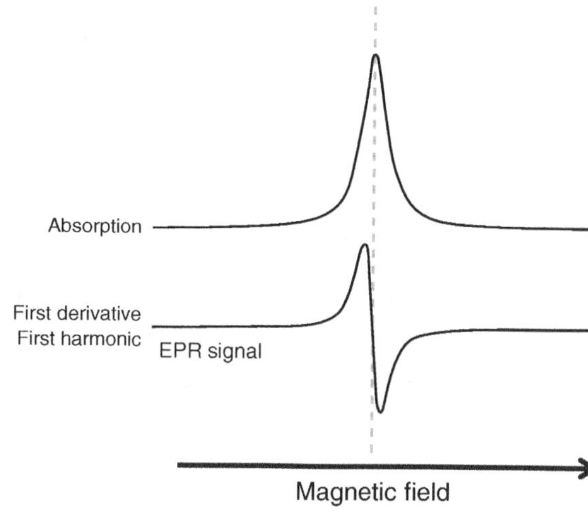

Figure 9.2 Absorption line of a free or isolated or not interacting unpaired electron with $S = 1/2$; first derivative or first-harmonic EPR signal in slow-passage condition.

used experimental conditions, corresponds to a Lorentzian function for the simplified case of free (isolated or not interacting) unpaired electron. Figure 9.2 shows an example of the absorption curve and of the detected cw EPR spectrum, which corresponds to the first derivative (named first-harmonic EPR signal) of the absorption function. In the rest of this chapter, we will use magnetic field instead of the more correct magnetic flux density. It has Tesla (T) and mT as SI units but Gauss (10 G = 1 mT) is still frequently used for historical reasons.

At thermal equilibrium, for a set of spins having the energy level scheme shown in the Figure 9.1, the ratio between the populations of the two levels is given by the Maxwell–Boltzmann law:

$$\frac{N_{\mathrm{ms}=1/2}}{N_{\mathrm{ms}=-1/2}} = \mathrm{e}^{-\frac{\Delta E}{kT}} = \mathrm{e}^{-\frac{g_e \beta B}{kT}} \tag{9.4}$$

where k and T are the Boltzmann constant and the absolute temperature, respectively.

According to Eq. (9.4), the ratio between the populations of the two levels is ~0.998 65 at 300 K with a magnetic field of 300 mT. Such a slight difference, related to the small energy separation (~0.28 cm^{-1}, 1 cm^{-1} = 1.986 · 10^{-23} J; 1 cm^{-1} = 1.239 · 10^{-4} eV) of the levels, is the reason for the absorption and it is much lower with respect to those involved in optical absorption experiments, evidencing a further difference between the two types of measurement. This is also the reason for the particular way of designing and building an EPR spectrometer with respect to other spectroscopic techniques based on absorption phenomena.

9.3 Anisotropy of *g* and Spectral Lineshape

In Section 9.2, we assumed that the magnetic moment was due only to the spin, but in a more realistic case, the spin Hamiltonian should consider the contributions due to the angular momentum and spin–orbit interaction. Furthermore, in almost all the realistic cases, the paramagnetic centers interact with the surrounding atoms. One of the main consequences, of these interactions and of the spin–orbit interaction, is that the g_e factor is replaced by a 3 × 3 matrix operator [2–4, 15]. A deep treatment of the **g** matrix can be found in [5] and in the following we will use the one obtained experimentally according to the procedure reported in [5]. In this type of approach, the g matrix can be diagonalized to obtain the three eigenvalues, which are known as principal components and which are usually indicated as g_x, g_y, g_z or g_1, g_2, g_3. The relation between them is used to identify the structural symmetry of the paramagnetic center (see the end of this paragraph and the next one).

To treat more in details, the relation between the spin–orbit coupling and anisotropic features, in the following we will assume that (i) ground state of the investigated system has no orbital degeneration, since the Jahn–Teller theorem makes this hypothesis quite general for nonlinear systems [3, 16] and (ii) the Zeeman and the spin–orbit interactions are smaller than the atomic Hamiltonian containing the electronic kinetic and potential energy terms. Then, applying the perturbative approach up to second-order corrections, the spin Hamiltonian [4]

$$\hat{\mathbf{H}}_{\text{spin}} = \beta \mathbf{B}^{\mathbf{T}} \cdot (\mathbf{L} + g_{\text{e}}\mathbf{S}) + \lambda \mathbf{L}^{\mathbf{T}} \cdot \mathbf{S} \tag{9.5a}$$

can be written [5]

$$\hat{\mathbf{H}}_{\text{spin}} = \beta \mathbf{B}^{\mathbf{T}} \cdot \hat{\mathbf{g}} \cdot \mathbf{S} + \mathbf{S}^{\mathbf{T}} \cdot \hat{\mathbf{D}} \cdot \mathbf{S} \tag{9.5b}$$

where $\mathbf{L}$ is the angular momentum operator in units of $\hbar$, λ is related to the spin–orbit coupling constant, $\hat{\mathbf{g}} = g_{\text{e}}\mathbf{1}_3 + 2\lambda\hat{\mathbf{\Lambda}}$, $\mathbf{1}_3$ is the 3×3 unit matrix, and $\hat{\mathbf{D}} = \lambda^2\hat{\mathbf{\Lambda}}$. $\hat{\mathbf{\Lambda}}$ is the 3×3 matrix expressed by

$$\mathbf{\Lambda} = \begin{vmatrix} \Lambda_{xx} & \Lambda_{xy} & \Lambda_{xz} \\ \Lambda_{yx} & \Lambda_{yy} & \Lambda_{yz} \\ \Lambda_{zx} & \Lambda_{zy} & \Lambda_{zz} \end{vmatrix} = -\sum_{n \neq G} \frac{\left\langle G \left| \mathbf{L} \right| n \right\rangle \left\langle n \left| \mathbf{L} \right| G \right\rangle}{E_n - E_0} \tag{9.6}$$

where $|G>$ is the ground state having energy E_0 and $|n>$ stands for a generic excited state with energy E_n. The term dependent on $\hat{\mathbf{D}}$ in Eq. (9.5b) takes into account the quadrupole interaction and is relevant for system with $S > 1/2$.

In the case of $S = ½$, $\hat{\mathbf{H}}_{\text{spin}} = \beta \mathbf{B}^{\mathbf{T}} \cdot \hat{\mathbf{g}} \cdot \mathbf{S}$, so in presence of a static magnetic field forming the angles $\theta_1, \theta_2, \theta_3$ with the principal axis of the $\hat{g}$ tensor of the paramagnetic defect, the energy separation of the levels can be written as:

$$\Delta E = \beta B \sqrt{g_1^2 \cos^2(\theta_1) + g_2^2 \cos^2(\theta_2) + g_3^2 \cos^2(\theta_3)} \tag{9.7}$$

so the resonance field can be expressed as:

$$B_{\text{r}} = \frac{h\nu}{\beta \sqrt{g_1^2 \cos^2(\theta_1) + g_2^2 \cos^2(\theta_2) + g_3^2 \cos^2(\theta_3)}}. \tag{9.8}$$

These equations show that for paramagnetic defects in crystalline matrix, the energy of the levels and their separation depend on the relative orientation of the external magnetic field with respect to their axis, highlighting the link between the paramagnetic defect and the surrounding environment.

Then, depending on the specific structure and symmetry of the center, in particular of the electronic distribution, the g can be isotropic with $g_1 = g_2 = g_3$ (cubic symmetry), axial with $g_1 = g_2 \neq g_3$ (usually indicated as $g_\perp$ and $g_\parallel$), or rhombic with $g_1 \neq g_2 \neq g_3$ [5, 15, 17]. In the first case, the lineshape of the absorption is Lorentzian (usually for homogeneous broadening), see Figure 9.2, Gaussian (typical of inhomogeneous broadening) or something similar, whereas in the other cases, the line of the absorption and as a consequence the one of the EPR signal is more complex, see Figure 9.3a and b for examples of the rhombic and axial cases.

9.4 The EPR Lineshape in Powder or in Amorphous

In this section, we will treat the EPR lineshape of paramagnetic systems consisting of powdered crystal or embedded in amorphous materials.

When a paramagnetic sample is a powdered crystal or an amorphous one, the cw EPR spectrum is due to the superposition of a multitude of single crystal signals, which are due to the different specific orientations of the principal symmetry axes of the paramagnetic center with respect to the direction of the external magnetic field. Despite this complication, some specific spectral features in the EPR line can be used to estimate the properties of the paramagnetic center [1, 5, 18, 19]. Indeed, some investigations have been focused on this topic [5, 11, 15, 17–23] and have shown that it is possible to evaluate the g_1, g_2, and g_3 principal values from the field positions of the first maximum, the zero cross, and the second minimum detected in the EPR derivative spectrum, and to eventually infer the symmetry of the paramagnetic center (see Figure 9.3).

Until now we did not any difference between crystal powders and amorphous samples, but relevant differences exist and they have to be at least mentioned. Indeed, in amorphous solids, the lack of the periodicity and order implies that bond angles and distances between atoms are statistically distributed; so, centers in different parts of the matrix can differ in principal values of the tensors g and A (the meaning of A will be reported in the next paragraph), resulting in a statistical distribution of the spin Hamiltonian parameters. These distributions originate an inhomogeneous

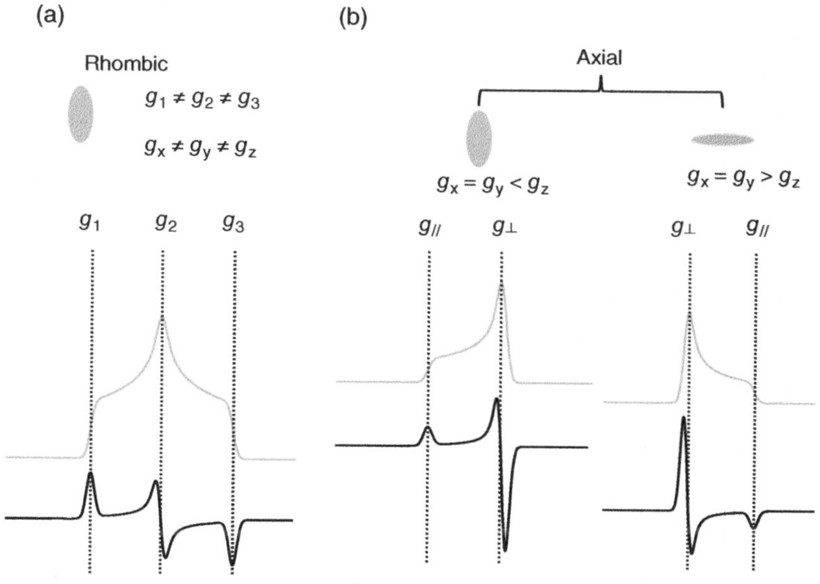

Figure 9.3 Absorption (top) and first derivative (bottom) EPR magnetic field scan for (a) rhombic symmetry; (b) axial symmetry centers, considering two possible relations for the main g values in the latter case. The dashed lines indicate the spectral positions used to evaluate the g values. The signals have been simulated using [26]. *Source:* Based on EPR simulator, www.eprsimulator.org.

broadening of the absorption curves. As a consequence, in some cases, it is necessary to perform computer simulations of the experimental spectra to take in to account the statistically distributed spin Hamiltonian parameters [11, 24, 25]. In other cases, the effect of the amorphous state on the EPR spectrum cannot be large enough to invalidate the observation of important spectroscopic features and the above reported procedure can be applied.

To conclude this section, we want to stress again the relation between the EPR signal and the surrounding matrix, and the possibility to use EPR spectra to investigate the microscopic structure of the material in which the paramagnetic centers are embedded.

9.5 Hyperfine Interactions

In many systems, the unpaired electron can be located near one or more atoms having nonzero nuclear spin. The interaction of these magnetic moments, that of the unpaired electron and those of the nuclei, is described by the hyperfine Hamiltonian [2–5]. When this interaction is present, it is important to distinguish the case in which the atom with nuclear magnetic moment is the one on which the electron is localized, from that in which the atom is just a neighbor of the former. It is also important to distinguish the states with s orbital from the others, which is relevant because only for the s orbitals the density of probability of the electron is not zero at the position of atomic nucleus on which the electron is localized. The s orbitals originate isotropic interaction (also called Fermi contact term) that can be described as [5]:

$$\hat{H}_{\mathrm{fc}} = A_0 \mathbf{S}^{\mathrm{T}} \cdot \mathbf{I} \tag{9.9}$$

where **I** is the nuclear spin and A_0 the isotropic hyperfine coupling constant, proportional to probability $|\psi(0)|^2$, with $|\psi(0)|$ being the electron wavefunction evaluated at the nucleus. Another hyperfine contribution has to be considered for not s electron orbitals, that is anisotropic and related to the dipolar interaction. Considering, for simplicity, that g and the nuclear g_n factors are isotropic, this term can be written as [5]:

$$\hat{H}_{\mathrm{dipolar}} = -\frac{\mu_0}{4\pi} g\beta g_n \beta_n \left[\frac{\mathbf{S}^{\mathrm{T}} \cdot \mathbf{I}}{r^3} - \frac{3(\mathbf{S}^{\mathrm{T}} \cdot \mathbf{r})(\mathbf{I}^{\mathrm{T}} \cdot \mathbf{r})}{r^5}\right] = \mathbf{S}^{\mathrm{T}} \cdot \hat{\mathbf{T}} \cdot \mathbf{I} \tag{9.10}$$

where β_n is the nuclear magneton, μ_0 the vacuum permeability and **T** matrix, using the distance r between dipoles with x,y,z components, can be expressed as:

$$\hat{\mathbf{T}} = -\begin{vmatrix} \left\langle \frac{r^2 - 3x^2}{r^5} \right\rangle & \left\langle -\frac{3xy}{r^5} \right\rangle & \left\langle -\frac{3xz}{r^5} \right\rangle \\ \left\langle -\frac{3xy}{r^5} \right\rangle & \left\langle \frac{r^2 - 3y^2}{r^5} \right\rangle & \left\langle -\frac{3yz}{r^5} \right\rangle \\ \left\langle -\frac{3xz}{r^5} \right\rangle & \left\langle -\frac{3yz}{r^5} \right\rangle & \left\langle \frac{r^2 - 3z^2}{r^5} \right\rangle \end{vmatrix}. \tag{9.11}$$

So, for a system with $S = ½$, the spin Hamiltonian considering the interaction with one atom with nonzero nuclear magnetic moment, not including nuclear quadrupole interaction ($I > 1/2$) and the nuclear Zeeman interaction, is given by

$$\hat{\mathbf{H}}_{\mathrm{spin}} = g\beta \mathbf{B}^{\mathrm{T}} \cdot \mathbf{S} + \mathbf{S}^{\mathrm{T}} \cdot \hat{\mathbf{A}} \cdot \mathbf{I} \tag{9.12}$$

with

$$\hat{\mathbf{A}} = A_0\mathbf{1}_3 + \hat{\mathbf{T}}.$$

In systems with $I > ½$, the nuclear quadrupole term should be considered. It is related to the interaction between the nuclear quadrupole moment and the gradient of electric field, caused by the asymmetric distribution of electron density, and its effects are often difficult to detect [15]. In addition, in EPR experiments, the nuclear Zeeman interaction is neglected being smaller than the other contributions. The presence of n nuclei with nonzero nuclear magnetic moment adds terms to the spin Hamiltonian; furthermore, until now, the g anisotropy has not been considered, so in a more general case Eq. (9.12) can be expressed as [2–5]:

$$\hat{\mathbf{H}}_{\text{spin}} = \beta\mathbf{B}^{\text{T}}\cdot\hat{\mathbf{g}}\cdot\mathbf{S} + \sum_{i=1}^{n}\mathbf{S}^{\text{T}}\cdot\hat{\mathbf{A}}_i\cdot\mathbf{I}_i. \tag{9.13}$$

One aspect that has to be noted is that the principal axes of the g and A tensors can be different; this can easily happen when the interaction is due to a nuclear magnetic moment of an atom different from that on which the electron is localized.

However, to evidence the effect of the hyperfine interaction on the cw EPR spectrum of a paramagnetic center, in the following we will consider simple cases. Assuming that (i) $S = 1/2$, (ii) the two tensors have the same axes, (iii) $A \ll g\beta B$, (iv) $kT \gg A$ [5], and (v) B is orientated along one of these axes (z for example), we can use the perturbation theory at the first order in B to evaluate the energy eigenvalues corresponding to the quantum numbers m_s and m_I of the **S** and **I** projections in the z direction [4]:

$$E_{m_S m_I} = \beta g_{zz} B m_S + A_z m_S m_I. \tag{9.14}$$

We note that each eigenstate m_s is separated into $2I + 1$ states. In Figure 9.4a, we report the energy levels scheme for the case $S = I = 1/2$. Regarding the spectrum, we underline that, the EPR line observed for $I = 0$ is substituted by other two lines (see Figure 9.4b) originated by the transitions that respect the selection rule $\Delta m_S = 1$, $\Delta m_I = 0$. The resonance condition is now $\beta g_{zz}B + A_z m_I$ and in first approximation the two lines appear, as shown in Figure 9.4b, with equal distance from the line observed when only the Zeeman interaction is present. In this way, an estimation of the

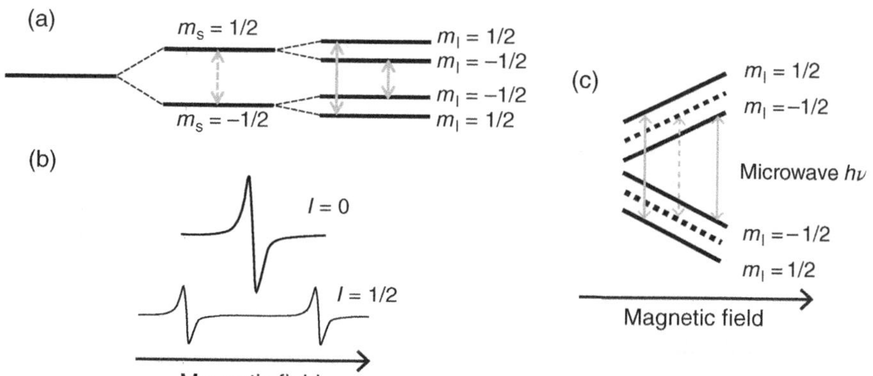

Figure 9.4 (a) Energy levels scheme for a system with $S = I = 1/2$, (b) EPR spectrum in absence of I and EPR spectrum in presence of $I = 1/2$; (c) splitting of the levels during typical EPR experiment in which the incident $h\nu$ is fixed (gray arrow) and B is swept to reach the resonance conditions for $I = 0$, dashed arrow, and $I = 1/2$, continuous arrows (see also panel a). The spectra are obtained for isotropic g and A.

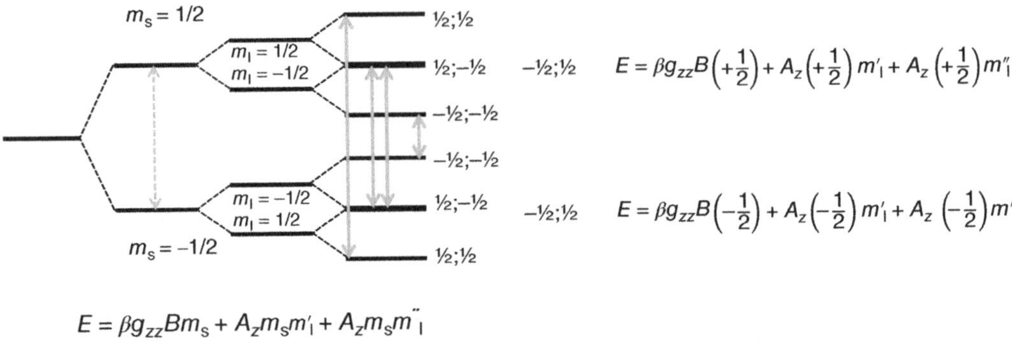

Figure 9.5 Energy levels scheme for a system with $S = 1/2$ and two equivalent magnetic nuclear moments with $I = 1/2$ (and quantum numbers m_I', m_I''). The equations used to evaluate the energy levels are also reported.

value of hyperfine interaction can be obtained by measuring the magnetic splitting between the two lines in an EPR spectrum. Figure 9.4c re-illustrates energy splitting in the concrete case of cw EPR experiment in which B is swept and the energy of the microwave photon, inducing the transition, is constant.

An interesting case is that of an unpaired electron interacting with two or more equivalent nuclear magnetic moments. Figure 9.5 illustrates the energy levels scheme for two equivalent nuclear magnetic moments. The lowest energy level state is the one with $m_s = -1/2$ and the two m_I equal to 1/2 whereas the highest energy level is the one with $m_s = 1/2$ and the two m_I equal to 1/2. We note that the transition between these levels is allowed when $\Delta m_s = 1$ and $\Delta m_I' = \Delta m_I'' = 0$. As reported in Figure 9.5, another allowed transition is the one due to the states in which both the m_I are $-1/2$. Then, when one m_I is 1/2 and the other is $-1/2$, the energy of the level returns to the one of the pure Zeeman term. These states are also responsible for allowed transitions and their EPR lines overlap, so that only three lines, with relative intensities 1 : 2 : 1, are observed. In the case of n equivalent magnetic moments, the number of EPR lines will be $2nI + 1$; Figure 9.6a shows an EPR spectrum in the case of 2 equivalent magnetic moments with $I = 1/2$ and the relative intensity of the hyperfine lines when a spin $S = 1/2$ interacts with n nuclei having $I = 1/2$.

In the case in which the two nuclear magnetic moments are not equivalent, the levels with m_I' and m_I'' equal to 1/2 and $-1/2$ and those with $-1/2$ and 1/2 will not have the same energy; so, four EPR lines will be observed. More in details, the spectrum is constituted by two doublets and the separation between the components of each doublet is related to the smaller hyperfine interaction, whereas the separation between the two doublets is related to the larger hyperfine interaction.

In agreement with what was above reported, when $I = 1$ each of the two levels of a system with $S = 1/2$ is split into three levels ($m_I = 1, 0, -1$) and the resonance lines become three with relative ratio 1 : 1 : 1 as illustrated in Figure 9.6b. This is for example the difference of having an unpaired electron interacting with a H or a D atom; so, it illustrates the potential of EPR spectroscopy in determining the structure of the defects and of its surrounding.

General and useful rules for the interpretation of EPR spectra in the case of the presence of several equivalent or inequivalent magnetic moments can be found in [5].

The above reported considerations are based on the use of the first-order contribution of hyperfine splitting. When second-order contributions become relevant, the interpretation of the data needs more accuracy. For example, in a system with unpaired electron with $S = 1/2$ and two nuclei

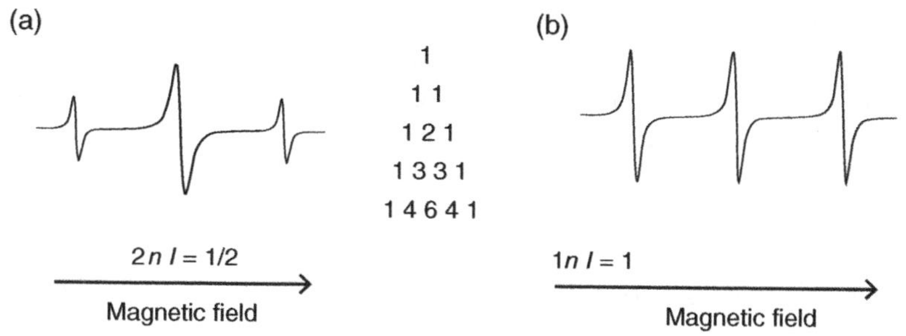

Figure 9.6 (a) EPR spectrum of a system with $S = 1/2$ and two ($n = 2$) equivalent magnetic nuclear moments with $I = 1/2$. The relative intensities of the lines for $n = 0, 1, 2, 3, 4$ equivalent nuclei are also reported by the triangle representing the Pascal binomial sequence; (b) EPR spectrum for a system with $S = 1/2$ and one nuclear magnetic moment with $I = 1$. Spectra are obtained assuming isotropic g and A.

with magnetic moments $I = 1/2$ (g and A are considered isotropic), the observed lines are four instead of three because all the lines, except one belonging to the central degenerate doublet, are shifted to lower magnetic field [5].

9.6 Paramagnetic Center with $S = 1$

In this section, we briefly treat the case of two electrons located at distance lower than 5 Å. In this circumstance, the spin Hamiltonian has to be written considering two additional terms, namely electron–electron dipole and electron-exchange interaction [2, 5]. First, we note that as a consequence of the second interaction, the two spins are coupled; so, there is a diamagnetic singlet state ($S = 0$, S being the total effective spin) and a paramagnetic state with $S = 1$ (triplet state). To simplify the treatment of such systems, we will assume that all hyperfine interactions are negligible, an isotropic $g \sim 2$, and that electron wavefunctions can be written as products of the orbital and of the spin components. Under these conditions, as reported in [5]:

$$\hat{\mathbf{H}}_{\text{spin}} = \beta g \mathbf{B}^{\mathrm{T}} \cdot \mathbf{S} + \mathbf{S}^{\mathrm{T}} \cdot \hat{\mathbf{D}}_{\text{dipolar}} \cdot \mathbf{S} + J_0 \left(\frac{1}{2} \mathbf{S}^2 - \frac{3}{4} \mathbf{1} \right) \tag{9.15}$$

with

$$\hat{\mathbf{D}}_{\text{dipolar}} = \frac{\mu_0}{8\pi} (g\beta)^2 \begin{vmatrix} \left\langle \frac{r^2 - 3x^2}{r^5} \right\rangle & \left\langle -\frac{3xy}{r^5} \right\rangle & \left\langle -\frac{3xz}{r^5} \right\rangle \\ \left\langle -\frac{3xy}{r^5} \right\rangle & \left\langle \frac{r^2 - 3y^2}{r^5} \right\rangle & \left\langle -\frac{3yz}{r^5} \right\rangle \\ \left\langle -\frac{3xz}{r^5} \right\rangle & \left\langle -\frac{3yz}{r^5} \right\rangle & \left\langle \frac{r^2 - 3z^2}{r^5} \right\rangle \end{vmatrix} \tag{9.16}$$

electron–electron dipole interaction

$$J_0 = -2 \left\langle \varphi_{\mathrm{a}}(1) \varphi_{\mathrm{b}}(2) \left| \frac{e^2}{4\pi\varepsilon_0 r} \right| \varphi_{\mathrm{a}}(2) \varphi_{\mathrm{b}}(1) \right\rangle \tag{9.17}$$

electron-exchange interaction

S is the total spin operator, resulting from the sum of the spins' angular momenta related to the two electrons, φ_a and φ_b are the spatial parts of the wavefunctions of the two electrons, e is the electron charge and ε_0 the vacuum permittivity. Then, we use the eigenstates of $\mathbf{S}^2$ and of the projection of **S** on the direction of **B** as basis set to describe the energy levels [5]:

$$E(S=0) = -\frac{3}{4}J_0 \tag{9.18}$$

$$E_{X,Y}(S=1) = -\frac{1}{4}J_0 + \frac{1}{2}\left\{D_Z \pm \left[4g^2\beta^2B^2 + (D_X - D_Y)^2\right]^{1/2}\right\} \tag{9.19}$$

$$E_Z(S=1) = \frac{1}{4}J_0 - D_Z \tag{9.20}$$

X, Y, and Z are the principal axes of the projection of the operator $\hat{\mathbf{D}}_{dip}$ in the states subspace of $S = 1$, D_i its diagonal values, and B is the modulus of the magnetic field along Z.

In some cases [5], the spin Hamiltonian is written using only the first two terms; this happens because the third one, related to the electron-exchange interaction, contributes to the triplet state energies with a constant value.

Some of the properties of the two interacting electrons system are listed in the following [5].

The singlet and the triplet states are separated by the exchange interaction as illustrated in Figure 9.7a; if $J_0 < 0$ and $|J_0| \gg kT$, only the paramagnetic triplet state is populated, whereas when $J_0 \gg kT$, the only populated one is the diamagnetic singlet state.

The states used to describe the system in Figure 9.7 are eigenstates of the Hamiltonian only when the term of Zeeman interaction dominates over the dipolar one, this condition occurs for sufficiently intense magnetic fields.

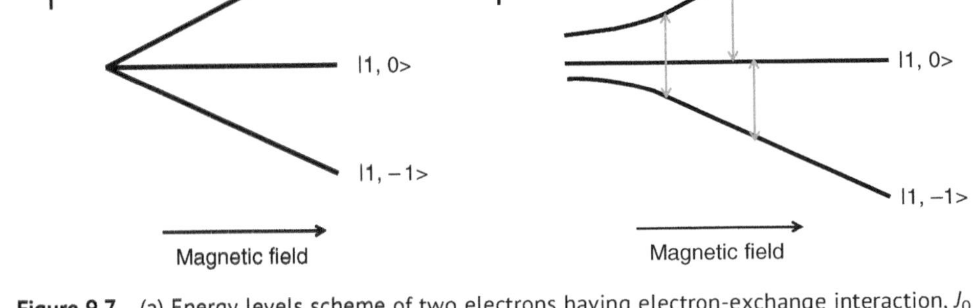

Figure 9.7 (a) Energy levels scheme of two electrons having electron-exchange interaction, $J_0 < 0$, for positive values of J_0 the singlet state has lower energy than the triplet state [5]. (b) Schematic representation of the energy levels of a system of two electrons considering also electron–electron dipole interaction as a function of the modulus of the magnetic field, when $J_0 < 0$. The arrows indicate the possible transitions detected in an EPR experiment. The quantum numbers are reported for high field states.

The energies of the levels for $S = 1$ do not, in principle, depend linearly on the amplitude of the magnetic field because of the dipolar interaction. The three states of such a system are not degenerate when the magnetic field is absent, and in presence of magnetic fields the energy separations of the states are dependent on the orientation of B with respect to the principal axis of the electron–electron dipole interaction ($\hat{\mathbf{D}}_{\text{dipolar}}$).

Finally, we make a few comments regarding the transitions that can be detected in the EPR spectrum and that are shown by arrows in Figure 9.7b. Indeed, the transition between the states |1,−1> and |1,0>, and that between |1,0> and |1,+1> are allowed transitions and they originate two lines with center of gravity at $g \sim 2$. In addition, another line at $g \sim 4$ (low-magnetic field component) can be detected in the EPR experiment. This line is due to the transition between the lowest and the highest energy levels of the triplet. Its presence is related to the fact that, at low magnetic field, the three triplet eigenstates of the system are linear combinations of |1,−1>; |1,0>; and |1,+1> and the selection rule $\Delta M = \pm 1$ does not apply (see Ref. [5] for a complete treatment).

9.7 Basics of Continuous Wave EPR Setup

In the EPR experiments, the samples are located inside a microwave resonant cavity (indicated by a rectangle named R. C. in Figure 9.8), which is positioned between the polar expansions of an electromagnet, dark gray rectangle in Figure 9.8. The resonant cavity is a metal box having high conductibility walls and it is employed to store the microwave energy at fixed frequency. A difference in the latter parameter implies a difference in magnetic field at which the resonance is observed. Common frequency values are ~10 GHz (X-band), ~1 GHz (L-band), ~3 GHz (S-band), ~20 GHz (K-band), ~35 GHz (Q-band), ~65 GHz (V-band), and ~94 GHz (W-band). The resonant cavity efficiency in accumulating microwave energy is given by its quality factor (Q); it depends on different factors and affects the sensitivity of the EPR spectrometer [1, 5]. Since the Q factor of a loaded cavity in absence of magnetic resonance absorption is a relevant parameter, we briefly remind the different origins for energy dissipation in a real cavity. One dissipation term is related to the heating of walls due to their resistivity and to the electrical currents that the microwaves

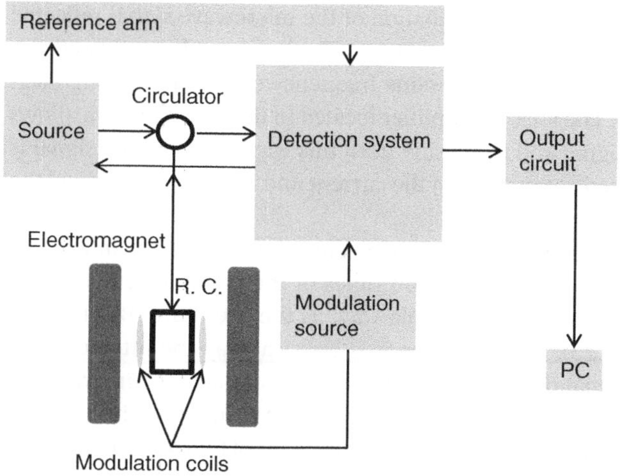

Figure 9.8 Simplified block scheme of a general EPR spectrometer.

generate in them. Other two origins of dissipation are the dielectric losses due to the sample energy absorption and the energy losses related to the holes, such as those required to insert the samples, to couple the cavity to the waveguide (employed to deliver to the cavity the microwave generated by the source), and to the iris.

The electromagnets illustrated in Figure 9.8 generate a homogenous and uniform magnetic field **B** that induces a Zeeman splitting discussed in the previous paragraph. As anticipated, a second magnetic field $\mathbf{B}_1$, oscillating at a microwave frequency and orthogonal to **B**, induces the resonant transition between two Zeeman splitted levels when **B** satisfies the resonance condition. We note that to maximize the sample absorption from the magnetic field $\mathbf{B}_1$ and limit the dielectric losses as much as possible, the sample position corresponds to the maximum amplitude of the magnetic field $\mathbf{B}_1$ and to the minimum amplitude of the electric field **E**, which are both determined by the normal modes of the cavity.

The EPR data are acquired fixing the frequency and the amplitude of the microwave field and measuring the microwave power reflected by the cavity while the amplitude of **B** is changed in a selected range. A further magnetic field, called modulation field and parallel to B, is applied to enhance the signal-to-noise ratio through a lock-in detection system.

It is also important to mention that the microwave sources in EPR spectrometers are equipped with a variable attenuator and with automatic frequency control system (AFC). As a consequence, the frequency of the microwave radiation is kept at the value of the resonance frequency of the cavity and the power losses related to the uncoupling are minimized. We remind that the detector receives the microwaves reflected by the cavity and usually generates a current that is proportional to the square root of the impinging microwave power on it. The detector output current then goes through an electronic system named lock-in amplifier. Since the best performances of the detector need a certain amount of microwave power, the system employs a reference arm that eventually supplies additional microwave power. A phase shifter in the reference arm guarantees that the additional microwaves are in-phase with the reflected signal from the cavity.

Then, the signal-to-noise ratio is increased using a RC filter and finally an analogical digital integrator, with selectable integration time, integrates the signal and the latter is displayed on the screen of the EPR spectrometer computer.

As above mentioned, a modulation field is employed and it is important to underline some aspects related to its presence. Indeed, as a consequence of this field, the total magnetic field amplitude oscillates at a certain frequency; this induces a modulation of the microwave signal reflected from the cavity, which is, then, converted in a periodic current $F(t)$. The latter can be expressed as a Fourier superposition of sinusoidal functions having the same frequency of the modulating magnetic field plus a number of its harmonics. The lock-in amplifier located in the detection arm allows to select the harmonic and phase of the signal and to amplify it. In this way, the recorded signal is proportional to the amplitude of the selected component of the current and as above mentioned the signal-to-noise ratio is increased.

To briefly discuss different acquisition modes and defects concentration estimation in an EPR experiment, it is necessary the introduction of the relaxation times of the paramagnetic centers. For those embedded in a solid and having $S = 1/2$ a "simple way" is the use of the Bloch equations [3, 27–29] (see Eqs. 9.21, 9.22 and 9.23). They describe in a classical way the interaction of the magnetic moments with the external magnetic field, taking into account also the interactions among themselves, and between them and the surrounding environment acting as a thermal reservoir. The latter interactions are introduced phenomenologically by means of two relaxation times, T_1 and T_2. The Bloch equations for an electronic paramagnetic system are the following:

$$\frac{dM_x(t)}{dt} = \gamma[\boldsymbol{M}(t) \times \mathbf{B}_T]_x - \frac{M_x(t)}{T_2} \tag{9.21}$$

$$\frac{dM_y(t)}{dt} = \gamma[\boldsymbol{M}(t) \times \mathbf{B}_T]_y - \frac{M_y(t)}{T_2} \tag{9.22}$$

$$\frac{dM_z(t)}{dt} = \gamma[\boldsymbol{M}(t) \times \mathbf{B}_T]_z - \frac{M_z(t) - M_0(t)}{T_1} \tag{9.23}$$

where γ is the gyromagnetic ratio of the electrons ($-1.76 \cdot 10^{11}$ s^{-1} T ^{1}); $\mathbf{B}_T$ is the whole magnetic field (static plus oscillating fields); $\boldsymbol{M}$ is the macroscopic magnetization (per unit of volume), resulting from the sum of microscopic magnetic moments and having M_x, M_y and M_z components; M_0 is the equilibrium value in the z direction (obtained in stationary conditions, since the static field along z is much more intense than the oscillating one), whereas T_1 and T_2 are the spin–lattice (or *longitudinal*) and spin–spin (*transversal*) relaxation times.

The solution of such equations is strictly dependent on the magnetic field time dependence and we will briefly consider two important cases. The first is called slow-passage condition, while the second is called rapid-passage [29, 30].

Basically, in the slow-passage condition, the rate of the variations of **B** plus $\mathbf{B}_m$ (**B** = the static magnetic field and $\mathbf{B}_m$= modulation field) is slow in comparison with the paramagnetic center relaxation rates; such a condition can be expressed as in Eq. (9.24). On the other hand, the rapid-passage (see Eq. 9.25) is the opposite condition

$$\frac{B_1}{\frac{d|\mathbf{B} + \mathbf{B}_m|}{dt}} \gg \sqrt{T_1 T_2} \tag{9.24}$$

$$\frac{B_1}{\frac{d|\mathbf{B} + \mathbf{B}_m|}{dt}} \ll \sqrt{T_1 T_2}. \tag{9.25}$$

We note that the slow-passage condition expresses the fact that at any time the magnetization, due to all the microscopic magnetic moments, reaches the stationary value associated to the total external magnetic field. We remind that $\mathbf{B}_1$ is the magnetic field that induces the transition; it oscillates at a microwave frequency and it is orthogonal to **B** and $\mathbf{B}_m$.

For slow-passage condition, it is possible to obtain an analytic solution; this is not the case for the rapid-passage, which is usually treated by empirical experimental characterizations.

In the slow-passage condition, from the Bloch equations, the absorption, of N paramagnetic centers for volume unit, is proportional to the square of the microwave field amplitude when the additional condition $\gamma^2 B_1^2 T_1 T_2 \ll 1$ is fulfilled (this condition is often called no-saturation condition). In the simpler case, the absorption curve has the shape of a Lorentzian function peaked at $B_r = \frac{h\nu}{g_e \beta}$ with a full width at half maximum equal to $1/\gamma T_2$. We note that when the condition $\gamma B_1^2 T_1 T_2 \ll 1$ is not fulfilled, EPR spectrum can be affected by distortions of the lineshape. Furthermore, it is worth noting that the no-saturation condition depends, through T_1 and T_2, on the features of the investigated system; so this condition has to be tested case by case.

In the large part of cases reported in EPR books and investigations [5, 11, 29, 31–37], the spectra are recorded in first-harmonic mode which means the acquisition of the component of the EPR signal, reflected from the resonant cavity, oscillating at the same frequency and in phase with the modulation magnetic field. This is often done in the slow-passage condition and the EPR signal reproduces the derivative of the absorption. In such a situation, taking care of the no-saturation condition and properly selecting the other parameters of measure, the recorded data allow to obtain undistorted spectra and the defect concentration. We also note that, in slow-passage conditions, the EPR signal in-phase with respect to the modulation magnetic field and having a double frequency (second harmonic) is proportional to the second derivative of the absorption curve [1]. The data that are reported in Figures 9.2 and 9.3 illustrate the lineshape of the EPR absorption and its first derivative, which corresponds to the first-harmonic signal. We note that the modulation magnetic field has to be much less than the width of the absorption profile to prevent lineshape distortion; some further suggestions regarding this aspect and how to select its value can be found in the next paragraph.

When the slow-passage and no-saturation condition are satisfied, the concentration of the defects can be estimated by evaluating the EPR spectrum double integral, normalized to the measurement conditions (sample mass or volume, modulation amplitude, detector gain, conversion time of the analogical digital integrator and square root of the microwave power), and comparing it with the one of a reference sample having a known concentration of defects. The use of the double integral is due to the fact that the EPR spectrum is the first derivative; so, its integration gives the absorption and the second integration gives a value that is proportional to the defects' concentration.

It is worth noting that when the slow-passage condition is satisfied, no EPR signals $\pi/2$-out-of-phase relative to the modulation magnetic field are observed independently from the frequency component (harmonic) selected by the lock-in amplifier [1]. On the other side, it is possible to observe $\pi/2$-out-of-phase EPR signals when the no-saturation condition is not fulfilled; for example, monitoring the first harmonic and using a phase of $\pi/2$, a signal similar to the correct EPR spectrum can appear. For this reason, a comparison of the signal recorded in-phase or with $\pi/2$-out-of-phase in first harmonic can be useful to evaluate the correctness of the experimental parameters of the measure.

Some investigations [11, 29, 35, 38–40], focused on centers having long relaxation times, have reported the detection of intense EPR spectra in the second-harmonic $\pi/2$-out-of phase signal under high microwave power. This result has been related to growth of the signal originating in the rapid-passage condition. Furthermore, it has been also observed that this second-harmonic signal is similar to the absorption profile for inhomogenously broadened EPR lines. This kind of measurement has proved a high detection sensitivity [11, 29, 35, 37–43], allowing to study signals with low amplitudes arising from a low amount of defects or from hyperfine interaction with isotopes of low natural abundance.

9.8 Parameters for EPR Signal Acquisition

In this section, we will comment on the different parameters that can be set in a EPR experiment and how to select them.

First of all, we consider the incident power (or the one in the resonant cavity) that must be set to avoid the effect of saturation ($\gamma^2 B_1^2 T_1 T_2 \ll 1$) of the absorption. As above anticipated, the saturation condition depends on the type of the investigated defect; Figure 9.9 illustrates the saturation curves of different Ge-related defects in Ge-doped silica [33, 44, 45]. Attention has to be given to the fact that these curves are temperature dependent [7] and the saturation (or the deviation from the linear behavior) is reached at lower incident power values by decreasing the temperature.

Then, it is necessary to select the modulation amplitude (A_m); such a parameter has to be adjusted so that the EPR signal reproduces the undistorted derivative of the signal. For this aim, the used value of A_m is usually less than 0.4 ΔB_{pp} [1] or even 0.1 ΔB_{pp} (the peak-to-peak width that is equal to the magnetic resonance field distance of adjacent maxima and minima). In this context, a good practice that should be applied is to compare the spectra recorded with different A_m. This allows

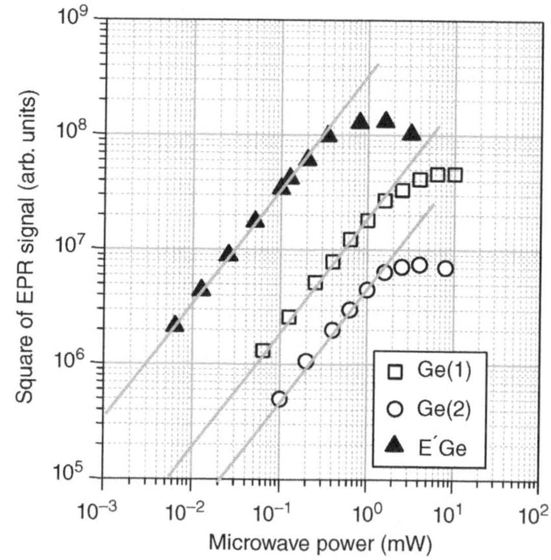

Figure 9.9 Saturation curve of Ge(1), Ge(2), and E′Ge defects in Ge-doped silica samples. Data recorded at room temperature. Gray lines indicate the linear trends.

to identify the maximum value that can be used without signal distortions or eventually select A_m values that increase the signal amplitude (and as a consequence the sensitivity) knowing the eventual impact on the EPR signal lineshape.

The other parameter of the modulation that has to be considered is its frequency. This is usually 100 kHz to avoid the appearance of sidebands for signal with ΔB_{pp} larger or equal to $3.6 \cdot 10^{-2}$ mT [1].

Regarding the magnetic field that induces the splitting of the levels, the operator can select a central value and an interval, ΔB_{sweep}, which have to be inside the maximum and the minimum B values allowed by the electromagnets. The ΔB_{sweep} should be selected in such a way that allows eventual correction of the baseline. Obviously, in any acquisition, the operator can also select the gain of the detection diode, high gains imply high sensitivity, but the selected gain has to be sufficiently low to avoid the saturation of the detection diode.

Finally, two time constants have to be selected by the operator, one is the conversion time T_{conv} of the analog-to-digital converter and the other is RC time constant (T_{const}). T_{conv} determines the time of the analog-to-digital conversion during the acquisition of each point. As a consequence, T_{conv} also determines the duration of the EPR measurement as well as the rate with which the magnetic field is changed. In some spectrometers, the parameter that can be selected is directly the total time of the measurement (T_{sweep}).

The T_{const} selection allows to cut noise frequency higher than $1/T_{const}$; so, increasing its value implies improving the signal-to-noise ratio. Simultaneously, in order to avoid distortions of the EPR line, the following empirical condition must be satisfied [1]:

$$T_{const} \leq 0.1 \frac{\Delta B_{pp}}{\Delta B_{sweep}} T_{sweep}. \tag{9.26}$$

An extended review of the experimental parameters and their impact on the acquired data can be found in [1, 5].

9.9 Cw EPR Case Studies

Among different possible cases of study, the $E_\gamma'Si$ (or simply E′-Si in the following) defect in silica has been selected to illustrate various aspects of the cw EPR spectroscopy. The signal attributed to this defect was observed for the first time by Weeks [32]; the structure of this defect is well accepted and it consists of an unpaired electron in a sp^3 orbital mainly localized on a silicon atom linked by three single bonds with three oxygen atoms [9, 11]. The structural model is reported in Figure 9.10a. In Figure 9.10b, we report also the spectrum (see the light gray or black curves recorded with all the correct experimental parameters) with $g_1 = 2.001\,80$, $g_2 = 2.000\,63$, $g_3 = 2.000\,37$, $\Delta g_{12} = 0.001\,17$, and $\Delta g_{13} = 0.001\,43$. Comparing this spectrum with that reported in Figure 9.3b, we note that the spectrum has a nearly axially symmetric lineshape. We also remind that this type of defects belongs to the set of paramagnetic defects that can be studied with the broken tetrahedron model [46–50] and that its analogies and differences with respect to the E'_1 in alpha quartz have been largely studied in the past by many research groups (more details can be found in [9]).

The first aspect that we treat in this paragraph is the effect of the modulation amplitude on the spectral lineshape. Indeed, the E′-Si is characterized by g values close to g_e and by small Δg; as a

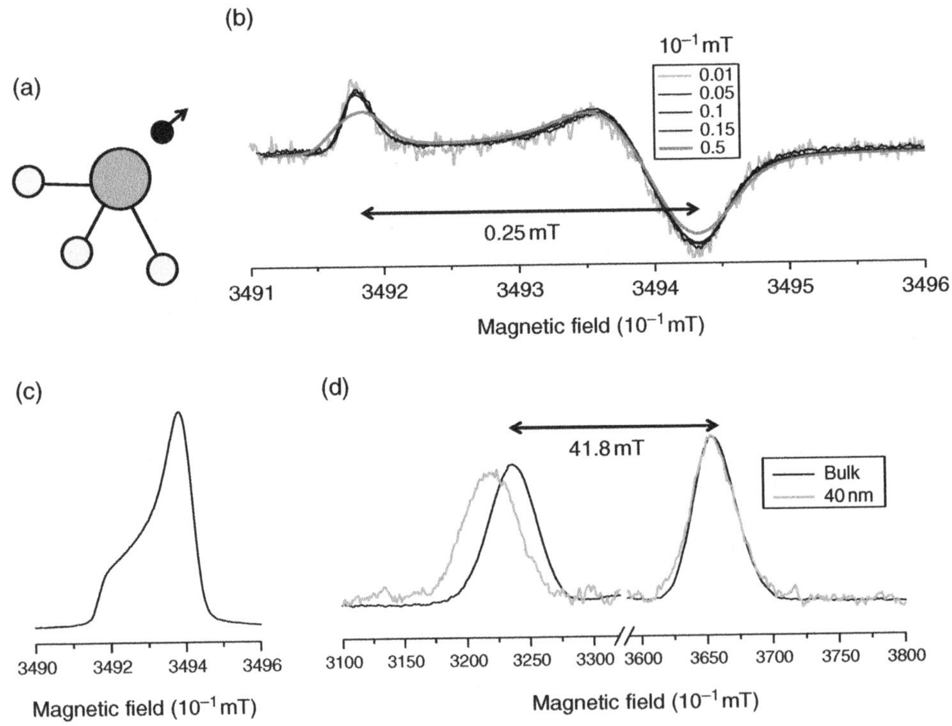

Figure 9.10 (a) Structural model of the $E_\gamma'Si$, the gray sphere represents the Si atom, the smaller light gray ones the O atoms, whereas the black sphere with the arrow the unpaired electron; (b) EPR signal of the E′-Si recorded with different values of the modulation amplitude (A_m) (—) 0.001 mT, (—) A_m = 0.005, 0.01 and 0.015 mT, (—) and A_m = 0.05 mT; (c) central EPR spectrum of E′-Si recorded with second-harmonic π/2-out-of phase measurements; (d) ^{29}Si-related hyperfine signal of E′-Si recorded with second-harmonic π/2-out-of phase measurements in bulk and nanoparticle samples (40 nm average diameter).

consequence of the latter, the EPR signal of these defects is narrow with the ΔB_{pp} being about 0.25 mT. So, it needs small modulation amplitudes to avoid distortions. In particular, from the spectra of Figure 9.10b, we can observe that the spectrum recorded with A_m of 0.05 mT features a relevant distortion of the first peak. Such a distortion implies a broadening of the first peak and the reduction of its amplitude; as a consequence of this, the peak-to-peak amplitude (amplitude difference between the first positive peak and the negative one), which can be often used to determine the defects concentration [5], is not reliable. On the other side, as a consequence of the broadening of the spectrum, the double integral of the curve can be less affected and can be still reliable for the evaluation of the defect concentration.

The spectra of Figure 9.10b have been recorded by modifying only the modulation amplitude, so they also allow to illustrate the impact of this parameter on the signal-to-noise ratio. Indeed, it can be seen that the spectrum recorded with the lowest A_m is the one with the higher level of noise. The signal-to-noise ratio increases by increasing the modulation amplitude and, as can be noted by the reported data, there are different A_m values for which the spectrum lineshape is not or poorly distorted. Considering the dependence of the EPR signal on the A_m, it is preferable to perform a study of the signal double integral and lineshape dependence on it. Indeed, in some cases it can be necessary to accept a compromise between detection and distortion.

All the spectra of Figure 9.10b have been recorded using an incident microwave power of 10^{-3} mW (53 dB of attenuation of a full power of 200 mW). Such a value is near the upper limit of incident microwave power for which the square of the EPR signal still depends linearly on the microwave power. This aspect has been tested experimentally even though the curve is not reported; this practice is always recommended because the power stored in the cavity can be slightly different in each spectrometer. As a consequence, further increase in such a parameter does not ensure an improvement of the detection limit (see Figure 9.9) and it implies further complication in the evaluation of the concentration.

As illustrated in the following, the intrinsic properties of the E′-Si give another efficient possibility to increase the detection limit. In fact, although the saturation regime for this defect is observed at low incident microwave powers, the features of the E′-Si allows to use second-harmonic $\pi/2$-out-of phase experiments to detect it. The procedure of this kind of measurements will be not presented, since it requires a deep study of the dependence of the recorded signal on all the experimental parameters (microwave power, modulation amplitude, phase). In fact, the impacts of each of them on the spectral shape and the concentration estimation have to be tested when one of the other experimental parameters is changed because of no trivial interdependence. An example of second-harmonic $\pi/2$-out-of phase signal of the E′-Si is reported in Figure 9.10c and the advantage in the detection limit is reported in [42].

The E′-Si is also responsible for a hyperfine structure constituted by two lines with separation of about 41.8 mT, which is related to the presence of the ^{29}Si (natural abundance 4.7%, $I = 1/2$) [11, 51]. The separation between the two lines has been related to structural features, such as the density, of the surrounding matrix [11, 31, 35, 52]. Such a relation has been used to investigate radiation-induced densification [31, 53–55] or the structural properties of silica nanoparticles [41, 56], often recording second-harmonic $\pi/2$-out-of phase data. In fact, because of the low abundance of the ^{29}Si and the fact that the signal is constituted by two lines, this EPR signature of the E′-Si has amplitude that is significantly lower than the one of the central line.

Figure 9.10d shows the spectra recorded in a bulk and in nanoparticles with diameter of about 40 nm; the spectra were shifted to anchor them to the peak of the high fields component to better show the difference in the lines separation, which is due to the higher density of the matrix surrounding the defects in the case of the nanoparticles.

We can find other defects related to Ge, Al, and P having structure similar to that of E′-Si. Such defects are usually called E′Ge [9, 11, 45, 57, 58], E′Al [59, 60], and P1 [61]. In these cases, the Si atom is replaced by the Ge, the Al, or by the P atoms.

As for the E′-Si, the lineshape of the EPR signal of the E′Ge is close to that of a defect with axial symmetry. However, the E′Ge (in silica) signal has a peak-to-peak distance ΔB_{pp} of about 12 Gauss, due to the different g and Δg values ($g_1 = 2.0012$, $g_2 = 1.9951$, $g_3 = 1.9941$; $\Delta g_{12} = 0.0061$ and $\Delta g_{13} = 0.0071$) [62]. We also note that these g values are close to those reported for the same defect but in pure GeO_2 [57]. In the comparison between these two E′ species, the authors of reference [57] also reported, that defining $\Delta g_3 = g_e - g_3$, the ratio $\Delta g_3(E'Ge)/\Delta g_3(E'\text{-}Si)$ is comparable to the ratio of the spin–orbit coupling of the Ge and Si.

While in the cases of the E′-Si and E′Ge, a minor part of the defects causes the hyperfine signal because of low natural abundance of the ^{29}Si and ^{73}Ge (natural abundance 7.76% and nuclear magnetic moment $I = 9/2$), for the Al and P cases all the defects contribute to the hyperfine signal. In fact, the ^{27}Al (nuclear magnetic moment $I = 5/2$) and the ^{31}P (nuclear magnetic moment $I = 1/2$) have natural abundance of about 100%. For this reason, the spectrum of the P1 in P-doped silica [61] is constituted by two lines with a separation of about 90 mT. We also note that for the studies of these hyperfine related signals in glass, the distribution of g and A have been considered to simulate the spectrum or to analyze it and that such investigations often were performed using second-harmonic $\pi/2$-out-of phase measurements.

9.10 Time-Resolved EPR Spectroscopy

In stationary EPR spectroscopy, the spin system is continuously subjected to the microwave oscillating magnetic field and is in a stationary condition in which the interaction of the system with external magnetic fields and relaxation interactions coexist. However, it is possible and very interesting to study the spin system in dynamic and nonstationary conditions, to obtain information not only on the interaction of the spin with the magnetic field but also on the characteristic times and the effectiveness of the relaxation interactions. In this dynamic regime, two timescales characterized by the times T_2 and T_1 can be distinguished ($T_1 \geq T_2$ and, in particular, in solid systems $T_1 \gg T_2$). At times $t \leq T_2$, the spins are driven by the microwave field and exhibit a coherent behavior, giving rise to so-called coherent transient effects that we will analyze in the next sections. At times $T_2 < t < T_1$, the spin system has already lost coherence but still evolves toward the steady state, in a regime in which the interaction with the magnetic field is still effective and the longitudinal relaxation is not yet completed (saturation transient regime).

9.10.1 Saturation Transients

To describe this regime, let us consider a two-level system subjected to an intense step-modulated microwave field. We expect that the energy absorption will vary from an initial maximum to a minimum value since the population difference between the up and down levels becomes minimal over time. The characteristic time of this decreasing trend will be related to the longitudinal relaxation time T_1 and the saturation of the transition between the two spin levels, which depends on the intensity of the oscillating field. A technique that allows to detect such transient regimes is the so-called second-harmonic (SH) spectroscopy [63, 64]. We will not describe here the experimental details of this technique, but we limit ourselves to recall that the SH signal is proportional to the absorption of microwave power and the resonance conditions are fulfilled for $\Delta E = 2h\nu$, where ΔE

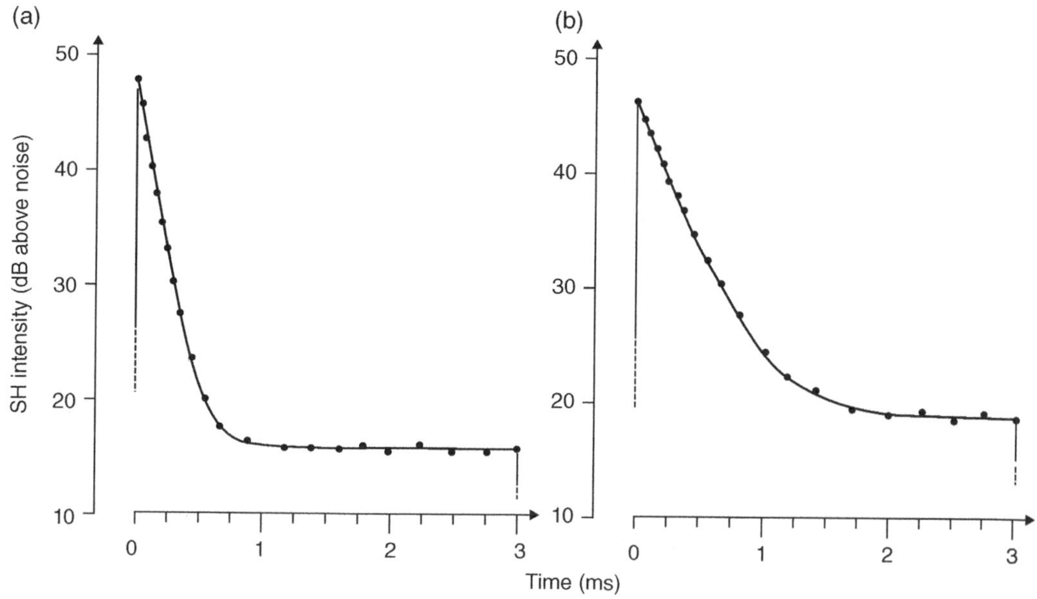

Figure 9.11 Pulse shapes of the SH intensity, emitted by a dilute ruby sample near the resonance, centered at H_0 = 107.5 mT, for a square pulse of input power P = 10 W, lasting 3 ms: (a) at resonance, $\Delta H = 0$; (b) off-resonance, ΔH = 1 mT. The full curves are the single-exponential laws which best fit the experimental points. *Source:* Boscaino and Gelardi [66]. © 1978, IOP Publishing.

is the splitting of the energy levels and ν is the frequency of the microwave field: so the resonance condition is characterized by double quantum (DQ) transitions, that is by absorption of two photons [65]. Two typical experimental curves of the SH intensity $I_{SH}(t)$, detected on a dilute ruby sample (Cr^{3+} : Al_2O_3) [66], are shown in Figure 9.11: curve a under resonance conditions, curve b with a detuning from the resonance of $\Delta H = 1$ mT.

Both curves, as shown by the continuous lines representing the best fit of the experimental points, follow a single-exponential decay law:

$$I_{SH}(t) = |A \exp(-t/\tau) + B|^2 \tag{9.27}$$

with

$$\left|\frac{A}{B}\right| = \frac{S}{1 + (2\omega - \omega_0)^2 T_2^2} \qquad \frac{1}{\tau} = \frac{1}{T_1}\left(1 + \frac{S}{1 + (2\omega - \omega_0)^2 T_2^2}\right) \tag{9.28}$$

where ω and ω_0 are, respectively, the frequency of the microwave magnetic field and the resonance frequency in rad s^{-1} and $S = (4\omega)^{-2} H_x^2 H_z^2 T_1 T_2$ is the saturation parameter of the DQ transition [65], with H_x and H_z components of the microwave field orthogonal and parallel to the static magnetic field, respectively. From the above expressions, confirmed by the experimental results, in resonance conditions, we have the maximum A/B ratio and the minimum value of the decay time τ. The parameters obtained by the best fit of the experimental decay curves allows us to determine both T_1 and S, obtaining information on the interactions of the spin system with its environment and on the intensity of the microwave field at the spin sites. Note that the observation of saturation transients in the timescale of Figure 9.11 requires a time $T_1 \sim ms$, a condition that for the system studied in [66] occurs at $T = 4.2\,K$ (liquid helium temperature). As last notation, we point out that the single-exponential decay law is valid in an homogeneous spin system. Otherwise, in an inhomogeneous system, a given spin, or a given packet of spectrally near spins, experiences a peculiar local field differing from the others, as regards both the resonance frequency and the saturation parameter. In this condition, the saturation transient shows a more complex decay than the single-exponential one, but also in this case it is possible, from the decay toward the stationary state, to obtain information on the spin dynamics and on the relaxation interactions within the system [67].

9.10.2 Spin Nutations

The spin nutations represent one of the coherent effects that can be observed in a timescale $t \leq T_2$, mentioned in the previous section. For a detailed description of the transient phenomena, it is necessary to solve the Eqs. (9.21)–(9.23) in nonstationary regime [68]. To this purpose, it is convenient to write the Bloch equations in a rotating system (RS) at the frequency ω of the microwave field. On the basis of the transformations from the laboratory system (LS) to the rotating one, the oscillating microwave field along x, $\boldsymbol{B} = \boldsymbol{i}2B_1\cos(\omega t)$ in LS, becomes static, as well as B_0, in RS and the equations take the following form:

$$\frac{dM_x}{dt} = (\gamma B_0 - \omega)M_y - \frac{M_x}{T_2} \tag{9.29}$$

$$\frac{dM_y}{dt} = -(\gamma B_0 - \omega)M_x + \gamma B_1 M_z - \frac{M_y}{T_2} \tag{9.30}$$

$$\frac{dM_z}{dt} = -\gamma B_1 M_y - \frac{M_z - M_0}{T_1}. \tag{9.31}$$

The above system of differential equations by the method of Laplace transforms and with the initial conditions $M_x(0) = M_y(0) = 0$ and $M_z(0) = M_0$ can be reduced to a system of algebraic equations in the variable s, whose determinant of coefficients is:

$$\left(s + \frac{1}{T_2}\right)^2 \left(s + \frac{1}{T_1}\right) + (\gamma B_0 - \omega)\left(s + \frac{1}{T_1}\right) + \gamma^2 B_1^2 \left(s + \frac{1}{T_2}\right) = 0. \tag{9.32}$$

By anti-transforming, in the time domain, the formal solution for the generic component of magnetization $M_i(t)$ $(i = x,y,z)$ is

$$M_i(t) = A\mathrm{e}^{-\alpha t} + B\mathrm{e}^{-\beta t}\cos\zeta t + \frac{C}{\zeta}\mathrm{e}^{-\beta t}\sin\zeta t + D \tag{9.33}$$

where α, β, and ζ are coefficients determined by equating (9.32) to the following factored expression

$$(s + \alpha)\left[(s + \beta)^2 + \zeta^2\right] \tag{9.34}$$

and A, B, C, and D are constants to be determined on the basis of the initial conditions and they depend on the component of $\boldsymbol{M}$ considered. Simple solutions are found in the resonance conditions $(\gamma B_0 - \omega = 0)$ with $1/T_2 \gg 1/T_1$, for which (9.32) becomes:

$$\left(s + \frac{1}{T_2}\right)\left[\left(s + \frac{1}{T_1}\right)\left(s + \frac{1}{T_2}\right) + \gamma^2 B_1^2\right] = 0. \tag{9.35}$$

One of the roots of (9.35) is $s_1 = \alpha = 1/T_2$ and the other two are

$$s_{2,3} = -\frac{1}{2}\left[\left(\frac{1}{T_1} + \frac{1}{T_2}\right) \pm \sqrt{\left(\frac{1}{T_1} - \frac{1}{T_2}\right)^2 - 4\gamma^2 B_1^2}\right]. \tag{9.36}$$

Note that if the condition $4\gamma^2 B_1^2 > \left(\frac{1}{T_1} - \frac{1}{T_2}\right)^2$ occurs, two roots of (9.36) will be complex and consequently $M_i(t)$ will exhibit damped oscillations (spin nutations). If the further hypothesis $1/T_2 \gg 1/T_1$ also occurs, the above condition reduces to $T_2 > \frac{1}{2\gamma B_1} \tilde{T}_0$ with T_0 being the nutation period. This means that the experimental detection of spin nutations requires a spin system characterized by a relatively long T_2. At variance, the occurrence of the opposite condition $\left(\frac{1}{T_1} - \frac{1}{T_2}\right)^2 \gg 4\gamma^2 B_1^2$ involves two real roots of (9.36) that generate two exponential decays with characteristic times $\tau_a \sim 1/T_2$ and $\tau_b \sim 1/T_1$.

In order to analyze in more detail the dependence of T_0 on parameters such as the intensity of the oscillating field and the detuning from resonance, we use a simplified version of Eqs. (9.29)–(9.31) where the relaxation interactions are neglected and, therefore, valid for a timescale $t \ll T_2, T_1$ [69]:

$$\frac{dM_x}{dt} = (\gamma B_0 - \omega) M_y \tag{9.37}$$

$$\frac{dM_y}{dt} = -(\gamma B_0 - \omega) M_x + \gamma B_1 M_z \tag{9.38}$$

$$\frac{dM_z}{dt} = -\gamma B_1 M_y. \tag{9.39}$$

The solutions for the components of the magnetization $M_i(t)$ taking into account the initial conditions

$M_x(0) = M_y(0) = 0$ and $M_z(0) = M_0$ are:

$$M_x(t) = \frac{\Delta \gamma B_1 M_0}{\beta^2}(1 - \cos \beta t) \tag{9.40}$$

$$M_y(t) = \frac{\gamma B_1 M_0}{\beta} \sin \beta t \tag{9.41}$$

$$M_z(t) = M_0\left(\frac{\Delta^2}{\beta^2} + \frac{(\gamma B_1)^2}{\beta^2} \cos \beta t\right) \tag{9.42}$$

where $\Delta = \gamma B_0 - \omega$ is the frequency detuning from resonance and $\beta = \sqrt{\Delta^2 + (\gamma B_1)^2}$ is the nutation frequency. Note that all the components $M_i(t)$ oscillate without damping, as expected, having ruled out the relaxation interactions in Eqs. (9.37)–(9.39).

As pointed out also in the previous section, these equations describe the time evolution of a homogeneous spin system. Actually, it is most probable that a system where the paramagnetic centers are diluted in a solid matrix is inhomogeneous. The inhomogeneity condition makes the effective time evolution of the magnetization more complicated than that reported in Eqs. (9.40)–(9.42), since it involves a convolution of $M_i(t)$ with a Gaussian representing the spectral distribution of the resonance frequencies of the spins around the central one $\omega_0 = \gamma B_0$. In particular, the absorption signal $V(t)$ is

$$V(t) = \frac{\gamma B_1 M_0}{\sqrt{\pi}\,\sigma} \int_{-\infty}^{+\infty} \frac{\sin \beta t}{\beta} \exp\left(-\frac{\Delta^2}{2\sigma^2}\right) d\Delta. \tag{9.43}$$

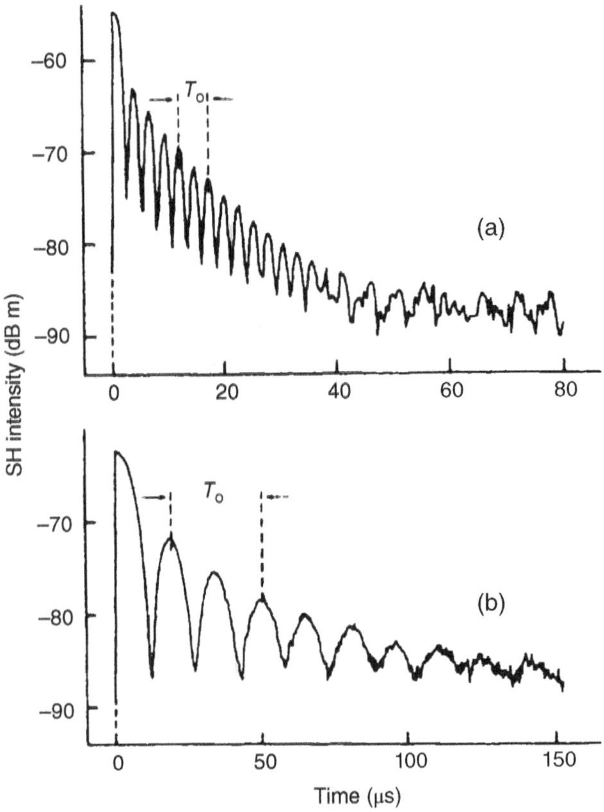

Figure 9.12 Experimental curves of the SH transient signal at the resonance in a sample of E′-Si centers in silica at $T = 4.2$ K, at two different power levels: (a) $P = 30$ W ($B_1 = 1.2$ mT), $T_o = 5.3$ μs; (b) $P = 4.8$ W ($B_1 = 0.49$ mT), $T_o = 31.0$ μs. *Source:* Boscaino et al. [69]. © 1986, American Physical Society.

This integral can be analytically resolved in case of high inhomogeneity, that is $\sigma \to \infty$, giving the result

$$V(t) = \frac{\gamma B_1 M_0 \sqrt{\pi}}{\sigma} J_0(\gamma B_1 t) \tag{9.44}$$

where J_0 is the zero-order Bessel function [69]. In Figure 9.12, an experimental curve detected by the SH spectroscopy technique in an inhomogeneous system of E'-Si centers in amorphous SiO_2 (silica) at $T = 4.2$ K is shown [69]. It is worth noting that the SH signal is proportional to $|V(t)|^2$ so that the signal is defined positive and the nutation period is the time lasting between two non-consecutive peaks. The oscillating behavior is well fit by the function $|J_0(\gamma B_1 t)|^2$ but the time decay is not described by a single exponential and it is not surprising because of the inhomogeneity of the spin system.

9.10.3 Free Induction Decay

The coherent dynamics of a spin system can be revealed not only when the system is driven by an oscillating magnetic field but also when this driving action ends, the microwave radiation is

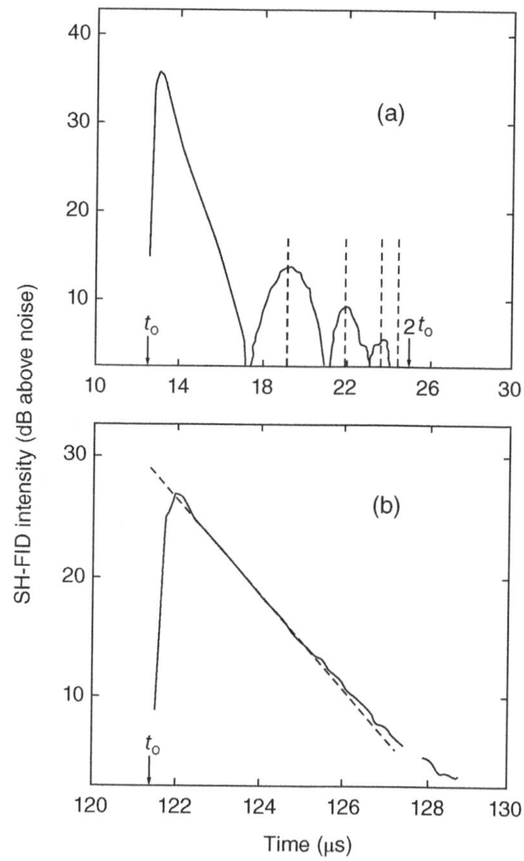

Figure 9.13 Experimental curves of the SH-FID signal of E′-Si centers in silica detected at the resonance: (a) after a pulse at maximum input power (P = 20 W), lasting t_0 = 12.5 μs; (b) after a pulse at P = 5.7 W lasting t_0 = 121.5 μs. The measured nutation period is T_n = 5 μs in (a) and T_n = 17.4 μs in (b). *Source:* Boscaino et al. [71]. © 1983, American Physical Society.

switched off and the only external magnetic field acting on the spin system is the static one. Obviously, in this case also, the coherent motions of the spin around the static field can subsist just for a short time, because of a dephasing effect due to the spin–spin interactions lasting for a time of the order of T_2. The typical coherent transient established in such conditions is the so-called free induction decay (FID) [70], well described, apart from the time decay, by the solutions of Eqs. (9.37)–(9.39) with the condition $\gamma B_1 = 0$:

$$M_x(t) = M_{x0}\cos\Delta t + M_{y0}\sin\Delta t \tag{9.45}$$

$$M_y(t) = -M_{x0}\sin\Delta t + M_{y0}\cos\Delta t \tag{9.46}$$

$$M_z(t) = M_{z0} \tag{9.47}$$

where M_{x0}, M_{y0}, and M_{z0} are no more the initial components of the magnetization but those at the ending time of the microwave pulse. Actually, the FID signal is characterized by damped oscillations when the preparative microwave pulse is short enough compared to T_2, whereas it exhibits an overdamped decay after a pulse whose duration is longer than T_2. In fact, in the former case, the dynamics of the spin system is still coherent when the radiation pulse is switched off and the oscillations persist; in the latter case, the dephasing interactions have been already effective during the pulse, lasting a time $t_w \gg T_2$, and the FID follows a single-exponential decay law. An example of these different behaviors of the FID signal is shown in Figure 9.13 reporting SH-FID signals detected in a system of E'-Si centers in silica at low temperature [71].

The comparison between the lasting time of the pulses with the corresponding nutation period and the measured spin–spin relaxation time in the system, $T_2 = 37$ μs, explains that: in case (a), the spin system is still in a coherent regime and exhibits oscillations until $t = 2t_0$; in case (b), at $t = t_0$ the spin system is in a steady state. In this latter case, the FID signal is characterized, after an initial abrupt increase, by a single-exponential decay law with a characteristic time $\tau_d = 2.2$ μs, as shown in Figure 9.13b by the fitting curve (dashed line).

9.10.4 Spin Echo

Among the coherent phenomena establishing in spin dynamics, the spin echo, firstly revealed by Hahn in a nuclear magnetic resonance experiment [72], is the most known, and experimental techniques based on its detection are largely used also in chemistry, biology and medicine [73–75]. Like the FID, the spin-echo phenomenon can also be described by Eqs. (9.45)–(9.47), since the echo signal occurs when the oscillating magnetic field is switched off. In order to understand the echo phenomenology, we remind that in the reference system rotating at frequency ω(RS), in resonance condition $\omega = \omega_0$, the components of the magnetic field are: $B_z = B_0 - \frac{\omega}{\gamma} = 0$ and $B_x = B_1$. So, in RS, the spins will rotate in the plane y-z around $\boldsymbol{B}_1$, spanning an angle $\theta_1 = \omega t_{w1}$ if t_{w1} is the lasting time of the microwave pulse. If $\theta_1 = \frac{\pi}{2}$, at the switching off of the radiation, all the spins will be aligned along y-axis and will rotate in the plane x-y around the direction of $\boldsymbol{B}_0$, which is the only magnetic field acting on the system. During this rotation, the spin–spin interactions affect the coherence in the spin motions and a spread in the angular velocity of the spins is expected. In RS, some spins having angular speeds $\omega_i \geq \omega$ appear to rotate in the given direction, other spins having $\omega_i \leq \omega$ appear to rotate in the opposite direction. When a second pulse rotates spins by an angle $\theta_2 = \pi$, all the spins will be again in the plane x-y but in specular positions with respect to the plane x-z. At the end of the second pulse, the spins will rotate around $\boldsymbol{B}_0$ but now the spins having $\omega_i \leq \omega$ will precede those having $\omega_i \geq \omega$ in the precession motion. Consequently, the faster

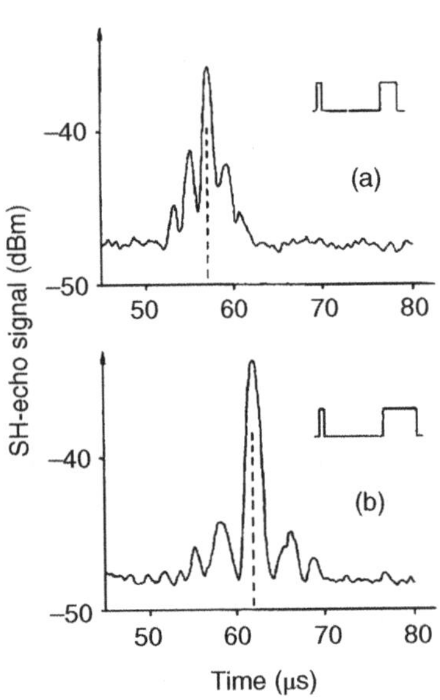

Figure 9.14 Echo signals of E′-Si centers in silica observed for the sequence of input pulses shown in the insets: $\varphi_1 = \frac{\pi}{2}, \tau = 25.5$ μs, $\varphi_2 = 2\pi$ (a) and $\varphi_2 = 4\pi$ (b). Time is measured from the leading edge of the former pulse. The dashed line marks t_e, at the center of the echo signal: $t_e = 56.9$ μs (a); $t_e = 61.5$ μs (b). *Source:* Boscaino et al. 76. © 1983, Elsevier.

spins will reach the slower ones, giving rise, at a given time, to a maximum of the magnetization in the plane *x-y*, that is the echo signal. If both pulses are very intense and their durations much shorter than the delay time τ between pulses, i.e. t_{w1}, $t_{w2} \ll \tau$, such a maximum value of $M_{xy}(t)$ will occur at $t \sim 2\tau$. Moreover, because of the spin–spin interactions affecting the coherence in the spin motion during the delay time τ between the pulses, one expects that the intensity of the echo signal decreases on increasing τ so that, as well as for the FID, the decay of the echo intensity follows a single-exponential decay law: $M_{xy}(2\tau) = M_{xy}(0)e^{-\frac{\tau}{T_2}}$. This property makes the detection of echo signal, as a function of delay time in a sequence of two pulses $\pi/2 - \pi$ (Hahn's sequence), a direct and reliable method for measuring the transverse relaxation time T_2 of a given spin system. There is a strict analogy between FID and echo signals, so that the echo can be considered as the junction of two mirrored FID signals; in particular, coherent oscillations can be observed in the spin-echo signal if FID is oscillatory [76]. This peculiar behavior was detected in the same spin system (E'-Si centers in silica) where nutations and oscillatory FID had been detected [69, 71] and it is evidenced in Figure 9.14.

Note that, the pulse sequences used in the experimental conditions of Ref. [76] and sketched in Figure 9.14 differ from the typical Hahn's sequence $\pi/2 - \pi$. Actually, the spin-echo signal can be generated by a variety of preparation sequences, aiming to evidence peculiar transient effects on the spin dynamics and to measure not only T_2 but also T_1 [5]. In conclusion, the detection of EPR (and nuclear magnetic resonance) time-dependent signals, in both coherent and non-coherent regime, allow to investigate in real time the dynamics of a spin system, obtaining information on the interactions occurring within the spin system, and between this and its environment, in a more detailed way than by the conventional stationary EPR spectroscopy.

References

1 Poole, C.P. Jr. (1967). *Electron Spin Resonance*. New York: Wiley.

2 Abragam, A. and Bleaney, B. (1970). *Electron Paramagnetic Resonance of Transition Ions*. Oxford, UK: Clarendon.

3 Pake, G.E. and Estle, T.L. (1973). *The Physical Principles of Electron Paramagnetic Resonance*. Reading, MA: WA Benjamin Inc.

4 Slichter, C.P. (1978). *Principles of Magnetic Resonance*. Berlin, Heidelberg, Germany: Springer-Verlag.

5 Weil, J.A., Bolton, J.R., and Wertz, J.E. (1994). *Electron Paramagnetic Resonance*, 2e. New York: Wiley.

6 Brückner, A. (2010). In situ electron paramagnetic resonance: a unique tool for analyzing structure–reactivity relationships in heterogeneous catalysis. *Chem. Soc. Rev.* 39: 4673–4684.

7 Naveed, K.R., Wang, L., Yu, H. et al. (2018). Recent progress in the electron paramagnetic resonance study of polymers. *Polym. Chem.* 9: 3306–3335.

8 Wang, B., Fielding, A.J., and Dryfe, R.A.W. (2019). Correction to electron paramagnetic resonance investigation of the structure of graphene oxide: pH-dependence of the spectroscopic response. *ACS Appl. Nano Mater.* 2: 19–27.

9 Lund, A. and Shiotani, M. (2014). *Applications of EPR in Radiation Research*. Heidelberg, Germany: Springer.

10 Tampieri, F. and Barbon, A. (2018, 1). Resolution of EPR signals in graphene-based materials from few layers to nanographites. In: *Frontiers in Magnetic Resonance* (eds. D. Savchenko and A.H. Kassiba), 36–66. Sharjah, UAE: Bentham Science.

11 Pacchioni, G., Skuja, L., and Griscom, D.L. (2000). *Defects in SiO_2 and Related Dielectrics: Science and Technology. Dordrecht, The Netherlands: Kluwer Academic.*

12 Schauer, D.A., Iwasaki, A., Romanyukha, A.A. et al. (2007). Electron paramagnetic resonance (EPR) in medical dosimetry. *Radiat. Meas.* 41: S117–S123.

13 Danhier, P. and Gallez, B. (2015). Electron paramagnetic resonance: a powerful tool to support magnetic resonance imaging research. *Contrast Media Mol. Imaging.* 10: 266–281.

14 Pan, Y. and Nilges, M.J. (2014). Electron paramagnetic resonance spectroscopy: basic principles, experimental techniques and applications to earth and planetary sciences. *Rev. Mineral. Geochem.* 78: 655–690.

15 Roessler, M.M. and Salvadori, E. (2018). Principles and applications of EPR spectroscopy in the chemical sciences. *Chem. Soc. Rev.* 47: 2534–2553.

16 Harris, D.C. and Bertolucci, M.D. (1978). *Symmetry and Spectroscopy an Introduction to Vibrational and Electronic Spectroscopy.* Oxford, UK: Oxford University Press Inc.

17 Hagen, W.R. (2006). EPR spectroscopy as a probe of metal centres in biological systems. *Dalton Trans.*: 4415–4434.

18 Kneubuhl, F.K. (1960). Line shapes of electron paramagnetic resonance signals produced by powders, glasses, and viscous liquids. *J. Chem. Phys.* 33: 1074–1078.

19 Sands, R.H. (1955). Paramagnetic resonance absorption in glass. *Phys. Rev.* 99: 1222–1226.

20 Weeks, R.A. and Nelson, C.M. (1960). Irradiation effects and short-range order in fused silica and quartz. *J. Appl. Phys.* 31: 1555–1558.

21 Siderer, Y. and Luz, Z. (1980). Analytical expressions for magnetic resonance lineshapes of powder samples. *J. Magn. Res.* 37: 449–463.

22 Poole, C.P. and Farach, H.A. (1979). Line shapes in electron spin resonance. *Bull. Magn. Res.* 1: 162–194.

23 Rowlands, C.C. and Murphy, D.M. (2016). EPR spectroscopy, theory. In: *Encyclopedia of Spectroscopy and Spectrometry* (eds. J.C. Lindon, G.E. Tranter and D.W. Koppenaal), 517–526. Oxford: Elsevier.

24 Taylor, P.C. and Bray, P.J. (1970). Computer simulations of magnetic resonance spectra observed in polycrystalline and glassy samples. *J. Magn. Res.* 2: 305–331.

25 Morin, G. and Bonnin, D. (1999). Modeling EPR powder spectra using numerical diagonalization of the spin hamiltonian. *J. Magn. Res.* 136: 176–199.

26 http://www.eprsimulator.org/ (accessed 23 February 2021).

27 Bloch, F. (1946). Nuclear induction. *Phys. Rev.* 70: 460–474.

28 Hyde, J.S. (1960). Magnetic resonance and rapid passage in irradiated LiF. *Phys. Rev.* 119: 1483–1492.

29 Harbridge, J.R., Rinard, G.A., Quine, R.W. et al. (2002). Enhanced signal intensities obtained by out-of-phase rapid-passage EPR for samples with long electron spin relaxation times. *J. Magn. Res.* 156: 41–51.

30 Weger, M. (1960). Passage effects in paramagnetic resonance experiments. *Bell. Syst. Tech. J.* 39: 1013–1112.

31 Devine, R.A.B., Duraud, J.P., and Dooryhée, E. (2000). *Structure and Imperfections in Amorphous and Crystalline Silicon Dioxide.* New York: Wiley.

32 Weeks, R.A. (1956). Paramagnetic resonance of lattice defects in irradiated quartz. *J. Appl. Phys.* 27: 1376–1381.

33 Agnello, S., Alessi, A., Gelardi, F.M. et al. (2008). Effect of oxygen deficiency on the radiation sensitivity of sol-gel Ge-doped amorphous SiO_2. *Eur. Phys. J. B* 61: 25–31.

34 Buscarino, G., Alessi, A., Agnello, S. et al. (2014). Isolation of the $CH_3^{\cdot}$ rotor in a thermally stable inert matrix: first characterization of the gradual transition from classical to quantum behaviour at low temperatures. *Phys. Chem. Chem. Phys.* 16: 13360–13366.

35 Griscom, D.L. (1984). Characterization of three E'-center variants in X- and γ-irradiated high purity a-SiO_2. *Nucl. Inst. Methods Phys. Res. B* 1: 481–488.

36 Todaro, M., Buscarino, G., Sciortino, L. et al. (2016). Decomposition Process of Carboxylate MOF HKUST-1 unveiled at the atomic scale level. *J. Phys. Chem. C* 120: 12879–12889.

37 Di Francesca, D., Girard, S., Agnello, S. et al. (2019). Combined temperature radiation effects and influence of drawing conditions on phosphorous-doped optical fibers. *Phys. Stat. Sol. A* 216: 1800553.

38 Galtsev, V.E., Galtseva, E.V., and Lebedev, Y.A.S. (1996). Optimal registration conditions for tooth EPR dosimetry at low accumulated dose. *Appl. Radiat. Isot.* 47: 1311–1315.

39 Galtsev, V.E., Grinberg, O.Y., Lebedev, Y.S., and Galtseva, E.V. (1993). EPR dosimetry sensitivity enhancement by detection of rapid passage signal of the tooth enamel at low temperature. *Appl. Magn. Res.* 4: 331–333.

40 Yan, B., Schultz, N.A., Efros, A.L., and Taylor, P.C. (2000). Universal distribution of residual carriers in tetrahedrally coordinated amorphous semiconductors. *Phys. Rev. Lett.* 84: 4180–4183.

41 Alessi, A., Agnello, S., Buscarino, G. et al. (2015). Silica nanoparticle core structure examined by the $E'Si_\gamma$ center ^{29}Si strong hyperfine interaction. *J. Non-Cryst. Solids* 423–424: 41–44.

42 Agnello, S., Boscaino, R., Buscarino, G., and Gelardi, F.M. (2004). Experimental evidence for two different precursors of E'_γ centers in silica. *J. Non-Cryst. Solids* 345–346: 505–508.

43 Buscarino, G., Agnello, S., and Gelardi, F.M. (2005). Delocalized nature of the E'_δ center in amorphous silicon dioxide. *Phys. Rev Lett.* 94: 125501.

44 Griscom, D.L. (2011). On the natures of radiation-induced point defects in GeO_2-SiO_2 glasses: reevaluation of a 26-year-old ESR and optical data set. *Opt. Mat. Expr.* 1: 400–412.

45 Giacomazzi, L., Martin-Samos, L., Boukenter, A. et al. (2015). Ge(2), Ge(1) and Ge-E′ centers in irradiated Ge-doped silica: a first-principles EPR study. *Opt. Mat. Expr.* 5: 1054–1064.

46 Cook, M. and White, C.T. (1989). Molecular electronic structure theory in the study of localised defects. *Semicond. Sci. Technol.* 4: 1012–1035.

47 Watkins, G.D. and Corbett, J.W. (1964). Defects in irradiated silicon: electron paramagnetic resonance and electron-nuclear double resonance of the Si-E center. *Phys. Rev. B* 134: A1359–A1377.

48 Feigl, F.J. and H. Anderson J. (1970). Defects in crystalline quartz: electron paramagnetic resonance of E' vacancy centers associated with germanium impurities. *J. Phys. Chem. Sol.* 31: 575–596.

49 Silsbee, R.H. (1961). Electron spin resonance in neutron-irradiated quartz. *J. Appl. Phys.* 32: 1459–1462.

50 Agnello, S., Buscarino, G., Gelardi, F.M., and Boscaino, R. (2008). Optical absorption band at 5.8eV associated with the E′γ centers in amorphous silicon dioxide: optical absorption and EPR measurements. *Phys. Rev. B* 77: 195206.

51 Griscom, D.L. (1979). E' center in glassy SiO_2. Microwave saturation properties and confirmation of the primary ^{29}Si hyperfine structure. *Phys. Rev. B* 20: 1823–1834.

52 Devine, R.A.B. and Arndt, J. (1987). Si-O bond-length modification in pressure-densified amorphous SiO_2. *Phys. Rev. B* 35: 9376–9379.

53 Buscarino, G., Agnello, S., Gelardi, F.M., and Boscaino, R. (2009). Polyamorphic transformation induced by electron irradiation in a-SiO_2 glass. *Phys. Rev. B* 80: 094202.

54 Douillard, L., Jollet, F., Duraud, J.P. et al. (1992). Radiation damage produced in quartz by energetic ions. *Radiat. Eff. Defects Solids* 124: 351–370.

55 Douillard, L., Dooryhee, E., Duraud, J.P. et al. (1993). Modifications of the atomic and electronic structure of quartz by high-energy ion irradiation. *Radiat. Eff. Defects Solids* 126: 237–241.

56 Stesmans, A., Clémer, K., and Afanas'ev, V.V. (2008). Primary ^{29}Si hyperfine structure of E′ centers in nm-sized silica: probing the microscopic network structure. *Phys. Rev. B* 77: 094130.

57 Tsai, T.-E., Griscom, D.L., and Friebele, E.J. (1987). Radiation induced defect centers in high-purity GeO_2 glass. *J. Appl. Phys.* 62: 2264–2268.

58 Watanabe, Y., Kawazoe, H., Shibuya, K., and Muta, K. (1986). Structure and mechanism of formation of drawing- or radiation-induced defects in SiO_2:GeO_2 optical fiber. *Jap. J. Appl. Phys.* 25: 425–431.

59 Griscom, D.L. (2011). Trapped-electron centers in pure and doped glassy silica: a review and synthesis. *J. Non-Cryst. Solids* 357: 1945–1962.

60 Trukhin, A.N., Teteris, J., Fedotov, A. et al. (2009). Photosensitivity of SiO_2–Al and SiO_2–Na glasses under ArF (193 nm) laser. *J. Non-Cryst. Solids* 355: 1066–1074.

61 Griscom, D.L., Friebele, E.J., Long, K.J., and Fleming, J.W. (1983). Fundamental defect centers in glass: electron spin resonance and optical absorption studies of irradiated phosphorus doped silica glass and optical fibers. *J. Appl. Phys.* 54: 3743–3762.

62 Alessi, A., Agnello, S., Gelardi, F.M. et al. (2011). Influence of Ge doping level on the EPR signal of Ge(1), Ge(2) and E'Ge defects in Ge-doped silica. *J. Non-Cryst. Solids* 357: 1900–1903.

63 Boscaino, R., Ciccarello, I., Cusumano, C., and Strandberg, M.W.P. (1971). Second-harmonic generation and spin decoupling in resonant two-level spin systems. *Phys. Rev. B* 3: 2675–2682.

64 Boscaino, R. and Tripo, A. (1978). Two-photon spectroscopy of a resonant three-level spin system. *J. Phys. C: Solid St. Phys.* 11: 365–376.

65 Alzetta, G., Arimondo, E., and Ascoli, C. (1968). Angular-momentum detection of many-photon transitions. *Nuovo Cim.* 54 B: 107–168.

66 Boscaino, R. and Gelardi, F.M. (1978). Pulse shape of the second harmonic generated by a two-level spin system. *J. Phys. C: Solid State Phys.* 11: L475–L478.

67 Boscaino, R. and Gelardi, F.M. (1980). Time evolution of inhomogeneous spin systems toward the saturated state. *J. Phys. C: Solid State Phys.* 13: 3737–3748.

68 Torrey, H.C. (1949). Transient nutations in nuclear magnetic resonance. *Phys. Rev.* 76: 1059–1068.

69 Boscaino, R., Gelardi, F.M., Messina, G. et al. (1986). *Phys. Rev. B.* 33: 3076–3082.

70 Hahn, E.L. (1950). Nuclear induction due to free Larmor precession. *Phys. Rev.* 77: 297–298.

71 Boscaino, R., Gelardi, F.M., and Messina, G. (1983). Second-harmonic free-induction decay in a two level spin system. *Phys. Rev. A* 28: 495–497.

72 Hahn, E.L. (1950). Spin echoes. *Phys. Rev.* 80: 580–594.

73 Tokuda, H., Hayamizu, K., Ishii, K. et al. (2005). Physicochemical properties and structures of room temperature ionic liquids. 2. variation of alkyl chain length in imidazolium cation. *J. Phys. Chem. B* 109: 6103–6110.

74 Jack, C.R. Jr., Garwood, M., Wengenack, T.M. et al. (2004). In vivo visualization of Alzheimer's amyloid plaques by magnetic resonance imaging in transgenic mice without a contrast agent. *Magn. Reson. Med.* 52: 1263–1271.

75 Johansson, A., Karlsson, M., and Nyholm, T. (2011). CT substitute derived from MRI sequences with ultrashort echo time. *Med. Phys.* 38 (5): 2708–2714.

76 Boscaino, R., Gelardi, F.M., and Messina, G. (1983). Oscillatory behaviour of two-photon echo signals. *Phys. Lett. A* 97: 413–416.

10

Nuclear Magnetic Resonance Spectroscopy

Alberto Spinella[1] and Pellegrino Conte[2]

[1] *Advanced Technologies Network Center (ATeN Center), University of Palermo, Palermo, Italy*
[2] *Department of Agriculture, Food and Forestry Sciences, University of Palermo, Palermo, Italy*

10.1 Introduction

Nuclear magnetic resonance (NMR) spectroscopy is a powerful technique used in many fields from basic to applied sciences in order to characterize both molecular and supramolecular structures of organic and inorganic compounds. Among the various spectroscopic techniques, NMR is one of the most versatile due to its ability to unveil three-dimensionality of molecular systems in each of the three different physical states (liquid, solid, and gas) and regardless of sample crystallinity.

Purcell, Torrey, and Pound at Harvard University [1] and Bloch, Hansen, and Packard at Stanford University [2] performed the first NMR experiments in condensed matter independently of each other in 1945. In particular, they obtained NMR signals from protons of paraffin wax and liquid water, respectively. However, since those pioneering days, NMR has become a well-established spectroscopic technique. In particular, both liquid and solid-state NMR spectroscopies are largely exploited for molecular structure determination and dynamics in several chemistry fields (e.g. protein chemistry, materials science, cultural heritage, food science, and so on), while NMR imaging is recognized as a very powerful tool in modern medicine in order to monitor biological tissues and living functions.

In the following, an overview of the basic principles of NMR spectroscopy and the main topics that may be useful to researches of different areas will be given. A detailed and in-depth introduction to NMR spectroscopy can be found in several textbooks [3–5].

10.2 NMR General Concepts

10.2.1 Nuclear Spin and Magnetic Moment

All elementary particles have an intrinsic property referred to as spin. This property has been introduced to explain the deflection of the elementary particles when they pass through an inhomogeneous magnetic field. In fact, according to the experiment by Stern and Gerlach, an elementary particle, such as an electron, being a moving electric charge, should behave like a small linear

Spectroscopy for Materials Characterization, First Edition. Edited by Simonpietro Agnello.

magnet when crossing a magnetic field gradient. Therefore, a continuous distribution should be revealed on the surface of a detector. Conversely, the image obtained by elementary particles crossing the magnetic field gradient is an ensemble of discrete points of accumulation. This suggests that all the elementary particles have an intrinsic spin angular momentum which can have only discrete values given by:

$$\hbar\sqrt{S(S+1)} \tag{10.1}$$

where S is the spin quantum number.

Electrons and photons are elementary particles having $S = ½$ and $S = 1$, respectively. Atomic nuclei are made up of subatomic particles (referred to as nucleons) which, in turn, are made up of quarks. Therefore, nuclear spin depends on the number of particles (i.e. protons and neutrons) forming the nuclei. For this reason, it is possible to have nuclei with integer, semi-integer, or zero spin quantum number. All the NMR techniques can only be applied to nuclei with $S \neq 0$.

Spin values of some of the most common nuclei are shown in Table 10.1 [6].

Atomic nuclei with a nonzero spin angular momentum also have a magnetic moment ($\overline{\mu}$) to which the former is parallel (Figure 10.1a).

The spin angular momentum is related to the magnetic moment by the following relationship:

$$\overline{\mu} = \gamma\overline{S} \tag{10.2}$$

where γ is a proportionality constant. It is characteristic for each nucleus and it is referred to as gyromagnetic ratio. The nuclear magnetic moment $\overline{\mu}$ can be oriented either along the same direction of the spin angular moment or in the opposite direction. This depends on the sign of γ. $\overline{\mu}$ magnitude, although very small (it is in the order of the mT), is still significant when compared to the magnitude of the Earth's magnetic field. In fact, the latter is approximately of 50 μT. In the volume of matter that is generally subjected to NMR experiments, the number of nuclear spins is in the order of 10^{20}. Noteworthy, the directions of all the nuclear magnetic moments are completely random in the absence of any external applied magnetic field. Therefore, the total magnetic moment of the nuclear system given by the vector sum of each i-th $\overline{\mu}$ component ($\sum_{i=1}^{N}\overline{\mu}_i$) is null.

Nuclear spins interact with each other within a single molecule (we can refer to these types of interactions as intramolecular spin interactions), while the intermolecular spin interactions are small. Nuclear spins in single molecules represent a spin-system; several spin-systems, given by different molecules, generate a spin-ensemble (Figure 10.1b).

Table 10.1 Examples of isotopes spin, natural abundance, gyromagnetic ratio, and Larmor frequency at 9.4 T.

Isotope	Spin	Abundance (%)	Gyromagnetic ratio γ (10^6 rad s^{-1} T^{-1})	Larmor frequency at 9.4 T (MHz)
^{1}H	1/2	99.97	267.522	−400
^{12}C	0	98.9	—	—
^{13}C	1/2	1.1	67.283	−100.602
^{14}N	1	99.6	19.338	−28.914
^{15}N	1/2	0.37	−27.126	+40.559

Source: Modified from Harris [6].

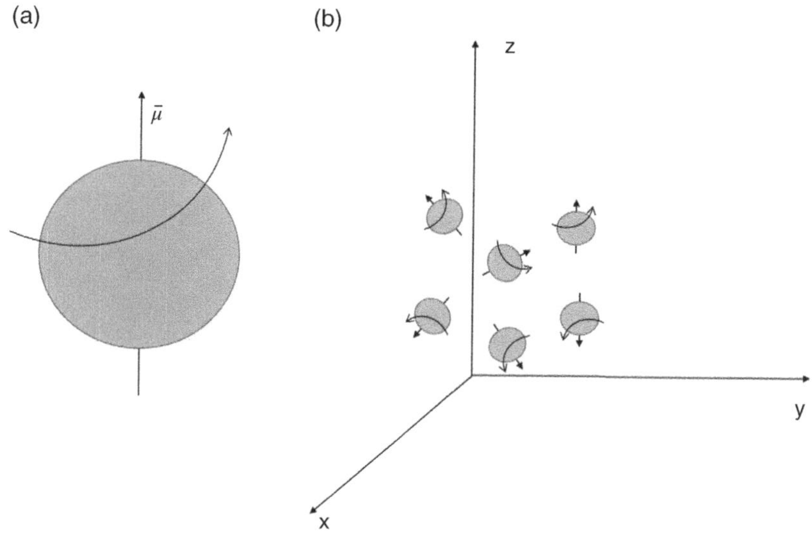

Figure 10.1 (a) The spin magnetic moment. (b) A spin ensemble.

10.2.2 Spin Precession and Larmor Frequency

When a static z-directed magnetic field ($\overline{B_0}$) is applied, the majority of the magnetic moments in a spin-system aligns with $\overline{B_0}$ and starts a precession motion around the direction of the applied magnetic field. In other words, the spin-system polarizes (Figure 10.2).

The angular velocity, ω_0, is indicated as the Larmor frequency (Figure 10.2).

The Larmor frequency and the applied static magnetic field, B_0, are related to each other by the proportionality relation:

$$\omega_0 = -\gamma B_0 \tag{10.3}$$

where B_0 and γ have been already defined. The minus sign indicates the direction of the rotation. In fact, the Larmor frequency can be either positive or negative according to the sign of the gyromagnetic ratio. As an example, $\gamma > 0$ for ^1H, while $\gamma < 0$ for the nuclide ^{15}N. The Larmor frequency can also be expressed either as an angular velocity in rad s^{-1} or as a frequency ($\nu_0 = \frac{\omega_0}{2\pi}$) in Hz.

10.2.3 Longitudinal Magnetization

The interaction between the nuclear spins and the z-oriented static magnetic field is referred to as Zeeman interaction. The torsional energy acting on the nuclear magnetic moment when B_0 is applied can be described by the scalar product:

$$E = -\bar{\mu} \cdot \overline{B_0}. \tag{10.4}$$

This magnetic energy assumes the lowest value when the magnetic moment $\bar{\mu}$ is parallel to the direction of vector $\overline{B_0}$. In the presence of B_0, the magnetic moments are no longer randomly oriented, but their orientation is in such a way that the vector sum of their contributions is a total net magnetization along the direction of B_0. This magnetization is referred to as bulk magnetization or longitudinal magnetization (M_0). To be slightly more precise, one can say that, within the observed sample, a thermal nonstatic equilibrium is generated where a slightly higher probability that a spin shows a component which is parallel to the magnetic field occurs.

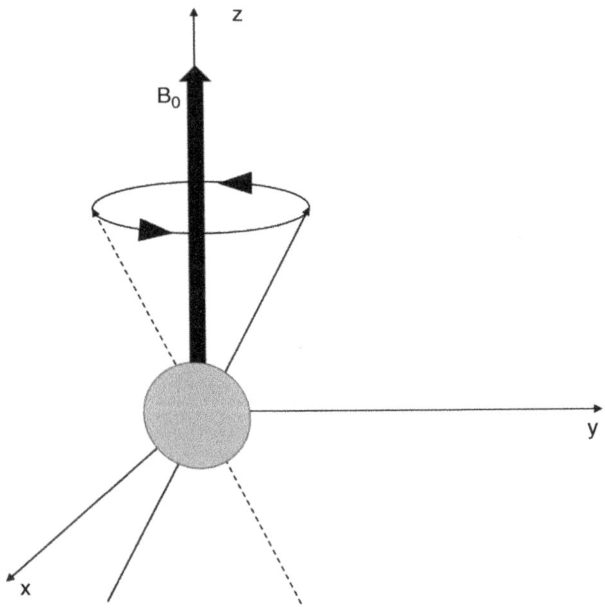

Figure 10.2 Precession motion of a magnetic moment around the magnetic field B_0 direction. ω_0 is the angular velocity.

10.2.4 Transverse Magnetization and NMR Signal

If a second magnetic field B_1 (less intense than B_0) is applied perpendicularly to the z axis, e.g. along the y axis, for a well-defined time, the result is that the M_0 vector flips by an angle θ and precesses around B_0 and B_1. In order to reduce the effect of B_0 on the motion of the bulk magnetization, a mathematical instrument referred to as the rotating frame must be introduced (see also Chapter 9).

Let us imagine that the magnetic moments, generating the magnetization vector M_0, precess at the Larmor frequency ω_0 in a reference system that is not static, but it rotates around the z axis at a frequency ω_{rot}. When $\omega_{\text{rot}} = \omega_0$, all the magnetic moments appear static, that is, the apparent Larmor frequency is zero.

The apparent Larmor frequency can be described by:

$$\Omega = \omega_0 - \omega_{\text{rot}} \tag{10.5}$$

where Ω is indicated as the offset. By combining relations (10.3) and (10.5), relation (10.6) comes out

$$\Omega = -\gamma \Delta B \tag{10.6}$$

where ΔB is the apparent magnetic field in the rotating frame which replaces B_0. If the offset is null, the apparent magnetic field is also zero.

From a practical point of view, B_1 must be not static. It must oscillate in a radio frequency (ω_{rf}) range. B_1 frequency can be modulated in such a way that it is resonant, that is, equal to or close to the Larmor frequency ω_0.

In summary, during an NMR experiment, two magnetic fields are used. B_0 is needed to generate the z-oriented bulk magnetization, while B_1 is applied perpendicularly to B_0 to tilt the magnetization. In particular, B_1 oscillates around ω_0, thereby reducing B_0 to $\Delta B = -\Omega/\gamma$. If $\omega_{rf} = \omega_{rot}$, from Eq. (10.5) it follows:

$$\Omega = \omega_0 \pm \omega_{rf} \tag{10.7}$$

therefore, the magnetization is subjected to an effective magnetic field (B_{eff}) which is the vector sum of ΔB and B_1:

$$B_{eff} = \sqrt{B_1^2 + \Delta B^2} \tag{10.8}$$

and the precession frequency of the magnetization around this field is given by:

$$\omega_{eff} = \gamma B_{eff}. \tag{10.9}$$

From Eq. (10.8), it appears that the appropriate ω_{rf} choice reduces the offset Ω around zero and, consequently, puts B_{eff} toward the xy plane. This means that M_0 is placed in the xy plane. In other words, a transverse magnetization (M_{xy}) is generated. This phenomenon is called resonance.

Once the transverse magnetization has been generated, the B_1 field is switched off. Therefore, M_0 resumes the precession in the xy plane around the direction of B_0. The evolution of the magnetization in the xy plane can be detected through a coil placed along the x axis where an oscillating electric current is induced by the magnetization motion. The induced current represents the NMR signal called free induction decay (FID, see also Chapter 9).

10.2.5 Spin Interactions

Nuclear spins interact with their environment and with each other. The nuclear spin interactions with the static magnetic field B_0 and with the oscillating RF magnetic field B_1 are called external interactions. The interactions with the spin environment and with the other spins are called internal interactions.

The relationship between the Larmor precession frequency of a nucleus, $\nu_L = \omega_0/2\pi$, its gyromagnetic ratio, and the strength of the effective magnetic field experienced by the nucleus is described in relation (10.10):

$$\nu_L = \gamma B_{eff}. \tag{10.10}$$

This relation accounts for the magnetic field B_{eff} actually experienced by the nucleus which is slightly different from B_0. This difference is due to nearby nuclei and electrons that generate secondary magnetic fields which contribute to the total magnetic field at the nucleus.

As moving electric charges, the electrons generate a magnetic field producing a diamagnetic effect (i.e. their magnetic field opposes to B_0). Therefore, the effect of the electrons is the shielding of a nucleus from the applied static magnetic field B_0. As a consequence, a given nucleus experiences a slightly different magnetic field depending on its chemical environment.

In the case of the NMR spectra of liquids, very narrow signals are observed, due to the dynamic mediation of the spin interactions caused by Brownian motions.

The fact that nuclei of the same element in a given molecule resonate in different areas of the spectrum depends, as aforementioned, on the differences in the shielding effect. These differences are caused by the electron donor or electron acceptor effect of the functional groups that surround them.

The electron donor or electron acceptor effect is proportional to B_0 intensity and the shielding magnetic field is given by relation (10.11):

$$B_s = -\sigma B_0 \tag{10.11}$$

or

$$B_0 + B_s = B_0 - \sigma B_0 = B_0(1-\sigma) \tag{10.12}$$

where σ is a proportionality factor called the shielding constant, which is characteristic of the shielding effect caused by the chemical environment of a given nucleus. Theoretically, σ is a constant, although this is not really true from an experimental point of view. In fact, the materials surrounding the sample have their own magnetic susceptibility. The latter causes distortions in the applied magnetic field B_0 in the region where the sample is located. An internal standard is used to provide a reference point for the different effective magnetic fields. This standard, in proton and carbon NMR spectroscopy experiments, is generally tetramethylsilane (TMS). The shielding constants have values of about 10^{-5} or less and the factor 10^6 transforms the σ units into parts per million (ppm). The chemical shift values are, therefore, expressed in ppm relative to the reference according to the relation (10.13):

$$\text{chemical shift (ppm)} = \frac{\nu_{\text{sample}} - \nu_{\text{ref}}}{\nu_{\text{ref}}} \cdot 10^6 \tag{10.13}$$

where ν_{sample} and ν_{ref} are the resonant frequencies of the sample and the reference, respectively. The total frequency range of the chemical shifts of a certain nucleus depends both on the applied magnetic field and on the isotope to be studied. The NMR spectra are represented with decreasing frequencies toward the right with zero coinciding with the resonance of the reference. If one nucleus is more shielded than another, its signal will be shifted to lower frequencies (or higher fields). The observation of an NMR spectrum of a specific molecule depends precisely on this chemical shielding property.

The importance of the chemical shift lies in the fact that it reveals changes in the chemical and physical surroundings of a molecule. The first advantage of the chemical shift is that the nuclei exhibit specific resonances which depend on their chemical nature. In addition to the characteristic frequencies of certain groups in IR or Raman spectroscopy, similar functional groups will have similar chemical shifts.

The spins can also interact with each other directly. If this interaction occurs through space, then it is called a direct dipole–dipole interaction. This interaction is important as it provides structural information since the intensity of the dipolar interactions depends on the internuclear distance.

Let us consider the general model of an isolated pair of spin ½ nuclei, which we will call μ_1 and μ_2, interacting through their own dipoles.

The dipole associated with μ_1 precesses around B_0 at its Larmor frequency, thus generating a static component along the direction of the field and a rotating component in the plane perpendicular to the direction of B_0. The static component of μ_1 produces a small static field at the μ_2 dipole site. The intensity of this last local magnetic field B_{loc} depends on the relative positions of the two spins and their orientations with respect to B_0. If a sample containing this isolated pair of nuclei is subjected to the static magnetic field B_0, the result will be that each nucleus will experience an effective magnetic field B_{eff}:

$$B_{\text{eff}} = B_0 \pm B_{\text{loc}} \left(\frac{\mu}{r_{ij}^3}\right)(3\cos^2\vartheta_{ij} - 1) \tag{10.14}$$

where θ_{ij} is the angle between the internuclear vector r_{ij} and the direction of B_0, while μ is the magnetic moment.

For very intense applied magnetic fields, only the parallel or the antiparellel components cause significant changes on the static magnetic field, thus causing a signal broadening.

Nuclear spins can also interact with each other indirectly through bonding electrons. This interaction is called scalar or *J*-coupling.

In the liquid state, the Brownian motions mediate the dipolar interactions between nuclei to zero. Furthermore, molecules have not fixed positions with respect to the magnetic field B_0. Therefore, the spectra of liquid samples are made of tight and well-resolved signals.

Nuclei with spin $I > ½$ reveal a nuclear electric quadrupolar moment (eQ). The interactions between such quadrupolar moment and the electric field gradient (EFG) result in the broadening of the NMR resonances. A tensor is usually used for the EFG description. This has three components (V_{xx}, V_{yy}, and V_{zz}) in its principal axis frame (PAF). The quadrupolar coupling constant (C_Q) describes the interaction intensity for a given nucleus and it is expressed (in Hz) by the following expression:

$$C_Q = \frac{eQV_{zz}}{h}. \tag{10.15}$$

An asymmetry parameter is used to express the EFG tensor deviation from the axial symmetry

$$\eta_Q = \frac{V_{xx} - V_{yy}}{V_{zz}} \tag{10.16}$$

where η_Q assumes values between 0 and 1.

The quadrupolar interaction can be large with respect to the Zeeman interaction. In such case, it can affect the Zeeman energy levels.

10.2.6 Fourier Transform NMR

The simplest NMR experiment consists in the application of a radio frequency (rf) pulse to the sample and the detection of the signal in a following period. The rf pulse is generated by a synthesizer. The latter is characterized by a precise amplitude, phase, and duration. The obtained signal is amplified and induces an oscillating magnetic field in the coil surrounding the sample. The NMR signal contains the Larmor frequencies of all spins in the sample.

As the rf pulse is switched off, the spins relax back to the equilibrium, thereby generating a signal indicated as FID. The frequency domain spectrum is obtained by applying a Fourier transform to the FID, see Figure 10.3.

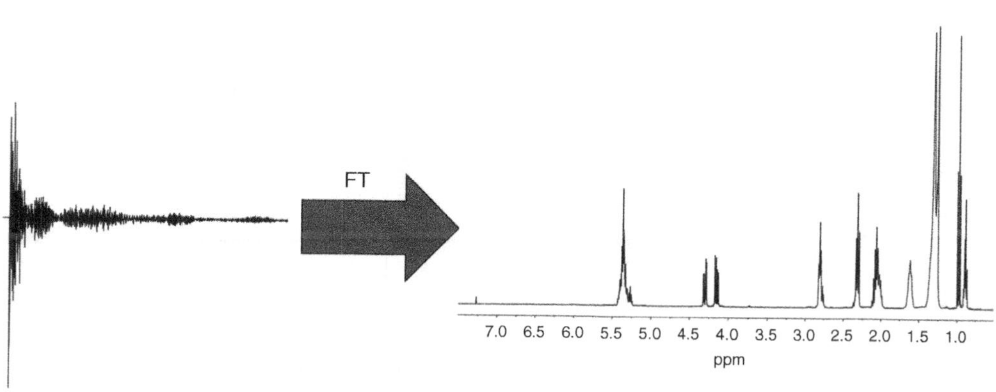

Figure 10.3 Fourier transform of a FID signal (time domain) to an NMR spectrum (frequency domain).

10.3 Liquid-State NMR

10.3.1 The NMR Spectrometer

In modern high-field NMR spectrometers, the static magnetic field $\boldsymbol{B}_0$ is generated by a superconducting magnet (Figure 10.4a).

The magnet is constituted by an external Dewar containing liquid nitrogen and an inner Dewar containing the coil (1) immersed in liquid helium to make it superconducting. A set of shim coils providing B_0 homogeneity adjustment is inside a room-temperature bore (2), the sample holder is rotated by a system located in a spinner device (3). The probehead (4) in the lower part of the magnet is connected to at least three cables that supply the ^{2}H lock (for liquid NMR experiments), ^{1}H and ^{Y}X frequencies (where ^{Y}X is any nuclide other than ^{1}H and ^{2}H). The probehead containing the rf coils transmits the rf to the sample and also detects the signal from it. A tuning of the rf coil is required at the specific frequency of the nucleus to be analyzed. Another important part of the spectrometer is the cabinet (Figure 10.4b).

Radio-frequencies and pulses generation are controlled by the computer. Typically, inside the cabinet, three rf channels, namely the observed, the lock, and the decoupling channel, are present. The frequencies are amplified and transmitted to the probehead. Being the output from the sample small, the further stage consists in the signal amplification by a radio receiver. Then, digitized signal is fed into the computer memory.

10.3.2 Sample Preparation

In NMR spectroscopy, the correct preparation of the sample plays the major role to obtain readable spectra where all the important information is not canceled or lost.

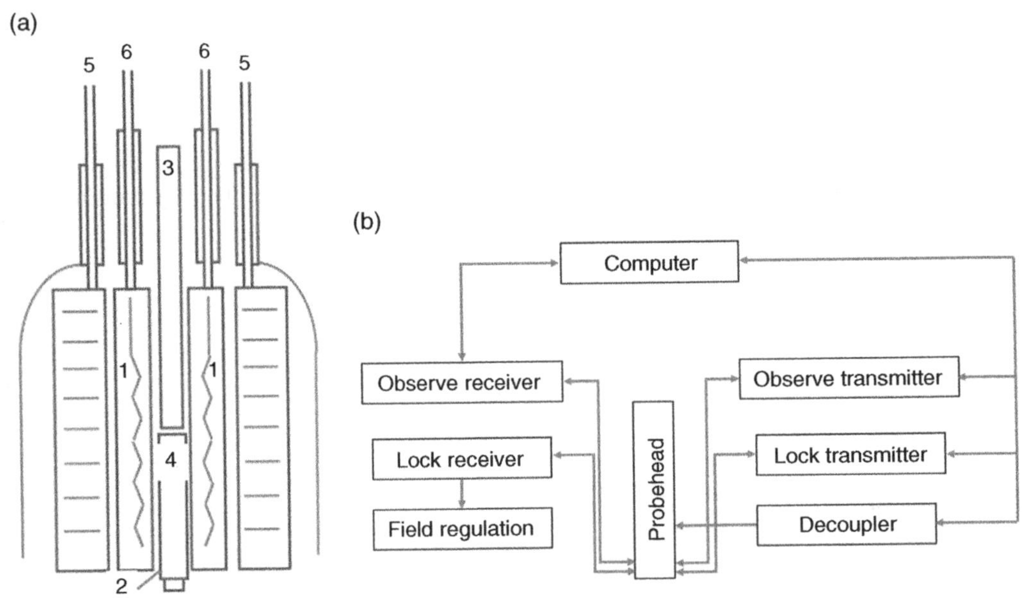

Figure 10.4 (a) Internal section of an NMR superconducting magnet: (1) Magnet coils in liquid He; (2) Shim coils system; (3) Spinner device; (4) Probehead; (5) Liquid N_2 ports; (6) Liquid He ports. (b) Diagram of an NMR cabinet.

In order to obtain a ^{1}H spectrum of an organic molecule, for instance, a quantity of sample ranging from 5 to 25 mg is required. The sample has to be dissolved in an appropriate deuterated solvent and transferred in an NMR glass tube which usually has a diameter of 5 mm. This is because the deuterium signal (lock) is used for the optimization of the magnetic field homogeneity around the sample three-dimensional region. The sample solution obtained must not contain any solid particle. A sample containing suspended particles shows a field homogeneity distortion around every single particle, thereby causing broad lines and indistinct spectra that cannot be corrected.

10.3.3 How to Set an Experiment

After the preparation of the sample discussed in the previous section, it is necessary to carry out the correct experimental setup.

The tube containing sample solution is placed inside the magnet after which the probe is tuned to the frequency of the nucleus to be observed. At this point, it is necessary to optimize the homogeneity of the magnetic field in the sample volume to be irradiated (shimming). The next operation consists in optimizing the duration of the excitation pulse in order to verify that it has the correct value of 90°. Finally, an appropriate spectral window to be irradiated is chosen (about 15 ppm for ^{1}H, and 250 ppm for ^{13}C) and the number of scans (i.e. the number of times a pulse sequence is used) to be acquired. Another important parameter to optimize the experiment is the delay time between one scan and another. It must be approximately five times longer than the T_1 of the nuclei to be observed (for the meaning of T_1, see next section).

10.3.4 Longitudinal Relaxation Time Measurement

T_1 (i.e. the longitudinal relaxation time) quantifies the time needed to recover the longitudinal component of the magnetization along the direction of the applied magnetic field (conventionally the z-axis). Its value is affected by the fluctuating local electric/magnetic fields that are generated by either unpaired electrons, or nuclear dipoles, or charged particles interacting with nuclear quadrupole moments for nuclei with spin number >½ (e.g. ^{14}N), or even anisotropy of the chemical shielding tensor, and finally fluctuating scalar coupling interactions and molecular rotations. Among the aforementioned factors, the molecular motions appear to be important in affecting the fluctuations of the local electromagnetic fields. Consequently, the evaluation either of the longitudinal relaxation time or the longitudinal relaxation rate (R_1) – which is the inverse of T_1 – can provide valuable information on molecular dynamics.

The classic rf pulse sequence for the T_1 measurement is the 180°–τ–90° inversion-recovery (IR) experiment. Three periods (preparation, evolution, and acquisition) are present in the sequence.

At the end of the preparation period (which is a delay time needed to let the magnetization completely regain its alignment along the direction of B_0), an inversion pulse is applied in order to flip the magnetization along $-z$. Then, the evolution period is set to be varied within a series of progressively increasing time intervals, τ. During each τ, the magnetization evolves toward the equilibrium condition along the direction of the applied magnetic field. Finally, at the end of each evolution time with the duration of τ, a 90° pulse is applied in order to generate the observable and acquire the FID signal.

Signal intensity depends on the τ values as described in Eq. (10.17),

$$I(\tau) = I_0\left(1 - 2e^{-\frac{\tau}{T_1}}\right) \tag{10.17}$$

where $I(\tau)$ is the signal intensity at the end of the evolution period with duration τ and I_0 is the signal intensity when $\tau = 0$.

The use of any fitting software where $I(\tau)$-vs-τ is reported allows the achievement of both I_0 and T_1.

10.3.5 Transverse Relaxation Time Measurement

The second relaxation process is due to the dephasing of the spins (also referred to as loss of coherence, Figure 10.5) in the rotating frame causing disappearance of the transverse magnetization. The time constant for this relaxation process is called T_2. T_2 relaxation is caused by three different mechanisms.

1) The inhomogeneity of the magnetic field. This alters the precessional frequency of the spins which differs from place to place within the sample, thereby allowing magnetization dephasing.
2) The dipolar coupling between spins which is referred to as spin–spin relaxation. This mechanism is sensitive to molecular motions and plays a fundamental role in the relaxation of most spin-1/2 nuclei.
3) The third mechanism is related to the anisotropic electron density distribution around the nucleus and is referred to as chemical shift anisotropy (CSA). This relaxation mechanism is important for spins having an anisotropic bonding environment and it is also sensitive to molecular motion.

The pulse sequence used for the T_2 measurement is the Carr, Purcell, Meiboom, Gill (CPMG) sequence. The CPMG sequence consists of an excitation pulse (90°), which flips the magnetization

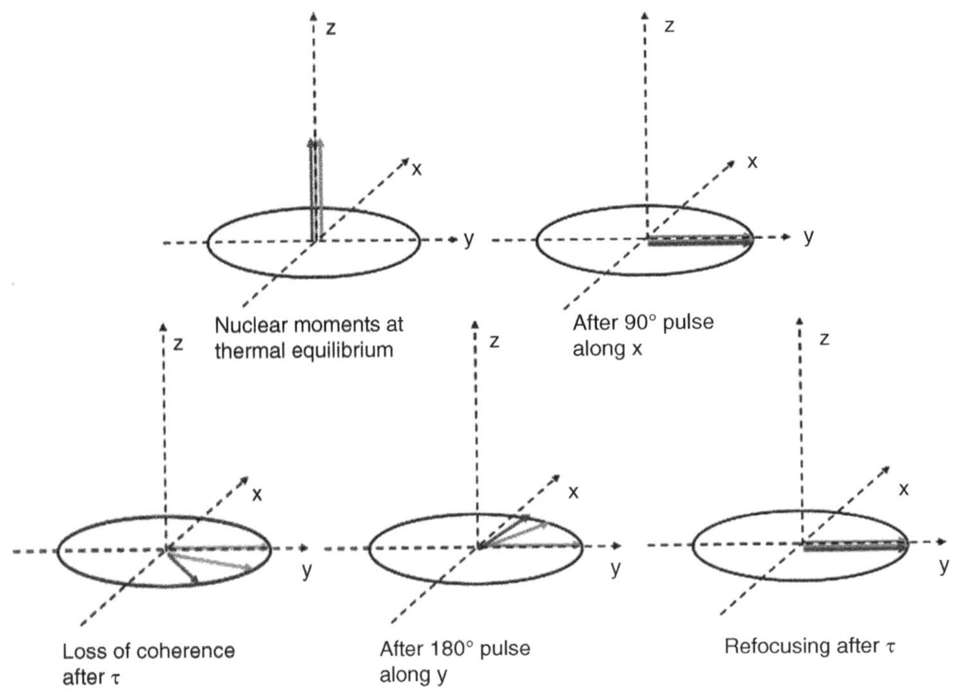

Figure 10.5 The effect of the CPMG pulse sequence on nuclear spins.

in the xy plane followed by refocusing 180° pulses. The two types of pulses are separated by a delay time (τ) of variable duration. The refocusing pulse generates an echo which reaches its maximum after 2τ. This is the point of maximal spin coherence (Figure 10.5), i.e. the point where the maximum M_{xy} intensity is achieved.

During a CPMG experiment a set of spectra are acquired by varying τ. The intensity obtained for each peak in the spectrum is plotted against τ. The obtained values are fitted by an exponential function where T_2 is the fitting parameter

$$I(\tau) = I_0 e^{-\frac{\tau}{T_2}}. \tag{10.18}$$

10.3.6 2D-Liquid-State NMR Techniques

A NMR spectrum (one-dimensional NMR or 1D NMR) is the graph where an intensity is reported against a frequency (ν). It is also possible to obtain NMR spectra where the intensity is a function of two frequencies ($\nu 1$ and $\nu 2$ or, more often, $F1$ and $F2$). In this case, a two-dimensional NMR (2D NMR) spectrum is obtained.

In a 2D spectrum, $F1$ is plotted versus $F2$. Both axes contain similar information. They refer to the chemical shift of the nuclei under investigation. When the two axes refer to the same nucleus, a homonuclear 2D NMR spectrum is acquired. Conversely, when the two axes refer to the chemical shift of different nuclei, the 2D NMR spectrum is indicated as heteronuclear [7]. As an example, the spectrum in Figure 10.6 reports a homonuclear 2D ^{1}H NMR spectrum. The figure reveals the presence of a diagonal. The latter is simply the mono-dimensional ^{1}H spectrum (e.g. the spectrum reported in Figure 10.3) as seen from above. On the left and right sides of the diagonal, stains (in the form of spots made by concentric circles) can be observed. The stains on both sides are correlated to each other and to the signals in the diagonal. The correlation can be found by drawing

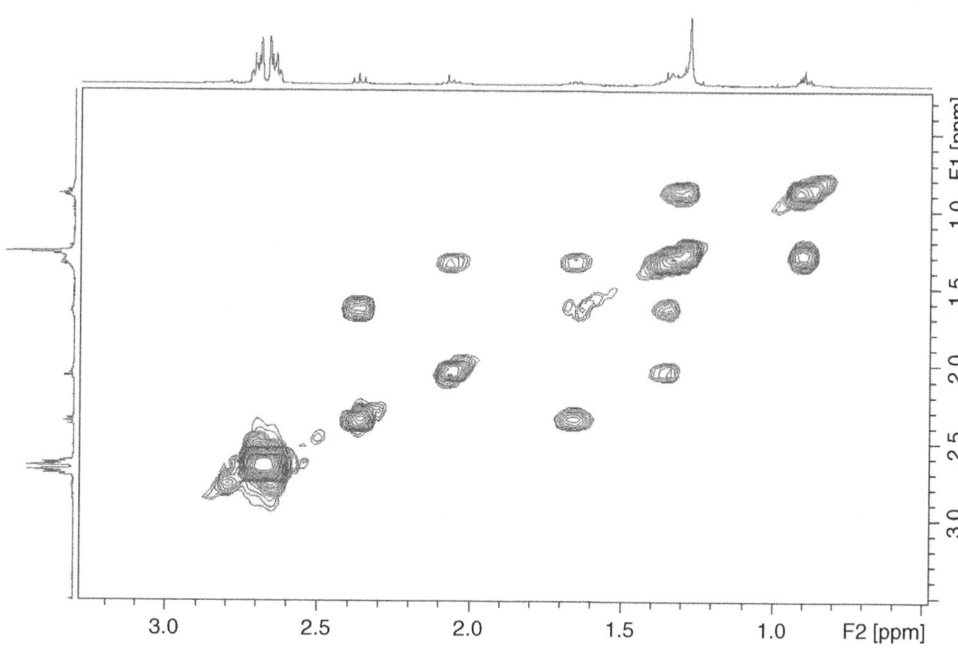

Figure 10.6 An example of H–H homonuclear correlation 2D NMR experiment (COSY).

lines which must be parallel to each axis. The aforementioned parallel lines must contain the stains. The parallel lines meet on the diagonal in the position where the resonances of different nuclei fall. When this occurs, it can be said that the two correlated nuclei "see" each other in the molecule. Following the example reported in Figure 10.6, the nuclei are correlated because of the nuclear interactions through the binding electrons. Depending on the pulse sequence applied, the nuclei can be spaced by three or more binding electron couples. Special sequences allow also to monitor the dipolar interactions among nuclei. In this latter case, the 2D NMR spectra reveal the so-called through-space nuclear interactions, thereby leading to very important information on the conformational arrangement of a given molecule.

10.3.7 Considerations on the Molecular Dynamics by NMR Spectroscopy

In order to describe a random motion of a molecule or a molecular fragment, it is necessary to analyze the temporal evolution of a function (referred to as correlation function) describing the effect of the magnetic field experienced at a certain time, t, by a nucleus belonging to the given molecule or fragment. Due to the randomness of the molecular motions, the field "felt" by the nucleus varies randomly over the time. The aforementioned correlation function, which has the form $F(t)F(t+\tau)$, is large and positive for small τ values. This is because the molecules move only lightly from the initial position, thereby altering only for a little the intensity of the applied magnetic field. As τ values increase, $F(t)F(t+\tau)$ tends to assume random values, positive or negative, depending on the stochastic memoryless characteristics of the molecular motions [3].

The correlation function can be seen as a "memory function" of the initial position of the molecules. Being real systems constituted by billions of molecules, it is necessary to account for a "macroscopic" function of correlation ($G(\tau)$), which is given by the sum of the correlation functions of the ith single spin:

$$G(\tau) = \sum_i F_i(t)F_i(t+\tau). \tag{10.19}$$

As for the microscopic function, also the macroscopic correlation function is large and positive when small values of τ are accounted for. $G(\tau)$ tends to zero for long τ values. The decay function, in the case of a single type of totally random motion, is exponential:

$$G(\tau) = \mathrm{e}^{-\frac{\tau}{\tau_c}}. \tag{10.20}$$

Here, τ_c is referred to as the correlation time and is an indication of how fast motion occurs. In other words, it is a kind of "period" of motion.

The Fourier transform of the correlation function is referred to as the spectral density function, $J(\omega)$. In the case of an exponential correlation function, $J(\omega)$ has the Lorentzian form (Bloembergen-Purcell-Pound model, BPP model) [8]:

$$J(\omega) = \frac{2\tau_c}{1+\omega^2\tau_c^2}. \tag{10.21}$$

The spectral density represents the "quantity" of the motion available at the different ω frequencies.

The correlation times τ_c depend on the temperature. In fact, as temperature increases, the thermal energy available for the motions also grows. As a consequence, the time needed for a molecule to rotate 1 rad or to move a distance corresponding to its length decreases, i.e. the correlation time becomes shorter.

The most common behavior τ_c-vs-T is described by the Arrhenius exponential law:

$$\tau_c(T) = \tau_c(\infty)e^{\frac{E_a}{RT}} \tag{10.22}$$

where E_a is the activation energy of the motion and R is the ideal gas constant (8.314462618 J (mol K)$^{-1}$).

The spectral properties that can be affected by molecular motions are essentially: signals' multiplicity, line shape, and width. The changes in the spectral properties depend on many factors. However, all of them are mainly related to the modification of the nuclear interactions generating the NMR spectrum (chemical shift, dipolar coupling, quadrupolar coupling, etc.).

It is possible to distinguish between three different types of dynamic studies: exchange processes in high-resolution spectra (liquid, solid), interference of motion with "static" line shapes (solid), and interference of motion with "instrumental" frequencies (solid).

The Lorentzian form of the spectral density function may not adequately describe molecular motions for several reasons. As an example, it does not apply when anisotropic and random motions are accounted for. Moreover, it does not apply either when a multitude of motions contribute to relaxation. In the first case, it is necessary to use more complex forms for the spectral density, which arise from theoretical considerations or, more often, from experimental equations already developed for other techniques (e.g. viscometry, dielectric relaxation, and so on). In the second case, it is necessary to write a more complex autocorrelation function or, with some approximations, perform a combination of spectral densities describing different motions. More complex spectral density models than BPP have been developed, in order to account for, e.g. of the possible distribution of correlation times or the correlation between motions that take place on different molecules or molecular fragments. These models are particularly used for motions occurring in the solid state and/or for high molecular weight molecules (e.g. polymers) [9].

On the basis of the equations obtained from the relaxation theory (Relaxation time, T vs J) and spectral density models (J vs. τ_c) and the behavior of τ_c with the temperature (e.g. Arrhenius), it is possible to predict the qualitative trend of relaxation times with the temperature. For example, if one assumes an Arrhenius-type behavior for τ_c, the condition of fast motion regime is encountered at the high temperature limit, $\omega_0\tau_c \ll 1$ and $T_1 = T_2$. By decreasing the temperature (slow motion regime), $\omega_0\tau_c > 1$ and T_1 becomes longer than T_2. In the rigid-lattice regime, i.e. $\omega_0\tau_c \gg 1$, the molecular motion is slower than T_2 and the spin–spin relaxation time becomes independent of the temperature (Figure 10.7).

10.4 Solid-State NMR

10.4.1 Powdered Samples

In a powdered solid, the anisotropic spin interactions are orientation dependent, they are different in each crystallite, and the signals are made by several frequencies, thereby resulting in a NMR spectrum with broadened signals.

The NMR signal of a powdered sample is a superposition of contributions from each crystallite, which are randomly oriented with respect to the external magnetic field. The broad powder line-shapes in the spectrum are mainly due to the anisotropic part of the chemical shift interaction. The spectral sensitivity and resolution of NMR spectra of powdered samples are generally low and this limits the application of this technique.

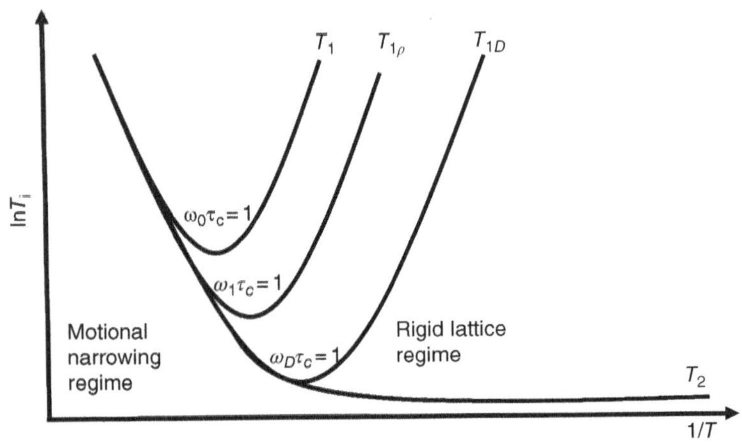

Figure 10.7 Dependence of the relaxation times on temperature. T_2 has an increasing trend with T, while, T_1, the spin lattice relaxation time in the rotating frame $T_{1\rho}$, and dipolar relaxation time T_{1D} show minima in correspondence of motion correlation times approximately equal to $1/\omega_0$ and $1/\omega_1$, respectively, which are the inverse of the reference frequencies associated to the motion under study. Solid-state (low T) T_2 reaches a limit value determined by residual interactions.

10.4.2 Cross-Polarization and Heteronuclear Decoupling

The cross-polarization (CP) technique [10] is applied to low natural abundance nuclei such as ^{13}C or ^{29}Si (1.1% for ^{13}C and 4.7% for ^{29}Si) and low gyromagnetic ratio, and thus having a very weak observable net magnetization. The NMR experiments on such nuclei are very time consuming since they require a high number of scans. The CP technique consists in transferring the polarization from abundant to dilute nuclei. The overall effect of the CP technique is the enhancing of the signal-to-noise ratio (S/N).

The polarization transfer between nuclei with different Larmor frequencies is obtained by varying the rf pulses (B_1) of the two nuclear species allowing the matching of their energy levels. This matching is called Hartman–Hahn condition [11]. As an example, in the case of proton and carbon systems, the Hartman–Hahn condition is $\gamma_C B_{1C} = \gamma_H B_{1H}$, where γ_C and γ_H are the carbon and proton gyromagnetic ratios, respectively. Since γ_H is four times γ_C, the match occurs when the strength of the applied rf field B_{1C} is four times the strength of the applied proton field B_{1H}. As a consequence, there is an enhancement of the dilute spin signal intensity by as much as the ratio of gyromagnetic ratios of the abundant and dilute spin. In the case of proton and carbon nuclei, as $\gamma_H/\gamma_C = 4$, the enhancement factor is 4. It is important to note that the Hartman–Hahn condition (i.e. the contact between the two nuclear species) must be maintained for a time (*contact time* t_C) that allows the maximization of the NMR signal. On the other hand, t_C cannot exceed a certain value, depending on the analyzed sample, above which relaxation processes occur, thus causing a loss of the signal intensity. A representation of the CP pulse sequence is reported in Figure 10.8.

In the presence of CP, as polarization is transferred from protons to carbons, the shorter T_1 relaxation time of protons dictates the recycle delay for signal averaging. As a result, the acquisition time for the carbon spectra is markedly shortened as compared to those acquired without any CP.

As aforementioned, the direct dipole–dipole interaction causes a signal broadening in solid samples. Let us now consider the case of a system consisting of the two nuclei ^{13}C and ^{1}H. ^{13}C will experience a force due to the z component of the magnetic field B_Z^1 generated by the ^{1}H nucleus.

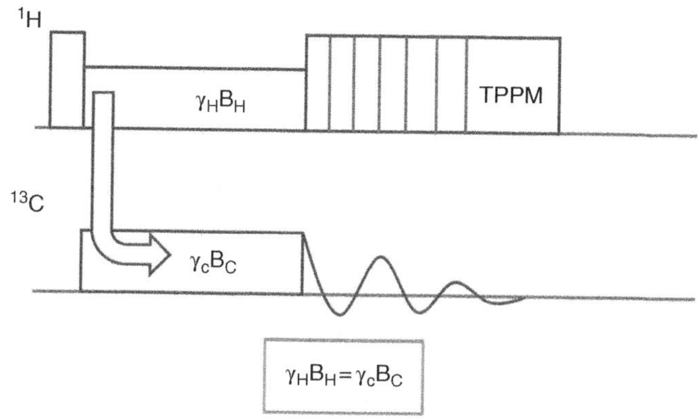

Figure 10.8 Cross-polarization pulse sequence. The equation describing the Hartman–Hahn condition is reported in the insert. Two Pulse Phase Modulation (TPPM) is a typical proton decoupling scheme.

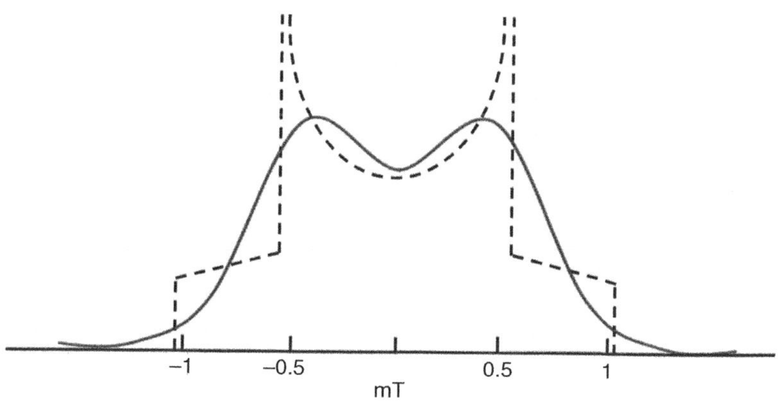

Figure 10.9 A two spin system spectrum and dipolar interactions. The dotted curve is the spectrum for two isolated nuclei. The continuous line shows the effect of the surrounding nuclei on the isolated system. The distance between the two lines of the doublet is determined by the length and orientation of the internuclear vector.

This component can be added or subtracted from B_0. Therefore, the ^{13}C NMR signal should appear as a doublet centered around its resonant frequency (Figure 10.9).

The distance between the two signals is given by

$$\Delta\nu_{CH} = \left(\frac{\gamma_c}{\pi}\right) B_z^1 \tag{10.23}$$

where

$$B_z^1 = \left(\frac{\mu_z^H}{r_{\text{CH}}^3}\right) \langle 3\cos^2\vartheta - 1\rangle \tag{10.24}$$

r_{CH} is the proton–carbon internuclear distance, B_Z^1 is the z component of the dipole of the H nucleus, μ_z^H is the component along the z axis of the dipole associated with the ^{1}H nucleus, and ϑ is the angle between the internuclear vector and the z axis. The angle brackets indicate the mean which depends on the molecular motions. For H–C dipolar couplings, values up to 40 kHz are possible.

This implies that the doublet we have discussed is only rarely observed. What is observed is generally a large signal that has the shape of a Gaussian. The dipolar spectrum is therefore not resolved due to the interactions between all spins.

The dipolar decoupling (DD) technique consists in irradiating abundant nuclei (such as protons) that have dipolar interactions with the rare nuclei (such as carbons) that we want to observe [12, 13]. This technique, through the zero mediation of heteronuclear dipolar interactions, allows the removal of the broadening of the signals and to obtain sharper signals. The DD technique causes the spins of protons to change their energy states at a rate that is higher compared to the frequency of the H–C dipolar interactions. Under these conditions, the local dipole fields on ^{13}C are canceled. To decouple the protons from the carbons in solids, the intensity of the decoupling field must be able to radiate all the transitions of the protons with a bandwidth of 40–50 kHz. In the most common experiments that involve DD, what is done is to irradiate only during acquisition. In this way, the total power consumption is lower and the sample does not overheat.

10.4.3 Magic-Angle Spinning

As we have already seen, the resonant frequency of a given nucleus is given by the shielding of the static magnetic field B_0 by the electrons surrounding it. When a magnetic field is applied to the sample, a secondary magnetic field is generated by the motions of the electrons and this secondary magnetic field partially shields the core from the applied magnetic field. In solid-state samples, this shielding due to electrons is anisotropic, i.e. it depends on the orientation of the sample with respect to the magnetic field.

In solution, random molecular motions are fast in the NMR time scale. The anisotropic part of the chemical shift is therefore averaged to zero and what is observed is only the isotropic part. In rigid solids where molecular motions are prevented, the spatial dependence of the shield determines the shape of the signals. CSA is the dependence of the chemical shift on the orientation of the bonds in the static magnetic field. The anisotropic nature of the chemical shift interaction reflects the local asymmetry of the electron cloud around a nucleus. Nuclei are shielded not only by their electrons but also by the polarized electronic clouds of neighboring chains. This causes the observed nucleus electron cloud to change by van der Waals interactions. Consequently, CSA depends on intermolecular distances as well as on intramolecular neighborhoods.

The directional nature of the chemical shift interaction can be described as follows.

The local magnetic field is given by

$$B_{\text{loc}} = \hat{\sigma} B_0 \tag{10.25}$$

where $\hat{\sigma}$ is a dimensionless second-rank tensor representing the anisotropic shift of the resonant frequency with respect to the totally unshielded core. Such a tensor can be written as the sum of an antisymmetric and a symmetric component. The antisymmetric component only affects the positions of the second-rank NMR lines and can therefore be neglected. In a system of coordinate axes, with x, y, and z relative to the structure of the molecule, the chemical shift tensor consists of nine elements of which only six are unique

$$\hat{\sigma} = \begin{bmatrix} \sigma_{xx} & \sigma_{xy} & \sigma_{xz} \\ \sigma_{yx} & \sigma_{yy} & \sigma_{yz} \\ \sigma_{zx} & \sigma_{zy} & \sigma_{zz} \end{bmatrix} \tag{10.26}$$

where $\sigma_{xy} = \sigma_{yx}$, $\sigma_{xz} = \sigma_{zx}$, and $\sigma_{yz} = \sigma_{zy}$.

At this point we can build a system of "molecular" axes x', y', z' called the PAF and in this axis system the chemical shift tensor is diagonal:

$$\hat{\sigma} = \begin{bmatrix} \sigma_{x'x'} & 0 & 0 \\ 0 & \sigma_{y'y'} & 0 \\ 0 & 0 & \sigma_{z'z'} \end{bmatrix} \tag{10.27}$$

where $\sigma_{x'x'}$, $\sigma_{y'y'}$, and $\sigma_{z'z'}$ are the principal or diagonal elements of the tensor. The six residual components can be geometrically interpreted as the axes of a 3D ellipsoid and the angles that define the orientation of the ellipsoid with respect to the laboratory reference system xyz (Figure 10.10). The eccentricity and the dimensions of the ellipsoid are determined by denoting the eigenvalues of (10.27) by σ_{11}, σ_{22}, σ_{33}.

In general, due to the symmetry properties of the Hamiltonian interactions, a representation in spherical polar coordinates is used. The resonant frequency observed for each single nucleus varies as a function of the electronic distribution (= magnetic screen) and is given by

$$\nu = \nu_{zz} = \nu_0 \left\{ \overline{\sigma} + \frac{\delta^*}{2} \left[3\cos^2\beta - 1\eta \sin^2\beta \cos(2\alpha) \right] \right\} \tag{10.28}$$

where α and β are the angles between the principal axis of the $\hat{\sigma}$ tensor and the laboratory reference system (see Figure 10.10) and

$$\overline{\sigma} = \frac{1}{3}\mathrm{tr}\hat{\sigma} = \frac{1}{3}(\sigma_{11} + \sigma_{22} + \sigma_{33}) \tag{10.29}$$

is the isotropic chemical shift,

$$\delta^* = \sigma_{33} - \overline{\sigma} = \frac{2}{3}\left[\sigma_{33} - \frac{1}{2}(\sigma_{11} + \sigma_{22})\right] \tag{10.30}$$

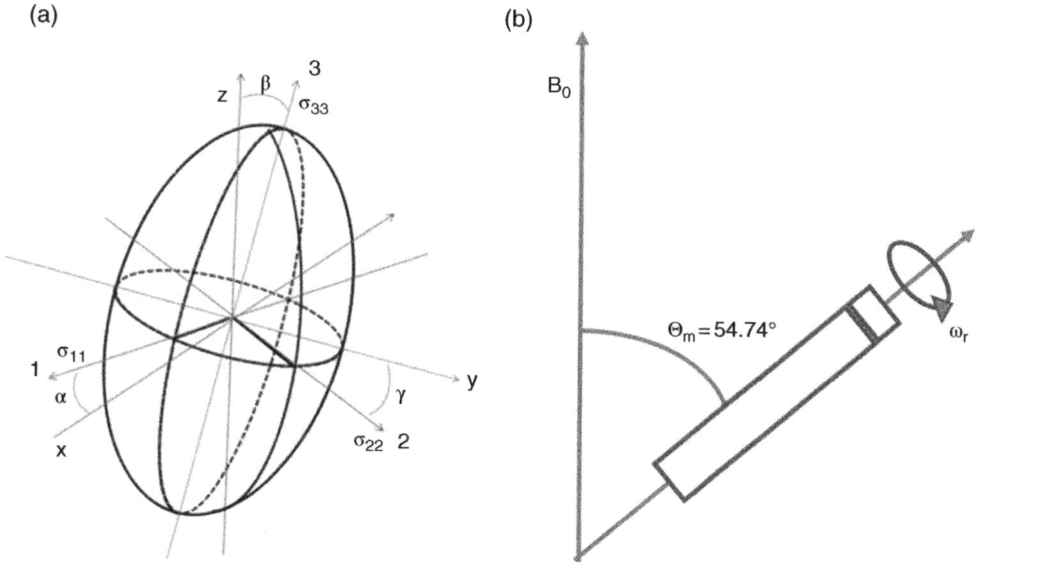

Figure 10.10 (a) Representation of the magnetic shielding tensor in terms of a rotation ellipsoid. (x, y, z): laboratory reference system; (1, 2, 3): reference system of the sample. (b) Magic angle rotation of a solid sample.

is the parameter of anisotropy, and

$$\eta = \frac{\sigma_{22} - \sigma_{11}}{\delta^*} = \frac{3(\sigma_{22} + \sigma_{11})}{2\sigma_{33} - (\sigma_{11} + \sigma_{22})} \tag{10.31}$$

is the asymmetry parameter.

In experiments on liquid samples, the Brownian motion to which the molecules are subject performs a temporal average operation of the space variables. The result is the isotropic chemical shift value.

The magic-angle spinning (MAS) technique [14–16] is potentially able to remove all magnetic interactions whose expressions contain the geometric term $3\cos^2\theta - 1$ (see Eqs. 10.14 and 10.24). These interactions include both the CSA and the dipolar interaction. The geometric term $3\cos^2\theta - 1$ takes into account the chemical shift and dipolar tensors with respect to the applied magnetic field B_0. It is possible to demonstrate that by rotating the sample to the magic angle, i.e. 54.74°, the geometric factor $3\cos^2\theta - 1$ is averaged to zero in the NMR time scale. This technique is then used to remove the effects of CSA and to facilitate the removal of the dipolar coupling. The MAS technique therefore consists in the high-speed rotation of the pulverized and packed sample in rotors at the magic angle with respect to the direction of the static magnetic field B_0, see Figure 10.10b.

However, there are limitations in the use of the MAS technique. First, the rotation speed must be greater than the interactions that must be removed. Due to the high intensity of some of these interactions, it is difficult to achieve sufficiently high rotational speeds. For example, the intensity of homonuclear dipolar interactions between protons can reach up to hundreds of kHz and generally such a speed of rotation is not practicable. Special pulse sequences are currently used to remove homonuclear dipolar interactions. For the ^{13}C nuclei, the dipolar interactions are of lower intensity due to the large distance between two nuclei of this species and because of the low isotopic abundance of ^{13}C. The MAS technique will therefore be able to remove the ^{13}C homonuclear dipolar interactions. The removal of anisotropic interactions in the NMR spectra of solids through the MAS technique allows to obtain the isotropic chemical shifts and fine structures of the spectra as in the high resolution for liquids. This method is used for solids to examine the structure in samples that are not soluble in suitable solvents or that do not melt without decomposing.

Since CSA has the same angular dependence as the dipolar interaction but is significantly lower in intensity, the MAS technique is effective for its removal. The chemical shift can be expressed in the following way: $\sigma_{zz} = \langle\sigma\rangle + \sigma_a$, where $\langle\sigma\rangle$ is the isotropic part which as seen is equal to 1/3Tr (σ) and σ_a is the anisotropic part.

For a uniaxial symmetric tensor, the anisotropic part can be written $\sigma_a = 1/3[\sigma_1 - \sigma_2(3\cos^2\theta - 1)]$ and when the sample is rotated to the magic angle ($\theta = 54.7°$) $\sigma_a = 0$.

Under these conditions, the CSA is eliminated and what remains is the isotropic average of the chemical shift.

However, there is another complication that must be taken into account for the MAS spectra. When the sample is rotated at a speed lower than CSA, rotational echoes or spinning sidebands (SSB) appear on the sides of the isotropic chemical shift peak. SSB are located at a distance from the isotropic peak equal to the rotation frequency. As the rotation speed increases, the sidebands are moved more and more toward the outside of the spectrum and become less intense. At very high speeds, the sidebands become negligible and the spectrum contains only narrow signals centered at the Larmor frequency.

The magic angle calibration is a very critical step for the correct acquisition of a solid-state NMR spectrum. If a calibration error of 0.1° is made for an anisotropy of 200 ppm (not uncommon for ^{13}C in solids), the result will be a broadening of the signal of 1.02 ppm. The MAS technique is the most effective in eliminating the effects of CSA. For aliphatic carbons, the CSA is small and low rotation speeds are sufficient. For carbons that form multiple bonds, the CSA is larger and higher rates must be used to mediate it. For aromatic carbons, the CSA is approximately 150 ppm. For a carbon Larmor frequency of 15 MHz, the CSA corresponds to 2.25 kHz and the sample must be rotated to this frequency to average the CSA. Very high rotation speeds are necessary for high applied magnetic fields and this is because the intensity of the CSA increases with increasing B_0.

10.4.4 Homonuclear Dipolar Decoupling

We have previously seen that heteronuclear DD is a technique currently used in the acquisition of spectra of rare nuclei such as ^{13}C, ^{29}Si, ^{15}N, etc. However, in the case of the acquisition of solid-state NMR spectra of nuclei such as ^{1}H or ^{19}F, their resonances are dominated by the strong homonuclear dipolar interaction that can reach values up to hundreds of kHz. The rotation to the magic angle achievable with common equipment is often not sufficient to obtain resolved spectra for such nuclei, although in some cases with MAS rotations of 70–80 kHz, it is possible to obtain good results.

For the acquisition of proton spectra and other nuclides characterized by strong homonuclear coupling, special pulse sequences known as combined rotation and multipulse sequence (CRAMPS) have been developed. Such pulse sequences use both the MAS rotation with synchronized rotations of the spins by means of rf pulses.

To this category of pulse sequences belong the frequency-switched Lee-Goldburg (FSLG) [17], the phase-modulated Lee-Goldburg (PMLG) [18] and the DUMBO (decoupling using mind-boggling optimization) [19] allow to obtain resolved spectra even at relatively low MAS spinning speed. These sequences can be windowed (i.e. they need an interruption of the RF for the acquisition of data points) or not windowed (i.e. with continuous RF irradiation) in the indirect dimension of a 2D experiment.

10.4.5 2D-Solid State NMR Techniques

In the last decades, several versatile two-dimensional solid-state NMR techniques have been developed. The correlation experiments in solid-state NMR can have multiple uses depending on the system to be investigated, the property to be studied, and the nuclei of the elements that constitute it.

As seen in the previous section, the most popular 2D pulse sequences are those designed to obtain an enhancement in the proton spectra resolution known as CRAMPS. These pulse sequences such as FSLG, PMLG, and DUMBO allow to obtain 2D spectra in which a sufficient resolved ^{1}H spectrum is present in the second dimension, thus allowing to correlate a certain X nucleus (such as ^{13}C) to the protons dipolar coupled with it. These experiments are of particular importance, for example, in supramolecular chemistry to study hydrogen bonds. As known, the anisotropic interactions are often a hindrance for the peak resolution in solid-state NMR spectra. Notwithstanding, these interactions, such as dipolar interaction, may contain important information, both structural and dynamic.

The dipolar interaction between two coupled spins is proportional to $1/r^3$ following the equation

$$D_{IS} = \left(\frac{\mu_0}{4\pi}\right)\frac{\gamma_I \gamma_S \hbar}{r_{IS}^3} \tag{10.32}$$

where D_{IS} is the dipolar coupling constant between nuclear spins I and S, r_{IS} is the internuclear distance I–S, μ_0 permeability of free space, γ_I and γ_S are the gyromagnetic ratio of spins I and S, respectively, $\hbar$ is the Planck's constant divided by 2π. Therefore, solid-state NMR experiments which provide access to this interaction to determine specific distances can be used. Heteronuclear dipolar couplings removed by MAS can be recoupled by applying pulses at proper intervals in order to counteract the effect of the sample spinning. For this purpose, a series of homonuclear and heteronuclear recoupling pulse sequences have been developed that combine a 2D approach with MAS and dipolar multiple-quantum NMR spectroscopy. In this way, the chemical shift resolution allows sites of interest to be distinguished and identified. Furthermore, since dipole–dipole couplings are sensitive to the distances between the nuclei as well as to the orientation of the internuclear vector, it is possible to obtain structural information. Information about the proximity or connectivity of different nuclei is achievable by ^{1}H double-quantum (DQ) MAS. In this kind of experiments, the observation of a DQ signal indicates the existence of a sufficient dipolar coupling between the respective nuclei. The 2D peaks are an evidence from which one can obtain information about internuclear proximities. Under MAS conditions, for the DQ coherence excitation, the use of a recoupling pulse sequence is required [20, 21]. The most used is the back-to-back (BABA) [22].

Distance measurements can also be performed by 2D heteronuclear correlation (HETCOR) spectroscopy through time-oscillatory magnetization build-up curves, rotational echo *d*ouble resonance dephasing curves or rotor-encoded SSB analysis. The combination of ^{1}H-chemical shift and H–X dipole–dipole coupling is exploited in several H–X dipolar recoupling pulse sequences, developed mainly for X = ^{13}C and ^{15}N [23, 24].

Another 2D NMR protocol to study the heteronuclear dipolar couplings is the separated local field spectroscopy (SLF) experiment which resolves for each chemically inequivalent site S in the sample its dipolar SSB manifolds that are correlated with heteronuclear dipolar couplings [25, 26].

In the case of quadrupolar interaction, signals are very distorted and often the center of gravity is changed. In order to separate the chemical shift and the quadrupolar coupling, a combination of pulse schemes (multiple quantum magic angle spinning [MQMAS]) was introduced [27]. Furthermore, in the case of supramolecular systems that were previously studied by conventional X-ray crystallography and neutron scattering, several 2D pulse sequences can be used to obtain structure and dynamic information [5]. This multiparameter, multinuclear approach represents a formidable tool for structural investigation since each NMR signal can be univocally assigned to the respective nucleus in the precise environment as a report at the molecular level.

10.4.6 Recoupling Techniques

Most of the applications of solid-state NMR in disordered solids exploit MAS to eliminate the effects of the anisotropic spin interactions (to a first-order approximation). However, anisotropic spin interactions contain information about the molecular structure, e.g. the direct dipolar coupling which depends on the internuclear distance. Therefore, to suspend the averaging effect of the MAS for a limited time interval, it is often useful to temporarily recouple certain anisotropic spin interactions by applying pulse sequences of resonant rf fields to the nuclear spins. These methods are called recoupling pulse sequences [28, 29].

The recoupling of dipolar couplings by rf pulse sequences is called dipolar recoupling. One of the possible applications of a rf pulse sequence, denoted RFDR [30, 31], recouples the dipolar couplings between spins of the same species. NMR spectra of multiply labeled compounds may contain several resonances which cannot usually be easily assigned. The spectra acquired by the RFDR pulse sequence display in both dimensions the isotropic chemical shifts of the molecular ^{13}C sites. If pairs of spins are close in space to each other, cross peaks in the 2D spectrum appear. Usually these are directly bonded ^{13}C ^{13}C pairs. This type of 2D spectrum is called a 2D chemical shift correlation spectrum and greatly assists the assignment of MAS NMR spectra.

10.4.7 Molecular Dynamics by Solid-State NMR Spectroscopy

In Section 10.3.6, correlation time, τ_c, to describe molecular motion has been defined. By studying the correlation time for a given system, one can obtain information about motional dynamics. In particular, small values of τ_c will indicate short times of changes of molecular rearrangements.

$\tau_c > 10^{-3}$ s can be studied by two-dimensional exchange methods. In such techniques, a particular nuclear spin interaction is monitored during the T_1 period. During the mixing time, molecular reorientation may occur. Finally, the change of molecular orientation/chemical caused by the new strength of the spin interaction is recorded in T_2. This correlation spectrum then correlates the strengths of the interaction during T_1 and T_2 and, from this, the angular reorientation produced by the motion can be obtained. The correlation time for the motion can be determined by recording different experiments with different mixing times.

If τ_c^{-1} is of the order of the nuclear spin interaction, the lineshape analysis can be applied. The specific nuclear spin interaction determining the powder lineshape (for a powder sample) is analyzed to study the molecular motion.

The powder lineshapes are sensitive to motions with τ_c^{-1} of the order of the width of the powder pattern, i.e. the anisotropy of the nuclear spin interaction which causes the powder lineshape. For chemical shift and dipolar interactions, these are motions with correlation times of 10^{-3} to 10^{-4} s, and smaller for quadrupolar interactions. The specific dynamic range will depend on the nucleus, its environment, and the interaction involved.

If the spin anisotropy is much greater than τ_c^{-1}, the powder pattern is the same as a normal static nucleus powder pattern. If τ_c^{-1} for the motion is much greater (approximately a factor of 50 greater) than the spin interaction anisotropy, the powder pattern lineshape motionally averaged reaches a fast motion limit; correlation time further decreases leave the lineshape unaltered.

Motions with lower correlation times (10^{-6} to 10^{-9} s) which are out of the dynamic range for lineshape analysis can be examined by relaxometric studies.

In particular, the spin-lattice relaxation times are affected by fluctuations of the spin interaction caused by molecular motions [32]. Thus, if the relaxation process is affected by a given nuclear spin interaction, the spin-lattice relaxation times can be calculated for different motions and compared with experimentally derived values to understand the motional behavior of the system. Other relaxation processes can also be used to study molecular motion.

10.5 Nonconventional NMR Techniques

The locution "nonconventional NMR techniques" refers to non-spectroscopic NMR approaches. The latter are mainly used to unveil molecular dynamics in complex systems when spectroscopy cannot provide suitable information [33]. As an example, in soil science, water dynamics is quite important to understand how to improve soil fertility. In this case, NMR spectroscopy cannot be helpful because water and, hence, nutrient dynamics encompass a wide variety of molecular motions which are not observable at the typical proton Larmor frequencies used in spectroscopy [34]. Another example is the monitoring of the variations occurring during food aging [35], conservation [36], and transformation [37] which are strictly related to the way how water molecules behave in the food matrix. Therefore, the following sections describe the most used nonconventional NMR techniques applied in fields which are traditionally far away from the typical NMR courses taken by the students of scientific faculties such as chemistry and physics.

10.5.1 Time Domain NMR

The mathematical models described above and used to achieve T_1 and T_2 values can be considered as a discrete representation of the longitudinal and transversal relaxation time distribution of a molecular system. They provide valuable information only when the relaxation times in a complex system are quite different between each other.

When the various components of the molecular dynamics in a multiphase frame are described by relaxation rates with values awfully close to each other, their T_1 or T_2 distributions can be more suitably obtained by applying an inverse Laplace transformation. It has the form reported in Eq. (10.33) when T_2 distribution must be accounted for, and the form in (10.34) when T_1 distribution must be achieved:

$$I(\tau) = \int_{T_2^{\min}}^{T_2^{\max}} D(T_2)\mathrm{e}^{-\left(\frac{\tau}{T_2}\right)} d(T_2) + \sigma \tag{10.33}$$

$$I(\tau) = \int_{T_1^{\min}}^{T_1^{\max}} D(T_1)\left(1 - 2\mathrm{e}^{-\left(\frac{\tau}{T_1}\right)}\right) d(T_1) + \sigma. \tag{10.34}$$

In the equations above, $I(\tau)$ is the signal intensity at the selected τ, the integral extremes are the suitable limits within which all the relaxation time values range; $D(T_2)$ and $D(T_1)$ are the relevant distribution functions that must be determined by solving either Eq. (10.33) or (10.34); σ is a parameter accounting for a suitable unknown noise component. The most likely distribution of relaxation time values may be obtained when some constraints, such as variance of the experimental data or smoothness of the solution, are kept into account.

Two algorithms have been developed to switch from $I(\tau)$ to $D(T_2)$ and $D(T_1)$. These are the continuous distribution, also referred to as CONTIN [38, 39], and the uniform penalty regularization, also referred to as UPEN [40–43]. The two algorithms differ in the smoothing procedure used. Nevertheless, it is worth noting that the two algorithms provide similar relaxation time distributions, also referred to as relaxograms, regardless of the procedure used to obtain the most probable distribution of relaxation times.

Figure 10.11 shows the T_1 distribution (upper part of the figure) of an extra virgin olive oil (EVOO) as obtained by applying the UPEN algorithm to the FID acquired at the proton Larmor frequency of 20 MHz.

The relaxogram consists of a broad band centered at 104 ms. This band can be deconvoluted (lower part of Figure 10.11) in order to reveal the number of relaxometric components which give rise to the overall relaxogram. Deconvolution reveals that the overall relaxogram is made by a slow (150 ms) and a fast (79 ms) relaxing component. According to the inverse micelle organization of the EVOO components [44], the former can be assigned to the apolar fatty acid tails into a kind of mobile hydrophobic sea, while the latter can be due to polar heads placed in the inner part of the inverse-micelle-like arrangement.

10.5.2 Fast Field Cycling NMR Relaxometry

Since the late seventies of the twentieth century [45], fast field cycling (FFC) NMR relaxometry has emerged as a useful technique to monitor molecular dynamics of systems which cannot be revealed

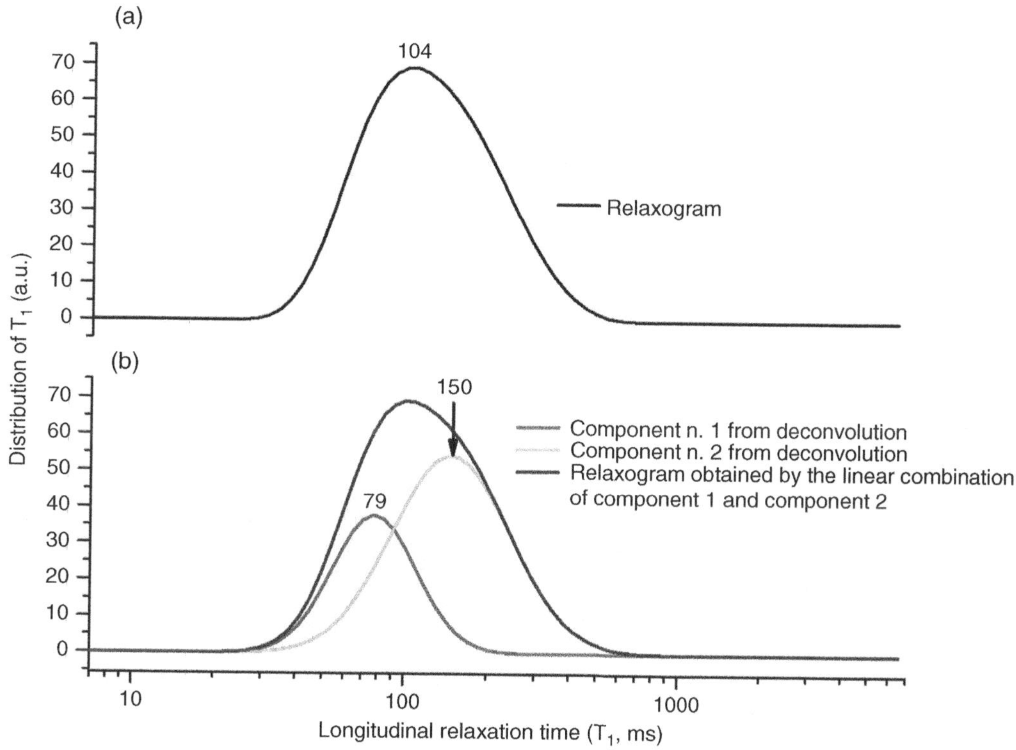

Figure 10.11 (a) Typical relaxogram of an extra virgin olive oil (black curve). (b) Combination (more intense curve) of the Gaussian curves (less intense ones) used to deconvolute the relaxogram on the top side of the figure.

by the more common NMR spectroscopy. In fact, the technique allows the evaluation of the molecular motions included in the time scale ranging between 10^{-8} and 10^{-3} s [34].

Figure 10.12a reports the time intervals for the molecular motions investigated when the proton Larmor frequencies on the top side of the scale are applied. In particular, molecular tumbling occurs in the time scale $<10^{-7}$ s, internal motions occur on a time scale comprised between $<10^{-9}$ s and slightly below 10^{-5} s, chemical exchanges take place in the time interval $\approx 10^{-5}$–10^{-3} s, and diffusion occurs on a time scale $>10^{-3}$ s. Spanning five orders of magnitude on the time scale reported in Figure 10.12a, FFC NMR relaxometry can monitor a wide set of molecular motions going from slow tumbling to slow diffusion. When compared to the other NMR techniques, the advantage of FFC NMR relaxometry is its larger sensitivity for the aforementioned molecular motions as a result of the special design of the hardware system [46].

The basic FFC NMR relaxometry sequence is given in Figure 10.12b. In this figure, the preparation time corresponds to the application (for a fixed period of time also referred to as polarization time, T_{POL}) of a polarization field (B_{POL}). Its value can be either non-null or null. In the first case, a pre-polarized (PP) sequence is achieved, while in the second case, a non-polarized (NP) sequence is retrieved.

The pre-polarization is needed to generate the magnetization that evolves to reach a new equilibrium condition (evolution period) under the action of a relaxation field (B_{RLX}) applied for a variable period τ. The acquisition time starts at the end of the evolution: the magnetic field intensity is

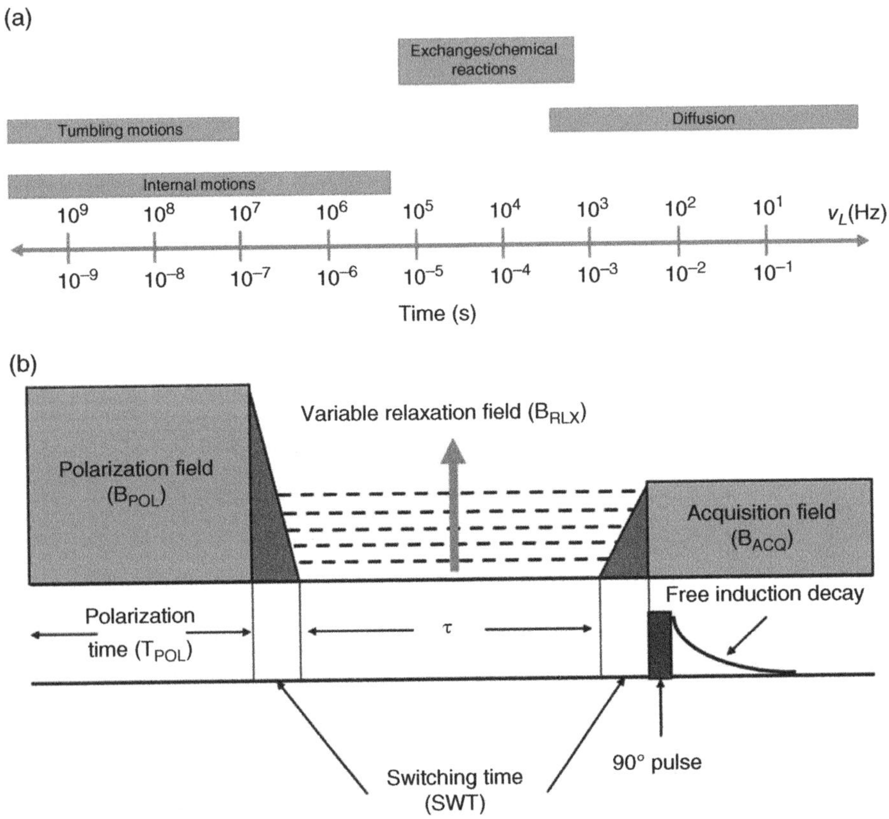

Figure 10.12 (a) Time scale of the molecular motions that can be investigated by different NMR techniques. The top side of the scale indicates the strength of the applied magnetic field in terms of proton Larmor frequency. (b) Typical fast field sequence applied to obtain the T_1 values in relaxometry experiments.

switched to a new value (indicated as B_{ACQ}), while a 90° pulse is applied to generate the observable magnetization, and the FID is finally acquired.

The use of the PP sequence is recommended whenever the intensity of the relaxation field is low; therefore, enhancement in sensitivity is needed. The crossover field between the PP and the NP sequences might be fixed empirically at $\nu_{RLX} = \frac{\nu_{POL}}{2}$, where ν_{RLX} and ν_{POL} are the B_{RLX} and the B_{POL} proton Larmor frequencies expressed in MHz, respectively.

In Figure 10.12b, a switching time (SWT) is also indicated, i.e. the time needed to switch between different magnetic field intensity values. SWT is usually set at ca. 3 ms. Shorter SWT values can be only applied providing that peculiar electronic precautions are adopted [47].

The shape of the curves acquired by applying either the NP or the PP sequence is given in Figure 10.13.

Different approaches can be applied to interpolate the curves in Figure 10.13. As an example, Eqs. (10.35) and (10.36) can be used when the number of different relaxing components in a mixture is known. Equation (10.35) is applied for the PP sequence, while Eq. (10.36) is used when the NP sequence is applied

$$I(\tau) = a + \sum_{i=1}^{N} b_i e^{-\frac{\tau}{T_{1i}}} \tag{10.35}$$

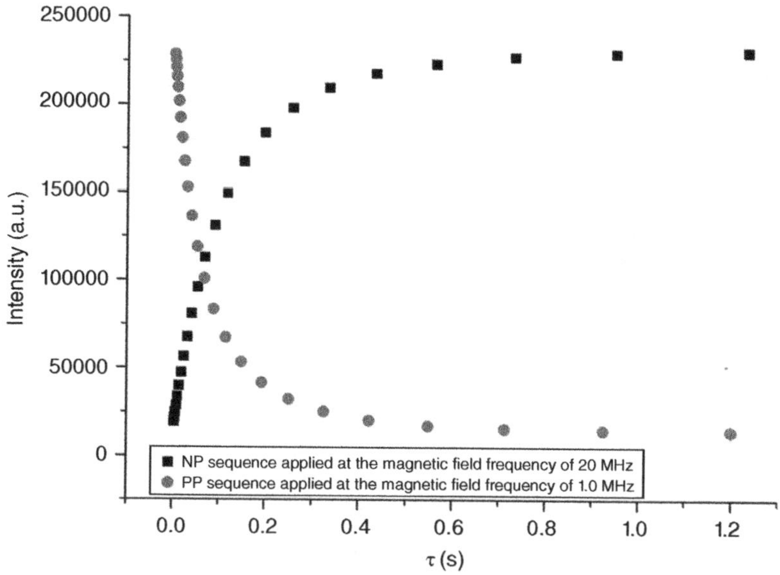

Figure 10.13 Decay (circles) and recovery (squares) curves obtained by applying the pre-polarized (PP) and non-polarized (NP) sequences, respectively.

$$I(\tau) = a + \sum_{i=1}^{N} b_i\left(1 - e^{-\frac{\tau}{T_{1i}}}\right). \tag{10.36}$$

In both equations, $I(\tau)$ is the signal intensity at the selected τ value, a is the offset, b_i is the magnetization intensity at the Boltzmann equilibrium of the ith component of the molecular motion at each fixed B_{RLX} intensity, and T_{1i} is the ith relaxation time.

When assumptions on the number of exponential components to be used for data fitting cannot be done, Eqs. (10.37) and (10.38) for the PP and NP sequences, respectively, must be applied

$$I(\tau) = a + b \times e^{-\left(\frac{\tau}{T_1}\right)^k} \tag{10.37}$$

$$I(\tau) = a + b \times \left(1 - e^{-\left(\frac{\tau}{T_1}\right)^k}\right). \tag{10.38}$$

In Eqs. (10.37) and (10.38), a and b have the same meaning as in Eqs. (10.35) and (10.36), and k is a heterogeneity parameter that is related to the stretching of the decay/recovery processes. The use of Eqs. (10.37) and (10.38) offers the advantage to deal with a large variety of different behaviors within a single model, thereby avoiding arbitrary assumptions about the number of exponentials to be applied in modeling the relaxometric results [34].

The $R_1 = 1/T_1$ values obtained via Eqs. (10.35)–(10.38) can be plotted versus the proton Larmor frequency in order to obtain a nuclear magnetic resonance dispersion (NMRD) profile. Its typical shape is a sigmoidal curve (Figure 10.14a) which is mathematically described by Eq. (10.39) [48]:

$$R_1 = \frac{2\tau_c}{1 + (\omega_L \tau_c)^2}. \tag{10.39}$$

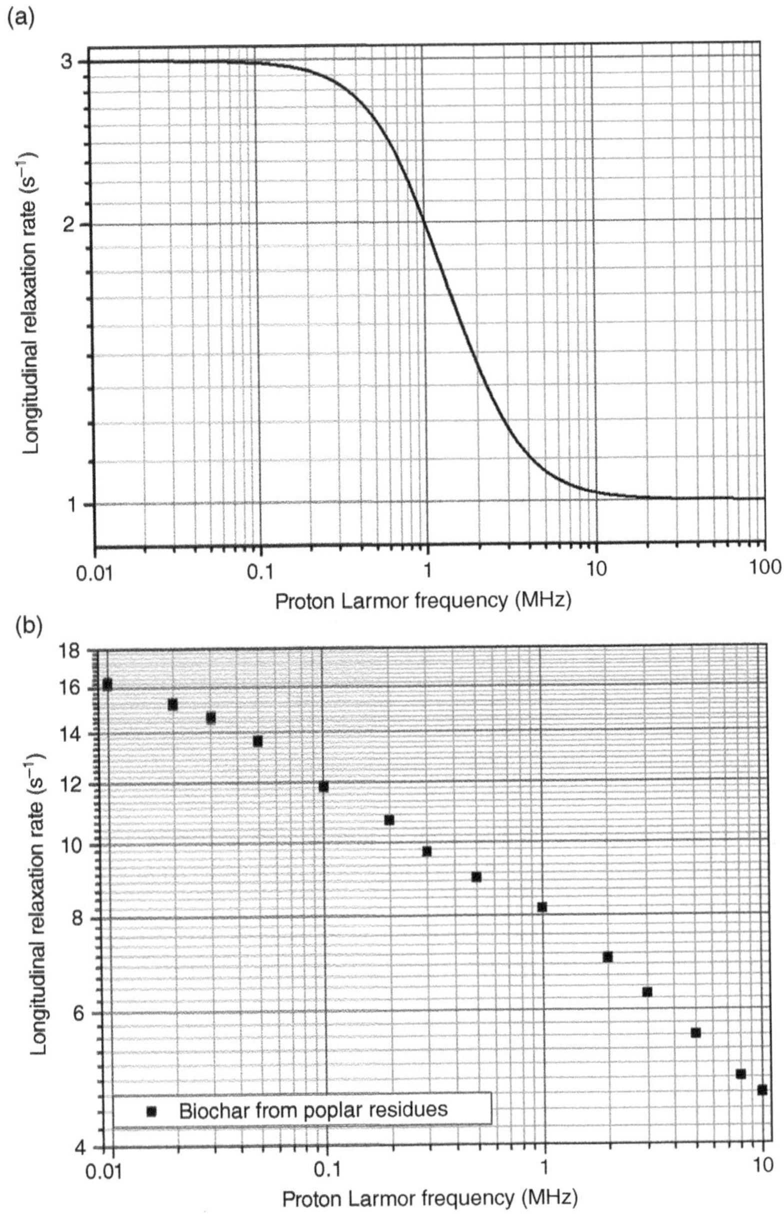

Figure 10.14 (a) Typical sigmoidal shape of a nuclear magnetic resonance dispersion (NMRD) profile. (b) Typical stretched NMRD profile acquired for a waterlogged biochar obtained by pyrolysis of poplar residues.

In Eq (10.39), R_1 is the longitudinal relaxation rate which represents the distribution of the motion frequencies in a molecular system, ω_L is the B_{RLX} intensity given as proton Larmor frequency, and τ_c is the relevant correlation time. The latter describes the random molecular motions of molecular systems either in solution or in porous media. In particular, the larger the τ_c value, the slower the molecular motions are, thereby accounting for the existence of restraints in the motional degrees of freedom.

Many models have been developed when systems showing different molecular characteristics are analyzed. As an example, Luchinat and Parigi [49] reported that protein dilute aqueous solutions can be investigated by the BPP model given in Eq. (10.40), which was developed for aqueous systems where bulk water is dominant [8]

$$R_1 = \alpha + \beta\left[\frac{0.2\tau_c}{1 + (\omega_L\tau_C)^2} + \frac{0.8\tau_c}{1 + (4\omega_L\tau_C)^2}\right]. \tag{10.40}$$

In Eq. (10.40), all the terms in the squared brackets have the same meaning as in Eq. (10.39), while α is the high-field relaxation rate and β is a parameter that relates to the dipolar interactions. The latter contains the Planck constant, the proton quantum-spin number, the gyromagnetic ratio, and the electron–nuclear hyperfine coupling constant (which in turn accounts for the interactions between resonant protons and unpaired electrons). The higher the β value, the stronger are the dipolar interactions responsible for the relaxation. Therefore, protein dynamics can be indirectly revealed by using water molecules as target for the relaxometric investigations.

It is well recognized that neither Eq. (10.39) nor (10.40) satisfactorily fit the experimental data points when a stretched dispersion is retrieved. The stretching depends on how much complex the re-orientational dynamics within the molecular systems is, as well as on the heterogeneous distribution of intermolecular dipole couplings and proton exchange rates [33]. An example of stretched dispersion is given in Figure 10.14b, where the NMRD profile of a water-saturated biochar from poplar residues is reported.

When a NMRD profile is stretched, the free model analysis proposed by Halle and coworkers can be applied [50]. It has the form:

$$R_1 = \sum_{i=1}^{N}\frac{C_i\tau_C^i}{1 + (\omega_L\tau_C^i)^2} \tag{10.41}$$

where the i index refers to the different components of the motion, τ_C^i is the correlation time of the ith component, and C_i is a fitting parameter. The sum of the C_i values constitutes the mean square fluctuation bearing the information on the equilibrium structure of the system and is not dependent on its dynamics.

For the sake of completeness, it must be noticed that the free model analysis is only a convenient mathematical approach to fit the experimental data for stretched NMRD profiles. The set of $\{C_i, \tau_C^i\}$ parameters from equation have no physical meaning, unless independent information suggests that the investigated system can be modeled by a fixed number of Lorentzians. In the latter case, a direct physical interpretation of the parameters can be attempted. According to Halle et al. [50], the $\{C_i, \tau_C^i\}$ ensemble can be applied to calculate a weight-averaged correlation time as in the following:

$$\langle\tau_C\rangle = \frac{\sum_{i=1}^{N} C_i\tau_C^i}{\sum_{i=1}^{N} C_i}. \tag{10.42}$$

The combination between the free model analysis given in Eq. (10.41) and the BPP model reported in Eq. (10.40) provides the BPP free model analysis [49]

$$R_1 = \alpha + \beta\left[0.2\sum_{i=1}^{N}\frac{C_i\tau_C^i}{1 + (\omega_L\tau_C^i)^2} + 0.8\sum_{i=1}^{N}\frac{C_i\tau_C^i}{1 + (\omega_L\tau_C^i)^2}\right]. \tag{10.43}$$

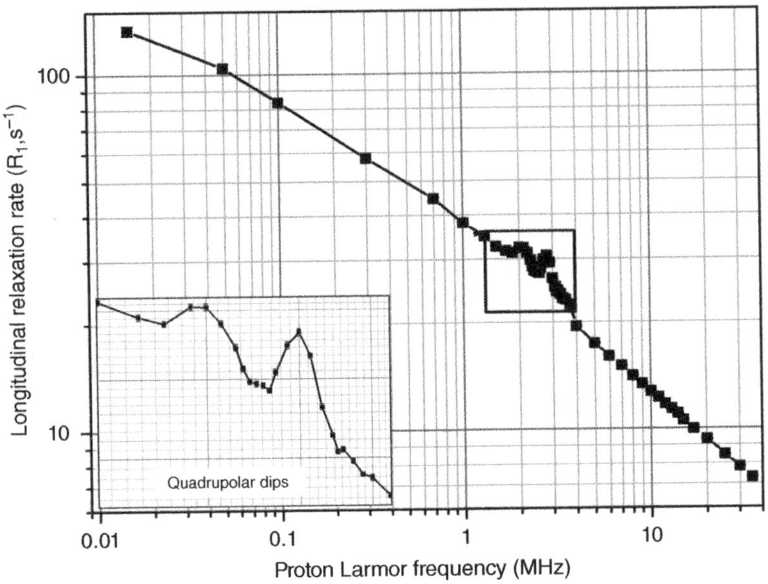

Figure 10.15 Relaxogram of Parmigiano Reggiano cheese. The square magnifies the quadrupolar dips discussed in the text.

All the parameters in Eq. (10.43) have been already defined. The set of $\{C_i, \tau_C^i\}$ parameters from the BPP free model analysis are used to obtain the weight-average correlation time given in relationship (10.42).

When dealing with rigid N-containing systems such as synthetic nanosponges [34], hyaluronan [51], and solid-state proteins [52], the NMRD profiles show quadrupolar dips (Figure 10.15).

The quadrupolar dips are due to the alteration of the longitudinal relaxation rate of proton nuclei in the proximity of quadrupolar centers, such as ^{14}N nuclei. When the proton Larmor frequency resembles the resonance frequency of the quadrupolar center, the excited protons exchange energy with the paramagnetic nuclei in the lattice at a relaxation rate faster than that occurring between ^{1}H and ^{1}H nuclei. The exchange preferentially arises when (both in dry and hydrated systems) molecular motions are restricted [48, 51, 54]. As a result, quadrupolar dips appear in the NMRD profiles of the investigated samples at the typical Larmor frequencies of the paramagnetic nuclei (Figure 10.15) [53].

A free model-based analysis accounting for quadrupolar dips is given by [52]:

$$\begin{aligned} R_1 = R_0 &+ \sum_{i=1}^{N} C_i^{\mathrm{HH}}\left(\frac{\tau_i}{1+(\omega_L\tau_i)^2}+\frac{4\tau_i}{1+(\omega_L\tau_i)^2}\right) \\ &+ C^{\mathrm{NH}}\left[\left(\frac{1}{3}+\sin^2\Theta\cos^2\phi\right)\left(\frac{\tau_Q}{1+(\omega_L-\omega_m)^2\tau_Q^2}+\frac{\tau_Q}{1+(\omega_L+\omega_m)^2\tau_Q^2}\right)\right. \\ &+\left(\frac{1}{3}+\sin^2\Theta\cos^2\phi\right)\left(\frac{\tau_Q}{1+(\omega_L-\omega_p)^2\tau_Q^2}+\frac{\tau_Q}{1+(\omega_L+\omega_p)^2\tau_Q^2}\right) \\ &+\left.\left(\frac{1}{3}+\cos^2\phi\right)\left(\frac{\tau_Q}{1+(\omega_L-(\omega_p-\omega_m))^2\tau_Q^2}+\frac{\tau_Q}{1+(\omega_L+(\omega_p-\omega_m))^2\tau_Q^2}\right)\right]. \end{aligned} \tag{10.44}$$

In this model, R_0 is a field-independent longitudinal relaxation rate that can be assigned to molecular components with the fastest motions in the complex mixture. The parameter C_i^{HH} is $\propto r_i^{-6}$ being r_i the ith H–H interspin distance of the ith molecular component of the motion described by the ith correlation time τ_i. C^{NH} is $\propto r_{NH}^{-6}$, where r_{NH} is the N–H distance, τ_Q is the correlation time measuring the time needed for the fluctuations of the N–H bond, ω_p, ω_m, and $(\omega_p - \omega_m)$ are the Larmor frequency values where the quadrupolar dips appear; Θ and Φ are the angles describing "the orientation of the ^{1}H–^{14}N dipole–dipole axis with respect to the principal axis system of the EFG at the position of ^{14}N."

Many other mathematical models can be used according to the chemical nature of the complex mixture which is investigated. The readers are addressed to more specific literature for more detailed features [33, 34, 48, 55].

10.5.3 Earth's Magnetic Field NMR

All the NMR techniques reported above need a very well-equipped lab. In fact, all of them need special technical measures such as the use of liquid gases in the high-field NMR spectroscopy or a connection to water tap to ensure refrigeration of the NMR-coil in the low-field FFC NMR technique. The aforementioned requirements make difficult the application of NMR in open fields which is, for example, needed in the investigation of rock porosity for petroleum industry [56], in the analysis of untraceable samples such as in cultural heritage [57], in the evaluation of plant metabolism [58], or, even, for food quality investigations [59]. Although different precautions have been developed for the aforementioned applications, the problem of the stability and the homogeneity of the magnetic field in the open fields still remains [60]. To this aim, Ganssle and coworkers realized an apparatus to use the most stable magnetic field worldwide [60]: the Earth's magnetic field. In particular, the authors were able to investigate the relaxometric properties (i.e. T_1 and T_2 evaluation) for some common solvents such as water, short chain alcohols, and hydrocarbons up to 10 carbon atoms. The technique for which special pulse sequences were designed, while the T_1 and T_2 calculations were done according to the equations reported above, resulted very promising for the development of compact, inexpensive devices which can take advantage of the Earth's highly homogeneous ambient magnetic field.

10.6 Case Studies

10.6.1 Polymers and Polymer-Based Composites

The study of polymer systems by NMR is often a challenging task, due to the presence of several dynamic processes on different timescales and structural disorder as well as dynamic heterogeneity.

Low values of relaxation times (T_2, $T_{1\rho}$) or low polarization transfer efficiencies, e.g. in CP, for experiments in solid-state NMR, due to molecular reorientations may lead to signal loss and inefficient detection. For this reason, the NMR experiment must be chosen in a way that it is representative of the entire sample.

The NMR experiments in the field of polymer-based materials can be aimed merely to confirm the chemical composition or to study specific structural or dynamic properties.

In liquid state, as already said, fast segmental motion and overall tumbling allow the acquisition of high-resolution well-resolved spectra by the most common 1D and 2D pulse sequences. In this case, the chains' conformation together with the isomerism and, in the case of copolymers, the monomer distribution can be studied by the chemical shift analysis. Furthermore, nuclear Overhauser effect (NOE) and J coupling-based experiments are used for structural assignments [61–63].

In the case of solid-state NMR, several techniques were exploited to study the molecular structure, conformation, and dynamics of polymer chains in various polymer systems. Solid-state ^{13}C NMR is applied for elucidation of cross-linking structures in thermoset systems, elastomers, and bonding structures in adhesives, for the study of polymer-based composites, conducting polymers, and so on.

A reactive suspension method and conventional mechanical mixing-prepared epoxy resin/TiO_2 nanocomposites are analyzed by ^{13}C CP-MAS NMR, relaxation times ($T_{1\rho}C$, $T_{1\rho}H$), and variable contact time (VCT) experiments in order to have a deeper insight into the interactions at the polymer/particle interface and the nanoparticles dispersion [64].

Specifically, solid-state NMR experiments showed that the nanoparticles, synthesized by reactive suspension method, can efficiently link the matrix by hydrogen bonds forming a hybrid organic–inorganic 3D network. Such evidences strongly confirm that the nanoparticles synthesized by reactive suspension method act not only as rigid filler, but also as actual cross-linking points, dramatically improving the mechanical properties of the polymeric matrix.

In another work, Ce:YAG-poly(methyl methacrylate) (PMMA) composites were prepared by using a melt compounding method, adding amounts of Ce:YAG in the range 0.1–5 wt.% [65]. The obtained materials were characterized by solid-state NMR and results have shown that the interaction between the two components is responsible for the observed emission shift of 15 nm.

No modification in the chemical shift and in the signal shape was observed in the presence of Ce:YAG in the ^{13}C solid-state spectra for the polymeric matrix. This confirmed that there were no chemical interactions and that only physical interactions occurred between the polymer and the Ce:YAG particles.

In order to investigate possible molecular interactions between the components, and to detect dynamic changes in the polymer induced by the presence of Ce:YAG, the proton and carbon spin lattice relaxation times in the rotating frame, $T_{1\rho}H$ and $T_{1\rho}C$, were determined through solid-state NMR measurements. Both the $T_{1\rho}H$ and $T_{1\rho}C$ values give information about molecular motions in the kHz range. These motions reflect the dynamic behavior of a polymeric chain in a range of a few nanometers. However, for natural-abundance experiments, the spin diffusion is less effective in the $T_{1\rho}C$ averaging, and this parameter is therefore a good probe for local mobility. For both composites, these values were strongly affected by the presence of the Ce:YAG particles. This indicates that a minor contribution to the $T_{1\rho}$ relaxation arises from each molecular group that modulates the ^{1}H–^{13}C dipolar coupling. These variations can account for specific interactions between the polymer and the surfaces of the Ce:YAG particles.

10.6.2 Mesoporous Materials

Solid-state NMR spectroscopy is widely used for the characterization of mesoporous materials. Structural information together with evidences on concentration, location of different functional groups bonded to these materials can be obtained by ^{1}H, ^{13}C, and ^{29}Si MAS NMR.

Moreover, two-dimensional ^{1}H–^{13}C NMR experiments can be performed to study in a deeper way the location of the functionalizing groups. The understanding of such functionalized systems cannot be achieved without a knowledge of pure materials.

Several silicon sites, with different Si–O–Si bond angles are present in the MCM-type materials, and their properties such as hydrogen bonding can be different depending on the synthesis strategies and sample treatment [66, 67].

The advanced solid-state NMR techniques, such as ultrafast MAS, are very useful for this purpose. At MAS rates approaching 45 kHz, one can obtain high-resolution ^{1}H spectra of MCM-41 surfaces in a routine fashion without CRAMPS sequences for homonuclear decoupling. This allows to perform ^{1}H–^{1}H homonuclear correlation methods, such as 2D exchange, RFDR, and DQ ^{1}H–^{1}H MAS NMR, as well as HETCOR NMR [66]. In the first case study [67], ^{29}Si CPMAS NMR was used for the characterization of MCM 41 mesoporous silica obtained with micelles as template and functionalized with neodymium ions by the incipient wetness method.

In spectra of the pure MCM 41 and of Nd-MCM 41 materials, three signals are present. The first peak, Q2, centered at around −90 ppm, is due to geminal silanols, the second peak, Q3, at around −100 ppm, is due to the silicon atoms with one hydroxyl group and the third, Q4, at around −109 ppm, is due to the silicon atoms without hydroxyl groups constituting the bulk. Significant changes in the spectra are observed as the neodymium content increases. A comparison between the relative contribution of the three different silicon atoms was performed assuming that Q2, Q3, and Q4 chemical shift positions do not change and that the CP is the same for all compositions. The deconvolution was performed to obtain the area of the three peaks. It is worth noticing that Q2 area does not change significantly, while Q4 and Q3 area increases and decreases, respectively. The MCM-41 structure modification is probably due to the conversion of the $(Si–O)_3–Si–OH$ groups in $(Si–O)_3–Si–O–Nd$ species.

The second example reported is the NMR characterization of a hexagonal mesoporous silica (HMS) functionalized with mercaptopropyl groups by co-condensation and grafting. ^{29}Si CPMAS NMR was employed to verify the presence of thiol groups in the HMS structure [68].

The pure HMS spectrum shows three main peaks due to the three different silicon atoms of the silica framework (Q2, Q3, and Q4).

The first and the second species are silicon atoms present at the silica surface, whereas the third type is present in the bulk. The ^{29}Si CPMAS NMR spectra of functionalized HMS show the peaks discussed above and new resonances in the range from −40 to −80 ppm (Tn species) that are an evidence of the HMS functionalization by the thiol groups.

The peaks correspond to three different environments for the siloxane groups: T1 units bound to one siloxane and represented at −47.5 ppm, T2 units bound to two siloxanes and represented at −57.5 ppm, and T3 units fully condensed and bound to three siloxanes and represented at −66.7 ppm. The most intense signals in the grafted sample are those related to T1 and T2, whereas in the co-condensed sample, the most intense peak arises from T3 and no signal attributable to T1 is present.

This is evidence that the mercapto groups in the co-condensed sample, in contrast with those in the grafted sample, are preferentially incorporated in the wall surfaces. Furthermore, in the co-condensed sample spectrum, the relative intensity of the Tn peaks with respect to the Qn peaks is much lower than that for the grafted spectrum. As observed before, the CP technique enhances the signals of those ^{29}Si nuclei closer to ^{1}H. The low intensity of the (Tn) peaks observed in the case of the co-condensed sample may then be indicative of a more homogeneous functionalization relative to the grafted sample, which gives rise to more distant mercapto groups and therefore less intense ^{29}Si signal.

10.6.3 Cultural Heritage

NMR is a widespread technique in many fields but its application to Cultural Heritage is rather recent. The main reason is that NMR has almost exclusively been a liquid-state technique so it was a problem for solid samples such as paper, pottery, glass, stone materials, and so on that could not be dissolved in an appropriate solvent.

Nevertheless, the characterization of soluble organic residues found in ancient artifacts can be carried out by NMR analysis in solution. The analysis of this kind of samples is a challenging task as they are often constituted by complex mixtures of compounds. Nowadays, by using NMR techniques based on difference in the molecular self-diffusion coefficients, complex mixtures of organic compounds may be successfully analyzed.

2D NMR techniques may be applied to identify the structure of each compound [69]. Recently, by using the cryogenically cooled NMR probes (cryoprobes), the required amount of material to be analyzed by NMR in solution has been minimized and it is possible to analyze quantities of sample down to the microgram scale [70, 71]. In the last decades with the solid-state 1D and 2D NMR techniques and multinuclear probes development that are able to investigate many nuclei of the element table, NMR has become a powerful technique in analytical and structural determinations both for organic and inorganic substances in the cultural heritage field.

In the first case study, a set of varnishes were prepared in laboratory with different linseed oil/colophony ratios to study those ones used by the great Italian violin maker A. Stradivari [72]. Many scientific researches on varnish samples of Stradivari's instruments revealed that they were generally made by a mix of linseed oil and colophony or metal rosinates in different ratios ranging between 75/25 (oil/resin) and 60/40 (oil/rosinate). However, it is not clear the relationship between the mixture composition and the structural and/or functional feature of the resulting varnish. Several NMR techniques were applied both in solid and in liquid state to obtain information about chemical–physical characteristics of the varnishes. Diffusion coefficients D_0 values obtained by diffusion ordered spectroscopy (DOSY) for linseed oil and colophony were found to clearly distinguish two components. This is suggested to be due to the fact that they have different motional behavior before the varnish drying process probably due to scarce interactions at the molecular level. This was confirmed by 2D-NOESY experiments. The cross peaks in the 2D NOESY spectrum are due to intramolecular dipolar correlations and are related only to fatty acids composing the linseed oil. No cross peaks related to signals of linseed oil and colophony can be observed in any of the analyzed mixtures, thereby indicating the absence of strong intermolecular interactions between linseed oil and colophony.

From the analysis by solid-state NMR of the $T_{1\rho}$H relaxation times, it was evidenced that all samples have different values for linseed oil and colophony except for the 75/25 mixture. This indicates that this latter sample can be considered a homogeneous material in a scale of tens of angstroms as $T_{1\rho}$H values determined for the resin and oil components are equivalent within the experimental error.

In the second case study, several solid-state NMR techniques were applied to assess the state of conservation of the wooden part of the roman Acqualadroni rostrum and to observe the interaction between holocellulose and lignin [73]. Qualitative information about the presence of the components of the wood was obtained by the ^{13}C CPMAS NMR spectra analysis. For this purpose, a sample of fresh wood of the same species (*Pinus pinaster*) was analyzed. Quantitative parameters such as the holocellulose/lignin ratio, the cellulose crystallinity index were determined by variable contact time (VCT) experiments and the state of conservation of the lignin was evaluated by non-quaternary suppression (NQS) spectra analysis.

Finally, $T_{1\rho}$H and T_1H gave information on the cellulose and lignin polymeric dynamics and on the interactions between them. In fact, it is possible to determine if wood behaves as a blend on a molecular scale (dynamically homogeneous) or as a mixture of noninteracting polymers, leading to distinct spin systems. The evidences obtained from the solid-state NMR analyses allowed to demonstrate that the degradation process of the wood has led to the progressive loss of polysaccharide components while the lignin remains intact. ^{13}C CPMAS NMR spectra show that in all the samples, except the inner layers, hemicelluloses are depleted. The quantitative analysis of the holocellulose/lignin ratio demonstrated that the degradation involves mainly the polysaccharides of the surface layers and then it proceeds to the inner layers. The analysis of the crystallinity degree of the cellulose has shown that the amorphous part is attacked preferentially with respect to the crystalline one.

10.6.4 Food

NMR techniques have been widely applied for the evaluation of food quality, origin, and transformations during cooking and conservation [74]. As an example, Boiani and coworkers integrated the data from NMR spectroscopy with those from low-field T_2 measurements to find that aging of mozzarella cheese is mainly related to a redistribution of water molecules inside the organic matrix of that dairy product [75]. These results appear to accord with those retrieved by Godefroy and coworkers [35] on the aging of mozzarella and gouda cheeses, and with those from Conte's group on Parmigiano Reggiano cheese ripening [76]. It is also worth noting that low-field T_2 measurements are, nowadays, routinely applied to monitor water state in lupin seeds [77], to monitor olive oil adulteration [78], or to evaluate fruit quality under different storage conditions [79].

In the recent years, also FFC NMR relaxometry has been applied for food evaluation. As an example, the technique has been able to reveal differences in molecular mobility of balsamic and traditional balsamic vinegars [80], to differentiate among wild eggplant cultivars [81], to unveil the tridimensional organizations of fatty acids in extra virgin olive oils [44], to discriminate among pistachio oils obtained from seed cultivated in different areas [82], to understand the transformations occurring during eggplant cooking [37] and aging of mozzarella and gouda cheeses [35], to reveal differences among honeys stored in different conditions [36], to discriminate between molten and cooled cocoa butter [83], and to explain the mechanisms of fat crystallization [84].

10.6.5 Environmental NMR: Rocks, Soils, Waters, Air

NMR techniques are worldwide recognized as powerful tools to unveil the chemical–physical characteristics of complex systems such as the environmental compartments [85, 86]. Just as an example, since the end of the seventies and the beginning of the eighties of the twentieth century, both liquid and solid-state NMR spectroscopies have been used to understand the role of natural organic matter in soil quality [87], although, with the technical improvements and better understanding, many limitations have been reported [88–91]. Certainly, the computer developments occurred in recent years, together with the amelioration of the miniaturization procedures, allow to apply, today, NMR spectroscopy in a way that could not be expected till a couple of years ago. Very recently, Simpson and coworkers designed a special pulse sequence in order to apply spectroscopy on small living organisms for the understanding of water, soil, and air contamination [84, 85, 92, 93].

As reported above, spectroscopy can only monitor molecular dynamics occurring at the high frequency fields used for that technique. When slow motions, such as those of molecular systems constrained in porous materials, must be investigated, fast-field NMR relaxometry is a more appropriate technique [33]. As an example, FFC NMR has been applied to monitor soil erosion [94–97], to measure the age of an Italian saltmarsh [95], and to acknowledge the mechanisms behind nutrient transport in soils [90].

10.6.6 NMR of "Exotic" Nuclei

Almost all the periodic table elements have a nonzero spin isotope, but not all these nuclei are easily observable with the NMR spectroscopy. Such isotopes are often called "exotic nuclei."

The problems arising studying exotic nuclei (such as ^{53}Cr, ^{59}Co, ^{61}Ni, ^{67}Zn, ^{73}Ge, ^{75}As, ^{87}Sr, ^{115}In, ^{119}Sn, $^{121/123}$Sb, ^{9}Be, ^{25}Mg, $^{135/137}$Ba, ^{33}S, $^{185/187}$Re, ^{39}K, ^{209}Bi, ^{43}Ca, $^{47/49}$Ti) are due to particular reasons such as low γ values and low natural abundances together with high quadrupolar constants (Q) and quantum spin numbers (I). In fact, most of these nuclei are quadrupolar with spin number $I > ½$ [98].

Furthermore, the sensitivity that is the main problem in NMR spectroscopy is still present and it represents the most significant challenge that NMR spectroscopy faces when compared with many other techniques.

Nonetheless, since the information that can be obtained from NMR analysis on such nuclei is of great importance in materials science, overcoming the problems related to their spectra acquisition is a challenge that NMR spectroscopists have to face. Depending on the problems that arise in the acquisition of the spectra of these nuclei, which depend on the specific properties of each of them, numerous techniques have been introduced for their observation.

For example, to overcome the low sensitivity due to low isotopic abundance, an effective method is isotope enrichment. Among others, some of the most important nuclei for which this technique is used are ^{17}O, ^{43}Ca, and ^{15}N which are characterized by isotopic abundances of 0.038, 0.135, and 0.37%, respectively. Furthermore, the MAS technique is largely used to eliminate totally or partially the CSA and the quadrupolar first-order interactions. Methods such as CP and dynamic nuclear polarization (DNP) are also used for nuclei characterized by low sensitivity.

Many exotic nuclei give rise to wide-spectra powder patterns of breadths up to ca. 300 kHz. Special techniques and hardware are required for such spectra whose pattern cannot be acquired by a single experiment.

Pulse sequences such as variable-offset cumulative spectra (VOCS) for the piecewise acquisition, together with techniques such as Carr-Purcell Meiboom-Gill (QCPMG) and Wideband Uniform Rate Smooth Truncated WURST-QCPMG, etc., belong to such methods applied in the field of the so-called wideline NMR.

Ultrawide solid-state NMR (UW SSNMR) was employed to distinguish three glycine polymorphic forms [99]. To this purpose, authors used the WURST-CPMG pulse sequence for the direct excitation of ^{14}N nuclei, to avoid problems arising from the breadths of the powder patterns. The glycine polymorphic forms differentiation was performed by the study of the ^{14}N C_Q and η_Q quadrupolar parameters together with the T_2^{eff} constants. Effectiveness of various methods to improve the ^{14}N powder pattern acquisition such as ^{1}H high-power decoupling, broadband CP, and variable temperature acquisition is also discussed. Furthermore, dynamical motion of the NH_3 group as a function of the temperature, and how this affects the signal-to-noise ratio of the ^{14}N CPMG and ^{1}H–^{14}N CP NMR spectra was highlighted. Finally, DFT calculations were used to the ^{14}N EFG tensor components' prediction by considering the glycine crystals' periodic nature.

In another work, O'Keefe et al. used the WURST-CPMG pulse sequence and the variable-offset cumulative spectrum (VOCS) to obtain ^{35}Cl SSNMR powder patterns with high signal-to-noise ratios (S/N) and short experimental times [100]. Such techniques were exploited in the study of chlorine ligands for several diamagnetic transition metal (TM) complexes. Two magnetic fields (9.4 and 21.1 T) were used to acquire the spectra. In particular, the use of the ultrahigh field allowed to increase the S/N ratios and to reduce CT patterns breadths. Both small and significant differences in molecular geometry and chemical environment were highlighted by the ^{35}Cl SSNMR due to the sensitivity of the EFG tensors. In addition, ^{35}Cl NQR experiments were performed to obtain sufficient resolution in the α-WCl_6 and β-WCl_6 polymorphic form differentiation. Differences in the molecular packing between $WOCl_4$ and $MoOCl_4$ were also studied by ^{35}Cl SSNMR. On the basis of the C_Q values, authors distinguished the chlorine environment in the dimeric pentahalide complexes $NbCl_5$ and $TaCl_5$. The influence of the size of the metal center on the value of C_Q was also observed in tantalum complex for all chlorine environments. A good agreement with experimental results especially in systems in which intermolecular interactions slightly affect the ^{35}Cl EFG with the DFT calculations was found.

References

1 Purcell, E.M., Torrey, H.C., and Pound, R.V. (1946). *Phys. Rev.* 69: 37.
2 Bloch, F., Hansen, W.W., and Packard, M. (1946). *Phys. Rev.* 69: 127.
3 Levitt, M.H. (2008). *Spin Dynamics Basics of Nuclear Magnetic Resonance*. John Wiley & Sons.
4 Keeler, J. (2002). *Understanding NMR Spectroscopy*. John Wiley & Sons.
5 Duer, M.J. (2002). *Solid-State NMR Spectroscopy Principles and Applications*. Blackwell Science.
6 Harris, R.K. (1996). *Encyclopedia of Nuclear Magnetic Resonance Nuclear spin properties & notation*. John Wiley & Sons.
7 Becker, E.D. (2000). *High Resolution NMR. Theory and Chemical Applications*. Academic Press.
8 Bloembergen, N.E., Purcell, M., and Pound, R.V. (1948). *Phys. Rev.* 73: 679–712.
9 Beckmann, P. (1988). *Phys. Rep.* 171: 85–128.
10 Pines, A., Gibby, M.G., and Waugh, J.S.J. (1973). *Chem. Phys.* 59: 569–590.
11 Hartmann, S.R. and Hahn, E.L. (1962). *Phys. Rev.* 128: 2042–2053.
12 Sarles, L.R. and Cotts, R.M. (1958). *Phys. Rev.* 111: 853–859.
13 Bloch, F. (1958). *Phys. Rev.* 111: 841–853.
14 Lowe, I. (1959). *J. Phys. Rev. Lett.* 2: 285–287.
15 Andrew, E.R., Bradbury, A., and Eades, R.G. (1959). *Nature* 183: 1802–1803.
16 Andrew, E.R., Bradbury, A., and Eades, R.G. (1958). *Nature* 182: 1659–1659.
17 Gerstein, B.C., Pembleton, R.G., Wilson, R.C., and Ryan, L.M. (1977). *J. Chem. Phys.* 66: 361–362.
18 Vinogradov, E., Madhu, P.K., and Vega, S. (2005). *New Techniques in Solid-State NMR*, vol. 246, 33. Berlin: Springer.
19 Madhu, P.K. (2009). *Solid State Nucl. Magn. Reson.* 35: 2–11.
20 Dusold, S. and Sebald, A. (2000). *Annual Reports on NMR Spectroscopy*, vol. 41, 185. San Diego, CA: Academic Press.
21 Levitt, M.H. (2002). *Encyclopedia of Nuclear Magnetic Resonance*, vol. 9, 165. John Wiley & Sons.
22 Sommer, W., Gottwald, J., Demco, D.E., and Spiess, H.W. (1995). *J. Magn. Reson. A.* 113: 131–134.
23 Schnell, I., Brown, S.P., Yee Low, H. et al. (1998). *J. Am. Chem. Soc.* 120: 11784–11795.
24 Brown, S.P. (2009). *Macromol. Rapid Commun.* 30: 688–716.
25 Munowitz, M.G., Griffin, R.G., Bodenhausen, G., and Wang, T.H. (1981). *J. Am. Chem. Soc.* 103: 2529–2533.
26 McElheny, D., De Vita, E., and Frydman, L.J. (2000). *Magn. Reson.* 143: 321–328.
27 Frydman, L. and Harwood, J.S. (1995). *J. Am. Chem. Soc.* 117: 5367–5368.
28 Griffin, R.G. (1998). *Nat. Struct. Biol.* 5: 508–512.
29 Ladizhansky, V. (2009). *Solid State Nucl. Magn. Reson.* 36: 119–128.
30 Bennett, A.E., Ok, J.H., Griffin, R.G., and Vega, S. (1992). *J. Chem. Phys.* 96: 8624–8627.
31 Bennett, A.E., Rienstra, C.M., Griffiths, J.M. et al. (1998). *J. Chem. Phys.* 108: 9463–9479.

32 Orrell, K.G. and Šik, V. (1993). *Annu. Rep. NMR Spectrosc.* 27: 103–171.
33 Conte, P. (2019). *Field-cycling NMR Relaxometry: Instrumentation, Model Theories and Applications*, 1ee., Issue 18 (ed. R. Kimmich), 229–254. The Royal Society of Chemistry.
34 Conte, P. and Lo Meo, P. (2020). *Agronomy* 10: 1040.
35 Godefroy, S., Korb, J.P., Creamer, L.K. et al. (2003). *Colloid Interface Sci.* 267: 337–342.
36 Cimò, G. and Conte, P. (2015). *Int. J. Spectrosc.* 2015 article ID 354327.
37 Lo Scalzo, R., Fibiani, M., Francese, G. et al. (2016). *Food Chem.* 194: 835–842.
38 Provencher, S.W. (1982). *Comput. Phys. Commun.* 27: 213–227.
39 Provencher, S.W. (1982). *Comput. Phys. Commun.* 27: 229–242.
40 Borgia, G.C., Brown, R.J.S., and Fantazzini, P. (1998). *J. Magn. Reson.* 132: 65–77.
41 Borgia, G.C., Brown, R.J.S., and Fantazzini, P. (2000). *J. Magn. Reson.* 147: 273–285.
42 Borgia, G.C., Brown, R.J.S., and Fantazzini, P. (2001). *Magn. Reson. Imaging* 19: 473–475.
43 Bortolotti, V., Brown, R.J.S., Fantazzini, P. et al. (2017). *Inverse Probl.* 33: article ID 3 015003.
44 Conte, P., Maccotta, A., De Pasquale, C., and Alonzo, G. (2010). *Fresen. Environ. Bull.* 19: 2077–2082.
45 Kimmich, R. (1979). *Bull. Magn. Reson* 1: 195–218.
46 Anoardo, E., Galli, G., and Ferrante, G. (2001). *Appl. Magn. Reson.* 20: 365–404.
47 Zampetoulas, V., Lurie, D.J., and Broche, L.M.J. *Magn. Reson.* 282: 38–46.
48 Kimmich, R. and Anoardo, E. (2004). *Prog. Nucl. Magn. Reson. Spectrosc.* 44: 257–320.
49 Luchinat, C. and Parigi, G. (2008). *Appl. Magn. Reson.* 34: 379–392.
50 Halle, B., Jóhannesson, H., and Venu, K.J. (1998). *Magn. Reson.* 135: 1–13.
51 Průšová, A., Conte, P., Kučerík, J., and Alonzo, G. (2010). *Anal. Bioanal. Chem.* 397: 3023–3028.
52 Kruk, D., Masiewicz, E., Borkowska, A.M. et al. (2019). *Biomolecules* 9: 652–665.
53 Fries, P.H. and Belorizky, E. (2015). *J. Chem. Phys.* 143: 044202.
54 Lo Meo, P., Mundo, F., Terranova, S. et al. (2020). *J. Phys. Chem. B* 124: 1847–1857.
55 Conte, P. and Alonzo, G. (2013). *EMagRes.* 2: 389–398.
56 Dunn, K.-J., Bergman, D.J., and Latorraca, G.A. (2002). *Handbook of Geophysical Exploration: Seismic Exploration*, 1ee, vol. 32 (eds. K.-J. Dunn, D.J. Bergman and G.A. Latorraca), 129–164.
57 Manz, B., Coy, A., Dykstra, R. et al. (2006). *J. Magn. Reson.* 183 (1): 25–31.
58 Ratcliffe, R.G. and Shachar-Hill, Y. (2001). *Annu. Rev. Plant Physiol.* 52: 499–526.
59 Capitani, D., Sobolev, A.P., Di Tullio, V. et al. (2017). *Chem. Biol. Technol. Agric.* 4: 1–14.
60 Ganssle, P.J., Shin, H.D., Seltzer, S.J. et al. (2014). *Angew. Chem. Int. Ed.* 53: 9766–9770.
61 Bovey, F.A. and Mirau, P. (1996). *NMR of Polymers*. San Diego: Academic Press.
62 Ibbett, R.N. (ed.) (1993). *NMR Spectroscopy of Polymers*, 63. Berlin: Springer.
63 Hatada, K. and Kitayama, T. (2004). *NMR Spectroscopy of Polymers*. Berlin: Springer.
64 Spinella, A., Bondioli, F., Nasillo, G. et al. (2017). *Colloid Polym. Sci.* 295: 695–701.
65 Armetta, F., Sibeko, M.A., Luyt, A.S. et al. (2016). *Phys. Chem.* 230: 1219–1231.
66 Trébosc, J., Wiench, J.W., Huh, S. et al. (2005). *J. Am. Chem. Soc.* 127 (9): 3057–3068.
67 Caponetti, E., Minoja, A., Saladino, M.L., and Spinella, A. (2008). *Micropor. Mesopor. Mat.* 113: 490–498.
68 Parola, V.L., Longo, A., Venezia, A.M. et al. (2010). *Eur. J. Inorg. Chem.*: 3628–3635.
69 Braun, S., Kalinowski, H.O., and Berger, S. (1998). *150 and More Basic NMR Experiments A Practical Course*. Germany: Wiley.
70 Spraul, M., Feund, A.S., Nast, R.E. et al. (2003). *Anal. Chem.* 75: 1546–1551.
71 Schlotterbeck, G., Ross, A., Hochstrasser, R. et al. (2002). *Anal. Chem.* 74: 4464–4471.
72 Spinella, A., Malagodi, M., Saladino, M.L. et al. (2017). *Pol. Sci. A.* 55: 3949–3954.
73 Bastone, S., Spinella, A., Chillura Martino, D.F. et al. (2016). *Microchem. J.* 124: 831–836.
74 Hatzakis, E. (2019). *Compr. Rev. Food Sci. Food Saf.* 18 (1): 189–220.

75 Boiani, M., Sundekilde, U., Bateman, L.M. et al. (2019). *Int. Dairy J.* 91: 155–164.
76 Conte, P., Cinquanta, P., Lo Meo, P. et al. (2021). *Food Res. Int.* 139: ID 109845.
77 Garnczarska, M., Zalewski, T., and Kempka, M.J. (2007). *Exp. Bot.* 58: 3961–3969.
78 Ok, S. (2017). *Grasas Aceites* 68: 1–13.
79 Gambhir, P.N., Choi, Y.J., Slaughter, D.C. et al. (2005). *J. Sci. Food Agric.* 85: 2482–2486.
80 Baroni, S., Consonni, R., Ferrante, G., and Aime, S.J. (2009). *Sci. Food Agric.* 57: 3028–3032.
81 Pellegrino, P., Bubici, S., Palazzolo, E., and Alonzo, G. (2009). *Spectrosc. Lett.* 42: 235–239.
82 Conte, P., Mineo, V., Bubici, S. et al. (2011). *Anal. Bioanal Chem.* 400: 1443–1450.
83 Ladd-Parada, M., Povey, M.J., Vieira, J., and Ries, M.E. (2018). *Mol. Phys.* 117: 1020–1027.
84 Ladd Parada, M., Povey, M.J., Vieira, J. et al. (2019). *Magn. Reson. Chem.* 57: 686–694.
85 Simpson, A.J., Simpson, M.J., and Soong, R. (2012). *Environ. Sci. Technol.* 46: 11488–11496.
86 Simpson, A.J., Simpson, M.J., and Soong, R. (2018). *Anal. Chem.* 90: 628–639.
87 Wilson, M.A. (1981). *J. Soil Sci.* 32: 167–186.
88 Berns, A.E. and Conte, P. (2010). *The Open Magnetic Resonance Journal* 3 (SPEC. ISS.1): 75–83.
89 Conte, P., Piccolo, A., Van Lagen, B. et al. (1997). *Geoderma* 80: 339–352.
90 Conte, P., Piccolo, A., Van Lagen, B. et al. (1997). *Geoderma* 80: 327–338.
91 Conte, P., Spaccini, R., and Piccolo, A. (2004). *Prog. Nucl. Magn. Reson. Spectrosc.* 44: 215–223.
92 Bastawrous, M., Jenne, A., Tabatabaei Anaraki, M., and Simpson, A.J. (2018). *Metabolites* 8: 1–24.
93 Simpson, A.J. (2015). *Magn. Reson. Chem.* 53: 633–634.
94 Conte, P., Di Stefano, C., Ferro, V. et al. (2017). *Environ. Earth Sci.* 76: 526.
95 Conte, P. and Ferro, V. (2018). *Hydrol. Process.* 2018 (32): 93–101.
96 Conte, P. and Ferro, V. (2020). *Magn. Reson. Chem.* 58: 41–50.
97 Maccotta, A., De Pasquale, C., Caruso, A. et al. (2013). *Environ. Sci. Pollut. R.* 20: 4847–4858.
98 Leroy, C. and Bryce, D.L. (2018). *Prog. Nucl. Magn. Reson. Spectrosc.* 109: 160–199.
99 Veinberg, S.L., Friedl, Z.W., Harris, K.J. et al. (2015). *Cryst. Eng. Comm.* 17: 5225–5236.
100 O'Keefe, C., Johnston, K.E., Sutter, K. et al. (2014). *Inorg. Chem.* 53: 9581–9597.

11

X-Ray Absorption Spectroscopy and X-Ray Raman Scattering Spectroscopy for Energy Applications

Alessandro Longo,[1,2] *Francesco Giannici*[3]*, and Christoph J. Sahle*[1]

[1] *The European Synchrotron, Grenoble, France*
[2] *ISMN-CNR, Palermo, Italy*
[3] *Department of Physics and Chemistry – Emilio Segrè, University of Palermo, Palermo, Italy*

11.1 Introduction

X-ray absorption spectroscopy (XAS) measures the X-ray absorption coefficient of a material as a function of incident X-ray energy. Immediately above an X-ray absorption edge, a fine structure is observed in the X-ray absorption coefficient. This fine structure, or X-ray absorption near edge structure (XANES), is sensitive to the direct electronic and/or chemical surrounding of the absorbing atom, its oxidation state, and chemical bonding with its nearest neighbors. Fine structure, however, extends far beyond an X-ray absorption edge, and is called extended X-ray absorption fine structure or EXAFS. This extended X-ray absorption fine structure can provide information on atomic arrangements, bond lengths, and coordination numbers.

Although EXAFS was first observed in the early 1930s by Kronig [1], it was not until the pioneering work of Sayers, Stern, and Lytle in 1971 that this fine structure was shown to contain useful structural information [2–4]. Initially, the phenomenon had largely been ignored due to two major constraints. Firstly, because the agreement between theory and experiment was quite poor and, secondly, the low intensity of X-ray sources available at the time rendered EXAFS measurements cumbersome. With the advent of third- and fourth-generation synchrotrons, this technique is omnipresent. The high flux available from these new electron storage rings allows drastically reduced collection times for an EXAFS data set, enabling EXAFS spectra to be collected quickly and adequately.

Moreover, the possibility to have fast data collection, like the quick-EXAFS (milliseconds region) opened new experimental opportunities such as in situ and *operando* measurements, which allow to monitor the evolution of the local environment of an absorber atom as a function of time during a thermal or chemical treatment.

In addition to the improvement of the experimental conditions, also the development of the theory, which started in the early 1970s, helped to spread application of this technique as a unique and valuable tool for materials characterization [5–11].

In a X-ray photoabsorption process, a photoelectron is emitted from the absorbing atom. The EXAFS signal then emerges from the constructive and destructive interference of the outgoing photoelectron with itself after scattering off of atoms in the immediate environment of the absorber. Therefore, measurement of the absorption coefficient's fine structure can provide atomic-scale

Spectroscopy for Materials Characterization, First Edition. Edited by Simonpietro Agnello.

information about the immediate surrounding of the absorbing atom: the atomic species, arrangements, and bonding mechanisms. Although the amount of information obtained from the absorption fine structure may be small compared to, for example, a typical X-ray diffraction experiment, EXAFS has some advantages over such more conventional methods. Since atoms absorb X-rays at characteristic energies corresponding to the electron binding energies of each element's different electronic shells, EXAFS is used to probe the structure around a particular atomic species element selectively. Due to its chemical specificity, the EXAFS technique is ideally suited for the study of materials where the atoms of interest form only a small proportion of the sample. As the probed volume is directly related to the electron mean free path, the effect is local and, therefore, EXAFS is extremely useful for the study of amorphous materials where other structural techniques are not applicable or fail.

11.2 The X-Ray Absorption Coefficient and the EXAFS Technique

The most straightforward means of obtaining an EXAFS spectrum is to measure the X-ray intensity before and after a sample as a function of X-ray's energy, this is the so-called transmission mode. The linear X-ray absorption coefficient is defined in terms of the transmitted and incident photon fluxes, Φ and Φ_0, according to:

$$\mu_{\text{lin}}(\omega) = \ln(\Phi_0/\Phi)/x \tag{11.1}$$

for a homogeneous sample of thickness x. It is usually the atomic X-ray absorption coefficient that is considered. This is related to μ_{lin} by, $\mu(\omega) = M\mu_{\text{lin}}/(\rho/N)$. M is the atomic weight of the element in question, ρ, the density, and N is Avagadro's number.

The atomic X-ray absorption coefficient is measured in barn (units of area). In Figure 11.1a, the X-ray absorption coefficient of an iron foil is plotted as a function of photon energy. From the figure, we can observe a number of general trends. The overall envelope of the X-ray absorption coefficient decreases monotonously with increasing photon energy, generally as $\sim\omega^{-3}$, except for sudden jumps, called absorption edges, at energies that correspond to the core binding energies of the elements constituting the sample.

Since the electron binding energies, for example for the K-shell electrons, grow monotonously with the atomic number Z (Figure 11.1b), every K-edge energy onset corresponds to a well-defined atomic species. In addition, each atomic species can exhibit excitations from its different electronic subshells, for example 1s, 2s, $2p_{1/2}$, and $2p_{3/2}$ states corresponding to the K, L_I, L_{II}, and L_{III} absorption edges, and so on, according to the established X-ray nomenclature (see e.g. [12]).

The spectral region between the edge onset and the sudden limit, the energy above which the photoelectron has enough kinetic energy so as to not feel the suddenly changed environment of the Coulomb potential it leaves behind in the form of a core hole, is called X-ray absorption near edge structure (XANES) (see Figure 11.1c). This region, for the first row of transition metals, usually spans approximately the first 30–50 eV above the edge onset. The spectral region beyond this point is usually referred to as the EXAFS region.

The X-ray absorption coefficient, in the EXAFS region, is usually written as the sum of two contributions: a part varying smoothly as a function of energy corresponding physically to the absorption coefficient of an atom embedded in a structure, where scattering of the electron wave is switched off, and an additional part containing the fine structure.

$$\mu(\omega) = \mu_0[1 + \chi(\omega)] \tag{11.2}$$

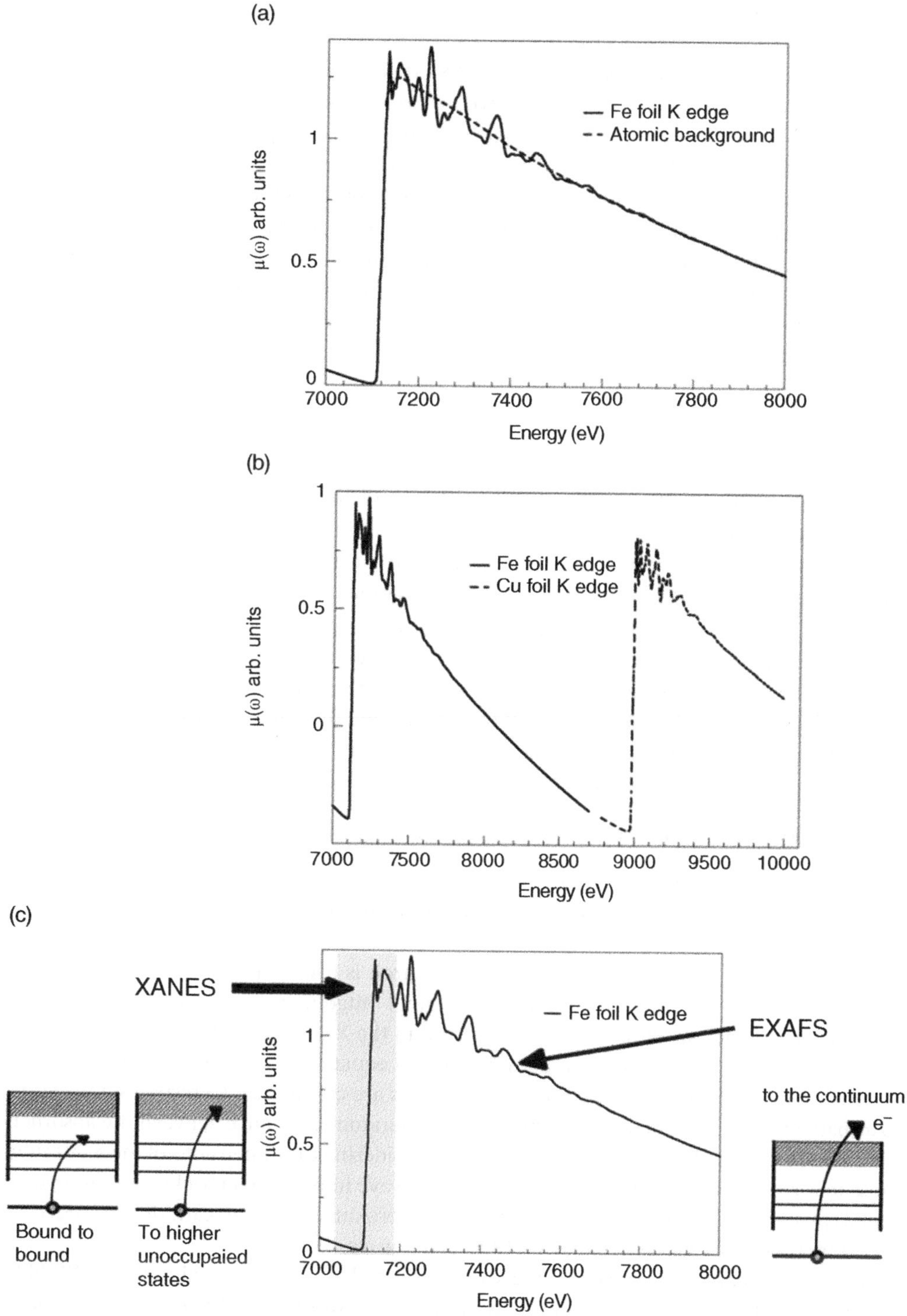

Figure 11.1 (a) Absorption coefficient as function of energy of a Fe foil measured at the Fe K-edge (7112 eV); (b) spectra of Fe and Cu foils measured at their K-edges, respectively (Cu 8979 eV); (c) schematic definitions of the XANES and EXAFS regions for the example of the spectrum of the Fe foil.

where μ_0 is the X-ray absorption coefficient of an isolated atom and χ is the EXAFS function which contains the information on the fine structure.

In the incident energy regime probed by most EXAFS experiments, typically 5–40 keV, the X-ray attenuation processes are dominated by photoelectric absorption. A photon of energy ω is absorbed by an atom and its energy is transferred to a single electron from a deep core orbital. The kinetic energy of this photoelectron is $\hbar\omega_k = (\hbar\omega - \mid \hbar\omega_0|) = (1/2)\hbar^2k^2 + E_F$, where ω_0 is the binding energy of the deep core orbital, k is the photoelectron wavenumber, and E_F is the thermodynamic Fermi energy of the material. k is defined as $k = \sqrt{(2m(\hbar\omega - E_b)/\hbar}$, where E_b is the core hole binding energy. Of course, in the process multiple electron excitations are also possible. These so-called shake-up and shake-off processes, will be briefly described in Section 11.2.1.2.2. For the sake of simplicity, we assume that only a single electron is excited, and may write the X-ray absorption coefficient from Fermi's golden rule as,

$$\mu(\omega) \sim \mid \psi_k\left(\vec{r}\right) \mid \hat{\varepsilon}\,\vec{r}\left|\psi_0\left(\vec{r}\right)\right|^2 \rho(\omega_k) \tag{11.3}$$

where $\rho(\omega_k)$ is the density of final states and $\hat{\varepsilon}$ is polarization unit vector.

Typically, this is approximated as the smooth variation of the unoccupied density of states. Eq. (11.3) $\hat{\varepsilon}\,\vec{r}$ describes the electron–photon interaction within the dipole approximation. In the dipole approximation, the X-ray wavelength is assumed to be small compared to the radius of the initial state ψ_0 it interacts with. This is a good approximation except for very large atomic numbers. As both $\hat{\varepsilon}\,\vec{r}$ and ψ_0 do not vary with energy, the only source of the observed oscillations in the absorption coefficient is the photoelectron final state $\psi_k(r)$. Therefore, EXAFS is a final state effect [7, 11]. Notably, the phenomenological theory used to interpret and fit the EXAFS data is based on electron scattering. The only difference between the EXAFS and the other electron diffraction processes is that the source of the EXAFS photoelectron lies within the atoms themselves.

11.2.1 The EXAFS Equation and the Key Approximations

As discussed in Section 11.1, the physical origin of EXAFS is due to a final state interference effect. In the absence of any neighboring atoms, the ejected photoelectron can be described by a spherical wave.

In condensed matter, however, the outgoing photoelectron is backscattered by the neighboring atoms. It is the resulting interference between the original outgoing wave and the backscattered wave that gives rise to the oscillatory structure observed in the X-ray absorption coefficient. Interpretation of EXAFS data can be based on a semiempirical equation first derived by Stern [2, 3].

In the derivation of the equation, two main assumptions are considered: (i) restricting the scattering to a spatial region very small with respect to the interatomic distance between the absorber and the backscatterer (small atom approximation); (ii) considering the dimensions of the scattering center close to zero, so that the curvature of the spherical wave impinging on backscatter atom can be neglected (plane wave approximation). Within these approximations, the scattering process can be described by the partial waves expansion of the plane wave.

Accordingly, this equation has since become known as the plane wave approximation to EXAFS and provides a robust parametrization of more complex forms for the fine structure used for most data analysis [4, 5, 10, 11]

$$\chi(k) = \sum_j S_0^2 e^{\frac{-2r_j}{\lambda}} \frac{N_j \mid f_j(k) \mid}{kr_j^2} e^{-2k^2\sigma_j^2} \sin\left(2kr_j + 2\delta_1\left(k, r_j\right) + \phi_j\right) \tag{11.4}$$

Equation (11.4) describes the extended X-ray absorption fine structure due to scattering of the photoelectron by shells of atoms surrounding the absorbing site. A typical shell consists of N_j neighbors located at a distance r_j from the absorbing atom. $f_j(k)$ is the backscattering amplitude from each of the N_j atoms of the jth type. The argument of the sine term contains the effective change of phase of the photoelectron as it travels to the scattering atom and back. The main contributions to the phase shift is a factor $2kr_j$ from the interatomic distance traveled, a factor $2\delta_1$ from a phase shift due to the excited central atom potential, and φ_j, which is an additional phase shift due to the potential at scattering site j.

It is this phase that controls the determination of the interatomic distance, r_j, and, since the (relative) phase can be measured very accurately, EXAFS is a good method for evaluating (relative) atomic distances, typically obtaining results at an accuracy of ±0.02 Å.

The amplitude of the EXAFS, however, is less well defined. It is proportional to the number of neighboring atoms N_j, but static and thermal disorder and many-electron processes all influence the amplitude significantly. These effects are less well understood and, typically, EXAFS amplitudes and hence coordination numbers, can be determined no more accurately than by ±10%. In the plane wave approximation, the amplitude effects are described by a number of semiempirical parameters. The Debye–Waller factor, $e^{\left(-2k^2\sigma_j^2\right)}$ in (11.2), allows for static and thermal disorder effects, where σ_j^2 is the mean square variation in atomic distance. The reduction factors, $e^{\left(-2r_j/\lambda\right)}$ and S_0^2, account for many-body processes, which contribute to the X-ray absorption coefficient but not to the fine structure, leading to an apparent decrease in the EXAFS amplitudes. $e^{\left(-2r_j/\lambda\right)}$ and S_0^2 model the losses in the EXAFS due to photoelectron mean free path effects and multi-electron excitations at the absorbing atom, respectively. The impact of these approximations will be discussed in more detail in the following sections.

11.2.1.1 Many-Body Effects

Equation (2.1) used in actual data analysis suffers from the problem that it is derived within the single-electron approximation. In fact, the EXAFS problem is inherently a many-body one. The system consists of many electrons, all of which interact via Coulomb forces and exchange interactions. The creation of a core hole–photoelectron pair results in a sudden change of the potential, which is, in general, extremely complex. This will obviously affect the behavior of the other electrons in the system, which are often referred to as passive electrons. However, the response of the passive electrons to the core hole–photoelectron system will in turn affect the original photoelectron via the strong electron–electron interaction, which, of course, is not negligible. Consequently, photoabsorption can induce transitions between any of the many-electron states ignored in a single-electron approximation. To model this complicated process exactly is practically impossible. Many-electron effects are generally approximated using additions to the effective single-electron scattering potentials. Ultimately, the many-body problem can be summarized by two effects. First, the Coulomb interactions between electrons modify the effective one-electron potential seen by the photoelectron. Second, the passive electrons in the atom may be excited due to the change in potential caused by the creation of the core hole and the subsequent photoelectron.

Two options have been proposed for approximating this potential [7]: either it can be derived from an atomic configuration with the passive electrons in their initial states or from the partially or fully relaxed final state. The latter is generally preferred and in the high-energy limit the unrelaxed Z (atomic number) approximation is generally considered appropriate. In this case, the atomic configuration is taken to be that of the Z atom without a contribution from the core orbital from which the photoelectron is ionized. For lower energy, it can be shown that the passive electron wavefunctions can be taken from the neutral $Z + 1$ atom [7, 9, 11].

11.2.1.2 Inelastic Effects

Inelastic processes involve the interaction between the photoelectron and the other electrons in the sample, both at the central atom and at the site of the scattering atoms. These processes are therefore completely neglected in the pure one-electron picture for the X-ray absorption coefficient (Eq. (11.3)) and can only be added to the EXAFS expression as ad hoc empirical factors designed to increase the agreement between theory and experiment.

The effect of inelastic processes is to diminish the EXAFS oscillations. Interference between the outgoing and incoming photoelectron waves can only take place if both waves are coherent; therefore, if the photoelectron has been inelastically scattered, it will not contribute to the EXAFS signal. Inelastic effects will, however, contribute to the total absorption coefficient. Inelastic effects are usually split into two kinds of processes: those in which the photoelectron is inelastically scattered as it propagates between the central and scattering atoms, known as extrinsic events, and those involving the creation of the core hole, known as intrinsic events.

11.2.1.2.1 Extrinsic Processes

The extrinsic effects are modeled in Eq. (11.4) by the mean free path term, $e^{(-2r_j/\lambda)}$. As the photoelectron propagates to and from the scattering atom, it may excite electron–hole pairs or cause collective excitations, such as plasmons. The losses lead to a decay of the final state, electrons are effectively removed from the elastically scattered beam giving rise to a weakening of the EXAFS signal. The energy dependence of the extrinsic losses is, however, important and is not included adequately in the simple, semiempirical, mean free path expression. Typically, in EXAFS calculations, the mean free path length is approximated by $\lambda = k/\text{VPI}$ where VPI is a constant imaginary part to the potential. Accordingly, in the calculations, the self-energy is most commonly approximated using the Hedin–Lundqvist potential [13]. This potential is obtained from uniform electron gas relations and was first applied to EXAFS calculations by Lee and Beni [14] using the local density approximation.

11.2.1.2.2 Intrinsic Losses

The intrinsic effects are those arising from the creation of a core hole. The core hole potential created upon photoexcitation is felt by the passive electrons, which can scatter off of this extra potential (shake-up) and thus transfer energy from the photoelectron. If the passive electrons are excited into the continuum, the process is referred to as shake-off. In this case, the possible range of secondary and hence final photoelectron energies is continuous and any interference between the photoelectron waves will tend to cancel. Events in which two or more electrons are excited into the continuum will therefore not contribute to the EXAFS. This effect is modeled using the amplitude reduction factor, S_0^2, in Eq. (11.4), representing the probability that each of the passive electrons remains in its initial state: it is assumed to be a constant in most cases, although this factor is actually energy-dependent.

Events are also possible where passive electrons are scattered into bound excited states. These shake-up processes result in photoelectrons with well-defined kinetic energy, which will cause oscillations in the X-ray absorption coefficient, although of a slightly different frequency than the primary EXAFS channel. However, the shake-up probability is small compared to that of shake-off events [10, 11] and is therefore usually ignored.

The core hole also has a finite lifetime. Eventually, the passive electrons will rearrange themselves so as to fill the deep core state whence the photoelectron came. This process, however, takes place on timescales greater than that of the photoelectron transit time. This value can be obtained from X-ray emission line widths, and the core hole lifetime can be measured fairly accurately. This finite lifetime effect can be added to the EXAFS expression (2.1) by effectively reducing the mean free path length. Such corrections are usually added automatically in the data analysis programs.

11.2.2 Multiple Scattering Theory: Basic Information

Equation (2.1) is based on the assumption that the X-ray photoexcited electron is scattered by just one atom before returning to the absorbing atom (e.g. Figure 11.2). However, the X-ray excited photoelectron can be scattered by two (or more) atoms before returning to the absorbing atom. Multiple scattering is particularly important at low k where the photoelectron has a very low kinetic energy and consequently a long mean free path. Multiple scattering (MS) is particularly strong if the two scattering atoms are nearly collinear since the photoelectron is strongly scattered in the forward direction (focusing effect). In this case, the EXAFS oscillations due to the multiple scattering pathway can be orders of magnitude stronger than those due to the single scattering pathway. Notably, failing to include multiple scattering contributions can lead to serious errors in both the amplitude and phase of EXAFS oscillations, with consequent errors in the coordination number and bond length estimation. An accurate treatment of the multiple scattering theory is out of the scope of

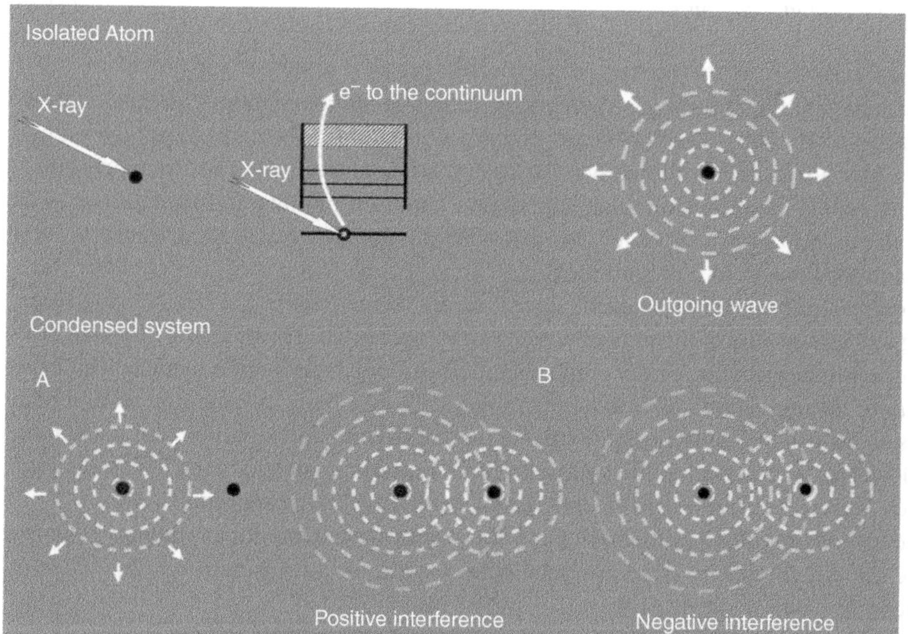

Figure 11.2 Schematic explanation of the EXAFS phenomenon for isolated atom (upper panel) and in a condensed system (lower panel). Upper panel: X-ray photon absorption (left); ejection of core electron to the continuum (center) and emission of photoelectron as spherical wave (right). Lower panel: X-ray photon absorption and emission of photoelectron spherical wave (a), backscattering of the photoelectron by neighboring atoms producing positive or negative interference effects (b).

this chapter, so in this section only the basic information of the theory will be outlined, which can be used only as guide for the reader.

According to the semiclassical treatment, photoelectric absorption is treated in time-dependent perturbation theory to first order in the operator $\vec{p}\cdot\vec{A}$ [15], where $\vec{A}$ is the potential vector of radiation field. Application of Fermi's golden rule yields the transition rate between the discrete initial and the many electron final states, Ψ_i and Ψ_f respectively, as

$$W_{i\rightarrow f} = 2\pi\hbar\sum_{f} \mid \Psi_f \mid \sum_{j} \vec{A}\left(\vec{r}_j\right)\cdot\vec{p}\,|\Psi_i|^2\rho(E_f) \tag{11.5}$$

where $\rho(E_f)$ is the density of the final states and the index j covers all the electrons in the system.

Calculation of the matrix elements is simplified extremely in the one-electron picture, in which it is assumed that only one electron is involved in the process. The other electrons are considered unaffected (sudden approximation); many-body corrections are added to the equation as described in Section 11.2.1. According to the single-electron approximation, it is assumed that one electron makes a transition between a localized core atomic orbital (e.g. the 1s orbital), described by a wave function Ψ_i, and a final state described by a wave function Ψ_f.

A further common simplification is the *dipole approximation*, which neglects the spatial variation of the vector potential of the radiation field (electron is confined in the small volume considering the variation of the vector potential), so this will only be a function of the time. The photoelectric absorption cross section for transitions to continuum final states can be written in these approximations as [15]

$$\sigma(\hbar\omega) = 4\pi^2\alpha\omega \mid \Psi_i \mid \hat{\varepsilon}\cdot\vec{r}\,|\Psi_f|^2, \tag{11.6}$$

where $\alpha = \frac{1}{137}$ is the fine structure constant whose correct physical dimensions are evident.

The selection rules are important consequences of the dipole approximation. These rules determine, on the basis of symmetry, the final states that are allowed for a given initial state; for a free atom, these are $\Delta l = \pm 1$; $\Delta_{m_l} = 0$ for linearly polarized radiation and $\Delta_{m_l} = \pm 1$ for circularly polarized radiation; here, l is the orbital angular momentum quantum number and m_l determines its projection on the quantization axis. Thus, for a K-absorption edge, the allowed transition will be to p states and for L-edges to d or s orbitals.

One must now treat the modification of the final state due to the presence of neighboring atoms corresponding to the scattering of the photoelectron by their potential. The outgoing photoelectron interacts mainly with the electron density of the neighboring atoms: this is often modeled using the very simple *muffin-tin approximation*. It consists in spherically averaging the potential around each atom, which is embedded in a constant interstitial potential. In this approximation, the calculation of the absorption cross section is simplified into a multiple scattering problem of the final state wave function, which is given by a collection of spherically symmetric scattering centers [16].

According to the assumptions illustrated so far, it was demonstrated that for a randomly oriented polycrystalline sample, the polarization averaged cross section for a transition to a final state of angular momentum l is [5, 7, 10, 11, 17]:

$$\sigma(\hbar\omega) = \frac{\sigma(\hbar\omega)}{(2l+1)\sin^2\left(\delta_l^0\right)}\mathrm{Im}\left(\sum_{m_l}\left[(I-TG)^{-1}T\right]_{0,0}^{L,L}\right) \tag{11.7}$$

where $\sigma(\hbar\omega)$ is the atomic cross section and δ_l^0 is the scattering phase shift centered on the absorber. Each matrix element in Eq. (11.7) is identified by four indices: i,j running over the sites of the atoms surrounding the central one, and L, L' (where $L = l, m_l$) being angular momentum indices. I is the unit matrix while T and G are the atomic scattering and propagator matrices in a local basis, respectively. The T matrix is diagonal in the atomic site and angular momentum indices, contains the phase shifts elements for the partial wavefunction of angular momentum l centered on the ith atom. The G matrix describes the free propagation of the electron from site i to site j. Accordingly, the T matrix depends on the atomic composition of the sample while the G matrix depends on the geometrical arrangement of the atoms. At sufficiently high energies above the edge (≈ greater than 50 eV), the formal matrix expansion:

$$T(I - TG)^{-1} = T[I + GT + GTGT + \cdots] \tag{11.8}$$

is convergent and it can be truncated. This formally allows to define the multiple scattering series (MS series). So, the cross section can be written as the sum of a small number of terms, each relative to a particular scattering path. Accordingly, the cross section can be written as:

$$\sigma(\hbar\omega) = \sigma_0(\hbar\omega)\left(1 + \sum_{i\neq 0}\chi_2^{0,i,0} + \sum_{i\neq j,\, i,j\neq 0,}\chi_3^{0,i,j,0} + \sum_{i\neq j\neq k,\, i,j,\,k\neq 0}\chi_4^{0,i,j,k,0} + \cdots\right) \tag{11.9}$$

where 0 identifies the central atom, the subscripts indicate the number of scattering segments and the superscripts indicate the scattering path, starting and ending on the central atom via atoms i and j. Scattering paths with two and three segments are pictorially represented in Figure 11.3; scattering paths with two segments are usually termed single scattering (SS) paths while those with three or more segments are called multiple scattering paths.

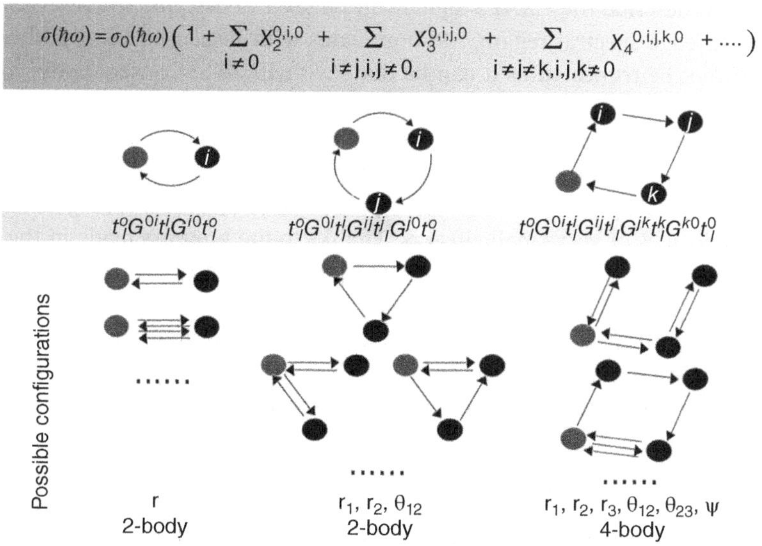

Figure 11.3 pictorially represents the single and multiple scattering paths for two-, three-, and four-body configuration respectively. The formal expression of multiple scattering series is expressed for the two-, three-, and four-body configurations as function of the atomic scattering T and propagator matrices G (see Section 2.1). Geometrical parameters needed to describe the configurations: r_i is the distance between the absorber and the ith scatter; θ_i represent the angle between the absorber and two scatters; and Ψ is the torsional angle used to define the four-body arrangement.

An alternative decomposition of the cross section in the EXAFS region into irreducible n-body signals was used by Filipponi et al. [6, 18]. Notably, this approach allows to perform the average for a defined n-body configuration, which becomes fundamental in the case of amorphous and nanostructured materials.

It can be shown that the MS contribution of order n identified by a path index $P = \vec{R}_i$ of a specific arrangement of atoms containing an absorber atom can be written as [7, 8, 11]:

$$\chi_n^P = A_n(k, P) \sin\left[kr_i + \varphi(k, P) + 2\delta\right] \tag{11.10}$$

The total EXAFS function will be the sum of all the SS and MS paths as indicated by Eq. (11.9) calculated using Eq. (11.10). In principle, the number of such paths is infinite, but in practice, it is limited by the rapid damping of the signal for long path lengths (it is found that paths whose length is greater than ≈8 A have negligible amplitude) and the fact that the amplitude of MS signal quickly decreases with increasing order n (usually paths with $n \geq 4$ can be neglected). MS paths are weak in the EXAFS region, but often they must be taken into account for an accurate structural determination. It can be shown that the multiple scattering series strongly depends on the scattering angle. For scattering angles less than ca. 150°, multiple scattering is weak and can often be neglected. However, for angles between 150° and 180°, multiple scattering must often be included. The angular dependence of multiple scattering means that EXAFS can, at least in principle, provide also a direct information about bond angles. Even when accurate angular information cannot be obtained, multiple scattering is still important since it is able to detect certain coordinating groups with unique EXAFS signatures.

11.2.3 XANES or Near-Edge X-Ray Absorption Fine Structure and Pre-Edge Region

The same MS formalism used to describe the EXAFS signal can be used to describe the XANES spectrum. However, in the near-edge spectral region, MS dominates with respect to SS, and the MS series of the Eq. (11.8) cannot be truncated as it can be done with EXAFS. Consequently, it is not possible to express the cross section as a sum of sine functions. Indeed, a simulation of the XANES region would require the inversion the matrix $(I - TG)$, which quickly becomes computationally difficult. Moreover, the muffin-tin approximation proves inadequate for the scattering potential in the XANES region, especially in cases where the local structure of the absorbing atom has a low symmetry, like in molecules or very small clusters. Thanks to the progress made in the theoretical modeling of XANES, software packages circumventing the limit of the muffin-tin approximation are now available [8, 18–22].

Hence, as a consequence of the sensitivity to MS, it is possible to extract information about the three-dimensional structure surrounding the absorber atom directly from the spectrum. Nevertheless, the full simulation and a complete optimization of the XANES spectrum still remains a cumbersome task.

It is worth noting that the ability to make even qualitative fingerprint-like comparisons of XANES spectra is still a very important tool (see Figure 11.4 Panel A). Indeed, if a representative library of reference spectra is available, spectral matching can be used to identify an unknown spectrum. To get more quantitative information on the composition of the spectrum, statistical tools, such as the principal component analysis (PCA) can be used. PCA is widely used in studies of element speciation in complex matrices (e.g. environmental chemistry, conservation science, geochemistry, and biology). It is based on a simple linear algebra: each reference spectrum (component) represents a vector, the data are reproduced by a vector sum. The algorithms automatically determine (statistics)

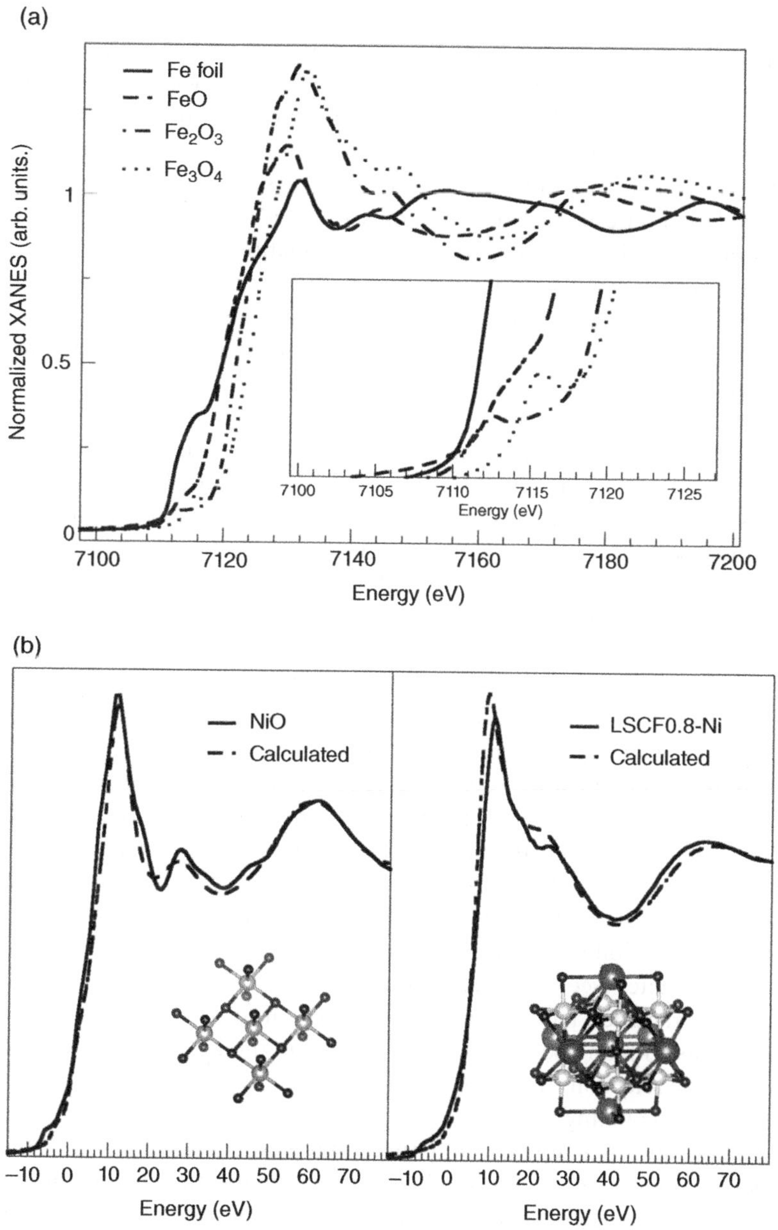

Figure 11.4 Panel A: XANES spectra of *Fefoil*, FeO Fe_2O_3 and Fe_3O_4. In the inset, an enlargement of the pre-edge is shown. Panel B: NiO and Ni inserted in a perovskite $La_{0.8}Sr_{0.2}Co_{0.7}Fe_{0.2}Ni_{0.1}O_3$ structure. Though, Ni is oxidized as Ni II in both materials, due to different MS paths, two different XANES are evident. Source: Data taken from Longo et al. [23].

the relevant (principal) components out of a given ensemble and reject the others. However, a priori knowledge on the basic composition of the sample is recommended.

Beyond these qualitative or semiquantitative applications, there are three main ways in which XANES spectra are used: to determine the oxidation state, to deduce three-dimensional structure, and as a probe of the electronic structure.

- Oxidation state:

  The energy of an absorption edge onset is not univocally defined. It can be taken as the energy at half-height or, more commonly, as the maximum in the first derivative with respect to the incident energy. However, as shown by Figure 11.4, edge spectra frequently have unresolved transitions superimposed on the rising edge. These will affect any attempt to define a unique *edge energy*. Despite such ambiguities, an assessment of the edge energies has been proven extremely useful in determining the oxidation state of the absorbing atoms for decades. It is well established that, in general, the energy of an edge increases together with the oxidation state of the absorber [11]. This observation is consistent with the fact that atoms with a higher oxidation state have a higher net charge (i.e. less electrons to screen the nucleus), thus requiring more energetic X-ray to eject a core electron. Notable exceptions however exist: gold, for example, does not follow this trend due to relativistic effects [22].
- Coordination geometry:

  As highlighted above, MS is particularly important in the XANES region. In principle, this means that it should be possible to determine the three-dimensional structure of the absorbing atom from analysis of the XANES features. Interestingly, the XANES region is quite sensitive to small variations in local geometry of the absorbing atom: nevertheless, two sites having very similar local atomic environment in the first coordination shells and equal oxidation state, could have very distinct XANES spectra (see Figure 11.4 Panel B) and vice versa.
- Electronic structure:

  Disregarding, for the sake of simplicity, the possible interactions between the core hole and partially filled d-bands (multiplet effects), in the vicinity of an absorption edge the X-ray absorption coefficient depends on the transitions of the photoelectron to those unoccupied states with lowest energy. The initial state is therefore a well-defined core level, insensitive to the chemical environment. In this respect, the edge structure reflects directly the modulations of the density of unoccupied electronic states that are available. In other words, the edge structure probes the density of final states of symmetry $\Delta l \pm 1$ with respect to the initial state symmetry, projected on the absorber site. However, besides the main XANES peak originating from the dipole allowed transition, weak pre-edge peaks are very often detected in a XANES spectrum, which often correspond to transitions into bound states (see Figure 11.5 Panel A).

  For the K-edge of a first row transition metal, pre-edge peaks arise from 1s → 3d transitions, and they are observed for metals with an open 3d shell [24]. Although the 1s → 3d transition is forbidden according to the dipole selection rules, it is nevertheless observed due both to 3d + 4p mixing and to direct quadrupolar coupling [22, 25].

  Breaking the symmetry of the absorber atom has an impact on the orbital mixing, so the pre-edge peaks become sharper and more evident. The sensitivity to 3d + 4p mixing implies that the intensity of the 1s → 3d transition can be used as a probe of geometry of the absorbed. As the site is progressively distorted from a centrosymmetric environment (i.e. octahedral ≤ square–pyramidal ≤ tetrahedral) [25], an increasing in the intensity is evidenced. This is also valid to distinguish between square–planar (i.e. centrosymmetric) and tetrahedral sites [11, 25]. Then, with a careful analysis and a correct use of references compounds, the details of the 1s → 3d transitions can be used to explore the electronic structure of the absorbing atom [22, 25, 26].

  The corresponding 1s → 4d transition for second transition series metals is generally not observed. These absorption edges occur at higher energy, where typical monochromator resolutions are worse and core hole lifetimes, determining the intrinsic line width of a transition, are much shorter [11]. This results in broad edges, which make the weak 1s → 4d transitions undetectable. However, for second-row transition metals, it is still possible to obtain information

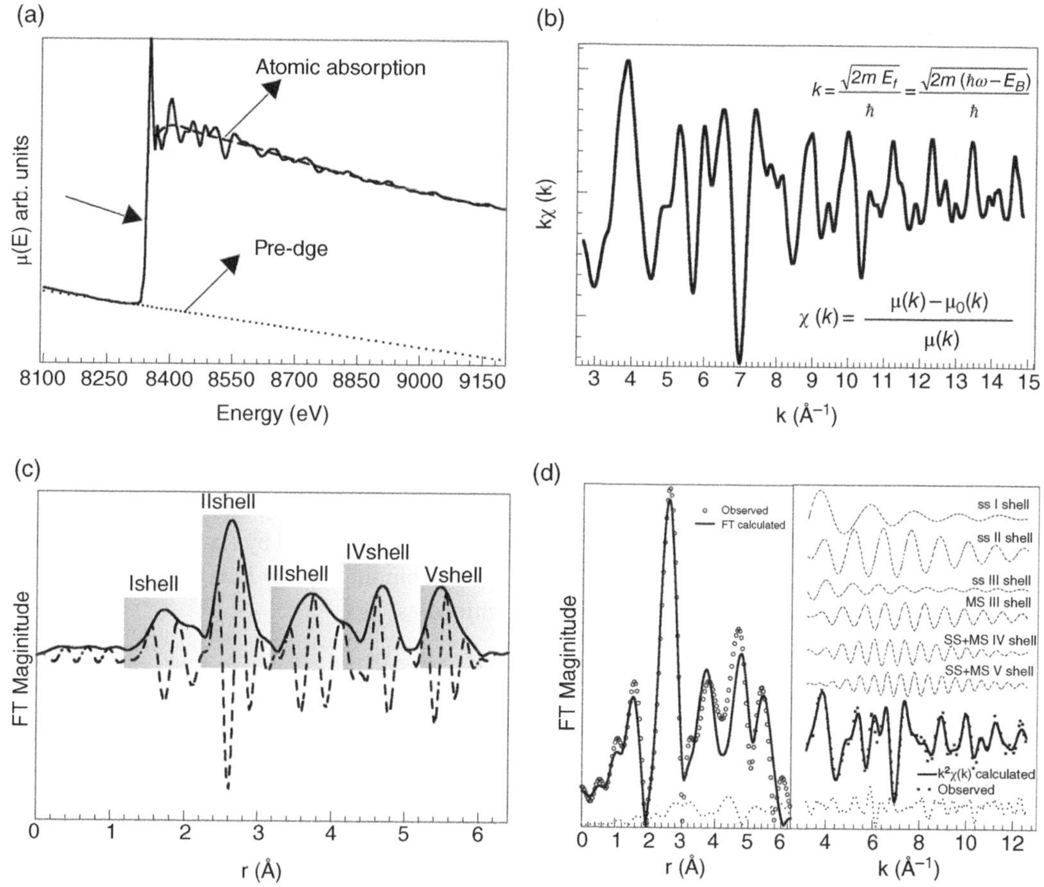

Figure 11.5 Different steps in data analysis for nickel oxide measured at 77 K: Panel A: pre-edge subtraction. Panel B: definition of EXAFS $\chi(k)$ (multiplied by k to amplify the oscillations at higher k values). Panel C: Fourier Transform of the $\chi(k)$. Panel D: example of fitting of NiO.

about the empty bound states by measuring data at the L_{III} and L_{II} edges, which probe 2p → 4d transitions [26]. The low energy of these edges makes the transitions relatively sharp, and the 2p → 4d transition is dipole allowed, thus making these transitions intense. Similar spectroscopic advantages (narrow lines, allowed transitions) are found for L-edge studies of the first transition series metals [25, 26]. However, in this case the very low edge energy is experimentally challenging, requiring the use of ultrahigh vacuum conditions for the samples, or X-ray Raman scattering spectroscopy (see Section 11.5).

11.3 EXAFS: Data Analysis Overview

The XAFS technique requires the measurement of the X-ray absorption coefficient as a function of photon energy. Synchrotron radiation is the ideal source for XAFS, thanks to the continuous spectrum of high intensity and strong collimation. Besides, the high degree of polarization is advantageous in several applications. A laboratory for X-ray spectroscopy with synchrotron radiation is generally composed of: (i) an optical apparatus, containing a monochromator, and one or more X-ray mirrors; (ii) a measurement apparatus, containing sample holders and detectors for measuring the absorption coefficient.

Let us consider the EXAFS signal at the K-edge of a given atomic species within a compound, measured using the transmission mode as described above. The output of measurement is:

$$\ln(I_0/I) = \ln(\Phi_0/\Phi) + C = \mu_t x + \text{const} \tag{11.11}$$

where I_0 and I are the detector signals, Φ and Φ_0 the X-ray photon fluxes before and after the sample, C is a constant depending on the detector efficiency. μ_t is the total absorption coefficient and x the sample thickness.

The first step of the analysis is the pre-edge removal, which consists in extracting the contribution $\mu_t x$ of the K-edge of the selected element from the experimental signal:

$$\mu_t x = \ln(I_0/I) - \mu_n x, \tag{11.12}$$

where μ_n is the contribution of all the other excitations of the selected element plus the excitations of the other elements of the compound. $\mu_n x$ includes also the constant C of Eq. (11.11). In the pre-edge energy region, $\mu_n x = \ln(I_0/I)$. Above the edge energy, $\mu_n x$ is estimated by extrapolating the pre-edge behavior (see Figure 11.5). In general, the core electron binding energy E_b is unknown, and the photoelectron wave vector k is experimentally determined as:

$$k = \sqrt{2m/\hbar^2(E - E_e)} = 0.263\sqrt{(E - E_e)}, \tag{11.13}$$

where E_e is a threshold energy, defined as the absorption edge energy (e.g. inflection point).

The last equality in Eq. (11.13) holds for wavenumbers measured in A^{-1} and energies in eV. The discrepancy between E_b and E_e is a priori unknown and is conventionally called E_0, so that $E_b = E_e + E_0$. The difference in E_0 is then treated as a free parameter to be optimized via a best fit.

The second step of the analysis, after the pre-edge removal mentioned above, is the normalization of the EXAFS signal with respect to the atomic background, calculated as $\chi(k) = (\mu - \mu_0)/\mu_0$. To evaluate μ_0, one looks for a curve, which averages the oscillations of the absorption coefficient. A frequently utilized approach is based on polynomial splines (see Figure 11.5, Panel A). This step of the analysis is generally achieved by trial and error.

Since the EXAFS function is the sum of sine functions, the argument of which is $k^r{}_i$, a Fourier transform (FT) of $\chi(k)$ will exhibit peaks in correspondence to the path length. The FT is usually plotted as a function of the half path length, which for the SS paths should correspond to the ith interatomic distance.

In addition, the FT operation also enables the different shell contributions which can easily be identified (see Figure 11.5, Panel B). Therefore, calculating the FT allows for an easy removal of the background signal, by selecting peaks below 1 A that cannot correspond to any meaningful interatomic correlation. In this respect, the FT represents a necessary visual representation of the distribution of atoms around the absorbing element and it provides guidance during the background removal process, helping to filter out the noise with low frequency.

The direct transformation from k- to r-space is given by:

$$F(r) = \int_{k_{min}}^{k_{max}} w(k)\chi(k)k^n e^{2ikr} dk, \tag{11.14}$$

where $w(k)$ is a window function whose aim is to reduce spurious oscillations induced by the finite k-range. The term k^n is used to balance the low- and high-k parts of the spectrum (typically $n = 1, 2,$

or 3). The integral limits k_{min} and k_{max} are chosen so as to exclude both the low-k signal, where the EXAFS formula is erroneous for the approximation used (typically $k_{min} \geq 2$–4 Å^{-1}), and the high-k signal, where the signal-to-noise ratio is too low.

$F(r)$ in Eq. (11.14) is a complex function, in principle extending over the entire r axis, with real and imaginary part Re[$F(r)$] and Im[$F(r)$], respectively. The modulus of $F(r)$ is given by:

$$|F(r)| = \sqrt{\left(\mathrm{Re}\left[F(r)^2\right] + \mathrm{Im}\left[F(r)^2\right]\right)} \tag{11.15}$$

and is characterized by peaks corresponding to the different coordination shells (Figure 11.5 Panel C). The peak positions do not correspond to the actual interatomic distances, because of the phase shifts $\varphi(k)$ present in the total phases of each coordination shell. In general, the peaks are shifted to lower r values by about 0.2–0.5 Å. Remarkably, the peak heights depend on the coordination number and on the Debye–Waller factor (both thermal and structural disorder).

The data analysis consists of optimizing the variable parameters, which are the shell structural parameters (N_i, σ_i, and R_i) in Eq. (2.1), so as to give the best fit to the observed signal using a non-linear least-squares fitting procedure.

In order to fit EXAFS data, it is first necessary to determine the parameters that define the scattering [$A_i(k)$, S_0^2, $\varphi(k)$, $\lambda(k)$] for all the multiple scattering paths. This can be done using ab initio calculations or from model compounds of known structure. Ab initio calculations are now relatively straightforward, with two main programs that are in wide use: FEFF [27] and GNXAS [18]. The different steps of data analysis are illustrated in Figure 11.5. The detailed description for an accurate data analysis is out of the scope of this chapter and, the interested reader should refer to manuals and textbook available in literature [18, 27, 28].

11.4 Experimental Setups

According to the definition of the absorption coefficient μ_t (Eq. (11.1)), $\mu_t = \ln(\varphi_0/\varphi)/x$ is proportional to the ratio between the photon fluxes impinging on and outgoing from the sample. Different instrumental configurations can be used to measure μ_t, depending on the sample dimension and composition as well as on the kind of information sought:

1) Direct measurement of μ_t through transmission experiments.
2) Indirect measurements of decay products, whose yield is proportional to μ_t.

- Detection of fluorescence photons (FLY, fluorescence yield)
- Detection of emitted electrons (AEY, Auger electrons yield; PEY, partial electrons yield; TEY, total electrons yield).

In this section, a brief overview of the transmission and fluorescence yield geometry will be outlined.

11.4.1 Transmission Geometry

The photon fluxes φ_0 and φ can be directly measured by two detectors immediately in front of and after the sample (see Figure 11.6).

In general, due to the high brilliance of synchrotron radiation sources, ionization chambers with plane parallel electrodes, some tens of centimeters long, are needed as detectors. The efficiency of ionization chambers can be adapted to different spectral regions by varying the atomic species of the filling gas and the corresponding gas pressure. The electrical signals I_0 and I generated by ionization chambers are low-intensity currents (typically 10^{-10} to 10^{-8} A). It is easy to see that $I_0/I = C(\varphi_0/\varphi)$ where C is a constant depending on the ionization chamber efficiency.

In order to have good signal/noise ratio on both I_0 and I, the sample needs to be relatively transparent to the beam, like in other absorption spectroscopies: for a metal foil, this corresponds to thickness around 10 μm.

When the sample is in powder form, compressed in a pellet or deposited on a thin film, one must be careful to avoid holes or inhomogeneities, which could cause spurious variations of the EXAFS amplitude. As a good rule of thumb, it is advised to check different pellets with different thicknesses. Transmission measurements are preferred, when possible, because they are easier and allow a better accuracy. In some cases, however, they are not suitable, for example for diluted samples or when surface sensitivity is needed.

11.4.2 Fluorescence Geometry

For dilute samples, i.e. when the absorbing species A constitutes only a tiny fraction of the entire sample, its contribution μ_A to the total absorption coefficient μ_{tot} can be exceedingly small. For concentrations typically lower than 1%, the XAFS structures at the absorption edge of the absorbing species μ_A may become comparable to the statistical noise of the total measured absorption coefficient μ_{tot}. In this case, transmission measurements are ineffective.

In general, fluorescence measurements are suitable for diluted samples. In this case, the intensity I_f of the X-ray fluorescence emitted at frequency ω_f by the absorbing species A, as a result of radiative de-excitation process, is measured as a function of incident energy $\hbar\omega$ as shown in Figure 11.6.

As a first approximation, fluorescence is one possible de-excitation mechanism following the absorption of the X-ray photon described in Figure 11.1. If the core hole created by the photoelectric absorption is filled by one of the outer-shell electrons, one X-ray photon is also emitted in the process, whose energy corresponds to the energy difference between the two atomic orbitals. A deep core hole can be filled by various electrons, so the emitted X-ray fluorescence spectrum will consist of several different lines, each with its own intensity. The fluorescence yield of a single edge is

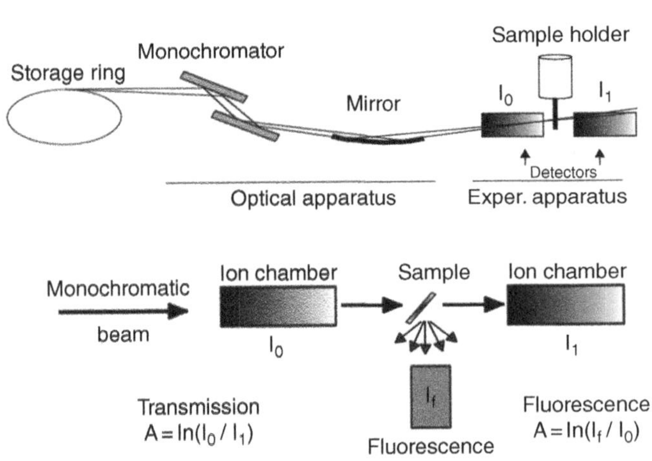

Figure 11.6 Top: typical experimental layout used in the EXAFS beamlines is shown. Bottom: experimental scheme used for transmission and fluorescence geometry.

defined as the probability that one core hole (e.g. the 1s core hole of the K-edge) decays through fluorescence de-excitation. Fluorescence yield η_f increases with atomic number, and is higher for K-edges than for L- and M-edges.

The fluorescence intensity is therefore:

1) proportional to the impinging flux Φ_0 at frequency ω, attenuated by absorption through the sample thickness z_n;
2) proportional to the absorption coefficient $\mu_A(\omega)$ of the atomic species A at frequency ω, and to the fluorescence yield η_f;
3) attenuated by the reabsorption of the fluorescence radiation traveling back through the sample thickness z_n;
4) proportional to the solid angle collected by the detector.

Two cases can be distinguished: thin and thick samples. For thin samples, only a small fraction of the impinging beam intensity is absorbed, and the fluorescence signal I_f is proportional to the absorption coefficient of the species A, independently of the degree of dilution. For thick samples, all of the incident beam intensity will be absorbed. This will produce aberration effects in the measured EXAFS signal μ, since basically all the fluorescence will be reabsorbed by the sample. Appropriate measures must be adopted in order to avoid spurious dampening of the EXAFS oscillations when measuring in fluorescence mode, usually through dilution of the absorbing element to minimize self-absorption.

However, it is worth noticing that the X-ray fluorescence signal originates from a two-photon process (excitation and decay), whereas description of the measured signal in transmission-mode EXAFS only requires a one-photon process. Care must be taken not to confuse these two modes and the theoretical descriptions behind them.

11.5 X-Ray Raman Scattering Spectroscopy

The last sections have outlined what a powerful tool the study of electronic excitations by X-ray absorption spectroscopy is. Numerous synchrotron- and even laboratory-based instruments exist and the XAFS techniques (XANES and EXAFS spectroscopy) are ubiquitous in the materials science research field. One aspect that remained without mentioning so far is the obvious dependence of the used X-ray energies on the investigated element. As such, the study of low-Z elements' K-absorption edges necessitate the use of soft X-rays; for example, the K-edges of first- and second-row elements can be found at energies between a few eV and about 1 keV. Soft X-ray absorption spectroscopy instruments of course exist and are routinely used; however, especially in the context of in situ materials science, the use of soft X-rays poses certain limitations to the possible studies. Furthermore, the information depth to be expected from a soft XAFS experiment is limited by the penetration depth of the used soft X-rays.

Here, we discuss the use of X-ray Raman scattering (XRS) spectroscopy to alleviate these shortcomings and provide a brief introduction to the underlying theory, instrumentation, and an instructive example.

11.5.1 Theoretical Background

X-ray Raman scattering spectroscopy is nonresonant inelastic X-ray scattering from bound electrons [29]. Figure 11.7A shows the typical outline of a nonresonant inelastic X-ray scattering

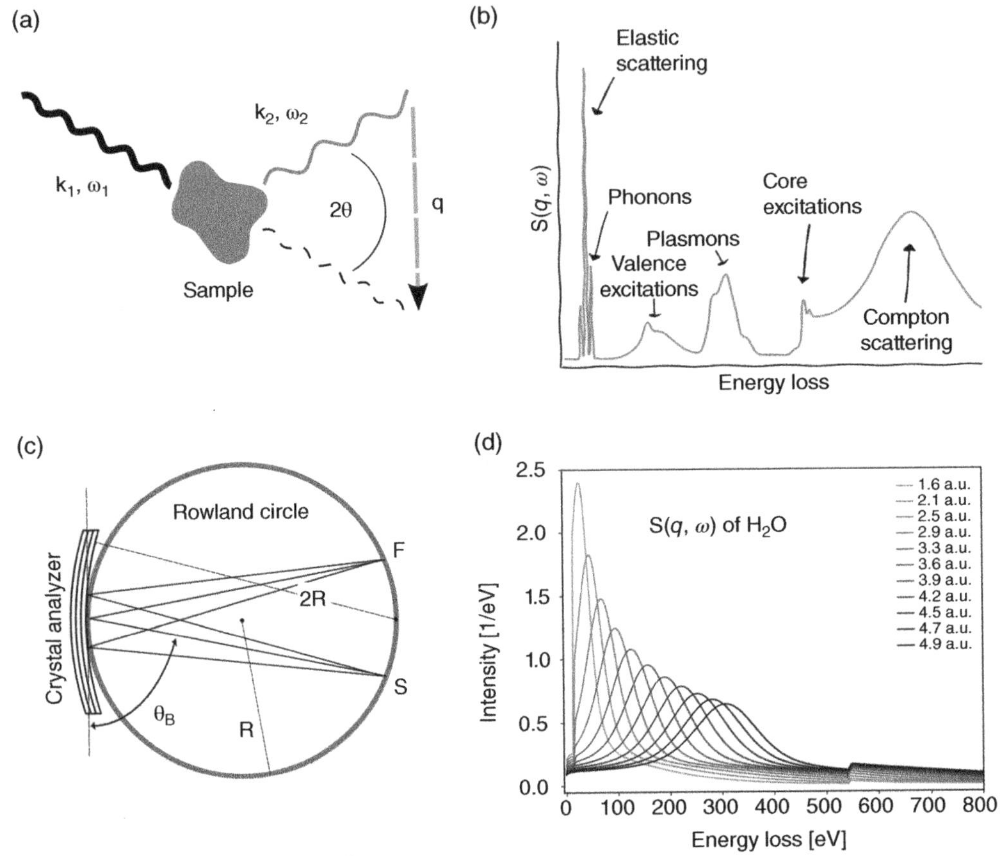

Figure 11.7 (a) Schematic drawing of a typical inelastic X-Ray spectroscopy experiment (IXS), an incident photon with wave vector k_1 and energy ω_1 is scattered off of a sample into a photon with wave vector k_2 and energy ω_2. The momentum transfer q is proportional to the scattering angle 2θ. (b) Caricature of an energy loss spectrum showing the different types of excitations measurable by nonresonant IXS (XRS is individuate in core excitations). (c) Schematic view of a Rowland circle spectrometer. Sample (S), the crystal analyzer, and the focus point (F) lie on a common circle of radius R, the crystal analyzer surface follows a circle with radius $2R$, the Bragg angle is θ_B. (d) Simulated $S(q, \omega)$ for liquid H_2O for different momentum transfers given in atomic units.

process. X-rays with well-defined energy $\hbar\omega_1$ and momentum k_1 impinge onto a sample and are scattered off of the sample under the loss of a portion of their energy $\hbar\omega$ and momentum q into photons of energy $\hbar\omega_2$ and momentum k_2. An illustrative caricature of a typical nonresonant inelastic X-ray scattering spectrum is shown in Figure 11.7B. At zero energy loss (often also referred to as energy transfer), elastic scattering dominates the spectrum. Energy transfers of a few to a few tens of meV result in phonon excitations [30], at a few eV valence electron excitations (for example crystal-field excitations), collective excitations such as plasmons and their dispersion can be measured. At the highest transferred energies and momenta, Compton scattering dominates the nonresonant inelastic loss spectrum [31, 32]. When the energy loss is, however, tuned across the binding energy of bound electrons, the observable spectra resemble those of XAS and we speak of XRS.

As for XAFS, we start off from Fermi's golden rule describing the transition probability of an initial state $|i\rangle$ to a final state $|f\rangle$ due to a perturbation H_i, which, of course, represents the interaction of the photon field of the incoming X-rays with the electrons of the sample system:

$$S(q,\omega) \sim |\langle f \mid H_i|i\rangle|^2 \delta(\hbar\omega_i - \hbar\omega_f - \hbar\omega). \tag{11.16}$$

Here, the δ-function ensures energy conservation. H_i derives from the famous Kramers–Heisenberg formula describing the coupling of electrons to an electromagnetic field and since here we are only interested in nonresonant scattering, it is proportional to the square of the vector potential A of the photon field $H_i \propto A^2 \propto e^{-iqr}$ [33]. $S(q, \omega)$ is the dynamic structure factor and, as expected, depends on both, the momentum and energy transfer. At small momentum transfers q, we can simply expand the transition operator e^{-iqr} in a Taylor series $e^{-iqr} \sim 1 + qr - (qr)^2/2 + \cdots$, and truncate the series after the second term. Since furthermore $|i\rangle$ and $|f\rangle$ are orthogonal, we recognize the similarity of XAFS and XRS at small momentum transfers immediately [34]:

$$\begin{aligned} S(q,\omega) &\sim |\langle f \mid qr|i\rangle|^2 \delta(\hbar\omega_i - \hbar\omega_f - \hbar\omega), \\ \sigma(\omega_0) &\sim |\langle f \mid er|i\rangle|^2 \delta(\hbar\omega_i - \hbar\omega_f - \hbar\omega_0). \end{aligned} \tag{11.17}$$

According to Eq. (11.17), the momentum transfer q in XRS takes on a similar role as the polarization vector $\hat{e}$ in XAFS (we have silently assumed throughout the text that $q = |\mathbf{q}|$ without loss of generality).

The Taylor series, of course, fails for finite q. However, an expansion of the transition operator e^{-iqr} in terms of spherical harmonics holds for all q and illustrates how the (magnitude of the) momentum transfer can be utilized to change the selection rules for the excitation process [35]: at low momentum transfer, the transition operator takes the shape of a dipole operator, with increasing q higher order multipole transitions (quadrupole, octupole, triacontadipole, etc.) contribute more and more to $S(q, \omega)$ [36].

An interesting consequence of the momentum transfer dependence of XRS follows from the formulation of $S(q, \omega)$ in terms of Green functions or propagators as was done for the case of XAS in Section 2.2 [37]

$$S(q,\omega) \sim \sum_L M(\hbar\omega)_L \rho(\hbar\omega)_L, \tag{11.18}$$

with the angular momentum channel $L = (l, m)$. Eq. (11.18) implies that, if the matrix elements $M(\hbar\omega)_L$ are known, measurements of $S(q, \omega)$ at different values of q allow for the experimental determination of the local core-excited electronic density of states $\rho(\hbar\omega)_L$ via the inversion of this equation [38]. Indeed, the $M(\hbar\omega)_L$ can be assumed to be atomic quantities and they can be calculated using, for example, the real space multiple scattering program FEFFq [37].

Summarizing this discussion on the theoretical background, XRS allows for the measurement of electronic excitation spectra of low-Z elements using hard X-rays. XRS therefore yields access to low-Z elements' absorption edges even if the samples are contained in complicated in situ or operando environments. Contrary to soft-XAS spectra, XRS spectra are practically free of artifacts such as self- or over-absorption, since the used X-ray energies are usually far away from resonances of the sample system. At low momentum transfer q, the measured quantity is directly proportional to soft-XAS spectra, at larger momentum transfer excitation channels other than dipole channels open and the entire unoccupied density of states (lDOS) are available.

An extensive overview of the theory of inelastic X-ray scattering in general and XRS in particular can be found in Refs. [29, 39], a shorter introduction to XRS can be found, for example, in Ref. [40].

11.5.2 Experimental Setup

11.5.2.1 Instrumentation

The evident benefits of XRS for the in situ study of shallow excitation spectra are clear from the discussion in the previous section. However, obviously these benefits arrive at a heavy price: the cross section of XRS is notoriously low and usually orders of magnitude smaller than that of photoelectric absorption. Therefore, today XRS is only routinely performed at dedicated and powerful instruments at third- and/or fourth-generation synchrotrons. Furthermore, since the overall requirements to energy resolution are unchanged compared to soft XAS, only high-resolution Bragg diffractive crystal spectrometers are to be considered for the analysis of the inelastically scattered photons.

Usually, spectrometers in so-called Rowland circle geometry are used. Sample, Bragg diffractive analyzer crystal, and detector are aligned on a common circle with radius R, the Rowland circle (see Figure 11.7C). The spherically curved crystal analyzer has a curvature radius of $2R$ for combined maximized focusing and minimized energetic aberrations (so-called Johann geometry) [41]. Since the absolute energy scales are irrelevant for XRS and only the energy transfer is important, the incident X-ray energy can be chosen relatively freely. Consequently, the Rowland spectrometer is operated as close as possible to the backscattering energy of the analyzer crystal ($\theta_B \sim 90.0°$), as many of the optical and energetic aberrations scale as $\cot(\theta_B)$ [42].

The detector is often chosen to be an energy dispersive detector, such as silicon drift diodes, in order to suppress spurious elastically scattered photons. In recent years, however, pixelated area detectors have gained popularity for the use in XRS spectrometers. In combination with the spherically bent analyzer crystals, which exhibit point-to-point focusing properties when operated close to backscattering, these two-dimensional detectors allow for spatially resolved XRS spectroscopy. This technique was coined direct tomography (DT) and is used to create three-dimensional tomographic images of heterogeneous samples with XRS as contrast mechanism [43–45].

Multiple dedicated instruments for XRS exist worldwide, all of which were built with the aim to combine high brilliance X-ray sources and efficient signal collection [42, 46–48]. Latter is usually achieved by using multiple Bragg analyzer crystals at the same time in order to increase the solid angle of detection.

11.5.2.2 Data Processing

The nonresonant nature of XRS leads to the fact that not only scattering from the target electrons (for example the two 1s-electrons in the case of the K-edge or the six 2p-electrons for the $L_{2,3}$-edge), but scattering from all electrons of the sample system contributes to the measured signal. This extra signal appears as unavoidable background and depends heavily on momentum transfer and energy loss. It is mostly due to the response of the valence electrons of the sample system and, to lesser extent, to other core electron excitations. At low momentum transfers, collective plasmon excitations dominate the background, electron–hole pair excitations and the Compton profile of the valence electrons dominate at intermediate and high momentum transfers, respectively. In this context, one way to understand the XRS signal of the desired core electron is that it represents the onset of the core Compton profile of the target electrons. In laboratory jargon, we say the XRS signal "rides" on the Compton profile. This implies that on the one hand, the Compton profile stemming from the valence electrons needs to be removed from the measured signal and, on the other hand, the signal-to-background ratio of the XRS signal can be tuned by adequate choice of the position of the Compton profile along the energy loss scale via the momentum transfer q [49]. If the maximum of the Compton profile coincides in energy loss with the desired edge onset, the subtraction of the Compton background becomes challenging, if the energy loss position of the Compton profile is too far away from the edge onset, the edge jump becomes small and the signal might be small even compared to the unavoidable general noise (spurious scattering from air or other parts of the sample container, detector noise, and spurious fluorescence).

Several recipes for the data extraction and processing exist [40, 50]. The subtraction of the valence electron response using parametrized functions, such as the PearsonVII function, seems to be most convenient. More elaborate schemes include the experimental determination of the valence Compton profile at a momentum transfer where edge onset and Compton profile are well separated and subsequent transfer to other momentum transfers by use of the definition of the Compton profile (see Refs. [40, 50] for details).

In a last step, the extracted XRS spectra have to be normalized. Ideally, Thomas–Reiche–Kuhn sum rules should be satisfied

$$\int_0^\infty S(q,\omega)\omega d\omega = N_{\text{el.}} q^2/2m, \tag{11.19}$$

where $N_{\text{el.}}$ is the number of electrons and m the electron mass. However, the small cross section of XRS often leads to experimentally available spectral ranges that are too short and consequently lead to faulty normalization integrals. Recent suggestions to alleviate this problem include the normalization to f-density, *i.e.* a small spectral region around the sudden limit [51].

11.6 Case Studies: Application of XAFS and XRS for Energy Materials

As it was highlighted so far, EXAFS specifically probes short range order and allows access to coordination numbers, bond distances, and chemical identity of nearest neighbors. Thus, it is an ideal technique for structural studies of poorly crystallized materials as for instance metal catalysts and solid oxides used in fuel cells devices. In this section, first an example of a gold catalyst used for CO oxidation will be discussed to describe the technique of EXAFS as a structural tool. Thereafter, results on dopant cations used in a solid oxide fuel cell (SOFC) material will be reported. Finally, an example on SOFCs, for which both EXAFS and XRS techniques have been used for the structural characterization, will be examined, highlighting the results obtained with XRS.

11.6.1 CO Oxidation Reaction: The Au/CeO_2 Catalyst

In chemistry, reactions very often do not take place because certain steps in the reaction, for instance the dissociation of molecules, require large amounts of energy. These steps can be facilitated by the presence of chemical species, the catalysts, which do not directly participate in the reaction. Although involved in the reaction, the catalysts are returned to their initial state when the reaction is completed. If the catalyst and the reactants are present in different phases, the specific scientific topic is called heterogeneous catalysis. In a such system, molecules are adsorbed to the surface of a catalytic particle, generally metal clusters of nanometer size, which are deposited onto an inert support. The interaction of the reacting molecules with the metal can influence the bonding in the reacting molecules, helping the desired reaction. It is evident that the size of the particles is an important parameter. This parameter is strongly influenced by the possible interaction occurring between the support and the metal particle itself. As a consequence, the nature of this interaction can strongly influence the catalytic properties of the system. Such small metal particles, used as catalyst, usually exhibit short range order and EXAFS is a unique tool used to study them and their variation during the chemical reaction, especially when in situ measurements are possible. Notably, in such extremely small nanoparticles (1–4 nm), a relatively large number of atoms is in contact with the support and EXAFS can provide information on the metal/support interface [52].

Catalytic CO oxidation has gained increasing attention in the last decade, primarily due to its demand and possible use in industrial processes such as: abatement of gaseous waste in petrochemical industries, synthesis of pure gases, ethanol or other fuel production, and pure hydrogen production for proton-exchange membrane fuel cells [53].

To achieve this aim, catalysts based on supported gold nanoparticles on ceria, owing to their exceptionally high activity for low-temperature CO oxidation, have been widely investigated.

Different aspects have been studied in order to explain such extraordinary catalytic behavior. Among the different factors including cluster size, preparation method, and different oxides used as support, the influence of support remains the most accredited and investigated. Indeed, one of the key factors, which makes ceria unique for this reactions is its ability to easily exchange oxygen atoms with the chemical environment and this process is strongly connected with the supported metal clusters and their size [54–57].

It is widely accepted that the high activity of gold/ceria catalysts is due to the enhanced electron transfer between defective ceria and partially charged gold nanoparticles via oxygen vacancies, with formation of a complex between partially charged gold clusters and the support [22, 56, 58]. Accordingly, the CO oxidation proceeds at the interface between small gold clusters in an intimate contact with the ceria support. To verify the intimate contact between the gold cluster and the ceria, the knowledge of the local structure and oxidation state of Au is essential. To this end, X-ray absorption spectroscopy, which is selective for the atomic species and probes the environment of the photo-absorber atom, plays a crucial role because it can highlight the existence of Au–Cerium bonds. Finally, the existence of this bond proves the establishment of the metal–support interaction. The volume of this interface is in the sub-nanometer region, so scattering bulk techniques fail in the identification of this structural modification. On the other hand, electron spectroscopies such as XPS, limited to the surface of the catalyst, are not able to detect the electronic modifications, which are fully averaged with the ceria surface.

The results shown in Figure 11.8 demonstrate the existence of a large interface between the gold particle and the support oxide, characterized by well-defined AuO and AuCe interactions extending up to 6.4 Å (see Figure 11.8 Panel A and B) [22]. The appearance of an interaction between gold and ceria is demonstrated by the stable AuO and AuCe distances observed in the samples evidenced by longer AuCe distance, associated with a AuAuCe three-body multiple scattering paths (see Figure 11.8 Panel C). This contribution is partially eclipsed during the reaction by a coarsening of the gold particles and an increase of the related disorder (Debye–Waller) factor. In conclusion, it is proposed that the gold–ceria interface structure involves a contact oxygen layer between the gold particle and the support, giving rise to roughly linear AuOCe bonds extending over a wide interface region.

11.6.2 Materials for Solid Oxide Fuel Cells

Solid oxide fuel cells (SOFCs) are high-temperature (≥600 °C) electrochemical devices that employ hydrogen or hydrocarbon fuel to produce electricity for stationary applications: the high operating temperature allows for a high thermodynamic efficiency, and less stringent requirements on fuel purity. On the other hand, thermal stress and long-term degradation of materials and their mutual compatibility are still key issues to be addressed. In addition, high ion conductivity in solid oxides is

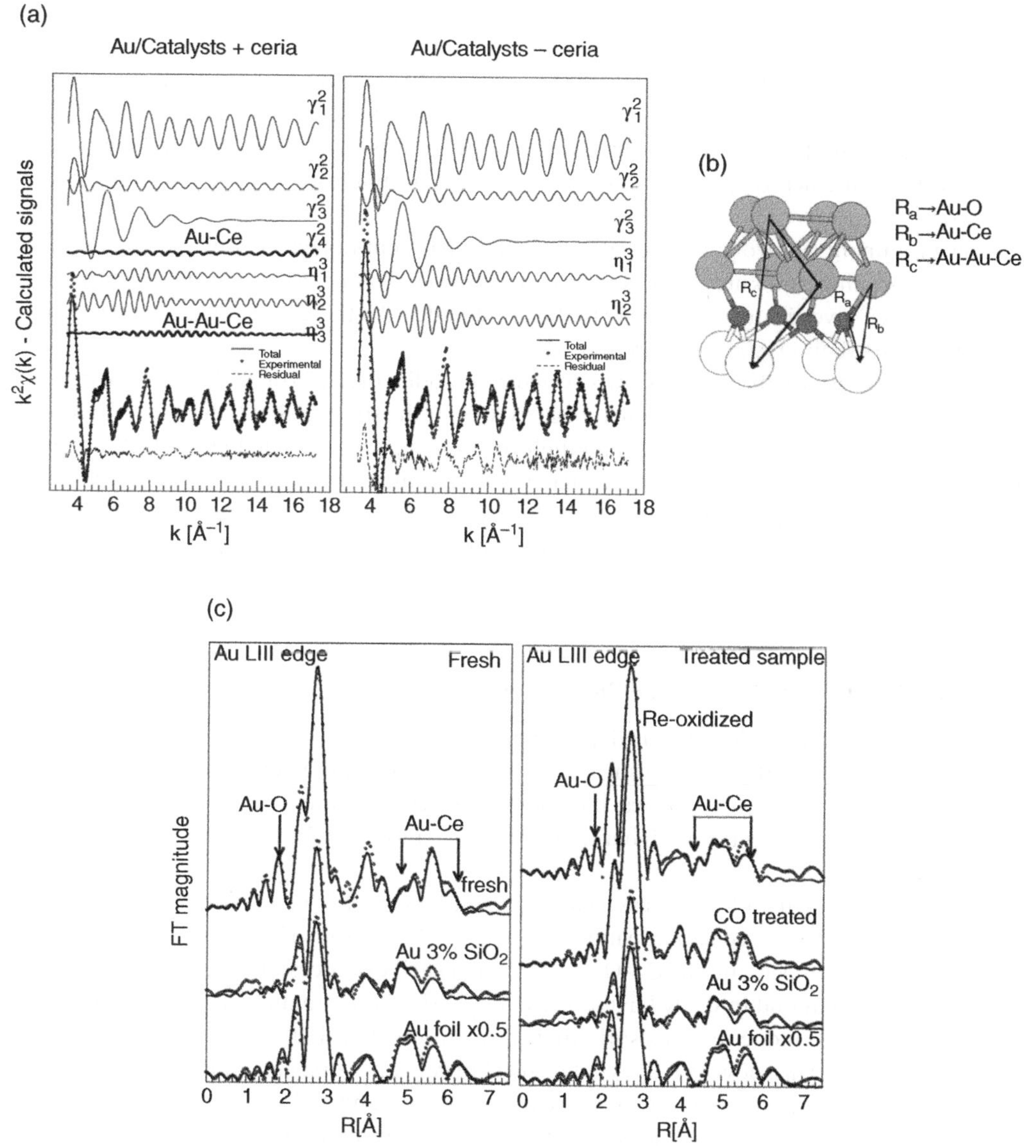

Figure 11.8 Panel A: observed and calculated EXAFS signals for the gold catalyst together with the single and multiple scattering paths used in the fitting procedure. Panel B: shows the model catalyst used to calculate the gold–ceria signals. Panel C: Fourier transform of the gold catalysts after CO and oxygen treatment compared with Au foil and a model gold catalyst supported on silica. Source: Figure Panel B with permission from Longo et al. [22]. Copyright 2012 American Chemical Society.

generally much lower than in liquid phases. SOFC devices and materials for SOFC electrolytes and electrodes have been the subject of many reviews [59, 60].

Most electrolyte and electrode materials for SOFC are doped metal oxides comprising often three, four, or more different cations, making them an ideal case study for XAFS techniques: in fact, in these structures the presence of different metal ions gives rise to a variety of local atomic structures, which are impossible to investigate by conventional structural probes such as X-ray diffraction.

In a typical XAFS experiment with a millimeter-sized X-ray beam, a very large number of absorbing atoms are probed simultaneously, on the order of 10^{17}. This ensures that the chemical and structural information derived from the analysis has statistical significance and that it represents a realistic picture of the material as a whole, regardless of surface effects or sampling artifacts. This fact is extremely important when critically evaluating the results vis-à-vis other techniques that do not possess the same degree of bulk averaging, e.g. electron microscopy or X-ray photoelectron spectroscopy.

Among prominent electrolyte materials for SOFC, trivalent-doped zirconia, trivalent-doped ceria, doped barium perovskites (cerates or zirconates), and alkaline earth-doped lanthanum niobate have all been investigated extensively in the last decades. All these oxides feature one or more doped sites inside their crystal structure [60–62].

In the following, we briefly outline a few selected case studies of XAFS on SOFC electrolytes. As a concluding remark, the application of microbeam XAFS for the study of SOFC devices is presented. The peculiar role of the dopant cations, and their relationship with mobile ionic defects and with the lattice as a whole, have been the subject of many XAFS studies. In particular, XAFS techniques have been extensively applied on several SOFC electrolyte materials (fluorites or perovskites): in such studies, the site selectivity of X-ray absorption spectroscopy was exploited to clarify the role of the dopant atoms in the lattice, and to correlate it with the ion transport properties.

11.6.3 Oxide-Ion Conductors: Dopants and Vacancies

In one such study, trivalent-doped ceria doped with either Yb^{3+}, Sm^{3+}, or Er^{3+} (or M^{3+} as a whole) with doping levels up to 30% was investigated: in these materials, the dopant ions are inserted in order to create mobile oxygen vacancies, but charged defects (dopants and oxygen vacancies) can also cluster together – especially at higher doping levels – into extended defect regions, eventually hindering the mobility.

Using a statistical model of defect association [63], it is possible to correlate the occurrence of different dopant–vacancy association patterns with the XAFS results (in particular, the Ce–O or M–O coordination number, and the Ce–O and M–O first shell distances). In this way, a few important differences were found in the behavior of seemingly similar lanthanide dopants: (i) on a local level, different dopants induce either a shrinking or expansion of the lattice, depending on their size difference with Ce^{4+}. This, however, does not result simply in an average modification of the crystal lattice as one would expect from Vegard's law [64]; (ii) Yb^{3+} and Er^{3+} attract oxygen vacancies, forming defect clusters, while Sm^{3+} does not: in Sm-doped ceria, therefore, the oxygen-ion conduction is not hindered by the dopant sites (see Figure 11.9); (iii) the Ce–O bonding states, probed by Ce XANES, correlate well with the activation energy for the oxygen-ion conductivity, and also in this case the lattice is less perturbed by Sm^{3+} cations, resulting in unimpeded ion mobility [65].

The highest oxygen ion conductivity of all oxide materials is observed in the disordered Bi_2O_3, whose structure could be better represented as a $BiO_{2-0.5}$ fluorite, with 1/2 oxygen vacancies per unit formula. In such a structure, every Bi^{3+} ion is coordinated by on average six oxygen atoms, but the coordination geometry is maintained cubic like in the fluorite aristotype. This high-symmetry disordered structure is thermodynamically favored over the low-symmetry one only at very high temperatures (e.g. 900 °C), but it can be stabilized at lower temperature by doping with a smaller hypervalent cation (e.g. Ta^{5+}). EXAFS and XANES on the Ta L-edge were used to show the formation of Ta_4O_{18} clusters and further aggregates like Ta_7O_{30} throughout the lattice: these clusters "freeze" oxygen vacancies in TaO_6 octahedra (as opposed to BiO_8 pseudo-cubes) [66].

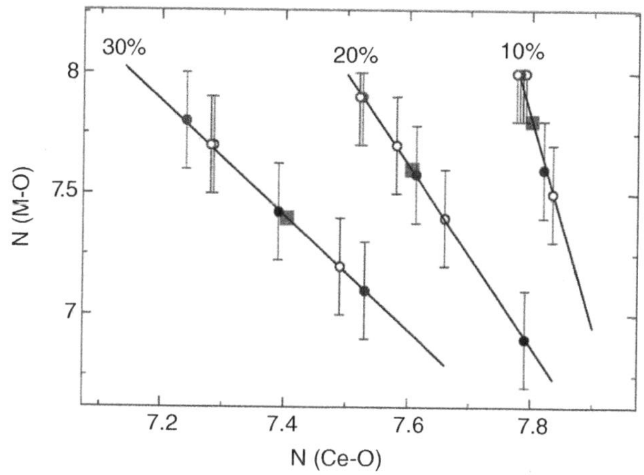

Figure 11.9 Combined EXAFS analysis of different edges is a powerful tool for understanding complex oxides. In this plot, the first shell coordination numbers around Ce (Ce–O) and dopants (M–O) in cerium oxide doped with Er^{3+}, Yb^{3+}, or Sm^{3+} derived from Ce and dopant L-edge spectra. Nanocrystalline samples sintered at 800 °C are empty circles, microcrystalline samples sintered at 1250 °C are full circles. The black lines represent the constraints due to oxygen stoichiometry for different dopant contents (10–30%), while the green squares correspond to a perfectly random distribution of dopants and vacancies. Relatively lower M–O and higher Ce–O values indicate association between the dopant and oxygen vacancy defects. Source: Reprinted with permission from Giannici et al. [65]. Copyright 2014 American Chemical Society.

11.6.4 Proton-Conducting Oxides

The systematic investigation of several dopant ions in barium perovskites ($BaCeO_3$ and $BaZrO_3$) was the subject of a number of studies. Different trivalent dopants affect ion conductivity in different ways, apparently uncorrelated to their ionic radii. Usual crystal structure investigations, based on X-ray diffraction or neutron diffraction techniques, did not help to clarify the issue, since the finer details of the local atomic structure around the dopants are lost in such methods. Also in this case, XAFS techniques proved successful in clarifying the peculiar chemical role of each dopant, and how different dopants interact with the surrounding lattice and with the other charged defects, and how the local structural effects in turn determine the ion conductivity [64].

In^{3+} is larger than Zr^{4+} and smaller than Ce^{4+}. Therefore, when In^{3+} is used as dopant in either cerates or zirconates, it causes a deformation of the lattice due to a size mismatch with the host cation. Such a strain affects a large area around the dopant, which encompasses several coordination shells, as witnessed by the EXAFS analysis of the In–O, In–Ba, and In–M distances. This kind of "strain dilution" is what most likely drives the complete miscibility of In^{3+} in the host lattice, so that both $BaCeO_3$ and $BaZrO_3$ form a solid solution with $Ba_2In_2O_5$ in any proportion. In fact, the effect of In^{3+} insertion affects a very large volume around the dopant even in the low-concentration regime (e.g. 2%), affecting the proton transfer activation energy [67, 68].

From the point of view of the local structure, different dopants display seemingly unpredictable behavior with respect to their ionic size, which is usually taken as the only parameter to interpret the doping process in a crystal lattice. In fact, the Y^{3+} cation, which is slightly larger than Ce^{4+} and definitely larger than Zr^{4+}, affects a very small volume around it. Several factors, including the polarizability of the dopant cation, and its match between dopant and host cation, have all been considered for the different behaviors of dopants. Another important point in the local structural analysis of proton conductors is based on the comparison of the EXAFS spectra on the absorption edges of both the dopant and host cation, between the protonated and deprotonated states, in order to evaluate any selective effect of the proton insertion on the local environment of the dopant. In this way, the possible attraction between the M^{3+} dopant (which carries a negative relative charge in a perovskite) and the protonic defect (positively charged) is measured. The association between these two defects is determined by the dopant/host couple, and it usually does not correlate unequivocally with the overall performance of the electrolyte in terms of proton conductivity. As an example, Y:$BaCeO_3$ and Gd:$BaCeO_3$ display significant association between defects, while Y:$BaZrO_3$ and In:$BaCeO_3$ do not, with the Y^{3+}-doped materials usually showing the highest proton conductivity [69–71].

11.6.5 The Role of Oxygen in Fuel Cell Cathodes

So far, we discussed the influence of dopant cations in the structure of oxides used as electrolyte for fuel cell applications. Besides the development of suitable electrolyte materials, the achievement of operative temperatures in the range of 500–700 °C, usually referred to as intermediate-temperature (IT) regime, also involves significant efforts to improve the electrocatalytic activity and the mixed electronic-ionic conductivity (MIEC) at the electrodes. These efforts concern, in particular, the cathodic materials that constitute the bottleneck toward the realization of efficient IT-SOFC devices.

Lanthanum strontium cobalt ferrites (LSCF), at different compositions, are still the most studied cathodic materials for IT-SOFCs and, as reported in many reviews, the conductivity of LSCF depends very much on the stoichiometric composition [72–75]. It is well recognized that the electronic conduction in LSCF is of p-type and that, when oxygen vacancies are created as a consequence of decreased O_2 partial pressure, electrons are injected into the Co_3d/O_2p hybridized state, thus affecting the hole conductivity [23, 76]. Recently, Ni-doped LSCF cathode have attracted interest because Ni^{3+} hinders the formation of oxygen vacancies near Co, thus stabilizing the higher oxidation state of Co [23, 76]. Moreover, Ni is able to stabilize oxygen vacancies as already reported for Ni-doped strontium titanate. In the latter case, the effect was attributed to the electronic structure of Ni^{3+}, which has one electron occupying one of the e_g orbitals geometrically directed toward the vacancy and, as observed by EPR spectroscopy, can form stable Ni^{3+}–Vo complexes [77, 78]. Ultimately, the stabilization of the oxygen vacancy and the better covalent interaction with the lattice oxygen strengthen the transition metal-oxygen (TM-O_6) configurations. The crucial role played by the TM_3d-O_2p band in the redox behavior of LSCF is highlighted by Mueller and coworkers [76]. Even though limited by the soft X-ray constrains, Mueller and coworkers demonstrated by operando EXAFS at the O K-edge, that the surface transition-metal (TM)-O_6 octahedron is the redox-active building block of LSCF and other perovskite oxides [23, 76]. In this respect, the knowledge of the oxygen K-edge is a unique tool to understand the electronic modification occurring during the O_2 absorption/insertion in the cathode especially if bulk information is involved. To this aim, as evidenced in the previous section, XRS, circumventing the soft X-ray limitations, plays a key role.

XRS measurements at the O K-edge for Ni:LSCF cathode are reported in Figure 11.10 Panel I. O K-edge excitations correspond to electronic transitions from the O 1s core level to 2p states hybridized with the TM3d states, thereby reflecting the extent of covalence between O2p and TM3d orbitals (TM=Ni, Fe, Co) [76]. Accordingly, the first double feature in the pre-edge region between 525 and 530 eV (peaks A and A′ in Figure 11.10) is attributed to bands of mixed O2p and TM3d character. The second peak (B) corresponds to bands deriving from Sr 4d/La 5d electronic states and the broad, intense peak at 544 eV (C) is attributed to bands of mixed O2p and TM 4s and 4p character.

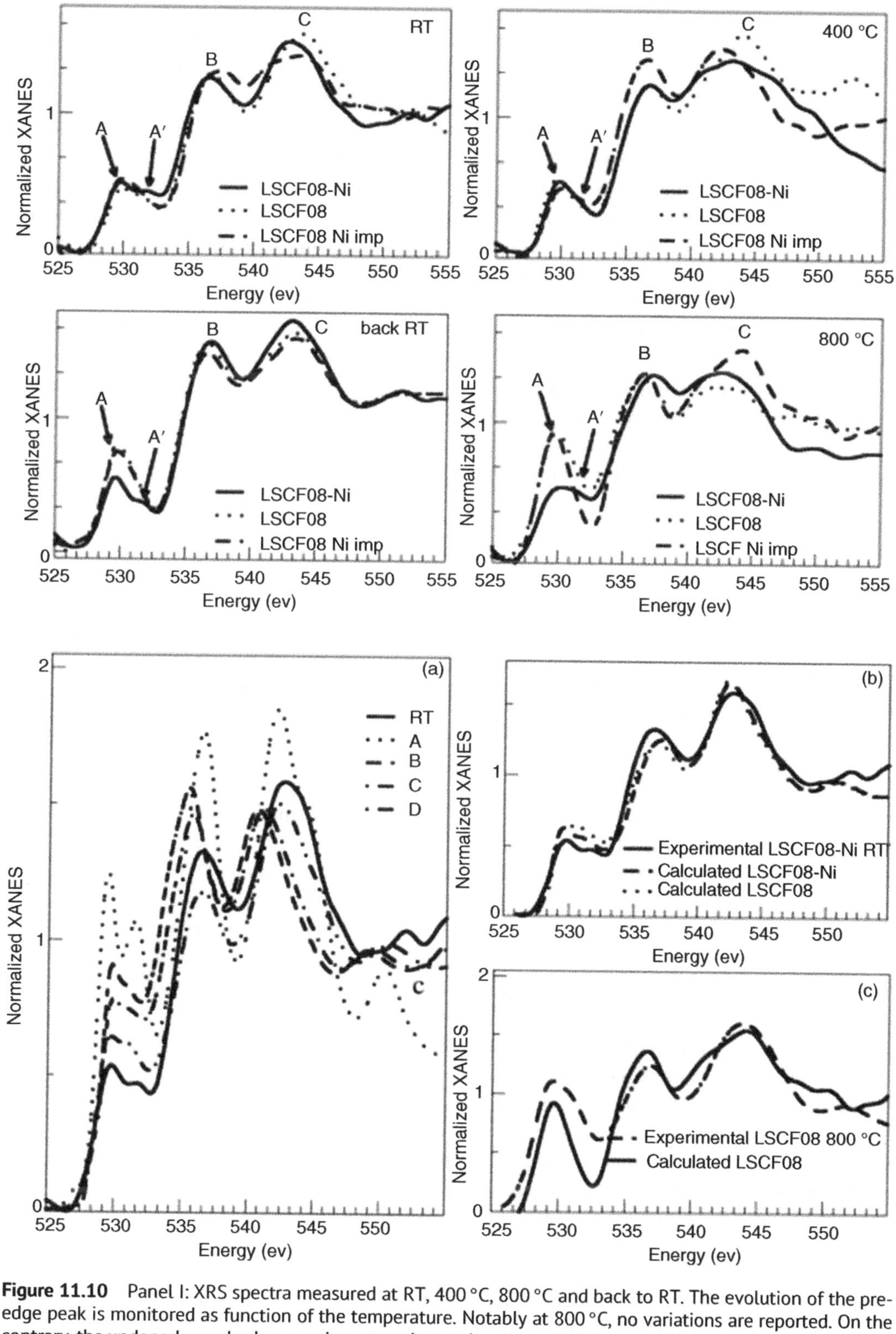

Figure 11.10 Panel I: XRS spectra measured at RT, 400 °C, 800 °C and back to RT. The evolution of the pre-edge peak is monitored as function of the temperature. Notably at 800 °C, no variations are reported. On the contrary, the undoped sample shows an important increasing connected to an emptying pf the states (see text). Panel II: Simulated XRS spectra with different density population; observed and calculated XRS spectra for RT b and Calculated XRS at 800 °C.

As evident in Figure 11.10, the spectra at RT and up to 400 °C are similar for all the samples. On the other hand, there is a clear and systematic change as a function of the thermal treatment, in particular for the pre-edge features, which are connected with the hybridization between the O2p and the TM3d states. At 800 °C, we observe an abrupt increase of intensity in the pre-edge of the samples. Notably, the pre-edge features in Ni-doped LSCF show only a modest variation with temperature and the initial RT spectrum is partially recovered after thermal treatment.

The pre-edge features can be associated to the σ^* (π^*) molecular orbital in LSCF, which originates from the antibonding mixture of TM (Fe/Co) 3d $e_{g\uparrow}$ ($t_{2g\downarrow}$) and O2p states. The variation in the vacancy concentration modifies the pre-edge features. An enhanced spectral weight of these is indicative of a depopulation of the corresponding electronic states near the Fermi level. The presence of Ni in the perovskite structure, as proved by calculations reported in Figure 11.10 (Panel II a, b, and c), retaining the electronic density on the $e_g\uparrow$ states during the treatment in air, inhibits the depopulation of the states, which instead are predominant in the undoped LSCF samples. Thus, during the thermal treatment in the LSCF–Ni sample, the lattice oxygen keeps its reduced state remaining bonded with the TM, so that the formation of oxygen vacancies is considerably hampered. Then, as reported elsewhere [23], the better electrochemical performances of Ni:LSCF are correlated to the ability of Ni to stabilize the lattice oxygen.

References

1 Kronig, R.d.L. (1931). Zur theorie der feinstruktur in de röntgenabsorptionsspektren. *Z. Phys.* **70**: 317–323.

2 Sayers, D.E., Stern, E.A., and Lytle, F.W. (1971). New technique for investigating noncrystalline structures: Fourier analysis of the extended X-ray—absorption fine structure. *Phys. Rev. Lett.* **27**: 1204–1207.

3 Stern, E.A. (1974). Theory of the extended X-ray-absorption fine structure. *Phys. Rev. B.* **10**: 3027–3037.

4 Lee, P.A. and Pendry, J.B. (1975). Theory of the extended X-ray absorption fine structure. *Phys. Rev. B.* **11**: 2785–2811.

5 Koningsberger, D.C. and Prins, R. *Principles, Applications and Techniques of EXAFS, SEXAFS and XANES*. New York: Wiley.

6 Filipponi, A., Di Cicco, A., and Natoli, C.R. (1995). X-ray-absorption spectroscopy and n-body distribution functions in condensed matter. I. Theory. *Phys. Rev B* **52**: 15122, 15134.

7 Rehr, J.J. and Albers, R.C. (2000). Theoretical approaches to X-ray absorption fine structure. *Rev. Mod. Phys.* **72**: 621–654.

8 Joly, Y. (2001). X-ray absorption near-edge structure calculations beyond the muffin-tin approximation. *Phys. Rev B* **63**: 125120-1–125120-10.

9 Mobilio, S. and Vlaic, G. (2003). Synchrotron radiation: fundamentals, methodologies and applications, conference proceedings. *Società Italiana di Fisica* **82**.

10 Bunker, G. (2010). *Introduction to XAFS: A Practical Guide to X-Ray Absorption Fine Structure Spectroscopy*. Cambridge University Press.

11 van Bokhoven, J. and Lamberti, C. (2016). *X-Ray Absorption and X-Ray Emission Spectroscopy Theory and Applications*. Wiley.

12 De Groot, F. and Kotani, A. (2008). *Core Level Spectroscopy of Solids*. CRC Press.

13 Hedin, L. and Lundqvist, S. (1969). Effects of electron-electron and electron-phonon interactions on the one-electron states of solids. *Solid State Phys.* **23**: 1–181.

14 Lee, P.A. and Beni, G. (1977). New method for the calculation of atomic phase shifts: application to extended X-ray absorption fine structure (EXAFS) in molecules and crystals. *Phys. Rev. B* **15**: 2862–2883.
15 Bransden, B.H. and Joachain, C.J. (2003). *Physics of Atoms and Molecules*, 2e. Prentice Hall.
16 Filipponi, A. (2001). Exafs for liquids. *J. Phys. Condens. Matter* **13**: R23.
17 k. Hakada, C., Natoli, R., and Benfatto, M. (1987). Self-consistent non-local optical potential for XANES. *J. Phys. Coll.* **48**: C9.
18 Filipponi, A. and Di Cicco, A. (1995). X-ray-absorption spectroscopy and n-body distribution functions in condensed matter. II. Data analysis and applications. *Phys. Rev. B* **52**: 15 135–15 149.
19 Bunau, O. and Joly, Y. (2009). Self-consistent aspects of X-ray absorption calculations. *J. Phys. Condens. Matter* **21**: 1–11.
20 Guda, S.A., Soldatov, M.A., Lomachenko, K.A. et al. (2015). *J. Chem. Theory Comput.* **11**: 4512–4521.
21 Gilmore, K., Vinson, J., Shirley, E.L. et al. (2015). Efficient implementation of core-excitation Bethe–Salpeter equation calculations. *Compt. Phys. Comm* **197**: 109–117.
22 Longo, A., Liotta, L.F., Pantaleo, G. et al. (2012). Structure of the metal–support interface and oxidation state of gold nanoparticles supported on Ceria. *J. Phys. Chem. C* **116**: 2960–2966.
23 Longo, A., Liotta, L.F., Banerjee, D. et al. (2018). The effect of Ni doping on the performance and electronic structure of LSCF cathodes used for IT-SOFCs. *J. Phys. Chem. C* **122**: 1003–1013.
24 Shadle, S.E., Penner-Hann, J.E., Schugar, H.J. et al. (1993). X-ray absorption spectroscopic studies of the blue copper site: metal and ligand K-edge studies to probe the origin of the EPR hyperfine splitting in plastocyanin. *J. Am. Chem. Soc.* **115**: 767–776.
25 Westre, T.E., Kennpohl, P., DeWitt, J.G. et al. (1997). A multiplet analysis of Fe K-edge 1s 3d pre-edge features of iron complexes. *J. Am. Chem. Soc.* **119**: 6297–6314.
26 Penner-Hahn, J.E. (2005). X-ray absorption spectroscopy. *Wiley Online Library* https://doi.org/10.1038/npg.els.0002984.
27 Reh, J.J., Kas, J.J., Vila, F.D. et al. (2010). Parameter-free calculations of X-ray spectra with FEFF9. *Phys. Chem. Chem. Phys.* **12**: 5503–5513.
28 Calvin, S. (2013). *XAFS for Everyone*. CRC Press.
29 Schülke, W. (2007). *Electron Dynamics by Inelastic X-Ray Scattering*, 7. Oxford University Press.
30 Krisch, M. and Sette, F. (2006). Inelastic X-ray scattering from phonons. In: *Light Scattering in Solid IX*, 317–370. Berlin, Heidelberg: Springer.
31 Eisenberger, P. and Platzman, P. (1970). Compton scattering of X-rays from bound electrons. *Phys. Rev. A* **2** (2): 415–423.
32 Cooper, M.J. (1985). Compton scattering and electron momentum determination. *Rep. Prog. Phys.* **48** (4): 415–481.
33 Kramers, H.A. and Heisenberg, W. (1925). Über die Streuung von Strahlung durch Atome. *Z. Phys.* **31** (1): 681–708.
34 Mizuno, Y. and Ohmura, Y. (1967). Theory of X-ray Raman scattering. *J. Phys. Soc. Jpn.* **22** (2): 445–449.
35 Haverkort, M., Tanaka, A., Tjeng, L., and Sawatzky, G. (2007). Nonresonant inelastic X-ray scattering involving excitonic excitations: the examples of NiO and CoO. *Phys. Rev. Lett.* **99** (25): 25740-1–257401-4.
36 Gordon, R., Seidler, G., Fister, T. et al. (2007). High multipole transitions in NIXS: valence and hybridization in 4f systems. *EPL (Europhys. Lett.)* **81** (2): 26004-p1–26004–p6.
37 Soininen, J.A., Ankudinov, A., and Rehr, J. (2005). Inelastic scattering from core electrons: a multiple scattering approach. *Phys. Rev. B* **72** (4): 045136-1–045136–10.
38 Soininen, J.A., Mattila, A., Rehr, J. et al. (2006). Experimental determination of the core-excited electron density of states. *J. Phys. Condens. Matter* **18** (31): 7327–7336.

39 Häamaäläainen, K. and Manninen, S. (2001). Resonant and non-resonant inelastic X-ray scattering. *J. Phys. Condens. Matter* **13** (34): 7539–7555.

40 Sahle, C.J., Mirone, A., Niskanen, J. et al. (2015). Planning, performing and analyzing X-ray Raman scattering experiments. *J. Synchrotron Radiat.* **22** (2): 400–409.

41 Johann, H.H. (1931). Die erzeugung lichtstarker Röntgenspektren mit Hilfe von Konkavkristallen. *Z. Phys.* **69** (3–4): 185–206.

42 Huotari, S., Sahle, C.J., Henriquet, C. et al. (2017). A large-solid-angle X-ray Raman scattering spectrometer at ID20 of the European Synchrotron Radiation Facility. *J. Synchrotron Radiat.* **24** (2): 521–530.

43 Huotari, S., Pylkkänen, T., Verbeni, R. et al. (2011). Direct tomography with chemical-bond contrast. *Nat. Mater.* **10** (7): 489–493.

44 Sahle, C.J., Mirone, A., Vincent, T. et al. (2017). Improving the spatial and statistical accuracy in X-ray Raman scattering based direct tomography. *J. Synchrotron Radiat.* **24** (2): 476–481.

45 Georgiou, R., Gueriau, P., Sahle, C.J. et al. (2019). Carbon speciation in organic fossils using 2D to 3D X-ray Raman multispectral imaging. *Sci. Adv.* **5** (8): eaaw5019-1–eaaw5019-9.

46 Fister, T., Seidler, G., Wharton, L. et al. (2006). Multielement spectrometer for efficient measurement of the momentum transfer dependence of inelastic X-ray scattering. *Rev. Sci. Instrum.* **77** (6): 063901-1063901–7.

47 Verbeni, R., Pylkkänen, T., Huotari, S. et al. (2009). Multiple-element spectrometer for non-resonant inelastic X-ray spectroscopy of electronic excitations. *J. Synchrotron Radiat.* **16** (4): 469–476.

48 Sokaras, D., Nordlund, D., Weng, T.-C. et al. (2012). A high resolution and large solid angle X-ray Raman spectroscopy end-station at the Stanford Synchrotron Radiation Lightsource. *Rev. Sci. Instrum.* **83** (4): 043112-1–043112-9.

49 Fister, T., Seidler, G., Hamner, C. et al. (2006). Background proportional enhancement of the extended fine structure in nonresonant inelastic X-ray scattering. *Phys. Rev. B* **74** (21): 214117-1–214117-7.

50 Sternemann, H., Sternemann, C., Seidler, G. et al. (2008). An extraction algorithm for core-level excitations in non-resonant inelastic X-ray scattering spectra. *J. Synchrotron Radiat.* **15** (2): 162–169.

51 Niskanen, J., Fondell, M., Sahle, C.J. et al. (2019). Compatibility of quantitative X-ray spectroscopy with continuous distribution models of water at ambient conditions. *Proc. Natl. Acad. Sci.* **116** (10): 4058–4063.

52 Bond, G.C., Louis, C., and Thompson, D.T. (2006). *Catalysis by Gold*, vol. **6**. London: Imperial College Press.

53 Al Soubaihi, R.M., Saoud, K.M., and Dutta, J. (2018). Critical review of low-temperature co oxidation and hysteresis phenomenon on heterogeneous catalysts. *Catalysts* **8**: 660-1–660-9.

54 Haruta, M., Tsubota, S., Kobayashi, T. et al. (1993). Lowtemperature oxidation of co over gold supported on tio$_2$, $- Fe_2O_3$*, and* Co_3O_4. *J. Catal.* **144**: 175–192.

55 Venezia, A.M., Pantaleo, G., Longo, A. et al. (2005). Relationship between structure and co oxidation activity of ceria-supported gold catalysts. *J. Phys. Chem. B* **109**: 2821–2827.

56 Zhang, C., Michaelides, A., King, A.D., and Jenkins, S. (2010). Positive charge states and possible polymorphism of gold nanoclusters on reduced ceria. *J. Am. Chem. Soc.* **132**: 2175–2182.

57 Trovarelli, A. (2002). *Catalysis by Ceria and Related Materials*, Catalytic Science Series, vol. **2**, 2175–2182. London: Imperial College Press.

58 Longo, A., Liotta, L.F., Di Carlo, G. et al. (2010). Structure and the metal support interaction of the Au/Mn oxide catalysts. *Chem. Mater.* **22**: 3952–3960.

59 Singhal, S.C. and Kendall, K. (2003). *High Temperature Solid Oxide Fuel Cells: Fundamentals, Design and Applications*. Elsevier.

60 Jacobson, A.J. (2010). Materials for solid oxide fuel cells. *Chem. Mater* **22** (3): 660–674.

61 Kreuer, K.-D. (2003). Proton-conducting oxides. *Annu. Rev. Mater. Res.* **33**: 333–359.

62 Tao, Z., Yan, L., Qiao, J. et al. (2015). A review of advanced protonconducting materials for hydrogen separation. *Prog. Mater. Sci.* **74**: 1–50.

63 Yashima, M., Ishizawa, N., and Yoshimura, M. (1992). Application of an ion-packing model based on defect clusters to zirconia solid solutions: I, modeling and local structure of solid solutions. *J. Am. Ceram. Soc.* **75**: 1541–1557.

64 Giannici, F., Longo, A., Kreuer, K. et al. (2010). Dopants and defects: local structure and dynamics in barium cerates and zirconates. *Solid State Ion.* **181**: 122–125.

65 Giannici, F., Gregori, G., Aliotta, C. et al. (2014). Structure and oxide ion conductivity: local order, defect interactions and grain boundary effects in acceptor-doped ceria. *Chem. Mater.* **26**: 5959–6006.

66 Gambino, M., Giannici, F., Longo, A. et al. (2015). Dopant clusterization and oxygen coordination in Ta-doped bismuth oxide: a structural and computational insight into the mechanism of anion conduction. *J. Phys. Chem. C* **119**: 26 367–26 373.

67 Giannici, F., Longo, A., Balerna, A. et al. (2007). Indium doping in barium cerate: the relation between local symmetry and the formation and mobility of protonic defects. *Chem. Mater.* **19**: 5714–5720.

68 Giannici, F., Longo, A., Balerna, A. et al. (2009). Proton dynamics in In:BaZrO3 insights on the atomic and electronic structure from X-ray absorption spectroscopy. *Chem. Mater.* **21**: 2641–2649.

69 Giannici, F., Longo, A., Deganello, F. et al. (2007). Local environment of barium, cerium and yttrium in BaCe1xYxO3 ceramic protonic conductors. *Solid State Ion.* **178**: 587–591.

70 Giannici, F., Longo, A., Balerna, A., and Martorana, A. (2009). Dopanthost oxide interaction and proton mobility in Gd:BaCeO3. *Chem. Mater.* **21**: 597–603.

71 Giannici, F., Shirpour, M., Longo, A. et al. (2011). Long-range and short-range structure of proton-conducting Y:$BaZrO_3$. *Chem. Mater.* **23**: 2994–3002.

72 Ivers-Tiffee, E., Weber, A., and Schichlein, H. (2003). *Handbook of Fuel Cells: Fundamentals Technology and Applications*, vol. **23**, 587600. New York: Wiley.

73 Ried, P., Bucher, E., Preis, W. et al. (2007). Characterization of La(0.6Sr0.4)(Co0.2Fe0.8)O_3 and Ba0.6Sr ECS. *TRANS Adv. Solid Oxide Fuel Cells VII* **7**: 1217–1224.

74 Itoh, T. and Nakayama, M. (2012). Using in situ X-ray absorption spectroscopy to study the local structure and oxygen ion conduction mechanism in (La0.6Sr0.4)(Co0.2Fe0.8)O_3. *J. Solid State Chem.* **192**: 3846.

75 Orikasa, Y., Ina, T., Nakao, T. et al. (2011). X-ray absorption spectroscopic study on La0.6Sr0.4CoO3 cathode materials related with oxygen vacancy formation. *J. Phys. Chem. C* **115**: 16 433–16 438.

76 Mueller, D.N., Machala, M.L., Bluhm, H., and Chueh, W. (2015). Redox activity of surface oxygen anions in oxygen-deficient perovskite oxides during electrochemical reactions. *Nat. Comm.* **6**: 7097-1–7097-7.

77 Muller, K.A., Berlinger, W., and Rubins, R.S. (1969). Observation of two charged states of a nickel-oxygen vacancy pair in SrTiO3 by paramagnetic resonance. *Phys. Rev.* **186**: 361–371.

78 Schie, M., Waser, R., and De Souza, R.A. (2014). A simulation study of oxygen-vacancy behavior in strontium titanate: Beyond nearest-neighbor interactions. *J. Phys. Chem. C* **118**: 15 185–15 192.

12

X-Ray Photoelectron Spectroscopy

Michelangelo Scopelliti

Department of Physics and Chemistry – Emilio Segrè, University of Palermo, Palermo, Italy

12.1 General Principles

X-ray photoelectron spectroscopy (XPS) – or, as it was called in its early years, ESCA (electron spectroscopy for chemical analysis) – is a spectroscopic technique in the family of electron spectroscopies, i.e. those methods whose probe (intended as a detected signal) consists of electrons instead of photons. Other electron spectroscopies are Auger electron spectroscopy (AES) and UV photoelectron spectroscopy (UPS); the latter will be discussed in Chapter 13.

These techniques measure the power of the electron beam, expressed as an electron count, as a function of their energy. The electron beam is produced by irradiating the sample with X-rays (in XPS), UV radiation (in UPS), or other electrons (in AES); in all these cases, in order to correctly detect and measure the electron beam, we need that the electrons expelled from the surface do not lose energy due to inelastic collisions, and thus we require ultrahigh vacuum conditions (see Section 12.2.1). Even if in the last years it has been possible to build detectors and systems able to operate in relatively high pressure (HP-XPS) or Near Ambient Pressure XPS (NAP-XPS) systems, their diffusion is currently rather low, and this chapter will not deal specifically with these systems; besides the vacuum conditions, though, general principles still apply.

In the case of XPS, a soft, usually monochromatic X-ray causes the emission of electrons by means of the photoelectric effect. Even if defined as "soft" (usually $\approx$1 keV; see Section 12.2.3), X-ray penetrating power enables the photon to enter a few micrometers (generally up to ten) in the analyzed material. Nonetheless, since electrons have to emerge outside the surface in order to be detected, the free mean path inside the material acts as the limiting factor of the analysis. In most cases, the electrons we are able to detect come from the uppermost 5–10 nm of the material, thus giving the XPS a surface-sensitive analysis character.

XPS is a powerful technique able to detect all elements of the periodic table (except for hydrogen and helium), and is able to give information about the chemical state (oxidation state, neighboring atoms, and electronic structure of the observed molecule) and can be applied successfully to solids, gases, and liquids. Due to the fact of being mainly a surface technique, however, the most important applications concern the superficial characterization of solids; liquids can be often frozen, though and there are no formal limitations to the temperature of spectral acquisition.

Besides, the technique is intrinsically nondestructive, and thus may be used to verify intermediate steps of preparation or manipulation without interfering with the next steps – useful, for example, when optimizing device manufacturing procedures.

Spectroscopy for Materials Characterization, First Edition. Edited by Simonpietro Agnello.

While the photoelectric effect is known since the nineteenth century, it was only in the 1950s that it was possible to develop such a technique, mainly thanks to the work of Kai Siegbahn and his work group in Uppsala, which recorded the very first XPS spectrum [1]. His pioneering work was awarded in 1981 with the Nobel Prize. The diffusion of this technique was hindered by the lack of the needed engineering for the instrument construction, but the development of vacuum and electronic technologies cause the spreading of this technique [2–4].

When monochromatic, soft X-rays are used to extract electrons from a (solid) surface *in vacuo*, expelled electrons possess a specific kinetic energy as a function of the difference between the incident X-ray energy, the work function (the minimum energy needed to remove an electron from a solid to a point in the vacuum immediately outside the solid surface, corrected by the instrument work function and contact potential), and the effective binding energy (BE):

$$E_K = h\nu - (\text{BE} + W) \tag{12.1}$$

where E_K is the kinetic energy, $h\nu$ the photon energy, and W the "complete" work function. Since W is a characteristic of the material, using a monochromatic X-ray, measuring the electron kinetic energies means to determine the BE and thus is possible to obtain chemical information from a surface.

Different atoms possess distinct BEs and cross sections; as a consequence, each element possesses a distinctive and thus identifiable pattern of emitted electrons.

Upon ionization, all orbital levels except ($l = 0$) originate a doublet, since the two possible states have different BE. This is called spin-orbit splitting (or coupling)[5]. p-type orbitals are then split in $p_{1/2}$ and $p_{3/2}$; d in $d_{3/2}$ and $d_{5/2}$; f in $f_{5/2}$ and $f_{7/2}$. The spin-orbit splitting ratio is usually fixed (1 : 2 for p; 2 : 3 for d; and 3 : 4 for f), but strong spin-orbit coupling may change such behavior, as coupling with unpaired electrons in the core with unpaired electrons in the shell (in such cases, very complex peak envelopes occur). Differences in chemical potential and/or polarizability of the molecules can cause an alteration of the BE (chemical shift); in addition to photoelectrons, an ordinary XPS spectrum may also contain Auger electrons (emitted because of the relaxation of the excited state produced by the X-ray). The sum of all these phenomena contributes to the final shape of the recorded spectrum. The spectral peaks arise from the electrons leaving the surface without energy loss; the electrons that undergo inelastic energy loss contribute to the background electron count (which, in some case, maybe annoyingly high). The relative positions (both in terms of energy and count, i.e. intensity) of the single peak can be translated in a complete qualitative and quantitative analysis, as we will see in the following sections.

In the next section, we will detail the instrumental setup of a typical laboratory-based XPS. While it is possible to perform XPS measurements at synchrotron facilities, this would go beyond the goal of this chapter. For sake of completeness, besides the different equipment, it is worth noticing that synchrotron radiation allows for higher X-rays fluxes, and wider energy ranges, letting lower density systems (like adsorbates) to be fully characterized. Principles and data analysis, however, share the same bases as the laboratory setup.

12.2 Instrumental Setup

Worldwide, there are currently about a dozen XPS systems manufacturers, each one with its specific (and patented) instrumental characteristics. The pricing, for a new system, ranges from a few hundred of thousand to a few millions euros (in year 2020), but all systems share the same main components: at least one X-ray source (Section 12.2.3), a sample holder/manipulator

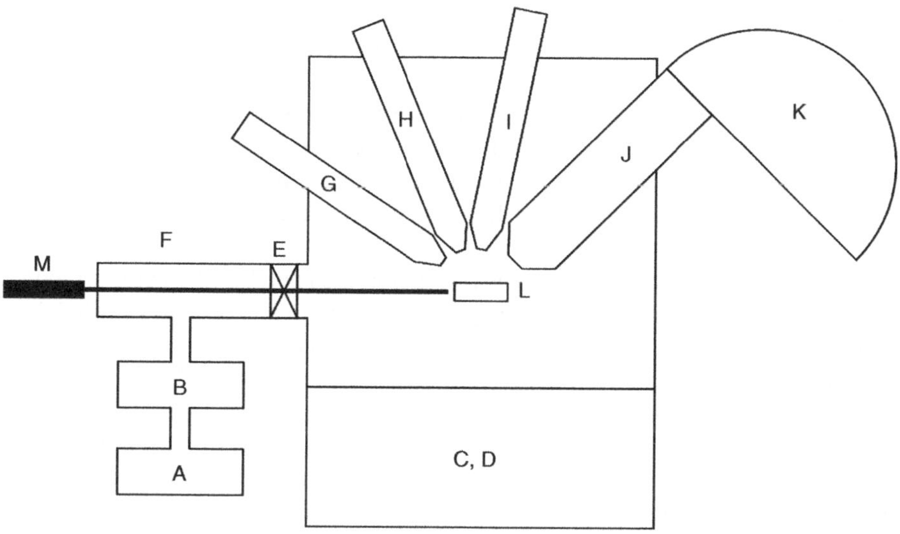

Figure 12.1 General scheme of a XPS instrument. (A) Roughing pumps (Section 12.2.1.1); (B) turbomolecular pump (Section 12.2.1.2); (C) ion pump (Section 12.2.1.3); (D) Titanium sublimation pump (TSP; Section 12.2.1.4); (E) interlock; (F) introductory chamber; (G) X-ray source (Section 12.2.3); (H) electron gun (Section 12.2.5.1); (I) ion gun (Section 12.2.5.2); (J) electron optics (Section 12.2.6); (K) hemispherical energy analyzer (Section 12.2.6); (L) sample (and manipulator; Section 12.2.4); (M) manipulator (Section 12.2.4).

(Section 12.2.4), an analyzer (Section 12.2.6), a detector, and a data elaboration system. And all share a common, (ultra) high vacuum environment (Section 12.2.1). A general instrument scheme is depicted in Figure 12.1.

12.2.1 Vacuum and Ultrahigh Vacuum, UHV

When working with electronic spectroscopies, a high level of vacuum is required, for two main reasons. The first one is bound to the effectiveness of the measurements: the collected electrons need to not lose energy, and any collision could lead to energy loss due to inelastic scattering. The second one, probably less evident, is the need to preserve the functionality of all filaments used internally to produce X-rays and/or electrons (mainly thermionic filaments). Such facilities can only work in the absence of contaminants, which could literally *burn* the filaments.

Both goals are achieved bringing the pressure between 10^{-7} and 10^{-10} Pa (i.e. 10^{-9} and 10^{-12} Torr). In such conditions, the mean free path of a gas is ca. 40 km, so almost all molecular interactions inside the vacuum chamber are negligible. In order to reach such pressure, and to maintain it, some conditions are needed:

- All the materials have to be low-outgassing (release of gas trapped or adsorbed in the material) and with small superficial area; to this extent, special steels are often used: carbon steel is to be avoided, and stainless steel with high chromium and nickel, and low carbon are the most popular choices. Besides that, some sequence of *bake-out* or *baking* is often required to prepare the main chamber. A "baking" is done by heating the whole system to temperatures usually over 100 °C while pumping the whole system. This way, the adsorbed gas is easily desorbed and pumped out. The high temperature is also required in cases where water or other slow desorbing materials are deposed on the chamber walls.

In order to avoid "gas pits" inside the chamber, using bolts, screws, and/or welding is discouraged. If some welding is needed, the surface roughness needs to be reduced by means of electropolishing.

- A rather high pumping speed is required, enough to remove gas excesses (especially during sample introduction, baking process or when electrons, ions, or neutral particles are produced, *e.g.* during depth profile analysis, Section 12.3.4.1, or when a neutralization system is active, Sections 12.2.5.1 and 12.2.5.2). To ease the pumping speed, high conductance tubing is preferred (short, fat tubes, and avoiding bends).
- Metal seals and flanges require particular care. The commonly used polybutadiene rubber, polytetrafluoroethylene (PTFE) and other "vacuum" polymers constitute a risk of outgassing in UHV environments; for this reason, when possible, dented flanges are used in conjunction with soft metal gaskets, typically copper, gold, or iridium. All these gaskets may be used only once.
- To enable some interaction between the chamber and the exterior (other chambers, or openings to exchange materials), one or more airlocks are used. To introduce the samples in the main (measurement) chamber, an introductory chamber is usually connected via an interlock and used as an intermediate stage to preserve the vacuum inside the main chamber. An introductory chamber ("pre-chamber") works at a higher pressure than the main one, and shuttles between ambient pressure and 10^{-4} Pa. Such changes are allowed only closing the interlocks, and allowing gas exchanges only when limits are reached. This, and the fairly smaller volume of the introductory chamber with respect to the main one, allows for a small pressure variation when the interlock is open.
- A constant monitoring of the pressure is required at all time. For such low pressures, vacuum is gauged via cold cathodes and hot cathodes.

  A hot cathode produces electrons while a current passes through it, while a cold cathode emits electrons without the need of thermionic effect. In both cases, the emitted electrons ionize the residual gas, and the vacuum is gauged by measuring the ionic current.

  As previously stated, a vacuum loss can severely damage the equipment.

To bring the pressure from ca. 1 atm to the required UHV, an XPS system uses a multistage set of pumping.

12.2.1.1 Roughing Pumps

The term "roughing pump" indicates a generic vacuum pump, usually a mechanical one, used in the first stage of vacuum system evacuation. As such, the use of roughing pumps is mainly as backing pumps, and supports the entire vacuum system. These pumps are usually able to reach pressure in the "rough vacuum" range (≥0.1 Pa). The most commonly used are the rotary pumps, where a rotor rotates inside a larger cavity. Rotor and cavity are not coaxial, and the ensuing eccentricity creates variable chambers, able to catch the fluid (gas) and push it toward an outlet. These pumps are usually oil-based (oil is used as lubricant and/or sealant) and such oil is exposed to the gas. This requires the use of low-degasing, low-adsorbing (and usually expensive) oils; demisters may be required to recover the oil after a separate degasing.

It is possible to chain multiple stages of rotary pumps to reach pressures of 10^{-4} Pa; however, lowering the pressure increases the risk for oil molecules to backstream, increasing the pressure or contaminating the chamber. For this reason, a second, different stage of pumping is used.

12.2.1.2 Turbomolecular Pumps

A turbomolecular pump is usually used as a "second stage" pumping in an XPS system. In such pumps, we abandon the concept of gas as fluid, and treat is as a set of discrete moving particles in a volume. As such, the basic principle is to give mechanical momentum to those particles, and to push them toward the backing pumps: a rapidly spinning fan achieves this goal by hitting the gas molecules and pushing them away.

Most of turbomolecular pumps use multiple sections, each consisting of a rotating blade and a stator blade. These sections are coaxial. Gas molecules captured by a stage are pushed to the next stage, and so on, until the exhaust is reached. The rotation frequency needed to accomplish such goal is in the kHz range, and the blade materials, angles, and relative sizes are carefully designed in order to avoid particle reflections and/or gas escapes in an undesired direction. Also, the different stages are designed to have decreasing volumes toward the outlet, in order to maintain the particle flux. To endure the extreme rotation conditions, very high-grade bearings are required, often magnetic bearing (oil is a great contamination hazard).

Turbomolecular pumps only work in molecular flow regimes, and thus require to be backed by other pumping systems (see Section 12.2.1.1). Operating a turbomolecular pump in atmospheric pressure, or at a non-designed flow, can stall the turbine; worse, it can put disrupting stress to the bearings, resulting in damages to the turbines and/or to the shaft. Despite those limitations, a turbomolecular pump is a very versatile device, able to reach up to 10^{-6} Pa (in some particular cases, and giving enough time, a pressure of 10^{-8} Pa can be reached).

12.2.1.3 Ion Pumps

With the ion pumps, we leave the field of mechanical pumps. As the name suggests, an ion pump operates ionizing the gas and accelerating these ions with a strong electric potential (tens of kV) onto a solid electrode. Small bits of the cathode are then sputtered into the chamber, and gases are removed by a chemical reaction with the freshly sputtered particles. Charged particles are then buried inside the electrode. Some designs use a double cathode: a highly reactive one, and a high inertia one, used intermittently. The high-inertia cathode (usually tantalum) is used to reflect and bury neutral particles, and is highly effective toward noble gases.

By construction, in an XPS system, the ion pump is built as integrated to the main chamber.

Electrons are used to produce ions; in order to be effective, these electrons have to survive free enough to collide with gas molecules. For this reason, the main element of an ion pump is a Penning trap.

Electric discharges are produced inside an ion pump, and the electrons are trapped as swirling clouds in one or more Penning traps, allowing them to ionize atoms or molecules of gas moving inside the chamber through the traps. The resulting ions are then accelerated toward the cathode (usually titanium [6]). Upon impact, these ions can be buried without consequences, or produce sputtered cathode material inside the chamber able to act as a "getter," removing the gas by means of physisorption or chemisorption. Ion pumps destined for XPS systems have to be high throughput by design in order to maintain optimal UHV conditions during the measurement operations (i.e. when electrons are expelled from the sample surface), when charge neutralization is active (extra flow of electrons and ions; see Section 12.2.5), when sputtered materials are produced during depth profiles (see Section 12.3.4.1), and when other techniques, sharing the instrumental setup, are operating – a possible example being a gas leak from a UPS source (see Chapter 13).

12.2.1.4 Titanium Sublimation Pumps

A last level of vacuum technology often used in common XPS systems is the titanium sublimation pump (TSP). As opposed to the previously described pumps, designed to work (almost) full time during the operations, TSPs are designed to operate only in special circumstances, and can be used instead of baking to "clean" a vacuum chamber.

A filament of titanium is sublimated when a high current (tens of amperes) is passed through it. The sublimated, high-reactive titanium covers the chamber walls, and the gases colliding with the walls chemically combine with the titanium, originating stable, solid compounds, effectively removing gas particles from the chamber.

The filament has a finite lifetime; since TSPs (as ion pumps) are integral to the main chamber, replacing such a component, as any other inside the main chamber, involves the opening and thus contamination of the measurement chamber. This, in turn, requires a sequence of "cleaning" (venting, pumping, and baking) before recovering operating conditions.

12.2.2 Magnetic Shielding

While it is very clear why we need a high vacuum in order to catch and measure the kinetic energy of the escaped electrons, it could be not so for the need to magnetically shield at least some parts of an XPS system.

Earth magnetic field, while not constant over the whole globe, is generally rather stable for short ranges, and an accurate calibration could take it into account when setting up an instrument. Usually, though, XPS are installed near other instruments – think about research facilities – and machines like NMR and EPR could cause major alterations in the electron trajectories, even controlling the relative distance of such apparatuses. Moreover, such instruments often require large amounts of high currents, and their use may be discontinuous. Current fluctuations, especially for high currents, may cause major magnetic field fluctuations. To avoid these disturbances, a magnetic shielding is often deemed necessary; mu-metal is the most common choice.

Mu-metal is a generic name for a nickel–iron alloy with very high magnetic permeability (μ), about two orders of magnitude larger than ordinary steel. Mu-metals possess low magnetic anisotropy, low magnetostriction, and low hysteresis; besides, it presents high malleability. These properties make it very effective as a magnetic shielding in material, in the form of thin sheets. Its effect, on low-frequency magnetic fields, is to concentrate the field lines in paths around the shielded area, thus diverting the effects of an external magnetic perturbation.

In an XPS system, the parts that may require shielding are mainly the measurement chamber and the analyzer.

12.2.3 Sources

Having reviewed the vacuum conditions required to perform an XPS experiment, let us proceed with the active part, i.e. the sources and the samples.

X-rays are generated (in laboratory instruments) bombarding a metallic anode with accelerated electrons. In turn, these electrons are produced by a current passing through a hot cathode (a thoriated filament, LaB_6...) by thermionic emission.

When high-energy particles (electrons, protons, and photons) collide upon a material, there is a finite probability that a particle strikes a bound electron in an atom. The target electron is then ejected from the shell, producing a vacancy in the energy level. Such core holes trigger a "fall" of outer shell electrons, which produces quantized photons, whose energy is given by the difference between the two energy levels. Such levels are characteristics of each atom, and thus the frequencies of produced X-rays is a unique set for each element (characteristic X-rays).

Such transitions are designed with a notation devised by M. Siegbahn [7] (Nobel Prize in Physics, 1924), father of K. Siegbahn (Nobel Prize in Physics, 1981): when an electron falls toward the innermost level (1s; K) from a 2p orbital (L), we have a Kα emission line, the strongest spectral X-ray. The line is in fact a doublet (Kα_1 and Kα_2, Kα_1 being higher in energy) with an intensity ratio of ca. 2 : 1.

While the frequency is characteristic of the used anode (in the case of electron bombardment), the intensity of the beam is a function of the current discharged at the anode.

Bombarding an anode with high-energy electrons transfers to the electrode large amounts of energy, which has to be dissipated to avoid damages to the anode. Water cooling is often used.

Early instruments usually worked with non-monochromatic X-rays: lower sensitivity did not require strict monochromaticity; Kα_1 and Kα_2 have very similar energies (examples: for Mg, Kα_1 = 1253.688 eV, Kα_2 = 1253.437 eV; for Al, Kα_1 = 1486.708 eV, Kα_2 = 1486.295 eV [8]); and building a monochromator could be expensive and/or complicated. Improvements in sensitivity made a monochromatic beam almost mandatory, and now non-monochromatic sources are often added as an option.

Besides beam width selection, specially crafted monochromators associated with a motion system allow for a limited beam displacement, useful sometimes for limiting the sample movement, during mapping procedures, and in some cases, for parallel data collection. With the right servomechanism, the motion of a monochromator could allow for source rastering, which can be useful when performing "shaped" data acquisition (different from maps; a shape, like a line or an area, gives *averaged* information about the scanned area, while an area gives information point-by-point).

Moreover, a monochromator can also focus the beam (usually in a 10–500 μm diameter) effectively determining the spatial resolution of the analysis itself.

At the moment, the most commonly used monochromatic radiation is Al Kα_1 (1486.7 eV, usually referred to as Al Kα; line width (full width at half maximum, FWHM) 0.85 eV [9, 10]; the most suitable monochromators are quartz crystals), while for non-monochromated sources, the most common is Mg Kα (1253.6 eV, FWHM 0.7 eV [9, 10]). While the energy resolution is limited to the natural line width for non-monochromatic sources, a well-monochromated Al Kα beam could be brought to a lower energy resolution, typically 0.25 eV (also taking into account analyzer broadening).

Besides Al and Mg, other sources are of course available, usually as an option. Recently, some XPS manufacturer started to introduce an additional Cr Kα_1 radiation (5415 eV, FWHM 1.9 eV [9, 11] as a standard feature).

Whatever the source is used, it has to be kept in mind that illuminating a surface means to discharge rather high power on a very small spot: depending on the instrument settings, a power of 100 W can be concentrated on a spot of 100 μm of diameter. For thermal sensitive samples, this could mean alteration or degradation.

The precision associated with the energy value of the sources can determine the quality of an XPS experiment. As previously stated (Section 12.1), in order to obtain useful information from an XPS measurement, according to Eq. (12.1), the BE cannot exceed the X-ray energy. Thus, a larger beam energy means a larger range of collected values. Most elements absorb in a BE range of 100–1100 eV, thus any of the previously indicated elements would suffice. There is, yet, another factor to account for.

Since we are collecting electrons emerging from a material, to evaluate the correct BE, it is important that those electrons do not lose energy before emerging from the surface. When a monochromatic electron beam interacts with a solid surface, most of the electrons lose their energy because of

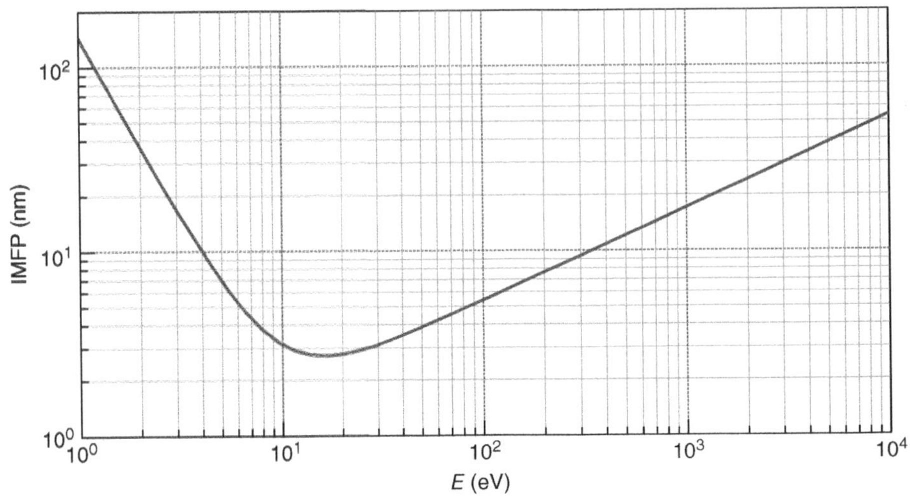

Figure 12.2 Universal curve for the electron inelastic mean free path, Eq. (12.3). Notice the log–log scale.

the strong interactions with matter, leading to plasmon excitation, electron–hole pair formation, and vibrational excitation.

The inelastic mean free path (IMFP) is an index of how far an electron travels through the material before its intensity decays to $1/e$ of its initial value. It depends on the material (nature and packing) but also on the energy of the colliding X-ray beam. It can be determined by the TPP-2M (Tanuma, Powell, and Penn 2nd predictive formula, Modified) equation [12]:

$$\lambda = \frac{E}{E_{\mathrm{p}}^2\left[\beta \ln(\gamma E) - \frac{C}{E} + \frac{D}{E^2}\right]} \tag{12.2}$$

where λ is the IMFP, E_{p} the free electron plasmon energy (in turn, depending on the valence electron number, the material density, and the molecular/atomic mass), E the kinetic energy, and C, D, and γ are material-dependent parameters (depending, e.g. on bandgap).

For practical purposes, a more used (and rough) equation is the "universal" relation [13, 14], valid for all materials:

$$\lambda = \frac{143}{E^2} + 0.054\sqrt{E} \tag{12.3}$$

where λ is the IMFP and E the electron kinetic energy. A plot is shown in Figure 12.2.

As it is shown in Figure 12.2, an increase in the energy, for (roughly) $E \geq 150$ eV, also increases the λ value, resulting in a deeper sampled volume. While in some occasion this feature could be desirable, it comes at some cost: since we are collecting *all* the electrons coming out from all the beamed volume (roughly, the cylinder described by the beamed surface and by the sampled depth), we hare homogenizing all the circumscribed materials, and we are no more able to discriminate between possible superficial material segregation and/or stratification.

12.2.4 Sample Manipulators

As in many other spectroscopic techniques, in order to avoid interactions between the probe and the excitation system, sources and analyzer are not collinear: in a common laboratory XPS system, source(s) and detector are fixed and usually positioned with a 90° (or less) angle. In such conditions, it is simpler to move the sample surface with respect to both source and analyzer. In order to do so, and in order to have reproducible positions (i.e. moving back and forth seeking exactly the same positions, within the spatial resolution of the instrument), at least two stages of manipulation systems (manipulators) are used.

The first and simpler one, rather coarse, is used to move the sample to be analyzed from the intro-chamber (Section 12.2.1) to the measurement chamber. This is usually limited to linear movements, either manual or automatic, and is mainly used to lock the sample holder in a reproducible position, fixed with respect to the second manipulator. Once the sample holder is locked, all the movements are referred to the locking position; in case of alignments or calibrations, usually a digitized system accounts for the local deviations and translates the movements to a "real" set of coordinates.

The second manipulator is placed inside the measurement chamber, in a position rather close to the focused X-ray beam, and is usually controlled by stepped motors, able to movements in the range of the micrometer; it is used to position the sample, aligning the spot to be analyzed with respect to X-ray beam and analyzer. This is accomplished via canonical axes movements (x, y, z) but also (often) with a rotatory movement, used to reach spots otherwise not accessible using the translation motors. A fifth motor is used to tilt the sample holder, in order to change the takeoff angle of the electrons leaving the sample surface. Tilting the surface has the goal selecting the sampling depth: according to both Eqs. (12.2) and (12.3), only the electrons within the IMFP are able to leave the surface; and named θ, the angle between the surface (assumed flat) and the analyzer, only the electrons of illuminated atoms up to $\lambda \sin \theta$ deep will be detected.

Such movement could be a way to select the kind of analysis, but could be also used to perform angle-resolved profiles (AR-XPS; Section 12.3.4.2).

Three translation axes, a rotation axis, and a tilting axis define the so-called "five axis" manipulators.

The reproducibility of all these movements is controlled and checked by means of visual correlations; for example, many systems use digital microcameras pointed inside the measurement chamber, aligning the movements with some external reference (digitally controlled) to synchronize servos and positions; other XPS manufacturers – namely ULVAC-PHI – uses a "secondary electrons" technology (SXI spectroscopy: a low-resolution, SEM-like system) to align positions, servos, and X-ray beam, all in one shot.

12.2.5 Charge Neutralization Systems

In any XPS experiment, the focus of the procedure is the expulsion of electrons from the surface.

It is easily imaginable that removing negative charges creates a charging effect, caused by the build-up of positive ions on the surface; this phenomenon is responsible for the increase in the contact potential. Without any correction, the observed BE value should drift toward increasingly higher values, progressively broadening the signal. Such an effect can be easily thwarted allowing the sample to "discharge" to ground the excess of positive charge. For this reason, it is important to ensure an electrical contact between the sample and the (grounded) instrument (the steel usually used to build the sample holders is conductive enough; sometimes, copper or other metal clamps may be needed). Nonetheless, such method is only useful when working with conductive materials, mainly metals; when analyzing semiconductors, insulators, or hybrid materials, in order to avoid signal drifts and broadening, some other equipment is required.

12.2.5.1 Electron Guns

An electron gun (e-gun, electron emitter) is a device able to produce a narrow, collimated electron beam, with a precise energy. Usually, a thermionic source is used to produce an electronic emission, which is then accelerated toward a holed cathode, allowing the electrons to escape unto the target. Electrostatic fields, magnetic fields, or a combination of both are used as electronic lenses to focus the beam.

In an XPS system, an electron gun is used to deliver a low-energy electron flux (in the 10^2 eV range) with a low current (nA scale) over the illuminated spot. A low current ensures a low collision probability, while maintaining the possibility for those electrons to be captured by the charged surface, thus reducing the contact potential to a constant value. The beam has to be angled in a way to avoid interference with the analyzer.

The thermionic source, after long periods of usage, undergoes massive deterioration, and the emission lowers with time. When a thermionic filament deteriorates, the only possible path is the substitution of the filament itself. A periodic monitoring of the e-gun status is required, both for emission and focusing. This is usually achieved by means of a Faraday cup used as a target (instead of the sample).

The use of such a compensation charge is required for XPS experiments involving semiconductors and/or insulators, but can be also used when dealing with metals. Some researchers prefer to always work with the same charge neutralization systems active for all the samples for sake of coherence and reproducibility of the experimental setup across different samples (even if this means to insulate the conductors from the grounded instrument).

A drawback associated with the e-gun neutralization is the formation of a negatively charged cloud over the analyzed surface, which can cause alteration to the measured BE (altering the contact potential). For this reason, the e-gun is often used in conjunction with an ion gun.

12.2.5.2 Ion Guns

An ion gun, as the name suggests, is a device that generates a beam of ions, with a definite energy distribution. An ion source, an extraction grid, and a collimator are the main parts of this device. The collimator can be also seen as a lens, and in many setups, is also used to allow a rastering of the beam.

In an XPS instrument, an ion gun may be used with two main purposes: the simpler one is to neutralize the electron excess over the analyzed area.

The most common material used in the charge neutralization system is argon gas: the ion source is a low-pressure argon ionized by electron ionization (these electrons can be produced by arc discharge). The generated radical ions ($Ar^{+\bullet}$) are then accelerated with a low intensity potential ($\approx 10^2$ eV) and "shot" over the analyzed surface. As in the case of electron guns, focusing and current have to be calibrated by means of a Faraday cup.

Due to the UHV conditions, and due to the risk of chamber contamination, a high-purity gas is required, usually N6.0 grade.[1]

When working with ion guns, the gas used as a source may leak into the main chamber; it is good practice to constantly monitor the vacuum, in order to avoid pressure jumps that can severely damage the equipment.

As a charge neutralization system, an ion gun can be operated in conjunction with an e-gun, or both can be operated independently, according to the required acquisition conditions. But, since an

1 Nines notation. N6.0 means 99.9999 % by volume pure (six nines).

ion gun is able to arbitrarily accelerate ions onto the surface, a second function is often associated with the same gun (or, in many cases, multiple ion guns are fitted on the same system).

Accelerating heavy ions onto a surface can cause the sputtering of the surface itself. A repeated sequence of etching/acquiring can be used to analyze the material beyond the "natural" surface, giving information about the composition through many layers. The price to pay, obviously, is that the analysis becomes destructive; the reward is the possibility to acquire depth profiles (see Section 12.3.4.1).

To sputter a surface, higher energy is required. In the case of $Ar^{+\bullet}$ radical ion, the potential goes up to 5–10 kV (higher energies could cause chemical damages to the interested material, or ion implantation; in case of organic polymers, graphitization processes could occur).

To etch a surface means to free into the chamber, many sputtered particles. Again, a close pressure checking is a must; between an etching and the subsequent data acquisition, vacuum conditions have to be recovered to read reliable data.

Depending on the size (i.e. the transferred kinetic energy) of the colliding ion, and depending on the nature of the analyzed material, a real crater forms onto the analysis surface. Such a crater needs to be centered with the X-ray beam, and to be as flat as possible at its bottom. It is then a good practice to raster the ion beam in a fairly large area (with respect to the X-ray beam size) to avoid the formation of irregular pits in the analysis spot. It could be also a good idea, when such technology is available, to rotate the etched spot – with respect to the ion beam – during the sputtering, thus assuring greater homogeneity to the crater bottom.

Another aspect to account for is the relative size of the accelerated ions when hitting the surface. A wrong combination of size/kinetic energy could alter the geometry and/or the packing of the superficial molecules – this effect becomes especially evident when dealing with organic polymers – leading to altered or useless results.

In such extent, additional ion guns, carrying different (larger) species, can help to reduce the damage upon etching.

One of the most used materials is C_{60} fullerene, whose larger mass/diameter is able to treat "gently" the etched surfaces. In order to transfer the same amount of kinetic energy, lower potential are required to accelerate the ionized C_{60} radical ions.

Some XPS manufacturers introduced the use of "gas cluster ion beam" – *very* gentle sputter agents: Ar is "clusterized" by adiabatic expansion, forming "particles" of ca. 40 000 atoms which are then ionized and accelerated to sputter a surface. To obtain the same sputtering results, only 1–2 eV acceleration is required. While fascinating, such a technology is though rather expensive.

12.2.6 Analyzers and Detectors

After illuminating the surface of a sample with an X-ray beam, the role of selecting the "correct" electrons to send to the detectors is delegated to the analyzer.

As shown in Figure 12.1, an analyzer is usually formed by two parts: a long cylinder, fitted with electron optics, and the "real" analyzer – the most common being the hemispherical one.

To measure electrons with a specific initial kinetic energy, the electron lenses focus the collected electrons and retard their velocity in order to match it with the pass energy (PE) of the hemispherical analyzer.

Inside the hemispherical analyzer, the electron beam is deflected by an electrostatic field, and the electrons move along a curved trajectory; the curve radius depends both on the electric field potential (the PE) and the kinetic energy of the electrons: using the lenses to vary the electron energy allows for selecting the energies sent to the detector.

A hemispherical analyzer can be operated in two different ways: Fixed Analyzer Transmission (FAT mode, also called CAE: Constant Analyzer Energy) or Fixed Retard Ratio (FRR; also known as CRR: Constant Retard Ratio). Most of XPS spectra are acquired in FAT mode, while FRR is mainly used in Auger spectroscopy.

In FAT mode, PE is held constant during the acquisition, and the lens system is responsible for selecting the kinetic energy in a range able to reach the detector. In these conditions, due to geometrical constraints, not all PE values are accessible to the system, and those available are generally multiple of a fixed value (function of the instrument design).

In FRR mode, both PE and lens system may variate in order to keep constant the initial electron energy:PE ratio.

The analyzer diameter and the PE can both influence the energy resolution available to the experiment. A wider aperture diameter can collect larger amount of electrons, thus allowing for greater electron counts (it affects the signal-to-noise ratio, SNR), but collects also electrons coming from a wider solid angle, needing careful refocusing. In terms of PE, a larger value allows for a coarser kinetic energy selection; this means that for higher PEs, a larger range of kinetic energies is available, while losing resolution – a large range of energies is sent to a limited number of detector channels; lower PE values allow for a finer resolution, with the usual trade-off resolution sensitivity (an increase in the resolution means an enhanced noise level). Since changing the PE value affects the SNR ratio, and since the high-frequency noise is mostly white noise, longer acquisition times can reduce considerably the noise level. Thus, larger PE requires less time to achieve the same spectrum quality (in terms of noise). In practice, PE values around 100 V or higher are used to collect fast, wide scans (≈1000 eV range), used to survey unknown samples, generally useful to identify the presence of specific atomic species; the used energy resolution, in these cases, is about 1 eV, and the time required for a full scan is usually few minutes. For higher resolutions (0.5–0.25 eV), a PE of 20–30 V is preferred, but the time increases to tens of minutes. Lower PE values (2–5 V) are seldom used, both to the time required and the high noise level. Such PE values are more useful when performing UPS with the same XPS equipment (see Chapter 13).

After analyzing (and thus selecting) the electrons, these are sent to the detectors.

The standard detector is an electron multiplier, using a discrete set of channels. The higher the channel number, the higher the detector resolution; digital discriminators may be present, with the goal of increasing the number of "virtual channels" available to the machine.

The electrons are counted as discrete events; the recorded spectrum is collected as a count (or a count per time unit) of events *vs.* the electron kinetic energy, and then presented as BE – after subtracting from the photon energy the kinetic energy and the work function, i.e. from Eq. (12.1),

$$\text{BE} = h\nu - (E_K + W). \tag{12.4}$$

12.3 Applications

From a practical standpoint, the laboratory XPS technique is used mainly to characterize materials and surfaces of various nature. Besides the elemental limiting factor (the inability to "read" hydrogen and helium atoms), the only hard limitations concern the ability of the sample to withstand such low pressure involved during the measurements and the physical size of the samples to be introduced into the measurement chamber. All those materials that require water to maintain specific molecular conformations (as biological molecules like DNA) or some highly outgassing materials (like some zeolites with high affinity toward water) may not be characterized without special preparations, which can sometimes alter some structural properties. It is possible, for some instruments, to offer a temperature control facility (both for cooling and heating) that can help to remove the water or to keep it frozen in place. Heating a sample *in vacuo* is rather easy – a Joule heating circuit is sufficient – while to cool down, a heat pipe is often used to transfer energy from the sample to a cold reservoir, typically liquid nitrogen.

Beyond those limitations, solids (bulky materials and powders) are characterized without particular problems, and with negligible or no damage at all – except for depth profile analysis (Section 12.3.4.1).

XPS is then routinely used to characterize organic and inorganic compounds and polymers, alloys, catalysts, biomaterials, and electronic devices; it is also useful in research fields like cultural heritage, etc., the only limits being those previously stated.

Analyzed materials are often solids, even powders. When preparing powdered samples, choosing the right support may be critical. By definition, a powdered sample is discontinuous, and for grain sizes lower than the analysis spot, the underlying material can show up as a constituent of the analysis. A homogeneous coating can limit or eliminate such effect: thus a rather thick layer of powder (with respect to the sampling depth), a drop cast or a spin-coat of the suspended powder are useful methods of preparation (assumed that the suspending medium does not alter the sample properties and can be easily removed).

A sample can (and in most cases, has to) be analyzed as is, i.e. with all the possible superficial contaminants it may carry (like thin layers of oxides and or adventitious carbon contamination), since usually those information help to characterize the material itself or, in some cases, can help to calibrate the spectrum scale (see Section 12.4).

If, however, these contaminants are not the main goal of the analysis, an ion gun sputter-etching may be used to remove those (superficial) impurities, allowing working with a "fresh" surface. Depending on the powder nature, a freshly ground sample, with highly reactive surfaces and with a high surface/volume ratio may drastically change the composition because of the contamination.

Once prepared and fixed onto the sample holder, and after the introduction in the chamber, the next steps are the identification of the area to investigate, and the research of an optimal height of the surface (with respect to the analyzer) to maximize the electron flux.

Since a typical XPS spectrum is the sum of the characteristic peaks of each of the analyzed elements, corresponding to the electron configuration within the atoms, a standard analysis of an unknown sample usually starts with a survey: since most of the XPS BE fall in the range of 0–1100 eV, a quick scan (2–4 minutes) of such a range, with low-energy resolution and high PE is useful to identify the presence of specific elements, which will be the subject of further analysis. An element can be identified by its specific absorption pattern; the presence of a single peak is not enough to confirm its presence (e.g. in the case of O, both O 1s, ca. 531 eV, and O KLL (Auger peak), ca. 745 eV, have to be present; also, O s2, ca. 21 eV may be observed, depending on the element concentration). The identification work is often delegated to software evaluation (automatic or semiautomatic), based on a simultaneous presence of different patterns. Since overlaps may occur, an operator overseeing is sometimes required.

Such a quick scan is also useful for two other goals: in case of strongly overlapping regions, it is possible to change the "standard" region to analyze an element (where *standard* refers to the spectral regions showing a better SNR ratio and/or carrying more XPS information: peak splitting, larger chemical shift sensitivity, and so on) to other cleaner regions. The other goal is to perform basic quantitative analysis on the sample.

12.3.1 Quantitative Analysis

For each peak in a spectrum, the number of detected electrons is directly related to the relative amount of the element within the XPS sampled volume. It is thus possible to have quantitative information about the sample composition, and such a technique is widely used to provide empirical formulas of (homogeneous) solids. Such formulas, however, exclude the hydrogen (it is not detected).

Quantitative data are reported as "at%" (atomic percent), and are simply an atomic mole fraction expressed as a percentage (in order to convert to "canonical" mass percentage, a conversion must account for atomic masses).

To evaluate the correlation between area and relative abundance, the most accurate method uses the peak area and the relative sensitivity factor (RSF).

In a homogeneous sample, for a specific peak, the signal intensity (as peak area) is given by

$$I = n\Phi\sigma\lambda A\eta_{\Phi}\eta_{\Theta}\eta_{D} \tag{12.5}$$

where n is the number of atoms per volume unit, Φ the X-ray flux in photons per area unit per time unit, σ the photoelectric cross-section for the given orbital (area unit), λ the IMFP, A the area and η_{Φ}, η_{Θ}, and η_{D} are specific efficiency factors (respectively of the photoelectric process in the formation of photoelectrons, the angular efficiency and the detection efficiency; those last two are characteristics of the instrumental setup) [15].

It is then possible to derive the dependency of the atomic abundance from the peak intensity:

$$n = \frac{I}{\Phi\sigma\lambda A\eta_{\Phi}\eta_{\Theta}\eta_{D}} = \frac{I}{S} \tag{12.6}$$

S, collecting all other parameters can be defined as the RSF, while λ and σ may vary from material to material, in a ratio in the form:

$$\frac{n_1}{n_2} = \frac{\frac{I_1}{S_1}}{\frac{I_2}{S_2}} \tag{12.7}$$

$\frac{\sigma_1}{\sigma_2}$ and $\frac{\lambda_1}{\lambda_2}$ are in fact nearly constant [15]. Since ηs are spectrometer specific, for a given instrument, it is possible to calculate (or better, to collect) a set of RSF that can in turn be used to assess relative atomic abundances.

As the relative abundance is presented as a mole fraction, for each j atomic species (except for hydrogen):

$$\chi_j = \frac{n_j}{n_{tot}} = \frac{n_j}{\sum n_i} = \frac{\frac{I_j}{S_j}}{\sum \frac{I_i}{S_i}} \tag{12.8}$$

over all i atomic species.

Values of S based on peak area measurements are based on empirical data [16], but need to be corrected for the spectrometer parameters.

The peak area is usually evaluated as a histogram (i.e. a summation of the height values in the sampled points after background removal – see Section 12.4.2) multiplied by the sampling pass. As such, no real shape evaluation is done, and a noise arising from the intrinsic resolution must be taken into account.

A practical outcome of the use of the RSF is the indication of the detection limit of the instrument for each element. As a matter of fact, while not rigorous, a rule of thumb suggests considering the lower limit of detection (LOD) in at% for jth element as:

$$\mathrm{LOD}_j = \frac{0.05}{S_j}. \tag{12.9}$$

It is nonetheless possible to further lower such a limit by increasing the acquisition time or in all the cases where a really high SNR can be achieved.

12.3.2 Qualitative Analysis

One of the most useful information an XPS spectrum can give is the chemical state of the observed atoms.

For each observed species, it is indeed possible to extract information about the local bonding environment: formal oxidation state, number and identity of the nearest-neighboring atoms. High-resolution data acquisition is needed; usually, a spectrum for region of interest.

At higher resolutions (ca. 0.5–0.1 eV), the technique is sensible enough to discriminate the different BEs associated with the various electron populations inside the single atomic orbitals. An example can be seen in the organic compound analysis: while the nominal BE of C 1s is 284.8 eV,[2] such value is shifted upon binding of other carbon atoms or heteroatoms (as a rule of thumb, electronegative species shift the BE to higher values; the higher the $\Delta\chi$, the larger the shift) allowing discriminating, for example, between alkanes, aromatic hydrocarbons, alcohols, carboxylic acids and so on. These general rules apply, of course, for all elements (specific sensitivity may vary).

Chemical shift detection, in combination with quantitative analysis, can lead to complete chemical characterization of the surface: a shift can be the result of semiconductor drugging, superficial reactions, or other chemical changes.

Besides the operator experience and chemical knowledge – always required to complete an analysis – some publicly accessible resources are available to the operators by instrument manufacturers [17], or by government institutions like the National Institute of Standards and Technology (NIST) [18].

12.3.3 Surface Maps

A standard spectrum is collected using (usually) a single spot of analysis. Location and beam size (and/or shape) are chosen with the goal of retaining most of the wanted features (details for the specific analysis) and for each spot, a spectrum is acquired. Each spectrum may contain several elemental regions, multiplexed in a single spectrum sharing the acquisition parameters. When collecting data for rastered areas, usually these spectra contain the *averaged* data for that area.

There is, though, another possibility. Once the sample is locked in a fixed position, using the monochromator ability to focus and raster the X-ray beam over a small area, it is possible to acquire multiple spectra, in a grid-like set of points, and digitally combine them to form an image. Each pixel is thus identified by its coordinates, and the color dimension is replaced by a value along a normalized scale representing the peak area values. These false color maps reproduce the relative abundance of elements in the scanned area. Collecting a map for each of the elements, such images can be recomposed to have a false color representation of the chemical composition of the surface.

2 By definition, the scale zero is fixed to the Fermi level. More on the energy scale in Section 12.4.

To collect a map, then, a small area – usually in the millimeter scale – is divided with a grid, and for each of the grid regions, a full spectrum is collected and analyzed. The analysis involves a data fitting procedure, or, more likely a background-level evaluation associated with an area gauging by means of histograms. The area values are then divided in discrete "bins" containing value ranges (typically (min – max)/*energystep*). Each of these bins is assigned to a color (a value in grayscale) which in turn can be represented in other color scales.

The limit of such a technique lays basically in the beam size of the focused X-ray. Since it is very rare to focus a beam below a diameter of 10 μm (the electron count would become very low, remarkably increasing both acquisition times and noise level), this value represents the maximum achievable lateral resolution of the image. This also implies that the grid size should not be chosen independently of the beam size: while a larger spot, averaging adjacent pixels, would result in a smoother image, a too large spot would create artifacts by averaging a large area and representing it a single data point. On the other hand, a beam too small with respect to the grid size would collect only partial information of the surface, leading to spotty or underestimated data: the partial collected data would be transformed in representations of larger areas, creating extrapolation errors.

An example of superficial map is reported in Figure 12.3, representing a bismuth island (impurity) over a natural, freshly exposed molybdenum sulfide (MoS_2) surface. The image is collected on 1.0 × 1.0 mm, 256 × 256 pixels, with a 9 μm diameter beam.

It has to be kept in mind that a surface map is an aggregate spectrum, i.e. each image pixel contains a full spectrum; as a consequence, it is possible to monitor how a species, or a species ratio, trends across a specific direction or along some portion of the surface. The drawback is, of course,

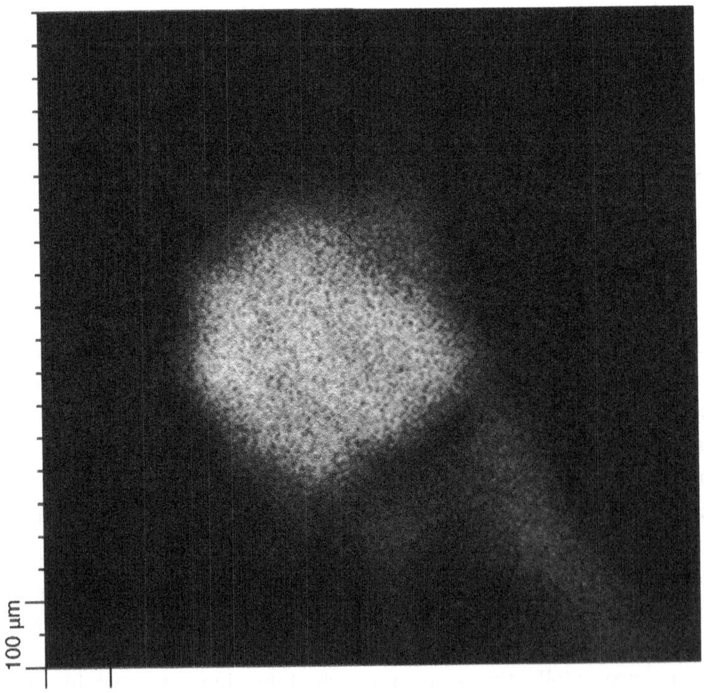

Figure 12.3 Example map. A metallic bismuth "island" naturally occurring in a just exposed molybdenum sulfide (MoS_2) mineral surface.

that a high-resolution image requires as much time to be acquired as the time required for a single spectrum multiplied by number of pixels composing the resulting image.

12.3.4 Profiles

Another very useful information that can be obtained by means of XPS is the depth of the observed element inside the sample. It is always possible to observe the presence or the absence of various energy losses, indicating whether the observed atom is in the bulk or on the surface. Also, some considerations can be made about the intensity loss of some peaks – the count of electrons emitted far from the surface get more attenuated than the ones closer to the surface. Those methods, though, are prone to errors, and require a great deal of calculation.

The profile techniques, both "real" depth profiles and the (less deep) angle profiles, result of far more practical use.

As for the maps, profiles are a collection of spectra, linked together; the acquisition coordinates (x, y), though, are the same for all the spectra, while the variable condition is the height (z) in depth profiles, and the tilting angle in angle profiles.

In the case of depth profiling, z is changed by ablating the superficial layer using ion beam (see Section 12.2.5.2). After every step of etching, the sample height needs to be recalibrated, optimizing again the electron emission toward the analyzer; then, spectrum is collected for each of the required elemental region.

Angle-resolved XPS (AR-XPS), on the other hand, correlates spectra acquired in the same (x, y, z) coordinates, but varying the takeoff angle. Such variation in sampling depth can then be correlated with the takeoff angle, providing information relative to only the topmost layers.

12.3.4.1 Depth Profiles

Depth profiling is accomplished by removing progressively the topmost layers of the material in a controlled way. For such reason, performing a depth profile means to destroy part of the sample, and such damage cannot be recovered.

The ion sputtering can be controlled both in energy and in exposure time, rather precisely. Yet, the real sputter rate is often unknown, and especially for organic materials, to convert sputtering time in nanometers requires calibrations specific for that sample. Preparation methods, drying, packing, and other factors can alter the sputter rate; and the few data available (usually for inorganic materials) are strictly dependent on the sputter technique details (energy, ion fluence, angle, impacted area size, and shape). Nonetheless, if the sputtering method is kept constant, a calibration can be obtained with the help of some external instrument (like a micro/nano indentometer) or whenever is possible to obtain data about the real ablated layer thickness (before and after the complete ablation).

Even in the absence of a full calibration, however, a depth profile can offer information about relative thickness of the various layers the instrument goes through, and – very useful when analyzing devices – can offer insights on the effective separation of the layers (e.g. it is possible to discriminate between plane and bulk heterojunctions, or to observe if some mixing occurs at the interface of two different layers). More on this in Section 12.5.

As stressed before (Section 12.2.5.2), the improper use of the ion gun, besides the layer removal, can cause chemical and mechanical alterations of the surface; especially sensitive samples, like organic polymers, may require a rather soft ion beam (<1000 V acceleration field) and delayed acquisition, to allow the surface to disperse the excess energy.

Acquiring a depth profile, depending on sample thickness and sturdiness, can be a very long process, involving the use of argon gas (or other sputtering medium) reservoir and producing large amounts of particles in the measurement chamber. To keep the instrument in operating conditions, special care is needed to remove those excesses (for example, by allowing more time between consecutive analysis in order to clean the chamber, or by using the acquisition time between two sputtering cycles to refill the gas reservoir).

12.3.4.2 Angle-Resolved Profiles

In a spectrum acquisition, the angle between the surface and the analyzer axis can make a difference in the observed relative abundances. For heterogeneous samples, a smaller takeoff angle (the angle between the plane surface and the analyzer axis) enhances the spectra of the components closer to the surface with respect to those in the bulk, while a takeoff angle of 90° (i.e. normal to the sample plane) tends to minimize them.

It is possible to exploit such effects to obtain information of ultra-thin films, without damaging the sample (as a difference with depth profiling). The maximum depth, though, is limited by $\lambda \sin \theta$ (λ = IMFP; θ = takeoff angle).

In a common AR-XPS, the sample is progressively tilted, and spectra are collected at different angles or sines (data may be reported as a linear function of the angle – nonlinear in depth – or as a linear function of the sine – linear in depth). With some information about involved IMFP, it is possible to map the distribution in a really thin layer ($<\lambda$; for Al Kα, <10 nm).

The major limitation of such method consists in the analyzer aperture angle. A large acceptance angle of the analyzer causes a "flattening" of the data, homogenizing many depths as one single, nominal angle. Specially designed, narrow acceptance angle analyzers are preferred for higher depth resolution. The drawback is, of course, that a too narrow angle lowers drastically the electron count, causing a diminution in SNR.

12.4 Data Analysis

Once data are acquired, the subsequent step is the data analysis. In all the acquisitions – single spot, area, profile – every spectrum is basically a collection of electron counts expressed as a function of the BE. Even in the case of correlated acquisition (maps, profiles), the basic analysis consists in the identification and quantification of the spectral peaks.

Many XPS vendors provide their own analysis software suite to collect and analyze the data; nonetheless, some hardware-independent software suite exists, the most used probably being CasaXPS [19, 20].

On the open source front, only few suites are available, and, while promising, often far from being complete: surfsci [21], XPS [22], SPECQP [23], and gxps [24] are just some of currently (at the moment of writing) actively developed projects. Besides these XPS-specific software, since we are dealing with the general problem of the peak fitting, some generic, very powerful tools are available: as open-source suites, fityk [25], and the more general-purpose scipy [26] are worth mentioning.

Two are the main limiting factors in the free access to XPS data analysis: the availability to the background calculation functions (mainly the Shirley background; see Section 12.4.2) and the virtual mutual incompatibility of the proprietary data format used by XPS vendors.

A common interchange data format exists (Versailles Project on Advanced Materials and Standards, VAMAS [27]) also as an ISO standard (ISO 14976:1998) but is seldom used, causing a *de facto* lock-in of the consumer onto vendor software.

Independently of the software used, due to the mechanics of data acquisition, before the effective peak fitting analysis, some preliminary operations are required: shift correction (that is, defining the energy scale; Section 12.4.1) and background subtraction (Section 12.4.2).

Moreover, often many researchers (and/or software suites) use, before any operation, some kind of data smoothing.

The use of data smoothing is a controversial topic, often shunned by many researchers since this means to alter the acquired, pristine data. The results, obviously, vary with the adopted smoothing method: smoothing splines, like Bézier or B-splines, while visually pleasant, are too vexing on data quality, and a filter approach should be preferred. The most common – and probably the less data-disturbing one – is the Savitzky–Golay filter (SG filter) [28–30].

The filter operates with a process known as convolution, by fitting a progression of subsets of the original data points (the larger the subset, the larger the distortion), with a polynomial of low order (usually 2–3) using the linear least square method. With equally spaced data points, an analytical solution can be found.

The result is a kind of moving average, able to smooth short-range fluctuations while maintaining long-range trends ("short" and "long" refer to the subset size). Problems may arise in the treatment of first and last data points, and coping strategies exist in the algorithm. The better strategy, from a spectroscopist perspective, is to acquire a few points (10 or so) beyond the range to be analyzed at both ends, and then discard the possible mathematical artifacts.

An example of how much data are distorted after filtering is shown in Figure 12.4.

Removing the short-term fluctuations means to remove the higher frequency noise – if the noise is independent of the signal – thus improving the SNR ratio; the noise reduction is paid by a slight signal distortion: the peak height will be reduced, and the FWHM will increase.

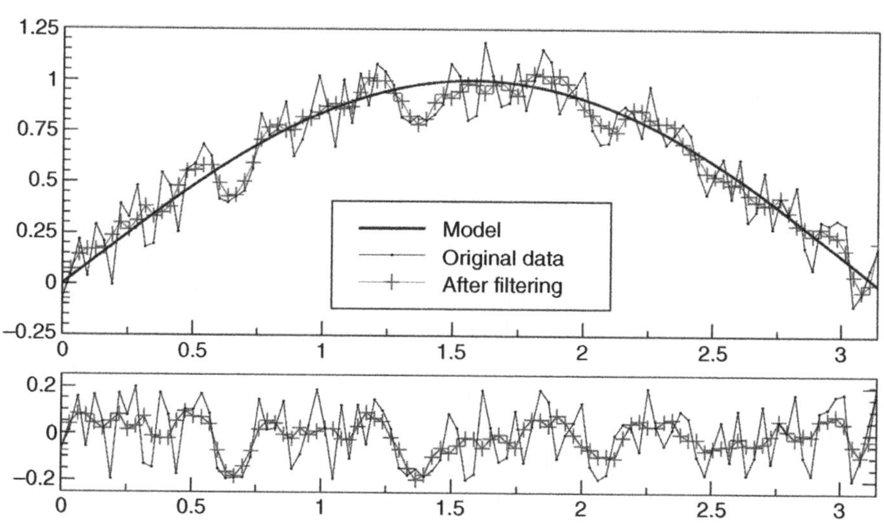

Figure 12.4 Effects of the SG filter. A white noise (black points) is superimposed to a sin function (continuous line). The noise reduction is evident in the "+" points and line (nine points subset, third order polynomials). The differences in noise level are shown below.

The SG filter can be used as a multipass filter, but the border distortion increases with the number of passes. Some improvement can be achieved by weighting the data according to their statistical error, e.g. modeling the noise as a function of the electron energy distribution, but such approach requires mathematical modeling beyond the scope of this chapter, and carries the pitfall of linking too strongly the data analysis to an ideal model, not necessarily responding to the studied case.

Since, however, the most useful parameters in a XPS data analysis are the relative positions of the peaks (i.e. the abscissa of peak maxima) and the relative area values, and since all FWHM are affected in a similar manner, the application of such a filter does not constitute a too large risk for the analysis result.

12.4.1 Shift Corrections

Besides the SNR ratio issue, one of the first operations performed on a data set before any analysis is a shift correction of the BE values.

By definition, the zero of BE scale is the Fermi level for the given sample at the given temperature. Ideally, for a grounded conductor, this should be directly observed without need of corrections (besides, of course, the instrument work function). In the real world, though, other factors, linked to the work function of the material and/or some spurious contact potential may interfere with the exact value. While the instrument work function may be determined (and assumed constant) and thus subtracted to the equations, the contact potential may change the real value. Besides, for insulators and semiconductors, or whenever a sample cannot be grounded, the shift becomes more significant. In the normal conditions of an XPS experiment for such samples (charge neutralization systems active), the contact potential reaches rapidly a steady state, and can be considered constant during the XPS experiment.

Once all the interfering components are constant, the overall effect on the spectrum is a shift for *all* BE values, and is usually enough to shift the whole spectrum of a fixed quantity in order to synchronize the read value with the real one. It is sufficient a single, well-known (and preferably narrow) peak value to perform such a shift. In the case of supported materials, or powder layered over a known background, the substrate material can be used as an "internal" reference (gold, silver, or silicon being the most frequent references in such cases). The reference values can be easily found in public repositories or in the instrument handbooks.

When "as-received" materials have to be analyzed, and it is not then possible to use those sharp references, the simpler way of correcting the data shift is to use the ubiquitous adventitious carbon.

Any real-life sample surface, exposed to the atmosphere, is able to collect the so-called "adventitious carbon," made of a really thin (1–2 nm) coating of hydrocarbon-based impurities. This is well evident, on a high-resolution spectrum, as a typical peak pattern: a major peak at 284.8 eV accompanied by two other peaks of lower intensity, namely at ca. 286 and 288.5 eV. These three peaks (attributable, respectively, to C—C and C—H sp^3 carbon, to alcoholic C—OH or etheric C—O—C carbons, and to C(O)=O carboxylic carbons) are so unavoidable to be used as calibration reference.

In very rare cases, for aging (or mistreated) instruments, some deviation from linearity may occur, and a single-point correction may not be enough. In such cases, the whole energy scale may be realigned by means of the recording of well-known pure element samples, and use the acquired peaks as a calibration. Good candidates for such operation are gold and copper.

It may be helpful to remind that a data shift does not alter the relative peak positions; thus, is not impossible to perform all the peak fitting operations beforehand and then use the fit-determined peak positions to shift the spectrum to energy values more accurate than the non-fitted data.

12.4.2 Background

In a spectrum, the observed defined peaks arise from the electrons which have not suffered inelastic energy loss during the emerging process. But those are not the only electrons generated in the photoelectron emission: the electrons losing energy while traveling in the bulk increase the background level at higher binding energies than the peak value. The energy loss processes are random, and, because of the large number of non-emerging electrons with respect to the emerging ones, those multiple events generate a continuous background. A Mg Kα-generated spectrum presents a higher background level due to Bremsstrahlung radiation of the non-monochromated X-ray; the peaks and their lineshapes add over such backgrounds.

To analyze a spectrum, a background function has to be calculated – beforehand or during the fit procedure – and may dramatically change the values of the produced data (mainly areas).

Standard functions, as horizontal, linear, or polynomial backgrounds, while computationally tempting, are seldom used: the data bracketing (left to operator) required to determine start- and end-point is somehow arbitrary, and may not account for a proper tail ending for the peak functions – depending on the operator point of view. As such, the repeatability and agreement between different sets of measurements are at risk. Other approaches are usually preferred; an interesting comparison has been made by Repoux [31].

Probably, a better and complete approach would be the use of the Tougaard background [32]. Such method requires to evaluate the primary excitation spectrum $F(E)$ from the flux of emitted electrons $j(E)$,

$$F(E) = j(E) - \lambda(E)\int_{E}^{\infty} K(E, E' - E)\, j(E')\, dE' \tag{12.10}$$

as usual, λ is the IMFP (as a function of energy), while $K(E, T)$ is the probability that an electron of energy E loses energy T per unit path length in the solid. Besides, the electron emitter's distribution has to be accounted for, as well as the emitter depth and the angle between the emitter and the surface. The equation may become computationally expensive, but the situation simplify itself knowing $K(T)$. Since λ is not very sensitive to E, an "universal" function may be used:

$$\lambda(E)\, K(E, T) \approx \frac{BT}{(C + T^2)^2} \tag{12.11}$$

where, empirically, $B \approx 2866\ \text{eV}^2$ and $C \approx 1643\ \text{eV}^2$.

Tougaard's approach, though, is not always easy to use routinely, since it requires a background data collection for about 100 eV below the peak. While it is possible to calculate the Tougaard background via theoretical methods, it is a computationally very expensive method.

A completely different approach is used in the Shirley background [33] that makes mostly "operating" consideration on the data analysis.

The background level is assumed to be proportional to the peak area above. Starting from a horizontal background, the procedure is iterated with better evaluations of the area for each cycle. This procedure can use both histogram-type areas (i.e. before the fitting) and the fully fitted area (the calculation of the background is performed during the fitting). A peculiarity of the method is the creation of a "step" under each of the observed peaks; these steps are linked by almost horizontal plateaus. While this is rather useful in case of overlapping peaks or when dealing with large differences in terms of background before and after the peak (from a data analysis standpoint), this could produce nonphysical (and thus wrong) backgrounds (often over the data baseline). An overview is shown in Figure 12.5. Nonetheless, Shirley background offers a high reproducibility, and

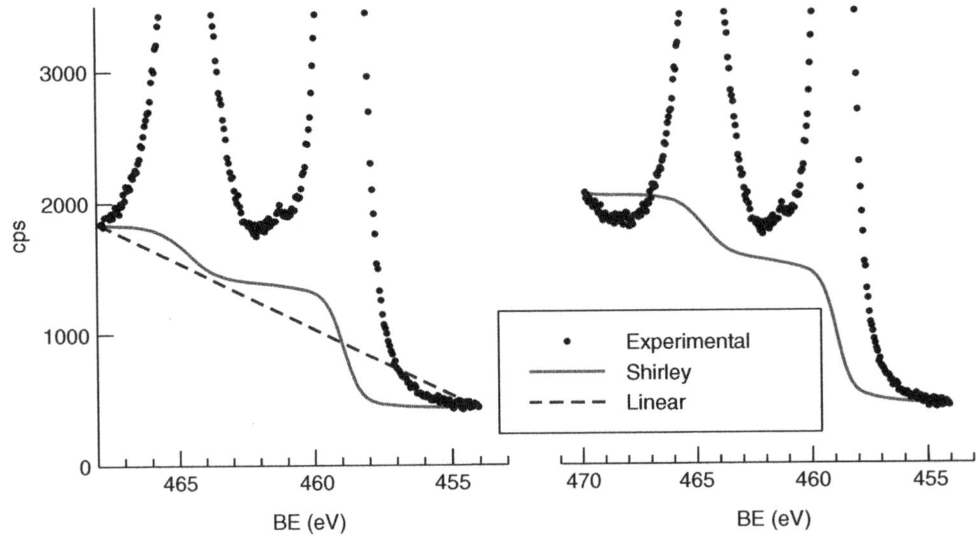

Figure 12.5 Left: direct comparison between a Shirley background (continuous line) and a linear one (dashed). Right: artifacts may arise without proper data bracketing: here the Shirley background has no physical meaning (notice the different BE ranges).

even without many collected points outside of the lineshape decay, offers comparable errors across different operators – even when different data intervals are chosen to calculate the background.

Interestingly, R language package "baseline" [34] implements Shirley background algorithm, and can be an easy reading to understand the involved calculations.

12.4.3 Line Shapes

Removed the background, for a "counting" spectroscopy as XPS, one would expect a Cauchy distribution, i.e. a Lorentzian function lineshape; in normalized form:

$$L(x) = \frac{1}{\pi\gamma}\left[\frac{\gamma^2}{(x-x_0)^2 + \gamma^2}\right] \tag{12.12}$$

where x_0 is the "location parameter" (the maximum coordinate) and γ the half width at half maximum (HWHM; FWHM $= 2\,$HWHM $= 2\gamma$).

The real outcome may differ: a real lineshape is broadened as a result of the convolution of the natural lineshape (Lorentzian profile) and other broadening factors (usually Gaussian in shape): the width of the X-ray line, instrumental effects, environmental fluctuations, and so on. Besides that, depending on the nature of the emitted electrons, some other phenomena (influencing the shape) may occur:

- In the case of *photoelectron lines*, the most intense lines are relatively symmetrical and rather sharp (usually FWHM < 1 eV); pure metal are an exception to the expected symmetry: the conducting electrons cause coupling phenomena and produce remarkable asymmetry, while retaining peak sharpness. Less intense photoelectron lines (at higher binding energies) are often wider (1–4 eV); insulators show wider lines of conductors. Another potential cause of asymmetry, in photoelectron lines, is the effective depth of the emitted electron: an electron that travels through the bulk may exhibit asymmetry due to overlap with energy loss peaks (see, for example, the second case study, Section 12.5); this phenomenon is usually negligible for homogeneous materials.

- An XPS spectrum may contain *Auger lines*. Such a topic is beyond the scope of this chapter; however, their complex pattern (nine or more lines) may be used in an XPS context to identify or to ascertain the presence of specific atomic species. Since the kinetic energies of Auger electrons do not depend on the ionizing X-ray, they may appear in different positions for different sources.
- Up to now, we assumed the photoelectric process produces ions in the ground state. There is, however, the possibility to produce ions in excited states, few electronvolts more energetic. This would result in satellite peaks a few electronvolt higher in BE. These are the so-called *shake-up lines*, corresponding to $\pi \rightarrow \pi^*$ and are most frequently observed for carbon in aromatic compound and in paramagnetic compounds, where the shake-up lines may be so intense to approach the main line intensity (e.g. in copper compounds).
- Sometimes, an apparent asymmetry is the result of a complex *splitting*. While a simple splitting scheme (Section 12.1) is expected for p-, d-, and f-type orbitals, strong spin-orbital coupling may generate complex patterns, especially in transition metals (see, for example, manganese, or chromium). Besides, spin-orbital coupling may be responsible for "violations" of the rules about fixed relative positions, or fixed intensity ratio in splittings (as in zinc).
- Some materials can show some additional peak at higher energy (20–25 eV) due to the interaction between the photoelectrons and other electrons in the sample. The effect is more marked in the case of metals. Those *energy loss* peaks are usually rare and of negligible intensity with respect to the main peak they refer to, but can cause difficult-to-diagnose problem when overlapping with other analyzed regions: an example of silicon interference is reported as a case study (see Section 12.5).

The peak fitting analysis requires recognizing (and dealing with) these spectral lines.

For symmetrical peaks, the ideal solution would be to use a Voigt lineshape, given by the convolution of a Lorentzian distribution with a Gaussian distribution (here normalized):

$$G(x) = \frac{1}{\sigma\sqrt{2\pi}} e^{-\frac{1}{2}\left(\frac{x-x_0}{\sigma}\right)^2} \tag{12.13}$$

where x_0 indicates the curve position, and σ is the standard deviation (notice that $\sigma \neq$ FWHM; for a Gaussian profile, FWHM $= 2\sigma\sqrt{2\ln 2}$).

A Voigt profile has the (normalized) form:

$$V(x;\sigma,\gamma) = \int_{-\infty}^{\infty} G(x';\sigma)\, L(x-x';\gamma)\mathrm{d}x' \tag{12.14}$$

where x is the shift from the line center, $G(x;\sigma)$ is a centered Gaussian, and $L(x;\gamma)$ is a centered Lorentzian. A good approximation [35] of FWHM for a Voigt profile (f_V) is given by the combination of the Lorentzian FWHM ($f_L = 2\gamma$) and the Gaussian one ($f_G = 2\sigma\sqrt{2\ln 2}$):

$$f_V \approx 0.5346 f_L + \sqrt{0.2166 f_L^2 + f_G^2}. \tag{12.15}$$

Since the convolution of many Gaussians produces still Gaussian profiles, the Gaussian in the equation is fit to represent all the (Gaussian) contributions to the line broadening. Such a curve would be ideal to fit all the symmetrical peaks; and, due to the nature of the convolution product, it would be ideally possible to calculate the "full Gaussian" and to remove it from the spectrum, sharpening the peaks, and considerably amplifying the peak separation. A visual comparison of the discussed curves is shown in Figure 12.6.

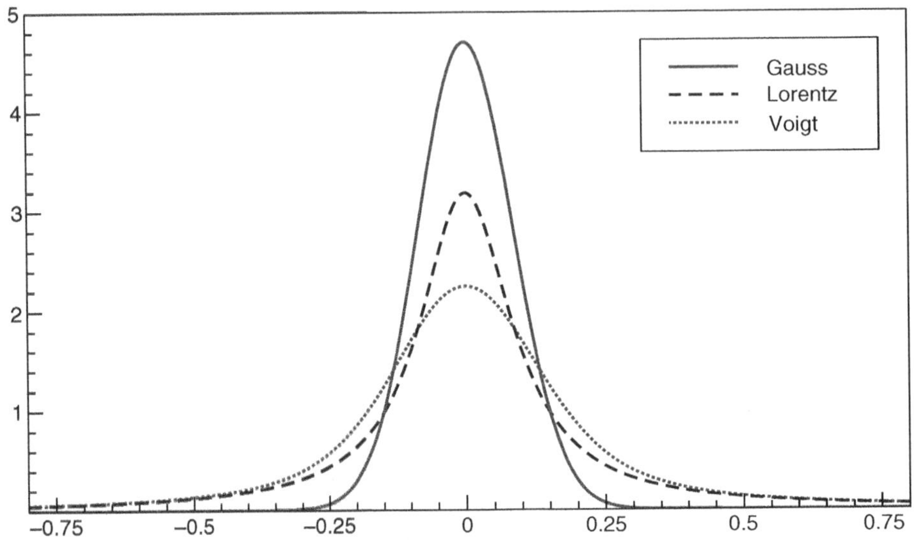

Figure 12.6 Visual comparison of normalized Gaussian, Lorentzian, and Voigt profiles. For these curves, $\sigma = \gamma = 0.1$.

The practice requires a "data training" over analogous spectra (spectra of analogous samples collected with the same instrument) in order to extrapolate the optimal Gaussian, and the possibility of generating mathematical artifacts has to be cared for. Many software suites distributed along with the XPS instrumentation offer such a feature.

Besides the deconvolution aspect, though, the real Voigt profile is a tough lineshape to be managed by most software. The integral form is not easily handled, and – especially during fitting procedures – may cause convergence errors or delays. For this reason, a linear combination (rather than a convolution) between a Lorentzian and a Gaussian is preferred. Those pseudo-Voigt may vary in their implementation [36–40], and the specific approximation is usually hidden in the peak fitting software. Whichever the approximation, though, the differences between different methods currently adopted are usually typically less than 1%.

For asymmetric peaks, various curves are possible (a summary is available in Ref. [41]), the most used being based on Doniach–Sunjic [42] profile:

$$D(x) = H\frac{\cos\left(\frac{\alpha\pi}{2} + (1-\alpha)\arctan\left(\frac{x-x_0}{w}\right)\right)}{\sqrt{\left(w^2 + (x-x_0)^2\right)^{(1-\alpha)}}} \tag{12.16}$$

where H indicates the height, x_0 the location parameter (*not* the center), w the width ($\neq$ FWHM), and α $(0 \leq \alpha < 1)$ the asymmetry of the curve. For $\alpha = 0$, the curve is symmetrical, and the asymmetry increases with α. A representation is shown in Figure 12.7.

While such a model was specifically designed to deal with metals in XPS spectroscopy, and thus is theoretically well defined, from a data fitting or statistical standpoint, there are many setbacks: depending on the value of α, x_0 is not the curve maximum; the H parameter is not really the value of the maximum (see Figure 12.7); the calculation of the curve area may require some adaption because of the asymmetry (the slower-sloping tail may require some "cut" to fall inside the fitted range); and the determination of FWHM may be tricky. Some implementations, moreover, to account for Gaussian broadening, often use a convolution with the usual Gaussian to better

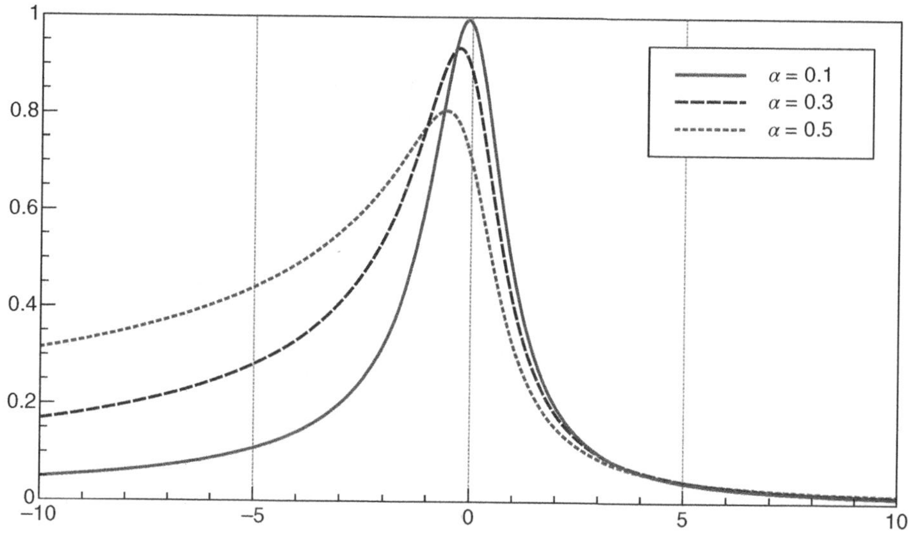

Figure 12.7 Representation of the Doniach–Sunjic profile. Used parameters: $H = 1$; $x_0 = 0$; $w = 0.5$; $\alpha = 0.1$ (continuous line), 0.3 (dashed), 0.5 (dotted). Note the scale (inverted with respect to the usual XPS BE scale), and the shifting maximum value, both in position and intensity, for the same $H = 1$ but varying α.

describe real situations, thus worsening the calculation part. In such cases, the uses of empirical (rather than analytical) parameters may result more useful, but correlating the formula with spectral parameters is not easy.

Another aspect of the use of the Doniach–Sunjic profile is the fact that an asymmetric curve over an asymmetric background (i.e. over a Shirley background) may overestimate the "after peak" baseline. A good suggestion would be to reduce α from $\approx$0.5 in case of linear background to less than 0.1 when using Shirley background.

12.4.4 Nonlinear Fitting

With these bases in mind, a fitting operation should be rather simple: use an *ad hoc* software, perform the shifts, apply the backgrounds, and fit the peaks.

Simple as it may seem, it has to be reminded that the analysis involves a nonlinear curve fitting procedure. This means that there is no unique solution to the general fitting problem, and the initial conditions of the analysis (that is, the operator choices) may drastically change the outcome.

Two factors are of the utmost importance, whether the fitting procedure concerns XPS data or not. The first one is of statistical nature. A nonlinear fit quality may be established by means of some goodness of fit parameter, usually the reduced χ^2, or χ^2_ν. Such value has to be taken for its real significance: a description of how well the used statistical model fits the experimental data. As a rule of thumb, $\chi^2_\nu \gg 1$ indicates a poor model; $\chi^2_\nu > 1$ means that the error variance has been underestimated (or the proposed model is incomplete in describing the data); $\chi^2_\nu \approx 1$ means a good fit of the model over the data; and, finally, $\chi^2_\nu < 1$ indicates an "over-fit" (the model treats the error as signal).

The adherence of the model to the experiment, however, is always delegated to the operator. And, here is the second factor: the operator is responsible for choosing a model physically significant, avoiding chasing a "perfect fit" (possibly leading to an over-fit) numerically correct but chemically irrelevant: with a high enough number of Gaussian peaks, it is possible to obtain a $\chi^2_\nu \approx 0$; but any of the peaks would not represent any atomic species.

It may be useful to remind that, unless explicitly considered as data weight, the errors associated with the data are not taken into account when performing least squares analysis; therefore, resolution and/or counting errors may have sensibly different magnitude from fit errors.

Once the peaks are fitted, the interpretation of the position of the single species and their quantitative and qualitative parameters is made by comparison to previous references; in some cases (like the correlation of the same species across a range of concentration, or a progressive degradation of a specific component as result of given procedures) may benefit of the superimposition of constraints on the fit (imposing a constant value for FWHM; forcing the area ratio; blocking the peak spacing). Such procedures may represent a good compromise between the "perfect fit" and the chemically sound interpretation of the data.

12.5 Case Studies

XPS measurements may be applied to various fields, and can be used as a tool to verify process results, contamination in synthesis, and the effect of depositions during device preparation. A good example, showing the ubiquitous presence of hydrocarbon contamination may be represented by the characterization of indium tin oxide (ITO) surfaces.

12.5.1 Hydrocarbon Contamination

ITO is a transparent conductor, largely used as an electrode in thin film solar cells and/or flexible photovoltaic devices. The manufacturing process involves the deposition of the photon-absorbing material over the ITO layer; superficial contamination and/or crack may cause spurious effects, generating irregularities in the adjacent layers.

A set of experiments of power generation using a specific batch of ITO was giving erratic results, independently of the cleaning technique used to prepare the surface (according to literature, a sequence of alcoholic solvent baths followed by ozone cleaning is sufficient to prepare for any kind of deposition). The deposition of the absorber layers was perfectly reproduced at every step, but the power generated suffered from deviation up to 50%.

XPS analysis of ITO surface showed a large carbon layer over the ITO itself, around 38 at% (Figure 12.8, continuous line; consider both relative intensity and carbon RSF. The carbon peak is found around 285 eV). In the same figure, the dashed line shows how much the residual carbon peak is still present after the literature used treatment (washing with a sequence of alcohols followed by ozone cleaning, 1 hour); the quantitative analysis shows a decrease from the previous value to about 18 at%. To remove almost the totality of the residual contamination (bringing C to less than 3 at%; dotted line), it was necessary an O_2 plasma treatment for 1 hour. Figure 12.8 also shows how such superficial cleanings (both ozone and O_2 plasma treatments are oxygen rich) enhance the overall SNR and how the increase in superficial oxygen makes evident the usually less visible O 2s region, ca. 24 eV.

12.5.2 Energy Loss

Sometimes, dubious interpretation may arise from unknown heterogeneous samples, especially when there is no prior knowledge about the single components. An example could be represented by some cultural heritage samples, like pigments in paintings.

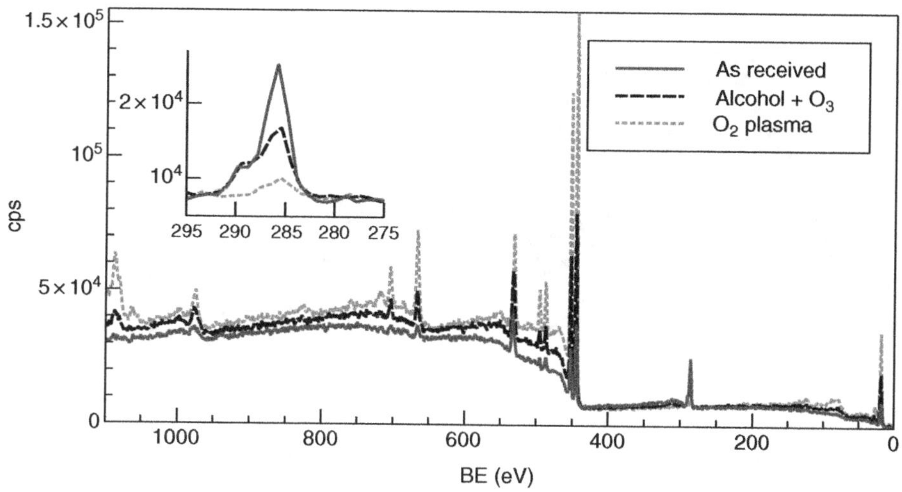

Figure 12.8 Comparison of survey spectra of the same sample, after different treatments. continuous line: as-received ITO sample. Dashed: solvent/O_3 treatment. Dotted: O_2 plasma treatment. Inset: C 1s region magnification.

In the case of paintings, only minimal chips are available, extracted from the original in a manner studied to cause the smaller amount of damage to the whole painting. These chips are multilayered mixes of substrates (inert minerals glued by some colloid, animal or vegetable in origin); XPS is useful in revealing the nature of these binding agents: an animal-based colloid (glue) will have a nitrogen-rich matrix (the glue being made by proteic residues), while a vegetable base, made by starch, will lack nitrogen signals and will be rich in alcoholic functions.

The pigment analysis is usually far more interesting (at least, for a spectroscopist): the single colors are in fact produced by very specific minerals, usually transition metal salts, and identifying the pigment may aid to ascertain the origin of the painting, or the painter, or even if different hands worked the same painting area.

An interesting case arose in the analysis of a red pigment, identified as vermilion (powdered cinnabar – mercury(II) sulfide, HgS) by other spectroscopic means. The sample was being analyzed to verify that no other red pigments were present, and to confirm the relative abundance of mercury sulfide.

The spectrum is shown in Figure 12.9. As expected for HgS, despite the low overall abundance, S shows an S $2p_{3/2}$ peak at 162 eV (metal sulfide) with the expected splitted value S $2p_{1/2}$ at +1.18 eVs. The real problem is the massive peak observed at 168.9 eV. This peak presents a larger Gaussian character (≈100%) and presents a too high value, attributable only to sulfate S. The problem was that there is no way to fit a doublet on the larger peak (too much symmetrical) and there is no other evidence of sulfate presence (no sulfates in FTIR nor from chemical analysis). Of course, these evaluations require a comparison between the various reduced χ^2 for each fit procedure. The high symmetry and the Gaussian character were determinant in the final attribution of the peak as an energy loss of silicon used in the substrate preparation of the painting (some aluminosilicates were present in the mixture); the presence of silicon was already evident in the Hg spectrum (not shown) where Si 2p region overlaps Hg $4f_{7/2}$ (typically the same range, 90–110 eV for both species); the difference in RSF between Si 2p and S 2p, and the differences in their relative abundance were the cause of the initial misinterpretation.

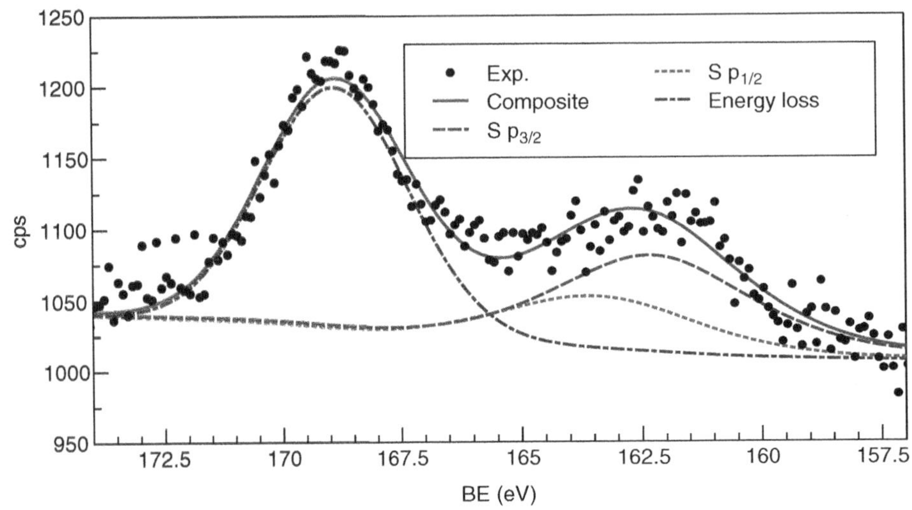

Figure 12.9 High-resolution S 2p region spectrum. Continuous line: cumulative fitted spectrum. Dashed and dotted lines: S 2p doublet (see text). Dash-dotted line: energy loss feature. The (Shirley) background was omitted for clarity.

12.5.3 Depth Profiles/1

A depth profile analysis (Section 12.3.4.1) consists of several spectra linked by the facts of being recorded in the same (x, y) coordinates, varying the z axis by removing the topmost layer via ion beams. It is thus possible to follow the concentration of atomic species and associate the values and the curve variation with the analyzed surface "passing through" the various layers constituting the sample.

Figure 12.10 reports two depth profile data sets, the S at% (as S 2p) showing the jumps in concentration values passing from the top toward the underlying ITO (where S at% drops to zero). The abscissa is already converted into nanometers. In both cases, it is possible to determine the discontinuity point by fitting a piecewise model function (e.g. with the use of SciPy [26]). The polygonal chain is a consequence of the fact that, while nominally "surface sensitive," the XPS technique really samples a finite depth ($\lambda \sin \theta$) and thus the species from an underlying layer shows even when the surface is not fully exposed because of the sputtering. In a way, it is possible to see the profile as a result of the convolution of the measuring window (a rectangular window function) with the step function representing the sharp concentration variation (such a function is a sharp step only when no mixing is appreciable; the curve smoothes with mixing). In perfectly formed planar heterojunctions, such convolutions assume the shape of the polygonal chain shown in Figure 12.10 for one layer change (bottom) or for multiple layers (top). It is possible to evaluate the layer thickness measuring the distances between the edge points, indicating the complete clearing of the layer: in the figure, while in "A" we observe the decrease of S 2p signal, it does not mean that we have completely cleared the superficial (P3HT) layer; this happens only when we reach "B" ($\approx$50 nm). This reasoning has to be applied also to multilayer systems. Layers thinner than the sampling depth may cause problems to this approach; in such cases, varying the takeoff angle may help lower the sampling depth; also, analyzing them with a combination of AR-XPS and depth profiles may result useful. Cases of alternating very thin layers (2–3 nm) of different materials may not be characterized by XPS.

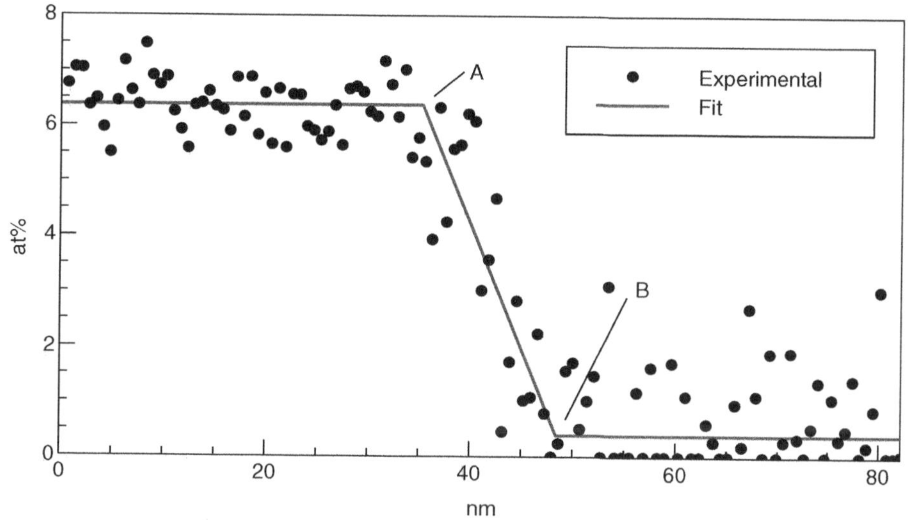

Figure 12.10 Depth profiles of a multilayer sample, following the relative abundance (at%) of S via S 2p. The sample is prepared deposing a P3HT [poly(3-hexylthiophene-2,5-diyl)] layer by spin-coating over ITO (indium tin oxide).

12.5.4 Depth Profiles/2

The last notable case still deals with depth profile analysis capabilities. Simply following the concentrations is usually enough. Yet, in some cases, extra information may be recovered.

Every point in a depth profile (like those in Figure 12.10) represents an area value extracted from an high-resolution spectrum for the observed species; if a single species changes its chemical properties in an automatically distinguishable way – for example, the change in oxidation/binding state leads to a well-resolved peak, fully separated from the original peak – it could be possible to follow across the profile not only the relative abundance, but also the chemical state.

A well-studied example is represented by silica-covered silicon surfaces: Si element shows a peak at 99.5 eV, while silica (SiO_2) has a peak at 103.5 eV. Both species should present a doublet each, but the p-type orbital splitting is so small to not pose a real problem here. But the low FWHM with respect to the silicon-silica energy vale difference allows to follow the disappearance of the silica peak (going from the top silica toward the underlying silicon) as a phenomenon separated from the appearing of the silicon peak.

Figure 12.11, top, shows the "usual" analysis: oxygen in the topmost silica layer (SiO_2), 66 at% O, disappears toward 0% when silicon layer is reached, while 33 at% silicon increases to 100 at% at the same time.

Separately analyzing the decrease and increase of the two peak families, however, produces the bottom plot: here, three species (O from silica, Si from silica, and Si from silicon) are clearly resolved (and a further linear least square optimization on the peak trend may be performed, lowering the noise level). Software suites able to deal with profiles usually offer both the fit-dependent (peak correlations across successive spectra) and fit-independent (i.e. just rough area evaluation via histograms) routines.

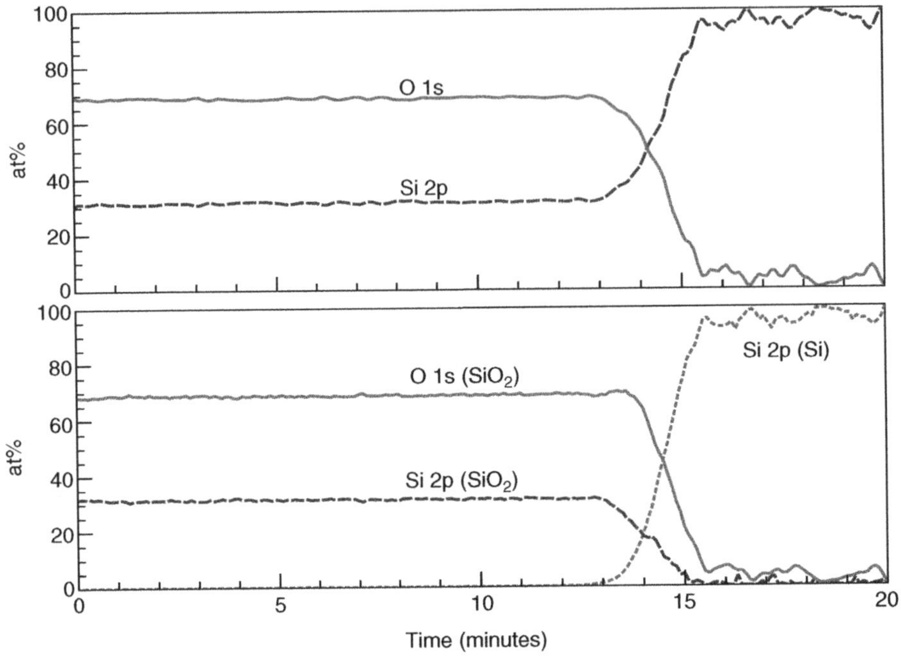

Figure 12.11 Depth profile of silica-covered silicon. Top: simple quantitative analysis, elements only. Bottom: differentiation between the relative abundances of Si 2p (Si element), Si 2p (silica), and O 1s (from silica).

References

1 Siegbahn, K. and Edvarson, K. (1956). β-ray spectroscopy in the precision range of $1:10^5$. *Nuclear Physics* **1** (8): 137–159. https://doi.org/10.1016/s0029-5582(56)80022-9.

2 Fahlman, A., Nordling, C., and Siegbahn, K. (1967). *ESCA: Atomic, Molecular and Solid State Structure Studied by means of Electron Spectroscopy*, volume 20 of *Nova acta Regiae Societatis Scientiarum Upsaliensis*. Uppsala, Sweden: Almqvist and Wiksell.

3 Siegbahn, K. (1976). Electron spectroscopy and molecular structure. *Pure and Applied Chemistry* **48** (1): 77–97. https://doi.org/10.1351/pac197648010077.

4 Siegbahn, K. (1982). Electron spectroscopy for atoms, molecules, and condensed matter. *Science* **217** (4555): 111–121. https://doi.org/10.1126/science.217.4555.111.

5 Briggs, D. and Grant, J.T. (eds.) (2003). *Surface Analysis by Auger and X-Ray Photoelectron Spectroscopy*. Chichester, West Sussex, U.K: IM Publications.

6 Weissler, G.L. and Carlson, R.W. (1979). *Vacuum Physics and Technology*. New York: Academic Press http://site.ebrary.com/id/10282314.

7 Siegbahn, M. (1916). Relations between the K and L series of the high-frequency spectra. *Nature* **96** (2416): 676–676. https://doi.org/10.1038/096676b0.

8 Schweppe, J., Deslattes, R.D., Mooney, T., and Powell, C.J. (1994). Accurate measurement of Mg and Al $kalpha_{1,\,2}$ X-ray energy profiles. *Journal of Electron Spectroscopy and Related Phenomena* **67** (3): 463–478. https://doi.org/10.1016/0368-2048(93)02059-u.

9 Bearden, J.A. (1967). X-ray wavelengths. *Reviews of Modern Physics* **39** (1): 78–124. https://doi.org/10.1103/revmodphys.39.78.

10 Citrin, P.H., Eisenberger, P.M., Marra, W.C. et al. (1974). Linewidths in X-ray photoemission and X-ray emission spectroscopies: what do they measure? *Physical Review B* **10** (4): 1762–1765. https://doi.org/10.1103/physrevb.10.1762.

11 Sorum, H. (1987). The k$\alpha_{1,2}$ X-ray spectra of the 3d transition metals Cr, Fe, Co, Ni and Cu. *Journal of Physics F: Metal Physics* **17** (2): 417–425. https://doi.org/10.1088/0305-4608/17/2/011.

12 Tanuma, S., Powell, C.J., and Penn, D.R. (1994). Calculations of electron inelastic mean free paths. V. Data for 14 organic compounds over the 50-2000 eV range. *Surface and Interface Analysis* **21** (3): 165–176. https://doi.org/10.1002/sia.740210302.

13 Seah, M.P. and Dench, W.A. (1979). Quantitative electron spectroscopy of surfaces: a standard data base for electron inelastic mean free paths in solids. *Surface and Interface Analysis* **1** (1): 2–11. https://doi.org/10.1002/sia.740010103.

14 Werner, W.S.M. (2001). Electron transport in solids for quantitative surface analysis. *Surface and Interface Analysis* **31** (3): 141–176. https://doi.org/10.1002/sia.973.

15 Wagner, C.D., Davis, L.E., and Riggs, W.M. (1980). The energy dependence of the electron mean free path. *Surface and Interface Analysis* **2** (2): 53–55. https://doi.org/10.1002/sia.740020204.

16 Wagner, C.D., Davis, L.E., Zeller, M.V. et al. (1981). Empirical atomic sensitivity factors for quantitative analysis by electron spectroscopy for chemical analysis. *Surface and Interface Analysis* **3** (5): 211–225. https://doi.org/10.1002/sia.740030506.

17 Thermo Fischer Scientific Inc (2013). XPS elements table. https://xpssimplified.com/periodictable.php (accessed August 2020).

18 Cedric Powell (1989). X-ray photoelectron spectroscopy database XPS version 4.1 NIST Standard Reference Database 20. http://srdata.nist.gov/xps (accessed August 2020).

19 Casa Software Ltd. (2020). CasaXPS (2.3.23). http://www.casaxps.com (accessed August 2020).

20 Walton, J., Wincott, P., Fairley, N., and Carrick, A. (2010). *Peak Fitting with CasaXPS: A Casa Pocket Book*. United Kingdom: Accolyte Science.

21 Kalliecharan, D. (2019). surfsci. https://gitlab.com/ddkn/surfsci (accessed August 2020).

22 Kalliecharan, D. (2019). XPS. https://gitlab.com/ddkn/xps (accessed August 2020).

23 Shiphilin, M. (2019). SPECQP. https://github.com/Shipilin/specqp (accessed August 2020).

24 Fischer, S. (2020). GXPS. https://github.com/schachmett/gxps (accessed August 2020).

25 Wojdyr, M. (2010). Fityk: a general-purpose peak fitting program. *Journal of Applied Crystallography* **43** (5): 1126–1128. https://doi.org/10.1107/s0021889810030499.

26 Virtanen, P., Gommers, R., Oliphant, T.E. et al. (2020). SciPy 1.0: fundamental algorithms for scientific computing in Python. *Nature Methods* **17** (3): 261–272. https://doi.org/10.1038/s41592-019-0686-2.

27 Dench, W.A., Hazell, L.B., and Seah, M.P. (1988). VAMAS surface chemical analysis standard data transfer format with skeleton decoding programs. *Surface and Interface Analysis* **13** (2-3): 63–122. https://doi.org/10.1002/sia.740130202.

28 Savitzky, A. and Golay, M.J.E. (1964). Smoothing and differentiation of data by simplified least squares procedures. *Analytical Chemistry* **36** (8): 1627–1639. https://doi.org/10.1021/ac60214a047.

29 Savitzky, A. (1989). A historic collaboration. *Analytical Chemistry* **61** (15): 921A–923A. https://doi.org/10.1021/ac00190a744.

30 Steinier, J., Termonia, Y., and Deltour, J. (1972). Smoothing and differentiation of data by simplified least square procedure. *Analytical Chemistry* **44** (11): 1906–1909. https://doi.org/10.1021/ac60319a045.

31 Repoux, M. (1992). Comparison of background removal methods for XPS. *Surface and Interface Analysis* **18** (7): 567–570. https://doi.org/10.1002/sia.740180719.

32 Tougaard, S. (1988). Quantitative analysis of the inelastic background in surface electron spectroscopy. *Surface and Interface Analysis* **11** (9): 453–472. https://doi.org/10.1002/sia.740110902.

33 Shirley, D.A. (1972). High-resolution X-ray photoemission spectrum of the valence bands of gold. *Physical Review B* **5** (12): 4709–4714. https://doi.org/10.1103/physrevb.5.4709.

34 Liland, K.H., Almøy, T., and Mevik, B.-H. (2010). Optimal choice of baseline correction for multivariate calibration of spectra. *Applied Spectroscopy* **64**: 1007–1016.

35 Olivero, J.J. and Longbothum, R.L. (1977). Empirical fits to the Voigt line width: a brief review. *Journal of Quantitative Spectroscopy and Radiative Transfer* **17** (2): 233–236. https://doi.org/10.1016/0022-4073(77)90161-3.

36 Wertheim, G.K., Butler, M.A., West, K.W., and Buchanan, D.N.E. (1974). Determination of the gaussian and lorentzian content of experimental line shapes. *Review of Scientific Instruments* **45** (11): 1369–1371. https://doi.org/10.1063/1.1686503.

37 Sánchez-Bajo, F. and Cumbrera, F.L. (1997). The use of the pseudo-Voigt function in the variance method of X-ray line-broadening analysis. *Journal of Applied Crystallography* **30** (4): 427–430. https://doi.org/10.1107/s0021889896015464.

38 Liu, Y., Lin, J., Huang, G. et al. (2001). Simple empirical analytical approximation to the Voigt profile. *Journal of the Optical Society of America B* **18** (5): 666. https://doi.org/10.1364/josab.18.000666.

39 Di Rocco, H.O. and Cruzado, A. (2012). The Voigt profile as a sum of a Gaussian and a Lorentzian functions, when the weight coefficient depends only on the widths ratio. *Acta Physica Polonica A* **122** (4): 666–669. https://doi.org/10.12693/aphyspola.122.666.

40 Ida, T., Ando, M., and Toraya, H. (2000). Extended pseudo-Voigt function for approximating the Voigt profile. *Journal of Applied Crystallography* **33** (6): 1311–1316. https://doi.org/10.1107/s0021889800010219.

41 Desimoni, E., Casella, G.I., Cataldi, T.R.I., and Malitesta, C. (1989). A comparison of some asymmetrical line shapes for XPS data analysis. *Journal of Electron Spectroscopy and Related Phenomena* **49** (2): 247–261. https://doi.org/10.1016/0368-2048(89)85012-1.

42 Doniach, S. and Sunjic, M. (1970). Many-electron singularity in X-ray photoemission and X-ray line spectra from metals. *Journal of Physics C: Solid State Physics* **3** (2): 285–291. https://doi.org/10.1088/0022-3719/3/2/010.

13

Ultraviolet Photoelectron Spectroscopy – Materials Science Technique

Dmitry A. Zatsepin and Anatoly F. Zatsepin

Russian Academy of Sciences, Ural Division, Yekaterinburg, Russia
Institute of Physics and Technology, Ural Federal University, Yekaterinburg, Russia

13.1 UPS History and Capabilities

Any narration about the set of photoelectron spectroscopy methods employed for materials science electronic structure analysis will be incomplete without taking into consideration ultraviolet photoelectron spectroscopy (UPS). Historically, this kind of spectroscopy was proposed and designed by Dr. David W. Turner (Imperial College of Science and Technology, London, Great Britain) jointly with his co-workers D. Frost, J.H.D. Eland, K. Kimura, and M.I. Al Jobory at the beginning of 1960 [1]. Independently from Dr. D.W. Turner and co-workers, the UPS was pioneered by Prof. Fedor I. Vilesov as well in 1961 (Leningrad State University, USSR) for studying photoemission spectra of free molecules in the gas phase [2]. In contrast to X-ray photoelectron spectroscopy (XPS), the UPS uses ultraviolet radiation range in order to excite and produce electron photoemission mostly from valence electronic shells, because both 21.2 eV (He I) and 40.8 eV (He II) photon sources employed in stationary laboratory spectrometers and even 100–150 eV-selected synchrotron advanced sources [3] are not giving any chance to pass inside electronic structure deeper than valence base-band area and low binding energy core-levels or semi-core levels (the latter predominantly used for material's electronic structure mapping – "fingerprinting") [4]. Such a restriction reasonably arises due to the known fact that it is impossible to ionize a sublevel with electrons having binding energy larger than excitation energy. As an additional stringent requirement, the need of an ultrahigh vacuum is considered in order to perform correct UPS-spectra acquisitions (the same is a standard for high-precision XPS measurements), and, with that, the contaminations sensitivity of UPS should be taken into account meaning the very accurate manipulation procedures with the sample before UPS probing. This does not mean the reasons to refuse at all an employment of ultraviolet photoelectron spectroscopy for electronic structure study, since all experimentally performed methods of materials inspection have their own certain limitations. Rather, it emphasizes the need for correct selection of research methods for concrete investigation, which is one of the reasons for the success of an ongoing experiment.

Despite the briefly mentioned UPS requirements and restrictions, an application of this kind of spectroscopy for the study of valence band electronic structure allows to obtain some advantages in

Spectroscopy for Materials Characterization, First Edition. Edited by Simonpietro Agnello.

comparison with the XPS method. Low-energy photons from He I – He II gas-discharge sources cannot damage most organic compounds (see e.g. Refs. [5–7]) in contrast to X-ray Mg $K\alpha$ (1253.6 eV), Al $K\alpha$ (1486.6 eV), and other XPS sources; so UPS is considered to be a nondestructive materials science research technique. Another significant advantage of UPS is the much higher energy resolution while recording valence band (VB) and valence base-band (VBB) spectra – typically, it is not more than a few millielectronvolts compared to ~ 0.15–0.2 eV in XPS. The higher energy resolution offered by UPS is important for detailed VB–VBB structure analysis and interpretation that is, sometimes, not so easy because of the valence states' essential hybridization (i.e. combination of orbitals originating from different neighboring atoms – components of a chemical compound). Usually hybridization effects exhibit in VB–VBB area as energy shifts of partial atomic components photoemission lines, their line-shape broadening and mutual overlapping thus yielding spectral ambiguity contributions that impedes standard mapping of electronic structure (especially when XPS technique had been applied) and does not allow to perform correct quantification [8, 9]. In this case, theoretical modeling of material atomic and electronic structures will be required as a rule.

When choosing a materials science spectral analysis technique, a researcher should realize that energy resolution and nondestructive excitation source are highly important but not the only features that should be taken into account as final key points for this choice. Spatial and depth resolution of spectral analysis technique and elements concentration sensitivity are of great significance as well. It is obvious, that beyond the physical origin of employed spectral analysis technique, these characteristics will be affected additionally by technical parameters related to implementation of a specific spectrometer (i.e. actual probe locality of excitation source, transmitting function of energy analyzer and possibilities of electron optics, signal-to-noise ratio of spectra recording channel and detector). Therefore, it seems reasonable to specify the parameters mentioned in relation to existing modern equipment rather than engage solely in appropriate theoretical background. Regarding the UPS method, most of the UPS spectrometers manufactured presently have ~0.1 at.% elements sensitivity, a few atomic monolayers depth resolution, and UPS probe diameter ~0.9 mm (spatial resolution). It is well enough for UPS to be employed in materials science as valuable day-by-day research technique for challenging application tasks.

13.2 Theory and Experimental Methodology of UPS

13.2.1 Physical Principles of UPS

Since ultraviolet spectroscopy is based on the well-known Einstein's photoelectric effect, the kinetic energy E_K of emitted photoelectron (directly measured by analyzer) can be written as

$$E_K = h\nu - E_I \tag{13.1}$$

where $h\nu$ is the energy of incident photons and E_I denotes energy needed to ionize an atom singly. Ionization energy is specific for each orbital and, according to Koopmans theorem, E_I may be identified approximately as binding energy of a particular orbital, but without accounting for intra- and extra-molecular relaxation effects [10]. The latter usually leads to the higher values of theoretically calculated binding energy when compared with experimentally measured values. In the frames of Koopmans theorem and energy conservation principle, it possible to determine binding energy E_B by re-writing expression (13.1):

$$E_B = h\nu - E_K - \phi. \tag{13.2}$$

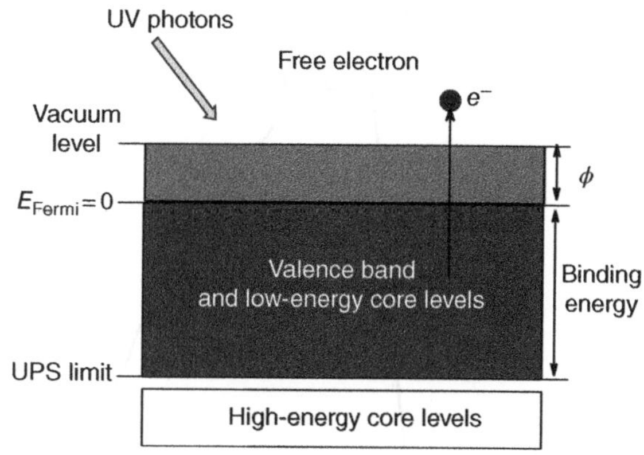

Figure 13.1 UV photons-stimulated photoemission and energy band diagram.

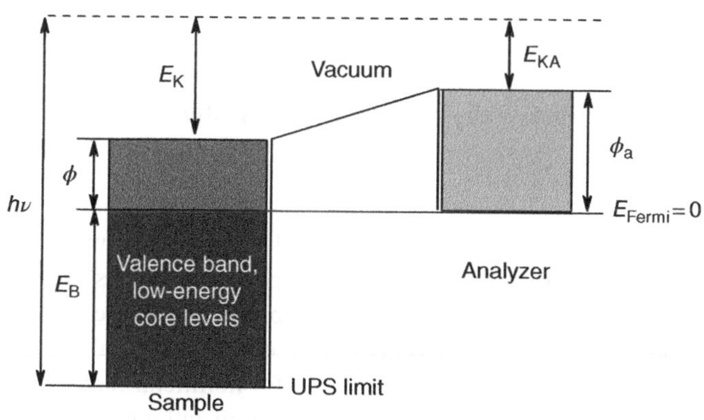

Figure 13.2 Diagram of relations between work functions of sample and energy analyzer, kinetic and binding energies due to contact potential.

Here, ϕ is the work function of the sample, which was added because the kinetic energy of an electron will be affected by electrostatic potential in vacuum nearby the sample surface which is losing electroneutrality due to electron UV-photons' stimulated photoemission (see Figure 13.1).

In general, when UPS measurements are performed, the kinetic energy of electron solely outside the sample will be modified additionally by the difference in work functions between the sample and energy analyzer of spectrometer, which is demonstrated in Figure 13.2.

In this regard, kinetic energy E_{KA} directly measured by the spectrometer and, hence, binding energy E_B can be written, respectively, as follows:

$$E_{KA} = E_K + \phi - \phi_a \tag{13.3}$$

$$E_B = h\nu - E_{KA} - \phi_a. \tag{13.4}$$

Generally speaking, ϕ is considered as a fixed characteristic of a material under study and, thus, additionally can be used for its identification. Ultraviolet photoelectron spectroscopy allows to obtain ϕ values directly from UPS spectra for different materials.

For pure metals, the work function is determined by simple expression $\phi_m = h\nu - \gamma$, where $h\nu$ denotes the energy of incident photons and γ is the width of UPS spectrum as shown in Figure 13.3. This simple procedure is derived from the fact that slowest loss electrons (contribute

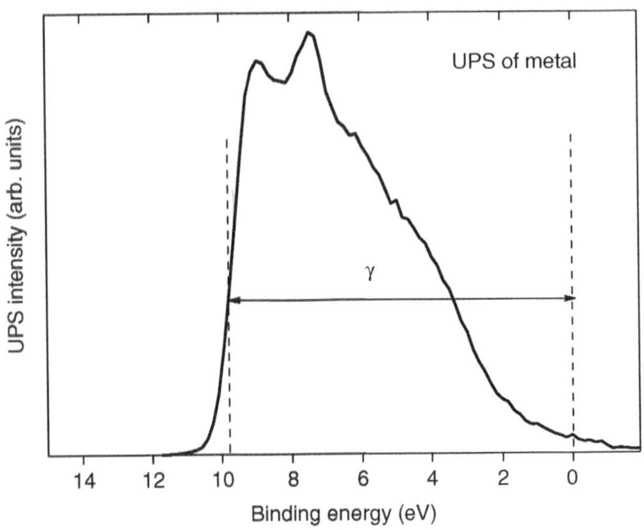

Figure 13.3 The scheme displaying how to measure the width γ of UPS calibrated spectrum from the onset of Fermi edge in the case of pure metal.

to the spectrum at highest binding energies cutoff edge) have nearly zero kinetic energy, whereas electrons near Fermi level have the highest energy. Actually, another procedure exists where energy inflection and intercept points are determined [4]; however, the simplified procedure shown here gives the final result with good-enough accuracy (differences between experimentally estimated value and calculated one in some cases are less than 0.07 eV). As for nonmetal molecular multicomponent compounds, interfaces, and wide-gap insulators, this procedure will not be as precise as expected because of possible interfacial chemical interactions known up to present for some organical coatings on metal substrate, near surface charge transfer, polarization and "push-back" effects [11, 12]. The latter means decrease in ϕ due to Coulomb repulsion between electronic subsystem of a material and near surface electrons, which occurs for the limiting case of weak hybridization between electronic states of adsorbate thin films and metal or metal-like substrate [13, 14]. Variations of ϕ are usually not less than about 0.2 eV in order of magnitude in most cases, which, nevertheless, make it possible to detect final ϕ value with enough confidence and precision, especially employing modern UPS spectrometers [15–17]. Chemisorption sensitivity of ϕ is quite important since ϕ-value variations can be really employed as an indicator of adsorbate contaminators of the surfaces of metals, metallic alloys, and metal-like bulk samples if surface purity control is needed. It is better to perform this kind of inspection on the basis of comparance with certified UPS standards, i.e. using pure appropriate substrate and the same substrate with known concentration of absorbate-contaminator or several, which will allow to make conclusions about surface purity even semiquantitatively. When measuring γ in UPS spectra, one should keep in mind the so-called "vacuum shift" Δ. It usually occurs at the lower kinetic energy side of spectrum (or at higher binding energies cutoff edge) due to the appearance of surface dipole momentum for the sample under study and band alignment in the energy range of highest occupied molecular orbitals (HOMOs) for semiconducting materials if referring to Au metal as a UPS standard (see e.g. Refs. [18–21] and Figure 13.4).

In order to interpret UPS spectra, it is necessary to decide first what kind of theoretical model to employ for electron photoemission process description. The well-known, simple three-dimensional

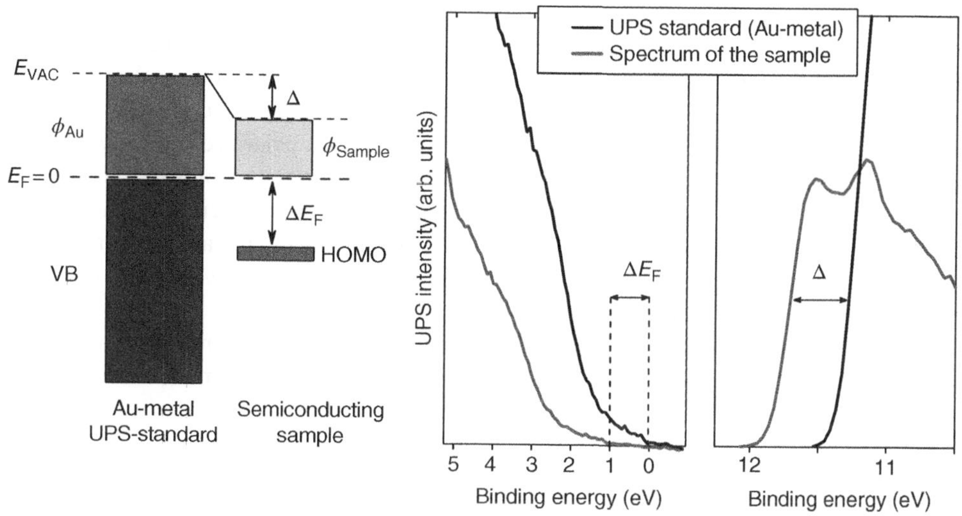

Figure 13.4 Electronic band alignment scheme: ΔE_F and "vacuum shift" Δ measurements procedure employing Au-metal UPS Standard (see explanation in the text). Right-hand side of the figure shows an example of E_F-vicinity area and highest binding energies cut-off edge of the same spectra.

one-step model can be used to predict and handle quantum efficiency with momentum distributions of electrons from the surface of ordered single-crystal and disordered correlated materials. It is giving relatively good conjunction with band-structure theory and most of existing wave-function theoretical calculations methods [22–24]. But, at the same time, electron–electron scattering interactions and kinematics effects are necessary to be considered for understanding and realistic description of the photoemission process. These requirements point to a known and experimentally proved Puff–Berglund–Spicer three-step photoemission model, which is well suited both for XPS and UPS [25, 26]. Subsequently, this model was developed and supplemented by Hüfner [27], Jensen [28], and Reinert and Hüfner [29].

In the framework of the three-step model, the photoemission process includes three distinct and independent sub-processes: (i) ionization of electronic shell due to photon absorption with onward excitation of an electron; (ii) traveling of excited electron through the solid to the surface; and (iii) emission of this electron from the solid to the vacuum where it is detected by kinetic energy analyzer. The second step in this model involves two separate approximations [27–29]. The first one – "sudden approximation" – is usually employed for the case of pure metals due to essentially suppressed electron–electron scattering effects, which follows from Pauli Exclusion Principle and Pauli Blocking (the lack of unoccupied final states). So electrons in metal have "undisturbed" possibility to travel to the surface and be emitted into the vacuum in comparance with other solids and electron energy distribution is easily described by Fermi–Dirac function:

$$N(E) = f_{\mathrm{F-D}}(E) = \frac{1}{1 + e^{(E - E_{\mathrm{F}})/k_{\mathrm{B}}T}} \tag{13.5}$$

where E_F is the Fermi energy, E denotes the energy of the state, and k_BT is electron gas thermal energy, k_B being the Boltzmann constant. For nonmetal solids, the many-body effects should be taken into consideration [27, 28], i.e. the electron–electron scattering effects and photo-hole lifetime during photoemission, especially for insulators, cannot be neglected. The reason for this is the nearly same timescale for electron emission process and electronic subsystem response to electron removal (polarization of nonconductive surrounding media and electron screening of created photo-hole) [27–31]. In this regard, one should realize the critical dependence of these responses on the chemical nature of the material under study. The most important point here is the fact that interaction of emitted electron and charge-coupling area of material's polarized media will affect both the escape process and traveling length, but electron kinetic energies vary only in the range of 0.1–0.4 eV [29–31]. This means that despite applying modern high-resolution energy analyzers, it will be very difficult to experimentally separate all these fine phenomena in the actual energy spectrum without employing state-of-the-art theoretical computational methods, which are, unfortunately, still machine-time costly [31, 32].

The three-step photoemission model supposes from general physical considerations that energy distribution function can be presented by a kind of proportionality relationship

$$N(E) \sim N_0(E)L(E)P \tag{13.6}$$

here $N_0(E)$ is the energy distribution of electrons in a material after photoexcitation of some electronic shell, $L(E)$ denotes characteristic traveling length of electron via solid to the surface (directly proportional to electron escape depth and inversely proportional to materials' absorption coefficient), and P is a probability for electron to be emitted into vacuum (can be assigned for simplicity as a constant for a given "perfect" excitation photon source) [33]. Experimentally recorded spectrum (formally represents energy distribution curve of all emitted electrons) includes contributions both from unscattered and scattered electrons; so, it is possible to write an intensity of experimental spectrum as:

$$I(E_B, h\nu) = N_0(E)T(E, h\nu)P + I_{BG}; \quad T(E, h\nu) = \frac{\lambda_e(E)/\lambda_p(h\nu)}{1 + \lambda_e(E)/\lambda_p(h\nu)} \tag{13.7}$$

where $T(E, h\nu)$ denotes electron transmitting function, which is significantly dependent on electron free mean path λ_e and absorbed photon attenuation depth λ_p in a homogeneous solid and I_{BG} is a contribution to the spectrum intensity from energy-loss electrons (or so-called energy degraded electrons) that are mostly responsible for background signal. We have to note that $T(E, h\nu)$ is a smooth-shape function thus it is not distorting spectrum shape. This means that UPS spectrum represents a convolution of occupied densities of states with contributed influence of emitted electron kinematic effects from surface and nearly surface layers with respect to material absorption ability. The probability P for electron to be emitted into vacuum is added in Equation (13.7) in order to account for stochastic effects, because it is well known that experimentally recorded spectra of even the same material with formally identical conditions of measurements being applied do not match each other in a perfect point-to-point manner [33].

P.-W. Huang and co-authors offered a modification of the three-step photoemission model where static and dynamic sub-cases for semiconducting materials had been included [34]. Comparing to original model they offered to take into considerations the contribution from acceptor levels located at the band gap and energy losses caused by scattering of electron with the crystal lattice, what is affecting electronic states near the surface, energy gap and, hence, electron emission process. From these additional considerations, an energy distribution $N_0(E)$ of electrons in a material after photoexcitation for the static sub-case will be given by expression

$$N_0(E, \nu) = N_A f_A(E_A) D_e(E) \delta(E - h\nu + E_g - E_A) + D_e(E) D_e(E - h\nu) H(h\nu - E_g - E) \tag{13.8}$$

where $N_A f_A(E_A)$ denotes the contribution from acceptor levels; f_A is the possibility of electrons to occupy acceptor level; $D_e(E)$ is the density at excited energy; and $\delta(E)$ and $H(E)$ are delta and Heaviside functions [34, 35], respectively. The electron transmitting function $T(E, h\nu)$ is now determined by modified Jensen $D(E)$ function [28, 34, 35]:

$$T(E, h\nu, s, x) = D\left((E - \Delta E)x^2 + E_a E_{\text{field}} E_{\text{bend}}\right). \tag{13.9}$$

This means that the static part of model after P.-W. Huang and co-authors allows to consider the transmission possibility of photoelectrons from depth s and direction x with energy E as a combination of semiclassical electron energy-loss ΔE, and near-surface layers' electric field E_{field} with bending distortion E_{bend} of initial band structure along with impurity levels and time-independent electron affinity E_a [34]. Recall that E_a is defined differently in solid state physics as opposed to physical chemistry and atomic physics, denoting here the energy difference for an electron to be at rest – outside the surface and the bottom of conduction band just – inside the surface. It is independent from dopants opposite to work function ϕ, thus serving exactly as a surface characteristic parameter. According to Ref. [34], energy loss ΔE can be found from equation

$$\Delta E = \frac{s}{x\sqrt{2E/m^*}} \lambda(E)\, E_{\text{PH}} \tag{13.10}$$

where $\lambda(E)$ is the electron scattering rate and E_{PH} is the phonon energy; $m^* = mE_g/R_\infty$ denotes electron's effective mass; R_∞ is the Rydberg energy ($R_\infty = 13.606$ eV); and s and x are the same parameters as in Equation (13.9). Taking into account (7.10), the intensity of UPS spectrum in this case can be written as:

$$I(E_B, h\nu) = N_0(E, \nu)T(E, h\nu, s, x) + I(\Delta E). \tag{13.11}$$

The dynamic sub-case of photoemission includes sample surface degradation during measurements because of residual gas absorption from vacuum media [36]. For this state of matter, Eq. (13.9) will include time-dependent electron affinity $E_a(t)$ instead of stationary value, additional time-dependent decay factor $T_{\text{decay}}(t)$, and $N_0(E, \nu, t) = \frac{dN_0}{dt}$ with all the consequences. Despite the fact that acceptor levels are treated both in static and dynamic sub-cases as an isolated monoenergy layer in the bandgap via employment of δ-function without temperature dependence, the modified three-step photoemission model made by P.-W. Huang and co-authors is believed to be the most improved up to the moment and considered as a suitable base for precise Monte-Carlo simulations of photoemission processes.

13.2.2 Angle-Resolved UPS

As in related XPS, the UPS can be employed in an angle-resolved mode (see Figure 13.5), i.e. photoelectrons are emitted from a solid at value of θ from the normal $\boldsymbol{n}$ to the sample surface due to angle adjustments of sample-holder relative to fixed angle position of energy analyzer entrance slit.

Thus angle-resolved ultraviolet photoelectron spectroscopy (ARUPS), or, as they sometimes call it, angle-resolved photoemission spectroscopy (ARPES), is a direct experimental UPS subtechnique allowing to obtain information about electronic excitations in reciprocal space of solids via spectroscopy band-mapping procedure [37]. From an experimental viewpoint, it is needed to determine the energy and momentum of emitted electron outside the sample under study and then to establish dispersion relations. Similar to conventional non-angular-resolved UPS, the negligible contribution of photon momentum can be ignored in comparison with electron momentum, and,

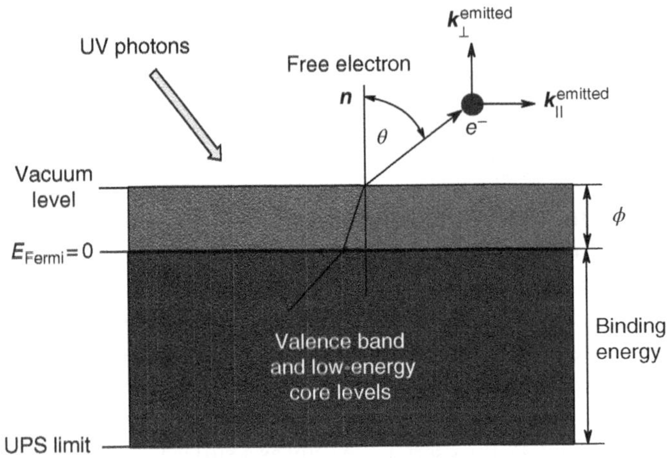

Figure 13.5 Angle-resolved ultraviolet photoemission (ARUPS) scheme.

with that, the component of electron momentum $\boldsymbol{k}_{\parallel}^{\text{initial}}$ in the plane of a solid should be conserved opposite to normal momentum component $\boldsymbol{k}_{\perp}^{\text{initial}}$ if translation symmetry exists:

$$\boldsymbol{k}_{\parallel}^{\text{initial}} = \boldsymbol{k}_{\parallel}^{\text{emitted}}. \tag{13.12}$$

Equation (13.12) is a fundamental base for ARUPS technique and formally means that ARUPS needs atomically flat surfaces for perfect conservation of surface-parallel electron momentum, which can be achieved by means of special sample preparations (high-precision surface sputtering, in situ cleaving of sample in corresponding crystallographic direction). Momentum components $\boldsymbol{k}_{\parallel}^{\text{emitted}}$ and $\boldsymbol{k}_{\perp}^{\text{emitted}}$ of photoelectron (see Figure 13.5) are determined by known expressions [37, 38]

$$\boldsymbol{k}_{\perp}^{\text{emitted}} = \frac{\sqrt{2\text{m}E_{\text{K}}}}{\hbar}\cos\Theta; \boldsymbol{k}_{\parallel}^{\text{emitted}} = \frac{\sqrt{2\text{m}E_{\text{K}}}}{\hbar}\sin\Theta \tag{13.13}$$

thus giving the possibility to obtain $\boldsymbol{k}_{\parallel}^{\text{emitted}}$ values via directly measured values of electron kinetic energy E_{K} at different values of θ. As a result, one can easily establish dispersion relation by means of simple variations in value of θ during ARUPS experiment:

$$\hbar\boldsymbol{k}_{\parallel}^{\text{emitted}} = \hbar\boldsymbol{k}_{\parallel}^{\text{initial}} = \sqrt{2m(E_{\text{initial}} + \hbar\omega - \phi)}\sin\Theta \tag{13.14}$$

$$\hbar\omega = E_{\text{f}} - E_{\text{initial}}\ (\text{Conservation law})\ \text{and}\ E_{\text{f}} = E_{\text{K}} + \phi \tag{13.15}$$

$$\hbar\boldsymbol{k}_{\parallel}^{\text{emitted}} = \hbar\boldsymbol{k}_{\parallel}^{\text{initial}} = \sqrt{2mE_{\text{K}}}\sin\Theta \tag{13.16}$$

where $\hbar\omega$ is the energy of incident UV-photons; E_{initial} denotes the energy of electron in the solid at the occupied state, which is then excited by UV-photons to the empty state with energy E_{f} and afterward emitted into vacuum with kinetic energy E_{K}; and ϕ is the work function (Figure 13.5). In order to determine the momentum component $\boldsymbol{k}_{\perp}^{\text{initial}}$, one should consider energy differences V_0 inside and outside the solid (so-called internal potential of a material):

$$\boldsymbol{k}_{\perp}^{\text{initial}} = \sqrt{\left(\boldsymbol{k}_{\perp}^{\text{emitted}}\right)^2 + \frac{2mV_0}{\hbar^2}}. \tag{13.17}$$

Formally, it is possible to obtain $E(\boldsymbol{k}_{\perp}^{\text{initial}})$ dispersion if photoelectron emission is measured in the direction normal to the sample's surface (i.e. $\boldsymbol{k}_{\parallel}^{\text{emitted}} = 0$) with a dependence on incident UV-photons' energy; but in this case, the dispersion of final state has to be exactly known. In fact, this implies an experimental limitation of ARUPS technique to study only 2D and quasi-2D systems (i.e. structure-layered materials such as graphite and its derivatives, some high-T_C superconductors, high-ordered surface frameworks on metal or semiconducting substrate and similar systems) [38]. At the same time, an employment of advanced synchrotron sources together with the latest generation of kinetic energy analyzers generally still extends informative possibilities of ARUPS measurements (see e.g. Refs. [39, 40]). If one takes into account that most of the physical and chemical features of a material are governed mainly by the density of states located in relatively narrow energy range around E_F-vicinity, it is clear why the UPS technique is now regarded as one of the most promising experimental methods of the solid state physics.

13.3 UPS Experiment and Factors of Influence

Structurally and technologically, the UPS experiment uses formally the same flowchart of equipment as the related XPS. Moreover, usually UPS modules can be installed as an upgrade to XPS spectrometer if the latter has good-enough high resolution kinetic energy analyzer with μ-metal protection and high-performance vacuum system based on turbomolecular pumps – exactly these parts of spectrometer will be jointly used both for UPS and XPS electronic structure inspections in the framework of upgraded "excitation source – electron passing-by channel" scheme. At least most of internationally accepted photoemission spectrometer manufactures (i.e. Thermo Scientific, ULVAC-PHI, Kratos Inc.) are doing just that and offer to the customer combined stationary UPS-XPS systems for Scientific, Research & Development, Industry & Research and other grades. As additionally installed units in this case only ultraviolet photon sources He I – He II and high-voltage gas-discharge ignition system are considered. We have to note that energy analyzer with μ-metal protection is exactly important requirement for precise UPS experiment due to essentially lower kinetic energies of UV-excited electrons compared to that of X-ray photoelectron spectroscopy. In general, a combined UPS-XPS spectrometer uses the same units and modules as any other conventional XPS spectrometer except the above-described UPS-additives; so, there is no reason to describe its flowchart in detail. As key-points the concrete technical implementation of particular UPS-XPS spectrometer and its specifications are accounted and since technical implementation depends on manufacturer engineering experience, it is rather useful to obtain a complete informative bundle of specifications exactly from manufacturer than to discuss known obvious UPS-XPS flow-charts which can be easily found in Chapter 12 or elsewhere. For this reason, we will discuss in brief the units and modules of UPS-XPS spectrometer that may negatively affect the accuracy of materials science UPS analysis.

13.3.1 Vacuum System and Pumping

When analyzing the electronic structure of some material, a vacuum rarefaction value in the analytical chamber of UPS-XPS spectrometer of 10^{-6}–10^{-7} Pa is usually considered as normal value; besides, the latter refers rather to spectrometer idle cycle (only pumping, preparations for run-up) than right to UPS measurements procedure. In general and from the very beginning, all surfaces facing high-vacuum media should be free from any possible types of contaminators (organic traces from human fingerprints, dust, traces of any oil and greases, absorbed water, etc.). At the same time, it is a well-known fact that vacuum hydrocarbons, oxygen lees, and other residual components of pumping deposit on and adhere to the surface of the material under study if an analytical chamber has rarefaction of ~8×10^{-6} Pa and worse. The deposition and adhesion of these residuals onto the sample occurs within four to five minutes, which is, generally speaking, comparable in order of magnitude to the registration time of single UPS spectrum. Since the time of deposition and adhesion of residuals is inversely proportional to the vacuum value in the first approximation, at 10^{-7} Pa, this process will take already about 50 minutes. Therefore, precise UPS materials science spectrometers should be equipped with a very high-performance and oil-free vacuum system in order to prevent vacuum fluctuations in analytical chamber during measurements, since unstable and/or "low" vacuum will lead directly to some distortion of the obtained data (the same is valid for XPS measurements [41]). Note, that electron optics of a spectrometer (entrance/exit slits, transfer channels, energy analyzer, plate of electron detector, etc.) as-well susceptible to hydrocarbon and other low-vacuum contaminators. As a result, undesirable distortions of working parameters or even short circuit may partially happen if the surface of electron optics unit is essentially contaminated; so, the results obtained in this state of matter cannot be accepted as valid experimental data. This means that in general, carbon-containing contaminations are even more harmful than others for proper run-up of a spectrometer. Thus, it is extremely important to exclude from the very beginning the source of "artificial" contaminations of the sample and spectrometer by means of correct and controllable management of vacuum system, especially when surface-sensitive UPS technique is be employed. From abovesaid the first simple rule is following: no stable high-vacuum media in analytical chamber during UPS measurements – no presice UPS data. Most of spectrometers manufacturers now already install low-vacuum lock system which does not allow to start UPS-XPS measurements at all if vacuum value in order of magnitude and stability have been violated in the analytical chamber. It is quite obvious that this feature of modern spectrometers essentially reduces the possibility to damage electron optics and other high-vacuum units; thus, warning indications from a low-vacuum lock system should not be ignored by an analytical scientist-operator.

Additionally, another simple recommendation exists in order to exclude vacuum system influence factor: use a scanning time-interval for one spectrum as short as possible, but do not forget about spectral signal-to-noise ratio and raw-data statistics accumulation. This means that it is better to employ short-time spectrum acquiring from several different target points on the sample surface with onward data summing and averaging (multi-point approach) than from one point but with long-time acquiring. Unfortunately, the multi-point spectra acquiring approach cannot be universally used in all states of matter because of the dependence on probe locality (spatial resolution, surface homogeneity) and actual sample surface area available for probing.

13.3.2 Sample and External Spectral Standard Preparation

As shown in previous sections of the current chapter, UPS is essentially sensitive to the surface electronic states and surface contaminations what is inflicting an extra accuracy with sample manipulations before UPS measurements. There are a lot of in situ contamination treatment procedures available for an analytical materials scientist in order to obtain valuable spectroscopy data. The main idea behind all these procedures is to remove contaminants without even minor modifications of sample surface if possible. In order to implement this idea successfully, it is the need to realize both the type and source of contaminant as well as possible cleaning strategy. The latter refers to sample preparation factors affecting UPS precision.

It is accepted to distinguish all contaminations into two formal groups: physical contaminators, which arise due to mechanical contaminations, and chemical contaminators (unwanted impurity). The first type refers to the so-called van der Waals contaminants and can be quite easily cleaned with the use of vacuum-neutral solvents and onward drying. Other methods (i.e. ultrasonic cleaning, soft heating-stage treatment, ion-beam sputtering, etc.) are also possible, but their employment is considered to be strictly individual for each sample due to physical-chemical origin. Contaminations of the second type are usually caused by technological purity violations during material synthesis and/or its storage and transportation conditions to the place of UPS-XPS certification. Notice that incorporation of chemical impurities into the sample structure at certain concentrations may be an integral part of synthesis process (so-called technological signature), as, for example, occurs in ball-milling chemistry methods when the material of chemical mixer-mill almost always contaminates the produced compound. These contaminators are usually distributed in the bulk of a material more or less uniformly and, unfortunately, cannot be removed by any cleaning procedures. Chemical surface and near-surface contaminants are the most difficult type of contaminants to remove; therefore, cleaving in high vacuum of spectrometer analytical chamber is possible directly before the measurements, if it is not necessary to save original geometric integrity of the sample. However, this procedure is useful only for bulk materials and cannot be employed at all for thin films on substrate or multilayered materials. Therefore, we will discuss those surface cleaning methods that are relatively simple and easy to employ, but at the same time they are effective and available in all laboratories of materials science spectral analysis.

Formally, an application of solvent cleaning method essentially depends on the type of molecular interaction between contaminant and employed solvent. In other words, cleaning dissolution successfully occurs only if a "solvent–contaminant" interaction is dominating in comparison with that for "contaminant–contaminant–sample surface" and, of course, "sample surface–sample" bonding structure is not destroyed. This method is quite simple, because the final results can be easily referred by the use of known Solubility Chart, theoretical thermodynamic estimations of "contaminant – interatomic surface forces" and chemical potentials of solvent and contaminant particles (details and solubility fundamentals are described well in Ref. [42]). Note that the principle of solvent cleaning method is based on solvent's ability to dissolve and remove only a certain contaminator. Of course, no residuals must remain on the sample surface after cleaning procedure. Simple solvents – the very high-grade methanol, ethanol, or other alcohols – are suitable for high-vacuum and UPS application and available in every laboratory of materials science analysis.

Ultrasonic solvent-bath cleaning is a variety of solvent method with essentially extended cleaning possibilities of surface (see e.g. Refs. [43, 44] or elsewhere). Here, ultrasonic waves with 7–110 μm bubbles in a liquid solvent media are produced by 20–40 kHz ultrasound generators, which allow to remove a wide range of heavy contaminators not only from flat areas of surface but also from micro-holes and scratches, blind clefts, etc. If the size and power of ultrasonic actuators are correctly chosen, there will be no damage even for thin-film on substrate material. Additionally, it is possible to combine solvent pre-cleaning with ultrasonic solvent-bath for better results. Final drying procedure is required both for solvent and ultrasonic solvent-bath cleaning and should be performed with accuracy and in fully controllable manner avoiding condensation of humidity from surrounding air media. Recall, that humidity is highly incompatible with high-vacuum techniques, resulting in vacuum instability or even inability to pump down intro-chamber of spectrometer at all.

Vacuum ion-beam cleaning of the surface (so-called sputtering, see e.g. Refs. [45–47]) allows to preserve sample initial geometric integrity and, in addition, has other valuable advantages:

- Selectively determined targeting of the process with high surface and angular accuracy.
- Possibility to employ ions of reactive gases in a vacuum environment instead of noble ones.

- Wide-range tunable variety of accelerated gas-ions energy, accuracy and simplicity of overall process management, high time-step discreteness of process (minimum sputtering time-step cannot be more than five seconds).
- Good reproducibility of sputtering results and compatibility with various materials.
- No need to transfer sample from spectrometer and back (most of modern UPS-XPS systems are already equipped with sputtering module from the very beginning).

In brief, the sputtering process produces a set of collision cascades with atoms of a contaminant at the sample surface. If remaining energy of collisions cascade is larger than contaminant's surface binding energy, an atom-contaminant will be ejected from surface (Figure 13.6). The number of ejected contaminant atoms per incident ion in the frame of theoretical Thompson' model is sputtering yield. If the type of sputtering-gun allows to employ relatively heavy and large ions as incident ion-beam and the material to be cleaned is dense enough, then many-body heat-spike sputtering occurs [48]. Usually, this case of sputtering results in essential increase of sputtering yield, thus should be employed with care in order not to damage sample surface. Note that application of possible minimum sputtering time-step is more preferable for that reason, since such modes can effectively clean the surface of even low-dimensional objects (i.e. thin-film coatings) without ion-induced sample structure destruction. As an alternative to noble and other gases sputtering solution, the fullerene C_{60} sputtering-gun can be used for the very gentle cleaning mode of organic samples or similar samples (serially manufactured by ULVAC-PHI, USA). Existing diversity of spectrometer built-in sputtering-gun designs and operating modes allows to consider this method as one of the most commonly and effectively used methods in spectroscopy. Note that total sputtering time is directly proportional to the depth of removed surface area. The latter can be easily measured using either optical interferometry or other techniques if an incident ion-beam had been sharply focused (depending on sputtering-gun features). An improper ion-beam sputtering mode may result in a solely analyzed surface, which sometimes is not a representative of sample chemical origin.

Vacuum-fresh surface of a material under study can be also achieved employing either a sharp tool for scraping or making vacuum cleavage (mechanical cleaning procedures). Unfortunately, these methods are too coarse since sample geometric integrity will be destroyed, and thus it would not be possible to use them for thin-film or nanostructured materials. In some cases, soft heating-stage treatment of van der Waals contaminations will give essentially effective and correct result of

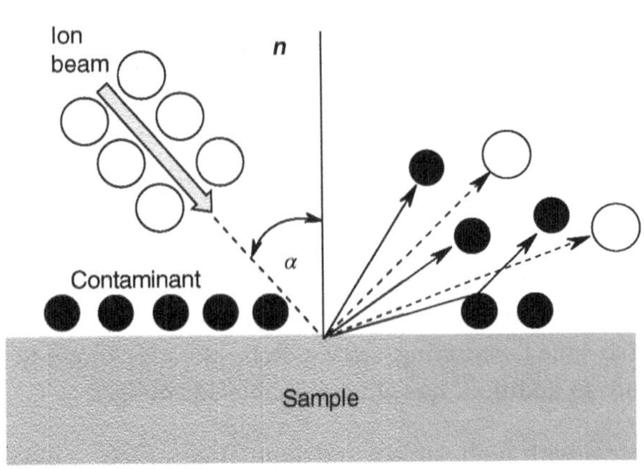

Figure 13.6 Ion-gun sputtering scheme of a sample surface.

surface cleaning compared to that of mechanical treatment. Most samples can tolerate 50 °C heating-stage and sometimes this kind of treatment is good-enough for onward UPS run-up.

Of course, above-mentioned surface cleaning methods are not absolutely universal and cannot be regarded as the only suitable receipts for everyday state of matter. In each case before cleaning a researcher should realize whether actually accidental contamination of the sample occurred and employment of cleaning procedures is really needed. This kind of decision should be made by accounting for synthesis, technological storage, and transportation history of the concrete sample. Also, one should not forget that all effects of surface cleaning will be naturally determined by the chemical origin of the sample and cleaning approach, both quantitatively and qualitatively.

Like other materials science spectroscopies, the UPS as well using external spectral standard method. In this case, the chemical compound or metal employed as a standard needs to be suitable following criteria from the very beginning:

- A fairly characteristic spectrum with well-resolved high-intensity line-shape (or a line).
- Durable chemical stability exactly under employed excitation probe and vacuum media that will be used in an experiment to study the electronic structure of a material.
- No chemical or catalytic interaction with a material under inspection.
- Well-known physical–chemical origin and proved purity.

External spectral standard means that it will be also probed by the same spectroscopy method, being loaded from air-media atmosphere as the sample under study. In this regard, only van der Waals contaminations are possible; if a researcher is using certified purity reference standard from Sigma-Aldrich, Alpha Aesar Fine Chemicals & Metal, or the like, all above reported cleaning procedures for van der Waals contaminants can be employed. Correct strategy and employment of cleaning methods for spectral standard and a material under inspection are considered the basis for obtained spectral data validity and reproducibility.

13.3.3 Ultraviolet Source

UPS spectroscopy uses ultraviolet-range sources in order to illuminate the sample with certain energy photons and produce electron photoemission. As a rule, commonly manufactured laboratory UPS spectrometers are equipped with well-known He I – He II gas-discharge compact sources giving $2\ ^1P \rightarrow 1\ ^1S$ (21.2 eV) and $2\ ^2P \rightarrow 1\ ^2S$ (40.8 eV) emission lines, respectively. Of course, there are other gas-discharge He-emission lines that formally can be employed in UPS (see Table 13.1.), but they are very sensitive to high-voltage ignition procedure and discharge operating conditions, thus have varying intensity yield essentially. Moreover, these lines do not provide any benefits

Table 13.1 Helium gas-discharge ultraviolet photon emission lines.

Emission line	Energy (eV)	Intensity and suitability for UPS
He 1α (He I)	**21.2**	**High-intensity, suitable**
He 1β	23.1	Medium-weak, unsuitable
He 1γ	23.7	Very weak, exactly unsuitable
He 2α (He II)	**40.8**	**High-intensity, suitable**
He 2β	48.4	Weak, unsuitable
He 2γ	51.0	Negligibly weak, useless

compared to He 1α – He 2α emission lines and in most cases usually cause an appearance of background or/and additional low-intensity "false" bands in UPS spectrum if UV-monochromator is not applied [49].

Not only helium can be used in UV-sources: hydrogen, neon, and argon gases as well are applicable due to H Lyman α (10.2 eV), Ne 1α (16.9 eV) – Ne 2α (26.9 eV), and Ar 1 (11.6 eV) having high-intensity and stable UV-emission lines. At the same time, it is obvious that energetics of these lines is not essentially different from that for He I – He II so their implementation as UV-source instead of helium rather depends on electronic structure specificity of a material under study and other particular experimental reasons than on any possible advantages. As for neon, the existence of many other emission lines with energy values close to Ne 1α (16.9 eV) – Ne 2α (26.9 eV) exactly allows to doubt the appropriateness of its employment as a valid UV-source for precise UPS. From above, the following criteria for an ideal UV-source can be formulated: (i) simple and fast gas-discharge ignition procedure; (ii) high intensity of yielding UV-emission and stable run-up during UPS measurements; (iii) a narrow single emission line with in wide-range adjustable photon flux. Unfortunately not possible to fulfill all these requirements at once in experimental reality for only one working gas, thus a universal multi-gas UV-source having separate gas-dedicated inlets, focusing-transfer capillary for UV-flux, and spot-size control with high-precision wavelength adjustable output monochromator will be really well-informed experimental decision (see the description of e.g. SPECS UVS-UVLS series, Henniker Scientific UVS 40A series or similar). Most of manufactured modern UV-sources are structurally and technologically different from initial primitive single gas-discharge UV-lamps, having now computer-controlled run-up preparation and operating facility in order to provide maximum output of UV-photons and exclude any possible negative influence upon working gas pressure, instability of gas-discharge current, and UV-source coolant flow. However, this does not mean that untrained staff will be able to operate it easily.

General spectroscopy requirements strongly recommend to apply the highest possible sample power-load mode of the source during measurements in order to achieve maximum signal intensity yield from the sample simultaneously using overall minimum scan-time of a spectrum; but this is not always universally suitable. Despite of the fact that UV gas-discharge sources much more gently affecting the sample in comparance with XPS X-ray probe, any selected mode can as well cause sample damage issues by means of UV-stimulated chemical bond-breaking, cross-linking and even sample decomposition especially in the case of weak-stable organic materials [49, 50]. Such phenomena usually result in visually-seen UPS spectrum-shape transformations: significant spectrum intensity drops, binding energy shifts, and appearance of "extra" and disappearance of "natural" spectral features. Also, the contribution of background may be increased as a negatively concomitant factor. In anyway, all these hand-made deviations exactly impede or even do not allow at all to perform correct spectral interpretation, so final selection of operating UV-power-load mode for UPS measurements of real sample should be made by analytical scientist carefully and reasonably.

A highly important procedure, an estimation of UV-source efficiency is considered before UPS-spectra acquiring will be started. This procedure involves intensity relations inspection of UPS external standard spectrum (Au or Ag) measured with all identical acquiring conditions (pass-energy of analyzer, power-load mode, detector sensitivity, etc.) for the same working gas but using different UV-emission lines and, hence, different energy regions. Normal intensity relations are usually listed by UV-source manufacturer for each possible working gas and recommended UPS-standard (i.e. typical maximum intensity value for Ag-reference under He II excitation should be not less than 20% of that for He I in the case of helium source).

Relatively simple procedure named multispectrum mode allows to inspect and establish whether sample damage occurred or not because of selected UV-power-load and UV-exposure time. Firstly, run-up 5–10 minutes idle-working cycle (i.e. without measurements) for UV-source in order to achieve stable operating temperature and discharge parameters. Then a spectra set of the same type should be acquired with onward mutual comparance. A difference between these spectra will exactly indicate that sample damage occurs due to abovementioned reasons; so it is necessary to repeat spectra acquiring with reduced sample power-load or/and exposure time parameters. Generally, this can be achieved by means of decreasing UV-flux or/and UV-probe spot size if possible, or otherwise by reducing the overall spectrum scan-time. As an additional sign of serious sample damage and decomposition, vacuum instability is considered. Mutual perfect spectra-shape and energy positions identity achieved during such inspecting measurements mean that selected UV-source mode and scan-time parameters are well-suited for this sample and indicate a good validity of analysis performed. At the same time, sample damage case should not be confused with that for initially heterogeneous material, which will always exhibit dissimilar spectra from different areas of inspected sample. Since the latter is the key-point difference, heterogeneity can be easily established by means of a multispectrum multipoint acquiring procedure, i.e. additional spectral inspection of as many different areas of the sample under study as possible. If a complexity of doubting interacting factors exists, always apply heterogeneity verification by other experimental techniques in order to achieve guaranteed reliable inspection result and avoid analysis distortion.

13.3.4 Charge Neutralizer

UPS allows to study semiconducting and insulating materials, but both of them are charging positively as photoelectrons have been emitted from the sample under UV-radiation. The same situation occurs in XPS but it can be easily fixed by employing charge neutralizer as opposed to UPS low-kinetic energy electron emission. Here, sample charging can cause essential distortions of spectrum-shape or even signal-losses, so it is necessary to minimize surface charge accumulation either via injection conductive pathways into insulating sample or employing charge neutralizer unit of a spectrometer. The first approach is not uniformly applicable because sometimes it is not allowed to modify surface composition and morphology of inspected sample. Alternatively, an increase of surface conductivity can be achieved by means of outer high-conductive surface grid mounting, sample heating, biasing, or using other ways, but these tricks as well will not work perfectly for particular materials. Second approach looks more universal, however, an effective minimizing of charge accumulation will be determined ceteris paribus mostly by construction design of charge-neutralizing unit and employed run-up parameters, which, sometimes, are not possible to vary in a needed range [51]. For this reason, post-acquisition binding-energy corrections of recorded spectra are usually applied with the use of verification databases especially when nonuniform surface charging occurs, because still no method exists that can guarantee elimination of charging effect. On the whole, surface charging and related phenomena are considered as a separate research area and recent successes in theoretical and experimental investigations (see e.g. Refs [51–54]) allow to achieve significant improvements in engineering and construction design of modern charge neutralizers suitable both for XPS and UPS analyses.

A starting idea for simple charge neutralizing procedure was to employ magnetic immersion lenses and low-energy electrons source located near energy analyzer entrance in order to trap and compensate negative charging between the "bottom" of these lenses and charged sample surface by means of magnetic field. Actually this scheme was able to compensate surface charge build-up only partially being incompatible with wide-gap insulating materials like SiO_2 or similar because of significant differential surface lateral and depth charging [55, 56], i.e. when time dependent spatial distribution of a charge is not uniform and several different local charge potentials distort photoelectron spectrum simultaneously. As a result, some spectral bands may be essentially shifted from their original binding energy location or/and broadened. Moreover, "spectral envelope" shape transformation occurs during measurements, thus making obtained data misleading and useless. For this reason, serious attempts were made to modify this simple scheme, but they did not yield any significant improvement; so, new principles of charge neutralization were needed in order to extend analytical possibilities of photoemission spectroscopy. These principles are valid up to the moment and imply two different approaches: charge neutralization and charge compensation. If the first approach assumes actual neutralization process in such way that zero surface charge will be achieved finally, the second one means only controllable reducing of surface charge value and distribution but not complete elimination. Despite of the difference mentioned, all systems are still named by manufacturers as charge neutralizers what is a little-bit fogging their real functionality. In order to simplify the situation, we will use onward the manufacturer's terminology while discussing concrete working schemes in brief.

At present, all modern charge neutralizers (CN) use a variety of designs, which are very different from what was manufactured at the very early development stages of XPS-UPS analysis. From technical-engineering point, they can be formally divided into static and dynamic groups. The first group performs only fixed charge compensation regardless of actual value and particular charge distribution on the surface of a sample (simple charge compensators). The second group allows to control surface charging dynamically by means of varying kinetic energy of slow electrons or/and ions according to the magnitude of appeared surface charge and in suitable cases can decrease it nearly up to zero, using manual or automatic operating modes (formally are as well using compensating principle). Both CN-groups can be manufactured either as a single-, dual-channel, or combined set-up units. At present the most commonly applied schemes are combined dual-channel and dual-source.

As a good example of dual-channel scheme, an ULVAC-PHI charge-neutralizer can be considered (see Figure 13.7). Here, low-energy (~1 eV) electrons from a "cold" cathode are used as a first channel and low-energy (~10 eV) ions covering larger sample surface are used as a second

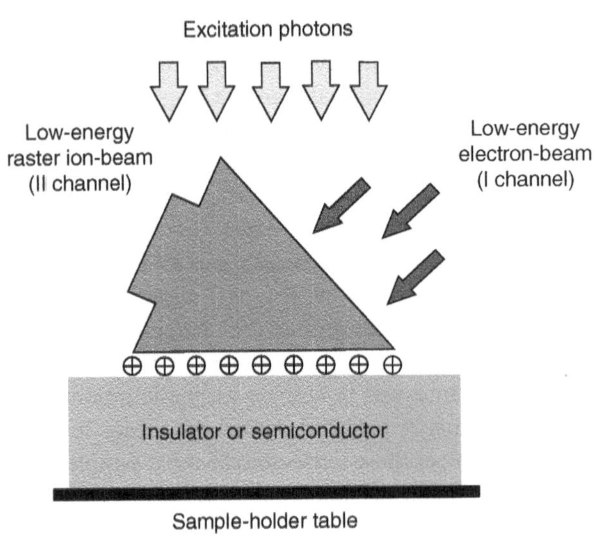

Figure 13.7 Dual-channel charge neutralizer working scheme. *Source:* Based on ULVAC-PHI, US Patent 5.990.476A, JP Patent P3616714B2, EP Patent 0848247B1.

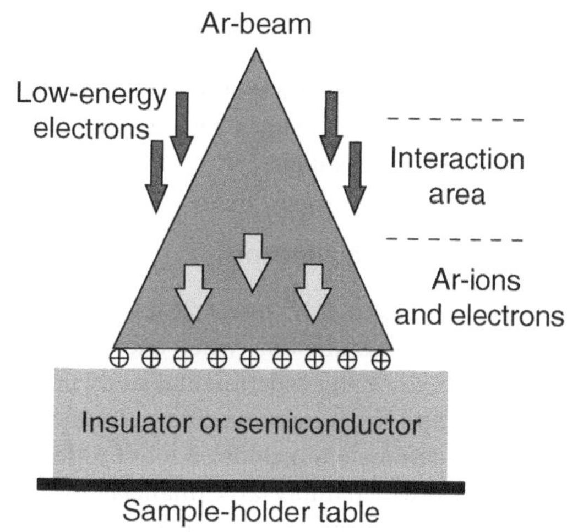

Figure 13.8 Combined dual-source charge neutralizer working scheme. *Source:* Based on GB Patent 2411763.

channel [57]. An optimum factory preset for combination of ion and electron beams allows effectively minimize surface charging in most obvious cases and separate independent management of Channels I–II formally makes possible customization of neutralizing process for damage-sensitive materials. At the same time, cases are known when the structure of some materials (i.e. nano-films, polymers, and bio-compounds) can be hardly destroyed by this type of charge neutralizing [57]. This means that extra accuracy during dual-channel neutralization is needed and, hence, additionally fine-tuned presets for this CN-type are highly preferable.

As an alternative to dual-channel neutralizing approach, the combined dual-source scheme is considered (see Figure 13.8 and Refs. [57, 58]). The main idea is to compensate surface positive charging because of photoemission by means of low-energy electron beam and simultaneously prevent negative charging deviations around probe area by employing Ar-ions. Thus low-energy electrons ("cold" cathode thermionic emission) and Ar-gas particles are applied in such way that they interact in order to produce Ar-ions. Then, both electrons and ions are additionally accelerated and transferred to the sample surface. Focusing and alignment of such combined electron-ion beam are essentially enhanced by extremely low residual outer magnetic fields present in the mu-metal protected analysis chamber. This type of neutralizer can be tuned in a wide range of electron and ion beam presets and their combinations (voltage and emission current for electron source, Ar-gas cell and extractor voltages, etc.) [57, 58], yielding single and dual flood-modes with electrostatic deflection for all sample types, standard (0–1000 eV) and ultralow (0–5 eV) energy dual-beam charge compensation for spectroscopy and imaging, manual charge compensation mode. Usually, no difficulties occur to select and apply optimum run-up parameters of this neutralizer type for most states of matter with minimum probability of sample damage and, thus, the combined dual-source approach is considered as a promising solution for day-to-day employment in high-precision XPS-UPS combined analyses. If remaining energy shift of spectrum or other small deviations are still present, they can be easily fixed using spectrum processing software tools bundled with any modern spectrometer after data collection is complete.

From the above, it is clear that surface charge neutralizing and compensating processes of semiconductors and insulators are important steps in spectral data reliable analysis. Practically, it is not possible to perform perfect elimination or compensation of surface charging at all even with the use of modern charge neutralizer units, which are rather essentially reducing an influence of this phenomenon on photoemission spectra than excluding it completely. For this reason, no universal "how to" recommendations still exist, because each situation of surface charging should be analyzed and finally considered separately in terms of sample structure care. Finally, it is exactly important to remember about test measurements with the use of reference sample or standard in order to verify neutralizing effectiveness of CN-unit before applying spectroscopy run-up routine to the samples under study.

13.3.5 Staff Requirements

From the aforesaid, it becomes clear that UPS methodology as a precise materials science tool can be applied successfully only when a lot of requirements are met strictly. This is not a negative feature of UPS since any experimental study primarily requires accuracy and precision from the very beginning and has its own specific and imperative requirements. In addition, the latest generation of UPS spectrometers includes a lot of different experimental sub-techniques and their technical components which imply large financial investments both for spectrometer installation, initial tuning and onward proper run-up operations. As a simple sequence of this it is supposed that an experienced specialist with a good experimental and technical background is needed in order to perform day-by-day procedures (standard servicing with regard to operating manual, the use and upgrade of driving software, system check-in routines prior measurements and long-term inspections, spectrometer baking, etc.). With no doubt, this specialist should also have high-vacuum equipment knowledge with proven skills. If one add to these requirements the need to obtain and analyze experimental data at a scientific level, then it becomes obvious that it is almost impossible to find such a specialist with encyclopedic knowledges. For this reason, all abovementioned activity is performed by an operating staff – system operator. This person is responsible for spectrometer's overall performance with manufacturer-provided servicing. In this case, any scientific activity will be performed by another staff member – an analytical scientist who as well should be familiar with current experimental methodology and equipment, but has solid state physics background at least in the area of experimental materials science analysis. Some additional knowledge about material stability under different treatments is highly desirable (i.e. ion-beam stability, vacuum stability), because analytical scientist is usually responsible both for sample preparations prior to UPS measurements (outside spectrometer), sputtering or other pre-run-up procedures (inside spectrometer) and final spectral data reliability, which includes correct data processing and primary interpretation. It is extremely wrong to rely totally on the high automation degree of modern spectrometers – this is only a set of support features that can be helpful in a limited case of electronic structure inspection. In this regard, an analytical scientist may find and apply suitable measurement procedures for some specific samples. But, of course, it will be possible only if system performance corresponds well to reliability conditions and peculiar possibilities of currently employed system are taken into consideration. To date, "system operator–analytical scientist" staff combination per experimental system is unofficially considered as the most preferred both from scientific and commercial viewpoints, thus being usually applied in most research laboratories. Other scientists (even from the same laboratory or group) employing this equipment to study their objects formally belong to the "user" group.

Finally, this means that a perfect materials science spectroscopy staff must be formed at least from two multi-skill specialists having relatively wide range of technical and scientific experience, which is difficult to implement in reality without additional special study and training. Usually, the manufacturer of a spectrometer provides familiarization training for the future staff, which basically includes "how to" run-up standard procedures, but, of course, it cannot be sufficient if an experimental background is completely absent in this area. In this regard, it is advisable to train future specialists in material science spectroscopy in advance and purposefully, providing a high-skilled technical and scientific background of material analysis. Perfectly designed experimental equipment is only a contributing part to overall reliability of yielding scientific data and their analytics.

References

1 Turner, D.W. and Al Jobory, M.I. (1962). Determination of ionization potentials by photoelectron energy measurement. *J. Chem. Phys.* 37 (12): 3007–3008.

2 Vilesov, F.I., Kurbatov, B.L., and Terenin, A.N. (1961). Electron distribution over energies in photoionization of aromatic amines in gaseous phase. *Sov. Phys. Dokl.* 6: 490–493.

3 Seki, K., Ishii, H., and Ouchi, Y. (2002). Functional organic material studied using UPS and NEXAFS. In: *Advanced Series in Physical Chemistry Vol.12: Chemical Applications of Synchrotron Radiation*. ISBN 978-981-4480-4 (ed. T.-K. Sham), 386–461. Singapore: World Scientific Publishing Co.

4 Zatsepin, D.A. (2014). Familiarization with XPS method and technique. In: *XPS Lectures Part I [Internet]* (eds. E.Z. Kurmaev, K. Jablonska and R. Berger). Warshaw: Institute of Physics – Polish Academy of Sciences, Poland. https://www.researchgate.net/publication/261586966_Lecture_Familiarization_with_XPS_method_and_technique (accessed 11 April 2020); doi: 10.13140/2.1.3354.9760.

5 Krause, S., Casu, M.B., Schöll, A., and Umbach, E. (2008). Determination of transport levels of organic semiconductors by UPS and IPS. *New J. Phys.* 10: 085001–085017.

6 Quinn, J.T.E., Zhu, J., Li, X. et al. (2017). Recent progress in the development of n-type organic semiconductors for organic field effect transistors. *J. Mater. Chem. C* 5: 8654–8681.

7 Li, P., Ingram, G., Lee, J.-J. et al. (2019). Energy disorder and energy level alignment between host and dopant in organic semiconductors. *Commun. Physiol.* 2 (1): 1–7.

8 Zatsepin, D.A., Boukhvalov, D.W., Gavrilov, N.V. et al. (2017). XPS-and-DFT analyses of the Pb 4f – Zn 3s and Pb 5d – O 2s overlapped ambiguity contributions to the final electronic structure of bulk and thin-film Pb-modulated zincite. *Appl. Surf. Sci.* 405: 129–136.

9 Zatsepin, D.A., Zatsepin, A.F., Boukhvalov, D.W., and Gavrilov, N.V. (2020). Bi-doped silica glass: a combined XPS – DFT study of electronic structure and pleomorphic imperfections. *J. Alloys Comp.* 829: 154459–154467.

10 Koopmans, T. (1933). Ordering of wave functions and eigenenergies to the individual electrons of an atom. *Physica* 1 (1): 104–113.

11 Ishii, H., Sugiyama, K., Ito, E., and Seki, K. (1999). Energy levels alignment and interfacial electronic structures at organic/metal and organic/organic interfaces. *Adv. Mater.* 11 (8): 605–625.

12 Kahn, A., Koch, N., and Gao, W. (2003). Electronic structure and electrical properties of interfaces between metals and conjugated molecular films. *J. Polym. Sci. Part B Polym. Phys.* 41 (21): 2529–2548.

13 Crispin, X., Geskin, V., Crispin, A. et al. (2002). Characterization of the interface dipole at organic/metal interfaces. *J. Am. Chem. Soc.* 124 (27): 8131–8141.

14 Bagus, P.S., Käfer, D., Witte, G., and Wöll, C. (2008). Work function changes induced by charged adsorbates: origin of the polarity asymmetry. *Phys. Rev. Lett.* 100 (12): 126101–126104.

15 Dudde, R. and Reihl, B. (1992). Complete electronic structure of oriented films of hexatriacontane. *Chem. Phys. Lett.* 196 (1–2): 91–96.

16 Bagus, P.S., Staemmler, V., and Wöll, C. (2002). Exchangelike effects for closed-shell adsorbates: interface dipole and work function. *Phys. Rev. Lett.* 89 (9): 096104–096108.

17 Grobosch, M. and Knupfer, M. (2007). Charge-injection barriers at realistic metal/organic interfaces: metals become faceless. *Adv. Mater.* 19 (5): 754–756.

18 Hill, I.G., Rajagopal, A., and Kahn, A. (1998). Energy-level alignment at interfaces between metals and the organic semiconductor 4,4^{l} -N,N^{l}-dicarbazolyl-biphenyl. *J. Appl. Phys.* 84 (6): 3236–3241.

19 Schlaf, R., Parkinson, B.A., Lee, P.A. et al. (1999). HOMO/LUMO alignment at PTCDA/ZnPc and PTCDA/ClInPc heterointerfaces determined by combined UPS and XPS measurements. *J. Phys. Chem. B* 103 (15): 2984–2992.

20 Liu, G., Jaegermann, W., He, J. et al. (2002). XPS and UPS characterization of the TiO_2/ZnPcGly heterointerface: alignment of energy levels. *J. Phys. Chem. B* 106 (23): 5814–5819.

21 Kashimoto, K., Yonezawa, K., Meissner, M. et al. (2018). The evolution of intermolecular energy bands of occupied and unoccupied molecular states in organic thin films. *J. Phys. Chem. C* 122 (22): 12090–12097.

22 Schattke, W. (1997). Photoemission within and beyond the one-step model. *Prog. Surf. Sci.* 54 (3): 211–227.

23 Dowell, D.H. and Schmerge, J.F. (2009). Quantum efficiency and thermal emittance of metal photocathodes. *Phys. Rev. ST Accel. Beams* 12: 074201–074210.

24 Karkare, S., Wan, W., Feng, J. et al. (2017). One-step model of photoemission from single-crystal surfaces. *Phys. Rev. B* 95 (7): 075439–075410.

25 Puff, H. (1964). Zur theorie der sekundärelektronenemission der transportprozeß. *Phys. Status Solidi B* 4 (1): 125–138.

26 Berglund, C.R. and Spicer, W.E. (1964). Photoemission studies of copper and silver: theory. *Phys. Rev.* 136 (4A): A1044–A1064.

27 Hüfner, S. (2003. ISBN 3-540-41802-4). Photoelectron spectroscopy and inverse photoelectron spectroscopy. In: *Photoelectron Spectroscopy: Principle and Applications*, 3ee, 14–20. Berlin: Springer-Verlag GmbH.

28 Jensen, K.L., Jensen, B.L., Montgomery, E.J. et al. (2008). Theory of photoemission from cesium antimonide using an alpha-semiconductor model. *J. Appl. Phys.* 104: 044907–044910.

29 Reinert, F. and Hüfner, S. (2005). Photoemission spectroscopy – from early days to recent applications. *New J. Phys.* 7: 97–131.

30 Koch, N., Heimel, G., Wu, J. et al. (2005). Influence of molecular conformation on organic/metal interface energetics. *Chem. Phys. Lett.* 413: 390–395.

31 Bröker, B. (2010). Electronic and structural properties of interfaces between electron donor & acceptor molecules and conductive electrodes. Dissertation zur Erlangung des Akademischen Grades Dr. rer. nat. im Fach Physik, Mathematisch-Naturwissenschaftlichen Fakultät I Humboldt-Universität zu Berlin.

32 Day, R.P., Zwartsenberg, B., Elfimov, I.S., and Damascelli, A. (2019). Computational framework *chinook* for angle-resolved photoemission spectroscopy. *Quantum Mater.* 4 (54): 1–10.

33 Maňa, C. (2017). *Probability and Statistics for Particle Physics*. ISBN 978-3-319-55737-3, 244pp. Springer International Publishing AG.

34 Huang, P.-W., Qian, H., Du, Y. et al. (2019). Photoemission and degradation of semiconductor photocathode. *Phys. Rev. Accel. Beams* 22: 123403–123412.

35 Jensen, K.L. (2003). Electron emission theory and its application: Fowler–Nordheim equation and beyond. *J. Vac. Sci. Technol. B* 21 (4): 1528–1544.

36 Chanlek, N., Herbert, J.D., Jones, R.M. et al. (2014). The degradation of quantum efficiency in negative electron affinity GaAs photocathodes under gas exposure. *J. Phys. D* 47 (5): 055110–055117.

37 Lv, B., Qian, T., and Ding, H. (2019). Angle-resolved photoemission spectroscopy and its application to topological materials. *Nat. Rev. Phys.* 1 (10): 609–626.

38 Sato, T. and Takahashi, T. (2011). Angle-resolved photoemission spectroscopy of graphene, graphite and related compounds. *Comprehensive Semiconductor Sci. Technol.* 1: 383–309.

39 Dudin, P., Lacovig, P., Fava, C. et al. (2010). Angle-resolved photoemission spectroscopy and imaging with a submicrometre probe at the SPECTROMICROSCOPY-3.2L beamline of Elettra. *J. Synchrotron Radiat.* 17 (4): 445–450.

40 Nakayama, Y., Hikasa, M., Moriya, N. et al. (2018). Anisotropic valence band dispersion of single crystal pentacene as measured by angle-resolved ultraviolet photoelectron spectroscopy. *J. Mater. Res.* 33 (20): 3362–3370.

41 Al-Ajlony, A.-M.B., Kanjilal, A., Harilal, S.S., and Hassanein, A. (2012). Carbon contamination and oxidation of Au surfaces under extreme ultraviolet radiation: an X-ray photoelectron spectroscopy study. *J. Vac. Sci. Technol. A* 30 (4): 041603–041606.

42 Israelachvili, J.N. (2011). *Intermolecular and Surface Forces*, 3ee ed. ISBN 978-0-12-375182-9. Amsterdam: Academic Press – Elsevier 674pp.

43 David, J. and Cheeke, N. (2012). *Fundamentals and Application of Ultrasonic Waves*, 2ee. ISBN-13: 978-1439854945. Boca Raton: Taylor & Francis Group – CRC Press 504pp.

44 Wang, J., Gao, Y., You, Z. et al. (2020). The effect of ultrasonic cleaning on the secondary electron yield, surface topography, and surface chemistry of laser treated aluminum alloy. *Materials* 13: 296.

45 Kelly, R. and Miotello, A. (1997). On the mechanisms of target modification by ions beams and laser pulses. *Nucl. Instrum. Meth.Phys. Res. Sect. B* 122 (3): 374.

46 Aumayr, F. and Winter, H. (2004). Potential sputtering. *Philos. Trans. R Soc. Lond. A* 362: 77.

47 Sricharoenchai, P., Panich, N., Visuttipitukul, P., and Wangyao, P. (2010). Effect of substrate temperature, biasing, and sputter cleaning on the structure and properties of nanostructured TiB_2 coatings on high speed steel. *Mater. Trans.* 51 (2): 246–252.

48 Jakas, M.M. and Bringa, E.M. (2000). Thermal-spike theory of sputtering: the influence of elastic waves in a one-dimensional cylindrical spike. *Phys. Rev. B* 62 (2): 824–830.

49 Boehm, A.M., Wieser, J., Butrouna, K., and Graham, K.R. (2017). A new photon source for ultraviolet photoelectron spectroscopy of organic and other damage-prone materials. *Org. Electron.* 41: 9–16.

50 Opitz, A., Frisch, J., Schlesinger, R. et al. (2015). Energy level alignment at interfaces in organic photovoltaic devices. *J. Electron Spectrosc. Relat. Phenom.* 190 (A): 12–24.

51 Baer, D.R., Engelhard, M.H., Gaspar, D.J. et al. (2002). Use and limitations of electron flood gun control of surface potential during XPS: two non-homogeneous sample types. *Surf. Interface Anal.* 33: 781–790.

52 Cazaux, J. (2000). About the charge compensation of insulating samples in XPS. *J. Electron Spectros. Relat. Phenom.* 113 (1): 15–33.

53 Baer, D.R., Windisch, C.F. Jr., Engelhard, M.H., and Zavadil, K.R. (2002). Practical aspects of charging phenomena in XPS as demonstrated in oxidized-Al films on Al and Al alloys. *J. Surf. Anal.* 9 (3): 396–303.

54 Sezen, H. and Suzer, S. (2013). XPS for chemical- and charge-sensitive analyses. *Thin Solid Films* 534: 1–11.

55 Yu, X. and Hantsche, H. (1993). Some aspects of the charging effect in monochromatized focused XPS. *Fresenius J. Anal. Chem.* 346: 233–236.

56 Tielsch, B.J. and Fulghum, J.E. (1996). Differential charging in XPS. Part I: demonstration of lateral charging in a bulk insulator using imaging XPS. *Surf. Interface Anal.* 24 (1): 28–33.

57 Baer, D.R., Artyushkova, K., Cohen, H. et al. (2020). XPS guide: charge neutralization and binding energy referencing for insulating samples. *J. Vac. Sci. Technol. A* 38 (3): 031204–031219.

58 Edwards, L., Mack, P., and Morgan, D.J. (2019). Recent advances in dual mode charge compensation for XPS analysis. *Surf. Interface Anal.* 51 (9): 925–933.

14

Transmission Electron Spectroscopy

Raffaele Giuseppe Agostino and Vincenzo Formoso

Dipartimento di Fisica, Università della Calabria, Rende, CS, Italy

14.1 Empirical Aspects of Electron–Matter Interaction

Exploring the nano-world is becoming an urgent necessity for scientists wishing to decipher the secrets of the stringent link between structure and physical properties of novel materials with unprecedented features.

Nowadays, a spectro-microscopic view of the intimate features of the nanomaterials is allowed by means of electron beams. This is achieved with the use of transmission electron microscopes (TEM, 30–300 keV) equipped with spectrometers such as electron or optical analyzers.

The experimental techniques we will introduce in the following mainly deal with the recovery of information with an extremely high spatial resolution by analyzing the electron spectra of the beam transmitted through an investigated specimen (electron energy loss spectroscopy – EELS) and the light emitted in the same conditions in different wavelength ranges from IR to X-rays (cathodoluminescence – CL and X-ray fluorescence – XRF). A schematic representation of these processes is reported in Figure 14.1.

The advantage of the spectro-microscopic approach consists in the allowed correlation between the morphological and structural features depicted by the TEM images with the information extracted from the acquired spectra such as stoichiometry, electronic structure, bonding order, and atomic-structure correlations.

14.1.1 Fast Electrons Interaction with a Solid

In a semiclassical picture of the processes occurring when a primary monoenergetic electron beam travels into a specimen, the associated electromagnetic field interacts with the condensed matter aggregate and a series of phenomena takes place. As a consequence, both the electron kinetic energy and their direction change with a strong dependence on the primary beam energy, the sample characteristics, and the penetration depth.

A typical electron spectrum is depicted in Figure 14.2, where we can distinguish three regions: the intense zero loss (ZL) peak, a low energy-loss region spanning few tens of electronvolts, and the large energy loss region ranging some hundreds of electronvolts.

The ZL peak is normally very intense due to the large electron mean free path for primary energies above 30 keV. Its width is mainly determined by the energy width of the incident beam,

Spectroscopy for Materials Characterization, First Edition. Edited by Simonpietro Agnello.

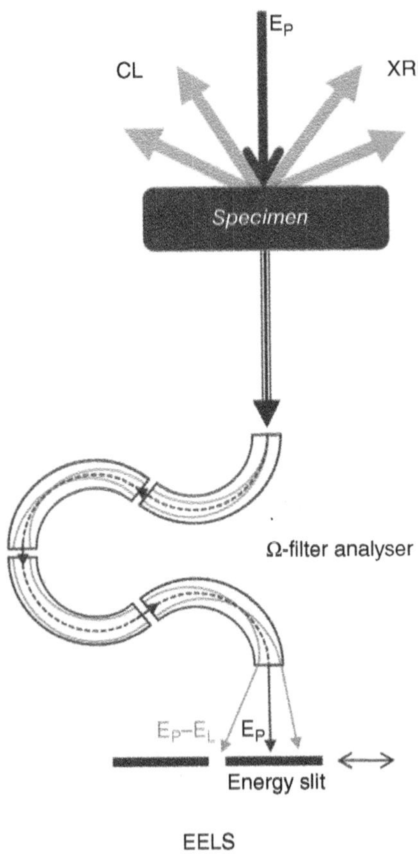

Figure 14.1 The main TEM-related conventional electron spectroscopies are the electron energy-loss spectrum (EELS) and the electron-induced light emission (CL, XRF). As the collimated electron beam (E_P = 30–300 keV) interacts with the sample, a series of energy-loss channels will be accessible, and the transmitted beam becomes polychromatic. The energy spectrum can be recovered by means of suitable analyzers (such as the Ω-filter depicted in figure). The radiative relaxation processes will give light emission ranging from IR to UV when valence electron bands are involved while X-rays are emitted when the de-excitation process involves core-holes.

which is 0.3 eV or even better in a field-emission gun. Monochromators can be used to improve the energy resolution down to 10 meV, allowing to unveil lower energy loss [1, 2]. These electrons are those giving rise to both the diffraction patterns and the Kikuchi lines and do not give any spectroscopic information on the sampled material [3]. We use their position in kinetic energy as a reference for calculating the electron energy loss.

14.1.2 Electron Energy Loss Spectroscopy (EELS)

The monoenergetic electron beam used in TEM is a minute and intense probe penetrating thin samples. This allows it to get analytical information from a limited interaction volume with a less-than-nm spatial resolution. Several review articles and textbooks are available on the argument giving the possibility to deepen each aspect of the TEM-related EELS spectroscopy [1, 2, 4–13].

A kinetic energy (KE) spectrum of the transmitted electrons is obtained by their dispersion and converted into an electron energy loss (E_L) spectrum by setting the zero to the elastic peak energy (E_P) and calculating

$$E_L = E_P - \text{KE}. \tag{14.1}$$

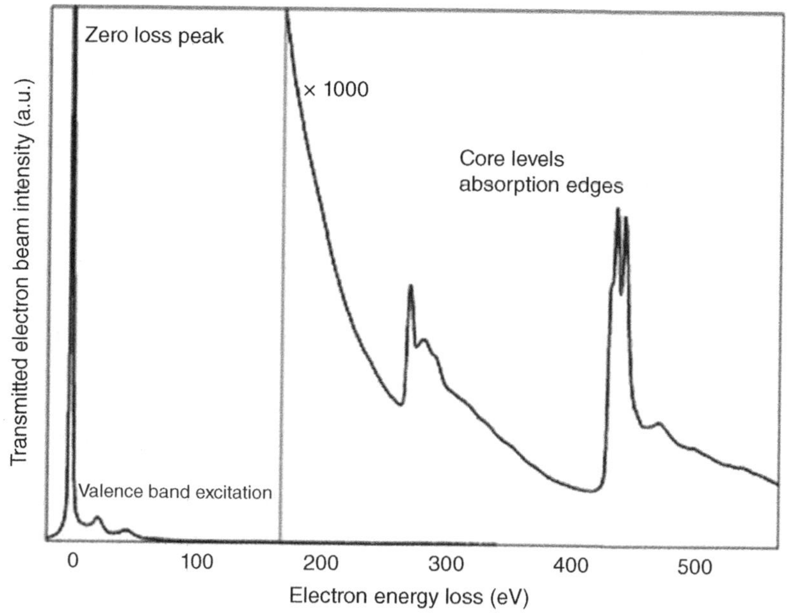

Figure 14.2 A schematic representation of electron energy loss spectrum. The zero loss peak, due to unperturbed and elastically scattered electrons, is the main feature and it is a reference for calculating the energy loss. The valence band excitation region, spanning some tens of electronvolts, is characterized by the presence of plasmons and valence-to-conduction band transitions. Finally, at the typical core electron binding energies, we find the faint features due to the absorption edges that are very similar to the X-ray absorption ones. A smooth background due to elastic/inelastic multiple events is present and gently declines in intensity with the energy loss.

When $E_P = 50$–300 keV, the spectral features that are meaningfully intense are distributed up to around 2 keV. For lower E_P, this range is reduced.

In an EELS spectrum, the energy resolution is given by the root sum squared of the energy spread on the incident beam and the energy resolution of the electron analyzer that is mainly due to aberration effects. As a thumb rule, the resolution can be increased by controlling the trajectories, i.e. the apertures and the field's homogeneity, with a consequent loss of intensity.

In general, the inelastic events involve a small-angle scattering when compared to elastic ones, giving signal in a cone centered on the incident electron beam direction and with an aperture angle [14]

$$\theta = \frac{E_L}{2E_P} \tag{14.2}$$

and thus, well below 10 mrad.

Differently from light, electrons can transfer a part of their kinetic energy to a solid for every single event. This translates into the possibility to cause excitations with energy between 0 and E_P, involving both solid core electrons and valence electrons. As we have seen, the energy spectrum is typically divided into energy ranges of energy loss E_L that typically involve a peculiar category of such excitations.

14.1.2.1 Inner Shell Excitations

At high energy loss, with E_L greater than 50 eV, we have the region of core electrons excitations that can bring the solid in a final neutral or ionized excited state characterized by the presence of a hole in a given internal orbital (see Figure 14.2).

A rather simple picture providing a good understanding of the physical processes involved in the electron inelastic scattering is given by the probability of transition between an initial core level $|i\rangle$ and a finale state $|f\rangle$ in the one-electron description [15]:

$$P_{i\to f} \propto \frac{1}{q^3} |\langle f \mid e^{i\,\boldsymbol{q}\cdot\boldsymbol{r}} \mid i\rangle|^2 \rho(E_f) \tag{14.3}$$

where $\boldsymbol{q}$ is the transferred momentum, $\boldsymbol{r}$ the electron position, and $\rho(E_f)$ is the final state density of unoccupied states.

Apart from the smooth dependence of the matrix element in Equation (14.3), the probability is modulated by the density of final states $\rho(E_f)$, whose variation is strong in the conduction band region, which is roughly limited between the Fermi (E_F) and the vacuum (E_V) levels.

In the limit of small transferred momentum, as in the case we are facing, we can neglect all the multipole terms in the exponential expansion, remaining, in a first approximation, with only the dipole term [16]. This is the basis of the correlation of the EELS spectral features with their UV-X-ray counterpart once the electron energy loss values in the former ones are treated as photon energies in the latter ones (Figure 14.3).

This allows us to interpret the EELS features close to binding energy of a specific core level in close similitude with the X-ray absorption near edge spectroscopy (XANES, see Chapter 11), which is widely used to probe the partial density of states (PDOS) for each element of a given sample [17].

A straight example of this powerful tool is given by the study of the carbon allotropes where the conduction band is strongly influenced by the 2s,p orbitals hybridization. Figure 14.4 reports diamond, graphite, and amorphous carbon EELS spectra at the C 1s edge [18]. The p-projected DOS $\rho_p(E)$ of the empty states above the Fermi energy is depicted by the spectral features once the proper energy shift is operated, $E = E_L - E_{C_{1S}}$ where $E_{C_{1S}} = 285.2\,\text{eV}$ is the C 1s electron binding energy [19].

The main differences between EELS and XANES come from:

a) A general dissimilarity between the electron and X-ray penetration depth.
b) The multipole terms in the matrix element of Eq. (14.3), which are absent in the X-ray absorption phenomenon.

Concerning the first point, the typical inelastic mean free path (IMFP) for 30–300 keV electrons ranges between 30 and 200 nm depending on the electron kinetic energy and the sample density. A detailed study of this dependence is found in literature [20, 21], while an empirical relationship, given by Eq. (14.4) [22], is widely used for a first estimation of the mean free path:

$$\lambda\,[\text{nm}] = \frac{538\,a}{E_{\text{kin}}^2} + 0.41\,a^{\frac{2}{3}}\,E_{\text{kin}}^{1/2} \tag{14.4}$$

where E_{kin} is the electron kinetic energy in eV and a is the mean atomic diameter evaluated as $a = \sqrt[3]{(M/1000\,\varrho A)}$ in nm with M molar mass (g/mol), ρ the density (kg/m^3), and A the Avogadro number (mol^{-1}) and where 538 has units eV^2 and 0.41 has units nm$^{1/3}$eV$^{-1/2}$. On the other hand, XANES is characterized by a lower spatial resolution, with the X-ray penetration depth being typically larger by two orders of magnitude [23].

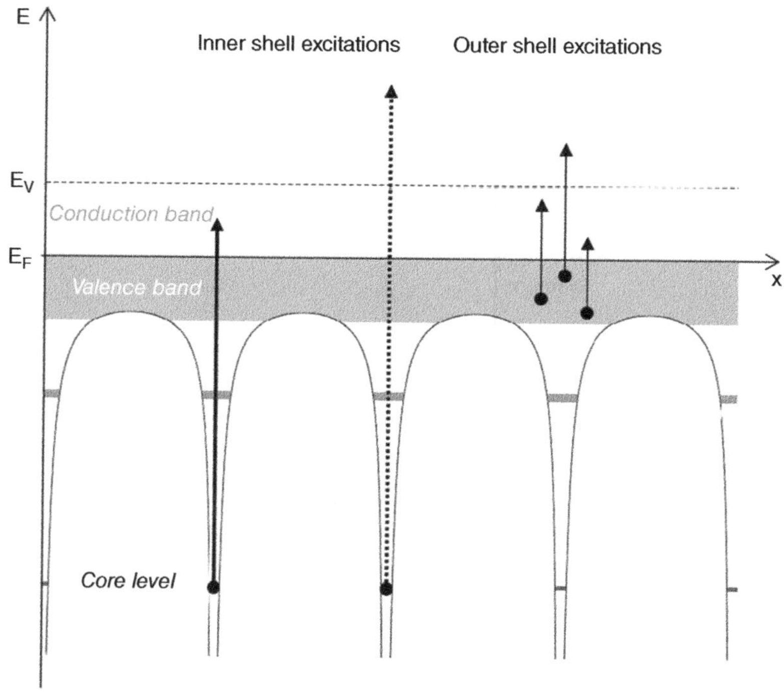

Figure 14.3 The electronic excitations involving "one-electron like" transitions depicted in a generic 1D energy-level diagram. The transition from a discrete inner core electron level to an unoccupied conduction band level is depicted with a tick and a dotted line distinguishing between the final states (below and above the vacuum level, respectively). On the right, we find the sketch of the possible transitions involving valence band electrons, which are characterized by a continuum of states both in the initial and final states. The delocalized valence band states cover a wide energy range below the Fermi level (E_F) while the inner shell levels are localized around the atomic positions. Note that each occupied state is coupled with an unoccupied state above E_F. The final states involving an electron in a delocalized orbital can give rise to a charge separation and, once their energy is above the vacuum level E_V, to the emission of the excited electron.

Concerning the effects of the multipole terms in the matrix element in Eq. (14.3), the interpretation of the EELS edge features deals with a quite subtle question: even faint multipolar matrix elements can be exalted by the presence of high partial density of states for dipole-forbidden final states, giving rise to a noncomparable EELS and XANES results [24, 25]. It is very useful, in those cases, to use numerical tools such as the FEFF package [17].

A further example of the limits afore exposed is depicted in Figure 14.5 where the C 1s edge is probed by EELS and XANES on the same sample. XANES spectrum shows the p-projected DOS above the Fermi level, E_F. In the 3–6 eV range above the edge (located at about 284 eV), a wide peak due to the COOH-states is present. This peak has no counterpart in the EELS spectrum due to the radiation-induced damage or evaporation of surface functional groups on the sample [26].

The presence of narrow peaks close to the edge is related to exciton states in which a neutral excited state is produced by an energy absorption close to the ionization energy in solids (semiconductors and insulators) and molecules where localized valence states are present. A typical example is shown in Figure 14.6. In a classical view, these states are those in which the electrostatic attraction bounds the core hole produced by the electron impact to the electron promoted in low energy states lying above the Fermi energy [27].

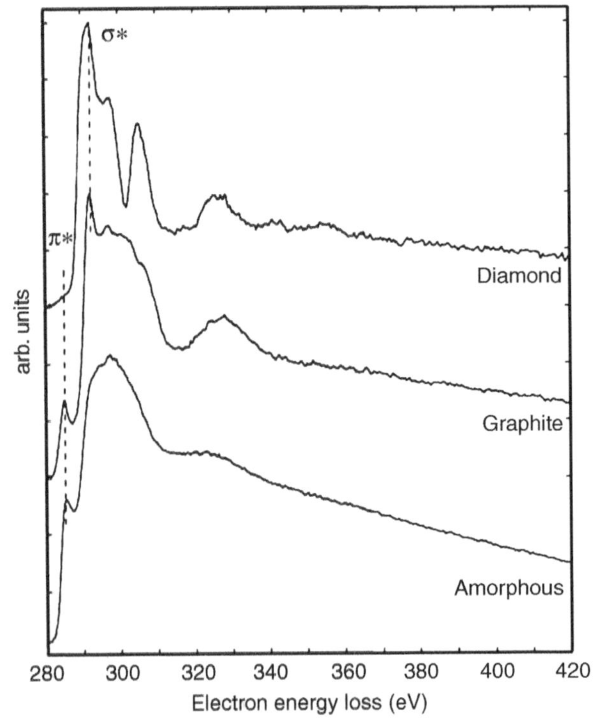

Figure 14.4 EELS spectra of diamond, graphite, and amorphous carbon core-loss (adapted from [18]). Note the presence of the π^*states just above the 285.2 eV edge in the graphite and amorphous carbon spectra.

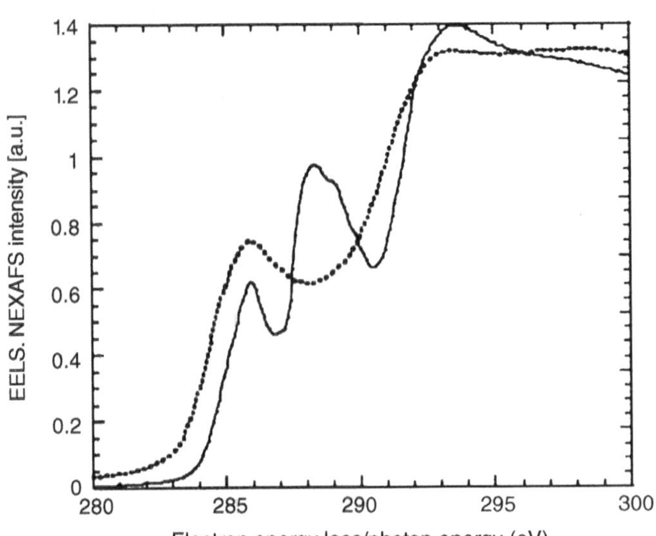

Figure 14.5 Comparison of near edge EELS (dotted line) and XANES (solid line) spectra of diesel soot [26]. A striking difference in the spectra is visible in the 287–291 eV energy range where COOH-group related states are present in the XANES spectrum. Source: Reprinted by permission from Elsevier.

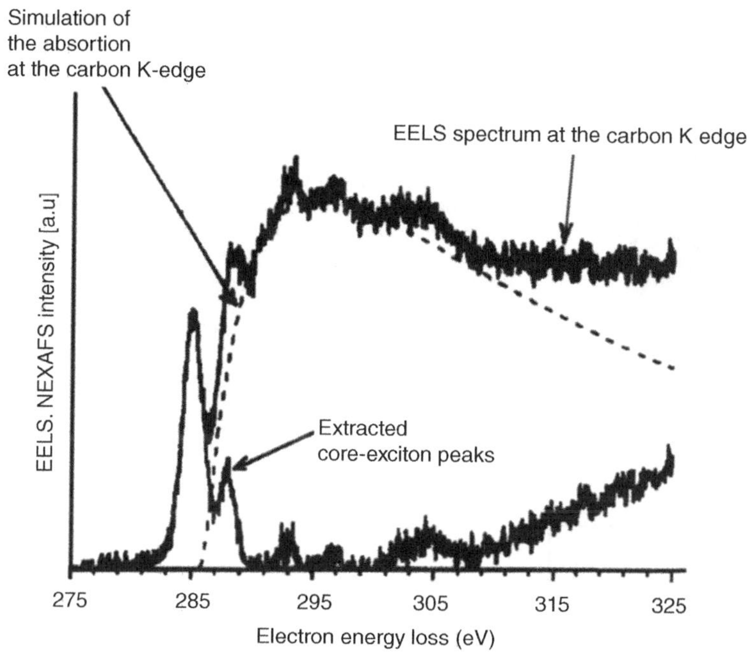

Figure 14.6 Core-exciton peaks from an EELS spectrum at the carbon K-edge for a PET sample [27]. Once a smooth function representing the C 1s-edge spectrum of an isolated C atom is removed, we can recover the peaks corresponding to excitons involving the $\pi*$ (C = C) and $\pi*$ (C = O) localized states. Source: Reprinted by permission from Elsevier.

A further general outcome is given by the fact that the elemental composition of the material under investigation can be recovered by looking for the presence of the specific core edges. Moreover, a quantitative analysis is possible by the evaluation of the edge intensities, each one of them being proportional to the element concentration.

14.1.2.2 Low-Loss Excitations

The excitations involved at low energy losses, below 50 eV, are mainly concerning the valence band electron and the shallow core states. The transitions involving the latter can be treated as those involving the inner core electrons, see the previous paragraph, once the energy distribution of the initial states is considered.

As long as the delocalized valence band states are concerned, the final state could result in a single electron transition or the excitation of plasmons, collective oscillation of the electron gas.

Generally, the collective excitations are by far more probable than the single-particle ones and predominate in the 0–50 eV energy-loss region. Here, once again, spectral features are exploited to identify the aggregates' nature by looking for the presence of the signature of specific structures. In the case of carbon allotropes, one can distinguish the sp^2 hybridization by the presence of both π and $(\sigma + \pi)$ plasmons whereas sp^3 hybridization shows only the $(\sigma + \pi)$ plasmons.

Together to the smooth and large peaks associated with plasmon excitation, in the low-energy-loss region, we can find weak features associated with intra- and inter-band transitions. These will give information on the joint density of states similar to optical UV-vis spectroscopies once the multipole effects are taken into account. As an example, direct measurement of the bandgap in

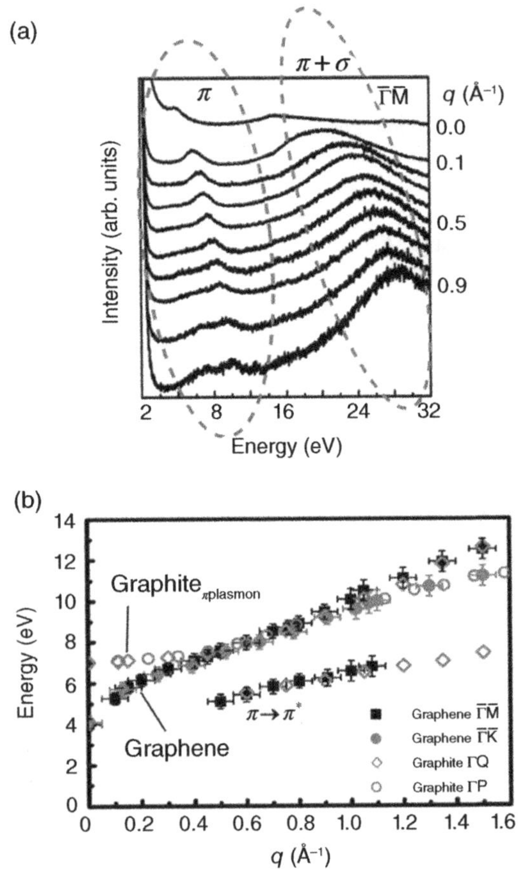

Figure 14.7 Plasmon dispersion in graphene and graphite depicted by EELS-TEM with a 100 keV electron beam. By recording the EELS spectra at different exit direction from the electron analyzer, it was possible to resolve the momentum transfer in the $0 < q < 0.9$ Å^{-1} range with an increment of 0.1 Å^{-1}. (a) Momentum-resolved valence band EELS spectra and (b) dispersion relation of π and $\pi/\pi*$ plasmon in 2D graphene (solid symbols) and 3D graphite (open symbols). The dispersion relation is extracted from the EELS peaks [29, 30]. Source: Reprinted with permission from Elsevier.

semiconductors and insulators can be extracted once the ZL peak contribution is removed from the spectrum.

A further possibility is given by the momentum-resolved EELS in which a 2D mapping of the electrons emerging from the analyzer is used to correlate the energy loss E_L to the transferred momentum q [28] as shown in Figure 14.7. An interesting case is given by the bidimensional systems such as graphene where the dispersion of the plasmon losses can be mapped with a 0.1 Å^{-1} resolution [29].

The main theoretical approach for the spectra interpretation involves the study of electronic behavior as depicted by the dielectric function $\varepsilon(q, \omega)$. In fact, the Energy Loss function can be expressed using the Full Penn Algorithm as [31]

$$Im\left[\frac{-1}{\varepsilon(q,\omega)}\right] = \int_0^\infty d\omega_p g(\omega_p) Im\left[\frac{-1}{\varepsilon^L(q,\omega;\omega_p)}\right] \tag{14.5}$$

where ε^L is given by the Lindhard model dielectric function of a free-electron gas with plasmon energy $\omega_p = \sqrt{4\pi n}$, n is the electron density, and $g(\omega_p)$ is a proper coefficient introduced to satisfy the $q = 0$ condition [20]. The classical Kramer–Kronig relation relates the real (ε_1) to the imaginary (ε_2) part of the dielectric function:

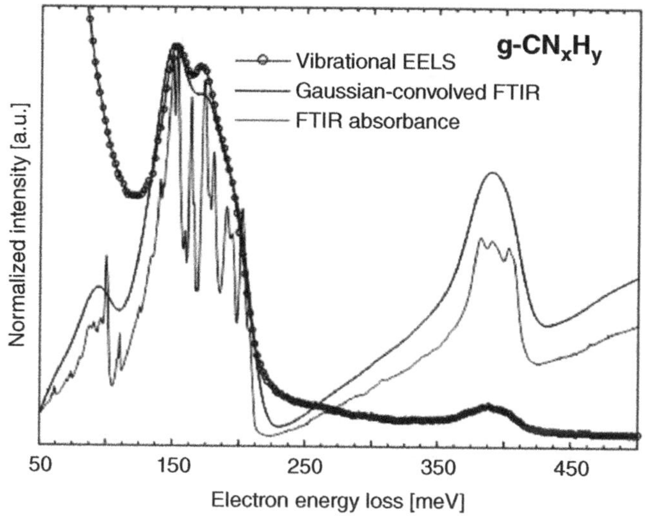

Figure 14.8 Vibrational EELS spectrum from g-CN$_x$H$_y$ aggregates recorded at 60 kV [32]. FTIR spectrum recorded from identical material and convolution of FTIR spectrum with response function corresponding to energy resolution of EELS are also displayed. Note the close similarity of the vibrational EELS spectrum with the optical one. Source: Reprinted with permission from Elsevier.

$$\varepsilon_1(\omega) = \frac{2}{\pi} P \int_0^{\infty} \frac{\omega' \varepsilon_2(\omega')}{(\omega\prime)^2 - \omega^2} \mathrm{d}\omega' \tag{14.6}$$

where P is the principal Cauchy value of the integral [20]. As already pointed out, the use of monochromators allows us to unveil very low energy losses such as those related to phonons and vibrations of the surface species, which span few hundred meV from the ZL peak. Exploiting both the high resolution of the TEM imaging and the delocalization of the valence electrons, it is possible to record vibrational spectra by recording the spectrum of a beam few nanometers away from the surface of a nanoparticle or an aggregate (Aloof beam EELS) [32]. A typical example is reported in Figure 14.8 for aggregates of g-CN$_x$H$_y$ where the opportune comparison with Fourier transform infrared spectra (FTIR) is shown to highlight the information related to vibrational modes.

14.1.2.3 Energy-Filtered Images

Employing the energy filter, it is possible to collect a spectrum from every pixel of a TEM image, to form a 3D spectral map where for each (x, y) position, the third dimension is the energy loss. This 3D map can be used, apart from the extraction of the individual spectrum from any real-space location, to get deeper information about the spatial distribution of each property evidenced by an EELS spectral feature. As shown in Figure 14.9, this is allowed by the extraction of the so-called energy-filtered TEM (EFTEM) images formed by extracting slices at a given energy loss, allowing a spectroscopic approach in the images analysis [33, 34].

This approach represents a significant advance in the EFTEM technique. The collection and treatment of an EFTEM images set allow quantitative elemental maps to be calculated. Moreover, the reliability of the obtained maps is strengthened by the analysis of the whole EELS spectral features.

The algebra applied to the EFTEM images has a close resemblance to the EELS spectrum treatment when edge intensities have to be recovered. A proper intensity background behavior has to be

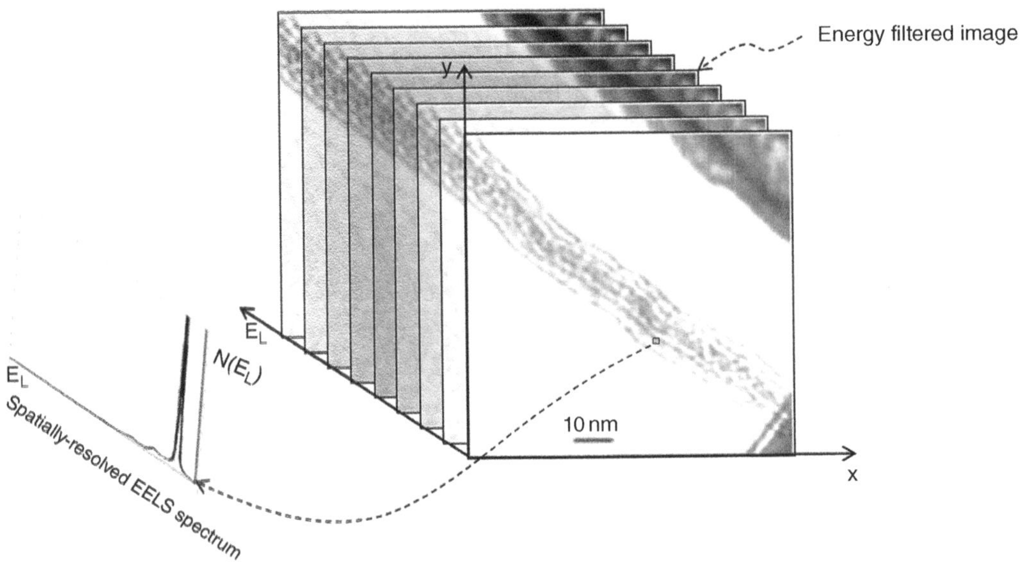

Figure 14.9 A schematic diagram showing how the 3D spectral map is built as a collection of EELS spectra, each at every (x, y) position. Individual spectra can be extracted as well as images filtered for a specific energy.

estimated and removed in such a way that the chemical map is recovered by plotting only the edge intensity variations. The whole process benefits from the conventional statistical methods for the determination of the signal-to-noise ratio and the reliability of the background-fitting procedure [33, 34]. An example of this treatment and the consequent analysis of the data is reported in Figure 14.10, where a compositional map is extracted after background subtraction.

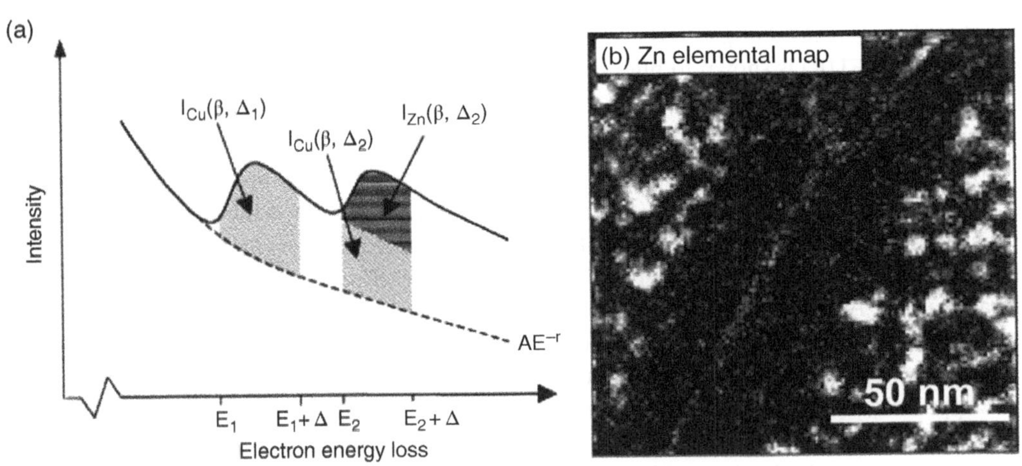

Figure 14.10 Method for the extraction of quantitative compositional maps in the presence of overlapping edges. (a) A background slow-varying curve is fitted for both the edges, allowing the measurement of the edges intensities. (b) Each single element map is depicted using the edge intensity (Zn L_{23} edge at 1020 eV in this case) [33]. Source: Reprinted with permission from Elsevier.

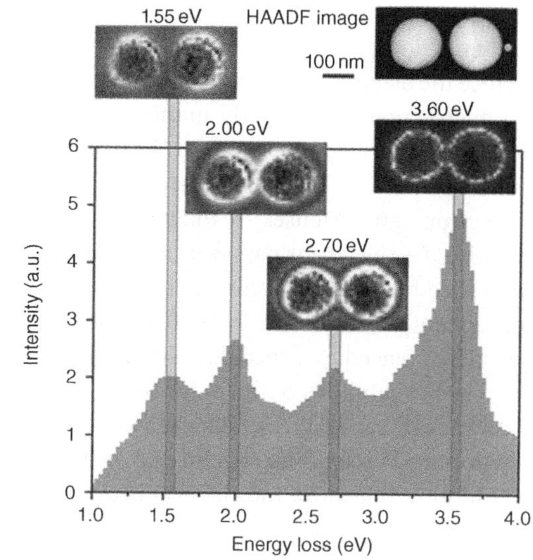

Figure 14.11 The localized surface plasmon resonance modes supported by a dimer of silver nanoparticles are mapped following the intensities of the plasmon peaks (loss energies of 1.55, 2.00, 2.70, and 3.60 eV) in the EEL spectrum. The beam position used for the spectrum acquisition is marked by the dot in the high-angle annular dark-field (HAADF) image [35]. Source: Reprint with permission from ACS publications.

The outer shell excitations of plasmons are usually linked to delocalized states whose typical span exceeds by orders of magnitude the bonding length. It becomes very interesting to map the EELS spectral features when nanoaggregates are taken into account. In these cases, in fact, the physical constraints due to the particles' borders give rise to peculiar forms for the spatial extension of the states [35]. An example of this result is shown in Figure 14.11. The mapping procedure is very similar to that used in the inner shell excitation case even though the background removal is less trivial owing to the peak overlapping.

14.2 Instrumental Setups

14.2.1 TEM in a Nutshell

A transmission electron microscope employs *monoenergetic* and collimated electrons as a source of radiation. As we depicted in the previous paragraph, detailed knowledge of the interaction between a primary electron beam and a sample target is very useful for material characterization: imaging, diffraction, and chemical analysis techniques can be used to investigate the properties of inorganic, organic, and biological materials. Thus, the field of investigation spans from mineralogy to solid state chemistry, from organic crystallography to protein structure determination [4–6, 8].

A conventional transmission electron microscope (CTEM) is a "simple" apparatus composed of five main important sections [5]:

- Electron source.
- Sample stage.
- Electromagnetic lens with their respective apertures.
- Projection chamber.
- Detector of the transmitted electrons.

There are two classes of electron sources: the thermionic and the field emission ones. The latter belong to two types: in the first type, electrons are extracted at room temperature (cold field-emission gun); in the second one, a low-temperature thermionic emission is combined with the field-emission process (thermal field emission or Schottky emission gun).

Once the electrons are extracted from the source surface, they are accelerated along the column of the microscope and properly manipulated by the electromagnetic lenses to get a collimated and monoenergetic electron beam. A TEM apparatus should be able to irradiate a sample with an electron beam having a uniform current density.

Electromagnetic lenses are made of magnetic material such as iron (polepieces) and a coil of copper wire. The electrons pass through a hole in the polepieces. The intensity of the magnetic field is controlled by the electric current in the coil. It is possible to modify the magnetic field strength and consequently control the focal length of the lens by changing the current in the coil. Using two polepieces placed at a specific distance, it is possible to obtain an important parameter governing the focus capabilities of the magnetic lens, the hole/gap ratio.

As we can see in the schematic view of the optical components of a commercial transmission electron microscopy instrument shown in Figure 14.12, different types of lens are necessary for a TEM to fully control the electron beam traveling from the electron source to the detector acquisition system.

Just below the electron source, we find an adaptable condenser system used to control the beam diameter and therefore the brightness of the microscope. The condenser system guarantees Köhler illumination in TEM mode (imaging, selected area diffraction), and spot illumination (spot analysis, scan mode, convergent beam electron diffraction) [5].

The condenser aperture system controls the beam intensity and selects only the electron within a certain angular aperture. Once we chose the electron beam energy, and, therefore, fix the wavelength λ of the traveling electrons, we can change the resolution r by changing the aperture according to the equation:

$$r = \frac{0.61\,\lambda}{\alpha} \tag{14.7}$$

In fact, several objectives with different aperture α can be inserted in the beam path, allowing to control the final resolution of the acquired image: increasing the aperture size, we can reach an improved image resolution. Besides, inserting the objective aperture permits to control the image contrast. High Z element scatters electrons at larger angle with respect to low Z elements for which, by inserting an aperture, we eliminate the electrons strongly scattered and the portion of material containing high Z elements appear darker than that with low Z elements.

The objective lens system is composed of two polepieces (upper and lower poles, objective prefield lens and objective lens). The specimen is inserted at the objective lens plane in the center of the polepiece gap (Figure 14.12).

This setup permits operation in different modes:

- High-resolution TEM mode.
- Spot mode, generating small electron probes for analysis.
- Selected area diffraction mode.
- Scan mode.
- Side-detectors optimized mode, secondary emission detectors arrangement in the scanning mode.
- Sample rotation (tilting) mode.

The objective lens plays an important role: in fact, because the two polepieces can be adjusted to vary the magnetic field, the objective lens can be used to regulate beam divergence varying from the

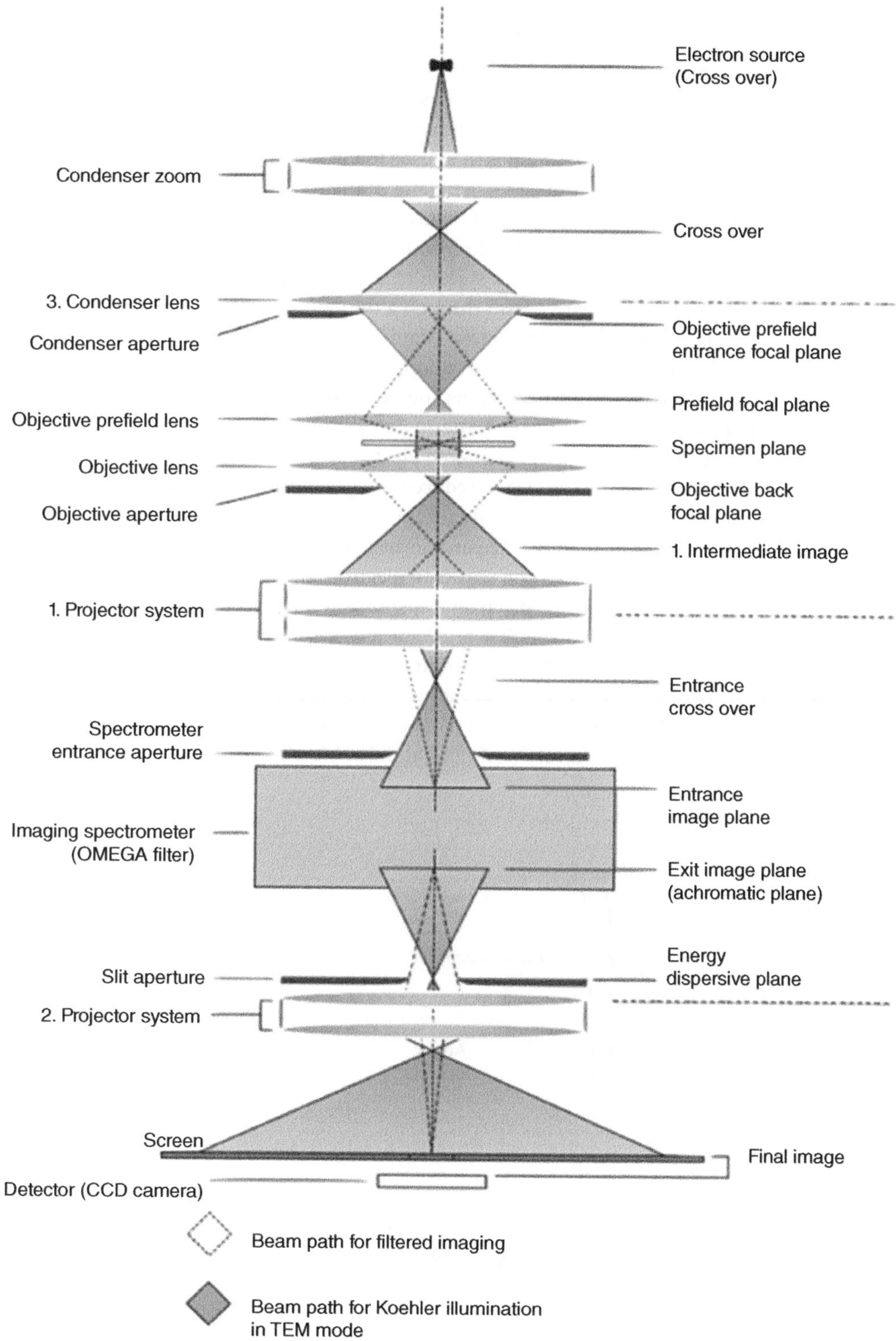

Figure 14.12 Schematic view of the focusing and acceleration lenses of a TEM. Note that the lower part is dedicated to the energy filtering of the transmitted electrons. Source: Courtesy of Assing SpA.

spot (convergent) to parallel illumination of the specimen placed in the objective lens plane. This electron-optics allows the magnification of the sample image, too.

The first projector lens system (labeled 1 in Figure 14.12) is located after the objective lens and it can operate in the image mode, where the lens focuses the beam on the image plane of the objective lens, or in diffraction mode in which the lens focuses the electrons on the back focal plane of the objective lens. The first projector lens system is then used to form an image of the first intermediate or diffraction pattern in the fixed entrance image plane of the following energy spectrometer.

The second projector lens system (labeled 2 in Figure 14.12) is a two-lens magnification arrangement able to image the filtered image or diffraction pattern or the electron energy-loss spectrum on the fluorescent screen or the data acquisition system of the microscope.

When the high-energy electron beam passes through a thin sample, elastic and inelastic scattering events can occur, and a poly-energetic electron beam is generated. Electrons passing through a sample lose energy depending on the energy of the primary beam, on the thickness of the target, and on the atomic number of the excited atoms. Besides, the scattered electrons change their momentum depending on the characteristic scattering event.

In conventional TEM, these inelastically scattered electrons are an annoying issue as they contribute to the diffuse background giving rise to reduced image contrast. One way to overcome this problem consists in inserting a small lens diaphragm aperture, as shown in Figure 14.13, which has to select only electrons falling in the aperture and to neglect electrons having a large scattering angle. Reducing the angular aperture of the scattered electron improves the diffraction contrast.

On the other hand, the most advanced electron transmission microscopes are equipped with an energy filter that can be inserted after the first projector lens system, allowing to discriminate the elastic and inelastic scattering portion of the transmitted beam as shown in Figure 14.14. This gives the so-called "energy loss capabilities" that together with the parallel illumination, the automated aperture selection, the quick switching between parallel to spot mode, and a field emission source transforms the TEM apparatus into a very advanced and flexible platform covering a strategic role

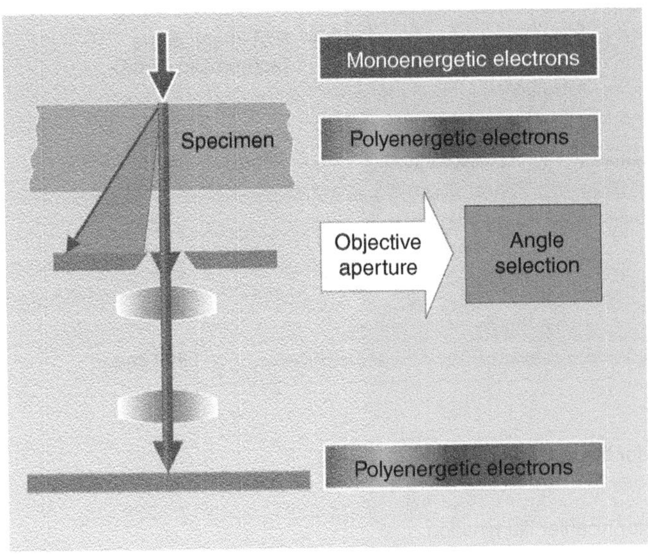

Figure 14.13 The insertion of a small diaphragm aperture allows the selection of the transmitted electrons and to filter out electrons having a large scattering angle. Source: Courtesy of Assing SpA.

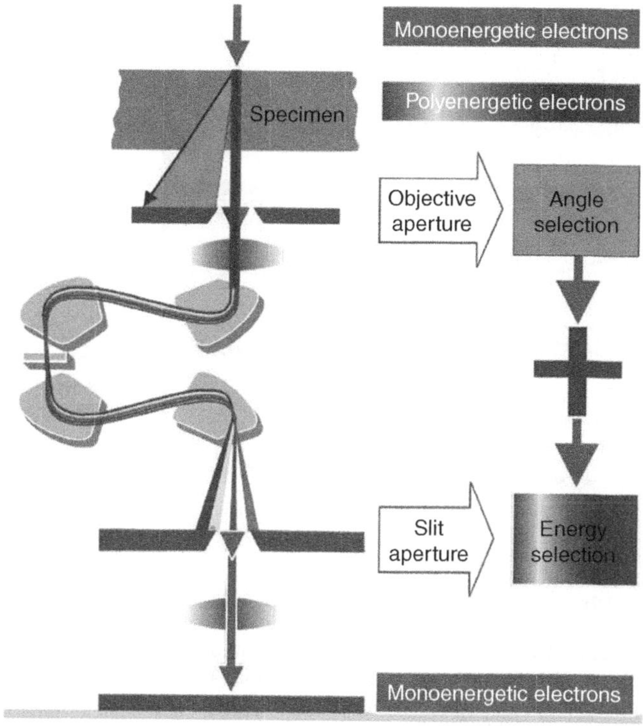

Figure 14.14 A simplified diagram to illustrate contrast generation in an energy-filtered transmission electron microscope. After the small-angle selection aperture, in the energy-filtered TEM, is possible to insert an additional energy selector resulting in an element-specific image contrast enhancement: the image is formed with only a monochromatic electron beam. Source: Courtesy of Assing SpA.

in the several fields of applied research. In fact, by dispersing the electron passing through the sample, we can form the sample image by selecting only a selected region of interest (ROI) in the energy dispersive spectrum.

Energy-filter TEM (EFTEM) can adopt two different experimental setups: the post-column and the in-column one as reported in Figure 14.15. A typical post-column imaging energy filter is composed of a magnetic prism generating a gradient field, dodecapole optics, a CCD camera, and an electrostatic shutter. This kind of filter is situated underneath the TEM viewing screen and can be attached to any TEM system. The spectrometer deflects the electron and transforms each pixel of the image into a spectrum. This spectrum can be filtered by an energy-selecting slit. The detector screen detects the selected electrons generating an energy-filtered image with additional contrast and improved resolution.

On the other hand, several different electron spectrometers (Mollenstedt, Wien, magnetic prisms combined with an electron mirror, Omega and Mandoline filter) have been constructed to be used as in-column electron energy analyzers. They are usually placed between the diffraction lens and the following intermediate lens. The use of an in-column energy-analyzer system has the advantage of a dual operational mode. On one side, it acts as energy spectrometer and, on another side, it enables the collection of an energy-filtered two-dimensional image or a diffraction pattern of the sample using electrons within a specified energy-loss range.

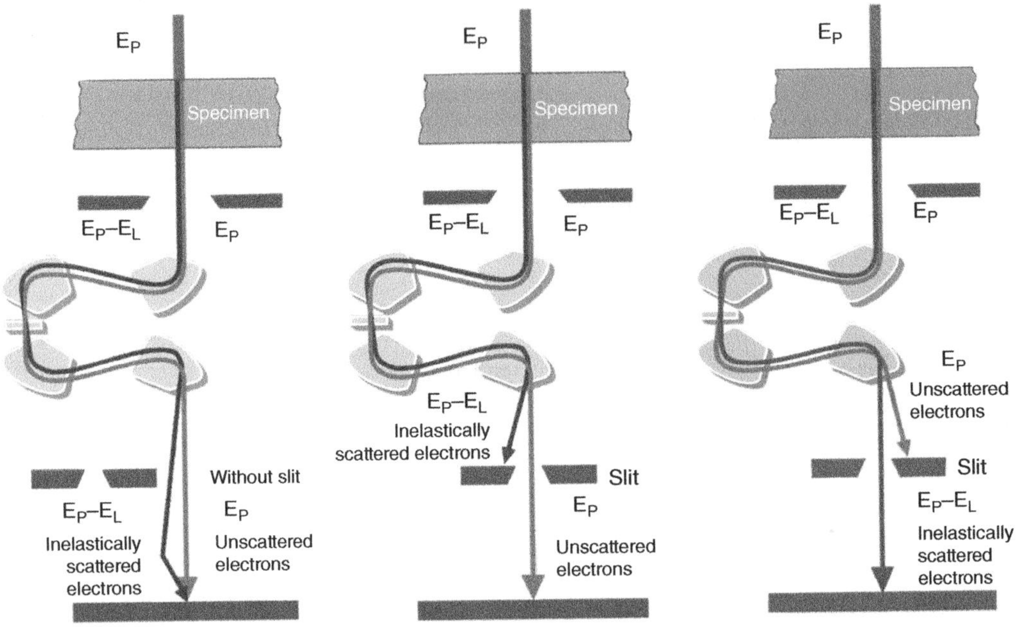

Figure 14.15 Comparison between conventional (CTEM) and an in-column integrated omega energy filter (EFTEM) transmission electron microscopy. In an EFTEM, we have an additional mode of operation able to extend the performances in imaging and spectroscopy by using unscattered and inelastically scattered electrons. Operational modes of an EFTEM: (left) An EFTEM can be used as a CTEM by simply removing the slit system from the beam path; (center) in the "elastic imaging" configuration, the sample image is formed with only the elastic electron: all the inelastic scattered electrons are filtered out; (right) in the "inelastic imaging" mode, only the inelastic electrons that have lost a specific energy (EL) are allowed to reach the detector. Source: Courtesy of Assing SpA.

This energy filter offers a typical energy resolution of 0.1–0.3 eV, depending on the used beam current. With some recent Omega-type monochromators, researchers were able to achieve an energy resolution of few tens of meV at 200 kV acceleration voltage [36].

By using only the zero loss electrons (Figure 14.15 left) to form the image, we gain contrast in bright-field, dark-field, and diffraction operation mode. A different way to improve the image contrast when we study a thin sample is presented in Figure 14.5 center. As we already pointed out, the inelastically scattered electrons reduce the image contrast considerably. In the elastic bright-field image mode, atoms with higher Z appear darker and, therefore, filtering out the inelastically scattered electrons, we gain higher contrast with respect to the unfiltered one. On the other hand, using only the inelastically scattered electrons on a narrow energy bandwidth (Figure 14.15 left), we obtain a dark field image that is element- and structure-sensitive.

Also, in the case of thick samples, it is possible to improve the resolution and enhance the contrast in comparison to a CTEM image, which is always blurred even when the image is acquired at high incident electron energy (>100 keV). In this case, the energy-loss spectra show an intense spectrum of inelastically scattered electrons compared to the zero loss peak. To improve both the image resolution and contrast, it is possible to filter the energy-loss electrons by choosing a small energy bandwidth. In this way, we collect only monoenergetic electron energy-loss to form the image, we reduce the chromatic aberration, and therefore the contrast is considerably enhanced.

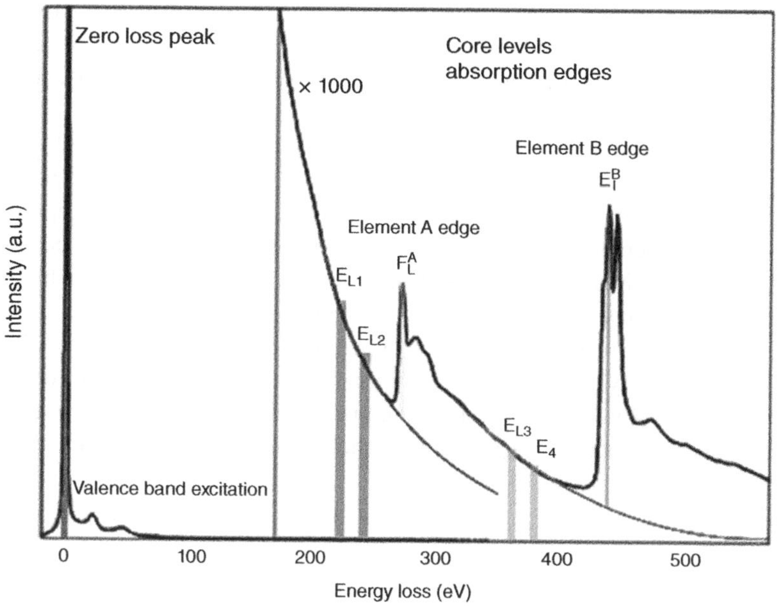

Figure 14.16 Element distribution images obtained with the three-windows method involves a careful inspection of the EELS spectrum. For each X element, a number of edges are present in the spectrum at the characteristic energy loss E_L^X. By evaluating the background behavior, the intensity of each edge can be extracted for each projected position in the image (see text). Elemental mapping is then recovered by the proper image combination.

The detectability of parts of the sample with different chemical compositions strongly depends on the imaging conditions, on the specimen thickness, and on energy-loss edge intensity. A typical procedure for elemental mapping involves the collection of two images (Figure 14.16) at energy losses E_{L1} and E_{L2} placed few tens of electronvolts from the selected element absorption edge E_L^A on the lower energy-loss side (evidenced by large vertical bars in Figure 14.16). These images are used to determine and subtract the background (continuous line) below the absorption edge at E_L^A. The third (A element-specific) image is then collected at the energy E_L^A where the maximum edge intensity is present (thin vertical bar in Figure 14.16). The difference between the A element-specific image and the calculated background intensity image corresponds to the chemically resolved map of the element A density in the specimen with high local resolution.

This procedure can be repeated for all the absorption thresholds present in the spectrum (such as for the B element edge in Figure 14.16), a map containing the elemental distribution is obtained for every single element present on the sample. By superimposing the chemical maps measured, the distribution of all the elements on the sample is obtained with a high lateral resolution sensitive to the chemical composition of the sample. In Figure 14.10, we find an example of elemental mapping with *nm* resolution.

As depicted, EFTEM microscopy can collect images with a narrow energy bandwidth. An images stack of the same region of the specimen can be collected as a function of the electron energy loss with a method called "Image EELS." By plotting the pixels' intensities corresponding to the same region of interest, an intensity profile is extracted as a function of the electron energy loss. At the limit, for each pixel of the sample acquired image, we can analyze an energy-loss spectrum, allowing a spectroscopic study with high spatial resolution, high detection element sensitivity, high energy resolution, and high-resolution chemical bond mapping. Very small and irregular objects can be properly investigated by this method, increasing the sensitivity of EELS analysis. The averaging of many pixels belonging to one object efficiently enhances the signal-to-noise ratio.

A different approach is called "Serial EELS" acquisition mode. The energy-selecting slit is inserted in the beam path, the whole energy distributed electrons (0–2500 eV) are recorded on a CCD camera having a large dynamic range. In such a way, the spectrum is recorded by sweeping the whole electron beam over the energy-selecting slit by changing the primary electron beam energy in very small steps. In this way, it is possible to achieve a good signal-to-noise ratio, which permits to acquire and evaluate the fine structures in the energy loss spectrum.

Finally, in the "Parallel EELS" mode, the image is acquired by removing the energy-selecting slit. This allows to acquire the image in a very short time and to release on the sample a very low dose compared to the serial EELS. This system is perfectly adequate in cases where the intensity of the transmitted electrons is very low such as in the case of a high-energy ionization threshold.

References

1 Mukai, M., Kim, J.S., Omoto, K. et al. (2014). The development of a 200kV monochromated field emission electron source. *Ultramicroscopy [Internet]* **140**: 37–43.

2 Lopatin, S., Cheng, B., Liu, W.T. et al. (2018). Optimization of monochromated TEM for ultimate resolution imaging and ultrahigh resolution electron energy loss spectroscopy. *Ultramicroscopy* **184**: 109–115.

3 Fundenberger, J.J., Morawiec, A., Bouzy, E., and Lecomte, J.S. (2003). Polycrystal orientation maps from TEM. *Ultramicroscopy* **96** (2): 127–137.

4 Egerton, R.F. (2011). *Electron Energy-Loss Spectroscopy in the Electron Microscope*. Boston, MA: Springer, ISBN 978-1-4419-9582-7.

5 Nellist, P.D. (2019). Scanning transmission electron microscopy. In: *Springer Handbook of Microscopy* (eds. P.W. Hawkes and J.C.H. Spence), 49–99. Switzerland AG: Springer Nature.

6 Polman, A., Kociak, M., and García de Abajo, F.J. (2019). Electron-beam spectroscopy for nanophotonics. *Nat. Mater.* **18** (11): 1158–1171.

7 MacLaren, I., Annand, K., Black, C., and Craven, A. (2018). EELS at very high energy losses. *Reprod Syst Sex Disord.* **67**: i78–i85.

8 Carter, C.B. and Williams, D.B. (2016). *Transmission Electron Microscopy: Diffraction, Imaging, and Spectrometry*. Switzerland: Springer International Publishing, ISBN 978-3-319-26649-7.

9 Krivanek, O.L., Dellby, N., Hachtel, J.A. et al. (2019). Progress in ultrahigh energy resolution EELS. *Ultramicroscopy* **203**: 60–67.

10 Reimer, L., Fromm, I., and Rennekamp, R. (1988). Operation modes of electron spectroscopic imaging and electron energy-loss spectroscopy in a transmission electron microscope. *Ultramicroscopy* **24** (4): 339–354.

11 Stroud, R.M. (2014). Transmission electron microscope-based spectroscopy. In: *Treatise on Geochemistry*, 2ee (eds. H.D. Holland and K.K. Turekian), 231–244. Elsevier Ltd.

12 Venkatraman, K., Levin, B., Crozier, P. et al. (2019). Vibrational spectroscopy at atomic resolution with electron impact scattering. *Nat. Phys.* **15**: 1237–1241.

13 Egerton, R.F. (2009). Electron energy-loss spectroscopy in the TEM. *Reports Prog Phys.* **72** (1): 016502.

14 Brydson, R. (2020). Electron energy loss spectroscopy. *Garland Sci.*

15 Paolucci, G., Santoni, A., Comelli, G. et al. (1991). M4,5 absorption edge of Ag, Pd, and Rh by reflection electron-energy-loss spectroscopy: role of nondipole transitions. *Phys. Rev. B.* **44**: 10888.

16 Santoni, A., Derossi, A., Finetti, P. et al. (1992). Nondipole transitions at the 4d edges of Ta, Pt, and Au: theory and experiment. *Phys. Rev. B.* **46** (24): 660–667.

17 Moreno, M.S., Jorissen, K., and Rehr, J.J. (2007). Practical aspects of electron energy-loss spectroscopy (EELS) calculations using FEFF8. *Micron* **38** (1): 1–11.

18 Turner S. (2010). Transmission electron microscopy characterisation of 0-D nanomaterials. PhD Thesis, University of Antwerp – Department of Physics.

19 Greczynski, G. and Hultman, L. (2020). X-ray photoelectron spectroscopy: Towards reliable binding energy referencing. *Prog. Mater Sci.* **107**: 100591.

20 Shinotsuka, H., Tanuma, S., Powell, C.J., and Penn, D.R. (2019). Calculations of electron inelastic mean free paths. XII. Data for 42 inorganic compounds over the 50 eV to 200 keV range with the full Penn algorithm. *Surf. Interface Anal.* **51** (4): 427–457.

21 Powell, C.J. (2020). Practical guide for inelastic mean free paths, effective attenuation lengths, mean escape depths, and information depths in x-ray photoelectron spectroscopy. *J. Vac. Sci. Technol. A* **38** (2): 023209-1–023209-14.

22 Briggs, D. and Seah, M.P. (eds.) (1993). *Practical Surface Analysis by Auger and X-ray Photoelectron Spectroscopy*. New York: Wiley.

23 Henke, B.L., Gullikson, E.M., and Davis, J.C. (1993). X-ray interactions: photoabsorption, scattering, transmission, and reflection at E=50-30000 eV, Z=1-92. *At. Data Nucl. Data Tables* **54** (2): 181–342.

24 Fuggle, J.C. and Inglesfield, J.E. (eds.) (1992). *Unoccupied Electronic States – Fundamentals for XANES, EELS, IPS and BIS*. Berlin, Heidelberg: Springer.

25 Braun, A., Huggins, F.E., Shah, N. et al. (2005). Advantages of soft X-ray absorption over TEM-EELS for solid carbon studies––a comparative study on diesel soot with EELS and NEXAFS. *Carbon N Y.* **43** (1): 117–124.

26 Braun, A., Kubatova, A., Wirick, S., and Mun, S.B. (2009). Radiation damage from EELS and NEXAFS in diesel soot and diesel soot extracts. *J. Electron Spectrosc. Relat. Phenomena* **170** (1–3): 42–48.

27 Varlot, K., Martin, J.M., Quet, C., and Kihn, Y. (1997). Towards sub-nanometer scale EELS analysis of polymers in the TEM. *Ultramicroscopy* **68** (2): 123–133.

28 Shekhar, P., Malac, M., Gaind, V. et al. (2017). Momentum-resolved electron energy loss spectroscopy for mapping the photonic density of states. *ACS Photonics* **4** (4): 1009–1014.

29 Ukhtary, M.S. and Saito, R. (2020). Surface plasmon in graphene and carbon nanotubes. *Carbon N Y.* **167**: 455–474.

30 Liou, S.C., Shie, C.-S., Chen, C.H. et al. (2015). π-plasmon dispersion in free-standing graphene by momentum-resolved electron energy-loss spectroscopy. *Phys. Rev. B.* **91** (4): 45418-1–45418-7.

31 Penn, D.R. (1987). Electron mean-free-path calculations using a model dielectric function. *Phys. Rev. B.* **35** (2): 482–486.

32 Crozier, P.A. (2017). Vibrational and valence aloof beam EELS: A potential tool for nondestructive characterization of nanoparticle surfaces. *Ultramicroscopy* **180**: 104–114.

33 Thomas, P.J. and Midgley, P.A. (2001). Image-spectroscopy – I. The advantages of increased spectral information for compositional EFTEM analysis. *Ultramicroscopy* **88** (3): 179–186.

34 Thomas, P.J. and Midgley, P.A. (2001). Image-spectroscopy – II. The removal of plural scattering from extended energy-filtered series by Fourier deconvolution. *Ultramicroscopy* **88** (3): 187–194.

35 Wu, Y., Li, G., and Camden, J.P. (2018). Probing nanoparticle plasmons with electron energy loss spectroscopy. *Chem. Rev.* **118** (6): 2994–3031.

36 Essers, E., Benner, G., Mandler, T. et al. (2010). Energy resolution of an Omega-type monochromator and imaging properties of the MANDOLINE filter. *Ultramicroscopy.* **110** (8): 971–980.

15

Atomic Force Microscopy and Spectroscopy

Gianpiero Buscarino

Department of Physics and Chemistry – Emilio Segrè, University of Palermo, Palermo, Italy

15.1 Introduction

Atomic force microscopy (AFM) is probably the most celebrated technique falling into the family of experimental methods known as scanning probe microscopy (SPM) [1]. Each of these methods is designed to characterize one or more specific properties of the surface of a solid sample with very high resolution. Such a result is achieved by raster scanning the surface of interest with a probe that typically consists of a long strip of a rigid material, the cantilever, with a sharp tip on top with apical radius of few nanometers. The interaction between the probe and the sample is usually confined to the tip. Different SPM techniques typically involve probes made of different materials and with a variety of geometries, to be able to sense different material properties (mechanical, electrical, magnetic, etc.). AFM is primarily designed to obtain the morphology of the surface of a solid material by using the force of interaction between an inert probe and the sample [2–7]. Moreover, as it will be discussed in Section 15.5, the AFM equipment can also be efficiently used to investigate many physicochemical properties of a material with sub-nanometer resolution [2–7]. A very distinctive property of AFM consists in the fact that it can be applied in many environments, such as air or controlled atmosphere of different gases, liquids, and vacuum. Due to its wide applicability and versatility, AFM is appreciated in physical, chemical, biological, and medical fields of research, and it is also involved in several important industrial applications.

The AFM microscope can generally be operated in many ways for morphology investigations, but they can be roughly classified into two main groups: static- and dynamic-deflection modes. In the former, the probe is not subjected to any external force, whereas in the latter it is forced to make a periodic motion. The most sensitive and diffused AFM acquisition method is a dynamic-deflection mode called the Tapping mode (also known as semicontact or intermittent contact), which consists in forcing the cantilever to oscillate at the frequency corresponding to its first flexural normal mode. Such oscillation is usually established when the probe is far above the sample, where its interaction with the sample is negligible. Subsequently, the surface of interest is approached until significant tip–sample forces are established, inducing a change in both the frequency of the first flexural mode and the amplitude of the cantilever oscillations. These changes are related to the amount of the tip–surface interaction and can then be used to determine the morphology of the sample surface.

Spectroscopy for Materials Characterization, First Edition. Edited by Simonpietro Agnello.

In fact, when the probe scans the surface, the intensity of the tip–sample interaction changes due to the surface roughness, as in some places it pushes the surface toward the tip (higher points), in other moves it away (lower points). Such variations induce continuous modification of the properties of the probe oscillation. A feedback circuit is then used to compensate for these changes by applying vertical movement of the probe, so as to restore the tip–surface interaction initially established by approaching the surface. There are two main feedback circuits that may be involved in an AFM equipment. One uses the amplitude of the forced oscillation of the probe at a fixed frequency as feedback parameter and it is the most diffused, whereas the other uses both the amplitude and the frequency of the normal mode of oscillation of the cantilever. The AFM technique involving the former feedback circuit is named Amplitude Modulation AFM (AM-AFM) and it is commonly used in all the environments but the vacuum (the reasons of this limitation will be discussed in Section 15.4.2), whereas the latter is known as Frequency Modulation AFM (FM-AFM) and it is usually reserved to measurements in vacuum. Due to its larger versatility, the AM-AFM is much more widespread in the research laboratories all over the world and is the subject of the present chapter.

15.2 The AFM Microscope

In Figure 15.1, a scheme is presented, showing the main components of an AM-AFM microscope. In the following, a detailed description of such components and their role in the measurement process are presented and discussed.

15.2.1 The Probe

The probe is the heart of the AFM microscope and its most common geometries are schematically represented in Figure 15.2. In general, it involves a sharp tip mounted on top of a strip of a rigid material (cantilever) and both these elements are connected to a large piece of silicon (substrate).

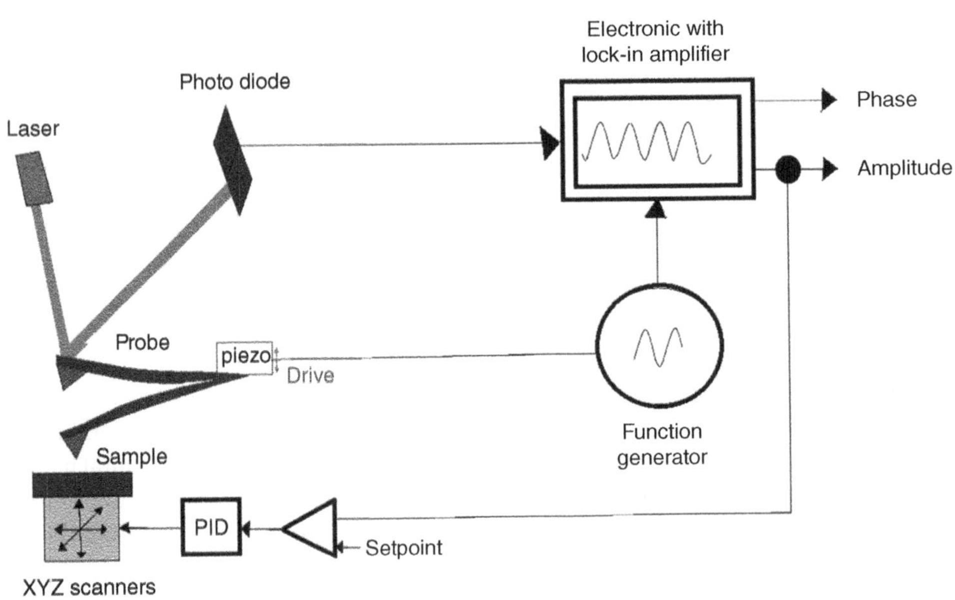

Figure 15.1 Typical scheme of an AM-AFM microscope.

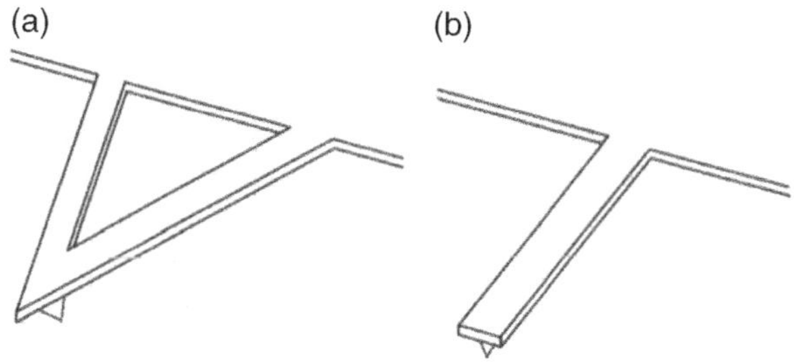

Figure 15.2 Schematic representation of (a) V-shaped and (b) rectangle-shaped probes.

Among the most important and obvious factors limiting the resolution and the quality of an AFM image are the tip apex radius and its aspect ratio. Commonly, the tips are made by silicon (Si) or silicon nitride (Si_2N_4) and are produced by using integrated circuit technologies. The standard tip is usually a 3 μm tall pyramid with approximately 2–5 nm end radius or more. Sharper tips can be obtained with more sophisticated methods such as chemical etching, ion milling, or addition of a carbon nanotube [4, 8–11].

The cantilevers are also made by Si or Si_2N_4 and fabricated with the same industrial techniques. They are usually V-shaped, Figure 15.2a, or rectangular, Figure 15.2b, with typical size: 100–300 μm (length), 10–30 μm (width), and 0.3–5 μm (thickness). Depending on the constituting material, geometry, and size, different spring constants and resonance frequencies of the first flexural mode can be obtained. Typical values of such parameters for AFM working in static-deflection mode are 0.01–1 N/m and 5–100 kHz, whereas for dynamic deflection mode 40 N/m and 300 kHz, respectively.

Finally, the substrate is just a macroscopic piece of Si or Si_2N_4 with standard size $3.5 \times 1.6 \times 0.5$ mm^3. It is used to grab the probe with tweezers and to put it in the AFM probe holder. Cantilever and substrate are usually different portions of a single crystal piece of material.

15.2.2 Harmonic Excitation of the Cantilever

As anticipated in the introduction, in dynamic AFM mode, the probe is forced to vibrate at a frequency close to its first flexural mode. In the most common case, this excitation is obtained by putting a piezoelectric material (actuator) just under the substrate of the probe [2, 3]. An oscillating voltage is then produced by a function generator and applied to the piezo element to obtain a periodic nearly harmonic oscillation of the whole probe (Figure 15.1). When the excitation frequency is nearly coincident with that of the first flexural mode of the cantilever, due to the phenomenon of resonance, it undergoes large amplitude oscillations. This method of acoustic excitation is very effective and has the advantage that it does not require to the cantilever to have any specific property. However, it has the disadvantage of mechanically exciting the whole probe, including the substrate, producing undesirable vibrations not related to the tip–sample interaction. In addition, when AFM measurements are performed in liquids, the environment is also put in oscillation by the actuator, seriously perturbing the motion of the probe [12]. For this reason, other methods have also been considered to drive the cantilever. The most diffused among them involves magnetic excitation, which consists in applying an oscillating magnetic field near the region of the tip [13]. Of course, this method requires the tip or part of the cantilever to be magnetic, making the production of the probes more complex and expensive. Furthermore, in some cases, the time-dependent magnetic fields involved may influence the properties of the system under study.

15.2.3 Scanning System

To scan the surface of a sample or, more generally, to put the probe on the point of interest over the sample with sub-nanometer precision, the AFM microscope makes use of piezoelectric actuators. They are used to change the relative position of the probe with respect to the sample in all the three directions of space, *X*, *Y*, and *Z*. When the voltage applied to a piezo element changes, a shift in one direction of the position of the tip with respect to the sample is produced. The ratio between these two quantities is known as sensitivity (volts/nm) of the piezo element along that direction. Two of these movements, *X* and *Y*, produce the raster scan, whereas the third, *Z*, follows the morphology of the sample. Of course, it is not relevant if the sample moves relative to the tip or vice versa and, in fact, both such movements can be found in different AFM equipments. The highest positioning precision is usually reached by the former, but it poses severe limitations on the sample size and weight, as the sample is usually placed just on the top of the piezoelectric actuator. The latter is in fact designed to investigate large size samples.

In most cases, the piezo actuator for *X*, *Y*, and *Z* movements consists in a hollow segmented tube, as schematically represented in Figure 15.3a [2]. These systems are in most cases preferred because they are compact and consequently have normal modes of resonance with high frequencies. This requisite is of fundamental importance to obtain reliable response of a piezo actuator to applied oscillating voltage. As shown in Figure 15.3b and c, the *Z* movement is obtained by applying a bias to inner and outer faces of the upper portion of the tube, whereas *X*–*Y* movements are generated by appropriate geometric distortions induced in the lower portion of the tube. The latter are induced by biasing the inner and outer faces of a couple of opposite segments (Figure 15.3c).

Piezoelectric actuators are very valuable in producing sub-nanometer movements, but they also have several limitations, such as nonlinearity, hysteresis, creep, cross coupling, aging, and thermal drift, that may produce severe distortions of the AFM images [2, 3].

Nonlinearity means that the geometry of a piezoelectric material is not linearly related to the applied voltage. Consequently, when a voltage ramp changing linearly in time is applied to the actuator, for example, its displacement as a function of time is not uniform. Furthermore, because of hysteresis, the actual elongation of a piezo element when it is subjected to a specific voltage depends on the details of the voltages previously applied to it. Both these effects are shown in Figure 15.4, where the positions of the piezo actuator in response to increasing and decreasing linear voltage ramps are compared.

The effects of nonlinearity and hysteresis in AFM scanners can be overcome by using closed-loop scanners. These scanners measure the actual displacement of the actuator in response to a change of the applied voltage with magnetic, capacitive, or optical systems; compare it with the nominal value; and, if necessary, apply a change to the sensitivity assigned to the piezo element in real time in order to make the actual displacement as reliable as possible. The main drawback of closed-loop scanners, apart from their higher costs, consists in a large inertia in the movements. This undesired effect can be observed, for example, in correspondence of the turning points of the raster scan movement produced during AFM image acquisition, where the effective change in the direction of movement takes place with a sizable delay. When a scanner does not involve closed-loop technology, it is classified as open-loop. In the latter case, the properties of the voltage ramps that have to be applied to the piezo element to obtain a defined movement are evaluated by the acquisition software on the basis of a number of fixed parameters, among which the sensitivity, that have to be determined periodically by the operator through boring and time-consuming calibration procedures.

The creep of a scanner is due to the fact that when a voltage is suddenly applied to a piezoelectric material, it usually undergoes a fast displacement followed by slower, undesired, displacement that

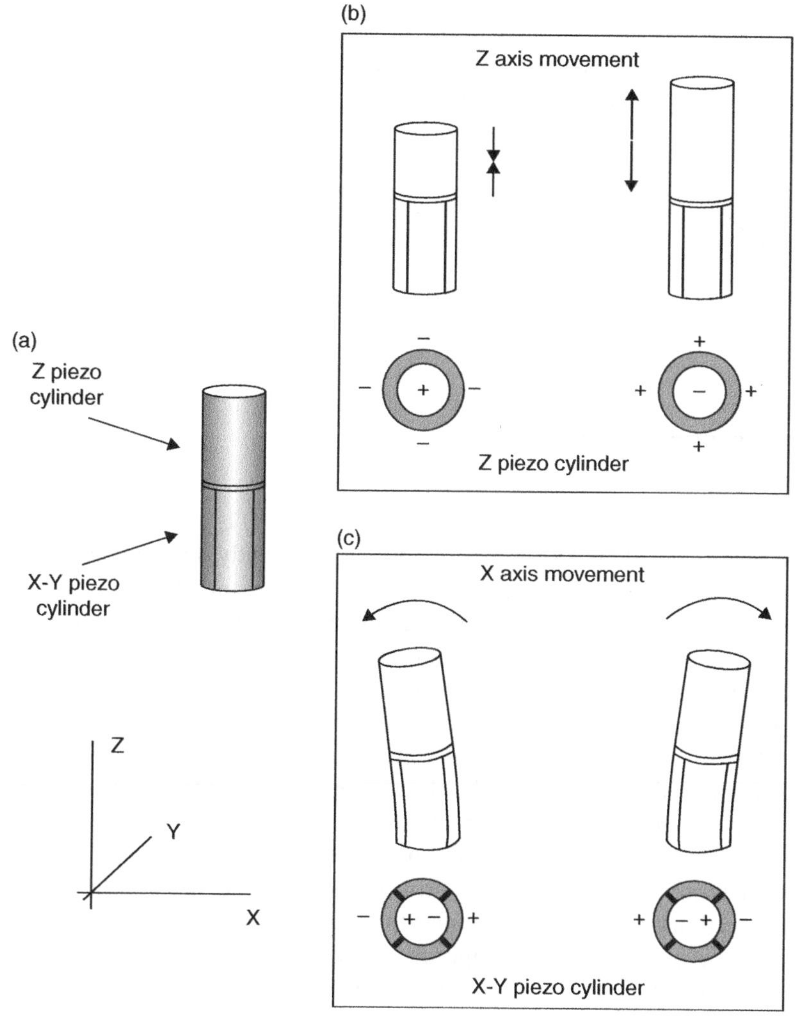

Figure 15.3 (a) Z and $X-Y$ piezo cylinders composing the segmented tube scanner. *Source:* Based on Connell and Smith [14]. Voltages applied to the piezo Z (b) and $X-Y$ (c) cylinders to generate scanner movements along Z and X, respectively.

reduces over time until it becomes undetectable [2]. Creep is usually observed for displacements in all the three directions of space. Such unideal response of the scanner, if not properly compensated, generates severe distortions of the image, especially when the probe is moved from one position of the surface to another or after zooming operations. This deleterious effect, inherently connected with the use of piezoelectric actuators, can be overcome by using closed-loop scanners.

Cross coupling consists in the observation of an undesired displacement along a direction nominally orthogonal to that of the intentional displacement. This effect is observed because the portions of the scanner devoted to the movements along the three mutually orthogonal directions are connected among them and consequently they are linked by geometrical constrains. Consider for example the tube scanner discussed above and shown in Figure 15.3. It is obvious that a distortion of the

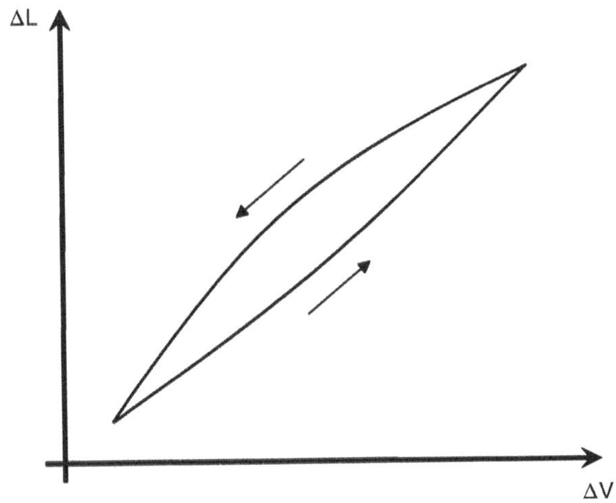

Figure 15.4 Displacement of a piezoelectric actuator in response to increasing and decreasing constant-rate voltage ramps, showing nonlinearity and hysteresis.

tube produced to obtain a displacement along *X* or *Y*, for example, concomitantly generates an undesired displacement along *Z*. Inversely, when the tube is placed in a generic distorted position, a displacement along *Z* concomitantly generates undesired displacement along *X* and/or *Y*. These effects are usually fixed in the post-acquisition analysis, by typically subtracting from the overall image a second (or even higher)-order polynomial surface. The fictitious curved surface, generated by cross coupling effects and lying under the actual surface morphology, is usually referred to as bow.

The response of a piezoelectric material to an applied voltage usually changes with time and usage. This aging effect is automatically compensated by closed-loop scanners, whereas periodical calibrations are required for open-loop ones. New scanners usually require to be recalibrated every three months the first year, every six months the second year, after which once a year will suffice.

All the AFM equipment inherently suffer from thermal drift effects, which may also contribute to generate undesired image distortions. They usually manifest themselves as variations with time of the measured value of the cantilever deflection (or amplitude of oscillation) or by a poor repeatability of the movements of the actuators during scanning. These deleterious effects are typically related to the warm-up of the components of the system and consequently they are very severe during the first hour from the power-up; after that they usually become much less evident and, in most cases, negligible. However, sudden changes of the temperature around the AFM microscope, due for example to the switch-on of a conditioning system or to the opening of the acoustic and vibration isolation enclosure which sometimes is used to isolate the equipment from the environment perturbations, may cause further severe thermal instabilities during measurements.

15.2.4 Measurement of the Cantilever's Deflection

The sensitivity of an AFM microscope is primarily related to that of the system used to measure the deflection of the cantilever, δ, quantitatively defined as the displacement of the free edge of the cantilever along *Z*. The success and the diffusion of the AFM technique is in fact mostly related to the fact that such a system is actually available [15, 16]. It consists in a laser beam produced by a solid state diode that is pointed to the free extremity of the cantilever, where it is reflected toward a position-sensitive photodetector consisting of four independent photodiodes (Figure 15.1). The extraordinary sensitivity of this detection system is related to the fact that the distance between cantilever and detector is generally three orders of magnitude longer than

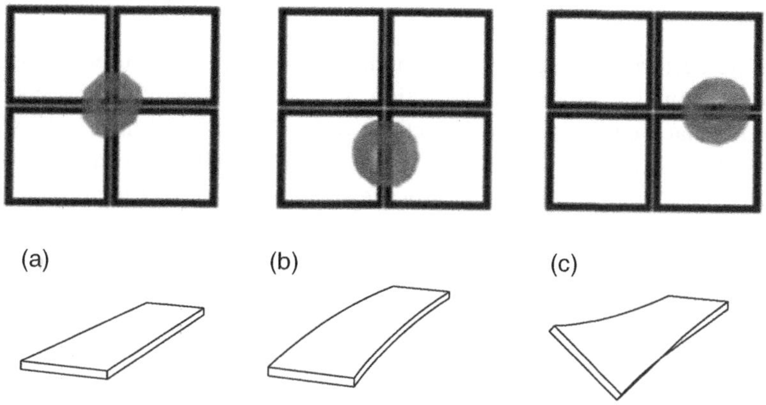

Figure 15.5 Scheme of the four-quadrants photodetector for three different geometries of the cantilever: (a) equilibrium (no distortions), (b) vertical deflection, and (c) horizontal torsion.

the length of the cantilever; as a result, the tiny deflection of the cantilever is greatly magnified before reaching the photodetector. When the probe is far from the sample, it does not interact and the laser spot falls at the center of the four quadrants photodetector (Figure 15.5a). This condition is usually obtained by moving the photodetector with appropriate screws. During scanning, when the cantilever bends due to tip–surface interaction, the reflected laser spot shifts vertically (Figure 15.5b). This change is promptly detected and measured by the system as a difference in the light intensity falling on the four distinct photodiodes. Similarly, horizontal torsions of the cantilever can also be detected, as they induce horizontal shifts of the reflected laser spot (Figure 15.5c).

The output of each photodiode involved in the detector is a photocurrent that is transformed into an output voltage. Therefore, the deflection of the cantilever is typically measured in Volts. In order to estimate the actual geometrical deflection of the cantilever, the optical sensitivity, OS, of the detection system has to be estimated, defined as the ratio between the measured voltage signal, ΔV, and the related geometrical deflection of the cantilever

$$\mathrm{OS} = \frac{\Delta V}{\delta}. \tag{15.1}$$

The OS of course depends on the geometry and on the components involved in the optical detection system. Furthermore, it is straightforward to verify by simple geometrical considerations that it also depends on the geometry of the probe and on the position of the laser on the top of the cantilever, and consequently on the initial procedure of laser and detector alignment. Therefore, a calibration procedure must be applied every time a quantitative estimation of the actual geometrical deflection of the probe is needed. The most common calibration procedure consists in putting the tip in contact with a very hard (ideally uncompliant) surface and moving the overall probe along Z by some amount ΔZ. Since the surface is assumed to be non-deformable, the vertical movement of the probe can be assumed to be equal to the geometrical deflection of the cantilever. The OS can then be simply obtained by dividing the deflection measured by the photodetector in Volt and ΔZ. Rigorously, the OS obtained with such procedure is the static one. Since different geometrical distortions of the probe are associated to the different eigenmodes of the cantilever, it follows that they have different OSs. The latter are usually determined based on the static value and by considering the specific geometry of the vibrational mode of interest [17, 18].

15.2.5 Feedback System

When AFM measurements are performed in static-deflection mode, the signal measured by the photodetector is simply proportional to the static deflection of the cantilever, which is just the parameter used by the feedback circuit to reconstruct the three-dimensional morphology of the surface. At variance, in most dynamic-deflection modes, as in Tapping, the cantilever is forced to oscillate at a fixed frequency and consequently the signal revealed by the photodetector is also oscillating. In these cases, as shown in Figure 15.1, the signal measured by the photodetector is sent to a lock-in amplifier. It selectively amplifies from the input signal just the component with the same frequency of a reference signal, which is equal to that exciting the cantilever oscillations. The main output of the lock-in amplifier consists in the amplitude of the signal oscillating at the same frequency of the reference signal, that means at the same frequency of the forced oscillations of the cantilever. This output signal is used by the feedback circuit to follow the sample morphology. Thanks to the lock-in amplifier, the noise components are strongly reduced and consequently the sensitivity of the technique is significantly enhanced. The phase-lag of the detected signal with respect to the reference one is also measured and stored. The latter output is also very important in AM-AFM, as it can be used to monitor the sample inhomogeneities [2, 3].

The role of the feedback circuit is to compare the value of the measured feedback parameter, usually the static deflection or the amplitude of the forced cantilever oscillations, with that settled by the operator, named the *set-point* value. An error signal is calculated as the difference between these two values and it is sent to a proportional-integral-differential (PID) system. The output of the PID is a voltage signal sent to the Z scanner to either increase or decrease tip–sample distance to minimize the error signal. As discussed in the introduction, such Z movements reproduce the profile of the sample morphology along the scanned line. The output voltage of the PID results from the weighted sum of three terms. The first (proportional) is an error component simply obtained by multiplying the last error received in input with a coefficient, P. The second (integral) is obtained by multiplying the sum of the last N (fixed by the manufacturer) errors received in input with a coefficient, I. Finally, the third (derivative) is obtained as a product of the difference between the last two errors received in input and a coefficient, D. The coefficients P, I, and D are empirically optimized by the operator and depend on many factors, such as the roughness of the surface, the scan rate, the value of set point, the size of the tip, and so on.

15.3 Tip–Surface Interaction Forces

Since atomic force microscopy makes use of tip–surface interaction to reconstruct the surface morphology of materials, it is of fundamental importance to fully understand the nature and properties of these forces. In particular, it is necessary to know with great precision the dependence of these forces on the distance between tip and surface. In fact, it is just this property that allows the AFM microscope to reconstruct the morphology of the surface and to study its physicochemical properties. Among the four fundamental forces of nature, the one at play during AFM image acquisition is the electromagnetic force, which manifests itself in different ways at different tip–sample distances. In air, the most relevant are: van der Waals, short-range repulsive, adhesion, and capillary. In liquids, other forces are usually present, such as typically the long-range electrostatic interaction, related to the formation of surficial charges.

Since AFM measurements are mostly performed in air, a detailed discussion of the forces acting in this environment is provided in the following. A brief description of the interactions acting in liquid is briefly presented in Section 15.3.5.

15.3.1 Van der Waals

The van der Waals interaction is a force of electrostatic origin acting between any pair of atoms and/or molecules and is linked to the presence of electric dipoles [19]. It comprises of three distinct components depending on the nature of the electric dipole involved: *permanent dipole–permanent dipole* interaction (Keesom), *permanent dipole–induced dipole* interaction (Debye), and *induced dipole–induced dipole* interaction (London) [19]. These three components have the same dependence on the distance between the atoms/molecules involved: the potential is proportional to r^{-6} and consequently the force is proportional to r^{-7} [20].

Obviously, to correctly describe the interaction between tip and surface, it is necessary to consider the fact that they are not single molecules. A process of integration of the van der Waals interactions between all the pairs of atoms/molecules present in a suitably extended volume portion of both the tip and the surface is necessary [21, 22]. This integration has two important consequences: (i) introduces a dependence on the geometries of tip and surface and (ii) changes the dependence of the interaction on their mutual distance. The most common and simple approximation consists in assimilating the tip to a sphere and the sample to a perfectly flat plane (sphere–fat geometry), as shown in Figure 15.6. This is a good approximation in the common case of tip radius of about 5–10 nm and samples with very limited roughness. For sphere–flat geometry, the van der Waals force is given by [22]:

$$F_{vdW} = -\frac{C_{\mathrm{H}}}{6}\frac{R}{d^2}, \tag{15.2}$$

where C_{H} is known as the Hamaker constant [23] and depends on the materials constituting tip and surface and on the environment, R is the radius of the sphere, and d is the effective distance between tip and surface. Equation (15.2) shows that the integration significantly reduces the dependence of the van der Waals interaction on the distance, as the potential becomes proportional to d^{-1} and the force to d^{-2}. This result points out that the van der Waals interaction in AFM experiments is a relatively long-range interaction. The distance between two interacting atoms/molecules is measured

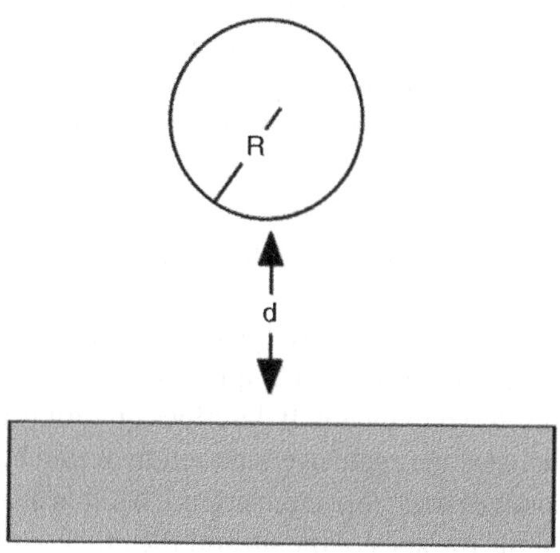

Figure 15.6 Sphere–flat geometry approximately describing tip and surface geometries.

by definition from their centers. The zero-distance condition is therefore not feasible (for the principle of impenetrability of bodies or for the Pauli principle). Experiments have shown that the zero "effective" distance is obtained for $d \simeq 0.16$ nm [22]. From Eq. (15.2), by assuming a tip radius $R \simeq$ 10 nm and a tip–sample distance of $d \simeq 0.16$ nm, it is possible to estimate in air an attractive force of about −4 nN, whereas for $d \simeq 1$ nm the force reduces (in absolute value) to about −0.3 nN.

The Hamaker constant has a value of the order of 10^{-20} J, that is, about an order of magnitude greater than the thermal energy at room temperature. If calculated in various situations, it indicates that in air and in vacuum the van der Waals interaction is always attractive, irrespective of the tip and sample shape and composition [22]. For different environments, as in liquids, it is already attractive if tip and sample are constituted of the same materials, whereas it can be attractive or repulsive depending on the properties of environment relative to tip and sample [22].

In some cases, the sphere–fat approximation is not appropriate and consequently other tip geometries must be evaluated, such as paraboloid of rotation, hyperboloid of rotation, cone, spherical-capped cone, or truncated cone [2]. By changing the tip geometry, both the dependence of the force on the distance and the intensity of the force at a fixed distance change significantly [2].

15.3.2 Short-Range Repulsive

When the distance between tip and surface falls nominally below 0.16 nm, a contact is established between them. To describe the contact interaction acting during AFM experiments, it is not possible to follow a procedure like that discussed previously for van der Waals interaction. In fact, as discussed in the following, a reliable description of the tip–surface interaction cannot be obtained by summing up the forces acting among all the couples of atoms/molecules involved in an appropriate volume of tip and surface [2]. When two atoms/molecules are forced to become closer than the sum of their radii, their electron clouds are forced to superimpose and consequently a strong electrostatic repulsive interaction is generated because of the Pauli exclusion principle. Many mathematical models have been proposed to describe such interaction, such as those of Morse [24] and Lennard-Jones (L-J) [25], both predicting a rapidly growing potential (exponentially and proportional to d^{-12}, respectively) on decreasing the interatomic distance. However, when a tip is pushed toward a rigid surface, the outermost atoms are not forced to superimpose their electronic clouds. Instead, tip and/or surface simply deforms, extending the surface of contact and increasing the number of interacting atoms involved in the physical contact. Furthermore, since the atoms constituting a solid are strongly bonded among them, the contact interaction force is also transmitted to many interior atoms located in proximity of the surface of both tip and surface. This mechanism of interaction is very different with respect to that occurring between two atoms/molecules and is expected to exhibit a potential dependence on the distance considerably mitigated with respect to those described by Morse and L-J models, for example.

Many analytical expressions have been reported to describe contact tip–-surface interaction under different hypotheses and approximations [26]. In most cases, the following assumptions are made: stress and strain are linearly related, the size of the contact is small with respect to the tip radius, axial symmetry is applicable in the region of contact, and shear forces and viscoelasticity of the materials can be neglected. One of the most important approaches working under these hypotheses and frequently used to describe the mechanical contact between tip and surface is the Hertzian model. It describes quantitatively the interaction between two rigid spheres and only includes the repulsive interaction acting between them within the contact area, just ignoring any possible attractive interaction [26]. It is a very useful model in AFM, as one sphere can be used to mimic the apex of the tip and the other, simply assuming infinite curvature radius, the flat surface.

In the Hertzian model, the relation between the applied force and the position of the tip is given by:

$$F = K\sqrt{R}\,h^{3/2} \tag{15.3}$$

where F is the force applied by the tip to the surface, h is the amount of penetration of the tip toward the sample with respect to the level of the unperturbed surface, usually named indentation, whereas R and K are given by

$$R = \left(\frac{1}{R_1} + \frac{1}{R_2}\right)^{-1} \tag{15.4}$$

$$\frac{1}{K} = \frac{3}{4}\left[\frac{(1-\nu_1^2)}{E_1} + \frac{(1-\nu_2^2)}{E_2}\right] \tag{15.5}$$

where R_1 and R_2 are the curvature radii of the two spheres, E_1 and E_2 their Young moduli and ν_1 and ν_2 their Poisson's coefficients, quantifying the ratio between the strain induced in the direction perpendicular to the force F and that induced in the direction parallel to it. Equation (15.3) shows that the force-indentation dependence, $F(h)$, characterizing the Hertzian model is a power law with exponent 3/2. The lateral resolution is defined by $2R_c$, where R_c is the radius of the circular area of contact, given by

$$R_c = \left(\frac{RF}{K}\right)^{1/3} \tag{15.6}$$

that fulfills the condition

$$\mathrm{R_c^2 = hR.} \tag{15.7}$$

Equation (15.6) suggests that a better lateral resolution (smaller value of R_c) is obtained with rigid tip and surface materials (E large, K large), tiny forces, and small tip radius. In using the equations reported above, it is important to keep in mind that the Hertzian model is valid only in the approximation $R_c \ll R$ and $h \ll R$. If these hypotheses are not satisfied, as it may occur for example by using sharp tips on compliant samples, other more sophisticated models must be considered, taking into account a larger portion of the actual probe shape than just the apical semispherical part [2].

In Figure 15.7, a representation of the tip–sample force is reported, as resulting from the superposition of van der Waals and short-range repulsive forces. This characteristic curve is of fundamental importance and it is frequently used to discuss the main properties of tip–sample interaction.

15.3.3 Adhesion

The Hertzian model must be considered just as a first approximation, surely lacking important components of the tip–surface interaction. First of all, when a tip comes into contact with a surface, two distinguishable adhesion contributions can be recognized: one acting within the surface of physical contact and one in the ring just outside it [27–29]. In general, both these attractive interactions should be contemporarily considered, but the high degree of complexity of the problem makes this objective very difficult to achieve [28, 30]. For this reason, two limiting cases are usually considered in which one of these contributions is dominant and consequently the latter is simply neglected. In both these cases, the repulsive forces are also considered by including an Hertzian term into the overall tip–surface force.

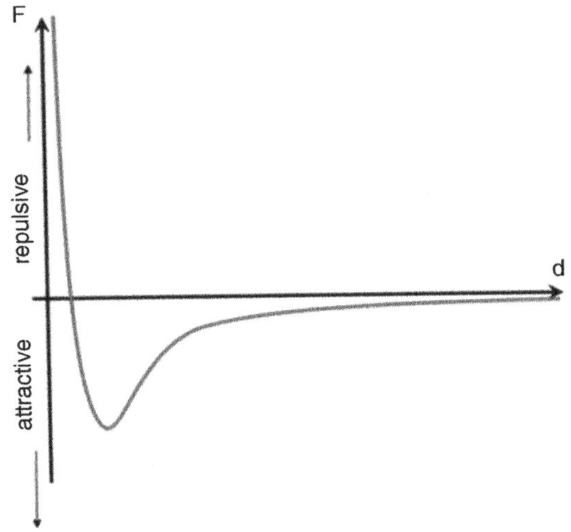

Figure 15.7 Qualitative representation of the interaction force between tip and sample as a function of their distance.

The Derjaguin–Muller–Toporov (DMT) model assumes that the only attractive interaction between tip and surface is that located in a ring just outside the contact region [31]. Additionally, it makes the simplifying assumption that such force is constant, i.e. independent of the tip–surface distance. Physically, it results from the integration of the van der Waals interaction between the tip (sphere) and the sample (flat plane) outside the contact region [31, 32] and it has the form

$$F = -2\pi R\gamma \tag{15.8}$$

where γ if the work of adhesion, defined as the energy necessary to separate a unit of area of the interface. The DMT model represents a good approximation for stiff materials, low adhesion, and small tip radii. The total force is then given by

$$F = \frac{K R_c^3}{R} - 2\pi R\gamma \tag{15.9}$$

where the first term comes from the Hertzian model. Due to the form of Eq. (15.9), the DMT model is also known as Hertz plus offset model. By reversing Eq. (15.9), one can express the contact radius as a function of the load force F

$$R_c = \left(\frac{R}{K}(F + 2\pi R\gamma)\right)^{1/3} \tag{15.10}$$

indicating that there is a nonzero contact region also for zero force, given by

$$R_{c0} = \left(\frac{2\pi R^2\gamma}{K}\right)^{1/3}. \tag{15.11}$$

The indentation depth has the same form as in the Hertz model

$$h = \frac{R_c^2}{R}. \tag{15.12}$$

The minimum permissible stable normal load, F_{adh}, can be found by minimizing F with respect to the radius of contact in Eq. (15.9). Since the parameter R_c is positive or zero by definition, the minimum value of F is obtained for $R_c = 0$, i.e. when the area of contact is zero, and we obtain

$$F_{adh} = -2\pi R\gamma \tag{15.13}$$

which is usually called pull-off force.

Another model, proposed by Johnson–Kendall–Roberts (JKR) [32], describes the opposite situation in which the only attractive term considered is that acting within the surface of contact. Any other adhesion contribution coming from outside such region is simply neglected. The JKR model represents a good approximation when low stiffness, high adhesion, and large tip radii are involved.

The equation for the force is

$$F = \frac{K R_c^3}{R} - \sqrt{6\pi\gamma K R_c^3} \tag{15.14}$$

where, together with the usual Hertzian repulsive term, an adhesive interaction is included that depends on the size of the contact area through the parameter R_c. In this model, the contact radius depends on the load force by the equation

$$R_c = \left[\frac{R}{K}\left(F + 3\pi R\gamma + \sqrt{6\pi R\gamma F + (3\pi R\gamma)^2}\right)\right]^{1/3}. \tag{15.15}$$

Also in this case, from Eq. (15.15), a finite contact region is predicted for zero applied force, given by

$$R_{c0} = \left(\frac{6\pi R^2\gamma}{K}\right)^{1/3}. \tag{15.16}$$

The indentation depth differs with respect to those of Hertz and DMT models, and is given by

$$h = \frac{R_c^2}{R} - \frac{2}{3}\sqrt{\frac{6\pi\gamma R_c}{K}}. \tag{15.17}$$

By using the same procedure discussed above for DMT, the following pull-off force is obtained for JKR model

$$F_{adh} = -\frac{3}{2}\pi R\gamma. \tag{15.18}$$

As anticipated, DMT and JKR describe just two opposite limiting cases. Of course, in general, a large variety of intermediate situations can be encountered. When a quantitative interpretation of tip–surface force data are needed, a proper analytical model must be selected. A practical method to find the most reliable model for a given real situation consists in calculating the elasticity parameter, λ, defined as

$$\lambda = \frac{2.06}{\xi_0}\left(\frac{R\gamma^2}{\pi K^2}\right)^{1/3} \tag{15.19}$$

where ξ_0 represents the interatomic distance between tip and surface atoms, whose typical value is 0.16 nm. For values of λ lower or equal to about 0.1, the DMT model applies, whereas for λ larger than 5 the JKR model is appropriate. When the elasticity parameter falls in the intermediate range $0.1 < \lambda < 5$, a more complete theoretical model has to be considered, which takes into account both the adhesive forces acting into and at the border of the contact region between tip and surface. An example of such a model is that of Maugis–Dugdale [2, 30]. This model, not further discussed here, is more complex than the previous ones and consequently rarely considered in ordinary AFM studies. Alternatively, when λ falls in the intermediate range, it is possible to force its value to increase by simply using probes with larger tip radii, loosing spatial resolution but easily falling in the JKR model regime. This method is actually widely used in spectroscopic application of AFM, where a reliable quantitative description of the tip–sample interaction is of fundamental importance.

15.3.4 Capillary

When a tip comes close to a surface beyond a certain limit, a meniscus of condensed water is formed between them. This characteristic distance, discussed quantitatively in the following, depends on the materials constituting tip and surface, their shape, and the relative humidity of the environment. The formation of meniscus is a manifestation of the capillary condensation, a thermodynamic process taking place because the actual vapor pressure of a substance and the curvature of its condensed form are intimately related between them. This relation is quantitatively described by the Kelvin equation [33, 34]

$$R_g T \ln\left(\frac{P}{P_0}\right) = \frac{\gamma_L V_m}{r_k} \tag{15.20}$$

where R_g is the universal gas constant, T the absolute temperature, P the actual vapor pressure, P_0 the vapor pressure at saturation, γ_L and V_m the surface tension and the molar volume of the liquid, respectively, whereas r_k, named Kelvin radius, describes the overall curvature of the meniscus and it is given by

$$\frac{1}{r_k} = \frac{1}{r_1} + \frac{1}{r_2}. \tag{15.21}$$

r_1 and r_2 are the principal radii of curvature of the meniscus assumed to have toroid shape [35–37] as shown in Figure 15.8, where the cross-sectional shape of meniscus at the interface of sphere and flat surface is represented. In particular, in Eq. (15.21), r_2 is positive (radius of a circle) whereas r_1 must be taken negative for concave and positive for convex surfaces. In AFM, a meniscus is formed between tip and surface whenever their distance becomes comparable or lower than the absolute

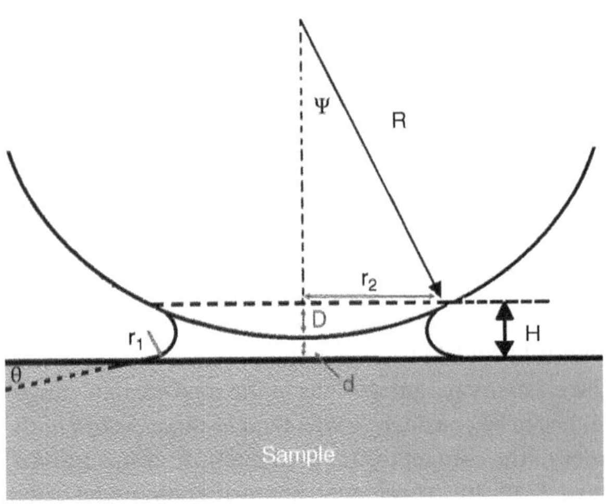

Figure 15.8 Cross-sectional shape of meniscus at the interface of sphere and flat surface. *Source:* Based on Israelachvili [22].

value of the Kelvin radius. Typical values of r_k for a spherical concave water meniscus at room temperature are $r_k = -0.8$ nm for 50% relative humidity (RH) and $r_k = -0.23$ nm for 10% RH.

Since water meniscus involves curved surfaces, the pressure observed inside it must be different with respect to that present outside, as described by the of Young–Laplace equation

$$P_{\text{inside}} - P_{\text{outside}} = \Delta P = \gamma_L \left(\frac{1}{r_1} + \frac{1}{r_2} \right). \tag{15.22}$$

Since r_1 is negative and usually its absolute value is lower than r_2, r_k is negative and consequently $P_{\text{inside}} < P_{\text{outside}}$, indicating an attractive interaction between tip and surface. When the angle ψ in Figure 15.8 is small, the resulting force can be approximately written as [22]

$$F(d) = \frac{4\pi R \gamma_L \cos\theta}{\left(1 + \frac{d}{D}\right)} \tag{15.23}$$

where θ and D are defined in Figure 15.8. When ψ cannot be assumed to be very small, another term in the tip–force interaction must be included, which is related to the surface tension acting at the tip–water and flat–water contact lines. This term significantly complicates the treatment and will not be further discussed here. For more details, see Ref. [2].

Capillary forces are usually undesirable, as they negatively interfere with the scan of the sample's surface. They are typically strong, reaching values of the order of 100 nN in ambient conditions. Capillary forces also exhibit hysteresis, which is still more undesirable: upon approach, the meniscus is formed when tip–surface distance becomes comparable with the Kelvin radius, whereas upon retract, it persists for distances some nanometers larger than r_k. Such tip–surface force hysteresis inherently leads to energy dissipation, indicating that a fraction of the mechanical energy associated to the cantilever is transferred to the sample as thermal energy.

Even if, in general, capillary forces are strongly undesirable, there are specific applications in which they are of fundamental importance, for example for nanolithography. It is possible, for example, to use the water meniscus spontaneously formed between a conducting tip and the surface of a semiconductor to selectively oxide the nanometric area under the tip when a voltage is applied between tip and surface [38, 39]. It is also possible to use the tip of an AFM probe to write with ink on a surface. In the latter application, named dip-pen nanolithography, the tip is first dipped in ink, where capillary forces induce a transfer of ink to the tip surface, and then scanned on the surface of a sample, where a new ink transfer is induced by capillary forces, in this case from tip to surface [40, 41].

15.3.5 Other Forces

The interactions presented in the previous sections are usually the most relevant when AFM measurements are performed in air. However, other forces may become also relevant when specific conditions apply. For example, long-range tip–surface electrostatic interactions are observed when tip and/or sample are electrically charged. This is of course the case when a finite voltage is intentionally applied between them. In fact, because of such voltage, a charge transfer takes place when they are put in contact, which generates net charges on both surfaces and consequently a measurable electrostatic attractive force is established. However, even in absence of such external voltage, net charges can be present on tip and sample surfaces. For example, when tip and sample are made of different conductive materials, a voltage is spontaneously established, named surface potential. It is because different metallic materials generally possess different values of the ionization potential (work function), even when they are both grounded. When tip and/or sample are made of dielectric material, this mechanism does not apply, but net charges can again be present on their surfaces, trapped during the scanning process of the sample because of the mechanical contact and shear forces.

When AFM measurements are performed in liquid, no capillary forces are present. Consequently, one may expect that the complexity of the description of the tip–surface interaction is significantly reduced. However, if the liquid is polar, then the surfaces of both tip and sample usually acquire net charges, as consequence of the processes of ionization, adsorption of ions from the liquid, or dissociation of surface atoms. Therefore, the electrostatic tip–surface interaction becomes very important and a reliable quantitative description of this long-range interaction must be considered. Furthermore, in liquids, other specific forces are also present, such as those due to solvation and hydration. A detailed description of such interaction goes beyond the scope of the present chapter and will not be further discussed. For more details on electrostatic interaction and on the forces acting in liquids, see refs [2, 3].

15.4 AFM Acquisition Modes

As discussed in the introduction, AFM images can be obtained in large variety of modes, falling into two main groups, static- and dynamic-deflection modes. Every mode has advantages and disadvantages, so it must be properly selected each time depending on the sample properties. The possibility to choose among many different modes gives the opportunity to optimize the acquisition conditions and makes the technique very powerful and versatile. The two most diffused static- and dynamic-modes are Contact and Tapping, respectively. Due to their wide diffusion and importance, a detailed description of these modes is presented in the following. Their main properties and their characteristic strengths and weaknesses are also discussed in detail.

15.4.1 Contact Mode

In AFM, the cantilever is described as an elastic object approximately following Hooke's law: the deflection, δ, is proportional to the restoring force, F, by a proportional constant, k, named Hooke constant or simply force constant [2–4]

$$F = -k\,\delta. \tag{15.24}$$

In Contact mode, the tip–sample interaction is initially established by approaching the probe to the surface until a measurable deflection of the cantilever due to repulsive forces is produced. Subsequently, the raster scan of the surface starts, which in the most common case proceeds at constant force. To fulfill such a requirement, during the scan, the feedback applies vertical displacements to the tip (head scanning) or to the surface (sample scanning) every time the measured deflection of the probe differs with respect to the set-point value. Since the deflection is maintained nominally constant during the scanning, the tip–sample force is also assumed to be constant. In Figure 15.9, the portion of the tip–sample force involved in Contact mode is indicated.

The two most important parameters involved in Contact mode are the force constant of the cantilever and the value of the deflection set point. The former is fixed by selecting the probe, the latter is settled by the operator during measurements. They both contribute to define the actual value of the force established between tip and surface during scanning: the higher the force constant and/or the deflection, the stronger the tip–sample interaction.

Contact mode was the first implemented AFM acquisition method and has several noticeable strengths. It allows higher scanning rates with respect to the other modes, in particular in

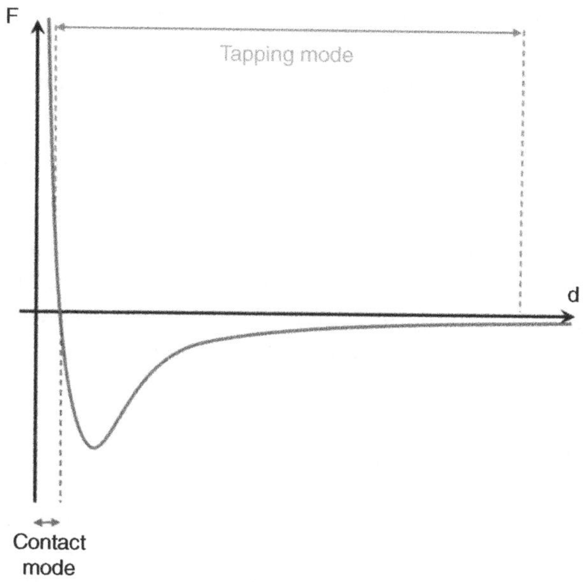

Figure 15.9 Portions of the tip–force curve involved in Contact and Tapping AFM modes.

comparison to dynamical ones that, as we will discuss in the following paragraph, have limitations in response time related to the physics of the forced motion of the cantilever. It may reach sub-nanometric resolution by using very sharp tips and limited tip–sample interaction forces. It has high sensitivity due to the use of long cantilevers ($\simeq$500 μm) with very low force constants (0.01–1 N/m), which undergo large deflections even in the presence of tiny forces (<1 nN). Furthermore, in Contact mode, the control of the tip–sample interaction force is straightforward, as the measured quantity, the cantilever deflection, is simply linearly related to it by the force constant. As we will see in the Section 15.5, reliable values of the effective Hooke constant can be experimentally estimated by well-established calibration methods [42]. The simple relation linking the measured deflection to the tip–sample interaction also greatly simplifies the process of optimization of the scanning parameters and confers on the operator a reliable control on the system.

Contact mode also has weaknesses, of course. Since the tip slides on the surface all the time during scanning, it is difficult to scan properly soft samples or nanosystems weakly adhered on the surface of the supporting substrate. In fact, even at the lowest possible tip–surface interaction forces, they can be damaged or dragged away by the tip. The tip itself may get dirty or broken during scanning. Furthermore, in Contact mode, notable shear forces are of course present, which usually further worsen such undesired effects.

However, it is worth noting that for sufficiently rigid samples, shear forces may represent a valuable source of information on the surface under study. In fact, a map of the surface inhomogeneities with sub-nanometric resolution can be obtained based on the different shear interactions that different materials generally establish with the tip [2]. To enhance the sensitivity of the AFM microscope with respect to these forces, the measurements are usually performed by using long rectangular cantilevers and by scanning perpendicularly to their direction of elongation. In these conditions, shear forces usually generate measurable horizontal torsions of the cantilever, which can be efficiently measured by the photodetector, as discussed in Section 15.2.4 and shown in Figure 15.5c.

The limitations pertaining to Contact mode discussed above are inherently related to the fact that the scan of the surface takes place with tip and surface constantly in contact. The necessity to surpass such drawbacks has inspired the development of new more powerful modalities, the dynamic-deflection ones.

15.4.2 Tapping Mode

In Tapping mode, the cantilever is typically forced to oscillate at a frequency near that of its first flexural normal mode. In AM-AFM, the feedback parameter is the amplitude of the forced oscillation, which is maintained constant to the set-point value during scanning by the feedback system. The vertical displacements imposed on the tip or the sample by the feedback reproduce the morphology of the sample, as usual.

Tapping mode is mostly applied in air, where a significant contribution of capillary forces is present. To moderate the disturbance effects related to them during scanning (see Section 15.3.4), the cantilever is excited to large values of amplitude (1–100 nm) [2–4]. This choice makes the tip oscillate from points very far from the surface, where poor or no interaction with the sample is detectable, down to the surface, where sizable contact repulsive forces are established, as indicated in Figure 15.9. These large-amplitude oscillations usually involve the formation and rupture of the water tip–sample meniscus at any cycle. However, large oscillation amplitudes imply large deflections of the cantilever from its equilibrium position, and consequently large restoring forces, which make the cantilever oscillation poorly influenced by the capillary forces. Another parameter typically considered to further stabilize the scan in the presence of capillary interaction is the force constant of the cantilever, as larger force constants give more intense restoring forces.

Tapping mode has comparable resolution with respect to Contact mode, but significantly higher sensitivity. This property is related to the fact that the signal revealed by the photodetector oscillates at a frequency of the order of kHz and consequently a lock-in amplifier can be used as a filter, appreciably improving the signal-to-noise ratio [2]. The higher sensitivity of Tapping mode with respect to Contact allows to obtain AFM images with considerably lower tip–surface interaction forces. Furthermore, due to the oscillation of the cantilever, the shear forces between tip and surface are greatly reduced. The properties mentioned above make the Tapping mode an powerful method, not only for hard stable surfaces, but also for soft samples or for nanosystems weakly adhered on a substate.

As usual, the advantages of an acquisition mode must be weighted with its weaknesses. As it will be discussed in the following, the oscillation of the probe imposes severe limitations on the response time, making the Tapping a slow scanning mode. In addition, the relationship between the observable, the oscillation amplitude, and the force of interaction between tip and surface is not straightforward. As a result, the origin of the contrast in Tapping mode AFM images is not fully understood and the identification of the optimal parameters of scanning is not trivial [14, 43, 44].

At the beginning of AM-AFM microscopy, it was found to be very easy to scan a large variety of different samples; therefore, it was assumed that there was no need to develop theoretical models of tip–sample interaction [3]. However, few years later, it became clear instead that it is mandatory to gain such knowledge, to acquire reliable AFM images of the sample morphology, and to avoid misleading interpretations. The main aim of such studies is the mathematical description of the motion of the cantilever–tip system, but a complete solution of this problem is very difficult to obtain. First, when the tip approaches a surface, many different interactions are established contemporarily, as different portions of the tip are located at different distances from the surface. When the tip apex is affected by repulsive contact forces, for example, its sides may contemporarily experience van der

Waals and capillary interactions. At the same time, since the cantilever is put in oscillation, it suffers forcing and hydrodynamic damping forces due to the environment. Furthermore, the relative weights of the forces acting on the cantilever–tip system depend on its geometry. Another important difficulty resides in the fact that since cantilever–tip system consists in an elastic three-dimensional object, it admits an infinite number of normal vibrational modes (eigenmodes), whose properties depend on its geometry, on the boundary conditions, and on the interaction it establishes with the surface. It is then clear that to obtain a quantitative, possibly analytical, treatment of the equation of motion of the cantilever–tip system, the problem needs to be strongly simplified. A common approximation consists in describing the cantilever–tip system by a damped ideal point-mass spring system driven by an external oscillating force, as shown schematically in Figure 15.10a. This oscillator is supposed to be located near the sample surface and consequently it is also affected by the tip–sample interaction. The motion of this simplified system can be mathematically described by the following second-order nonlinear differential equation [2, 3]

$$\frac{d^2\delta}{dt^2} + 2\gamma\frac{d\delta}{dt} + \omega_0^2\delta = \frac{F_0}{m^*}\sin(\omega t) + \frac{F_{ts}(d)}{m^*} \tag{15.25}$$

where F_0 and ω are the amplitude and angular frequency of the driving force, m^* and ω_0 the effective mass and characteristic frequency of the ideal oscillator, whereas γ is the damping coefficient associated to the motion in the viscous environment (Figure 15.10a). The effective mass m^* depends on the geometry of the cantilever and of the excited oscillation mode. In Eq. (15.25), δ and d are the deflection of the cantilever and the tip–surface distance, respectively, as defined in Figure 15.10b. Note that Eq. (15.25) is a differential equation of δ with respect to time, consequently d has to be written in terms of δ. As indicated in Figure 15.10b, the simple relation $d = \delta + Z_c$ holds, where Z_c is the mean position of the probe with respect to the surface and it is assumed to keep a constant value during the scanning process.

To solve Eq. (15.25), it is necessary to define an analytical description of the tip–sample interaction force, $F_{ts}(d)$. However, since it is a nonlinear function, it prevents Eq. (15.25) from having analytical solutions. Consequently, two main possibilities are usually considered: to solve the equation numerically or to find its analytical solutions by taking a linear approximation of the function $F_{ts}(d)$

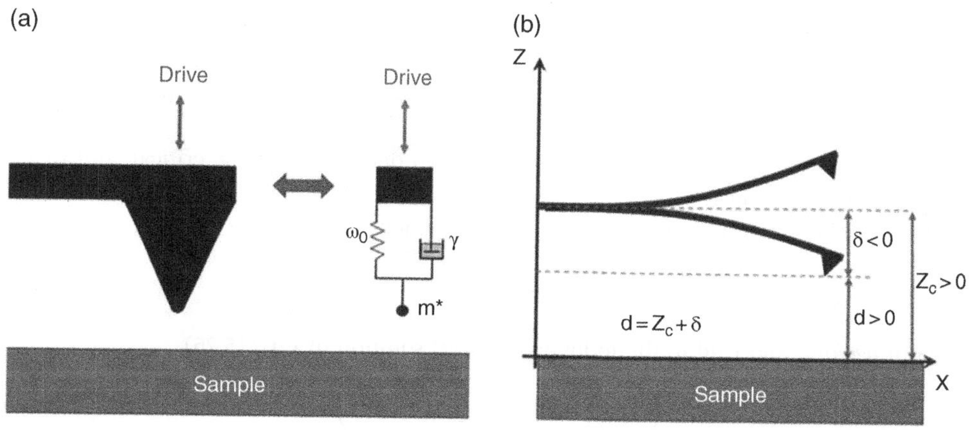

Figure 15.10 (a) Schematic representation of the damped ideal point-mass spring system used to approximate the cantilever–tip system. (b) Scheme of the tip oscillating upon a surface. The quantities δ (deflection of the cantilever), d (tip–sample distance), and Z_c (distance of the probe from the surface in static condition) are indicated.

with respect to δ. These methods give often complementary precious information on the properties of the cantilever dynamics during the acquisition of AFM images [45–49]. In the following, the latter approach is further discussed, leading to a problem known as weakly perturbed harmonic oscillator.

Equation (15.25) is a nonhomogeneous differential equation, due to the terms of the driving force, $F_0 \sin(\omega t)$ and tip–sample interaction, $F_{ts}(d)$. The mathematical strategy used to find the complete set of solution consists in solving first the related homogeneous equation, giving a general parametric solution of the restricted problem. Subsequently, a particular solution of the complete nonhomogeneous differential equation is just added to it, to find the general parametric solution of the whole problem. The homogeneous differential equation associated to Eq. (15.25) is

$$\frac{d^2\delta}{dt^2} + 2\gamma\frac{d\delta}{dt} + \omega_0^2\delta = 0. \tag{15.26}$$

This equation is also physically interesting, as it describes how the cantilever restores its equilibrium position after encountering a perturbation, when no driving and tip–sample forces are present. By assuming a solution of Eq. (15.26) in the form $\delta(t) = e^{\alpha t}$, as usual, we find

$$\frac{d^2}{dt^2}(e^{\alpha t}) + 2\gamma\frac{d}{dt}(e^{\alpha t}) + \omega_0^2(e^{\alpha t}) = 0 \tag{15.27}$$

and

$$\alpha^2(e^{\alpha t}) + 2\gamma\alpha(e^{\alpha t}) + \omega_0^2(e^{\alpha t}) = e^{\alpha t}\left(\alpha^2 + 2\gamma\alpha + \omega_0^2\right) = 0\,. \tag{15.28}$$

The last equation indicates that a function in the form $\delta(t) = e^{\alpha t}$ is a solution of Eq. (15.26) only if the following condition, known as characteristic equation, is satisfied

$$\alpha^2 + 2\gamma\alpha + \omega_0^2 = 0 \tag{15.29}$$

or equivalently if the exponent α has one of the following values

$$\alpha = -\gamma \pm \sqrt{\gamma^2 - \omega_0^2}. \tag{15.30}$$

The latter equation gives solutions with considerably different characteristics in the three following cases

$$\gamma^2 > \omega_0^2, \gamma^2 = \omega_0^2, \gamma^2 < \omega_0^2 \tag{15.31}$$

that depend on the physical properties of the oscillator and define the strong, critical, and weak damping regimes, respectively.

In the strong damping regime, the solutions for the exponent α are

$$\alpha_1 = -\gamma + \sqrt{\gamma^2 - \omega_0^2}, \alpha_2 = -\gamma - \sqrt{\gamma^2 - \omega_0^2} \tag{15.32}$$

which are both *negative* and lead to the following general solution of Eq. (15.26)

$$\delta(t) = A_0\,e^{\alpha_1 t} + B_0\,e^{\alpha_2 t} = e^{-\gamma t}\left(A_0\,e^{t\sqrt{\gamma^2-\omega_0^2}} + B_0\,e^{-t\sqrt{\gamma^2-\omega_0^2}}\right). \tag{15.33}$$

In the critical damping regime, one has

$$\alpha_1 = \alpha_2 = -\gamma \tag{15.34}$$

and consequently the general solution is

$$\delta(t) = e^{-\gamma t}(A_0\, t + B_0). \tag{15.35}$$

Equations (15.33) and (15.35) indicate that in strong and critical damping regimes, when no driving and tip–sample forces are present, no oscillation of the probe is possible, as an eventual initial deflection of the cantilever is damped by the viscous forces represented by the parameter γ.

In the weak damping regime, the solutions for α are

$$\alpha_1 = -\gamma + i\sqrt{\omega_0^2 - \gamma^2} = -\gamma + i\,\omega_r \tag{15.36}$$

$$\alpha_2 = -\gamma - i\sqrt{\omega_0^2 - \gamma^2} = -\gamma - i\,\omega_r \tag{15.37}$$

where $\omega_r = \sqrt{\omega_0^2 - \gamma^2}$. The general solution then becomes

$$\delta(t) = A_0\, e^{\alpha_1 t} + B_0\, e^{\alpha_2 t} = e^{-\gamma t}\left(A_0\, e^{i\omega_r t} + B_0\, e^{-i\omega_r t}\right). \tag{15.38}$$

By using the Euler's formula

$$e^{\pm i\omega_r t} = \cos(\omega_r t) + i \sin(\omega_r t) \tag{15.39}$$

we obtain

$$\delta(t) = e^{-\gamma t}[(A_0 + B_0)\cos(\omega_r t) + i(A_0 - B_0)\sin(\omega_r t)]. \tag{15.40}$$

Since the solution must be a real function, the constants A_0 and B_0 have to be complex conjugated numbers

$$A_0 = a_0 + ib_0, B_0 = a_0 - ib_0 \rightarrow A_0 + B_0 = 2a_0, A_0 - B_0 = i2b_0 \tag{15.41}$$

so that Eq. (15.40) becomes

$$\delta(t) = e^{-\gamma t}[2a_0 \cos(\omega_r t) - 2b_0 \sin(\omega_r t)] \tag{15.42}$$

which, by using trigonometric addition formulas, can be equivalently written in the following more compact form

$$\delta(t) = C_0\, e^{-\gamma t} \sin(\omega_r t + \varphi_0) \tag{15.43}$$

where C_0 and φ_0 are constants depending on the initial conditions of motion.

Equation (15.43) indicates that in the conditions of weak damping, the cantilever makes oscillations at a frequency given by

$$\omega_r = \sqrt{\omega_0^2 - \gamma^2} < \omega_0 \tag{15.44}$$

which differs from that of the simple harmonic oscillator ω_0. Furthermore, such oscillations are damped, with a characteristic damping time equal to γ^{-1}. A graphical representation of the deflection of the cantilever as a function of time in the weak damping regime is shown in Figure 15.11.

Once the general solution of the associated homogeneous differential equation (Eq. 15.26) has been found, it is now time to find a particular solution of the nonhomogeneous one (15.25), which requires the inclusion of two terms: the driving force and the tip–sample interaction. We first consider the effects induced by the inclusion of the former term. The differential equation becomes the following

$$\frac{d^2\delta}{dt^2} + 2\gamma\frac{d\delta}{dt} + \omega_0^2\delta = \frac{F_0}{m^*}\sin(\omega t). \tag{15.45}$$

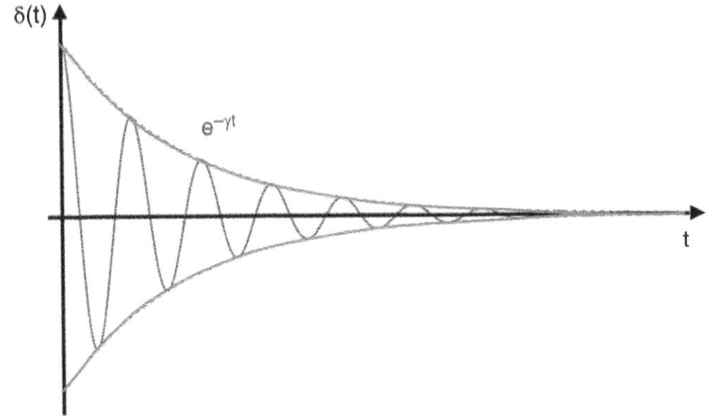

Figure 15.11 Graphical representation of the deflection of the cantilever as a function of time in the weak damping regime.

Since the new term represents harmonic oscillation with frequency ω, in order to find a particular solution of the nonhomogeneous equation, it is natural to try a function of the form $\delta(t) = A\sin(\omega t + \emptyset)$. Putting a solution with this form in Eq. (15.45), we obtain

$$-\omega^2 A\sin(\omega t + \emptyset) + 2\gamma\omega A\cos(\omega t + \emptyset) + \omega_0^2 A\sin(\omega t + \emptyset) = \frac{F_0}{m^*}\sin(\omega t). \tag{15.46}$$

By using again the trigonometric addition formulas

$$\left[(\omega_0^2 - \omega^2)A\cos\emptyset - 2\gamma\omega A\sin\emptyset\right]\sin(\omega t) + \left[(\omega_0^2 - \omega^2)A\sin\emptyset + 2\gamma\omega A\cos\emptyset\right]\cos(\omega t) = \frac{F_0}{m^*}\sin(\omega t). \tag{15.47}$$

Since the equality must apply for every value of t, the following separate equations must be satisfied

$$(\omega_0^2 - \omega^2)A\cos\emptyset - 2\gamma\omega A\sin\emptyset = \frac{F_0}{m^*} \tag{15.48}$$

$$(\omega_0^2 - \omega^2)A\sin\emptyset + 2\gamma\omega A\cos\emptyset = 0 \tag{15.49}$$

from which we finally have

$$A = \frac{F_0}{m^*}\frac{1}{\sqrt{(\omega_0^2 - \omega^2)^2 + 4\gamma^2\omega^2}} \tag{15.50}$$

and

$$\tan\emptyset = -\frac{2\gamma\omega}{\omega_0^2 - \omega^2}. \tag{15.51}$$

In conclusion, we have found that a function of the form $\delta(t) = A\sin(\omega t + \emptyset)$ is a particular solution of the differential equation of Eq. (15.45) (including the driving force) only if the parameters A and Φ satisfy the conditions in Eqs. (15.50) and (15.51). Note that the parameters A and Φ are therefore determined by the physical properties of the system and not by the initial conditions of the oscillator. The particular solution we have found indicates that the oscillation of the cantilever has the same frequency of the driving force, ω, and not that of the unperturbed oscillator, ω_0.

Furthermore, in general, this oscillation is not in-phase with respect to that of the driving force. The amplitude of the oscillation and its phase-lag with respect to the driving force depend on the physical properties of the oscillator (cantilever–tip system) as described by Eqs. (15.50) and (15.51), respectively. In Figure 15.12, some representative curves of the amplitude A as a function of the driving frequency are presented for different values of the damping constant γ. As shown, some curves of $A(\omega)$ exhibit a peak. From Eq. (15.50), it can be proved that this peak is present only for $2\gamma^2 < \omega_0^2$(weak damping), otherwise $A(\omega)$ decreases monotonically, as for the two lowest curves in Figure 15.12. When the peak is present, it is observed at a frequency different with respect to that of the unperturbed harmonic oscillator and it is given by

$$\omega_{\text{peak}} = \sqrt{\omega_0^2 - 2\gamma^2} < \omega_0 \tag{15.52}$$

and its amplitude is

$$A_{\text{peak}} = A(\omega_{\text{peak}}) = \frac{F_0}{2m^*\gamma\sqrt{\omega_0^2 - 2\gamma^2}} > A(\omega_0). \tag{15.53}$$

Equations (15.52) and (15.53) indicate that in the limit of negligible damping ($\gamma \ll \omega_0$), the peak position of the resonance falls approximately at ω_0 and its amplitude tends to infinity: it is the phenomenon of the *resonance*.

To obtain information on the energy transfer during the forced oscillation of the cantilever, it is necessary to consider the power provided by the driving force to the cantilever. For the AFM systems where the driving force is applied to the free end of the tip, it can be simply obtained by multiplying the driving force, $F_0 \sin(\omega t)$, to the velocity of its application point, v, obtained by taking the derivative of the deflection $\delta(t)$ with respect to time

$$P(t) = F_0 \sin(\omega t)\frac{d\delta}{dt} = \omega A F_0 \sin(\omega t)[\cos(\omega t)\cos\emptyset - \sin(\omega t)\sin\emptyset]. \tag{15.54}$$

By averaging over a period, we have

$$P_{\text{med}} = \frac{F_0^2\,\omega_0^2\,\gamma}{m^*}\,\frac{1}{|\omega_0^2 - \omega^2|^2 + 4\gamma^2\omega^2} = \gamma\, m^*\omega^2 A^2. \tag{15.55}$$

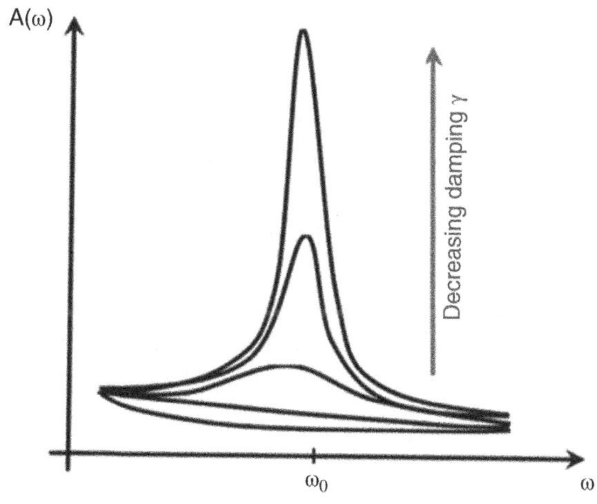

Figure 15.12 Graphical representation of the function $A(\omega)$ of Eq. (15.50) for different values of the damping constant γ. The peak becomes more pronounced on decreasing the damping.

It can be easily proved by setting the derivative of Eq. (15.55) with respect to ω to zero that the maximum transferred power is obtained for $\omega = \omega_0$. It is also useful to evaluate the width of the curve $P_{med}(\omega)$ by finding the frequencies ω_1 and ω_2 such that the mean transferred power is one-half that of the maximum, $P_{med,\ peak}$, pertaining to resonance

$$P_{med}(\omega_1) = P_{med}(\omega_2) = \frac{1}{2} P_{med,peak}. \tag{15.56}$$

From Eqs. (15.55) and (15.56), we obtain

$$\omega_1 = -\gamma + \sqrt{\gamma^2 + \omega_0^2}, \omega_2 = \gamma + \sqrt{\gamma^2 + \omega_0^2} \tag{15.57}$$

giving a width

$$\Delta\omega = \omega_2 - \omega_1 = 2\gamma. \tag{15.58}$$

Such a result allows to define the quality factor Q as

$$Q = \frac{\omega_0}{\Delta\omega} = \frac{\omega_0}{2\gamma}. \tag{15.59}$$

This latter quantity plays a key role in AM-AFM, as it quantifies the width of the resonance relative to its characteristic frequency. High values of the quality factor indicate a strong resonance, which means very large oscillation amplitudes of the cantilever in response to the driving force. This is the physical situation usually encountered for AFM measurements performed in air with a standard probe, as the typical value of the quality factor is about 400 (in absence of tip–sample interaction). Furthermore, since the typical unperturbed resonance frequency is $\omega_0 \simeq 2 \times 10^6$ rad s^{-1}, a value of about 2.5×10^3 rad s^{-1} can be estimated for γ based on Eq. (15.59), indicating a weak damping regime.

The complete general solution of Eq. (15.45) is obtained by summing the particular solution just discussed in the form $\delta(t) = A \sin(\omega t + \emptyset)$, with A and Φ given by Eqs. (15.50) and (15.51), with that in parametric form pertaining to the homogeneous differential equation, Eq. (15.26), to give

$$\delta(t) = C_0\, e^{-\gamma t} \sin(\omega_r t + \varphi_0) + A \sin(\omega t + \emptyset) \tag{15.60}$$

where the solution in the weak damping regime has been considered because, as discussed above, it represents the most commonly encountered regime for AM-AFM measurements performed in air.

To obtain the complete solution of the differential equation in Eq. (15.25), we have now to consider the last term: the tip–surface interaction. As discussed above, to obtain an analytical solution of the problem, we assume that the cantilever undergoes small deflection, so that the force $F_{ts}(\delta)$ can be approximately described by a linear function with respect to δ [43]

$$F_{ts}(\delta) = F_{ts}(0) + \left(\frac{dF_{ts}}{d\delta}\right)_0 \delta. \tag{15.61}$$

In this approximation, the tip–sample interaction force is just summarized by the coefficient multiplying δ, which represents the gradient of the force felt by the probe during its oscillations. By defining an effective spring constant as follows

$$k_{ts} = -\left(\frac{dF_{ts}}{d\delta}\right)_0 \tag{15.62}$$

and including the linearized tip–surface interaction force in Eq. (15.25), one obtains

$$\frac{d^2\delta}{dt^2} + 2\gamma\frac{d\delta}{dt} + \omega_0^2\delta = \frac{F_0}{m^*}\sin(\omega t) + \left[\frac{F_{ts}(0)}{m^*} - \frac{k_{ts}}{m^*}\delta\right]. \tag{15.63}$$

By neglecting the constant term $\frac{F_{ts}(0)}{m^*}$ that, as is well known, just shifts the equilibrium position of the oscillator, and by grouping together the terms that multiply δ, we have

$$\frac{d^2\delta}{dt^2} + 2\gamma\frac{d\delta}{dt} + \left(\omega_0^2 + \frac{k_{ts}}{m^*}\right)\delta = \frac{F_0}{m^*}\sin(\omega t). \tag{15.64}$$

Since by definition $\omega_0 = \frac{k}{m^*}$, it is possible to define an effective frequency as follows

$$\omega_{eff}^2 = \omega_0^2 + \frac{k_{ts}}{m^*} = \frac{k}{m^*} + \frac{k_{ts}}{m^*} = \frac{k + k_{ts}}{m^*} = \frac{k_{eff}}{m^*} \tag{15.65}$$

where $k_{eff} = k + k_{ts}$ is the effective force constant, then Eq. (15.64) becomes

$$\frac{d^2\delta}{dt^2} + 2\gamma\frac{d\delta}{dt} + \omega_{eff}^2\delta = \frac{F_0}{m^*}\sin(\omega t). \tag{15.66}$$

Comparison of Eqs. (15.45) and (15.66) shows that, as far as the tip–sample interaction is described as a linear function of δ, its effect on the cantilever dynamic can be easily interpreted just as a shift of the unperturbed oscillation frequency ω_0 to a new value ω_{eff}. Furthermore, for $k_{ts} \ll k$ the shift can be approximated as a linear function of k_{ts}, as $\omega_{eff} - \omega_0 \approx \frac{\omega_0 k_{ts}}{2k}$.

Summarizing, the solution of the complete problem of the weakly perturbed harmonic oscillator is that already reported in Eq. (15.60), but the parameters A and $\emptyset$ are given by

$$A = \frac{F_0}{m^*}\frac{1}{\sqrt{\left(\omega_{eff}^2 - \omega^2\right)^2 + 4\gamma^2\omega^2}} \tag{15.67}$$

$$\tan\emptyset = -\frac{2\gamma\omega}{\omega_{eff}^2 - \omega^2} \tag{15.68}$$

which are similar to those of Eqs. (15.50) and (15.51), but now the characteristic frequency of the ideal oscillator, ω_0, is replaced by the effective one, ω_{eff}, which includes the effect of tip–sample interaction. The first term of the solution in Eq. (15.60) is transitory and depends on the initial conditions, whereas the second one is stationary and is related to the driving force and to the tip–sample interaction. To obtain proper scanning of a surface in AM-AFM, it is mandatory to be sure that the transitory part of Eq. (15.60) is exhausted. This term, in fact, is not directly related to the morphology of the sample and it is also inherently time dependent. It is also important to keep in mind that during scanning, any perturbation of the cantilever due to the roughness of the surface is mathematically equivalent to the imposition of a new set of initial conditions to the motion of the cantilever described by Eq. (15.60). Consequently, a new transitory damped oscillation is expected to start in correspondence of any of such perturbative event. These considerations indicate that, to obtain a faithful reproduction of the surface morphology, the scanning speed must be slow enough to be sure that the time necessary to travel between two relevant features on the surface is significantly longer than the characteristic time of the transitory term of Eq. (15.60). In these conditions, the measured oscillation amplitude of the cantilever is due to the stationary term of the cantilever motion, which is strictly related to the tip–sample interaction and consequently to the surface morphology. These considerations explain why, as anticipated above, severe limitations on the scanning speed apply to Tapping mode at variance to other modes, such as the Contact mode. Moreover, again due to the transitory part of Eq. (15.60), AM-AFM measurements are practically inapplicable in vacuum, as in such environment, the damping coefficient becomes negligible and consequently the duration of the transitory becomes very long.

By assuming that the transitory term is exhausted, the equation of motion of the cantilever is described by

$$\delta(t) \cong A \sin(\omega t + \emptyset). \tag{15.69}$$

If we assume, for example, a tip–sample interaction involving the attractive part of the force–distance curve due to van der Waals interaction (see Figure 15.7), then locally the gradient of F_{ts} with respect to δ, $\frac{\partial F_{ts}}{\partial \delta}$, is positive (tip–sample distance, d, and δ are related by the equation $d = \delta + Z_c$, where Z_c is a constant, see Figure 15.10b) and consequently $k_{ts} < 0$ (Eq. (15.62)) and $\Delta\omega < 0$ (Eq. (15.65)). Conversely, if the dominant tip–sample interaction is repulsive, then $\frac{\partial F_{ts}}{\partial \delta} < 0$, $k_{ts} > 0$ and $\Delta\omega > 0$. These results indicate that within the linear approximation, the effects of establishment of attractive and repulsive tip–sample interactions simply shift the $A(\omega)$ curve toward lower and higher frequencies, respectively, as shown in Figure 15.13.

These results give the opportunity to obtain a first rough understanding of the characteristic reduction of the cantilever oscillation amplitude observed when the tip interacts with the sample's surface. In AM-AFM, the cantilever is driven by a force oscillating at a frequency near that of its first flexural normal mode. This frequency is selected before the tip interacts with the surface and it is maintained fixed during the entire scanning process. The value of the amplitude of the cantilever oscillation corresponding to the selected frequency is established by the curve $A(\omega)$, of course. Suppose, for example, when the probe is far from the surface to select a driving frequency equal to ω_0, then the cantilever oscillates with an amplitude approximately corresponding to that of the maximum of the curve $A(\omega)$. When the tip approaches the surface, a shift of the curve $A(\omega)$ is induced, toward higher or lower frequencies for repulsive or attractive forces, respectively. Whatever the sign of the shift, it determines a reduction of the oscillation amplitude corresponding to the exciting frequency ω_0, as shown in Figure 15.13. When the amplitude of cantilever oscillations reduces to the set-point value selected by the operator, then the tip–sample approach process ends and the scanning starts. Since the repulsive portion of the force–distance curve is steeper than the

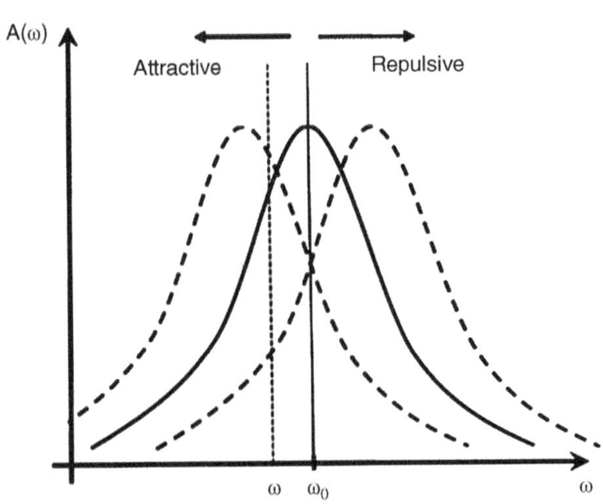

Figure 15.13 Representation of the effects induced by tip–sample interaction on the function $A(\omega)$ within the weakly perturbed harmonic oscillator approximation.

attractive one, more stable conditions of measurement are usually obtained in the former regime. It is possible to impose such a condition by selecting a driving frequency slightly lower than that of the peak, as indicated by dashed vertical line in Figure 15.13. In fact, during the approaching process, attractive forces induce an increase of the amplitude of the cantilever oscillation, which is not considered by the AM-AFM microscope. The approach then proceeds further until the repulsive forces produce a shift in the curve $A(\omega)$ toward frequencies higher enough to reduce the oscillation amplitude of the cantilever to the set-point value.

Before concluding the discussion on the weakly perturbed harmonic oscillator model, it is important to underline its limitations. First of all, it has been supposed that the tip–surface interaction function can be approximately assumed to be liner with respect to δ. As it can be easily recognized by looking at Figure 15.9, this assumption is inapplicable to Tapping mode, as by definition the cantilever oscillates between points located very far from the surface, where the interaction is almost negligible, to points where sizable repulsive contact forces are established. The portion of the force–distance curve explored during such oscillations is evidently nonlinear. Furthermore, in the theory above discussed, no energy dissipation related to tip–surface interaction has been included at all. These drawbacks of the model add up with those discussed at the beginning of Section 15.4.2, related to the finite size of cantilever and tip. Due to all these difficulties, the weakly perturbed harmonic oscillator must be considered just a first simple point of reference to understand the properties of AFM images and to optimize the acquisition conditions, but not a comprehensive satisfying theory of AM-AFM [50]. When a reliable quantitative analysis is required, alternative theoretical and computational approaches must be considered [3].

15.5 AFM Spectroscopy

The AFM microscope can also be used to evaluate some relevant physicochemical properties of the material under study [2, 5, 21]. This type of application is known as AFM spectroscopy. In principle, both static and dynamic modes can be used, even if the former is indubitably the most diffused one and it is discussed in more detail in the present paragraph.

AFM spectroscopy consists in the experimental determination and analysis of the curve representing the interaction force between tip and sample as a function of their mutual distance. The typical experiment consists in the measurements of the deflection of the cantilever during an entire path of approaching (withdrawing) of the probe to (from) the surface. This measurement gives a curve relating the deflection of the cantilever, δ, to the mean position of the cantilever, Z_c (Figure 15.10b). To obtain the curve relating the interaction force, F_{ts}, to the tip–sample distance, d, the force has to be calculated from the deflection, and the quantity Z_c has to be related to d [2]. The latter is straightforward, as the simple relation $d = \delta + Z_c$ holds, as shown in Figure 15.10b, whereas the former requires a calibration protocol. As discussed in Section 15.2.4, the deflection of the cantilever is first measured as a voltage, but it can be easily transformed into the physical deflection of the cantilever, δ, by a simple calibration procedure. The second important step is to determine the force constant associated to the excited mode of the cantilever. A variety of methods have been proposed to date [42, 51–59], but the most widespread is that of thermal noise [56, 58]. This method is founded on the thermodynamic equipartition theorem, implying that when a cantilever is in thermodynamic equilibrium with the environment, its mean square deflection caused by thermal vibrations, $\langle\delta^2\rangle$, must satisfy the following relation

$$\frac{1}{2}k\left\langle\delta^2\right\rangle = \frac{1}{2}k_B T \tag{15.70}$$

where k is the force constant of the cantilever, whereas k_B and T are the Boltzmann constant and the temperature, respectively. Based on this property, it is possible to determine the force constant of the cantilever by measuring the spectrum of the cantilever deflections due to thermal fluctuations and by fitting the main peak found in this graph with a Lorentzian profile [56, 58]. This procedure is today well established and included in the acquisition software of the most diffused AFM microscopes.

In Figure 15.14a and b, the deflection vs. Z_c, $\delta(Z_c)$, and force vs. distance, $F_{ts}(d)$, curves are shown, respectively, as continuous lines for the idealized case of an infinitely hard sample without distance-dependent surface forces [21]. As shown, before the contact is established, no deflection or force is measured, whereas when the tip touches the surface, the deflection changes of the same amount as Z_c [$\Delta\delta = \Delta(Z_c)$] and the force goes to infinity. The point of physical contact between tip and surface, corresponding to $d = 0$, is evaluated experimentally by the same curves as the point corresponding to the abrupt change of the slope.

Also shown in the same figures, as dashed and dotted lines, are the curves observed when repulsive and attractive distance-dependent surface forces are present, respectively. The regions of the curves pertaining to contact and distance-dependent forces are easily distinguishable. Many relevant information on the properties of the material can be obtained by fitting the portions of the tip–sample force curve associated to the latter interactions with appropriate analytical models, such as those presented and discussed in the Section 15.3. When distance-dependent forces are present, the evaluation of the contact point is less obvious. Usually, it is defined as the intersection point between the two straight lines observed at the two extremes of the curves in Figure 15.14a and b.

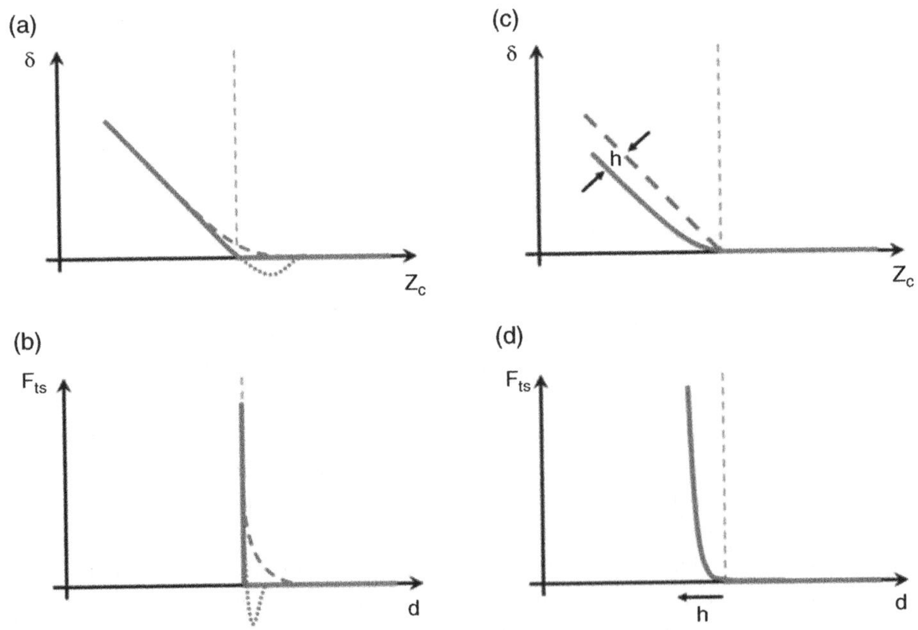

Figure 15.14 (a) $\delta(Z_c)$ and (b) $F_{ts}(d)$ curves for the idealized case of an infinitely hard surface without distance-dependent surface forces. Dashed and dotted lines refer to cases where repulsive and attractive surface forces are also present, respectively. (c) $\delta(Z_c)$ and (d) $F(d)$ curves for the case of a deformable surface, but without including distance-dependent surface forces.

Another instructive case is that of deformable surfaces. The curves $\delta(Z_c)$ and $F_{ts}(d)$ associated to this situation obtained under the assumption of negligible distance-dependent forces are reported in Figure 15.14c and d, respectively [21]. The portions of the curves deviating from the horizontal null reference line are associated to the physical contact established between tip and surface. As shown, due to deformation, both the curves are not linear in this region. Such deviations from the linear trend in the contact region define the indentation h, as indicated in Figure 15.14c and d. For mathematical convenience, the force curve is usually studied as a function of h instead of d, and it is fitted with appropriate models, such as the Hertzian model discussed in Section 15.3. This method gives the opportunity to determine many relevant physicochemical parameters with sub-nanometer resolution, such as the Young modulus of the surface.

In most real situations, both deformation of the contacting surfaces and distance-dependent forces are observed. Furthermore, if the measurements are performed in air, as usual, then the formation of the meniscus between tip and surface strongly affects both $\delta(Z_c)$ and $F_{ts}(d)$ curves, increasing significantly the complexity of the problem and making more difficult the proper analysis of the experimental data. These situations are usually encountered when soft samples are studied under the action of relevant attractive and/or adhesive forces.

A further characterization of the sample can be obtained by comparing the force–distance curves obtained by approaching and withdrawing the probe to/from the surface. An example of such typical curves is given in Figure 15.15 as dashed and continuous lines, respectively. As shown, they usually do not coincide, except at very large distances where no interaction takes place, and indicate that smaller adhesion forces act on tip during the approaching process. For measurements performed in air, most of this effect is related to capillary forces. To obtain a measure of such adhesion, the value of the force corresponding to the minimum of the withdrawing force curve, F_{adh}, is taken. The hysteretic trend shown in Figure 15.15 indicates that energy is dissipated during each approaching–withdrawing cycle. The value of the dissipated energy is by definition equal to the work done by tip–surface forces during a cycle and consequently it can be easily calculated from the force vs. Z_c curve. It can be easily shown that it is represented by the area between approaching and withdrawing curves. Since the values of the adhesion forces and of the related dissipated energy

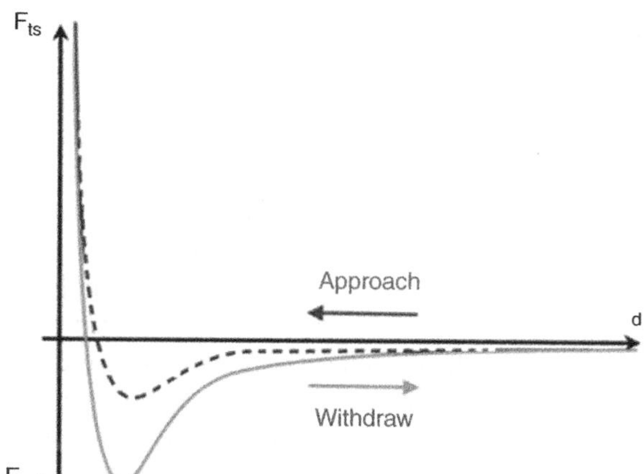

Figure 15.15 Typical $F_{ts}(d)$ curves obtained when the probe approaches (dashed) and withdraws (continuous) the probe to/from a surface.

depend on the nature of the adhesion forces acting between tip and surface, their experimental estimation gives a valuable contribution to the physicochemical characterization of the material under study.

15.6 Case Studies

In the following, Section 15.6.1 and 15.6.2, we present two reference examples to show how in practice the AFM technique can be used to obtain relevant morphological information on a sample. In the first study, the characterization of the roughness of an optically polished surface of amorphous silicon dioxide (SiO_2) is discussed, whereas in the second the focus is on the determination of the size distribution of a collection of latex nanoparticles distributed over a flat substrate. The proposed case studies are representative of important concrete applications of AFM microscopy, both in industrial and in research fields. Since SiO_2 is a hard material, the measurements were acquired in the fast and simple Contact mode, whereas to study the latex nanoparticles that are just put on the substrate surface, we use the more sensitive Tapping mode.

15.6.1 Roughness of a Flat Surface

Here, an experimental estimation of the surface roughness of a sample of SiO_2 of commercial origin with polished surface for optical applications and size $5 \times 5 \times 1\ \text{mm}^3$ is presented. To clean the surfaces of the sample before measurements, it was sonicated for one hour in an equal mixture of acetone and ethanol and then dried in air. The measurements were performed in air with a Bruker Multimode V working in Contact mode. The probes used were V-shaped (silicon nitride) with nominal spring constant of $0.3\ \text{N m}^{-1}$. The tip velocity on the surface and the PID gains were optimized to give the higher scan rate compatibly with a reliable scan of the surface. The optimal tip velocity was found to be $25\ \mu\text{m s}^{-1}$.

The roughness of a surface, R_q, is defined as root mean square vertical deviations of the surface points with respect to the main local plane. Practically, it is obtained by acquiring an AFM image of the surface, by subtracting from it a mean background surface and then calculating the following parameter

$$R_q = \sqrt{\frac{1}{n}\sum\nolimits_{i=1}^{n} z_i^2} \tag{15.71}$$

where z_i are the values of heights associated to the points of the surface and n is their total number. It is known that the value of the roughness estimated experimentally by AFM depends on the tip size and consequently we have considered supersharp tips with nominal tip radius of 2 nm [60]. Furthermore, it is also known that the estimated roughness increases on increasing the scan size, until a saturation value is reached [60]. Of course, the most reliable value of the surface roughness is that obtained at saturation, as it reflects the properties of a larger, more representative portion of the surface. These considerations indicate that to properly estimate the roughness of a surface, many scans with different sizes must be acquired and we have considered the following: $0.25 \times 0.25, 0.5 \times 0.5, 1 \times 1, 2 \times 2, 4 \times 4, 6 \times 6\ \mu\text{m}^2$. To obtain a reliable estimation of the experimental uncertainty, for each scan size, 10 measurements were acquired in different points of the surface and the experimental uncertainty was evaluated statistically from the fluctuations of the estimated values of roughness.

In Figure 15.16a, a 2 μm × 2 μm representative AFM image of the sample surface is shown. It was obtained by subtracting a background polynomial surface of third order to the as-acquired AFM image. In Figure 15.16b, the vertical profile obtained along a line of the image is also reported, showing the characteristic vertical fluctuations of the points of the surface with respect to the main line.

The results of the quantitative estimation of the roughness obtained for the images with different size are collected in Figure 15.16c. As shown, the roughness increases with scan size up to 2 μ m × 2 μm, whereas for larger images it maintains a constant saturation value, which was evaluated as $R_q = 0.5 \pm 0.05$ nm.

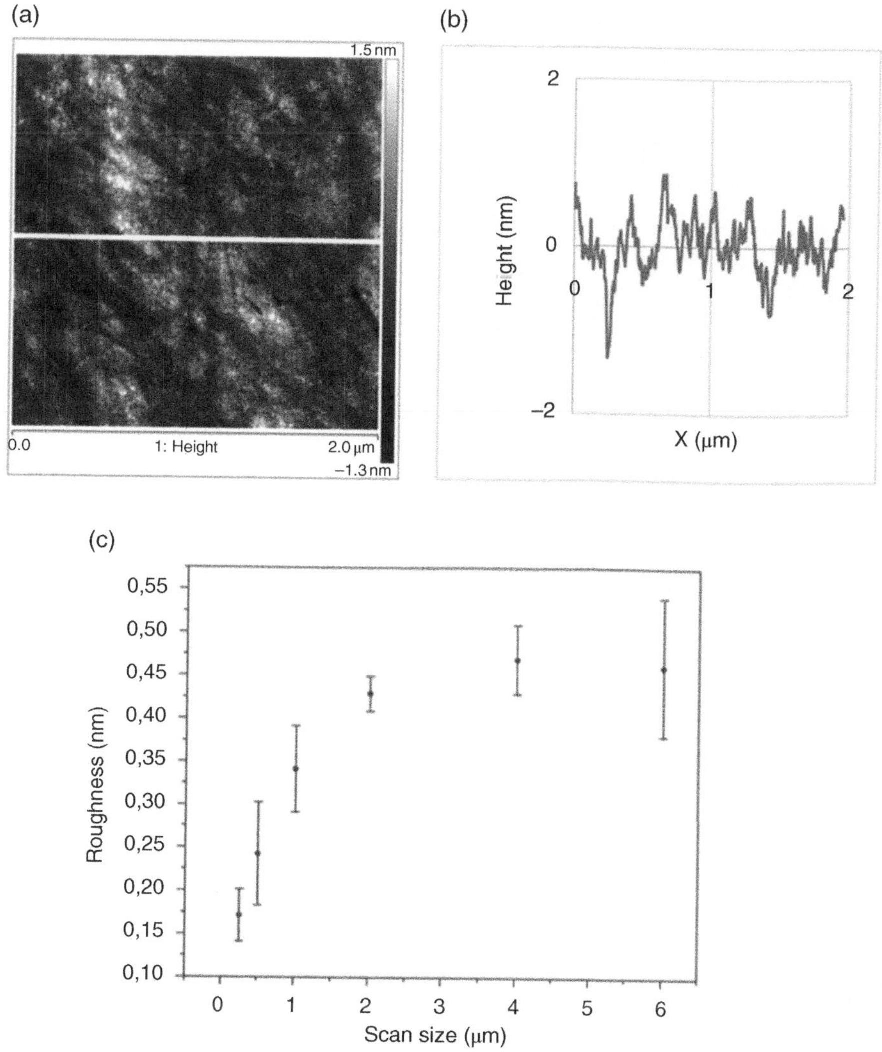

Figure 15.16 (a) 2 μm × 2 μm AFM image of the SiO_2 sample surface obtained after subtraction of a background polynomial surface of third order. (b) Vertical profile of the surface obtained along the white line indicated in (a), showing the characteristic vertical fluctuations of the surface. (c) Values of roughness estimated in the AFM images obtained with different values of the scan size.

15.6.2 Size Distribution of Nanoparticles

Here, a simple example is presented showing the procedure used to determine the size distribution of a system of nanoparticles. The sample considered consists in latex nanoparticles of commercial origin. They were preliminarily dispersed in water and then a drop of the mixture was put on a flat and clean surface of mica. The latter is an important mineral that has become a standard substrate due to its easy cleavage along the {001} planes by Scotch Tape exfoliation. The measurements were performed in air with a Bruker Multimode V working in Tapping mode. The probes used were

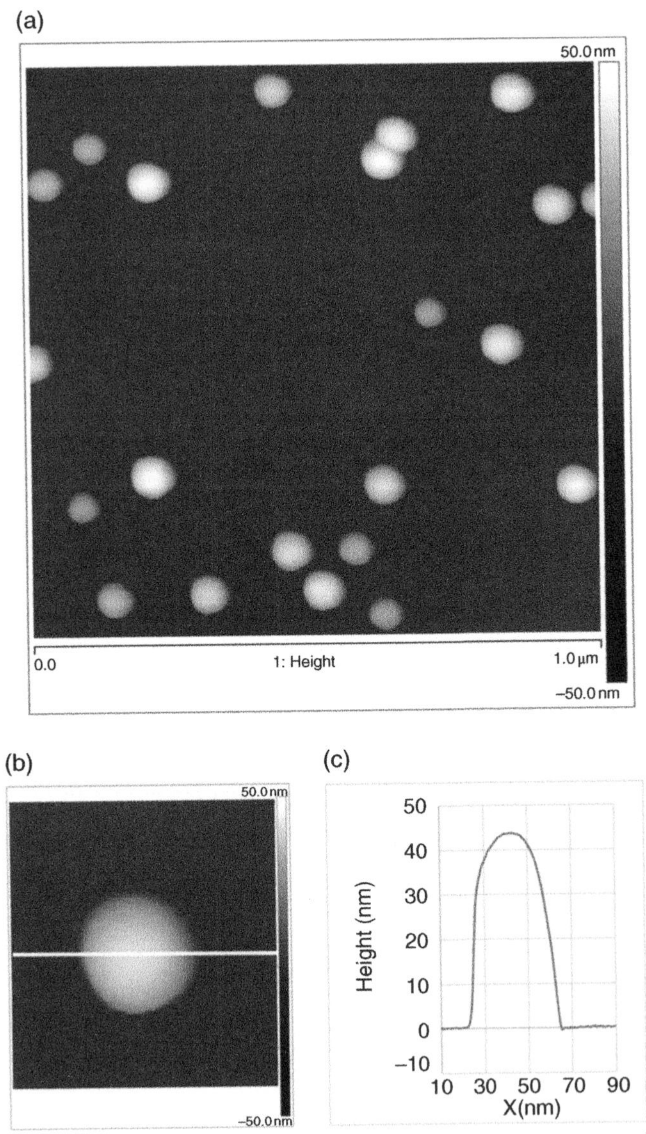

Figure 15.17 (a) 1 µm × 1 µm AFM image of the system consisting of latex nanoparticles deposited on mica. (b) AFM image of a single nanoparticle extracted from the image in (a). (c) Vertical profile of the surface estimates along the line shown in white in (b).

rectangle-shaped (silicon) with nominal tip radius of 5 nm and nominal spring constant of 37 N m^{-1}. The tip velocity on the surface and the PID gains were optimized to give the higher scan rate compatibly with a reliable scan of the surface. The optimal tip velocity was found to be 2 μm s^{-1}. The scan size was fixed at a value of 1 μm × 1 μm, to limit the total duration of each measurement and consequently the undesired thermal drift effects.

In Figure 15.17a, a typical AFM image of the nanoparticles on mica is reported. As shown, they are well distinguishable with respect to the substrate plane and mostly unaggregated. To obtain a reliable estimation of the size distribution of the nanoparticles, a large number of them were extracted from the images and singly analyzed. The analysis consisted in the estimation of the height of the particle with respect to the background plane and consequently only well-isolated nanoparticles were considered. In particular, the height was estimated along a line passing through the highest point (the top) of the nanoparticle, as outlined in Figure 15.17b. Even though the nanoparticles we examined have spherical shape, as guaranteed by the commercial supplier of the sample, the profile of the nanoparticle shown in Figure 15.17c, for example, indicates a lateral size (~70 nm) significantly larger than the vertical one (~40 nm). This well-known effect is observed because the AFM profile inherently results from the convolution between the shape of tip and that of the surface. If the actual shape of the nanosystem is unknown, then the true shape must be reconstructed by geometrical and/or numerical methods [61, 62]. In our case, since we already know that the particles are spherical, we can estimate the diameter of the particles by simply evaluating the height from the extracted profile, without taking care of the convolution effects. In fact, it is straightforward to verify that the convolution between tip and surface shapes cannot affect the measured height of the profile, but just its width.

By accurately measuring the diameters of more than 200 nanoparticles, we have obtained the size distribution reported in Figure 15.18. Also shown in the same figure is the Gaussian profile obtained by fitting curve to the data. Based on these data, we have estimated a mean nanoparticle diameter $D = 42 \pm 1$ nm.

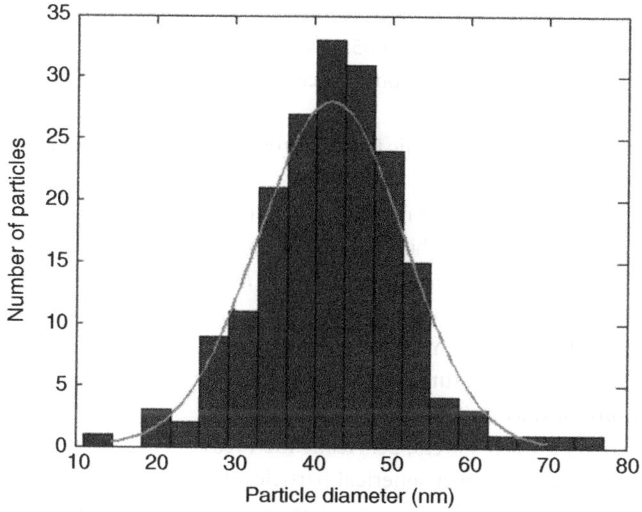

Figure 15.18 Size distribution of the nanoparticle's diameter estimated by AFM measurements. The Gaussian profile obtained by fitting the data is also shown.

References

1 Friedbacher, G. and Fuchs, H. (1999). Classification of scanning probe microscopies (technical report). *Pure and Applied Chemistry* 71 (7): 1337.

2 Haugstad, G. (2012). *Atomic Force Microscopy: Understanding Basic Modes and Advanced Applications*, ISBN:9780470638828. John Wiley & Sons, Inc.

3 Garcia, R. (2010). *Amplitude Modulation Atomic Force Microscopy*, ISBN:9783527408344. Wiley-VCH Verlag GmbH & Co. KGaA.

4 Eaton, P. and West, P. (2010). *Oxford, Atomic Force Microscopy*, ISBN-10: 0199570450. OUP Oxford.

5 Seo, Y. and Jhe, W. (2008). Atomic force microscopy and Spectroscopy. *Reports on Progress in Physics* 71: 016101.

6 Braga, P.C. and Ricci, D. (eds.) (2004). Methods in molecular biology. In: *Atomic Force Microscopy: Biomedical Methods and Applications*, vol. 242. Totowa, NJ: Humana Press Inc.

7 Giessibl, F.J. (2003). Advances in atomic force microscopy. *Reports on Progress in Physics* 75 (3): 949.

8 Martinez, J., Yuzvinsky, T.D., Fennimore, A.M. et al. (2005). Length control and sharpening of atomic force microscope carbon nanotube tips assisted by an electron beam. *Nanotechnology* 16: 2493.

9 Klinov, D. and Magonov, S. (2004). True molecular resolution in tapping-mode atomic force microscopy with high resolution probes. *Applied Physics Letters* 84: 2697.

10 Snow, E.S., Campbell, P.M., and Novak, J.P. (2002). Single-wall carbon nanotube atomic force microscope probes. *Applied Physics Letters* 80: 2002.

11 Cheung, C.L., Hafner, J.H., and Lieber, C.M. (2000). Carbon nanotube atomic force microscopy tips: direct growth by chemical vapor deposition and application to high-resolution imaging. *Proceedings of the National Academy of Sciences USA* 97: 3809.

12 Putman, C.A.J., Van der Werf, K.O., De Grooth, B.G. et al. (1994). Tapping mode atomic force microscopy in liquid. *Applied Physics Letters* 64 (18): 2454.

13 Han, W., Lindsay, S.M., and Jing, T. (1996). A magnetically driven oscillating probe microscope for operation in liquids. *Applied Physics Letters* 69: 4111.

14 Connell, S.D. and Smith, D.A. (2006). The atomic force microscope as a tool for studying phase separation in lipid membranes (review). *Molecular Membrane Biology* 23 (1): 17–28.

15 Alexander, S., Hellemans, L., Marti, O. et al. (1989). An atomic-resolution atomic-force microscope implemented using an optical-lever. *Journal of Applied Physics* 65: 164.

16 Meyer, G. and Amer, N.M. (1988). Novel optical approach to atomic force microscopy. *Applied Physics Letters* 53: 1045.

17 Schaffer, T.E. (2005). Calculation of thermal noise in an atomic force microscope with a finite optical spot. *Nanotechnology* 16: 664.

18 Schaffer, T.E. and Fuchs, H. (2005). Optimized detection of normal vibration modes of atomic force microscope cantilevers with the optical beam deflection method. *Journal of Applied Physics* 97: 083524.

19 Landau, L.D. and Lifshitz, L.M. (1981). *Quantum Mechanics: Nonrelativistic Theory*, 3e. Butterworth-Heinemann.

20 Baym, G. (1969). *Lectures on Quantum Mechanics*. New York: Benjamin.

21 Butt, H.J., Capella, B., and Kappl, M. (2005). Force measurements with the atomic force microscope: technique, interpretation and applications. *Surface Science Reports* 59: 1.

22 Israelachvili, J.N. (2010). *Intermolecular and Surface Forces*, 3e. Academic Press.

23 Hamaker, H.C. (1937). van der Waals attraction between spherical particles. *Physica* 4 (10): 1058.

24 Morse, P.M. (1929). Diatomic molecules according to the wave mechanics. II. Vibrational levels. *Physics Review* 34: 57.

25 Lennard-Jones, J.E. (1931). Cohesion. *Proceedings of the Physical Society* 43: 461.

26 Johnson, K.L. (1985). *Contact Mechanics*. New York: Cambridge University Press.

27 Ciavarella, M., Joe, J., Papangelo, A., and Barber, J.R. (2019). The role of adhesion in contact mechanics. *Journal of the Royal Society Interface* 16 (151): 20180738.

28 Grierson, D.S., Flater, E.E., and Carpick, R.W. (2005). Accounting for the JKR-DMT transition in adhesion and friction measurements with atomic force microscopy. *Journal of Adhesion Science and Technology* 19 (3–5): 291.

29 Greenwood, J.A. (1997). Adhesion of elastic spheres. *Proceedings Mathematical Physical & Engineering Sciences* 453: 1961.

30 Maugis, D. (1992). Adhesion of spheres – the JKR-DMT transition using a dugdale model. *Journal of Colloid and Interface Science* 150: 243.

31 Derjaguin, B.V., Muller, V.M., and Toporov, Y.P. (1975). Effect of contact deformations on the adhesion of particles. *Journal of Colloid and Interface Science* 53: 314.

32 Johnson, K.L., Kendall, K., and Roberts, A.D. (1971). Surface energy and the contact of elastic solids. *Proceedings of the Royal Society of London A* 324: 301.

33 Adamson, A.W. (1990). *Physical Chemistry of Surfaces*. New York: John Wiley & Sons, Inc.

34 Atkins, P.W. (1990). *Physical Chemistry*, 4e. New York: W.H. Freeman and Co.

35 Mayer, R.P. and Stowe, R.A. (2005). Nodoids and toroids: comparison of two geometries for the meniscus profile of a wetting liquid between two touching isolated spheres and extensions to the model of a collection of packed spheres. *Journal of Colloid and Interface Science* 285: 781.

36 Bisschop, F.R.E. and Rigole, W.J.L. (1982). A physical model for liquid capillary bridges between adsorptive solid spheres: the Nodoid of Plateau. *Journal of Colloid and Interface Science* 88 (1): 117.

37 Orr, F.M., Scriven, L.E., and Rivas, A.P. (1975). Pendular rings between solids: meniscus properties and capillary force. *Journal of Fluid Mechanics* 667: 723.

38 Garcia, R., Martinez, R.V., and Martinez, J. (2005). Nano-chemistry and scanning probe nanolithographies. *Chemical Society Reviews* 35 (1): 29.

39 Day, H.C. and Allee, D.R. (1993). Selective area oxidation of silicon with a scanning force microscope. *Applied Physics Letters* 62: 2691.

40 Braunschweig, A.B., Huo, F., and Mirkin, C.A. (2009). Molecular printing. *Nature Chemistry* 1: 353.

41 Salaita, K., Wang, Y., and Mirkin, C.A. (2007). Applications of dip-pen nanolithography. *Nature Nanotechnology* 2: 145–155.

42 Cook, S.M., Schaffer, T.E., Chynoweth, K.M. et al. (2006). Practical implementation of dynamic methods for measuring atomic force microscope cantilever spring constants. *Nanotechnology* 17 (9): 2135–2145.

43 Garcia, R. and Perez, R. (2002). Dynamic atomic force microscopy methods. *Surface Science Reports* 47 (6–8): 197.

44 Lantz, M., Liu, Y.Z., Cui, X.D. et al. (1998). Dynamic force microscopy in fluid. In: *3rd Conference on Development and Industrial Application of Scanning Probe Methods (SXM-3)*, 354–360. Switzerland: John Wiley.

45 Paulo, A.S. and Garcia, R. (2002). Unifying theory of tapping-mode atomic force microscopy. *Physical Review B* 66 (041406(R)): 1–4.

46 Garcia, R. and San Paulo, A. (1999). Attractive and repulsive tip–sample interaction regimes in tapping-mode atomic force microscopy. *Physical Review B* 60: 4961.

47 Nony, L., Boisgard, R., and Aime, J.P. (1999). Nonlinear dynamical properties of an oscillating tip–cantilever system in tapping mode. *The Journal of Chemical Physics* 111: 1615.

48 Chen, J., Workman, R.K., Sarid, D., and Hoper, R. (1994). Numerical simulations of a scanning force microscope with a large-amplitude vibrating cantilever. *Nanotechnology* 5: 199.

49 Wang, L. (1998). Analytical descriptions of the tapping-mode atomic force microscopy response. *Applied Physics Letters* 73: 3781.

50 Holscher, H., Schwarz, U.D., and Weisendanger, R. (1999). Calculation of the frequency shift in dynamic force microscopy. *Applied Surface Science* 140: 344.

51 Clifford, C.A. and Sheah, M.P. (2005). The determination of atomic force microscope cantilever spring constants via dimensional methods for nanomechanical analysis. *Nanotechnology* 16: 1666.

52 Sader, J.E., Pacifico, J., Green, C.P., and Mulvaney, P. (2005). General scaling law for stiffness measurement of small bodies with applications to the atomic force microscope. *Journal of Applied Physics* 97: 124903.

53 Burnham, N.A., Chen, X., Hodges, C.S. et al. (2003). Comparison of calibration methods for atomic-force microscopy cantilevers. *Nanotechnology* 14 (1): 1–6.

54 Levy, R. and Maaloum, M. (2002). Measuring the spring constant of atomic force microscope cantilevers: thermal fluctuations and other methods. *Nanotechnology* 13: 33.

55 Sader, J.E., Chon, J.W.M., and Mulvaney, P. (1999). Calibration of rectangular atomic force microscope cantilevers. *The Review of Scientific Instruments* 70: 3967.

56 Butt, H.J. and Jascke, M. (1995). Calculation of thermal noise in atomic force microscopy. *Nanotechnology* 6 (1): 1–7.

57 Sader, J.E., Larson, I., Mulvaney, P., and White, L.R. (1995). Method for the calibration of atomic-force microscope cantilevers. *The Review of Scientific Instruments* 66: 3789.

58 Hutter, J.L. and Bechhoefer, J. (1993). Calibration of atomic-force microscope tips. *The Review of Scientific Instruments* 64: 1868.

59 Cleveland, J.P., Manne, S., Bocek, D., and Hansma, P.K. (1993). A non-destructive method for determining the spring constant of cantilevers for scanning force microscopy. *The Review of Scientific Instruments* 64: 403.

60 Sedin, D.L. and Rowlen, K.L. (2001). Influence of tip size on AFM roughness measurements. *Applied Surface Science* 182: 40.

61 Canet-Ferrer, J., Coronado, E., Forment-Aliaga, A., and Pinilla-Cienfuegos, E. (2014). Correction of the tip convolution effects in the imaging of nanostructures studied through scanning force microscopy. *Nanotechnology* 25: 395703.

62 Bukharaev, A.A., Berdunov, N.V., Ovchinnikov, D.V., and Salikhov, K.M. (1998). Three-dimensional probe and surface reconstruction for atomic force microscopy using a deconvolution algorithm. *Scanning Microscopy* 12 (1): 225.

Index

Spectroscopy for Materials Characterization, First Edition. Edited by Simonpietro Agnello.